YONG-JICK KIM.
김 용 직.

1983. Winter.

CONTROL
AND DYNAMIC SYSTEMS

CONTROL
AND DYNAMIC SYSTEMS

YASUNDO TAKAHASHI
University of California, Berkeley, California

MICHAEL J. RABINS
Polytechnic Institute of Brooklyn

DAVID M. AUSLANDER
University of California, Berkeley, California

ADDISON-WESLEY PUBLISHING COMPANY
Reading, Massachusetts
Menlo Park, California · London · Amsterdam · Don Mills, Ontario · Sydney

Second printing, November 1972

Copyright © 1970 by Addison-Wesley Publishing Company, Inc.
Philippines copyright 1970 by Addison-Wesley Publishing Company, Inc.

All rights reserved. No part of this publication may be reproduced, stored in a retrieval system, or transmitted, in any form or by any means, electronic, mechanical, photocopying, recording, or otherwise, without the prior written permission of the publisher. Printed in the United States of America. Published simultaneously in Canada. Library of Congress Catalog Card No. 77-87045.

ISBN 0-201-07440-0
HIJKLMNOPQ-HA-898765432

PREFACE

Prefaces invariably are written last. In our case, too, this seems appropriate, since the subject of our book is automatic controls and dynamic systems, but the subject of the preface is the book.

From our combined teaching experience we judge that there are three important needs in the field of automatic control theory: (1) bridging the theory-applications gap; (2) emphasizing the essential unity of control theory rather than dealing only with specialized areas or stressing modern versus classical approaches; and (3) taking advantage of the most recent developments in the field. We have directed our main efforts in writing this book towards these three considerations.

Automatic control theory, as we see it, is a coherent body of knowledge that acts as a bridge between mathematical abstraction and physical reality. Our common inclination has been to emphasize practical considerations at the expense of mathematical rigor. The generality and elegance of the mathematical approach has great appeal to engineers; however, engineers must deal with the real world. Therefore, we have taken a heuristic approach, explaining the underlying meaning of theorems and illustrating the concepts with simple but, we hope, meaningful examples. Mathematical rigor, however, is preserved where appropriate, and the more theoretically inclined reader can refer to the many cited references. We hope that this approach will contribute to a closing of the often-mentioned "gap" between current theory and practice.

The rapid growth of the control field all too often emphasizes limited areas of interest. We have chosen the wide approach because we believe that broadness of knowledge and imagination is a crucial qualification for systems-oriented engineers and researchers. To successfully carry through design or research projects requires the effective integration of the classic and the modern, advanced and basic techniques, and theoretical and computational approaches. This same rapid growth makes it imperative that each new book be as up-to-date as possible. We have tried to include as much from the wealth of new ideas on the scene as could be incorporated without displacing the necessary introductory material. A particular example is the last chapter, on switching control. We feel that the importance of including this material has not yet been recognized by most American control engineers.

We have used the state-space approach as the main vehicle for most of the theoretical developments in this book. Though there is an obvious chronological distinction between "modern" (or state-space) theory and "classical" control theory, we feel that the schism that has developed between the modernists and the classicists is, in reality, artificial. There is much in common between the two approaches and many areas where they can be profitably joined. Thus, the importance and usefulness of classical theory is enhanced by the generality of the fundamental state-space formulation.

We have intended this book as a senior or graduate-level text and as a reference for practicing engineers. For the teacher of a one-semester senior level course, we recommend the following sequence: chapters 1, 2, 3, 4 (omitting parts of the Lyapunov theory coverage), 5, 8, and 9. We have also found that seniors enjoy the material on switching control (Chapter 15) and it seems to whet their appetites for further work in this area. We thus urge that teachers of undergraduate courses consider *starting* with material from the last chapter.

For use on the graduate level we assume that the student has already had a course on the fundamentals of control theory; if not, the usual accelerated graduate pace should be slowed for the first five chapters, and Chapters 8 and 9 will have to be presented in more detail than they ordinarily would be. For a one-semester (or equivalent) graduate course, then, we recommend covering Chapters 1 and 2 rapidly, 3, 4, 5, and 6, the first half of Chapter 13, and then, depending on the inclination of the class, either 7, 10, or 15. For a two-semester or three-quarter offering, it should be possible to cover anywhere from eleven chapters to the complete book. As a guideline, the material in the first fourteen chapters has been covered in three ten-week quarters, with the class meeting for two eighty-minute sessions per week.

Acknowledgments should be printed in large letters on the cover of every book! To say that we could not even have undertaken this work without the help of many people is an understatement. We particularly thank Professor J. Lowen Shearer, Dr. Rolf Wagner, Roger Bakke, Dr. Walter Loscutoff, Professor Herman Thal-Larsen, Professor Robert Donaldson, Dr. M. Iwama, and Professor René Perret. To our many students, teachers, and colleagues who have inspired, helped, and criticized, to Mrs. Barbara von der Meden, a tireless, cheerful, and accurate typist, to Jim Piles, our helpful editor, and, most of all, to our wives whose contributions are beyond measure, we extend our thanks.

Berkeley, California Y. T.
June 1969 M. R.
  D. A.

CONTENTS

Chapter 1	**Introduction**	1
1–1	Dynamic Systems and Control (history of control)	1
1–2	Basic Concepts and Terminology (system modeling, problem formulation)	5
1–3	Static Behavior and Linearization (causality, steady-state relations, linear approximation)	8
1–4	Response of a First-Order System (the integrator, differential equations, difference equations, free response, convolution)	12
1–5	Signal Flow in a First-Order System (Laplace and z-transforms, block and signal flow diagrams, the zero-order hold)	18
PART 1	**LINEAR SYSTEMS THEORY**	33
Chapter 2	**Mathematical and Graphical Representations of Lumped-Parameter Dynamic Systems**	35
2–1	The State Vector and Vector Differential Equation (the state and output equations, linear and nonlinear, vector-block and signal-flow diagrams)	35
2–2	Linear Relations in the s-Domain (Laplace domain solution, matrix transfer function, the characteristic equation)	41
2–3	State Models from Transfer Functions (signal flow diagram for unfactored transfer functions, realizability condition)	44
2–4	Diagonalization of the A-Matrix (partial fraction form of the transfer function, diagonal form of the system matrix, and the Schwartz form)	48
2–5	Discrete-Time Linear Lumped Systems (difference equations for state and output, z-domain relations, and the pulse transfer function)	50
Chapter 3	**Response of Linear, Lumped-Parameter Stationary Systems**	57
3–1	The Solution Matrix (matrix exponential, time-domain series solution of vector state equation)	57

viii   Contents

	3–2	Solution via the Inverse Laplace Transform 61
		(closed analytical form of the solution matrix, output as a convolution of the impulse response and the input)
	3–3	Diagonalization of the A-Matrix 65
		(modal matrix as a linear transformation, eigenvalues, eigenvectors, modified canonical form for oscillatory system, the Schwartz form)
	3–4	Controllability and Observability 75
		(modal domain relations, analytical conditions for controllability and observability, geometric interpretations)
	3–5	Motion in the State Space 87
		(the eigenvector as a straight-line trajectory, the isocline method, various trajectory patterns)
	3–6	Discrete-Time Systems 101
		(time-domain solution, machine computation of the solution matrix, z-domain solution, Jordan canonical form)

Chapter 4   Stability of Linear Lumped-Parameter Systems 114

	4–1	Trajectory Patterns and Stability 115
		(equilibrium state, stability concept, stability of motion, Lyapunov functions, positive- and positive-semi-definite functions, energy state, geometry in state space)
	4–2	The Lyapunov Theorems 120
		(stability, asymptotic stability, instability, sharpness of the stability test, construction of test functions)
	4–3	Stability of Linear, Stationary, Lumped Systems 125
		(quadratic form, Sylvester condition, hyperellipsoids, necessary and sufficient condition for stability, integrated squared error, time-varying case)
	4–4	The Lyapunov Approach to Discrete-Time Systems 132
		(difference form for test function increment, stability theorem, unit circle condition in the z-plane, squared-error summation)
	4–5	The Routh Test 137
		(characteristic polynomial, two conditions for stability, constructing the array, handling singular situations, transformation for discrete-time system test)

PART 2   FORMULATION OF DYNAMIC SYSTEMS 151

Chapter 5   Mathematical Modeling of Engineering Systems 153

	5–1	The Systems Concept 153
		(definition of a system, interactions, modeling, lumping, causality, linearization)
	5–2	Resistance, Capacitance, and Inductance 162
		(potential and flow variables in energetic interactions, electrical, liquid-flow, gas-flow, thermal, and mechanical systems)

	5-3	Some Examples of *RC*-Systems	127
		(systems with negative real eigenvalues, unilaterally coupled systems)	
	5-4	Reaction Processes and Stability.	178
		(autocatalytic processes, chemical reactors, nuclear reactors, an epidemic process, fermentation processes)	
	5-5	Controllers and Control Laws	187
		(the feedback principle, analog control laws, pneumatic controllers, hydraulic controllers, the human operator)	
Chapter 6		**Bilaterally Coupled Systems**	206
	6-1	Energy Ports and One-Port Elements	206
		(energy bonds, one-port systems, causality, source elements, the bond graph)	
	6-2	Ideal Junction: Series Impedance, Shunt Admittance. . . .	213
		(zero- and one-junctions, two-port elements, the transfer matrix, analogies for mechanical systems, transfer functions)	
	6-3	Transmission Lines and Iterative Chains	226
		(Z-, Y-, ZY- and YZ-chains, uniform fluid lines, state and output equation, overall transfer matrices)	
	6-4	Transducers and Transformers	236
		(two-port energy converter, ideal transformers and transducers, gyrators)	
	6-5	Multiport Systems.	245
		(three-port systems, modulators, amplifiers, unilateral measurement, linkage and gear trains, state-space formulation)	
Chapter 7		**Distributed-Parameter Systems**	266
	7-1	Some Characteristics and Examples of Distributed-Parameter Systems	266
		(mathematical modeling, partial differential equation of YZ-chain, uniform beam, percolation processes, heat exchangers, distillation columns, and highway traffic flow)	
	7-2	Uniform, Two-Port Transmission Line	278
		(solution of vector partial differential equation, wave scattering model, block diagram and transfer function, wave reflection)	
	7-3	A Generalization of the Transfer Matrix Approach	286
		(vector equation with symmetric matrix and its solution, bending-beam problem, diagonalization of nonsymmetric matrix)	
	7-4	Processes with Carrier Flow	291
		(parallel-flow heat exchangers, counterflow heat exchangers, hybrid systems of lumped- and distributed-parameter elements)	
	7-5	Percolation Processes and Some Remarks on Distributed-Parameter Systems	296
		(second-order partial differential equations, example of exact solution, irrational transfer functions, models for lumped approximation, transfer function for distributed input and output)	

x Contents

PART 3   CONTROL OF LINEAR SYSTEMS 315

Chapter 8   Scalar Input–Output Systems and Feedback Control 317

8–1   Structure of a Feedback Control System 317
(single-loop control systems, block diagrams, Laplace domain relations, open-loop transfer function, closed-loop transfer function, ideal control, closed-loop system stability)

8–2   Response of Feedback Control Systems 321
(step-input response pattern, time-response criteria, dominant oscillation, steady-state error, sensitivity, servomechanisms, time-domain solution by the convolution integral)

8–3   The Root-Locus Method 329
(closed-loop pole locations and their corresponding response patterns, general relations for the root locus and basic construction rules, root-locus patterns, stability limit condition, systems with dead time, quarter-wave decay)

8–4   Quality of Control and Its Improvement by Compensation . . 343
(Ziegler-Nichols rules, approximate second-order response, quadratic performance index, parameter optimization, series compensation, feedforward compensation, pole cancellation)

Chapter 9   Frequency Response 356

9–1   Sinusoidal Input and Output 356
(complex representation of sinusoidal steady state, Nyquist plots, Bode diagrams and their construction, nonminimal phase systems, irrational transfer functions)

9–2   The Nyquist Stability Theorem 372
(motion of a phasor, derivation of the theorem, modified Nyquist criterion, examples, singular point, Dzung criterion)

9–3   Open-Loop Approach to Control-System Design 382
(gain and phase margin, lead-and-lag networks, pole cancellation, complex frequency response)

9–4   Frequency Response of Closed-Loop Systems 391
(closed-loop system bandwidth, M_c circle on the Nyquist plane, Nichols chart, response approximation by a Fourier series)

9–5   Frequency Response in State Space 398
(state and output equation for frequency response, trajectory for sinusoidal response, Lissajous figure, invariant axis of rotation)

Chapter 10   Multivariable Control Systems 412

10–1   Transfer-Function Approach to Multivariable Systems . . . 413
(open- and closed-loop relations in terms of the matrix transfer function, classical noninteracting control, decouplability theorem, feedforward control)

10–2	State-Space Formulation 421
	(vector feedback, open- and closed-loop characteristic polynomial, matrix polynomial for resolvent, algorithm to determine control law for completely specified closed-loop poles)
10–3	Modal Control of Lumped-Parameter Objects. 430
	(controller matrix for the purely ideal case, design in the modal domain, mode analyzer and synthesizer, independent modal control, interaction due to nonideality of measurement and control)
10–4	Modal Control of Distributed-Parameter Systems. 446
	(space eigenfunction, measurement of distributed response, application of distributed input, independent modal control of band-limited system, manipulator design)
10–5	Linear Optimal Feedback Control For Quadratic Criteria . . 452
	(performance index, optimal condition in terms of a Lyapunov function, matrix Riccati equation for linear systems, geometric interpretation of the optimal condition)

Chapter 11 **Linear Digital Control**. 467

11–1	Introduction to Digital Control 467
	(Direct Digital Control, discrete-time signals in the frequency domain, sampling theorem, noise aliasing and filter design, quantization error, DDC algorithms)
11–2	Single-loop Digital Control Systems 479
	(parameter tuning, state-space formulation, stability, classical finite-time-settling controller design)
11–3	Control Algorithm for Finite-Time Settling 486
	(state-vector feedback control for time-optimal settling, control law for time-optimal single-loop systems)
11–4	Some Multivariable and Optimal-Control Systems 491
	(multivariable finite-time-settling control law, use of pseudo-inverses for modified finite-time-settling algorithms, modal-domain approach)

PART 4 **NONLINEAR, STOCHASTIC, OPTIMAL CONTROL, AND LOGIC SYSTEMS** 507

Chapter 12 **Nonlinear Systems** 509

12–1	On–Off Control Systems in the State Plane 510
	(piecewise linear objects, switching functions, phase-plane trajectories of switching control, the analytical determination of symmetric limit cycles, improvement of performance, inverse hysteresis, and limit cycles in discrete-time switching control)

xii   Contents

12–2   Oscillation of Nonlinear Systems 523
(the describing function, some examples for switching control systems, amplitude and frequency loci, velocity saturation as an incidental nonlinearity, stability of limit cycles, stability boundary in state-space, jump phenomenon)

12–3   Stability of Nonlinear Systems 536
(the first method of Lyapunov, Krasovskii method, Lure and Letov method, upper bound of quantization error, Popov criterion, circle criterion)

12–4   Positive Use of Nonlinear Action 548
(time-optimal position servo, its realization, Posicast control, adaptive control, system matrix identification)

12–5   Automatic Hill-climbing Control 558
(optimalizing control, Perret–Rouxel's extremal computer, its switching logic, stability, and limit cycles)

Chapter 13   Linear Stochastic Systems 575

13–1   Statistical Considerations. 575
(stochastic variables, probability functions, expected value, Gaussian distribution, ergodic hypothesis, correlation functions, pseudorandom binary noise, properties of Gaussian noise)

13–2   Statistical Relationships in the Frequency Domain 584
(auto- and cross-power density spectra, stochastic input–output relations in the frequency domain, mean squared error)

13–3   Statistical Relationships in the State Space. 591
(Gaussian distribution of random vector noise, covariance matrix, evaluation of mean squared value, shaping filter for colored noise)

13–4   The Principle of Optimum Filtering. 597
(optimum estimate of state, filtering algorithm for first-order discrete-time systems, its derivation, optimum prediction, filter for continuous-time first-order systems)

13–5   Optimum Filter Algorithms 607
(discrete-time systems, continuous-time systems, systems with deterministic input, continuous-time systems with colored measurement noise)

Chapter 14   Optimal Control 624

14–1   Problem Statement and Properties of Optimal Trajectories . . 625
(typical problem statement, the principle of optimality, dynamic programming approach, the maximum principle, property of an optimal trajectory)

14–2   The Maximum Principle 635
(theorem statement, covariant vector and cone of attainability, relation between the maximum principle and dynamic programming)

Contents xiii

14–3	Some Applications of the Maximum Principle.	. . . . .	646
	(mode of optimal control, final time specified case, nonautonomous case, transversality condition, constraint in state-space)		
14–4	Linear Optimal Feedback Control		657
	(derivation of time-varying optimal gain for first-order system, generalization to the n-th order case and its variation)		
14–5	Discrete-Time Optimal Control		666
	(discrete-time maximum principle, linear optimal feedback control of discrete-time linear object, matrix Riccati equation)		
14–6	Optimal Feedback Control of Linear Stochastic Systems. . .		675
	(control algorithm for discrete-time and continuous-time system, derivation of the algorithm for first-order discrete-time system, algorithm to obtain performance, linear optimal control system design)		

Chapter 15 Switching Systems 699

15–1 Logic Elements 699
(electromagnetic relay, diode gate element, transistor gate, fluidic elements, piston logic elements, elementary logical statements and truth tables)

15–2 Elements of Switching Algebra 706
(Boolean algebra and coding, duality, fundamental operations, functions of a single variable, Venn diagrams, functions of two variables and their generalization to n variables, DeMorgan theorem, simplification by a Karnaugh map, numerical representation, minterms and maxterms)

15–3 Combinational Systems 719
(combinational system design, application of the Karnaugh map for simplification, prime implicants, tabulation method, realization of logic function, hazards, bond-graph representation)

15–4 Sequential Systems 736
(synchronous *vs.* asynchronous system, flip-flops, system design)

15–5 Theory of Finite-State Systems 747
(state and transition probability, vector equation for Markov processes, flow diagrams, solution of the equation, ergodic sets, and transient sets)

Appendix Matrix Algebra 759

A–1 Definitions 759
(vector, matrix, addition, and subtraction)

A–2 Multiplication 760
(scalar product of vectors, matrix product)

A–3 Matrix Inversion 762
(condition for existence of inversion, solution of algebraic equations, inverse matrix)

Contents

	A–4	Minimum Right Inverse	765
		(problem statement, minimal-norm solution, proof of the minimal-norm condition)	
	A–5	Minimum Left Inverse	768
		(problem statement, minimal-distance solution, proof of the property)	

MAJOR SYMBOLS 773

INDEX 779

CONTROL
AND DYNAMIC SYSTEMS

1
INTRODUCTION

The field of dynamic systems and controls has recently matured from an art to a science, and is expanding rapidly. We begin this chapter, and the book, with a brief historical review of this growth. Then, to enable us to study this field without overwhelming the reader with definitions, we will introduce some basic concepts and terms on a heuristic basis.

The foundation of our study will be the basic theory of linear lumped systems. To arrive at this point, we shall first consider the technique of linearizing the static characteristics of a device. The concept of the integrator, the basic element necessary for studying the dynamics of physical systems, will follow. While we shall be concerned in this chapter only with first-order systems (that is, systems involving only one integration), we shall briefly introduce the concept of "lumping," which is indispensable in formulating models of higher-order physical systems with a finite number of integrators.

To study the response of single-integrator systems, we will begin by solving the resulting linear and scalar ordinary differential equation of the first order for both the stationary and nonstationary cases. As part of the solution, the convolution integral will be derived. A numerical solution in finite difference form will also be discussed. The chapter will close with an introduction to transformation methods: the Laplace transformation for continuous-time systems and the z-transform method for discrete-time or sampled-data systems.

1-1 DYNAMIC SYSTEMS AND CONTROL

Figure 1-1 spotlights some trends and events in the field of control during the last few centuries. It attempts to place in perspective how and why the field has developed as it has. We have chosen to use a nonlinear vertical time scale in Fig. 1-1, because our natural tendency in looking back to our scientific heritage is to place more weight on recent developments, and also because the growth rate in various branches of science and technology appears to be exponential. Although the three periods shown in the figure occupy roughly equal amounts of space because of the nonlinear time scale, in reality they have a duration ratio of approximately 9 : 3 : 1.

2　Introduction

Period	Year	Control Theory	Control Application	Background
I. Art	1750		Watt Governor and its improvement	Windmill Steam Engine
	1800			Progress in some machinery
	1850			
II. Transition		Maxwell's paper Analysis of telescope control system	Regulators related to power generation and transmission	Electric power
	1900		Autopilot for airplane	
		Book on speed control Use of differential equations and Routh–Hurwitz criteria on some simple systems	Instrument and regulators for process and power industries Controls for communication Servomechanism	World War I Progress in industry
III. Science		Ziegler-Nichols' method	Controls for weapon	World War II
	1950	Laplace domain approach and frequency response method Root locus	Electronic controllers Plant and processes with controls as essential part	Nuclear power Computers
		z-transform method	Data-logging	
	1960	State space approach— Lyapunov concept, optimum control theory, and mathematical theory of control processes	Digital computer for computing control Direct digital control	Automation Space projects Systems and control concept in biomedical and various other fields
		Detailed analyses of optimal controls	Progress towards dynamic optimization Software developments	
				Man on the Moon
	1970			

Fig. 1-1　Some events in the history of control.

The first period extends from the late eighteenth to the early twentieth century. During this time most of the progress in the field of control was empirical in nature. The Watt governor applied to steam engine control was perhaps the single most important development in this era of control. Because of its proportional-control action (see Chapter 8), the Watt governor resulted in static error of the engine speed. Attempts to eliminate this error led to the *art* of integrating control action, which, however, turned out to be unstable when applied to steam engines. Without current theory and the insight it affords, early experimenters had to rely entirely on art and intuition. Accordingly, progress in obtaining solutions to early problems was slow. One exception is an early paper by J. C. Maxwell, entitled "On Governors" [1],* which represents the beginning of theoretical development.

Period II extends from about 1900 to the start of World War II, encompassing an age when industrial progress was gaining momentum. This era was characterized by large-scale power generation and transmission, the birth of the aeronautics industry, rapid growth in the processing and chemical industries, and the advent of communications and electronic engineering. Such technological progress created an increasing demand for instruments and regulators to which a new industry was quick to respond. Some of the resulting designs for pneumatic process controllers and hydraulic and electric regulators that appeared in the middle of Period II are basic even today.

The increasing complexity of engineering systems both required and justified the use of theory in conjunction with intuition for the design of control systems. The publication of Max Tolle's book on speed regulation [2] was one of the first attempts in this direction. The rate of publication of books and papers on control gradually increased, although the total number of publications over the entire Period II was less than the present annual output. The use of theoretical methods was limited mostly to solutions of ordinary differential equations and to applications of the Routh–Hurwitz stability criteria (see Section 4–5). The application of such theoretical methods (with manual computations) was limited to simple and low-order systems. The control theorists engaged in such analysis were a minority group compared to the "handbook" and "steady-state" engineers. To some of the group, however, it was nevertheless evident by the end of Period II that a new field of theory was emerging which unified all feedback control applications [3].

The third and present period began with the initial development of this new field of theory and with the fresh insight into the area of automatic controls as a potential science. Both the development and the application of theory experienced a phenomenal growth rate during World War

---

* Numbers in brackets are keyed to the references at the end of each chapter.

II. We therefore date the beginning of Period III from the early 1940's, even though the use of the frequency-response method (see Chapter 9) predated the war by several years. This method, which characterized the initial stage of Period III, was originally developed in the communications field, but was heavily applied to weaponry servomechanism (position-control system) design during World War II. Frequency-response techniques bypassed the main difficulty caused by high-order differential equations, and, with their applications, feedback control theory as a science came of age. The boundaries of this science were vastly enlarged by Dr. N. Wiener's concept of "Cybernetics" [4], incorporating new problem areas to which the feedback control point of view could be applied.

The development of the frequency-response techniques was soon followed by the introduction of Evans' root-locus method ([5, 6, 7, 8] and Chapter 8). Basic feedback control theory quickly advanced to include such problems as sampled-data control, random-signal systems, and some phenomena caused by system nonlinearities and nonlinear control actions ([9 through 16] and Chapters 8 through 12). Throughout Period II many engineers believed that control devices were relatively unimportant and would yield only slight performance improvement when added to an existing system. This shortsightedness disappeared during Period III, to the extent that most engineering companies now include separate groups which are concerned primarily with system controls and dynamic response. Since World War II, automatic control systems, with and without feedback, have become an integral part of the overall design of dynamic systems. Progress in this area has been accelerated by the use of analog and digital computers and switching logic for direct control purposes [17, 18, 19].

The late 1950's marked the rediscovery of the state-space point of view and the adoption of an entirely new approach to control problem solutions. Originally developed in the field of classical mechanics by Poincaré, Lyapunov, Gibbs, and others fifty to eighty years earlier, the method was found to have great utility in the analysis and design of control systems [20, 21, 22]. A particularly important aspect of this new approach was the development of the theory of static and dynamic optimization (see Chapter 14) resulting from an ever-increasing demand for improved quality of control. This new state-space viewpoint, based on old techniques, acquired the label of "modern control theory." Besides the possibility of becoming inappropriate a generation from now, the use of this expression has the unfortunate effect of relegating what had come before to old-fashioned and second-class status. This will certainly not be our intention, even though we shall occasionally refer to work in the state space as "modern" for want of a more widely accepted short description. The theory initially developed during Period III, in the frequency and Laplace domains, is often referred to as "classical" or "conventional" control theory. We shall retain this nomenclature with the clear understanding that in certain applications the classical approach is still to be preferred over the modern approach. Throughout this book we shall endeavor to use the two different approaches interchangeably, as each individual application demands. In the process, it should become clear that there is still much valuable classical control theory (some probably as yet undiscovered) which is valid and of current utility.

With the application of the state-space point of view to the control area, there has been a growing interest in the purely theoretical. In fact, the mathematical level of sophistication has reached a stage where some recent literature treats completely abstract and nonreal systems. This is a far cry from the humble beginnings of the controls area with Watt's experimentally developed governor. To continue to perform an engineering function, the engineer working with the new and elegant theory must keep in mind the goal of *application*.

1-2 BASIC CONCEPTS AND TERMINOLOGY

We learn from history that both theory and practice are important in many control- and systems-engineering problems. We attack a typical problem by defining an objective, which helps us to determine the most feasible theoretical approach. For instance, a system consisting of a spring attached to a mass may be treated as a *point mass* if the system is thrown out into space and if we are interested only in its flight trajectory. On the other hand, if the spring is attached to a fixed frame and a force acting on the mass causes an oscillation, we may focus our attention on this particular motion. Several possible approaches which appear at this stage of mathematical modeling are illustrated in Fig. 1-2.

The first step in many cases is to decide whether a system can be assumed to be linear or nonlinear (Fig. 1-2a). In the spring-mass system of the figure, the relation between force $u(t)$ and displacement $x(t)$ is *linear* if $x(t)$ obtained for forces $u_1(t)$ and $u_2(t)$ acting together $[u(t) = u_1(t) + u_2(t)]$ is the sum of displacements $x_1(t)$ and $x_2(t)$ obtained for forces $u_1(t)$ and $u_2(t)$ acting independently. Such superposition does not hold when the system is *nonlinear*. As an example, the spring-mass system becomes nonlinear if its spring constant varies with deflection, as shown at the right-hand side of Fig. 1-2a.

A causal relation of variables is assumed at this stage. In the spring-mass system, force is assumed to cause a change in displacement. In other words, force $u(t)$ is an *input*, and displacement $x(t)$ is an *output*. We shall return to this concept of cause and effect in subsequent chapters.

Another important aspect of modeling is the *lumping* of system parameters. The mass can experience elastic deformation, and the spring may have some inertia. The process of separating the mixed effects and considering the system as composed of idealized elements, such as an ideal

6 Introduction

mass (no elasticity) and an ideal elasticity (no inertia), is called *lumping*. The opposite extreme to lumping is a model represented by an elastic rod consisting of infinitely small inertia and elastic elements combined in series. In such a model, which illustrates a *distributed parameter system* (Fig. 1–2b), partial rather than ordinary differential equations are used.

A system parameter may change as a function of time, as in the case of a spring supporting a leaking water tank. A system whose parameters change with time is called *nonstationary* or *time variant*. Otherwise, the system is called *stationary* (Fig. 1–2c).

Let us turn our attention to the pattern of variable changes. This pattern may be either *deterministic*, as in the case of a step or sinusoidal change,

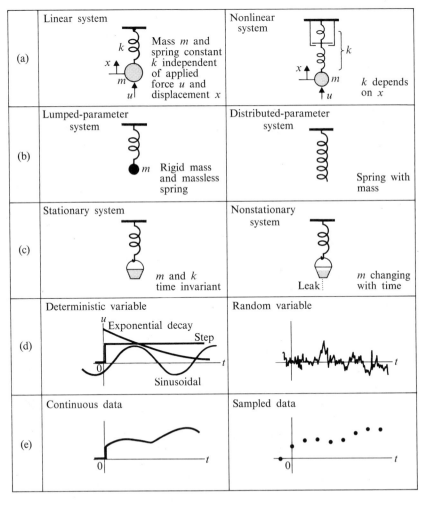

Fig. 1–2 Some classifications for mathematical modeling.

or *random*, where the value of a variable at each instant is given on a probabilistic basis (Fig. 1–2d). The change of a variable, which usually occurs *continuously* with time, can be made intermittent (discrete) by *sampling* (Fig. 1–2e). Also, a variable may take any value over a given range, or the value may be limited to some quantized, distinct points, as is the case for digital signals or binary-coded data. As we shall see in Chapter 15, there are only two values in switching logic: ONE or ZERO.

Whether a theoretical system will be simple or complex very often depends on the purpose of the investigation. For example, a simple dynamic mathematical model (linear, lumped, stationary, and deterministic) may apply to a system of molten metal in a pot if we merely seek the overall cooling trend of the liquid metal over some limited temperature range. On the other hand, it is obvious that there are many other possibilities and requirements of the investigation which might make the model of the system very complex, even beyond the capabilities of existing methods and computing techniques.

The complexity of a problem may also depend on the purpose of the project. The design of an adequately stable feedback control system for a simple object can be achieved by manual computation. However, if we want to optimize the system based on some kind of performance index, then the use of a computer may become necessary. Moreover, a meaningful performance index is sometimes very difficult to define. Engineers are

Table 1-1 SYSTEM RELATIONS

u → [Filter or controller] — α, J, t → (Plant: x, ẋ) → y
v →

Problem	α	v	u	x	y	J	t
Transfer function			○——▶○				
Observability				○◀——○			
Controllability			○——▶○				
Invariance			○——▶○--▶○				
Parameter sensitivity	○——————▶○--▶○						
Optimization	○◀--------○◀——					○◀--○	

often forced to be satisfied with incomplete solutions to problems which cannot be adequately defined.

The above discussion should lead us to the conclusion that the simplest possible problem formulation is not only a logical first step in learning but is also a valuable tool when properly applied in treating some specific aspects of a problem. Thus, in the design of the components of a complex system, or even in the overall design of the whole system, a first approximation of the problem in simplest possible terms often is the initial step in arriving at a satisfactory solution.

As noted in the previous paragraphs, the problem formulation and the purpose(s) of the investigation are intimately related. Table 1–1 has been constructed to offer some insight into the various possible relationships. In this table, it is assumed that there are some yet-to-be-determined overall system relations of the form

$$F(y, x, \dot{x}, u, v, \alpha, J, t) = 0,$$

where

F denotes a matrix relationship,
y an output (measured) vector,
x a state (internal variable description of the system) vector,
u a control (input) vector,
v a disturbance (noise) vector,
α system parameters,
J the performance index, and
t is time.

In a first-order, single-variable system the vectors become scalars. The state vector x and the terms *transfer function*, *controllability* and *observability* in Table 1–1 will be discussed in detail in Chapters 2 and 3. Shown in this table are the forms of the desired relation F for the various problems indicated.

1-3 STATIC BEHAVIOR AND LINEARIZATION

Our introduction to the dynamic analysis of systems and components will begin with a brief discussion of static input–output relationships and static elements. Let us consider, for example, a thermometer. Its temperature reading $x(t)$, which represents the state of the system, is the variable of interest. This temperature reading results from the imposition of an ambient temperature, $u(t)$, on the system; thus, $u(t)$ is the input and $x(t)$ is the output (or response) of the thermometer. The causal relationship can be described graphically by the two kinds of diagrams shown in Fig. 1–3(a), where the direction of causality is indicated by the arrowhead. In the block-diagram representation, signals appear as lines, and functional relations as

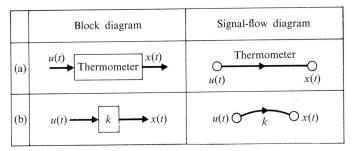

Fig. 1-3 Input and output.

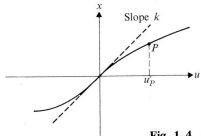

Fig. 1-4 A static characteristic.

blocks; while in the signal-flow diagram, signals are represented by small circles and functional relations by labeled lines.

In the expressions $u(t)$ and $x(t)$, (t) means that both the input u and the output x are functions of time t. A calibration chart for a thermometer presents a relationship between the steady-state values of the input (u_{ss}) and output (x_{ss}) rather than the full, time-dependent relationship between input and output. In this steady-state relationship,

$$x_{ss} = f_s(u_{ss}). \tag{1-1}$$

In Eq. (1-1), $f_s(u_{ss})$ is a single-valued function of u_{ss}, as shown in Fig. 1-4, where the origin is taken at any specific temperature (arbitrarily set equal to zero here). Equation (1-1) applies to the class of dynamic systems and components for which such *static* relations hold.

If the temperature $u(t)$ to be measured is changing with time, the thermometer response $x(t)$ lags the input, and thus the static relation does not apply. However, if the input $u(t)$ is changing slowly enough, or if the thermometer has a fast enough response, a static equation gives an approximate relation between a changing input $u(t)$ and a changing output $x(t)$:

$$x(t) = f_s(u(t)). \tag{1-2}$$

The error due to this static approximation depends on the time scale; perhaps t must be measured in hours or days for a slowly responding thermometer.

An *ideal* component, for which Eq. (1–2) holds for all t, is called a *static-no-memory* component. Zero memory in this case means that no hysteresis or backlash type phenomena are present, so that for each $u(t)$ there is a unique $x(t)$ that does not depend on past inputs.

Let us suppose that the time scale of interest in the study of a dynamic system is such that the static relationship of Eq. (1–2) applies with sufficient accuracy. We can further simplify the input–output relationship by assuming a proportionality between the variables:

$$x(t) = ku(t), \qquad (1\text{–}3)$$

where k is the proportionality constant, often called the *gain constant*. (The term "gain" as used in the context of the frequency-response method of Chapter 9 has a broader meaning; the gain defined here must be referred to as "zero-frequency gain" or "d-c gain.") A geometric interpretation of this gain idealization is shown in Fig. 1–4. An experienced engineer may choose the gain as the mean slope of the static characteristic over the range of interest of input u. Mathematically, it is given by the slope of the tangent at the origin of the diagram if the curve $f_s(u)$ is smooth through the origin:

$$k = \left.\frac{df_s(u)}{du}\right|_{u=0}. \qquad (1\text{–}4)$$

The process of replacing Eq. (1–2) by Eq. (1–3) is one form of linearization. The relation linearized by means of Eq. (1–4) holds only in the vicinity of the origin. If our interest is focused on the vicinity of a point P removed from the origin (Fig. 1–4), we take P as a new origin and apply Eq. (1–4) with $u = u_P$.

The proportional relationship of Eq. (1–3) can be represented by the diagrams in Fig. 1–3(b). The informal description of Fig. 1–3(a) is thus reduced to a constant k. This is the simplest idealized relation between an input and an output. With this proportionality relationship, the response pattern becomes an exact replica of the input pattern with a scale factor k; there is no delay or distortion of the signal pattern between the input and output. Typical units of the scale factor k might be cm/°C in this case.

Mathematically, it is possible to write the inverse relationships of Eqs. (1–2) and (1–3):

$$u(t) = f_s^{-1}(x(t)) \qquad (1\text{–}5)$$

and

$$u(t) = k^{-1}x(t). \qquad (1\text{–}6)$$

However, since the input–output concept is based on a physically realizable causality, its inverse form does not exist in a physical system unless the causality is reversible. For instance the causal relationship in a thermometer is definitely irreversible, whereas the ideal link of Fig. 1–5 may have a reversible causality. In this figure $k = b/(a + b)$ for small deflections.

We can generalize the concept discussed so far for one input and one output to r inputs and n outputs. For a nonlinear, static, multivariable system the relations that correspond to Eq. (1–2) become:

$$x_i(t) = f_{si}(u_1(t), \ldots, u_r(t)), \quad i = 1, \ldots, n. \quad (1\text{-}7)$$

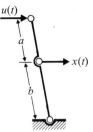

Fig. 1–5 A link.

This equation describes n hypersurfaces (not representable graphically if $r > 2$) in an $(r + 1)$-dimensional space. If the surface (which may be warped due to nonlinearity) is smooth, and is replaced by a tangent hyperplane at the origin, a linearized relation results:

$$x_i(t) = \sum_{j=1}^{r} k_{ij} u_j(t), \quad i = 1, \ldots, n, \quad (1\text{-}8)$$

where

$$k_{ij} = \left. \frac{\partial f_{si}(u_1, \ldots, u_r)}{\partial u_j} \right|_{u_1=0, \ldots, u_r=0}. \quad (1\text{-}9)$$

Example 1–1 *A control valve* [23]. Turbulent flow through a valve can be described by the equation

$$Q = \eta A \sqrt{\Delta P}, \quad (1\text{-}10)$$

where Q is the total flow rate through the valve, A is the area of the valve opening, ΔP is the pressure drop across the valve, and η is a coefficient. It is obvious that A is an input and Q is the output. In the case at hand we may have another input: the supply pressure to the valve. For simplicity let us assume the downstream pressure to be constant.

In order to linearize Eq. (1–10), small deviations u_1, u_2, and x must be introduced such that

$$u_1 = (A - A_0), \quad u_2 = (\Delta P - \Delta P_0), \quad x = (Q - Q_0),$$

where A_0, Q_0, and ΔP_0 are a set of variable values corresponding to some nominal conditions called the operating point. Substituting the deviations in Eq. (1–10), we get a specific form of Eq. (1–7),

$$x(t) = \eta(A_0 + u_1(t))\sqrt{(\Delta P_0 + u_2(t))} - Q_0. \quad (1\text{-}11)$$

Applying Eq. (1–9), we can linearize the relation to

$$x(t) = k_1 u_1(t) + k_2 u_2(t), \quad (1\text{-}12)$$

where

$$k_1 = \left. \frac{\partial Q}{\partial A} \right|_{A_0, Q_0, \Delta P_0} = \left. \frac{\partial x}{\partial u_1} \right|_{x=0, u_1=0, u_2=0} = \eta \sqrt{\Delta P_0}$$

and

$$k_2 = \left. \frac{\partial Q}{\partial (\Delta P)} \right|_{A_0, Q_0, \Delta P_0} = \left. \frac{\partial x}{\partial u_2} \right|_{x=0, u_1=0, u_2=0} = \frac{1}{2} \frac{\eta A_0}{\sqrt{\Delta P_0}}.$$

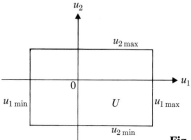

Fig. 1-6 Admissible range U.

The gains k_1 and k_2 include values of variables at the nominal operating point and hence are constants.

The limiting effect of saturation is one of the most important factors to be considered in dealing with control valves and other control components. For example, the range of signal (air pressure, electric current, etc.) used to operate a valve as well as the valve stroke itself is bounded. Thus one must recognize that the magnitude of any admissible signal is limited. When, for example, two controlling elements (such as incremental area and pressure in Example 1-1) are used in a system, the admissible input signal range may be defined by a rectangular zone U as in Fig. 1-6. Such constraint is symbolized by

$$(u_1, u_2) \in U, \qquad (1\text{-}13)$$

where the symbol $\in$ is read "must be contained in."

There is a class of controlling components in which U can take on end values only: open or shut, ON or OFF, or plus or minus. Such a construction or mode of operation results in "bang-bang" or switching control. The linearization method discussed above does not apply to these components, and we will have to develop special techniques for handling such nonlinearities (see Chapters 12 and 15).

1-4 RESPONSE OF A FIRST-ORDER SYSTEM

As was stated in Section 1-3, static behavior is a special case of dynamic behavior, the latter being the basis of control theory. In this section we will introduce the integrator, and then derive the first-order differential equation for a single-integrator system. The general solution of a first-order linear differential equation will be given, followed by discussion of output response patterns for step and impulse inputs. A sampled-data input–output relation, suitable for digital computer solution, will then be briefly introduced.

The *integrator* is the basic building block used in developing the mathematical model of a physical system. Its output $x(t)$ is the integral of the input

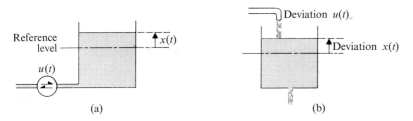

Fig. 1-7 A liquid tank as integrator.

$u(t)$. The equation relating $x(t)$ and $u(t)$ is:

$$\frac{d}{dt}x(t) = Bu(t), \quad (1\text{-}14)$$

where B is a constant or time-dependent coefficient. The water tank in Fig. 1-7(a) is a physical model of an integrator which, because of its physical simplicity, we will often use in examples throughout this book. The input $u(t)$ in this system is a volumetric-supply flow rate of liquid. It may take both positive and negative values if a reversible positive-displacement pump is used as the input device. The output $x(t)$ represents the tank level measured from an arbitrary reference point. A mass balance taken over a differential time increment dt immediately yields Eq. (1-14), where B is equal to the reciprocal of the free surface area. To obtain a linear relationship for a lumped mass balance and a tank with vertical sides, we must assume that the surface is perfectly flat. By ignoring any local level differences caused by liquid motion we may consider the storage action of the tank as a single lump.

In order to derive the general form of a first-order ordinary differential equation, let us consider a tank with a free discharge (Fig. 1-7b). Equation (1-14) is now generalized to the form

$$\frac{d}{dt}x(t) = f(x(t), u(t), t), \quad (1\text{-}15)$$

which means that the rate of change of the tank level is a function of $u(t)$, $x(t)$, and time t. The relation can be linearized in the vicinity of a normal equilibrium point [where $dx(t)/dt = 0$] of the system by means of the technique delineated in Section 1-3. For small deviations of the signals $u(t)$ and $x(t)$ from an equilibrium state,

$$f(x(t), u(t), t) \approx A(t)x(t) + B(t)u(t),$$

where

$$A(t) = \left.\frac{\partial f(x, u, t)}{\partial x}\right|_{u=0, x=0}, \quad B(t) = \left.\frac{\partial f(x, u, t)}{\partial u}\right|_{u=0, x=0}.$$

14   Introduction

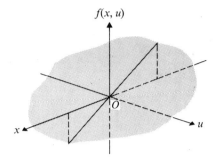

Fig. 1-8 A plane $f(x, u) = Ax + Bu$, $A < 0$, $B > 0$.

Geometrically, the surface $f(x, u, t)$ in three-dimensional space is thus approximated by a tangent plane in the vicinity of the origin (Fig. 1–8). The surface must be smooth in order for this approximation to have meaning. With this linearization, Eq. (1–15) becomes

$$\frac{d}{dt} x(t) = A(t)x(t) + B(t)u(t) . \tag{1-16}$$

A further simplification can be made by assuming both A and B to be constant. The equation then becomes linear and stationary:

$$\frac{d}{dt} x(t) = Ax(t) + Bu(t) . \tag{1-17}$$

Applied to the tank of Fig. 1–7(b), $Ax(t)$ in the equation represents the discharge flow rate. Therefore A must be negative for this system.

The solution of Eq. (1–17) for general $u(t)$ is

$$x(t) = e^{At}x_0 + \int_0^t e^{A(t-\tau)} Bu(\tau) \, d\tau , \tag{1-18}$$

where x_0 is the initial value of $x(t)$, and τ is a dummy variable of integration. Equation (1–18) can be derived by starting with the equality

$$\frac{d}{dt}(e^{-At}x) = e^{-At}\left(\frac{dx}{dt} - Ax\right).$$

Substituting Eq. (1–17) in the right-hand side of this equality, we get

$$\frac{d}{dt}(e^{-At}x) = e^{-At} Bu(t) .$$

After integration, the latter equation becomes

$$e^{-At}x - x_0 = \int_0^t e^{-A\tau} Bu(\tau) \, d\tau ,$$

which reduces to Eq. (1–18) (see [24] for a rigorous derivation).

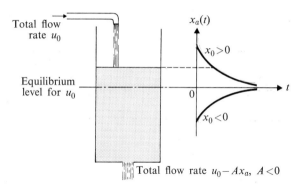

Fig. 1-9 Free response $x_a(t)$.

The first term $e^{At}x_0$ of the solution is called the free response. This is the response when the system is not forced, that is, when $u(t) = 0$. The second term is the forced response with a zero initial value,

$$x_f(t) = \int_0^t e^{A(t-\tau)} Bu(\tau)\, d\tau, \qquad x_f(0) = 0, \qquad (1\text{-}19)$$

and is called the convolution integral. The total response is given by the sum of the two components. Such a summation, valid only for linear systems, is called a linear superposition. It also applies when the input $u(t)$ is given as a sum of components $u_1(t), \ldots, u_k(t)$,

$$u(t) = \sum_{i=1}^{k} u_i(t).$$

The forced response is then

$$x_f(t) = \sum_{i=1}^{k} x_{fi}(t),$$

where

$$x_{fi}(t) = \int_0^t e^{A(t-\tau)} Bu_i(\tau)\, d\tau.$$

The pattern of the free response

$$x_a(t) = e^{At}x_0 \qquad (1\text{-}20)$$

is an exponential decay when A is negative. An initial deviation x_0, either positive or negative, of the tank level from an equilibrium level exponentially decreases with time (Fig. 1-9).

Figure 1-10 shows the forced response for a step and an impulse input. If $u(t)$ is a step of a magnitude h,

$$u(t) = h \text{ for } t > 0, \qquad u(t) = 0 \text{ for } t < 0,$$

and $u(t)$ is discontinuous and undefined at $t = 0$. For this input, Eq. (1-19)

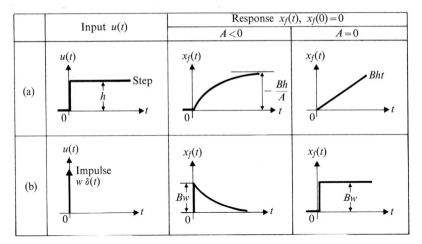

Fig. 1-10 Some patterns of forced response.

gives the response:

$$x_f(t) = e^{At} \int_0^t e^{-A\tau} Bh \, d\tau = \frac{Bh}{-A}(1 - e^{At}). \tag{1-21}$$

When $A = 0$, the forced response becomes a pure integration of the step input or a ramp output as shown in Fig. 1-10(a) on the right. This may be corroborated by applying l'Hôpital's rule to Eq. (1-21).

If a finite amount of liquid is instantaneously added at $t = 0$, we expect the level (in the ideal situation of no agitation) to go up immediately, followed by a free response for $t > 0$ (Fig. 1-10b). An idealized impulse signal of this sort occurring at $t = 0$ is described by a Dirac delta function $\delta(t)$ of unit total area (for a unit volume of liquid in our example):

$$\int_{-\varepsilon}^{+\varepsilon} \delta(t) \, dt = 1, \qquad \varepsilon > 0$$

$$\delta(0) = \infty \tag{1-22}$$

$$\delta(t) = 0 \qquad \text{for all} \quad t \neq 0.$$

(See [22] for a rigorous definition of the delta function.) The impulse input for w units of liquid is

$$u(t) = w\delta(t),$$

and the forced response is

$$x_f(t) = e^{At} \int_{0^-}^t e^{-A\tau} Bw\delta(\tau) \, d\tau = e^{At} Bw, \tag{1-23}$$

where 0^- means the integral is to start at $-\varepsilon$ to include the delta function.

1-4   Response of a first-order system

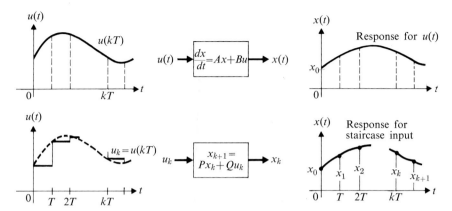

Fig. 1-11 Differential equation vs. difference equation.

Bw is the instantaneous level change at $t = 0$, but it is equivalent to x_0 in Eq. (1-20) in the final form of the free solution. Therefore, an impulse input at $t = 0$ has the effect of introducing an initial condition into an unforced mathematical model.

With the ever-increasing application of digital computers to data-processing and direct control, sampled-data theory is gaining in importance. Instead of considering the input–output signals over a continuous-time domain, we now focus our attention on sets of samples of the signals taken at discrete intervals of time. For instance, we replace a signal $x(t)$ by sampled data $x_0, x_1, \ldots, x_k$, as shown in Fig. 1-11, where the time interval (called the sampling period) is assumed constant. We shall show that a differential equation in the continuous-time domain becomes a difference equation in the discrete-time domain.

In order to derive a difference equation which corresponds to Eq. (1-17), we assume that an input $u(t)$, which is usually smooth, can be approximated by a series of steps (a "staircase") without causing any serious error (Fig. 1-11). Hence,

$$u(t) = u_k \text{ (const)} \qquad \text{for} \qquad kT \leqslant t < (k+1)T, \qquad (1\text{-}24)$$

where $u(t)$ is not defined at the points of discontinuity. We now apply Eq. (1-18) over a sampling period from kT to $(k+1)T$ and determine

$$x_{(k+1)} = x[(k+1)T],$$

$$x_{(k+1)} = e^{AT} x_k + \int_{kT}^{(k+1)T} e^{A[(k+1)T-\tau]} Bu_k \, d\tau$$

$$= e^{AT} x_k + (e^{AT} - 1) A^{-1} Bu_k$$

18 Introduction

or
$$x_{k+1} = Px_k + Qu_k, \tag{1-25}$$
where
$$P = e^{AT} \quad \text{and} \quad Q = (P-1)A^{-1}B.$$

The computational scheme associated with Eq. (1–25) is graphically depicted in Fig. 1–14(a), which will be further discussed along with z-transform methods in the next section.

The response x_k at time $t = kT$ can be computed by repeated application of Eq. (1–25) in the following way:

$$\begin{aligned} x_1 &= Px_0 + Qu_0 \\ x_2 &= Px_1 + Qu_1 = P^2 x_0 + PQu_0 + Qu_1 \\ &\vdots \\ x_k &= P^k x_0 + \sum_{i=0}^{k-1} P^{k-1-i} Qu_i. \end{aligned} \tag{1-26}$$

The staircase approximation of $u(t)$ is the only assumption in the derivation of the solution. (See [25] for more accurate approximations.) If the system is free,

$$x_k = P^k x_0 \tag{1-27}$$

completely agrees with the free response $x_a(t)$ of Eq. (1–20) at all sampling instants.

1-5 SIGNAL FLOW IN A FIRST-ORDER SYSTEM

For convenience we denote the derivative operation d/dt by the operator D; for example, Eq. (1–14) would be written

$$Dx(t) = Bu(t).$$

Denoting the integral $\int_0^t dt$ by $1/D$, we may write an integrator response

$$x(t) = x_0 + \int_0^t Bu(t)\, dt$$

in the form

$$x(t) = x_0 + \frac{1}{D} Bu(t). \tag{1-28}$$

The block and signal-flow diagrams of Fig. 1–12(a) describe the causal relation of an integrator using this operator notation. The relation can be generalized for the case of Eq. (1–17). We write

$$x(t) = x_0 + \frac{1}{D}\left(Ax(t) + Bu(t)\right) \tag{1-29}$$

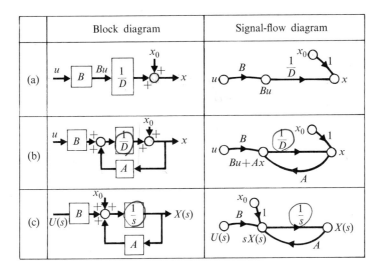

Fig. 1-12 Graphical representations of a first-order, linear, stationary, continuous-in-time system.

and obtain the diagrams of Fig. 1-12(b). Note that the signal-flow diagram in the figure can be easily converted into an analog computer program, where x_0 is introduced as an initial charge on a feedback capacitor. We delete x_0 from the equations and diagrams if $1/D$ is defined for an indefinite integral, $\int (\ )\, dt$.

The Laplace transform of a function of time $f(t)$, written $\mathscr{L}[f(t)] = F(s)$, converts a differential equation into algebraic form. The result of the transformation is somewhat similar to the form arrived at by use of the derivative operator D. The Laplace transform of a signal $f(t)$ is defined by

$$F(s) = \int_0^\infty f(t) e^{-st}\, dt, \qquad (1\text{-}30)$$

where the signal $f(t)$ is assumed to exist only for positive time, $0 \le t < \infty$, and where $s = \sigma + j\omega$ is complex-valued. The signal pattern with which we are concerned must be included in this range. If, for example, $f(t)$ is a unit step, we start the integration immediately after the discontinuous change that occurs at $t = 0$, and obtain [assuming $\mathscr{R}e(s) > 0$]

$$\mathscr{L}[\text{unit step}] = \int_{0^+}^\infty e^{-st}\, dt = \frac{e^{-st}}{-s}\Big|_{0^+}^\infty = \frac{1}{s}. \qquad (1\text{-}31a)$$

Laplace transformation thus converts a function in the time domain to one in the s-domain. The integration must start at $t = 0^-$ if $f(t)$ is a unit impulse defined by the delta function of Eq. (1–22). Thus,

$$\mathscr{L}[\text{unit impulse}] = \mathscr{L}[\delta(t)] = \int_{0-}^{\infty} \delta(t)e^{-st}\,dt$$

$$= \int_{-\varepsilon}^{+\varepsilon} \delta(t)\,dt = 1 \qquad (1\text{--}31\text{b})$$

because $e^{-st} = 1$ for $-\varepsilon \leqslant t \leqslant \varepsilon$ when $\varepsilon \to 0$.

In general, we allow s to be complex with positive real part so that the integral of Eq. (1–30) has meaning; it must, in particular, satisfy the condition $f(t)e^{-st} \to 0$ as $t \to \infty$. With this assumption satisfied, we have (integrating by parts)

$$\mathscr{L}\left[\frac{d}{dt}f(t)\right] = \int_0^{\infty} f'(t)e^{-st}\,dt$$

$$= f(t)e^{-st}\Big|_0^{\infty} + s\int_0^{\infty} f(t)e^{-st}\,dt = -f(0) + sF(s). \qquad (1\text{--}32)$$

The Laplace transform can be used to derive the solution in the s-domain for $t \geqslant 0$ of a constant-coefficient linear differential equation. For Eq. (1–17), with initial condition $x(t) = x_0$ when $t = 0$, letting

$$\mathscr{L}[x(t)] = X(s) \quad \text{and} \quad \mathscr{L}[u(t)] = U(s),$$

we have $\frac{dx(t)}{dt} = Ax(t) + Bu(t)$

$$sX(s) - x_0 = AX(s) + BU(s),$$

so that

$$X(s) = (s - A)^{-1}(x_0 + BU(s)). \qquad (1\text{--}33)$$

The relation is shown graphically in Fig. 1–12(c). Note the similarity of s to D in Fig. 1–12(b) and (c). Also, note that the initial value x_0 in Fig. 1–12(c) can be looked upon as an impulse input signal of magnitude x_0, because $\mathscr{L}[x_0\delta(t)] = x_0$ in the s-domain. If $x_0 = 0$, the input–output relation of a system can be conveniently represented by the ratio

$$\frac{X(s)}{U(s)} = G(s).$$

This is called the system-transfer function. For example,

$$G(s) = \frac{B}{s - A} \qquad (1\text{--}34)$$

for the first-order system of Fig. 1–12(c).

The inverse process of the Laplace transform, $f(t) = \mathscr{L}^{-1}[F(s)]$, is defined by a contour integral in the complex s-plane [24, 25]. Use of a table (Table 1–2) in which Laplace transform pairs are listed (see [26, 27] for

1-5 Signal flow in a first-order system

Table 1-2 LAPLACE AND $\mathscr{Z}$-TRANSFORMATION PAIRS

	$F(s)$	$f(t) = \mathscr{L}^{-1}[F(s)]$	$f_{(kT)} = f_k$	$\mathscr{Z}[f_k] = F(z)$
1	$\dfrac{1}{s}$	1	1	$\dfrac{z}{z-1}$
2	$\dfrac{1}{s^2}$	t	kT	$\dfrac{Tz}{(z-1)^2}$
3	$\dfrac{1}{s^3}$	$\dfrac{1}{2}t^2$	$\dfrac{1}{2}(kT)^2$	$\dfrac{T^2}{2}\dfrac{z(z+1)}{(z-1)^3}$
4	$\dfrac{1}{s+a}$	e^{-at}	$c^k, \quad c = e^{-aT}$	$\dfrac{z}{z-c}$
5	$\dfrac{1}{(s+a)^2}$	te^{-at}	$(kT)c^k, \quad c = e^{-aT}$	$\dfrac{cTz}{(z-c)^2}$
6	$\dfrac{1}{(s+a)^3}$	$\dfrac{1}{2}t^2 e^{-at}$	$\dfrac{1}{2}(kT)^2 c^k, \quad c = e^{-aT}$	$\dfrac{T^2}{2}\dfrac{cz(z+c)}{(z-c)^3}$
7	$\dfrac{a}{s(s+a)}$	$1 - e^{-at}$	$1 - c^k, \quad c = e^{-aT}$	$\dfrac{(1-c)z}{(z-1)(z-c)}$
8	$\dfrac{a}{s^2(s+a)}$	$\dfrac{1}{a}[at-(1-e^{-at})]$	$\dfrac{1}{a}[kaT-1+c^k], \quad c = e^{-aT}$	$\dfrac{Tz}{(z-1)^2} - \dfrac{(1-c)z}{a(z-1)(z-c)}$
9	$\dfrac{s}{(s+a)^2}$	$(1-at)e^{-at}$	$(1-kaT)c^k, \quad c = e^{-aT}$	$\dfrac{z^2 - c(1+aT)z}{(z-c)^2}$
10	$\dfrac{a^2}{s(s+a)^2}$	$1 - (1+at)e^{-at}$	$1 - (1+kaT)c^k, \quad c = e^{-aT}$	$\dfrac{z}{z-1} - \dfrac{z}{z-c} - \dfrac{caTz}{(z-c)^2}$

(Cont.)

Table 1-2 (CONTINUED)

	$F(s)$	$f(t) = \mathscr{L}^{-1}[F(s)]$	$f_{(kT)} = f_k$	$\mathscr{Z}[f_k] = F(z)$
11	$\dfrac{b-a}{(s+a)(s+b)}$	$e^{-at} - e^{-bt}$	$c^k - d^k$, $c = e^{-aT}$, $d = e^{-bT}$	$\dfrac{(c-d)z}{(z-c)(z-d)}$
12	$\dfrac{(b-a)s}{(s+a)(s+b)}$	$-ae^{-at} + be^{-bt}$	$-ac^k + bd^k$, $c = e^{-aT}$, $d = e^{-bT}$	$\dfrac{(b-a)z^2 - (bc-ad)z}{(z-c)(z-d)}$
13	$\dfrac{ab}{s(s+a)(s+b)}$	$1 + \dfrac{be^{-at} - ae^{-bt}}{a-b}$	$1 + \dfrac{bc^k - ad^k}{a-b}$, $c = e^{-aT}$, $d = e^{-bT}$	$\dfrac{z}{z-1} + \dfrac{b}{a-b}\dfrac{z}{z-c} - \dfrac{a}{a-b}\dfrac{z}{z-d}$
14	$\dfrac{\beta}{s^2 + \beta^2}$	$\sin \beta t$	$\sin k\beta T$	$\dfrac{z \sin \beta T}{z^2 - 2z \cos \beta T + 1}$
15	$\dfrac{s}{s^2 + \beta^2}$	$\cos \beta t$	$\cos k\beta T$	$\dfrac{z^2 - z \cos \beta T}{z^2 - 2z \cos \beta T + 1}$
16	$\dfrac{\beta}{s^2 - \beta^2}$	$\sinh \beta t$	$\sinh k\beta T$	$\dfrac{z \sinh \beta T}{z^2 - 2z \cosh \beta T + 1}$
17	$\dfrac{s}{s^2 - \beta^2}$	$\cosh \beta t$	$\cosh k\beta T$	$\dfrac{z^2 - z \cosh \beta T}{z^2 - 2z \cosh \beta T + 1}$
18	$\dfrac{s+\alpha}{(s+\alpha)^2 + \beta^2}$	$e^{-\alpha t} \cos \beta t$	$c^k \cos k\beta T$, $c = e^{-\alpha T}$	$\dfrac{z^2 - cz \cos \beta T}{z^2 - 2cz \cos \beta T + c^2}$
19	$\dfrac{s+\alpha}{(s+\alpha)^2 + \left(\dfrac{\pi}{T}\right)^2}$	$e^{-\alpha t} \cos \dfrac{\pi t}{T}$	$(-c^k)$, $c = e^{-\alpha T}$	$\dfrac{z}{z-c}$
20	$\dfrac{\beta}{(s+\alpha)^2 + \beta^2}$	$e^{-\alpha t} \sin \beta t$	$c^k \sin k\beta T$, $c = e^{-\alpha T}$	$\dfrac{cz \sin \beta T}{z^2 - 2cz \cos \beta T + c^2}$

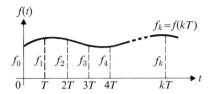

Fig. 1-13 Sampled data.

example) is the easiest way to find an inverse transform for a given $F(s)$. Leaving all details of the process to these references, here we shall study the case of an exponential function of time, since this function results in an especially important pair. Let

$$f(t) = e^{\lambda t}, \quad t \geq 0, \tag{1-35}$$

where λ is a constant which may be real or complex. The Laplace transform of this function is

$$F(s) = \mathscr{L}[e^{\lambda t}] = \int_0^\infty e^{\lambda t} e^{-st}\, dt = \frac{e^{(\lambda-s)t}}{\lambda - s}\bigg|_0^\infty = \frac{1}{s - \lambda}. \tag{1-36}$$

Therefore,

$$\mathscr{L}^{-1}\left[\frac{1}{s - \lambda}\right] = e^{\lambda t}. \tag{1-37}$$

Note that λ, which characterizes the exponential pattern of $f(t)$, appears as the root of the denominator (or characteristic) equation of $F(s)$, $s - \lambda = 0$. Such a root (or roots) is often called the pole(s) of $F(s)$, for $F(s)$ goes to infinity at $s = \lambda$.

The z-transform of a function $f(t)$ is useful when the function is available in the form of sampled data $f_0, f_1, f_2, \ldots, f_k$ taken at fixed time intervals T, as shown in Fig. 1-13. It is defined by

$$\mathscr{Z}[f_n] = \sum_{k=0}^\infty f_k z^{-k} = F(z), \tag{1-38}$$

where the symbol $\mathscr{Z}$ denotes the z-transform. The variable z, which is in general complex, must satisfy the convergence condition of the series:

$$|z| > (\text{radius of convergence of the series})^{-1}.$$

For example, we find the z-transform of a unit step as follows:

$$\mathscr{Z}[\text{unit step}] = 1 + z^{-1} + z^{-2} + \cdots = \frac{1}{1 - z^{-1}} = \frac{z}{z - 1}. \tag{1-39}$$

We take the value $f_0 = 1$ after the jump at $t = 0$. Generally, if $f(t)$ has a jump discontinuity at $t = kT$, we shall interpret f_k as $f(kT^+)$. The z-transform of an exponential function is:

$$\mathscr{Z}[e^{\lambda t}] = 1 + z^{-1}e^{\lambda T} + \cdots = \frac{1}{1 - z^{-1}e^{\lambda T}} = \frac{z}{z - e^{\lambda T}}. \qquad (1\text{--}40)$$

In a sampled-data system, the shifting theorem,

$$\mathscr{Z}[f(t + T)] = z(F(z) - f(0^+)), \qquad (1\text{--}41)$$

where

$$F(z) = \mathscr{Z}[f(t)],$$

is analogous to the Laplace transform of the derivative operation (Eq. 1–32) for a continuous-time system. We derive Eq. (1–41) by the following procedure:

$$\begin{aligned}
\mathscr{Z}[f(t + T)] &= \sum_{n=0}^{\infty} f_{n+1} z^{-n} = z \sum_{n=0}^{\infty} f_{n+1} z^{-(n+1)} \\
&= z \left\{ \left(f_0 + \sum_{n=0}^{\infty} f_{n+1} z^{-(n+1)} \right) - f_0 \right\} \\
&= z \sum_{m=0}^{\infty} f_m z^{-m} - z f_0, \qquad \text{where } m = n + 1.
\end{aligned}$$

Applying the shifting theorem to the difference equation (Eq. 1–25, restated),

$$x_{k+1} = P x_k + Q u_k,$$

we obtain

$$z X(z) - z x_0 = P X(z) + Q U(z),$$

where

$$\mathscr{Z}[x] = X(z) \quad \text{and} \quad \mathscr{Z}[u] = U(z).$$

Therefore,

$$X(z) = z^{-1}(P X(z) + Q U(z)) + x_0, \qquad (1\text{--}42)$$

from which follows the graphical representation shown in Fig. 1–14(b). The signal flow diagram shown is a graphical representation of the sequence of operations that take place in a digital computer to compute the response sequence. The initial value x_0 is first registered in the right-hand cell (the small circle) labeled $X(z)$ in the figure. It is then multiplied by P, while the first sample of the input u_0 is multiplied by Q. The sum of $P x_0$ and $Q u_0$, which first appears in the left-hand cell, is then shifted into the right-hand cell when x_0 is cleared from it. This is x_1, and the sequential process continues in the same way. Comparing Fig. 1–14 with Fig. 1–12, we note that an integrator in a continuous-time system is replaced by a delay operator in a sampled-data system.

1-5 Signal flow in a first-order system

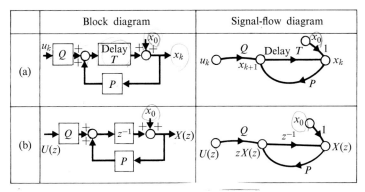

Fig. 1-14 Graphical representations of a first-order, linear, stationary, sampled-data system.

In order to relate the z-transform to the Laplace transform (see [15] for rigorous details), let us consider the Laplace transform of a time-shifted signal. If

$$\mathscr{L}[f(t)] = F(s) = \int_0^\infty f(t)e^{-st}\,dt\;,$$

$$f(t) = 0 \quad \text{for} \quad t < 0\;,$$

then it follows that

$$\mathscr{L}[f(t+T)] = \int_0^\infty f(t+T)e^{-st}\,dt$$

$$= e^{sT}\int_0^\infty f(t+T)e^{-s(t+T)}\,dt\;, \quad f(t) = 0 \text{ for } t < T$$

$$= e^{sT}\int_T^\infty f(\tau)e^{-s\tau}\,d\tau\;, \quad \text{where } \tau = t+T\;,$$

$$\qquad\qquad\qquad\qquad\qquad\qquad f(\tau) = 0 \text{ for } \tau < T$$

$$= e^{sT}F(s)\;. \qquad (1\text{-}43)$$

In other words, e^{sT} is an operator for a shift T, and

$$z = e^{sT} \qquad (1\text{-}44)$$

for the equivalence of z-domain shifting (Eq. 1-41) and s-domain shifting (Eq. 1-43).

Since $\mathscr{L}[\delta(t)] = 1$, z^{-n} is a delayed impulse:

$$z^{-n} = \mathscr{L}[\delta(t - nT)]\;. \qquad (1\text{-}45)$$

Therefore, although it is a mathematical fiction, a sequence

$$f_0, z^{-1}f_1, \ldots, z^{-n}f_n, \ldots$$

is a train of impulses with time spacing T, as shown at the left in Fig. 1-15(b).

26 Introduction

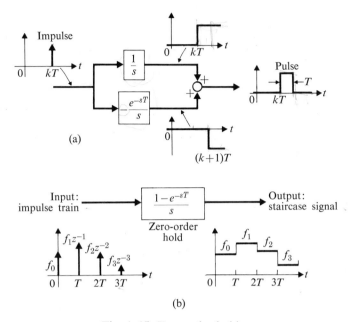

Fig. 1-15 Zero-order hold.

In many cases a control signal, generated by a discrete-time system and applied to a control object, appears as a staircase pattern rather than an impulse train. Since an integrator $1/s$ produces a step for an impulse input, and the output of a delayed negative integrator $-e^{-sT}/s$ for the same impulse input is a delayed negative step, the sum of the two step responses becomes a pulse of a width T (Fig. 1-15a). Therefore, the conversion of an impulse train into a staircase pattern can be accomplished by the transfer function

$$G_h(s) = \frac{1 - e^{-sT}}{s}. \tag{1-46}$$

This operation is called a zero-order hold (Fig. 1-15b).

REFERENCES

1. Maxwell, J. C., "On governors." *Proc. Roy. Soc.*, **16**, 1868, pp. 270–283.
2. Tolle, M., *Regelung der Kraftmaschinen*. Berlin: Springer, 1921.
3. Oppelt, W., "Vergleichende Betrachtung verschiedener Regelaufgaben hinsichtlich der geeigneten Regelgesetzmässigkeit." *Luftfahrtforschung*, **16**, 1939, pp. 447–472.
4. Wiener, N., *Cybernetics*. New York: Wiley, 1948.

5. James, H. M., Nichols, N. B., Phillips, R. S., *Theory of Servomechanisms*. New York: McGraw-Hill, 1947.
6. Oldenbourg, R. C., Sartorius, H., *Dynamik selbsttätiger Regelungen*. Munich: Oldenbourg, 1944. (English translation published by ASME.)
7. Chestnut, H., Mayer, R. W., *Servomechanisms and Regulating System Design*. New York: Wiley, Vol. 1, 1951, Vol. 2, 1955.
8. Evans, W. R., *Control-System Dynamics*. New York: McGraw-Hill, 1954.
9. Truxal, J. G., *Automatic Control System Synthesis*. New York: McGraw-Hill, 1955.
10. Young, A. J., *An Introduction to Process Control System Design*. London: Longmans, 1955.
11. Laning, J. H., Battin, R. H., *Random Processes in Automatic Control*. New York: McGraw-Hill, 1956.
12. Truxal, J. G. (ed.), *Control Engineers Handbook*. New York: McGraw-Hill, 1958.
13. Smith, O. J. M., *Feedback Control Systems*. New York: McGraw-Hill, 1958.
14. Cosgriff, R. L., *Nonlinear Control Systems*. New York: McGraw-Hill, 1958.
15. Jury, E. I., *Sampled Data Control Systems*. New York: Wiley, 1958.
16. Gille, J. C., Pelegrin, M. J., Decaulne, P., *Feedback Control Systems*. New York: McGraw-Hill, 1959.
17. Oppelt, W., *Kleines Handbuch technischer Regelvorgänge*. Verlag Chemie (Germany), 1953, fourth edition, 1960.
18. Ledley, R. S., *Digital Computer and Control Engineering*. New York: McGraw-Hill, 1960.
19. Oldenbourger, R., "Automatic control, a state-of-the-art report." *Mech. Eng.*, April 1965, pp. 38–45.
20. LaSalle, J., Lefschetz, S., *Stability by Liapunov's Direct Method with Applications*. New York: Academic Press, 1961.
21. Pontryagin, L. S., Boltyanskii, V. C., Gamkrelidze, R. V., Mishchenko, E. F., *The Mathematical Theory of Optimal Processes* (English Translation). New York: Wiley, 1962.
22. Zadeh, L. A., Desoer, C. A., *Linear Systems Theory: The State Space Approach*. New York: McGraw-Hill, 1963.
23. *Control Engineering*, January 1957, p. 69, December 1955, p. 46, April 1959, p. 123, March 1959, p. 103, etc.
24. Kaplan, W., *Ordinary Differential Equations*. Reading, Mass.: Addison Wesley, 1961.
25. Kaplan, W., *Operational Methods for Linear Systems*. Reading, Mass.: Addison-Wesley, 1962.
26. Nixon, F. E., *Handbook of Laplace Transformation*. Englewood Cliffs, N. J.: Prentice Hall, 1960.
27. Tou, J. T., *Digital and Sampled Data Systems*. New York: McGraw-Hill, 1959, pp. 588–592.

28 Introduction

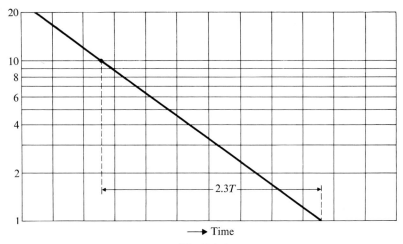

Fig. P1-2

PROBLEMS

1-1 The number N of systems-oriented engineers in an industry has increased as follows:

Year "t"	1951	1961	1968
N	5	28	60

Determine T and t_0 in the empirical formula

$$N = \exp\left(\frac{t - t_0}{T}\right).$$

1-2 When an exponential decay (or increase) is plotted on semilog paper, a straight line results. (a) Show that the time constant T can be determined, as shown in Fig. P1-2, by measuring the time for $\frac{1}{10}$ drop (or rise). (b) Determine k and T of the system function $k/(TD + 1)$ whose unit step input response is given by the following sampled data:

Time t	0	2	4	6	8	10	12	14	16
Response x	0	2.3	5.5	7.3	8.5	9.4	10.1	10.6	11.0

1-3 A square-root law was given in Section 1-3 for turbulent flow through a valve. Define dimensionless variables $x = \Delta Q/Q$, $u_1 = \Delta A_1/A_1$, $u_2 = \Delta A_2/A_2$, and $u_3 = \Delta P/P$, where total normal values are $Q = $ flow rate, A_1 and A_2 are areas of valve opening, P is the pressure drop, and their deviations are expressed by Δ. For the parallel and series arrangements shown in Fig. P1-3, determine $k_1, k_2,$ and k_3 of the linearized relation

$$x = k_1 u_1 + k_2 u_2 + k_3 u_3.$$

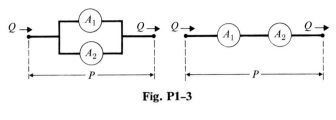

Fig. P1-3

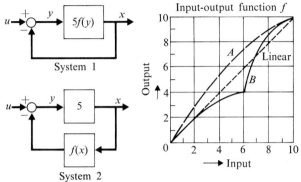

Fig. P1-4

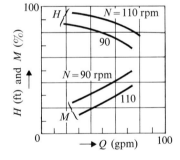

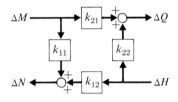

Fig. P1-5

1-4 Two different static feedback systems are shown in Fig. P1-4. Compute the static relation between u and x, for characteristics A and B of function f, and plot the results in a ux-diagram (see Section 8-2 for a more general discussion). Which of systems 1 and 2 is more effective in linearizing the overall characteristics?

1-5 A pump characteristic is shown at the left of Fig. P1-5, where H, M, N, and Q are head, torque, shaft speed, and discharge, respectively. Linearize the relation at $Q = 50$ gpm, $N = 100$ rpm, and determine $k_{11}, k_{12}, k_{21}, k_{22}$ defined by the block diagram at the right of the figure.

1-6 Outflow from the bottom of a tank is $\eta \sqrt{x}$, where x is the total head (measured from the tank bottom) and η is assumed to be constant. The area of the free surface C is constant, the initial head is x_0, and there is no inflow. Determine the transient response $x(t)$ and the time t_1 when the tank is completely emptied. Does this mathematical model hold all through the transient, $0 \leqslant t \leqslant t_1$?

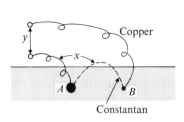

Fig. P1-7   Fig. P1-8

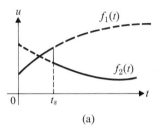   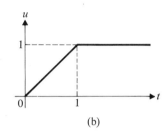

(a)   (b)

Fig. P1-10

1-7 Let k be the emf per unit temperature change, in mV/°F, of a copper-constantan thermocouple (plus on copper side). Junction A in Fig. P1-7 has a time constant T, while junction B has practically no time lag. Obtain equations for x and y (see the figure), and then relate input temperature u and output emf y by a transfer function. Sketch the response pattern for $x(0) = 0$ and a unit step of u.

1-8 In the analog computer circuit of Fig. P1-8, initial charge on the capacitor is $+10$ V. Determine the response $y(t)$ for $t \geqslant 0$, and sketch the pattern.

1-9 The total mass m [kg/(m/sec^2)] of a rocket decreases due to a constant rate of fuel consumption w (kg/sec), so that the mass at time t is given by

$$m = m_0 - \frac{w}{g_0} t,$$

where m_0 is the initial mass. The velocity v (m/sec) of the jet relative to the rocket is proportional to w:

$$v = kw,$$

where k is a constant. The velocity x (m/sec) of the rocket is zero at $t = 0$. There is neither resistance nor gravitational acceleration. Find $x(t)$. Assume that the rocket is a mass point that moves on a straight line. [*Hint:* Thrust = rate of momentum change of jet.]

1-10 A system is described by

$$\frac{dx}{dt} = Ax + Bu, \quad x(0) = 0.$$

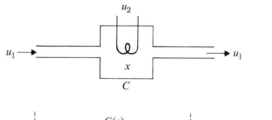

Fig. P1-11

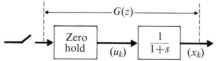

Fig. P1-12

(a) Input is [as shown in Fig. P1-10(a)],

$u(t) = f_1(t)$ for $0 \leqslant t < t_s$,
$u(t) = f_2(t)$ for $t_s \leqslant t$.

Obtain $x(t)$.

(b) $f_1(t) = u_0 = $ const, and $f_2(t) = 0$. Determine the response.

(c) $A = -1$ and $B = 1$ in the system's equation. Compute the response for $u(t)$ shown in Fig. P1-10(b).

1-11 The heat capacitance of a fluid in a completely agitated space is C. The container wall has negligible heat capacitance, and it is a complete insulator against heat loss. The product of flow rate and specific heat of the liquid flowing through the system is an input u_1, and its inlet temperature is zero. Heat input u_2, into the container is another input (see Fig. P1-11). (a) Obtain the differential equation for temperature, x, in the tank. Is this equation linear, and/or stationary? (b) Solve the equation for the following conditions:

$x(0) = x_0$, $u_1 = 1 - \cos \omega t$, $u_2 = 0$.

(c) Assuming that u_1 and u_2 remain constant during sampling intervals, obtain a difference equation for the system.

1-12 Problems on discrete-time systems: (a) Using Eq. (1-38), obtain $F(z)$ for $f(t) = t$ (a ramp function). (b) Apply Eq. (1-40) and obtain $F(z)$ for

$$f(t) = \sin \omega t = \frac{e^{j\omega t} - e^{-j\omega t}}{2j}.$$

(c) Determine the difference equation and $G(z)$ for the system shown in Fig. P1-12.

PART 1

LINEAR SYSTEMS THEORY

Chapter 2

Mathematical and Graphical Representations of Lumped-Parameter Dynamic Systems

Chapter 3

Response of Linear Lumped-Parameter Systems

Chapter 4

Stability of Linear Lumped-Parameter Systems

2
MATHEMATICAL AND GRAPHICAL REPRESENTATIONS OF LUMPED-PARAMETER DYNAMIC SYSTEMS

The state equation, which was introduced as a scalar first-order differential equation in Chapter 1, is in this chapter generalized to a vector differential equation to describe nth-order dynamic systems. For such a system, which may be multivariable, the dynamic state, the inputs, and the outputs are represented by column vectors. The state equations are transformed into the Laplace domain. Then for linear stationary systems, the transformed equations can be solved for a set of transfer functions describing the input–output relations. The reverse procedure of deriving state equations from transfer functions is also discussed.

Block diagrams and signal-flow graphs, which are graphical representations of dynamic systems, are both complementary to the mathematical description. They clearly demonstrate the separation of static and dynamic actions. The signal-flow graph is closely related to an analog computer program through its dynamics, a set of n integrators. Further, state equations may be written by inspection of the signal-flow graph, and vice versa.

Discrete-time systems are represented by signal-flow graphs in which a set of n time delays replace the integrators. The vector differential equation of continuous-time systems becomes a vector difference equation for discrete-time systems.

2-1 THE STATE VECTOR AND VECTOR DIFFERENTIAL EQUATION

The state of a system is characterized by a set of dynamic variables, called state variables. These variables are not unique, and several different methods of choosing them will be presented. We can define the system output (the observed response of a system) as a linear function of the chosen state variables and the system inputs. Consider, for example, the first-order linear system discussed in Chapter 1. The canonical form of the differential equation for that system is written in terms of the (single) state variable $x(t)$ and the input to the system $u(t)$:

$$\frac{d}{dt} x(t) = ax(t) + bu(t). \tag{2-1}$$

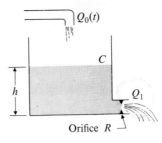

Fig. 2-1 A leaking tank.

The system response $y(t)$ is a linear function of $x(t)$ and $u(t)$:

$$y(t) = cx(t) + du(t), \qquad (2\text{-}2)$$

where a, b, c, and d are constant coefficients.

A differential equation in $y(t)$ can be established by eliminating $x(t)$ from the two equations. Thus,

$$\frac{d}{dt} y(t) = ay(t) + d \frac{d}{dt} u(t) + (bc - ad)u(t). \qquad (2\text{-}3)$$

This equation shows that there are three independent quantities which must be fixed to determine an input–output relation: a, d, and $(bc - ad)$. For a given input–output relationship, only the product bc is fixed; the system state variable is established by specifying either b or c.*

Example 2-1 *The leaking tank.* A flow $Q(t)$, an independent function of time, flows into a tank which has straight sides (Fig. 2-1). The orifice at the bottom of the tank has a linear head-flow relation, so that the flow through the orifice is related to the height of liquid in the tank as follows:

$$Q_1 = \frac{1}{R} h. \qquad (2\text{-}4)$$

The output, or system response, we are looking for is the net flow into the tank, designated by Q_n. The net inflow is the difference between the inflow $Q_0(t)$ and the outflow Q_1. Although not explicitly written as such, Q_n, Q_1, and h are all functions of time:

$$Q_n = Q_0(t) - Q_1. \qquad (2\text{-}5)$$

The outflow Q_1 is related to h by Eq. (2-4):

$$Q_n = Q_0(t) - \frac{1}{R} h.$$

This equation can be differentiated with respect to time:

$$\frac{d}{dt} Q_n = \frac{d}{dt} Q_0(t) - \frac{1}{R} \frac{d}{dt} h. \qquad (2\text{-}6)$$

---

* A similar result may be found for an nth-order multivariable system; namely the (**CB**) matrix product is invariant (see Problem 3-21).

2-1 State vector and vector differential equation

The rate of change of the liquid height, however, is related to Q_n by

$$\frac{d}{dt}h = \frac{Q_n}{C},$$

where C is the cross-sectional area. Substituting, we obtain

$$\frac{d}{dt}Q_n = \frac{d}{dt}Q_o(t) - \frac{1}{RC}Q_n,$$

or

$$\frac{d}{dt}Q_n + \frac{1}{RC}Q_n = \frac{d}{dt}Q_o(t). \tag{2-7}$$

The overall transfer function for the system is $sY(s) - y_0 + Y(s) = sU(s) - y_0$

$$G(s) = \frac{Y(s)}{U(s)} = \frac{s}{s + 1/RC} = \frac{RCs}{RCs + 1}, \tag{2-8}$$

where $Y(s) = \mathscr{L}[Q_n]$, $U(s) = \mathscr{L}[Q_o(t)]$, and the initial state is zero.

For sufficiently small values of the time constant RC, this transfer function represents an approximate differentiator with a gain of RC. For example, if the flow input is a unit step,

$$U(s) = \frac{1}{s},$$

the response is found by inverse Laplace transformation:

$$Q_n = \mathscr{L}^{-1}[Y(s)] = \mathscr{L}^{-1}\left(\frac{RC}{RCs + 1}\right) = e^{-t/(RC)}. \quad \mathscr{L}\left\{\frac{1}{s+\frac{1}{RC}}\right\}$$

If this were an ideal differentiator, the response to a unit-step input would be a unit impulse. But when RC is small, $e^{-t/(RC)}$ is a good approximation of an impulse. Note that Eq. (2-8) relates the net inflow to flow input, and the liquid level h does not appear. In the "black-box" approach of the classical transfer function method we thus lose track of an important intermediate physical measurement in the system.

A state variable for this system can be found by comparing the input-output relation (Eq. 2-7) with the standard form of input–output relation (Eq. 2-3). Matching terms, we find that the coefficients are

$$a = -\frac{1}{RC}, \quad d = 1, \quad bc - ad = 0.$$

Either b or c can be chosen arbitrarily; we will take $b = 1/C$. Substitution into Eq. (2-1) yields the following differential equation:

$$\frac{d}{dt}x(t) = -\frac{1}{RC}x(t) + \frac{1}{C}Q_o(t). \quad = ax(t) + bu(t)$$

Examination of the system-constitutive relations shown above reveals that the state variable $x(t)$ is the liquid height h.

38 Representations of lumped-parameter systems

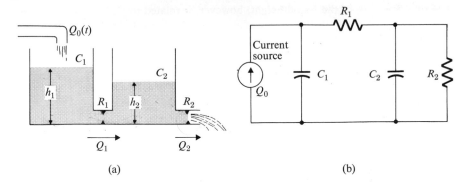

Fig. 2-2 Two-tank liquid-level system and its electrical analogy.

The relations expressed by Eqs. (2–1) and (2–2) can be generalized to apply to higher-order systems. An nth-order linear, lumped, stationary system, for example, has n integrators and requires n state variables, $x_1, x_2, \ldots, x_n$ to designate a dynamic state. A set of column vectors can be conveniently used to represent the system state, input, and output. For a system with r inputs $(u_1, u_2, \ldots, u_r)$ and m outputs $(y_1, y_2, \ldots, y_m)$, we have the following vectors:

$$\text{input vector } \mathbf{u}(t) = \begin{bmatrix} u_1 \\ u_2 \\ \vdots \\ u_r \end{bmatrix}, \quad \text{state vector } \mathbf{x}(t) = \begin{bmatrix} x_1 \\ x_2 \\ \vdots \\ x_n \end{bmatrix},$$

$$\text{output vector } \mathbf{y}(t) = \begin{bmatrix} y_1 \\ y_2 \\ \vdots \\ y_m \end{bmatrix}.$$

For multivariable systems, the scalar equation (2–1) is replaced by a vector equation:

$$\frac{d}{dt}\mathbf{x}(t) = \mathbf{A}\mathbf{x}(t) + \mathbf{B}\mathbf{u}(t), \qquad (2\text{–}9)$$

where $\mathbf{A}$ is an $n \times n$ matrix of constants and $\mathbf{B}$ is an $n \times r$ matrix of constants. Similarly, the general form of Eq. (2–2) is

$$\mathbf{y}(t) = \mathbf{C}\mathbf{x}(t) + \mathbf{D}\mathbf{u}(t), \qquad (2\text{–}10)$$

where $\mathbf{C}$ and $\mathbf{D}$ are, respectively, $m \times n$ and $m \times r$ matrices of constants.*

---

* The reader who is not familiar with matrix methods may refer to the Appendix, and to Bellman and Kreider et al. [1].

Example 2-2 *The two-tank system.* The two-tank liquid level system shown in Fig. 2-2(a) [or its electrical analogue, Fig. 2-2(b)] is similar to the leaking-tank system examined above. If we ignore the effects of fluid inertia and assume that the system elements are linear, we can write the mass-continuity equations in terms of the liquid levels h_1 and h_2:

$$C_1 \frac{dh_1}{dt} = -\frac{h_1 - h_2}{R_1} + Q_0(t),$$

$$C_2 \frac{dh_2}{dt} = \frac{h_1 - h_2}{R_1} - \frac{h_2}{R_2},$$

where C_1 and C_2 are the cross-sectional areas of tanks 1 and 2. Rearranging terms, we have

$$\frac{d}{dt} h_1 = -\frac{1}{R_1 C_1} h_1 + \frac{1}{R_1 C_1} h_2 + \frac{1}{C_1} Q_0(t),$$

$$\frac{d}{dt} h_2 = \frac{1}{R_1 C_2} h_1 - \left(\frac{1}{R_1 C_2} + \frac{1}{R_2 C_2}\right) h_2. \tag{2-11}$$

Equations (2-11) must be equivalent to the matrix equation (2-9). If we equate corresponding terms, we find that

$$\mathbf{x}(t) = \begin{bmatrix} h_1 \\ h_2 \end{bmatrix}, \quad \mathbf{A} = \begin{bmatrix} -1/R_1 C_1 & 1/R_1 C_1 \\ 1/R_1 C_2 & -(1/R_1 C_2 + 1/R_2 C_2) \end{bmatrix},$$

$$\mathbf{B} = \begin{bmatrix} 1/C_1 \\ 0 \end{bmatrix}, \quad u(t) = Q_0(t). \tag{2-12}$$

If we are interested in the flow Q_1 through the connecting pipe as the system response, we have

$$Q_1 = \frac{h_1 - h_2}{R_1}. \tag{2-13}$$

In standard matrix form the output is expressed as a function of the state vector and the input vector (cf. Eq. 2-10). In matrix form, (2-13) becomes

$$Q_1 = [1/R_1 \quad -1/R_1] \cdot \begin{bmatrix} h_1 \\ h_2 \end{bmatrix} + 0 \cdot Q_0(t).$$

The coefficient matrices, $\mathbf{C}$ and $\mathbf{D}$, are thus given by

$$\mathbf{C} = [1/R_1 \quad -1/R_1], \quad \mathbf{D} = 0,$$

The matrix elements are indicated explicitly in the signal-flow diagram Fig. 2-3.

Matrix notation can be extended to signal-flow diagrams and block diagrams to permit clearer visualization of the functional relationships for multivariable systems. For example, contrast the signal-flow diagram of Fig. 2-3, which is for a second-order system, with the matrix signal-flow diagram of Fig. 2-4(a), which depicts an *n*th-order system as described by Eqs. (2-9) and (2-10). Figure 2-4(b) shows the matrix block diagram for

40   Representations of lumped-parameter systems

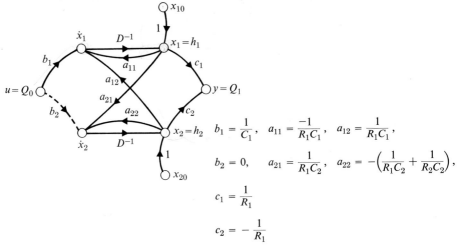

Fig. 2–3 Signal-flow diagram for two-tank liquid-level system.

the same system. Note that the only dynamic operator is $(1/D)\mathbf{I}$, where $\mathbf{I}$ is the identity matrix; the operators $\mathbf{A}$, $\mathbf{B}$, $\mathbf{C}$, and $\mathbf{D}$ are all static.

If the system is time varying, but still linear, any of the elements in the matrices $\mathbf{A}$, $\mathbf{B}$, $\mathbf{C}$, or $\mathbf{D}$ could become time varying. In practice, however, as we will see in the next Chapter, it is the $\mathbf{A}$-matrix that has the most important effect on system behavior.

A system which is nonlinear and time varying has a functional relationship of the form

$$\frac{d}{dt} x_i = f_i(\mathbf{x}, \mathbf{u}, t), \qquad i = 1, 2, \ldots, n,$$

$$y_j = h_j(\mathbf{x}, \mathbf{u}, t), \qquad j = 1, 2, \ldots, m.$$

For nonlinear but time-invariant systems, the reference to "t" in the functions f and h should be deleted. These equations can be expressed in vector form:

$$\frac{d}{dt} \mathbf{x}(t) = \mathbf{F}(\mathbf{x}, \mathbf{u}, t), \tag{2–14}$$

$$\mathbf{y}(t) = \mathbf{H}(\mathbf{x}, \mathbf{u}, t), \tag{2–15}$$

where

$$\mathbf{F}(\mathbf{x}, \mathbf{u}, t) = \begin{bmatrix} f_1(\mathbf{x}, \mathbf{u}, t) \\ \vdots \\ f_n(\mathbf{x}, \mathbf{u}, t) \end{bmatrix}, \qquad \mathbf{H}(\mathbf{x}, \mathbf{u}, t) = \begin{bmatrix} h_1(\mathbf{x}, \mathbf{u}, t) \\ \vdots \\ h_m(\mathbf{x}, \mathbf{u}, t) \end{bmatrix}.$$

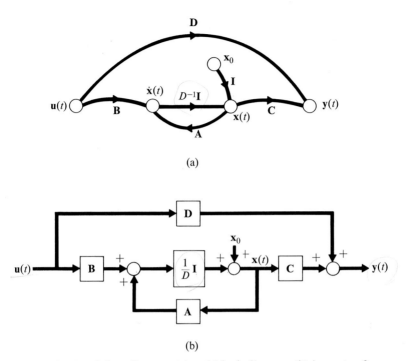

(a)

(b)

Fig. 2-4 Signal-flow diagram (a) and block diagram (b) in vector form.

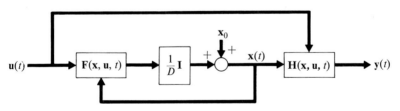

Fig. 2-5 Vector block diagram for a nonlinear system.

The block diagram for a system satisfying the state equation (2-14) and output equation (2-15), shown in Fig. 2-5, indicates the separation of the dynamic operator $(1/D)\mathbf{I}$ from the static, nonlinear operators.

2-2 LINEAR RELATIONS IN THE s-DOMAIN

As shown in Chapter 1, ordinary differential equations can be transformed into algebraic equations by using the Laplace transformation to go from the time domain to the s-domain. Since the Laplace transformation is a scalar operator, it can be applied to matrix equations. We can thus take the Laplace transformation of Eqs. (2-9) and (2-10) to obtain the following *s*-domain equations:

$$sX(s) - x_0 = AX(s) + BU(s), \quad (2\text{--}16)$$

and

$$Y(s) = CX(s) + DU(s), \quad (2\text{--}17)$$

where

$$\mathscr{L}[x(t)] = X(s), \quad \mathscr{L}[y(t)] = Y(s), \quad \mathscr{L}[u(t)] = U(s),$$

and

$$x_0 = \lim_{t \to 0} x(t).$$

These relations can be expressed graphically in a block diagram (Fig. 2–6). Note its strong structural similarity to the block diagram for the untransformed equations, shown in Fig. 2–4(b). The difference of location of the x_0 input in the two figures is due to the shift from one domain to another. In Fig. 2–4(b), x_0 is a constant bias in the *t*-domain, while in Fig. 2–6, in the *s*-domain, it is an impulsive input which "charges" the integrators.

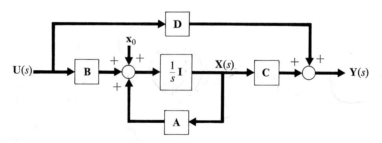

Fig. 2–6 Vector block diagram in the *s*-domain.

The algebraic equations (2–16) and (2–17) can be solved for $X(s)$ and $Y(s)$ (making sure to observe the rules of matrix algebra):

$$X(s) = (sI - A)^{-1}(x_0 + BU(s)), \quad (2\text{--}18)$$

$$Y(s) = C(sI - A)^{-1}(x_0 + BU(s)) + DU(s). \quad (2\text{--}19)$$

The transfer function, as defined below by Eq. (2–20), relates input to output in the *s*-domain for systems *having zero initial state*. For linear systems, it can be defined—without loss in generality—as the relation between changes in input state and the resulting changes in output:

$$Y(s) = G(s)U(s). \quad (2\text{--}20)$$

We can solve Eqs. (2–19) and (2–20) for $G(s)$ with $x_0 = 0$:

$$G(s) = C(sI - A)^{-1}B + D. \quad (2\text{--}21)$$

$\mathbf{G}(s)$, the matrix transfer function, is an $m \times r$ matrix which relates the r-input vector to the m-output vector. Each element of $\mathbf{G}$ represents one component of the input–output relation:

$$G_{ij}(s) = \frac{Y_i(s)}{U_j(s)}.$$

The complete expression for the ith output $Y_i(s)$ is

$$Y_i(s) = G_{i1}(s)U_1(s) + G_{i2}(s)U_2(s) + \cdots + G_{ir}(s)U_r(s). \tag{2-22}$$

Examining Eqs. (2-18) through (2-21), we see that the inverse matrix $(s\mathbf{I} - \mathbf{A})^{-1}$ plays an important role in the s-domain solution of the system equations and in determining the transfer function. In fact, it is this matrix that will determine the system characteristic equation. We must therefore consider the characteristics of inverse matrices.

The inverse of a square matrix $\mathbf{M}$ is defined as a matrix $\mathbf{M}^{-1}$ such that

$$\mathbf{M} \cdot \mathbf{M}^{-1} = \mathbf{M}^{-1} \cdot \mathbf{M} = \mathbf{I}.$$

We obtain $\mathbf{M}^{-1}$ from the adjoint matrix of $\mathbf{M}$ and determinant of $\mathbf{M}$ (see Appendix A-3)

$$\mathbf{M}^{-1} = \frac{\operatorname{adj} \mathbf{M}}{\det \mathbf{M}}. \tag{2-23}$$

Since the determinant $|s\mathbf{I} - \mathbf{A}|$ appears in the denominator of all the $G_{ij}(s)$, the equation

$$|s\mathbf{I} - \mathbf{A}| = 0 \tag{2-24}$$

is the characteristic equation from which can be determined the poles of the transfer function. These poles are the eigenvalues of $\mathbf{A}$.

Example 2-3 In Example 2-2, Section 2-1, let us assume the following set of numerical values:

$$C_1 = \tfrac{1}{6}, \qquad C_2 = \tfrac{3}{2}, \qquad R_1 = 4, \qquad R_2 = \tfrac{1}{2}.$$

Then it follows that

$$\frac{d}{dt}\begin{bmatrix} x_1 \\ x_2 \end{bmatrix} = \begin{bmatrix} -\tfrac{3}{2} & \tfrac{3}{2} \\ \tfrac{1}{6} & -\tfrac{3}{2} \end{bmatrix}\begin{bmatrix} x_1 \\ x_2 \end{bmatrix} + \begin{bmatrix} 6 \\ 0 \end{bmatrix} u, \qquad y = \begin{bmatrix} \tfrac{1}{4} & -\tfrac{1}{4} \end{bmatrix}\begin{bmatrix} x_1 \\ x_2 \end{bmatrix}.$$

Therefore,

$$s\mathbf{I} - \mathbf{A} = \begin{bmatrix} s+\tfrac{3}{2} & -\tfrac{3}{2} \\ -\tfrac{1}{6} & s+\tfrac{3}{2} \end{bmatrix}, \qquad (s\mathbf{I} - \mathbf{A})^{-1} = \frac{1}{\Delta}\begin{bmatrix} s+\tfrac{3}{2} & \tfrac{3}{2} \\ \tfrac{1}{6} & s+\tfrac{3}{2} \end{bmatrix},$$

where

$$\Delta = \begin{vmatrix} s+\tfrac{3}{2} & -\tfrac{3}{2} \\ -\tfrac{1}{6} & s+\tfrac{3}{2} \end{vmatrix} = (s+1)(s+2).$$

44   Representations of lumped-parameter systems

Hence

$$G(s) = \begin{bmatrix} \frac{1}{4} & -\frac{1}{4} \end{bmatrix} \frac{1}{\Delta} \begin{bmatrix} s + \frac{3}{2} & \frac{3}{2} \\ \frac{1}{6} & s + \frac{3}{2} \end{bmatrix} \begin{bmatrix} 6 \\ 0 \end{bmatrix} = \frac{1.5s + 2}{(s+1)(s+2)}.$$

The poles of the system are

$$s = -1 \quad \text{and} \quad s = -2.$$

2–3 STATE MODELS FROM TRANSFER FUNCTIONS

We must often find a set of state variables and the A-matrix for a system for which we know the transfer function. Such a system is usually a single-input, single-output system. However, the material given below is also applicable to multiple input–output systems where one component, $G_{ij}(s)$, of the matrix transfer function is given and the ith output and jth input meet certain conditions of observability and controllability (see Chapter 3 for the discussion of these conditions).

The number of state variables (which defines the order of the system or the number of energy storage modes in the system) can be found by an inspection of $G(s)$. For linear, lumped, stationary systems, $G(s)$ can be expressed as a ratio of polynomials:

$$G(s) = \frac{N(s)}{D(s)}. \tag{2-25}$$

The order of the system is equal to the order of the denominator polynomial $D(s)$. Therefore, for example, a system with a transfer function

$$G(s) = \frac{b_0}{a_3 s^3 + a_2 s^2 + a_1 s + a_0}$$

will have three state variables, as will the system described by

$$G(s) = \frac{b_1 s + b_0}{a_3 s^3 + a_2 s^2 + a_1 s + a_0}.$$

We can most easily see why this is true by looking at Eq. (2–21). Substituting the relation for $(s\mathbf{I} - \mathbf{A})^{-1}$, we get

$$G(s) = \frac{1}{|s\mathbf{I} - \mathbf{A}|} \mathbf{C}[\text{adj}\,(s\mathbf{I} - \mathbf{A})]\mathbf{B} + \mathbf{D}. \tag{2-26}$$

Since **A**, **B**, **C**, and **D** are constant matrices, the denominator of all elements of **G**(s) must be equal to the determinant $|s\mathbf{I} - \mathbf{A}|$. For **A**, an $n \times n$ matrix of constants, this determinant will be a polynomial of order n. A system with n state variables has an $n \times n$ **A** matrix and thus has a transfer function with an nth-order polynomial in the denominator, while the order of the numerator polynomial is equal to or less than n.

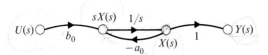

Fig. 2-7 Signal-flow diagram for $G(s) = b_0/(s + a_0)$.

Signal-flow diagrams offer us a convenient means of finding a state model without reversing the matrix operations implied by Eq. (2–26). A signal-flow diagram for a transfer function with an nth-order denominator requires n integrators; we can choose a set of state variables by designating the output of each integrator as a state variable. The state equations are found directly by writing an equation for each summing node that is at the input to an integrator. Note that, because there are many possible signal-flow diagrams for a given transfer function, there are many possible state variable sets for a given transfer function.

Consider first the first-order transfer function

$$G(s) = \frac{b_0}{s + a_0}. \qquad (2\text{--}27)$$

Since a signal-flow diagram is made up of integrators, we divide the numerator and denominator by s to transform the transfer function into integration ($1/s$)-operator form:

$$G(s) = \frac{b_0/s}{1 + a_0/s}.$$

A possible signal-flow diagram is shown in Fig. 2–7. The numerator b_0/s is realized by a feedforward path, and $1 + a_0/s$ in the denominator is realized by a feedback path. With this choice of state variable (X in Fig. 2–7), the system state equation is

$$\frac{d}{dt}x = -a_0 x + b_0 u,$$

and the output equation is

$$y = x.$$

The transfer function of a first-order system can also have first-order terms in the numerator:

$$G(s) = \frac{b_1 s + b_0}{s + a_0}. \qquad (2\text{--}28)$$

To simplify the task of drawing the signal-flow graph, rearrange this by division into the form,

$$G(s) = b_1 + \frac{b_0'}{s + a_0}, \qquad b_0' = b_0 - b_1 a_0.$$

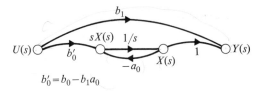

Fig. 2-8 Signal-flow diagram for $G(s) = (b_1 s + b_0)/(s + a_0)$.

The graph of Fig. 2–7 can be modified for this transfer function by adding a direct feedforward path from U to Y (Fig. 2–8). The state equation for the new signal-flow diagram is

$$\frac{d}{dt} x = -a_0 x + b_0' u ,$$

with an output equation

$$y = x + b_1 u .$$

In general, the direct path from input to output (hence D in the output equation) appears only when the order of the numerator polynomial $N(s)$ of a transfer function is equal to the order of its denominator polynomial $D(s)$. This is the limit for a physically realizable system, since the output cannot depend on future input. This point will be discussed further in Chapter 3.

Signal-flow diagrams for second-order systems are constructed in much the same manner. Consider the simple (constant-numerator) second-order transfer function

$$G(s) = \frac{b_0}{s^2 + a_1 s + a_0} . \tag{2-29}$$

In integration-operator form, this equation becomes

$$G(s) = \frac{b_0/s^2}{1 + a_1/s + a_0/s^2} .$$

Two signal-flow diagrams for this transfer function are shown in Fig. 2–9. The matrix equations obtained from the graphs in Fig. 2–9(a) and (b) are, respectively:

$$\frac{d}{dt}\begin{bmatrix} x_1 \\ x_2 \end{bmatrix} = \begin{bmatrix} -a_1 & 1 \\ -a_0 & 0 \end{bmatrix} \begin{bmatrix} x_1 \\ x_2 \end{bmatrix} + \begin{bmatrix} 0 \\ b_0 \end{bmatrix} u, \quad y = \begin{bmatrix} 1 & 0 \end{bmatrix} \begin{bmatrix} x_1 \\ x_2 \end{bmatrix}, \tag{2-30}$$

and

$$\frac{d}{dt}\begin{bmatrix} x_1 \\ x_2 \end{bmatrix} = \begin{bmatrix} 0 & 1 \\ -a_0 & -a_1 \end{bmatrix} \begin{bmatrix} x_1 \\ x_2 \end{bmatrix} + \begin{bmatrix} 0 \\ 1 \end{bmatrix} u, \quad y = \begin{bmatrix} b_0 & 0 \end{bmatrix} \begin{bmatrix} x_1 \\ x_2 \end{bmatrix}. \tag{2-31}$$

2-3 State models from transfer functions   47

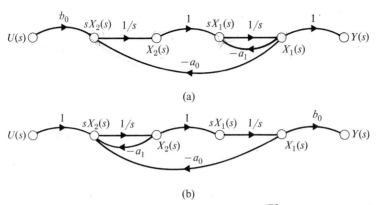

Fig. 2-9 Signal-flow diagrams for $G(s) = b_0/(s^2 + a_1 s + a_0)$.

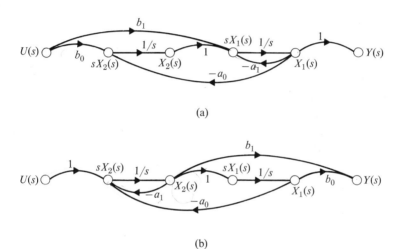

Fig. 2-10 Signal-flow diagrams for $G(s) = (b_0 + b_1 s)/(s^2 + a_1 s + a_0)$.

Finally, let us look at the second-order transfer function

$$G(s) = \frac{b_1 s + b_0}{s^2 + a_1 s + a_0}. \qquad (2\text{-}32)$$

Two signal-flow diagrams are shown in Fig. 2-10. The matrix equation corresponding to the first form (Fig. 2-10a) is

$$\frac{d}{dt}\begin{bmatrix} x_1 \\ x_2 \end{bmatrix} = \begin{bmatrix} -a_1 & 1 \\ -a_0 & 0 \end{bmatrix} \begin{bmatrix} x_1 \\ x_2 \end{bmatrix} + \begin{bmatrix} b_1 \\ b_0 \end{bmatrix} u, \quad y = \begin{bmatrix} 1 & 0 \end{bmatrix} \begin{bmatrix} x_1 \\ x_2 \end{bmatrix}. \qquad (2\text{-}33)$$

The alternative form (Fig. 2-10b) is not recommended if the numerator factor $b_1 s + b_0$ is nearly or exactly canceled by a factor hidden in the denominator, that is, if one of the poles is very close or equal to $-b_0/b_1$. To see the reason for this, consider the scalar differential equation that corresponds to Eq. (2–32),

$$\frac{d^2 y}{dt^2} + a_1 \frac{dy}{dt} + a_0 y = b_1 \frac{du}{dt} + b_0 u.$$

When such a cancellation occurs, the initial state, $x_1(0)$ and $x_2(0)$, in the second state model (Fig. 2–10b) will go to infinity for a prescribed set of nonzero initial conditions, y and dy/dt at $t = 0$, in the scalar differential equation (see Problem 2–6).

The above principle of state model construction can be extended to transfer functions of any arbitrary order.

2–4 DIAGONALIZATION OF THE A-MATRIX

In the above examples, it was assumed that the transfer function was given in terms of unfactored polynomials. If the denominator polynomial can be factored (that is, if the roots of $|s\mathbf{I} - \mathbf{A}| = 0$ are known), a set of state variables that leads either to a diagonal or Jordan canonical form of the A-matrix can be chosen.

The simplest case occurs where the poles of $G(s)$ are all real and distinct. We can then expand $G(s)$ into partial-fraction form:

$$G(s) = \frac{N(s)}{D(s)} = \frac{K_1}{s - p_1} + \frac{K_2}{s - p_2} + \cdots + \frac{K_n}{s - p_n}, \qquad (2\text{–}34)$$

where the expansion constants are

$$K_i = \frac{N(p_i)}{D'(p_i)}, \qquad D'(p) = \frac{dD(p)}{dp}, \qquad i = 1, \ldots, n.$$

K_i is called the residue of $G(s)$ at p_i and is real when p_i is real* (for all i). Proceeding as we did in the preceding section, we construct a signal-flow graph for $G(s)$. A parallel form of the graph can be constructed (Fig. 2–11) and leads to the decoupled vector state equation

$$\frac{d}{dt}\begin{bmatrix} x_1 \\ x_2 \\ \vdots \\ x_n \end{bmatrix} = \begin{bmatrix} p_1 & 0 & \cdots & 0 \\ 0 & p_2 & 0 & 0 \\ 0 & 0 & p_3 & \\ \vdots & & & \vdots \\ 0 & & \cdots & p_n \end{bmatrix} \begin{bmatrix} x_1 \\ x_2 \\ \vdots \\ x_n \end{bmatrix} + \begin{bmatrix} K_1 \\ K_2 \\ \vdots \\ K_n \end{bmatrix} u, \qquad (2\text{–}35)$$

$$y = [1 \quad 1 \cdots 1]\mathbf{x}.$$

---

* See [2] for a more detailed explanation of the partial-fraction expansion.

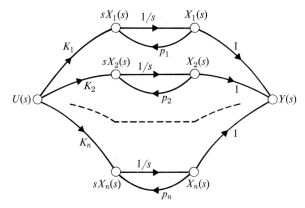

Fig. 2-11 Signal-flow diagram of a system with distinct real poles.

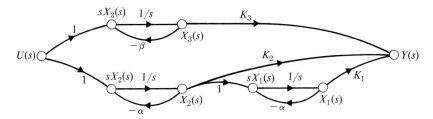

Fig. 2-12 Signal-flow diagram of a system with a double pole.

A diagonalized system matrix **A** thus results when the system characteristic equation has been factored. Conversely, diagonalizing **A** yields the system poles or eigenvalues; but more on this in the next chapter. The distribution matrix **B** is made up of the residues K_i.

Unfortunately, if $G(s)$ has any multiple poles, we cannot achieve the simple partial-fraction expansion of $G(s)$. The partial-fraction expansion for a transfer function with a multiple pole p_m of multiplicity k,

$$G(s) = \frac{H(s)}{(s - p_m)^k},$$

where $H(s) = N(s)/D(s)$ and $D(s)$ is a polynomial having distinct roots, is

$$G(s) = \frac{H(p_m)}{(s - p_m)^k} + \frac{H'(p_m)}{1!\,(s - p_m)^{k-1}} + \cdots + \frac{H^{(k-1)}(p_m)}{(k-1)!\,(s - p_m)}$$
$$+ \left(\text{expansion of } G(s) \text{ on poles other than } p_m\right). \quad (2\text{--}36)$$

Note that a kth-order multiple pole has k terms and k residues associated with it [2].

The parallel signal-flow diagram of Fig. 2-11 must be modified to the parallel-series form of Fig. 2-12 to represent a transfer function with

50 Representations of lumped-parameter systems

multiple poles. Figure 2-12 applies to the following example:

$$G(s) = \frac{b_0 + s}{(s + \alpha)^2(s + \beta)}.$$

We have

$$H(s) = \frac{b_0 + s}{s + \beta}, \qquad H'(s) = \frac{\beta - b_0}{(s + \beta)^2}$$

and

$$G(s) = \frac{K_1}{(s + \alpha)^2} + \frac{K_2}{s + \alpha} + \frac{K_3}{s + \beta},$$

$$K_1 = \frac{\alpha - b_0}{\alpha - \beta}, \qquad K_2 = \frac{\beta - b_0}{(\alpha - \beta)^2}, \qquad K_3 = \frac{b_0 - \beta}{(\alpha - \beta)^2}.$$

The system state equation has the modified diagonal form of A-matrix, the Schwartz form of the Jordan canonical matrix for the first two state variables (see Chapter 3 for general discussion):

$$\frac{d}{dt}\begin{bmatrix} x_1 \\ x_2 \\ x_3 \end{bmatrix} = \begin{bmatrix} -\alpha & 1 & 0 \\ 0 & -\alpha & 0 \\ 0 & 0 & -\beta \end{bmatrix} \begin{bmatrix} x_1 \\ x_2 \\ x_3 \end{bmatrix} + \begin{bmatrix} 0 \\ 1 \\ 1 \end{bmatrix} u. \qquad (2\text{-}37)$$

The output equation for this system is

$$y = [K_1 \quad K_2 \quad K_3]\mathbf{x}.$$

Because the above development was limited to transfer functions with all real poles, the residues K_i and the coefficients in the signal-flow diagrams are all real. An analog computer circuit can be constructed almost immediately from the signal-flow diagrams; we need only add inverters in the proper places to taken care of the sign change introduced by the analog operational amplifiers. An example will be described at the end of Chapter 3.

Looking back at the mathematics, we find that the method applies with equal validity even to system transfer functions which have complex poles. In this case, however, the convenience of real coefficients in the A-matrix and the similarity between the signal-flow diagram and an analog computer program are lost. A modified set of state equations will be described in the next chapter that eliminates the need to work with complex numbers for oscillatory systems.

2-5 DISCRETE-TIME LINEAR LUMPED SYSTEMS

Sampled-data systems are specified only at discrete times, and in place of the state differential equations (2-9 and 2-10) we write vector difference equations

$$\mathbf{x}_{k+1} = \mathbf{P}\mathbf{x}_k + \mathbf{Q}\mathbf{u}_k, \qquad (2\text{-}38)$$

$$\mathbf{y}_k = \mathbf{C}\mathbf{x}_k + \mathbf{D}\mathbf{u}_k, \qquad (2\text{-}39)$$

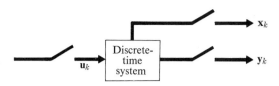

Fig. 2-13 Input–output representation of a discrete-time system.

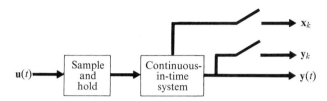

Fig. 2-14 Continuous-in-time system with input through sampler/holder.

where $k = 0, 1, 2, \ldots$ is a running index,

$$\mathbf{x}_k = \begin{bmatrix} x_1(kT) \\ \vdots \\ x_n(kT) \end{bmatrix}, \quad \mathbf{y}_k = \begin{bmatrix} y_1(kT) \\ \vdots \\ y_m(kT) \end{bmatrix}, \quad \mathbf{u}_k = \begin{bmatrix} u_1(kT) \\ \vdots \\ u_r(kT) \end{bmatrix}$$

are the state, response, and input vectors, respectively, and T is the sampling period (assumed constant).

Such a system is shown in Fig. 2-13. The three sets of switches indicate that the system operates on sampled (or pulsed) data, and that all variables are discrete in time; the switches are assumed to close simultaneously for a short time every sampling period. A digital control computer is an example of this kind of system.

The difference equations also apply to systems which are continuous in time, but have inputs which are sampled by a set of zero hold elements. In order to indicate that we are considering the output and state of the system only at discrete times, two sets of fictitious samplers are added to the system block diagram (Fig. 2-14).

A sampled data formulation for a first-order continuous-time system is given by Eq. (1-25). The relation between the difference equation constants, $\mathbf{P}$ and $\mathbf{Q}$, and the original system constants, $\mathbf{A}$ and $\mathbf{B}$, is derived in Chapter 3.

The Laplace transformation, from the time domain to the s-domain, was used to reduce differential equations to algebraic equations. The z-transform, which we defined in Section 1-5, can be used to reduce difference equations to algebraic equations in the z-domain. Let

$$\mathbf{U}(z) = \mathscr{Z}[\mathbf{u}_k]; \quad \mathbf{X}(z) = \mathscr{Z}[\mathbf{x}_k]; \quad \mathbf{Y}(z) = \mathscr{Z}[\mathbf{y}_k].$$

Using Eq. (1-41), restated in the form,

$$\mathscr{Z}[\mathbf{x}_{k+1}] = z\mathbf{X}(z) - z\mathbf{x}_0,$$

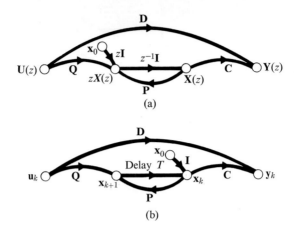

Fig. 2-15 Signal-flow diagrams for sampled-data systems.

we take the z-transform of Eqs. (2-38) and (2-39):

$$z\mathbf{X}(z) - z\mathbf{x}_0 = \mathbf{P}\mathbf{X}(z) + \mathbf{Q}\mathbf{U}(z), \qquad (2\text{-}40)$$

$$\mathbf{Y}(z) = \mathbf{C}\mathbf{X}(z) + \mathbf{D}\mathbf{U}(z). \qquad (2\text{-}41)$$

The z-domain signal-flow diagram is shown in Fig. 2-15(a). Since we know from Section 1-5 that $\mathbf{I}z^{-1}$ is the transform of a set of time-delay operators, the time-domain signal-flow diagram (Fig. 2-15b) is easily constructed from the z-domain version.

A transfer function in the z-domain, $G(z)$, called the pulse-transfer function, is derived by performing the same matrix operations that were used to find the s-domain transfer function. Again assuming zero initial conditions, the pulse transfer function is

$$G(z) = \mathbf{C}(z\mathbf{I} - \mathbf{P})^{-1}\mathbf{Q} + \mathbf{D}, \qquad (2\text{-}42)$$

and the characteristic equation is

$$|z\mathbf{I} - \mathbf{P}| = 0. \qquad (2\text{-}43)$$

As we might expect, the rules developed in Section 2-3 for the construction of state models from s-domain transfer functions also apply to z-domain transfer functions if $1/s$ is replaced by $1/z$. The following example shows this process for a particular transfer function.

Example 2-4 The given transfer function is

$$G(z) = \frac{2z + 6}{4z^3 - z}.$$

In the canonical form,

$$G(z) = \frac{b_2 z^2 + b_1 z + b_0}{z^3 + a_2 z^2 + a_1 z + a_0},$$

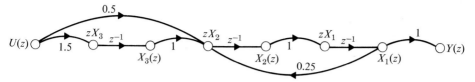

Fig. 2–16 Signal-flow diagram for $G(z) = (2z + 6)/(4z^3 - z)$.

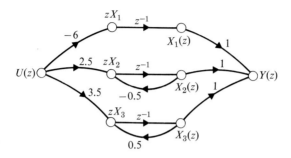

Fig. 2–17 Signal-flow diagrams for the expanded form of $G(z) = (2z+6)/(4z^3 - z)$.

the coefficients are

$$a_0 = 0, \quad a_1 = -0.25, \quad a_2 = 0; \quad b_0 = 1.5, \quad b_1 = 0.5, \quad b_2 = 0.$$

Applying the rule presented in Section 2–3, and replacing s by z, we get the signal-flow diagram shown in Fig. 2–16. The resulting state and output equations are

$$\mathbf{x}_{k+1} = \begin{bmatrix} 0 & 1 & 0 \\ 0.25 & 0 & 1 \\ 0 & 0 & 0 \end{bmatrix} \mathbf{x}_k + \begin{bmatrix} 0 \\ 0.5 \\ 1.5 \end{bmatrix} u_k, \quad y_k = \begin{bmatrix} 1 & 0 & 0 \end{bmatrix} \mathbf{x}_k.$$

The poles of $G(z)$ are

$$z = 0, -\tfrac{1}{2}, +\tfrac{1}{2},$$

and its partial-fraction expansion is

$$G(z) = \frac{-6}{z} + \frac{2.5}{(z + \tfrac{1}{2})} + \frac{3.5}{(z - \tfrac{1}{2})}.$$

We use this form of $G(z)$ to get the parallel signal-flow diagram Fig. 2–17, and from that the state equation

$$\mathbf{x}_{k+1} = \begin{bmatrix} 0 & 0 & 0 \\ 0 & -\tfrac{1}{2} & 0 \\ 0 & 0 & \tfrac{1}{2} \end{bmatrix} \mathbf{x}_k + \begin{bmatrix} -6 \\ 2.5 \\ 3.5 \end{bmatrix} U_k$$

and the output equation

$$y_k = \begin{bmatrix} 1 & 1 & 1 \end{bmatrix} \mathbf{x}_k.$$

Note that the states of Fig. 2–16 and 2–17 are not the same.

Representations of lumped-parameter systems

REFERENCES

1. Bellman, R., *Introduction to Matrix Analysis*. New York: McGraw-Hill, 1960. Kreider, D. L., Kuller, R. G., Ostberg, D. R., Perkins, F. W., *An Introduction to Linear Analysis*. Reading, Mass.: Addison-Wesley, 1966.

2. Kaplan, W., *Operational Methods for Linear Systems*. Reading, Mass.: Addison-Wesley, 1962.

PROBLEMS

2-1 In Example 2-1 the state variable is the total height h as shown in Fig. 2-1. Show that all the relations in this example hold for a small perturbation Δh about a nominal level taken as the state variable x when the orifice resistance is nonlinear.

2-2 Two leaking tanks are cascaded, as shown in Fig. P2-2. Obtain the matrices **A**, **B**, **C**, and **D** of the state and output equations.

2-3 In the system of Fig. P2-2, the parameters are

$$C_1 = 1, \quad C_2 = 1, \quad R_1 = 1, \quad R_2 = \tfrac{1}{2}.$$

Determine the matrix transfer function for the inputs and outputs shown in the figure.

2-4 Three identical tanks are connected via three equal resistors in a delta form, as shown in Fig. P2-4. There is no leakage. Determine the A-matrix. For $C = 1$ and $R = 1$, find the eigenvalues. For the said parameter values, obtain the transfer function for the input and output shown in the figure. Does this transfer function represent the dynamics of the system without a serious loss of information?

2-5 For the third-order transfer function

$$G(s) = \frac{b_2 s^2 + b_1 s + b_0}{s^3 + a_2 s^2 + a_1 s + a_0},$$

construct a signal-flow diagram in the first canonical form (Fig. 2-9a) and obtain **A**, **b**, and **c** for the state model.

2-6 Consider a second-order ordinary differential equation

$$\frac{d^2 y}{dt^2} + a_1 \frac{dy}{dt} + a_0 y = b_1 \frac{du}{dt} + b_0 u, \quad b_1 \neq 0,$$

with $u(t)$ given for $0 < t$, and with initial conditions

$$y(0) = c_1, \quad \left(\frac{dy}{dt}\right)_{t=0} = c_2.$$

Moreover, the poles of the system are $-\alpha$ and $-\beta$, both real, and $b_0/b_1 = \gamma$. (a) Determine $\mathbf{x}_0$ in terms of c_1 and c_2 for the two canonical state models given in Fig. 2-10. (b) Does the second model (Fig. 2-10b) apply in the special case when $\alpha = \gamma$ or $\beta = \gamma$?

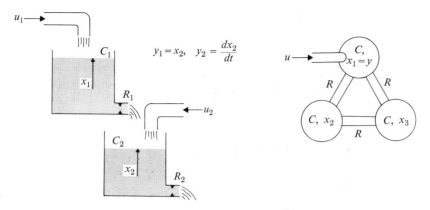

Fig. P2-2   Fig. P2-4

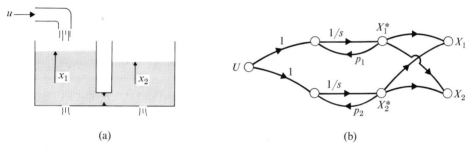

Fig. P2-7

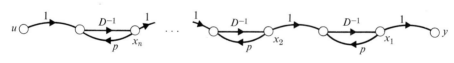

Fig. P2-8

2-7 In the two-leaking-tank liquid-level system shown in Fig. P2-7(a), the system parameters (C's and R's) are so chosen that

$$\frac{d}{dt}\begin{bmatrix} x_1 \\ x_2 \end{bmatrix} = \begin{bmatrix} -2 & 1 \\ 2 & -3 \end{bmatrix} \begin{bmatrix} x_1 \\ x_2 \end{bmatrix} + \begin{bmatrix} 1 \\ 0 \end{bmatrix} u.$$

By expanding $G_1(s) = X_1(s)/U(s)$ and $G_2(s) = X_2(s)/U(s)$ into partial fractions, and assigning new state variables $X_1^*(s)$ and $X_2^*(s)$ for each fraction as indicated in part (b) of the figure, determine the transformation matrix **T** such that

$$\mathbf{X}(s) = \mathbf{TX}^*(s).$$

2-8 In the chain shown in Fig. P2-8, p is real. Obtain (a) the state-space equations, and (b) the transfer function. Also determine (c) the unit-impulse input response for zero initial state.

Representations of lumped-parameter systems

2-9 In the system of Fig. P2-7(b), $u(t)$ is an input fed from a sampler–holder, so that it takes a staircase pattern of values $u_0, u_1, \ldots, u_k, \ldots$ (a) Determine $\mathbf{P}^*$ and $\mathbf{q}^*$ in the matrix difference equation:

$$\mathbf{x}^*_{k+1} = \mathbf{P}^* \mathbf{x}_k + \mathbf{q}^* u_k .$$

(b) Applying the transformation matrix $\mathbf{T}$ determined in Problem 2-7, obtain $\mathbf{P}$ and $\mathbf{q}$ in the difference equation

$$\mathbf{x}_{k+1} = \mathbf{P}\mathbf{x}_k + \mathbf{q} u_k .$$

2-10 A second-order sampled-data system is given by

$$\mathbf{x}_{k+1} = \begin{bmatrix} p_{11} & p_{12} \\ p_{21} & p_{22} \end{bmatrix} \mathbf{x}_k + \begin{bmatrix} q_1 \\ q_2 \end{bmatrix} u_k .$$

Determine the pulse-transfer functions $G_1(z) = X_1(z)/U(z)$ and $G_2(z) = X_2(z)/U(z)$.

3

RESPONSE OF LINEAR, LUMPED-PARAMETER STATIONARY SYSTEMS

We continue the analysis of linear lumped-parameter systems by examining the system response from several points of view. The general solution is derived for a first-order vector differential equation, which represents a set of n first-order linear differential equations or, indirectly, an nth-order differential equation. The solution involves the matrix exponential function and, except for certain simple cases, is not amenable to manual computation. The discrete-time-domain solution, however, provides a technique for efficient digital computer computation of the response in continuous-time systems.

An analytic approach leads to the concept of a set of modes inherent in a given dynamic system. The modes are characterized by the exponential time functions of the eigenvalues, or poles, of the system. Each element of a system's solution matrix is a sum of constants times each of the modes. A simplification of the solution is obtained by performing a coordinate transformation which decouples a system into individual modes. With this transformation, which is equivalent to a diagonalization, a new set of decoupled state variables is established which has a one-to-one correspondence with the system's modes.

We define the *state space* as an n-dimensional Euclidian space whose coordinate axes are the state variables of an nth-order system. For high-order systems this space must remain a mathematical abstraction, but for second-order systems and some third- and higher-order systems, graphical techniques allow visualization of system responses in the state space. For a second-order system, when one of the state variables is a derivative of the other, the resulting state plane is the well-known phase plane.

In the case of complex systems, we are not always sure whether a particular control input is really capable of completely controlling the system, or whether some particular output is a sufficient observation of the complete state of the system. The concepts of controllability and observability are introduced, and techniques for the determination of the suitability of control inputs and observations are discussed.

3-1 THE SOLUTION MATRIX

In Chapter 2 we saw that a linear, lumped-parameter, stationary system of any order could be described by the vector differential equation

$$\frac{d}{dt}\mathbf{x}(t) = \mathbf{A}\mathbf{x}(t) + \mathbf{B}\mathbf{u}(t).$$

Since this is an ordinary differential equation, we seek its general solution as the sum of a particular solution and the homogeneous solution. The homogeneous solution is found by solving the differential equation with the forcing term set equal to zero:

$$\frac{d}{dt}\mathbf{x}(t) = \mathbf{A}\mathbf{x}(t) .$$

The solution is written first as an infinite series with undetermined coefficients:

$$\mathbf{x}(t) = \mathbf{S}(t)\mathbf{x}(0) = (\mathbf{I} + \mathbf{C}_1 t + \mathbf{C}_2 t^2 + \cdots + \mathbf{C}_n t^n + \cdots)\mathbf{x}(0), \quad (3\text{-}1)$$

where $\mathbf{S}(t)$ is the solution matrix and $\mathbf{x}(0)$ is the initial-state vector. (There is no loss of generality by starting the series with the unit matrix $\mathbf{I}$; this solution is equally valid when premultiplied by any constant matrix.) To determine the coefficients, we substitute (3-1) into the homogeneous differential equation,

$$\frac{d}{dt}\mathbf{S}(t)\mathbf{x}(0) = (\mathbf{C}_1 + 2\mathbf{C}_2 t + 3\mathbf{C}_3 t^2 + \cdots + n\mathbf{C}_n t^{n-1} + \cdots)\mathbf{x}(0)$$

$$= \mathbf{A}\mathbf{S}(t)\mathbf{x}(0) = (\mathbf{A} + \mathbf{A}\mathbf{C}_1 t + \mathbf{A}\mathbf{C}_2 t^2 + \cdots + \mathbf{A}\mathbf{C}_n t^n + \cdots)\mathbf{x}(0)$$

and equate the coefficients of like power terms. This yields

$$\mathbf{C}_1 = \mathbf{A},$$

$$\mathbf{C}_2 = \frac{1}{2}\mathbf{A}\mathbf{C}_1 = \frac{1}{2}\mathbf{A}^2,$$

$$\mathbf{C}_3 = \frac{1}{3}\mathbf{A}\mathbf{C}_2 = \frac{1}{3\cdot 2}\mathbf{A}^3, \ldots, \mathbf{C}_n = \frac{1}{n!}\mathbf{A}^n,$$

or

$$\mathbf{S}(t) = e^{\mathbf{A}t} = \mathbf{I} + \mathbf{A}t + \frac{1}{2!}\mathbf{A}^2 t^2 + \frac{1}{3!}\mathbf{A}^3 t^3 + \cdots + \frac{1}{n!}\mathbf{A}^n t^n + \cdots \quad (3\text{-}2)$$

The similarity of the series in Eq. (3-2) to the series expansion of the scalar exponential leads us to define $\mathbf{S}(t)$ as the matrix exponential $e^{\mathbf{A}t}$ (see any text on matrix calculus, for instance Bellman [1], for a discussion of matrix functions). The exponential solution obtained in Chapter 1 for the first-order scalar differential equation can now be seen as a special case of the solution of the first-order vector differential equation. The solution matrix $e^{\mathbf{A}t}$ is also called the state-transition matrix. Note that it is an $n \times n$ matrix for an $n \times n$ matrix $\mathbf{A}$, and that the matrix exponential operation is defined only for square matrices (because $\mathbf{M}^n$ is not defined for nonsquare matrices).

Several important properties of the state-transition matrix must be demonstrated to show the extension of properties of the scalar exponential to the matrix exponential, or vice versa.

The differentiation relation

$$\frac{d}{dt} e^{At} = Ae^{At} = e^{At}A \tag{3-3}$$

can be proved by direct differentiation of the series:

$$\frac{d}{dt} e^{At} = A + \frac{2}{2!} A^2 t + \frac{3}{3!} A^3 t^2 + \cdots + \frac{n}{n!} A^n t^{n-1}$$

$$= A\left(I + At + \frac{1}{2!} A^2 t^2 + \cdots + \frac{1}{n!} A^n t^n\right) = Ae^{At} = e^{At}A.$$

Equation (3-1) expresses the homogeneous solution in terms of an initial-state vector at time zero. We could also write the solution in terms of an initial condition at time τ:

$$x(t + \tau) = S(t)x(\tau). \tag{3-4}$$

However, by Eq. (3-1), $x(\tau)$ is

$$x(\tau) = S(\tau)x(0), \tag{3-5}$$

and therefore

$$x(t + \tau) = S(t)S(\tau)x(0). \tag{3-6}$$

We can also apply Eq. (3-1) directly to $x(t + \tau)$,

$$x(t + \tau) = S(t + \tau)x(0), \tag{3-7}$$

and thus show that

$$S(t + \tau) = S(t)S(\tau)$$

or

$$e^{A(t+\tau)} = e^{At}e^{A\tau}. \tag{3-8}$$

Note that by setting $\tau = -t$, we get $I = S(t)S(-t)$ or $S(-t) = S^{-1}(t)$. Since $S(t + \tau) = S(\tau + t)$, we also note the commutative property

$$S(t)S(\tau) = S(\tau)S(t). \tag{3-9}$$

We now turn our attention to the full, nonhomogeneous, differential equation

$$\frac{d}{dt} x(t) = Ax(t) + Bu(t).$$

For the particular solution, we try

$$x_p(t) = S(t)p(t) = e^{At}p(t),$$

where **p**(t), an unknown vector function, has the property

$$\mathbf{p}(t) = \mathbf{0} \quad \text{when} \quad t = 0.$$

Differentiation of $\mathbf{x}_p(t)$ with respect to time yields

$$\frac{d}{dt}\mathbf{x}_p(t) = \frac{d}{dt}\mathbf{S}(t)\mathbf{p}(t) + \mathbf{S}(t)\cdot\frac{d}{dt}\mathbf{p}(t);$$

but $(d/dt)\mathbf{S}(t) = \mathbf{A}\mathbf{S}(t)$ and thus

$$\frac{d}{dt}\mathbf{x}_p(t) = \mathbf{A}\mathbf{S}(t)\mathbf{p}(t) + \mathbf{S}(t)\cdot\frac{d}{dt}\mathbf{p}(t)$$

$$= \mathbf{A}\mathbf{x}_p(t) + \mathbf{S}(t)\cdot\frac{d}{dt}\mathbf{p}(t). \tag{3-10}$$

Since $\mathbf{x}_p(t)$ must also be a solution to the original differential equation, we have

$$\frac{d}{dt}\mathbf{x}_p(t) = \mathbf{A}\mathbf{x}_p(t) + \mathbf{B}\mathbf{u}(t). \tag{3-11}$$

Equations (3-10) and (3-11) can be solved for **p**(t):

$$\frac{d}{dt}\mathbf{p}(t) = \mathbf{S}^{-1}(t)\mathbf{B}\mathbf{u}(t)$$

or

$$\mathbf{p}(t) = \int_0^t \mathbf{S}^{-1}(\tau)\mathbf{B}\mathbf{u}(\tau)\,d\tau, \tag{3-12}$$

where τ is a dummy variable of integration. By Eq. (3-8)

$$\mathbf{S}(t)\mathbf{S}^{-1}(\tau) = \mathbf{S}(t-\tau),$$

and therefore

$$\mathbf{x}_p(t) = \mathbf{S}(t)\int_0^t \mathbf{S}^{-1}(\tau)\mathbf{B}\mathbf{u}(\tau)\,d\tau = \int_0^t \mathbf{S}(t-\tau)\mathbf{B}\mathbf{u}(\tau)\,d\tau; \tag{3-13}$$

and the total solution is given by the sum of the homogeneous solution (3-1) and the particular solution:

$$\mathbf{x}(t) = \mathbf{S}(t)\mathbf{x}(0) + \int_0^t \mathbf{S}(t-\tau)\mathbf{B}\mathbf{u}(\tau)\,d\tau. \tag{3-14}$$

This solution is applicable to stationary systems. If the state equation involves time as an explicit variable, for example if $\mathbf{A} = \mathbf{A}(t)$, the form of the solution changes somewhat [2].

Because of its concise notation, the matrix form of the solution (Eq. 3-14) is ideally suited to mathematical description of multivariable systems and to digital computation. However, because of the matrix exponential, expansion of Eq. (3–14) into a set of scalar equations is almost impossible unless the system is no more than second or third order or if the A-matrix is composed mostly of zero elements. The simplest nontrivial form of the A-matrix is the purely diagonal form:

$$\mathbf{A} = \mathbf{\Lambda} = \begin{bmatrix} p_1 & & & 0 \\ & p_2 & & \\ & & \ddots & \\ 0 & & & p_n \end{bmatrix}.$$

When the A-matrix is in this form, the solution matrix $\mathbf{S}(t)$ is

$$\mathbf{S}(t) = e^{\mathbf{\Lambda} t} = \begin{bmatrix} e^{p_1 t} & & & 0 \\ & e^{p_2 t} & & \\ & & \ddots & \\ 0 & & & e^{p_n t} \end{bmatrix}.$$

3-2 SOLUTION VIA THE INVERSE LAPLACE TRANSFORM

In Chapter 2 we found the solution to the matrix differential equation in the Laplace domain. By taking the inverse Laplace transform of that solution (which is repeated below)

$$\mathbf{X}(s) = (s\mathbf{I} - \mathbf{A})^{-1}\mathbf{x}_0 + (s\mathbf{I} - \mathbf{A})^{-1}\mathbf{B}\mathbf{U}(s), \qquad (3\text{–}15)$$

we must arrive by a different route at the same solution we found above. By inverse transformation, we find:

$$\mathbf{x}(t) = \mathscr{L}^{-1}[(s\mathbf{I} - \mathbf{A})^{-1}]\mathbf{x}_0 + \mathscr{L}^{-1}[(s\mathbf{I} - \mathbf{A})^{-1}\mathbf{B}\mathbf{U}(s)],$$

which must be equivalent to Eq. (3–14), so that

$$\mathbf{S}(t) = e^{\mathbf{A} t} = \mathscr{L}^{-1}[(s\mathbf{I} - \mathbf{A})^{-1}],$$

and

$$\int_0^t \mathbf{S}(t - \tau)\mathbf{B}\mathbf{u}(\tau)\, d\tau = \mathscr{L}^{-1}[(s\mathbf{I} - \mathbf{A})^{-1}\mathbf{B}\mathbf{U}(s)]. \qquad (3\text{–}16)$$

Example 3–1 As will be illustrated in this example, Eq. (3–16) can be used to find all the elements of the solution matrix $\mathbf{S}(t)$. In order to do this, we must be able to find the inverse matrix $(s\mathbf{I} - \mathbf{A})^{-1}$.

Consider the system having the A-matrix:

$$\mathbf{A} = \begin{bmatrix} -\frac{3}{2} & \frac{3}{2} \\ \frac{1}{6} & -\frac{3}{2} \end{bmatrix},$$

$$(s\mathbf{I} - \mathbf{A}) = \begin{bmatrix} s + \frac{3}{2} & -\frac{3}{2} \\ -\frac{1}{6} & s + \frac{3}{2} \end{bmatrix}, \qquad (s\mathbf{I} - \mathbf{A})^{-1} = \frac{1}{\Delta}\begin{bmatrix} s + \frac{3}{2} & \frac{3}{2} \\ \frac{1}{6} & s + \frac{3}{2} \end{bmatrix}$$

where $\Delta = (s+1)(s+2)$. By Eq. (3-16), we have

$$S(t) = \mathscr{L}^{-1} \begin{bmatrix} \dfrac{s+\frac{3}{2}}{(s+1)(s+2)} & \dfrac{\frac{3}{2}}{(s+1)(s+2)} \\ \dfrac{\frac{1}{6}}{(s+1)(s+2)} & \dfrac{s+\frac{3}{2}}{(s+1)(s+2)} \end{bmatrix}.$$

According to matrix theory, the inverse Laplace transform of a matrix is a matrix whose elements are the inverse transforms of the corresponding elements of the original matrix. Thus

$$\mathscr{L}^{-1}\left[\frac{s+\frac{3}{2}}{(s+1)(s+2)}\right] = \tfrac{1}{2}(e^{-t} + e^{-2t}),$$

$$\mathscr{L}^{-1}\left[\frac{\frac{3}{2}}{(s+1)(s+2)}\right] = \tfrac{3}{2}(e^{-t} - e^{-2t}),$$

$$\mathscr{L}^{-1}\left[\frac{\frac{1}{6}}{(s+1)(s+2)}\right] = \tfrac{1}{6}(e^{-t} - e^{-2t}),$$

and

$$S(t) = \begin{bmatrix} \tfrac{1}{2}(e^{-t}+e^{-2t}) & \tfrac{3}{2}(e^{-t}-e^{-2t}) \\ \tfrac{1}{6}(e^{-t}-e^{-2t}) & \tfrac{1}{2}(e^{-t}+e^{-2t}) \end{bmatrix}. \tag{3-17}$$

Note that the coefficients in the exponential, -1 and -2, are the roots of the characteristic equation $|sI - A| = 0$. This can be verified by direct computation. Also, the constants in each term are the residues at each pole.

Now, by Eq. (3-16), we find that

$$S(t) = \exp(At) = \exp\left(\begin{bmatrix} -\tfrac{3}{2} & \tfrac{3}{2} \\ \tfrac{1}{6} & -\tfrac{3}{2} \end{bmatrix} t\right) = I + At + \frac{1}{2!}A^2 t^2 + \cdots,$$

$$A^2 = \begin{bmatrix} -\tfrac{3}{2} & \tfrac{3}{2} \\ \tfrac{1}{6} & -\tfrac{3}{2} \end{bmatrix}\begin{bmatrix} -\tfrac{3}{2} & \tfrac{3}{2} \\ \tfrac{1}{6} & -\tfrac{3}{2} \end{bmatrix} = \begin{bmatrix} \tfrac{5}{2} & -\tfrac{9}{2} \\ -\tfrac{1}{2} & \tfrac{5}{2} \end{bmatrix}, \quad A^3 = \begin{bmatrix} -\tfrac{9}{2} & \tfrac{21}{2} \\ \tfrac{7}{6} & -\tfrac{9}{2} \end{bmatrix},$$

and hence

$$S(t) = \begin{bmatrix} 1 - \tfrac{3}{2}t + \tfrac{1}{2!}\tfrac{5}{2}t^2 - \tfrac{1}{3!}\tfrac{9}{2}t^3 + \cdots & \tfrac{3}{2}t - \tfrac{1}{2!}\tfrac{9}{2}t^2 + \tfrac{1}{3!}\tfrac{21}{2}t^3 + \cdots \\ \tfrac{1}{6}t - \tfrac{1}{2!}\tfrac{1}{2}t^2 + \tfrac{1}{3!}\tfrac{7}{6}t^3 + \cdots & 1 - \tfrac{3}{2}t + \tfrac{1}{2!}\tfrac{5}{2}t^2 - \tfrac{1}{3!}\tfrac{9}{2}t^3 + \cdots \end{bmatrix}.$$

We can show that this is equivalent to Eq. (3-17) by summing each of the series; for example,

$$1 - \tfrac{3}{2}t + \tfrac{1}{2!}\tfrac{5}{2}t^2 - \tfrac{1}{3!}\tfrac{9}{2}t^3 + \cdots = \tfrac{1}{2}\left(2 - 3t + \tfrac{1}{2!}5t^2 - \tfrac{1}{3!}9t^3 + \cdots\right)$$

$$= \tfrac{1}{2}\left[\left(1 - t + \tfrac{1}{2!}t^2 - \tfrac{1}{3!}t^3 + \cdots\right)\right.$$

$$\left.+ \left(1 - 2t + \tfrac{1}{2!}(2t)^2 - \tfrac{1}{3!}(2t)^3 + \cdots\right)\right] = \tfrac{1}{2}(e^{-t} + e^{-2t}), \text{ etc.}$$

In the above example, the roots of the characteristic equation

$$|s\mathbf{I} - \mathbf{A}| = 0$$

appear as coefficients in the exponentials that describe the system response. These roots, which are also poles of the transfer function $G_{ij}(s)$, are called eigenvalues of the matrix $\mathbf{A}$ and play a dominant role in the mathematical description of a dynamic system.

In the simplest case, where the eigenvalues (roots) p_i, $i = 1, \ldots, n$, are all real and distinct (that is, there are no repeated roots), the exponential components (modes) $e^{p_i t}$, $i = 1, \ldots, n$, appear in the elements of the solution matrix $\mathbf{S}(t)$. The exponentials appear because in essence we are solving an nth-order linear differential equation, which, as differential equation theory tells us, has as its general solution a sum of exponentials. The case of a repeated root gives the following modes (see Problem 2–8):

$$e^{p_i t}, \quad t e^{p_i t}, \quad \frac{t^2 e^{p_i t}}{2!}, \quad \ldots, \quad \frac{t^{k-1} e^{p_i t}}{(k-1)!},$$

where k is the multiplicity of the root. A pair of conjugate complex poles,

$$p_i = -\sigma + j\omega, \qquad p_{i+1} = -\sigma - j\omega,$$

where σ is real and ω is real and nonzero, combine to form an oscillatory mode:

$$e^{-\sigma t} \sin(\omega t + \phi),$$

where ϕ is a phase angle.

The second equality in Eq. (3–16) relates the input function and its transform via a convolution integral. The convolution integral is well known in Laplace transform theory [3] and appears whenever the inverse transform is taken of a product of two functions of s. A similar situation arises in writing the relation of the output function $y(t)$ or its transform $Y(s)$ to the input $u(t)$ and its transform $U(s)$. The transfer function $G(s)$ has been defined by the equation

$$Y(s) = G(s)U(s). \tag{3-18}$$

The inverse transform gives the time-domain solution in the form of a convolution integral:

$$y(t) = \mathscr{L}^{-1}[Y(s)] = \int_0^t g(t - \tau)u(\tau)\, d\tau, \tag{3-19}$$

where $g(t) = \mathscr{L}^{-1}[G(s)]$. The convolution integral of Eq. (3–19) can be interpreted in terms of the system's impulse response. The time function $g(t)$ is the response to a unit impulse input $u(t) = \delta(t)$ at time $t = 0$. We can verify this by noting that the Laplace transform of the unit impulse is $U(s) = 1$. If we substitute it for $U(s)$ in Eq. (3–18) and then take the inverse transform, we get

$$Y(s) = G(s), \qquad y(t) = g(t).$$

As shown in Fig. 3-1, $g(t - \tau)$ in Eq. (13-19) represents the signal amplitude at time t in response to a unit impulse input at time τ. The input $u(\tau)\,d\tau$ is considered an impulse input of strength $u(\tau) \cdot d\tau$ at time τ. Its contribution to the response at time t $(t > \tau)$ is $g(t - \tau) \cdot u(\tau)\,d\tau$. The integral of (3-19), then, is the response of the system to a train of impulses acting over the time $t = 0$ to t; this arbitrary impulse train, when summed via an integral, represents the continuous input function $u(t)$. Note that τ is a dummy variable of integration ranging from 0 to t (the time at which the response is to be evaluated).

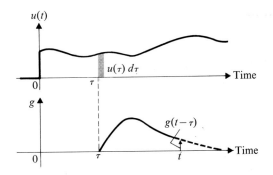

Fig. 3-1 An interpretation of the convolution integral.

The matrix transfer function $\mathbf{G}(s)$ was defined in Chapter 2 in terms of the coefficients of the system differential equation, restated:

$$\mathbf{G}(s) = \mathbf{C}(s\mathbf{I} - \mathbf{A})^{-1}\mathbf{B} + \mathbf{D}.$$

Since by Eq. (3-16)

$$\mathscr{L}^{-1}[(s\mathbf{I} - \mathbf{A})^{-1}] = \mathbf{S}(t),$$

the vector impulse response $\mathbf{g}(t)$ becomes

$$\mathbf{g}(t) = \mathscr{L}^{-1}[\mathbf{G}(s)] = \mathbf{CS}(t)\mathbf{B} + \mathbf{D}\delta(t),$$

where $\delta(t)$ is the unit impulse function (Dirac delta function). Note that the output involves the delta function only when $\mathbf{D}$ exists. This implies that the order of the numerator polynomial is equal to the order of the denominator polynomial in some of the elements of $\mathbf{G}(s)$ and that there is a direct connection between input and output on the signal-flow diagram.

3-3 DIAGONALIZATION OF THE A-MATRIX

We have already seen the idea of diagonalization used in Chapter 2. An input–output transfer function was expressed in factored form and then expanded into the partial-fraction representation. This yielded a signal-flow graph with a parallel pattern and a state model with a diagonalized A-matrix. In this section, instead of starting with a transfer function, we will start with a state equation in nondiagonal form and use matrix theory to arrive at the diagonalized form of the state equation.

Let us consider the problem of converting the A-matrix of the homogeneous state equation

$$\frac{d}{dt}\mathbf{x}(t) = \mathbf{A}\mathbf{x}(t), \qquad \mathbf{x}(0) = \mathbf{x}_0,$$

into diagonal or Jordan canonical form. We can perform a linear transformation from the state vector $\mathbf{x}(t)$ to a new state vector $\mathbf{x}^*(t)$ by means of the constant square matrix $\mathbf{T}$:

$$\mathbf{x}(t) = \mathbf{T}\mathbf{x}^*(t), \qquad \mathbf{x}^*(t) = \mathbf{T}^{-1}\mathbf{x}(t). \tag{3-20}$$

If we replace $\mathbf{x}(t)$ by $\mathbf{x}^*(t)$ in the differential equation, we get the transformed equation

$$\frac{d}{dt}\mathbf{T}\mathbf{x}^*(t) = \mathbf{A}\mathbf{T}\mathbf{x}^*(t)$$

or

$$\frac{d}{dt}\mathbf{x}^*(t) = \mathbf{T}^{-1}\mathbf{A}\mathbf{T}\mathbf{x}^*(t). \tag{3-21}$$

Suppose that the matrix $\mathbf{T}^{-1}\mathbf{A}\mathbf{T}$ is diagonal:

$$\mathbf{T}^{-1}\mathbf{A}\mathbf{T} = \boldsymbol{\Lambda} = \begin{bmatrix} p_1 & & & 0 \\ & p_2 & & \\ & & \ddots & \\ 0 & & & p_n \end{bmatrix}. \tag{3-22}$$

It is obvious that the eigenvalues of a diagonal matrix are equal to the diagonal elements (prove this by simply substituting into the defining equation for the eigenvalues and then carrying out the indicated operations); thus the eigenvalues of $\mathbf{T}^{-1}\mathbf{A}\mathbf{T}$ are p_1 through p_n. However,

$$s\mathbf{I} - \mathbf{T}^{-1}\mathbf{A}\mathbf{T} = \mathbf{T}^{-1}(s\mathbf{I} - \mathbf{A})\mathbf{T},$$

and thus the characteristic equation for $\mathbf{T}^{-1}\mathbf{A}\mathbf{T}$ can be written

$$|\mathbf{T}^{-1}(s\mathbf{I} - \mathbf{A})\mathbf{T}| = 0$$

or, canceling **T** and its inverse,

$$|s\mathbf{I} - \mathbf{A}| = 0,$$

so that the eigenvalues p_1 through p_n are also the eigenvalues of **A**.

The T-matrix is called the _modal matrix_ because of the decoupling of the modes that is accomplished when the differential equation is transformed. To find out more about the nature of the modal matrix, we first go back to Eq. (3-22) and premultiply both sides by **T**:

$$\mathbf{AT} = \mathbf{T}\begin{bmatrix} p_1 & & 0 \\ & \ddots & \\ 0 & & p_n \end{bmatrix}.$$

If we carry out the multiplication on the right-hand side, we get

$$\mathbf{AT} = \begin{bmatrix} p_1 T_{11} & p_2 T_{12} & \cdots & p_n T_{1n} \\ p_1 T_{21} & p_2 T_{22} & & \\ \vdots & & & \vdots \\ p_1 T_{n1} & p_2 T_{n2} & \cdots & p_n T_{nn} \end{bmatrix}. \tag{3-23}$$

For convenience, we designate each column of **T** by a column vector $\mathbf{v}^i$, that is,

$$\mathbf{v}^i = \begin{bmatrix} T_{1i} \\ T_{2i} \\ T_{3i} \\ \vdots \\ T_{ni} \end{bmatrix} = \begin{bmatrix} v_1^i \\ v_2^i \\ v_3^i \\ \vdots \\ v_n^i \end{bmatrix}.$$

By the rules of matrix multiplication, we note that each column of the matrix **AT** on the left-hand side of Eq. (3-23) can be represented by the column matrix

$$\mathbf{A}\mathbf{v}^i = \mathbf{A}\begin{bmatrix} T_{1i} \\ T_{2i} \\ \vdots \\ T_{ni} \end{bmatrix}.$$

If the left-hand side of Eq. (3-23) is equated to the right-hand side, column by column, a set of n equations for the $\mathbf{v}^i$ is established:

$$\mathbf{A}\mathbf{v}^i = p_i \mathbf{v}^i,$$

or

$$(p_i\mathbf{I} - \mathbf{A})\mathbf{v}^i = 0, \quad i = 1, \ldots, n. \tag{3-24}$$

Equation (3–24) is equivalent to n sets of homogeneous linear algebraic equations with each set having n unknowns, $v_1^i, v_2^i, \ldots, v_n^i$. A set of homogeneous equations has either one solution, the trivial solution with all the v_n^i equal to zero, or an infinite number of solutions. In the latter case only the direction of the vector $\mathbf{v}^i$ is determined; any vector having the proper direction, regardless of its magnitude, is a solution. Nontrivial solutions are obtained only if the coefficient determinant is equal to zero; that is, we must have

$$|p_i\mathbf{I} - \mathbf{A}| = 0. \tag{3-25}$$

Since p_i is an eigenvalue, Eq. (3–25) must be satisfied and a nontrivial solution is obtained. For each value of p_i the vector $\mathbf{v}^i$ determined according to Eq. (3–24) has a fixed direction and arbitrary magnitude (in general, the vector is written with one of its nonzero elements arbitrarily set equal to one; all the other elements are then uniquely determined). The $\mathbf{v}^i$ vectors are known as *eigenvectors*.

The modal matrix $\mathbf{T}$ thus has as its columns the eigenvectors of the A-matrix. However, the derivation which led to this result used the inverse matrix $\mathbf{T}^{-1}$ in several places. The diagonalization process is possible only if $\mathbf{T}$ is not singular, that is, if it actually has an inverse. It therefore follows that diagonalization is possible only if the eigenvectors of the A-matrix are all linearly independent. A set of vectors $\mathbf{v}^1, \ldots, \mathbf{v}^n$ is said to be linearly dependent if there exists a set of scalars c_i (not all zero) such that

$$\sum_i c_i \mathbf{v}^i = 0.$$

For instance, a set of two-dimensional vectors is linearly dependent if they are parallel to each other. According to matrix theory, any matrix that has two columns (or rows) which are linearly dependent will have a zero determinant and thus be singular.

Diagonalization is possible for all systems with distinct eigenvalues. However, in most of the systems that have any repeated eigenvalues, it becomes impossible to choose n linearly independent eigenvectors when the system is nth order. It then becomes necessary to use the Jordan canonical form in place of a diagonal form.

Example 3-2 Use the above method to find the eigenvectors and the modal matrix of the matrix

$$\mathbf{A} = \begin{bmatrix} -\frac{3}{2} & \frac{3}{2} \\ \frac{1}{6} & -\frac{3}{2} \end{bmatrix},$$

which has eigenvalues

$$p_1 = -1 \quad \text{and} \quad p_2 = -2.$$

68 Response of linear, lumped-parameter systems

The eigenvectors $\mathbf{v}^1$ and $\mathbf{v}^2$ are established by Eq. (3-24). For $\mathbf{v}^1$

$$\left(-1\begin{bmatrix} 1 & 0 \\ 0 & 1 \end{bmatrix} - \begin{bmatrix} -\frac{3}{2} & \frac{3}{2} \\ \frac{1}{6} & -\frac{3}{2} \end{bmatrix}\right)\begin{bmatrix} v_1^1 \\ v_2^1 \end{bmatrix} = \mathbf{0}.$$

This matrix equation corresponds to the pair of algebraic equations

$$\tfrac{1}{2}v_1^1 - \tfrac{3}{2}v_2^1 = 0 \quad \text{and} \quad -\tfrac{1}{6}v_1^1 + \tfrac{1}{2}v_2^1 = 0.$$

Both of these equations give

$$v_1^1 = 3v_2^1.$$

We can arbitrarily select $v_2^1 = 1$ and thus get

$$\mathbf{v}^1 = \begin{bmatrix} 3 \\ 1 \end{bmatrix}.$$

The second eigenvector is found in the same manner:

$$\mathbf{v}^2 = \begin{bmatrix} -3 \\ 1 \end{bmatrix}.$$

The modal matrix is then

$$\mathbf{T} = \begin{bmatrix} 3 & -3 \\ 1 & 1 \end{bmatrix} \quad \text{and} \quad \mathbf{T}^{-1} = \frac{1}{6}\begin{bmatrix} 1 & 3 \\ -1 & 3 \end{bmatrix}.$$

We confirm this solution by checking to see that $\mathbf{T}^{-1}\mathbf{A}\mathbf{T}$ is indeed diagonal:

$$\mathbf{T}^{-1}\mathbf{A}\mathbf{T} = \frac{1}{6}\begin{bmatrix} 1 & 3 \\ -1 & 3 \end{bmatrix}\begin{bmatrix} -\frac{3}{2} & \frac{3}{2} \\ \frac{1}{6} & -\frac{3}{2} \end{bmatrix}\begin{bmatrix} 3 & -3 \\ 1 & 1 \end{bmatrix} = \begin{bmatrix} -1 & 0 \\ 0 & -2 \end{bmatrix}.$$

The normalization,

$$|\mathbf{v}^1| = 1, \quad |\mathbf{v}^2| = 1,$$

is possible if it is desired for some reason. In our example, $\mathbf{v}^1$ can be normalized by first letting $v_2^1 = c$, $v_1^1 = 3c$ and fixing c by the relation

$$c^2 + (3c)^2 = 1, \quad c^2 = \tfrac{1}{10}.$$

Although all elements of the **A**-matrix are real for engineering systems, imaginary numbers will appear in the diagonalization process for systems with conjugate complex (or imaginary) eigenvalues. For instance, a second-order system with eigenvalues $-\sigma \pm j\omega$, σ and ω both real and $\omega > 0$, must have a diagonal matrix of the form

$$\Lambda = \begin{bmatrix} -\sigma + j\omega & 0 \\ 0 & -\sigma - j\omega \end{bmatrix}.$$

A pair of conjugate complex eigenvectors composed of real vectors $\boldsymbol{\alpha}$ and $\boldsymbol{\beta}$,

$$\mathbf{v}^1 = \boldsymbol{\alpha} + j\boldsymbol{\beta}, \quad \mathbf{v}^2 = \boldsymbol{\alpha} - j\boldsymbol{\beta},$$

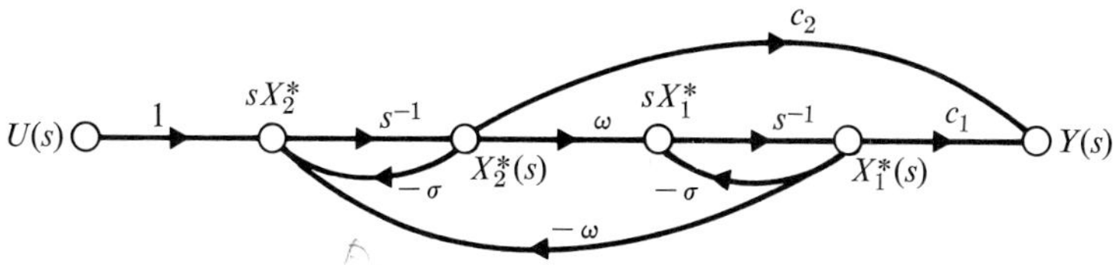

Diagonalization of the A-matrix

$$\frac{d}{dt}\begin{bmatrix} x_1^* \\ x_2^* \end{bmatrix} = \begin{bmatrix} -\sigma & \omega \\ -\omega & -\sigma \end{bmatrix}\begin{bmatrix} x_1^* \\ x_2^* \end{bmatrix} + \begin{bmatrix} 0 \\ 1 \end{bmatrix}u, \quad y = [c_1 \ c_2]\begin{bmatrix} x_1^* \\ x_2^* \end{bmatrix},$$

$$G(s) = \frac{Y(s)}{U(s)} = \frac{b_1 s + b_0}{(s+\sigma)^2 + \omega^2}, \quad c_1 = \frac{b_0 - b_1 \sigma}{\omega}, \quad c_2 = b_1$$

Fig. 3-2 Canonical structure for a second-order oscillatory system.

will satisfy the equality,

$$AT = T\Lambda, \quad T = [\mathbf{v}^1, \mathbf{v}^2].$$

In some engineering applications, it is desirable to completely avoid imaginary quantities in manipulation. In such cases, we make use of real vectors $\boldsymbol{\alpha}$ and $\boldsymbol{\beta}$ and apply the following transformation:

$$\mathbf{x} = T_m \mathbf{x}_m^*, \quad T_m = [\boldsymbol{\alpha}, \boldsymbol{\beta}].$$

Since

$$A\mathbf{v}^1 = (-\sigma + j\omega)\mathbf{v}^1 = (-\sigma + j\omega)(\boldsymbol{\alpha} + j\boldsymbol{\beta})$$
$$= (-\sigma\boldsymbol{\alpha} - \omega\boldsymbol{\beta}) + j(\omega\boldsymbol{\alpha} - \sigma\boldsymbol{\beta}),$$

or

$$A\boldsymbol{\alpha} = -\sigma\boldsymbol{\alpha} - \omega\boldsymbol{\beta}, \quad A\boldsymbol{\beta} = \omega\boldsymbol{\alpha} - \sigma\boldsymbol{\beta},$$

it follows that

$$A[\boldsymbol{\alpha}, \boldsymbol{\beta}] = [\boldsymbol{\alpha}, \boldsymbol{\beta}]\begin{bmatrix} -\sigma & \omega \\ -\omega & -\sigma \end{bmatrix}$$

or

$$AT_m = T_m \Lambda_m, \quad \Lambda_m = \begin{bmatrix} -\sigma & \omega \\ -\omega & -\sigma \end{bmatrix}. \tag{3-26}$$

The signal-flow diagram for the modified canonical form,

$$\frac{d}{dt}\mathbf{x}_m^* = \Lambda_m \mathbf{x}_m^*,$$

is given in Fig. 3-2, where input and output are added to form a second-order transfer function. We shall see in the next section that an undistorted (or normalized) spiral pattern appears in the state space for $\mathbf{x}_m^*$. An application of the modified canonical form for modal control will be discussed in Chapter 10.

Example 3-3 The matrix,

$$\mathbf{A} = \begin{bmatrix} 0 & 1 & 0 \\ -1 & -2 & 1 \\ -2 & 0 & 0 \end{bmatrix}$$

has for its characteristic equation

$$s^3 + 2s^2 + s + 2 = 0$$

and eigenvalues $p_1 = j$, $p_2 = -j$, $p_3 = -2$. The first eigenvector for $p_1 = j$ is found from the equation

$$(\mathbf{A} - j\mathbf{I})\mathbf{v}^1 = \begin{bmatrix} -j & 1 & 0 \\ -1 & -2-j & 1 \\ -2 & 0 & -j \end{bmatrix} \begin{bmatrix} v_1^1 \\ v_2^1 \\ v_3^1 \end{bmatrix} = \begin{bmatrix} 0 \\ 0 \\ 0 \end{bmatrix},$$

that is,

$$-jv_1^1 + v_2^1 = 0,$$
$$-v_1^1 - (2+j)v_2^1 + v_3^1 = 0,$$
$$-2v_1^1 - jv_3^1 = 0.$$

Taking $v_1^1 = 1$, we obtain

$$\mathbf{v}^1 = \begin{bmatrix} 1 \\ j \\ 2j \end{bmatrix} = \begin{bmatrix} 1 \\ 0 \\ 0 \end{bmatrix} + j \begin{bmatrix} 0 \\ 1 \\ 2 \end{bmatrix} = \boldsymbol{\alpha} + j\boldsymbol{\beta},$$

where

$$\boldsymbol{\alpha} = \begin{bmatrix} 1 \\ 0 \\ 0 \end{bmatrix}, \quad \boldsymbol{\beta} = \begin{bmatrix} 0 \\ 1 \\ 2 \end{bmatrix}.$$

The second eigenvector is then $\mathbf{v}^2 = \boldsymbol{\alpha} - j\boldsymbol{\beta}$. The third one, for the real eigenvalue $p_3 = -2$, is easily found to be

$$\mathbf{v}^3 = \begin{bmatrix} 1 \\ -2 \\ 1 \end{bmatrix},$$

where the first element was chosen to be $v_1^3 = 1$.

The modal matrix $\mathbf{T}$ is made up from the eigenvectors:

$$\mathbf{T} = [\mathbf{v}^1 \ \mathbf{v}^2 \ \mathbf{v}^3] = \begin{bmatrix} 1 & 1 & 1 \\ j & -j & -2 \\ 2j & -2j & 1 \end{bmatrix}.$$

We can check the result without having to find the inverse matrix $\mathbf{T}^{-1}$. We expect that the diagonal matrix will have the eigenvalues of $\mathbf{A}$ as its diagonal elements. Thus

$$\mathbf{T}^{-1}\mathbf{A}\mathbf{T} = \begin{bmatrix} j & 0 & 0 \\ 0 & -j & 0 \\ 0 & 0 & -2 \end{bmatrix} = \boldsymbol{\Lambda}.$$

Left-multiplying **T** on both sides, we must have

$$\mathbf{AT} = \mathbf{T}\mathit{\Lambda}.$$

Indeed, we find that

$$\mathbf{AT} = \begin{bmatrix} 0 & 1 & 0 \\ -1 & -2 & 1 \\ -2 & 0 & 0 \end{bmatrix} \begin{bmatrix} 1 & 1 & 1 \\ j & -j & -2 \\ 2j & -2j & 1 \end{bmatrix} = \begin{bmatrix} j & -j & -2 \\ -1 & -1 & 4 \\ -2 & -2 & -2 \end{bmatrix},$$

and

$$\mathbf{T}\mathit{\Lambda} = \begin{bmatrix} 1 & 1 & 1 \\ j & -j & -2 \\ 2j & -2j & 1 \end{bmatrix} \begin{bmatrix} j & 0 & 0 \\ 0 & -j & 0 \\ 0 & 0 & -2 \end{bmatrix} = \begin{bmatrix} j & -j & -2 \\ -1 & -1 & 4 \\ -2 & -2 & -2 \end{bmatrix}.$$

Hence the result is confirmed. Although the second eigenvector $\mathbf{v}^2$ could be chosen independently from the first eigenvector, for example

$$\mathbf{v}^2 = \begin{bmatrix} j \\ 1 \\ 2 \end{bmatrix},$$

it is usually simpler to find $\mathbf{v}^2$ by only changing the sign of the imaginary part of $\mathbf{v}^1$. We use

$$\mathbf{T}_m = [\boldsymbol{\alpha} \quad \boldsymbol{\beta} \quad \mathbf{v}^3] = \begin{bmatrix} 1 & 0 & 1 \\ 0 & 1 & -2 \\ 0 & 2 & 1 \end{bmatrix}$$

to obtain the modified canonical form,

$$\mathit{\Lambda}_m = \begin{bmatrix} 0 & 1 & 0 \\ -1 & 0 & 0 \\ 0 & 0 & -2 \end{bmatrix}.$$

The result can be confirmed as before by the following equality:

$$\mathbf{AT}_m = \mathbf{T}_m \mathit{\Lambda}_m = \begin{bmatrix} 0 & 1 & -2 \\ -1 & 0 & 4 \\ -2 & 0 & -2 \end{bmatrix}.$$

If **A** has multiple poles (repeated eigenvalues) it may be impossible to find a set of n linearly independent eigenvectors, and the diagonalization becomes impossible. It is always possible, however, to find a form

$$\mathbf{T}^{-1}\mathbf{AT} = \begin{bmatrix} p_1 & q_1 & 0 & \cdot & \cdot & \cdot & 0 \\ 0 & p_2 & q_2 & 0 & \cdot & \cdot & 0 \\ \cdot & \cdot & \cdot & \cdot & \cdot & \cdot & \cdot \\ \cdot & \cdot & \cdot & \cdot & \cdot & \cdot & \cdot \\ 0 & \cdot & \cdot & \cdot & 0 & p_{n-1} & q_{n-1} \\ 0 & \cdot & \cdot & \cdot & \cdot & 0 & p_n \end{bmatrix}, \qquad (3\text{-}27)$$

where the q_i, $i = 1, \ldots, n-1$ are either one or zero. This is the Jordan canonical form which includes the diagonal form as a special case. The canonical form can be divided into blocks, $L_1, \ldots, L_k$:

$$T^{-1}AT = \begin{bmatrix} L_1(p_1) & & 0 \\ & \ddots & \\ 0 & & L_k(p_k) \end{bmatrix}$$

where

$$L_j(p_j) = \begin{bmatrix} p_j & 1 & & & \\ & p_j & 1 & & 0 \\ & & \ddots & & \\ & 0 & & p_j & 1 \\ & & & & p_j \end{bmatrix}, \quad j = 1, 2, \ldots, k.$$

The simplest case is that in which there is only one pole, p_1, and where only one eigenvector can possibly be chosen for this pole:

$$T^{-1}AT = \begin{bmatrix} p_1 & 1 & & & \\ & p_1 & 1 & & 0 \\ & & \ddots & & \\ & 0 & & p_1 & 1 \\ & & & & p_1 \end{bmatrix}. \tag{3-28}$$

The Schwartz form is the name for this form of the Jordan canonical matrix. To determine the column vectors of the T-matrix for this case, we will again compute the T-matrix column by column, where each column is represented by a column vector v^i. First, we premultiply both sides of Eq. (3-28) by T and then perform the indicated matrix operations. On the left-hand side, the first column of AT is

$$Av^1 = p_1 v^1$$

or

$$(p_1 I - A)v^1 = 0 \tag{3-29}$$

which is the same as Eq. (3-24). Equation (3-29), however, applies only for the first column of T; the 1's in the Jordan matrix causes coupling in the relations for the remaining columns:

$$Av^2 = v^1 + p_1 v^2,$$
$$\vdots$$
$$Av^n = v^{n-1} + p_1 v^n,$$

or

$$(A - p_1 I)v^2 = v^1,$$
$$\vdots$$
$$(A - p_1 I)v^n = v^{n-1}. \tag{3-30}$$

The vector $\mathbf{v}^1$ is the eigenvector of p_1 and is determined first, from Eq. (3–29). Since there are no more linearly independent eigenvectors, the set of linearly independent vectors $\mathbf{v}^2$ through $\mathbf{v}^n$ (which are not eigenvectors) are then derived from Eqs. (3–30). The sets of vectors for $L_1, \ldots, L_k$ in the general form can be computed in a similar way.

Example 3-4 The system A-matrix

$$\mathbf{A} = \begin{bmatrix} -1 & 0.5 \\ -2 & -3 \end{bmatrix}$$

has a double pole $p_{1,2} = -2$. The eigenvector is found from Eq. (3–29):

$$\left(-2 \begin{bmatrix} 1 & 0 \\ 0 & 1 \end{bmatrix} - \begin{bmatrix} -1 & 0.5 \\ -2 & -3 \end{bmatrix} \right) \mathbf{v}^1 = 0$$

or

$$v_2^1 = -2v_1^1; \qquad \mathbf{v}^1 = \begin{bmatrix} 1 \\ -2 \end{bmatrix}.$$

Since this is the only independent eigenvector, we must turn to Eq. (3–30) to find the vector describing the second column of $\mathbf{T}$, $\mathbf{v}^2$:

$$(\mathbf{A} - p_1 \mathbf{I}) \mathbf{v}^2 = \mathbf{v}^1.$$

Substituting, we have

$$\begin{bmatrix} 1 & 0.5 \\ -2 & -1 \end{bmatrix} \begin{bmatrix} v_1^2 \\ v_2^2 \end{bmatrix} = \begin{bmatrix} 1 \\ -2 \end{bmatrix},$$

which yields

$$v_1^2 = 1 - 0.5 v_2^2, \qquad \mathbf{v}^2 = \begin{bmatrix} 1 \\ 0 \end{bmatrix},$$

and

$$\mathbf{T} = \begin{bmatrix} 1 & 1 \\ -2 & 0 \end{bmatrix}, \qquad \mathbf{T}^{-1} = \frac{1}{2} \begin{bmatrix} 0 & -1 \\ 2 & 1 \end{bmatrix}.$$

We verify this solution by checking that

$$\mathbf{T}^{-1} \mathbf{A} \mathbf{T} = \begin{bmatrix} -2 & 1 \\ 0 & -2 \end{bmatrix}.$$

The case of a third-order system with a double pole and one simple pole follows. A third-order system is still simple enough for the matrix operations to be done by hand; for higher-order systems it is usually necessary to use a computer to obtain numerical results.

Example 3-5 Consider a third-order system represented by

$$\mathbf{A} = \begin{bmatrix} -3 & 1 & 0 \\ 0 & -3 & 1 \\ -4 & 0 & 0 \end{bmatrix}.$$

The characteristic equation is

$$s^3 + 6s^2 + 9s + 4 = (s+1)^2(s+4) = 0.$$

The eigenvalues are

$$p_1 = -1, \quad p_2 = -1, \quad \text{and} \quad p_3 = -4.$$

We fix the first eigenvector by

$$(\mathbf{A} - p_1\mathbf{I})\mathbf{v}^1 = (\mathbf{A} + \mathbf{I})\mathbf{v}^1 = \begin{bmatrix} -2 & 1 & 0 \\ 0 & -2 & 1 \\ -4 & 0 & 1 \end{bmatrix} \begin{bmatrix} v_1^1 \\ v_2^1 \\ v_3^1 \end{bmatrix} = \mathbf{0}.$$

The first equation is $-2v_1^1 + v_2^1 = 0$, and we choose $v_1^1 = 1$, $v_2^1 = 2$. From the second equation $-2v_2^1 + v_3^1 = 0$ we get $v_3^1 = 4$ and confirm that the third equation is satisfied since $v_3^1 = 4v_1^1$.

Since we could fix only one eigenvector for $p_1 = -1$, we must find another vector by

$$(\mathbf{A} - p_1\mathbf{I})\mathbf{v}^2 = \mathbf{v}^1,$$

that is,

$$\begin{bmatrix} -2 & 1 & 0 \\ 0 & -2 & 1 \\ -4 & 0 & 1 \end{bmatrix} \begin{bmatrix} v_1^2 \\ v_2^2 \\ v_3^2 \end{bmatrix} = \begin{bmatrix} 1 \\ 2 \\ 4 \end{bmatrix}.$$

For the first condition $-2v_1^2 + v_2^2 = 1$ we choose $v_1^2 = 0$, $v_2^2 = 1$. The second equation $-2v_2^2 + v_3^2 = 2$ gives $v_3^2 = 4$. We confirm that the third condition $-4v_1^2 + v_3^2 = 4$ is satisfied by the chosen values of v_1^2, v_2^2, v_3^2.

Similarly, we take $v_1^3 = 1$, $v_2^3 = -1$, and $v_3^3 = 1$ for $p_3 = -4$. The canonical form is

$$\mathbf{T}^{-1}\mathbf{A}\mathbf{T} = \begin{bmatrix} -1 & 1 & 0 \\ 0 & -1 & 0 \\ 0 & 0 & -4 \end{bmatrix},$$

where

$$\mathbf{T} = [\mathbf{v}^1, \mathbf{v}^2, \mathbf{v}^3] = \begin{bmatrix} 1 & 0 & 1 \\ 2 & 1 & -1 \\ 4 & 4 & 1 \end{bmatrix}.$$

A pair of identical first-order systems, without mutual coupling, can be considered as a second-order system. In this case, the system matrix will be in diagonal form,

$$\frac{d}{dt}\mathbf{x} = \begin{bmatrix} p_1 & 0 \\ 0 & p_1 \end{bmatrix} \mathbf{x},$$

even though the two eigenvalues are the same. In most second-order systems that we will encounter, however, if the system has a double pole, a modal transformation will yield the Schwartz form. Application of Eq. (3–29) to find the eigenvector for p_1 gives the interesting result

$$0\mathbf{v}^1 = \mathbf{0}.$$

Since *any* finite vector will satisfy this relation, we can choose any two linearly independent vectors to form the transformation matrix **T**. That is to say, since the modes are already decoupled, we would expect that any nonsingular square matrix is a satisfactory transformation matrix. This is easily confirmed:

$$\mathbf{T}^{-1}\mathbf{A}\mathbf{T} = \mathbf{T}^{-1}\begin{bmatrix} p_1 & 0 \\ 0 & p_1 \end{bmatrix}\mathbf{T} = p_1\mathbf{T}^{-1}\mathbf{I}\mathbf{T} = p_1\mathbf{I}.$$

This seemingly trivial case of second-order system can also come in the form of decoupled modes which are embedded in the system equations for higher-order systems (see [4], pp. 256–257, for a test method to determine the mode decoupling). Such is the case in the third-order system example that follows below.

Example 3-6 The matrix

$$\mathbf{A} = \begin{bmatrix} -1.8 & 0.4 & -0.4 \\ 0 & -1 & 0 \\ -0.4 & 0.2 & -1.2 \end{bmatrix}$$

has eigenvalues $p = -1, -1, -2$. In solving for the first eigenvector instead of finding two equations relating the three variables v_1^1, v_2^1, and v_3^1, we find only one equation

$$2v_1^1 - v_2^1 + v_3^1 = 0. \tag{3-31}$$

Because we have only one equation and three unknowns, we can pick two linearly independent vectors satisfying Eq. (3-31) and eliminate the need to use Eq. (3-30). The eigenvector corresponding to $p = -2$ can be found in the usual manner, using Eq. (3-29). Three possible eigenvectors are:

$$\mathbf{v}^1 = \begin{bmatrix} 0 \\ 1 \\ 1 \end{bmatrix}, \quad \mathbf{v}^2 = \begin{bmatrix} 1 \\ 2 \\ 0 \end{bmatrix}, \quad \mathbf{v}^3 = \begin{bmatrix} 2 \\ 0 \\ 1 \end{bmatrix}.$$

Since we were able to choose two linearly independent vectors for the double pole without assuming the presence of any off-diagonal terms in the canonical matrix [as we were forced to do in deriving Eq. (3-30)], the transformed **A**-matrix is purely diagonal:

$$\mathbf{T}^{-1}\mathbf{A}\mathbf{T} = \begin{bmatrix} -1 & 0 & 0 \\ 0 & -1 & 0 \\ 0 & 0 & -2 \end{bmatrix}.$$

3-4 CONTROLLABILITY AND OBSERVABILITY

The modal transformation leading to a diagonalized **A**-matrix can be used to illustrate the concepts of controllability and observability. To examine the controllability of a system, we ask the question, "Can the complete state of the system be controlled by a particular input u_i?" The observability of a system tests the possibility of observing the complete state of the system by

76 Response of linear, lumped-parameter systems

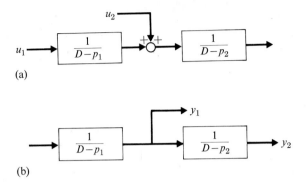

Fig. 3-3 A second-order system (a) uncontrollable by u_2 and (b) unobservable by y_1.

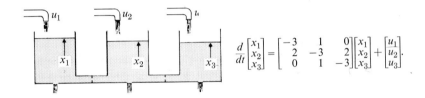

$$\frac{d}{dt}\begin{bmatrix} x_1 \\ x_2 \\ x_3 \end{bmatrix} = \begin{bmatrix} -3 & 1 & 0 \\ 2 & -3 & 2 \\ 0 & 1 & -3 \end{bmatrix}\begin{bmatrix} x_1 \\ x_2 \\ x_3 \end{bmatrix} + \begin{bmatrix} u_1 \\ u_2 \\ u_3 \end{bmatrix}.$$

Fig. 3-4 A three-tank system.

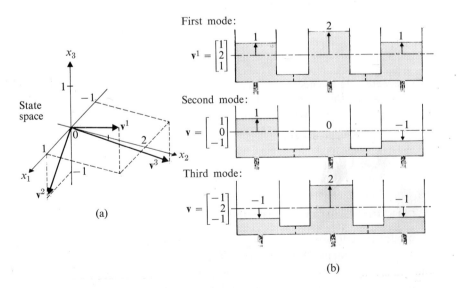

Fig. 3-5 Characteristic modes of the three-tank system.

noting a particular output y_i. The formalization of the concepts of controllability and observability is not necessary for some systems of simple structure; the questions involved can easily be answered by inspection. For example, the forcing input u_2 in Fig. 3-3(a) cannot reach the first mode, $e^{p_1 t}$; thus the system is said to be uncontrollable by u_2. In Fig. 3-3(b) it is obvious that the second mode $e^{p_2 t}$ will not appear in the output y_1; the system is unobservable with y_1. For more complex systems, however, as we will see in the next example, it is not always obvious whether a control input can completely control the state of the system or a particular output can observe it completely.

Consider the three-tank liquid-level system shown in Fig. 3-4. As control system designers, for example, we might be interested in finding out which of the three possible inputs, u_1, u_2, or u_3, and three possible observations, x_1, x_2, or x_3, could be used in designing the control for this system. The state equation will take the following form (with numbers already substituted for the resistances and surface areas):

$$\frac{d}{dt}\begin{bmatrix} x_1 \\ x_2 \\ x_3 \end{bmatrix} = \begin{bmatrix} -3 & 1 & 0 \\ 2 & -3 & 2 \\ 0 & 1 & -3 \end{bmatrix} \begin{bmatrix} x_1 \\ x_2 \\ x_3 \end{bmatrix} + \begin{bmatrix} u_1 \\ u_2 \\ u_3 \end{bmatrix}. \qquad (3\text{-}32)$$

We solve the characteristic equation $|p\mathbf{I} - \mathbf{A}| = 0$ and find that the eigenvalues are

$$p_1 = -1, \qquad p_2 = -3, \qquad \text{and} \qquad p_3 = -5$$

and a corresponding set of eigenvectors is

$$\mathbf{v}^1 = \begin{bmatrix} 1 \\ 2 \\ 1 \end{bmatrix}, \qquad \mathbf{v}^2 = \begin{bmatrix} 1 \\ 0 \\ -1 \end{bmatrix}, \qquad \mathbf{v}^3 = \begin{bmatrix} -1 \\ 2 \\ -1 \end{bmatrix}.$$

The modal matrix is constructed from the eigenvectors:

$$\mathbf{T} = \begin{bmatrix} 1 & 1 & -1 \\ 2 & 0 & 2 \\ 1 & -1 & -1 \end{bmatrix}.$$

The eigenvalues give the response characteristics of a system's modes (the exponential terms described in Section 3-2). The eigenvectors are the link that relate these response characteristics to particular changes in the state of the system as measured by the state variables x_1, x_2, and x_3. It is often convenient to view the state of the system as a point (or vector) in the space having x_1, x_2, and x_3 as coordinate axes. This space is called the state space. The eigenvectors are three nonparallel direction vectors in the state space (Fig. 3-5a); it is motion at different rates along these vectors that is described by the characteristic modes. (Note that since the eigenvectors are not par-

Response of linear, lumped-parameter systems

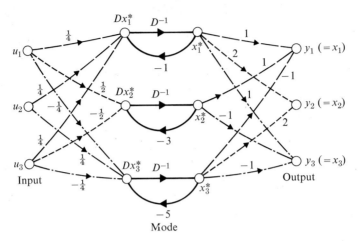

Fig. 3–6 Signal-flow diagram of the three-tank system in the modal domain.

allel to each other, any point in the state space can be represented in terms of components in each of the eigenvector directions.) The physical meaning of the modes is shown in Fig. 3–5(b). The first mode (corresponding to $\mathbf{v}^1$) describes a state of motion in which the level is simultaneously rising (or falling) in all the tanks. The second mode describes a state in which the first and third tank levels are moving in opposite directions and the second tank level is not moving at all. In the third mode the levels in tanks one and three are going in the direction opposite to the level in tank two. The decay rate of each of the modes is given by the exponential term $e^{p_i t}$, and since the modal motion represents a motion in a diagonalized domain, there is no coupling between the modes. In this case, the third mode decays fastest ($p_3 = -5$) and the first mode decays the slowest ($p_1 = -1$).

The state equation in the diagonalized (modal) domain is

$$\frac{d}{dt}\mathbf{x}^* = \mathbf{T}^{-1}\mathbf{A}\mathbf{T}\mathbf{x}^* + \mathbf{T}^{-1}\mathbf{B}\mathbf{u} \tag{3-33}$$

or

$$\frac{d}{dt}\begin{bmatrix} x_1^* \\ x_2^* \\ x_3^* \end{bmatrix} = \begin{bmatrix} -1 & 0 & 0 \\ 0 & -3 & 0 \\ 0 & 0 & -5 \end{bmatrix} \begin{bmatrix} x_1^* \\ x_2^* \\ x_3^* \end{bmatrix} + \frac{1}{4}\begin{bmatrix} 1 & 1 & 1 \\ 2 & 0 & -2 \\ -1 & 1 & -1 \end{bmatrix} \begin{bmatrix} u_1 \\ u_2 \\ u_3 \end{bmatrix}. \tag{3-34}$$

We can establish the ability of inputs u_1, u_2, or u_3 to control the system by examining Eq. (3–34). For an input to be able to control the system, it must be able to affect all the modes of the system. The system at hand is controllable by either the first or third inputs, because we see that the matrix $\mathbf{T}^{-1}\mathbf{B}$ has no zeros in its first or third rows; it is not controllable by the second input, since the zero in the second column of the second row of $T^{-1}B$ shows that the input u_2 has no effect on the second mode (see Fig. 3–6).

The outputs can be tested in a similar manner. Consider the output vector **y**,

$$\mathbf{y} = \begin{bmatrix} x_1 \\ x_2 \\ x_3 \end{bmatrix}.$$

The output equation is

$$\mathbf{y} = \mathbf{Cx} + \mathbf{Du} = \begin{bmatrix} 1 & 0 & 0 \\ 0 & 1 & 0 \\ 0 & 0 & 1 \end{bmatrix} \mathbf{x} + 0\mathbf{u}.$$

In the modal domain this equation becomes

$$\mathbf{y} = \mathbf{CTx^*} = \begin{bmatrix} 1 & 1 & -1 \\ 2 & 0 & 2 \\ 1 & -1 & -1 \end{bmatrix} \begin{bmatrix} x_1^* \\ x_2^* \\ x_3^* \end{bmatrix}. \tag{3-35}$$

The zero that appears in the second column of the second row of **CT** shows that the second mode is not observable by means of the output y_2 (see Fig. 3–6).

What is the physical significance of these findings? If, in order to achieve some control objective, this system is forced by introducing liquid into the middle tank (input u_2) of the system, it would be found that no manipulation of the input would have any effect on the second mode;[†] it would only affect the amplitude of the first and the third modes. It is also clear that no reading on the second mode can be obtained by making observation at its node, the level of the second tank.

To make a general statement on controllability and observability for a system that has distinct real eigenvalues, we apply the above reasoning to the general form of the modal-domain state and output equations for scalar input and output:

$$\frac{d}{dt}\mathbf{x^*} = \mathbf{\Lambda x^*} + \mathbf{b^*}u, \qquad y = \mathbf{c^*x^*}, \tag{3-36}$$

where

$$\mathbf{b^*} = \mathbf{T^{-1}b}, \qquad \mathbf{c^*} = \mathbf{cT}.$$

For a system to be controllable (or observable) by input u (or output y), the

---

† The uncontrollability of the second mode in this example is not only due to the presence of its node in the second tank. The pattern of the first and the third modes must be symmetric relative to the middle tank.

Table 3-1 A SYSTEM WITH POLE CANCELLATION

	System I	System II
Block diagram	$U(s) \rightarrow \boxed{\dfrac{s-1}{s+2}} \rightarrow \boxed{\dfrac{1}{s-1}} \rightarrow Y(s)$ Compensation network — Original system	$U(s) \rightarrow \boxed{\dfrac{1}{s-1}} \rightarrow \boxed{\dfrac{s-1}{s+2}} \rightarrow Y(s)$ Original system — Compensation network
Signal-flow diagram	(signal-flow graph with nodes $D^{-1}x_2$, $D^{-1}x_1$; gains -3, 1, -2, 1, 1)	(signal-flow graph with nodes $D^{-1}x_2$, $D^{-1}x_1$; gains 1, -3, -2, 1, 1)
System's matrices	$\mathbf{A} = \begin{bmatrix} 1 & 1 \\ 0 & -2 \end{bmatrix}$, $\mathbf{b} = \begin{bmatrix} 1 \\ -3 \end{bmatrix}$, $\mathbf{c} = [1\ 0]$.	$\mathbf{A} = \begin{bmatrix} -2 & -3 \\ 0 & 1 \end{bmatrix}$, $\mathbf{b} = \begin{bmatrix} 0 \\ 1 \end{bmatrix}$, $\mathbf{c} = [1\ 1]$.
Modal matrix	$\mathbf{T} = \begin{bmatrix} 1 & 1 \\ 0 & -3 \end{bmatrix}$, $\mathbf{T}^{-1} = \dfrac{1}{3}\begin{bmatrix} 3 & 1 \\ 0 & -1 \end{bmatrix}$.	$\mathbf{T} = \begin{bmatrix} 1 & 1 \\ -1 & 0 \end{bmatrix}$, $\mathbf{T}^{-1} = \begin{bmatrix} 0 & -1 \\ 1 & 1 \end{bmatrix}$.
Modal-domain relation	$\mathbf{T}^{-1}\mathbf{b} = \mathbf{b}^* = \begin{bmatrix} 0 \\ 1 \end{bmatrix}$, $\mathbf{cT} = \mathbf{c}^* = [1\ 1]$. (signal-flow diagram with $D^{-1}x_1^*$, $D^{-1}x_2^*$; gains 1, 1, 1, -2) Uncontrollable	$\mathbf{T}^{-1}\mathbf{b} = \mathbf{b}^* = \begin{bmatrix} -1 \\ 1 \end{bmatrix}$, $\mathbf{cT} = \mathbf{c}^* = [0\ 1]$. (signal-flow diagram with $D^{-1}x_1^*$, $D^{-1}x_2^*$; gains -1, 1, 1, -2) Unobservable

column matrix $\mathbf{b}^*$ (or row matrix $\mathbf{c}^*$) must not contain zeros. If zeros appear in the matrix, their number is equal to the number of uncontrollable (or unobservable) modes. However, there are some exceptions (see Examples 3-8, 3-9, and the end of Section 3-5).

Example 3-7 In the classical control technique a compensation network was sometimes applied with the intention of replacing a system's undesirable pole ($p_1 = 1$ in Table 3-1) by a desired pole (for instance $p_2 = -2$ in the table). The pitfall in this approach can be uncovered by the modern control theory. For this purpose we first construct the signal-flow diagram by the method given in Chapter 2, and obtain the matrices $\mathbf{A}, \mathbf{b},$ and $\mathbf{c}$. It is then possible to diagonalize the relation and obtain the

3-4 **Controllability and observability** **81**

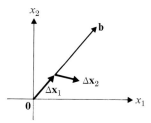

Fig. 3-7 Forcing the state of a second-order system from the origin.

modal-domain expressions:

$$\frac{d}{dt}\begin{bmatrix} x_1^* \\ x_2^* \end{bmatrix} = \begin{bmatrix} 1 & 0 \\ 0 & -2 \end{bmatrix} \begin{bmatrix} x_1^* \\ x_2^* \end{bmatrix} + \mathbf{b}^*u, \qquad y = \mathbf{c}^*\mathbf{x}^*.$$

The modal-domain structure given in the table reveals that the system is made either uncontrollable or unobservable. In system I (left-hand side) control $u(t)$ acts only on the second mode ($p_2 = -2$). This means that the state $\mathbf{x}^*$ is forced by u only in the direction of the x_2^*-axis in the $x_1^* x_2^*$ modal state plane. Although the two modes are both forced in System II (right-hand side in the table), the observation is made only in the direction of the x_2^*-axis in the state plane. So far as the theoretical input-output relation for zero initial state is concerned, classical control theory, according to which the overall transfer function $G(s)$ is given by the product of $G_1(s) = 1/(s-1)$ and $G_2(s) = (s-1)/(s+2)$, is correct:

$$G(s) = G_1(s)G_2(s) = G_2(s)G_1(s) = \frac{1}{s+2}.$$

Since, however, the original system's eigenvalue $p_1 = 1$ produces an unstable mode e^t (an exponential rise), a nonzero initial value $x_1^*(0)$, however small, may eventually build up in System I. In System II the unstable mode is forced by input u, but it is ignored in the observation. The pole cancellation scheme will not work in either case.

A geometric interpretation of controllability and observability in the state space leads to criteria which do not require diagonalization and hence are also applicable to systems with complex or multiple eigenvalues. The uncontrollability of the first mode by the input u in System I of Example 3-7 (Table 3-1) means that in the original state space [the $(x_1 x_2)$-plane] the input cannot force the state $\mathbf{x}$ of the system in the direction of the first eigenvector $\mathbf{v}^1$. To derive the controllability condition from this viewpoint, let us consider a second-order system described by

$$\frac{d}{dt}\mathbf{x} = \mathbf{A}\mathbf{x} + \mathbf{b}u, \qquad (3-37)$$

and an initial state at the origin of the state space. If a control input $u = u_1$ is applied for a short time interval Δt, the state of the system will initially move away from the origin in the direction of vector $\mathbf{b}$ (Fig. 3-7) since the first term of Eq. (3–37) will be very small:

$$(\Delta \mathbf{x})_1 \approx \mathbf{b} u_1 \cdot \Delta t .$$

However, the magnitude of the x-vector soon becomes appreciable, and for a second short time interval, with control $u = u_2$, the motion can be approximated by the equation

$$(\Delta \mathbf{x})_2 \approx \mathbf{A}(\Delta \mathbf{x})_1 \Delta t + \mathbf{b} u_2 \Delta t = \mathbf{A} \mathbf{b} u_1 (\Delta t)^2 + \mathbf{b} u_2 \Delta t .$$

If the second move is in a direction different from the first, then the system will be capable of reaching any region in the state space (a plane in this case), and thus the system will be controllable. From the last expression we see that controllability (the capability of u to move the state from the origin to any point in the state space) is assured if the vectors $\mathbf{b}$ and $\mathbf{Ab}$ are linearly independent, or if the matrix whose columns are vectors $\mathbf{b}$ and $\mathbf{Ab}$ has a nonzero determinant:

$$|\mathbf{b}, \mathbf{Ab}| \neq 0 .$$

For a third-order system we consider three steps of control, $u = u_1, u_2$, and u_3, to reach a point in a three-dimensional space, and require for its controllability the condition

$$|\mathbf{b}, \mathbf{Ab}, \mathbf{A}^2 \mathbf{b}| \neq 0 .$$

For the general case of an nth-order system this condition becomes

$$|\mathbf{b}, \mathbf{Ab}, \mathbf{A}^2 \mathbf{b}, \ldots, \mathbf{A}^{n-1} \mathbf{b}| \neq 0 . \tag{3-38}$$

The controllability conditions for the systems of Example 3–7 are given in Table 3–2. For system I the two vectors

$$\mathbf{b} = \begin{bmatrix} 1 \\ -3 \end{bmatrix}, \quad \mathbf{Ab} = \begin{bmatrix} -2 \\ 6 \end{bmatrix}$$

are linearly dependent; thus

$$|\mathbf{b}, \mathbf{Ab}| = \begin{vmatrix} 1 & -2 \\ -3 & 6 \end{vmatrix} = 0 ,$$

and the system is uncontrollable. The rank* of this determinant is 1, meaning that one mode is controllable: the state at the origin can be forced by u in the direction of $\pm \mathbf{v}^2$ (and only in that direction). In general, the rank of the determinant (3–38) is equal to the number of controllable modes. When

---

* The rank, r, of a matrix is a measure of the largest nonzero $r \times r$ determinant that can be formed from the elements of the matrix.

3-4 Controllability and observability

Table 3-2 CONTROLLABILITY AND OBSERVABILITY OF SYSTEMS I AND II IN THE PRECEDING TABLE

	System I	System II
Controllability	(a) Uncontrollable	(b) Controllable
Observability	(c) Observable	(d) Unobservable

this number is less than the order of the system n, the system is said to be uncontrollable.

We follow the same line of argument to derive the necessary and sufficient condition for the observability of a system whose output (or observation) matrix is represented by a row vector $\mathbf{c}$ such that

$$y = \mathbf{cx}.$$

In order for the system to be observable the requisite condition is:

$$|\mathbf{c}', \mathbf{A}'\mathbf{c}', \ldots, (\mathbf{A}')^{n-1}\mathbf{c}'| \neq 0, \tag{3-39}$$

where all the component matrices are transposed because $\mathbf{c}$ is a row matrix. Let us again consider a second-order system to see the geometric interpretation of this condition. Suppose we make measurements of a free response at time t_1 and t_2:

$$y_1 = \mathbf{cx}(t_1), \qquad y_2 = \mathbf{cx}(t_2).$$

For a short time interval between the measurements, $\Delta t = t_2 - t_1$, the following will hold:

$$\mathbf{x}(t_2) \approx \mathbf{x}(t_1) + \mathbf{Ax}(t_1) \cdot \Delta t.$$

Thus

$$y_2 \approx y_1 + \mathbf{cAx}(t_1) \cdot \Delta t.$$

Hence

$$\begin{bmatrix} y_1 \\ y_2 - y_1 \end{bmatrix} \approx \begin{bmatrix} \mathbf{c} \\ \mathbf{cA}\Delta t \end{bmatrix} \mathbf{x}(t_1),$$

and we can fix the state $\mathbf{x}(t_1)$ of a second-order system so long as $\mathbf{c}$ and $\mathbf{cA}$ are linearly independent (so that the inverse of the right-hand side matrix will exist). The condition (3–39) is a generalization of this argument.

The scalar product $\mathbf{cx}$ ($\mathbf{c}$ is a row vector and $\mathbf{x}$ is a column vector) represents the projection of the state vector $\mathbf{x}$ onto the observation vector $\mathbf{c}$. As shown in Table 3–2(d), $\mathbf{c}$ in System II of Example 3–7 (and Table 3–1) is orthogonal to $\mathbf{v}^1$ (that is, the inner product of $\mathbf{c}$ and $\mathbf{v}^1$ is zero—the two vectors are perpendicular). This is why the first mode is unobservable in this system. Although

$$|\mathbf{c}', \mathbf{A}'\mathbf{c}'| = \begin{vmatrix} 1 & -2 \\ 1 & -2 \end{vmatrix} = 0,$$

the determinant has rank 1, meaning one out of the two modes is observable. In general, the rank of (3–39) is equal to the number of observable modes. When this number is less than the order of the system, the system is said to be unobservable.

The above discussions can be extended to systems with multiple inputs or outputs. A system whose input matrix is $\mathbf{B}$ is said to be controllable if and only if

$$\text{Rank } [\mathbf{B}, \mathbf{AB}, \ldots, \mathbf{A}^{n-1}\mathbf{B}] = n. \tag{3–40}$$

A system with an output matrix $\mathbf{C}$ is called observable when

$$\text{Rank } [\mathbf{C}', \mathbf{A}'\mathbf{C}', \ldots, (\mathbf{A}')^{n-1}\mathbf{C}'] = n. \tag{3–41}$$

Equation (3–40) implies that every mode is reached by at least one input, and Eq. (3–41) has the meaning that every mode is connected to at least one output. For u_j to control all modes, then the jth column of $\mathbf{T}^{-1}\mathbf{B}$ must contain no zeros. If a zero is in the kth row of the jth column, then the kth mode is not controlled by u_j. Similarly, for y_j to observe all modes (states), then the jth row of $\mathbf{CT}$ must not contain any zeros. If a zero is in the kth column of the jth row, then the kth mode is unobservable by y_j.

Example 3–8 A second-order system in Jordan canonical form (the Schwartz form),

$$\frac{d}{dt}\mathbf{x}^* = \begin{bmatrix} p & 1 \\ 0 & p \end{bmatrix}\mathbf{x}^* + \begin{bmatrix} b_1 \\ b_2 \end{bmatrix}u, \quad y = [c_1 \quad c_2]\mathbf{x}^*,$$

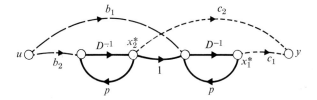

Fig. 3-8 A second-order system in Jordan canonical form.

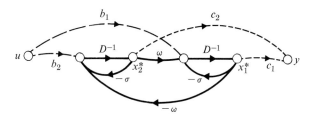

Fig. 3-9 A second-order oscillatory system in the modified canonical form.

has a chain structure, as shown in Fig. 3-8. The system loses controllability when

$$|\mathbf{b}, \mathbf{Ab}| = \begin{vmatrix} b_1 & pb_1 + b_2 \\ b_2 & pb_2 \end{vmatrix} = b_2^2 = 0 .$$

The system is controllable when $b_2 \neq 0$, regardless of b_1. We see in Fig. 3-8 that the first link of the chain (state variable x_2) becomes unreachable by u when $b_2 = 0$. Similarly, we anticipate that $c_1 \neq 0$ will be the only condition for the system to be observable. This can be confirmed:

$$|\mathbf{c}', \mathbf{A}'\mathbf{c}'| = \begin{vmatrix} c_1 & pc_1 \\ c_2 & c_1 + pc_2 \end{vmatrix} = c_1^2 = 0 .$$

Example 3-9 A second-order system with conjugate complex eigenvalues expressed in the modified canonical form

$$\frac{d}{dt}\mathbf{x}^* = \begin{bmatrix} -\sigma & \omega \\ -\omega & -\sigma \end{bmatrix}\mathbf{x}^* + \begin{bmatrix} b_1 \\ b_2 \end{bmatrix} u , \quad y = [c_1 \quad c_2]\mathbf{x}^* , \quad \omega > 0 ,$$

has a signal-flow diagram with a feedback loop (Fig. 3-9). The system will become uncontrollable when

$$|\mathbf{b}, \mathbf{Ab}| = \begin{vmatrix} b_1 & -\sigma b_1 + \omega b_2 \\ b_2 & -\omega b_1 - \sigma b_2 \end{vmatrix} = \omega(b_1^2 + b_2^2) = 0 .$$

This condition is satisfied only when *both* b_1 and b_2 are zero, which means a complete isolation of the system from u. The rank of the determinant is either 2 or 0. A similar condition holds on observability because of the following determinant:

$$|\mathbf{c}', \mathbf{A}'\mathbf{c}'| = \begin{vmatrix} c_1 & -\sigma c_1 - \omega c_2 \\ c_2 & \omega c_1 - \sigma c_2 \end{vmatrix} = \omega(c_1^2 + c_2^2) .$$

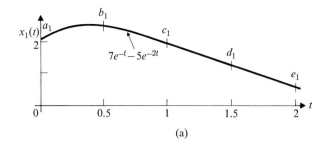

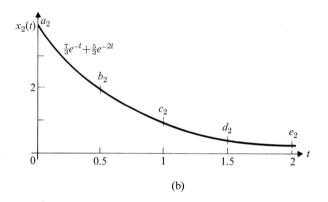

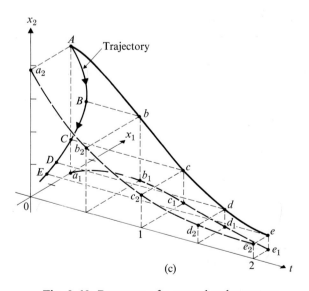

Fig. 3–10 Response of a second-order system.

3-5 MOTION IN THE STATE SPACE

The response of a system starting from an initial point $\mathbf{x} = \mathbf{x}(t_0)$, $t = t_0$, traces a curve in the space which has the n state variables and time as its coordinate axes [an $(n + 1)$-dimensional space]. The projection of the response curve onto the n-dimensional space having the n state variables $x_1, \ldots, x_n$ as coordinate axes does not include time explicitly but has proved to be of great value in systems analysis. The curve is called a trajectory, and, as we noted in Section 3-4, the space is a state space.* For second-order systems the state space is a plane, in which case graphical interpretation of system responses is particularly easy; for systems of order higher than three, the concept and mathematics of the trajectory is still used, but of course graphical representation cannot be easily visualized.

Example 3-10 Consider the free response $[u(t) = 0]$ of the system whose **A**-matrix is

$$\mathbf{A} = \begin{bmatrix} -\frac{3}{2} & \frac{3}{2} \\ \frac{1}{6} & -\frac{3}{2} \end{bmatrix}. \quad p = -1, -2$$

The solution matrix for this system was found in (3-17); and, for example, for an initial state of

$$\mathbf{x}_0 = \begin{bmatrix} 2 \\ 4 \end{bmatrix}$$

the response is given by

$$\mathbf{x}(t) = \mathbf{S}(t)\mathbf{x}_0 = \begin{bmatrix} 7e^{-t} - 5e^{-2t} \\ \frac{7}{3}e^{-t} + \frac{5}{3}e^{-2t} \end{bmatrix}.$$

$$S(t) = \mathcal{L}^{-1}\{[sI - A]^{-1}\}$$

The response is represented in Fig. 3-10 by the projection of the response curve onto the planes $x_1 t$, $x_2 t$, and by the space curve in the space (x_1, x_2, t). The (x_1, x_2)-plane in Fig. 3-10(c) and Fig. 3-11 is the state space for this system. The arrows on the trajectory in the state space show the direction along the trajectory for increasing time, and may be determined by examining any one of the state variables as is found convenient. Thus $x_2(t)$ in the example must be less than $x_2(0)$, and the trajectory must drop.

If the initial state happens to be either

$$\mathbf{x}_0 = \begin{bmatrix} 3 \\ 1 \end{bmatrix} \quad \text{or} \quad \mathbf{x}_0 = \begin{bmatrix} 3 \\ -1 \end{bmatrix},$$

then either

$$\mathbf{x}(t) = \mathbf{S}(t)\begin{bmatrix} 3 \\ 1 \end{bmatrix} = \begin{bmatrix} 3e^{-t} \\ e^{-t} \end{bmatrix}$$

---

* The state space is sometimes called the phase space, especially when the state variables are related according to the equation

$$\frac{d}{dt}x_1 = x_2, \ldots, \frac{d}{dt}x_{n-1} = x_n.$$

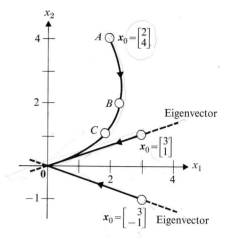

Fig. 3-11 Trajectories for a second-order system for various initial states.

and thus $x_1(t)/x_2(t) = 3$ for all t, or in the latter case,

$$\mathbf{x}(t) = \mathbf{S}(t)\begin{bmatrix} 3 \\ -1 \end{bmatrix} = \begin{bmatrix} 3e^{-2t} \\ -e^{-2t} \end{bmatrix},$$

and $x_1(t)/x_2(t) = -3$ for all t.

Since the state space for this system is the $x_1 x_2$-plane, the trajectories for both of these initial conditions are straight lines (Fig. 3-11) on the state plane.

Why should these particular starting points lead to straight-line trajectories? In seeking to generalize this result, we refer to Section 3-3 (where the eigenvectors for this system were calculated) and note that these starting points are each on one of the system's eigenvectors. The general conclusion to be drawn from this observation is that any free-response trajectory that starts on an eigenvector will follow along that eigenvector to the origin. We will now show that this conclusion is correct.

A point in the state space can be expressed vectorially as

$$\mathbf{x} = \begin{bmatrix} x_1 \\ x_2 \\ \vdots \\ x_n \end{bmatrix} = x_1 \mathbf{e}_1 + x_2 \mathbf{e}_2 + \cdots + x_n \mathbf{e}_n, \qquad (3\text{-}42)$$

where the $\mathbf{e}_i$ are unit vectors in the coordinate directions (also called basis vectors):

$$\mathbf{e}_1 = \begin{bmatrix} 1 \\ 0 \\ 0 \\ \vdots \\ 0 \end{bmatrix}, \quad \mathbf{e}_2 = \begin{bmatrix} 0 \\ 1 \\ 0 \\ \vdots \\ 0 \end{bmatrix}, \quad \ldots, \quad \mathbf{e}_n = \begin{bmatrix} 0 \\ 0 \\ 0 \\ \vdots \\ 0 \\ 1 \end{bmatrix}.$$

3-5 Motion in the state space

This set, however, is not the only possible set of basis vectors. Any set of n linearly independent vectors can serve as basis vectors for the general column vector **x**. In particular, we have already established that an nth-order system has n linearly independent eigenvectors (if the eigenvalues are all distinct and real) and they can serve as a set of basis vectors;

$$\mathbf{x} = x_1^* \mathbf{v}^1 + x_2^* \mathbf{v}^2 + \cdots + x_n^* \mathbf{v}^n, \tag{3-43}$$

where, assuming that all eigenvectors are normalized (their lengths are made unity):

$$\|\mathbf{v}^i\| = 1, \qquad i = 1, \ldots, n.$$

The differential equation for the free response (that is, homogeneous equation) is

$$\frac{d}{dt}\mathbf{x}(t) = \mathbf{A}\mathbf{x}(t) = x_1^*(t)\mathbf{A}\mathbf{v}^1 + \cdots + x_n^*(t)\mathbf{A}\mathbf{v}^n.$$

However, by the defining equations for the eigenvectors,

$$(\mathbf{A} - p_i\mathbf{I})\mathbf{v}^i = 0 \qquad \text{or} \qquad \mathbf{A}\mathbf{v}^i = p_i\mathbf{v}^i.$$

Therefore,

$$\frac{d}{dt}\mathbf{x}(t) = p_1 x_1^* \mathbf{v}^1 + \cdots + p_n x_n^* \mathbf{v}^n. \tag{3-44}$$

If, for instance, $\mathbf{x} = x_i^* \mathbf{v}^i$ (that is, the state of the system, as a point in the state space, is on the ith eigenvector), then

$$\frac{d}{dt}\mathbf{x} = p_i x_i^* \mathbf{v}^i,$$

which means that the velocity vector (the instantaneous tangent to the trajectory) is oriented along the eigenvector. Interpreted geometrically, this result shows that any trajectory that crosses (or starts on) an eigenvector will remain on that eigenvector.

As we have seen, when a trajectory starts on an eigenvector, an analytic expression for it can be derived easily because the solution involves only a single mode. In the general case, for instance in the example problem we just solved, the solution involves all of the modes:

$$x_1(t) = 7e^{-t} - 5e^{-2t}, \qquad x_2(t) = \tfrac{7}{3}e^{-t} + \tfrac{5}{3}e^{-2t}.$$

We can eliminate the variable t from the above equations by solving first for e^{-t} and e^{-2t}:

$$e^{-t} = \frac{x_1 + 3x_2}{14}, \qquad e^{-2t} = \frac{-x_1 + 3x_2}{10}.$$

We square the first equation and equate:

$$\left(\frac{x_1 + 3x_2}{14}\right)^2 = -\frac{x_1 - 3x_2}{10}$$

to get the analytic expression for the trajectory that is shown in Fig. 3–11.

Although it is often possible to trace a trajectory by deriving an analytic expression for it, this approach is usually not advisable because it is slow and cumbersome. The *isocline method* is a commonly used graphical technique for rapid determination of trajectories for second-order systems. Briefly, the method involves finding a family of curves, called *isoclines*, in the state plane. These curves have the property that a given trajectory will always cross the same isocline with the same slope dx_2/dx_1. For linear stationary second-order systems, the isoclines turn out to be straight lines. Once the isoclines are determined for a given system, it is possible to determine the trajectory from any initial point quickly and easily (note that if the trajectories are found by deriving the analytic expression, a new expression must be derived for every initial condition).

The actual use of the isocline method is best illustrated with an example. We will continue with the system we have been considering in this section and find its isoclines and then use the isocline method to find the trajectories.

We recall that

$$\mathbf{A} = \begin{bmatrix} -\frac{3}{2} & \frac{3}{2} \\ \frac{1}{6} & -\frac{3}{2} \end{bmatrix}.$$

The state equation can be written as a pair of first-order differential equations,

$$\frac{d}{dt} x_1 = -\tfrac{3}{2} x_1 + \tfrac{3}{2} x_2,$$

$$\frac{d}{dt} x_2 = \tfrac{1}{6} x_1 - \tfrac{3}{2} x_2.$$

To find dx_2/dx_1, we divide the second equation by the first:

$$\frac{dx_2/dt}{dx_1/dt} = \frac{dx_2}{dx_1} = \frac{\tfrac{1}{6} x_1 - \tfrac{3}{2} x_2}{-\tfrac{3}{2} x_1 + \tfrac{3}{2} x_2} = \frac{\tfrac{1}{3} - 3r}{-3 + 3r}, \qquad (3\text{--}45)$$

where $r = x_2/x_1$. Each value of r determines a straight line through the origin and, by Eq. (3–45), determines the slope dx_2/dx_1 of trajectories crossing that straight line. The line is, of course, an isocline. We now put in the isoclines corresponding to various values of r and draw the slope lines indicating the angle at which a trajectory will cross each isocline (Fig. 3–12). The arrow on the slope lines is determined by inspection of the original dif-

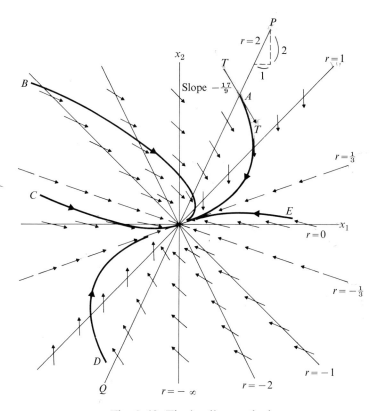

Fig. 3-12 The isocline method.

ferential equation. On the isocline $r = 2$ (*PQ* in Fig. 3-12), for instance, with $2x_1 = x_2$,

$$\frac{dx_1}{dt} = -\tfrac{3}{2}x_1 + \tfrac{3}{2}x_2 = \tfrac{3}{2}x_1,$$

and x_1 increases with time when x_1 is positive; the arrow on slope line *TT* in the figure is on the right-hand end.

As we should expect from previous considerations, the slope lines coincide with the isoclines along the eigenvectors. Therefore, the eigenvectors can be fixed by the following condition for r:

$$\frac{dx_2}{dx_1} = \frac{\tfrac{1}{3} - 3r}{-3 + 3r} = r, \quad \text{or} \quad r = \pm\frac{1}{3}.$$

Using these slope lines as guides, we can easily construct trajectories from arbitrarily chosen initial states (*A, B, C*, etc., in Fig. 3-12). The state-plane solution illustrates that solutions to the differential equation are unique since none of the trajectories cross each other. Further examination of the

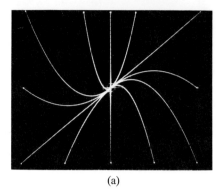

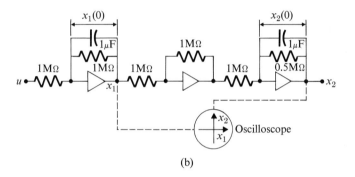

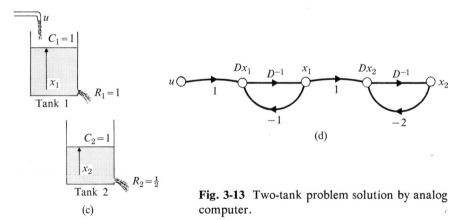

Fig. 3-13 Two-tank problem solution by analog computer.

trajectories shows that near the origin they are all tangent to one of the eigenvectors; the eigenvectors correspond to the modes e^{-t} and e^{-2t}, and since the e^{-2t} mode decays the fastest, near the origin all of the trajectories will be tangent to the eigenvector of the first or slow mode.

Example 3-11 Figure 3-13(a) is a photograph of a set of trajectories generated, one at a time, on an oscilloscope screen by the analog computer circuit shown in Fig.

3-13(b) (an *xy*-plotter may be used for the same purpose). Each trajectory corresponds to a different pair of initial conditions $x_1(0)$ and $x_2(0)$ set on the integrators, where various starting points appear as dots on the photograph. The system that has been simulated in Fig. 3-13(b) is shown in Fig. 3-13(c), and its signal-flow diagram is indicated in Fig. 3-13(d). Negative tank levels do not actually occur, of course, and must be interpreted as deviations about some positive equilibrium (datum) levels. It is instructive to note the similarity in form between the analog computer circuit and the signal-flow diagram (in the computer circuit, the middle operational amplifier is necessary to account for sign inversion on the computer). The details of choosing the circuit elements of the simulation or of scaling the solution will not be discussed here, but the resulting trajectories are worthy of a few more comments.

For the physical constants shown in Fig. 3-13(c), it is a straightforward process to derive the following:

$$\mathbf{A} = \begin{bmatrix} -1 & 0 \\ 1 & -2 \end{bmatrix}, \quad \mathbf{b} = \begin{bmatrix} 1 \\ 0 \end{bmatrix}, \quad p_1 = -1, \quad p_2 = -2$$

and

$$\mathbf{T} = [\mathbf{v}^1 \quad \mathbf{v}^2] = \begin{bmatrix} 1 & 0 \\ 1 & 1 \end{bmatrix}.$$

Thus the "slow" eigenvector, associated with the e^{-t} mode, has a slope of $+1$. Similarly the "fast" eigenvector, associated with the e^{-2t} mode, has an infinite slope, the vector being coincident with the x_2-axis. It is clear that, starting at any arbitrary initial state in the state plane, the resulting motion along a trajectory is influenced by both modes of response. The exceptions, as we have already discussed, are the cases where the initial state falls on an eigenvector. In these cases, linear trajectories are generated along the eigenvectors. Consider an initial state at the lower right-hand corner of Fig. 3-13(a). Initially the fast mode (e^{-2t}) predominates, and the state moves rapidly upward in the general direction of the fast eigenvector (pointed upward along the negative vertical axis). As the slow mode (e^{-t}) takes precedence, the trajectory bends over to the left and approaches the origin (equilibrium point) along the $+45°$ line.

Two parenthetical comments will be made before we return to the general discussion. Observe the *gravity* feed of fluid from tank 1 to tank 2, Fig. 3-13(c). The zero in the upper right-hand corner of the **A**-matrix and the lack of feedback loops in Fig. 3-13(b) or (d) which would couple x_2 to x_1 are caused by this physical decoupling of tank 1 from tank 2. Thus x_2 exerts no return effect on x_1, and in reality the second-order system amounts to two separate first-order systems. If the input u were introduced into tank 2, there would be no control over x_1 ("uncontrollable").

Our second comment bears on the general utility of analog simulation. One immediate advantage of analog simulation is that the effect of parameter variations on response patterns [such as Fig. 3-13(a)] may be detected at a glance. Perhaps more important is the fact that changes in system structure can be dictated by the results of analog studies. If it were not obvious to us by such studies that x_1 could not be controlled by a u introduced to tank 2, it would certainly be apparent from a simulation study. For high-order systems, each of the state variables would have to be examined separately on a time-based axis as in Fig. 3-10(a) and (b), or on pro-

Table 3-3 MOTION OF SECOND-ORDER SYSTEMS WITH DISTINCT REAL POLES

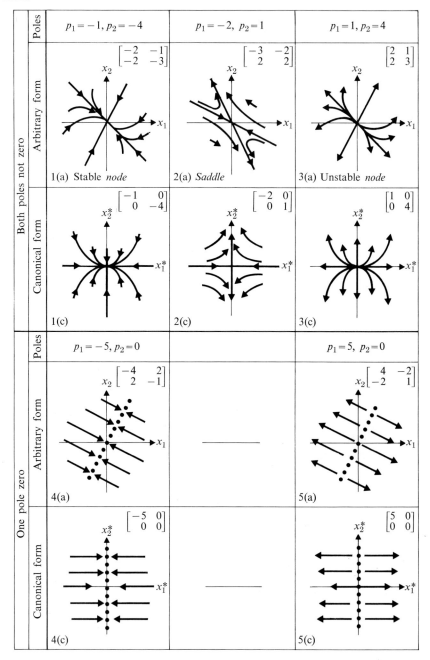

Table 3-4 MOTION OF SECOND-ORDER SYSTEMS WITH A DOUBLE POLE

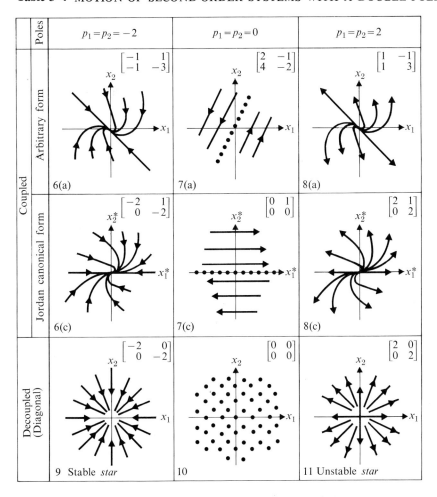

jections of trajectories on various coordinate planes (where two projected curves may cross each other).

We now resume our general discussion.

Various free trajectory patterns for second-order systems with two distinct real poles are shown in Table 3-3. Item 1(a) of the table is similar to the example we just discussed; the center of this pattern is called a *node*. The two eigenvectors become coordinate axes in a diagonalized system in 1(c). If the poles are both positive, we get the diverging node pattern of 3(a); and if one pole is negative and the other positive, we get a saddle as in 2(a).

If one of the poles is zero, the velocity component along that eigenvector becomes zero. As a consequence, an infinite number of equilibrium states are possible for the system, as shown by the dots along the second eigenvector in items 4 and 5 of Table 3-3.

Table 3-4 shows the behavior of second-order systems which have double poles. The systems 6 and 8 have only one eigenvector but otherwise have patterns similar to systems 1 and 3 of Table 3-3. When the two first-order components of a second-order system are completely decoupled, all the trajectories become straight lines, giving the star patterns 9 and 11. If we consider a purely inertial system, with x_1-displacement and x_2-velocity, Newton's equation of motion becomes

$$\frac{d^2}{dt^2} x_1 = \text{force} = 0,$$

and the trajectory pattern of 7(c) results. Pattern 10 shows a degenerate system; since its A-matrix is the null matrix (all zeroes), it is really a static system and any initial point is a stable equilibrium point.

If a second-order system has a pair of conjugate complex poles, its eigenvectors will have complex values and there will be no eigenvectors appearing on the state plane. We already know that a pair of complex conjugate eigenvalues results in an oscillatory mode $e^{-\sigma t}[\sin(\omega t + \phi)]$, which must result in a trajectory that encircles the origin on the state plane. Since there are no real eigenvectors, there are no "uncrossable" lines in the state plane as there are for systems with real poles, and it is possible for a trajectory to encircle the origin. The center of the trajectory pattern for a damped oscillatory system (complex poles) is called a *focus* [Table 3-5, 12(a)]. The trajectory pattern for an undamped oscillatory system (pure imaginary poles) is a set of ellipses about a *center* [13(a) of Table 3-5].

Example 3-12 The system whose A-matrix is

$$\mathbf{A} = \begin{bmatrix} -1 & 2 \\ -5 & 1 \end{bmatrix}$$

has poles $p_1 = 3j$ and $p_2 = -3j$.

A form of the general solution for a second-order undamped oscillatory system is

$$x_1 = C \cos(3t + \phi),$$

where C and ϕ are arbitrary constants. Substituting this expression into the first state equation $dx_1/dt = -x_1 + 2x_2$, we get

$$x_2 = \tfrac{1}{2} C \cos(3t + \phi) - \tfrac{3}{2} C \sin(3t + \phi).$$

To eliminate t, we solve for $\sin(3t + \phi)$ and $\cos(3t + \phi)$ separately,

$$\sin(3t + \phi) = \frac{x_1 - 2x_2}{3C}, \qquad \cos(3t + \phi) = \frac{x_1}{C},$$

Table 3-5 MOTION OF SECOND-ORDER SYSTEM WITH A PAIR OF CONJUGATE COMPLEX POLES

Poles	$p_1, p_2 = -1 \pm 3j$	$p_1, p_2 = \pm 3j$	$p_1, p_2 = 1 \pm 3j$
Arbitrary form	$\begin{bmatrix} 2 & -3 \\ 6 & -4 \end{bmatrix}$ 12(a) Stable *focus*	$\begin{bmatrix} -1 & 2 \\ -5 & 1 \end{bmatrix}$ 13(a) *Center*	$\begin{bmatrix} -2 & 3 \\ -6 & 4 \end{bmatrix}$ 14(a) Unstable *focus*
Modified canonical form	$\begin{bmatrix} -1 & 3 \\ -3 & -1 \end{bmatrix}$ 12(c)	$\begin{bmatrix} 0 & 3 \\ -3 & 0 \end{bmatrix}$ 13(c)	$\begin{bmatrix} 1 & -3 \\ 3 & 1 \end{bmatrix}$ 14(c)

and then square both terms in order to use the Pythagorean relation:

$$\sin^2(3t + \psi) + \cos^2(3t + \psi) = 1 = \frac{x_1^2 + \frac{1}{9}(x_1 - 2x_2)^2}{C^2}.$$

Therefore

$$\tfrac{10}{9} x_1^2 - \tfrac{4}{9} x_1 x_2 + \tfrac{4}{9} x_2^2 = C^2 .$$

Since this is the equation for an ellipse (see the next chapter), the trajectories must be ellipses. Since the eigenvectors are complex-valued, we are not interested in the diagonal form of the canonical system matrix. Instead, we use the symmetric (or modified canonical) form (Section 3-3):

$$\frac{d}{dt} \mathbf{x}^*(t) = \begin{bmatrix} -\sigma & \omega \\ -\omega & -\sigma \end{bmatrix} \mathbf{x}^*(t) .$$

For the undamped system, the "center" pattern becomes a set of circles [Table 3-5, 13(c)] in canonical form. The "focus" pattern becomes a set of logarithmic spirals [12(c) and 14(c)] in the canonical form.

The characteristic equation for a third-order system is a cubic equation, and thus at least one of the eigenvalues must be real and at least one straight line will appear in the trajectory pattern for a third-order system (in a three-dimensional state space). Some trajectory patterns are shown in Table 3-6.

Table 3-6 TRAJECTORY PATTERNS OF THIRD-ORDER OSCILLATORY SYSTEMS

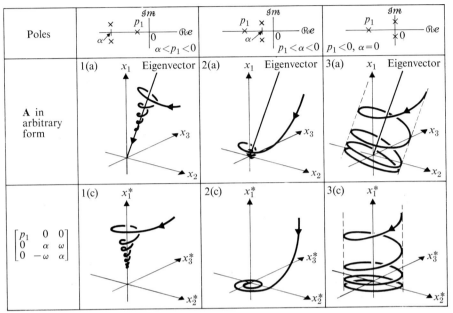

for systems with a real pole p_1 and a pair of complex conjugate poles p_2 and p_3. Depending on which modes are more strongly damped, the trajectories will follow the patterns 1, 2, or 3 of Table 3-6. If the oscillatory mode is more heavily damped than the exponential mode, the trajectory approaches the eigenvector and follows the eigenvector to the origin [Table 3-6, 1(a)]. If, on the other hand, the damping of the oscillatory mode is not as great as that of the exponential mode, then, as time increases, the response trajectory approaches a damped oscillation which takes place on a plane through the origin [Table 3-6, 2(a)]. The trajectory lies on a cylinder when the oscillatory mode has no damping [3(a) and 3(c)]. The trajectories for the canonical forms are shown in Table 3-6, 1(c), 2(c), and 3(c).

There is no known simple graphical method for the determination of three-dimensional trajectories. Although it is possible to obtain projections onto the coordinate planes, the projected trajectories may have crossings, and it is often difficult to get an idea of the behavior of a three-dimensional trajectory from its projections. Analog simulation may be helpful here with the third state fed into the z-axis of the oscilloscope to offer depth perception of the trajectory.

A geometric description of a trajectory becomes impossible for systems of order higher than three. If, however, a system has some poles with large negative real parts, the modes for these poles disappear quickly and the rest of the response can be investigated in a lower-dimensional space.

3-5 Motion in the state space

A double oscillator (Fig. 3–14) is a fourth-order system with all imaginary poles; therefore, the modes do not decay and there is no possibility of neglecting one or more of the modes to enable us to examine the trajectories geometrically. Nevertheless, in this case a decoupling is possible, and then the projections of the trajectory become simple. Taking x_1 and x_3 as the displacements of the masses, we write the following equations of motion:

$$\frac{d}{dt} x_1 = x_2 \quad \text{(velocity of } m_1\text{)},$$

$$m_1 \frac{d}{dt} x_2 = -k_1 x_1 + k_2(x_3 - x_1) \quad \text{(inertial force of } m_1\text{)},$$

$$\frac{d}{dt} x_3 = x_4 \quad \text{(velocity of } m_2\text{)},$$

$$m_2 \frac{d}{dt} x_4 = k_2(x_1 - x_3) \quad \text{(inertial force of } m_2\text{)},$$

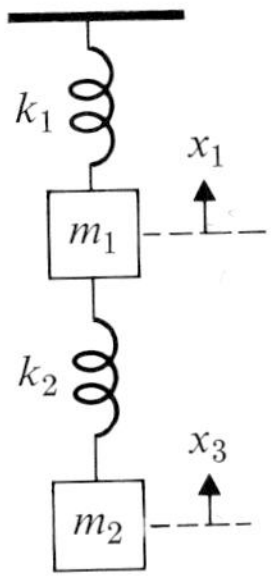

where k_1 and k_2 are spring constants. If m_1, m_2, k_1, and k_2 are all 1, then

Fig. 3–14 A double oscillator.

$$\frac{d}{dt} \begin{bmatrix} x_1 \\ x_2 \\ x_3 \\ x_4 \end{bmatrix} = \begin{bmatrix} 0 & 1 & 0 & 0 \\ -2 & 0 & 1 & 0 \\ 0 & 0 & 0 & 1 \\ 1 & 0 & -1 & 0 \end{bmatrix} \begin{bmatrix} x_1 \\ x_2 \\ x_3 \\ x_4 \end{bmatrix}. \tag{3-46}$$

The characteristic equation $|s\mathbf{I} - \mathbf{A}| = 0$ is

$$s^4 + 3s^2 + 1 = 0;$$

thus

$$s^2 = -\tfrac{3}{2} \pm \tfrac{1}{2}\sqrt{5} = -0.382, \quad -2.618,$$

$$s = \pm j\omega_1, \quad \pm j\omega_2, \quad \text{where} \quad \omega_1 = 0.618, \quad \omega_2 = 1.618.$$

The free motion of the system consists of two undamped oscillations. Since the ratio of the two periods is an irrational number, the superimposed pattern of the two sinusoids never repeats itself in a transient response. However, by using the modified canonical form for systems with complex roots, the two oscillations can be written in decoupled form:

$$\frac{d}{dt} \mathbf{x}^* = \begin{bmatrix} 0 & \omega_1 & 0 & 0 \\ -\omega_1 & 0 & 0 & 0 \\ 0 & 0 & 0 & \omega_2 \\ 0 & 0 & -\omega_2 & 0 \end{bmatrix} \mathbf{x}^*. \tag{3-47}$$

The projection of the trajectory onto the $x_1^* x_2^*$-plane is a circle on which

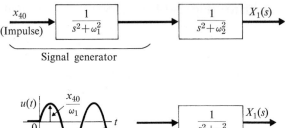

Fig. 3-15 Block diagrams for a double oscillator.

a point (a projection of the state) moves with an angular velocity ω_1, and the projection onto the $x_3^* x_4^*$-plane is another circle on which a point moves with an angular velocity ω_2.

Up to this point we have considered only free motions. Sometimes, however, it is possible to look on a free motion as a forced motion, and vice versa. As an example, let us look further into the double oscillator we discussed before. Consider an initial state in which all the state variables except x_4 are zero, and $x_4(0) = x_{40}$. If we write the equations in the Laplace domain, we have

$$sX_1 = X_2, \quad sX_2 = -2X_1 + X_3,$$
$$sX_3 = X_4, \quad sX_4 = x_{40} + X_1 - X_3.$$

We solve for X_1 and obtain

$$X_1(s) = \frac{x_{40}}{(s^2 + \omega_1^2)(s^2 + \omega_2^2)}.$$

As we can see from the system block diagram (Fig. 3-15a), the response is equivalent to a second-order system's output for a sinusoidal forcing input (Fig. 3-15b).

An insight into trajectory patterns in state space provides us with a guide for making judgments on controllability and observability. If the vector **b** for input u (Eq. (3-37)) is linearly dependent on some eigenvector(s), the system is uncontrollable for the modes other than those eigenvectors. All modes whose eigenvectors are orthogonal to vector **c** of output equation $y = \mathbf{cx}$ are unobservable. Examples are shown for third-order systems in Table 3-7. The star pattern trajectory (#9 of Table 3-4) yields a special case, where a given vector **b** can move a state only in that direction, and there always exists a trajectory which is orthogonal to the given vector **c**; hence the system is uncontrollable and unobservable.

Table 3-7 CONDITIONS FOR LOSS OF CONTROLLABILITY OR OBSERVABILITY IN THIRD ORDER SYSTEMS WITH DISTINCT EIGENVALUES

System	Controllability	Observability
Linearly independent eigenvectors (straight line trajectories) $\mathbf{v}^1$, $\mathbf{v}^2$ and $\mathbf{v}^3$ for poles p_1, p_2 and p_3.	\multicolumn{2}{c}{one mode, p_3, is not:}	
	controllable if k_1, k_2 (one of which is nonzero) exists such that $$\mathbf{b} = k_1\mathbf{v}^1 + k_2\mathbf{v}^2$$ (**b** is in the $\mathbf{v}^1\mathbf{v}^2$-plane)	observable if $$\mathbf{c}'\mathbf{v}^3 = 0$$ (**c** is orthogonal to $\mathbf{v}^3$)
	two modes, p_2 and p_3, are not:	
	controllable if $k \neq 0$ exists such that $$\mathbf{b} = k\mathbf{v}^1$$ (**b** and $\mathbf{v}^1$ forming one line)	observable if $$\mathbf{c}'\mathbf{v}^2 = 0 \text{ and } \mathbf{c}'\mathbf{v}^3 = 0$$ (**c** orthogonal to $\mathbf{v}^2\mathbf{v}^3$-plane)
Eigenvector (straight line trajectory) **v** for real pole p and vectors $\boldsymbol{\alpha}$ and $\boldsymbol{\beta}$ for conjugate complex poles (see Eq. (3-26))	real mode, p, is not	
	controllable if k_1, k_2 (one of which is nonzero) exists such that $$\mathbf{b} = k_1\boldsymbol{\alpha} + k_2\boldsymbol{\beta}$$ (**b** is in the $\boldsymbol{\alpha\beta}$-plane)	observable if $$\mathbf{c}'\mathbf{v} = 0$$ (**c** is orthogonal to **v**)
	oscillatory mode is not:	
	controllable if $k \neq 0$ exists such that $$\mathbf{b} = k\mathbf{v}$$ (**b** and **v** forming one line)	observable if $$\mathbf{c}'\boldsymbol{\alpha} = 0 \text{ and } \mathbf{c}'\boldsymbol{\beta} = 0$$ (**c** orthogonal to the oscillation plane $\boldsymbol{\alpha\beta}$)

3-6 DISCRETE-TIME SYSTEMS

A sampled-data system, or any other system which can be observed or forced only at discrete times, is described by a difference equation rather than a differential equation. A useful application of the relations that will be derived in this section is to the digital computer solution for the time response of linear, lumped, continuous-time systems. By the nature of digital computation, the time-domain solution must be done on a step-by-step basis;

that is, the solution is actually obtained only for a finite number of points in time. Therefore, the computational model of the continuous-time system is itself a discrete-time system, and all the solutions presented below can be applied directly. A matrix difference equation was presented without derivation in Chapter 2 [Eqs. (2–38) and (2–39)] for linear lumped discrete-time systems. We begin our derivation of this difference equation by applying Eq. (3–14) to a time interval $kT < t \leq (k+1)T$, where T represents the sampling interval:

$$\mathbf{x}_{k+1} = e^{\mathbf{A}T}\mathbf{x}_k + e^{\mathbf{A}T}\int_0^T e^{-\mathbf{A}\tau}\mathbf{B}\mathbf{u}(\tau)\,d\tau\,, \qquad \tau = t - kT\,. \qquad (3\text{–}48)$$

We consider that the input $\mathbf{u}(t)$ is constant over the time interval in question and that changes in the input function can be made only at the end of the time interval. With this assumption we can integrate Eq. (3–48):

$$\begin{aligned}\mathbf{x}_{k+1} &= e^{\mathbf{A}T}\mathbf{x}_k + e^{\mathbf{A}T}(\mathbf{I} - e^{-\mathbf{A}T})\mathbf{A}^{-1}\mathbf{B}\mathbf{u}_k \\ &= e^{\mathbf{A}T}\mathbf{x}_k + (e^{\mathbf{A}T} - \mathbf{I})\mathbf{A}^{-1}\mathbf{B}\mathbf{u}_k\,,\end{aligned} \qquad (3\text{–}49)$$

where it is further assumed that $\mathbf{A}^{-1}$ exists. We now have an equation in the form given by Eq. (2–37):

$$\mathbf{x}_{k+1} = \mathbf{P}\mathbf{x}_k + \mathbf{Q}\mathbf{u}_k\,,$$

where

$$\mathbf{P} = e^{\mathbf{A}T} = \mathbf{S}(T)\,, \qquad \mathbf{Q} = (\mathbf{S}(T) - \mathbf{I})\mathbf{A}^{-1}\mathbf{B}\,. \qquad (3\text{–}50)$$

In order to find the time-domain solution starting from some initial condition $\mathbf{x}_0$, we apply Eq. (3–49) repeatedly, that is, we use the equation once to find $\mathbf{x}_1$ and then substitute the newly found $\mathbf{x}_1$ into the right-hand side of the equation to find $\mathbf{x}_2$, and so on, up to $\mathbf{x}_n$. The result thus obtained is equivalent to

$$\mathbf{x}_k = \mathbf{P}^k\mathbf{x}_0 + \sum_{i=0}^{k-1}\mathbf{P}^{k-1-i}\mathbf{Q}\mathbf{u}_i\,. \qquad (3\text{–}51)$$

Note that the form of this solution is the same as that of the solution to the first-order case which was presented in Chapter 1. But since the solution now involves matrices rather than scalars, the computation process is usually too tedious to be done by hand computation: A possible technique for solving this equation with the aid of a digital computer is given below.

To use Eq. (3–51) to find the time response of a system, we must first calculate the matrices $\mathbf{P}$ and $\mathbf{Q}$, which are related to the system's $\mathbf{A}$- and $\mathbf{B}$-matrices by Eq. (3–50). Since only the sampling interval T is involved in computing $\mathbf{P}$ and $\mathbf{Q}$, these matrices need only be calculated once for any given system. The sampling interval is either known from the characteristics of the system or, if the "system" is a digital model of a continuous-time system, it is chosen to give the desired resolution in the calculated time response. A convenient and simple way to calculate **P** and **Q** is to use the series expansion for **S**(T) (Eq. 3-2). We can avoid the matrix inversion, (**A**$^{-1}$) in the relation for **Q** by expanding the quantity inside the parentheses into a power series:

$$(S(t) = I + At + \frac{1}{2!} A^2 t^2 + \cdots)$$

$$(\mathbf{S}(T) - \mathbf{I}) = \left(\mathbf{I} + \mathbf{A}T + \frac{1}{2!}(\mathbf{A}T)^2 + \cdots\right) - \mathbf{I}$$

$$= \mathbf{A}T + \frac{1}{2!}(\mathbf{A}T)^2 + \cdots.$$

If we now postmultiply this series by **A**$^{-1}$, we get the following expression, which does not involve any matrix inversions for **Q**:

$$\mathbf{Q} = T\left(\mathbf{I} + \frac{1}{2!}\mathbf{A}T + \frac{1}{3!}(\mathbf{A}T)^2 + \cdots\right)\mathbf{B}. \tag{3-52}$$

The following procedure has been suggested by H. M. Paynter for digital computation:

1. Let $q = \max |A_{ij}T|$, where A_{ij} is an element of the **A**-matrix.
2. Find a p that approximately satisfies the equation

$$\frac{1}{p!}(nq)^p e^{nq} = 0.001,$$

where n is the order of the system.

3. Compute **P** from Eq. (3-2), truncating the series after the pth term.
4. Compute **Q** from Eq. (3-52), truncating the series also after p terms.
5. Calculate the time response from Eq. (3-51).

Example 3-13 *Discrete-time model of a continuous-time system.* We are often interested in obtaining numerical solutions for the time response of continuous-time systems. Digital computer solutions are fast, simple, and inexpensive; they can be used for highly complex systems, and often to obtain accurate approximations to the solution of nonlinear and nonstationary (time-varying parameter) systems. The analog computer solution for the two-tank system (Fig. 3-13c) is given in Section 3-5; we will now obtain a digital computer solution for the same system.

The digital computer solution (or any numerical solution) is a discrete-time model of the actual continuous-time process. In formulating the discrete-time model, we follow the steps outlined above and produce an *algorithm*, a sequence of numerical and logical operations, which is then translated into a computer program. In constructing the algorithm, we do not include any specific data. The algorithm is a general set of instructions that can be used to solve a *class* of problems. After the algorithm has been obtained and a computer program is written, the data are introduced to fit specific problems. The computer program can be used repeatedly to solve many specific problems. The algorithm for formulating the discrete-time model and

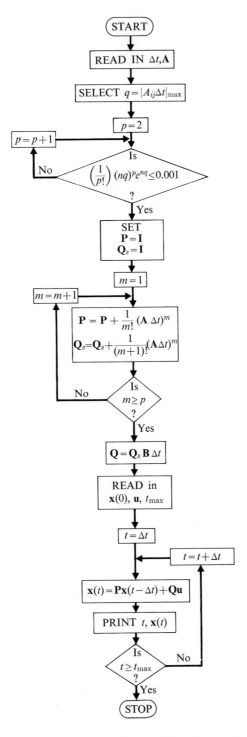

Fig. 3-16 Flow chart for finding discrete-time model and computing system response.

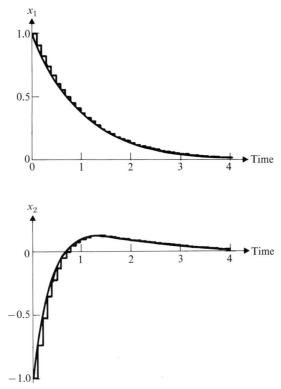

Fig. 3-17 Response of discrete-time model of two-tank system; smooth curves are analog results, presented for comparison.

calculating its time response is given in Fig. 3-16 in the form of a flow chart, a graphical representation of an algorithm. The boxes labelled "READ" are points in the program at which data are introduced; other boxes can be compared directly to the original statement (in engineering-oriented rather than computer-oriented language) of the procedure above.

The system response for the two-tank system is shown in Fig. 3-17 with analog results shown for comparison. Initially there is liquid in both tanks [$x_1(0) = 1$ and $x_2(0) = -1$], and the system forcing is zero (that is, there is no flow into the upper tank). Because we are using a discrete-time model to represent the original system, the response remains at the same value for an entire sampling period. At the end of that period it jumps to the value it will take for the next sampling interval. This discrete jump behavior shows up in the state plane (Fig. 3-18), as a series of points; the state vector remains at a point for a sampling period and then jumps to the next point. Since the sampling period is a constant, the distance between adjacent points on the state plane is proportional to the mean speed along the trajectory.

The theoretical relations developed for linear, lumped-parameter, stationary continuous-time systems in Sections 3-2 through 3-5 apply to linear,

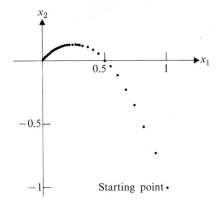

Fig. 3-18 State-plane representation of discrete-time model response.

lumped-parameter, stationary, discrete-time systems, with some modifications. In particular, the **P**-matrix replaces the **A**-matrix of the continuous-time system. Some of the basic relations for the discrete-time system will be reviewed next.

As shown in Chapter 2, we use $z = e^{sT}$ in the Laplace domain. The difference equation

$$\mathbf{x}_{k+1} = \mathbf{P}\mathbf{x}_k + \mathbf{Q}\mathbf{u}_k,$$

solved in the Laplace (or the z) domain,

$$\mathbf{X}(z) = (z\mathbf{I} - \mathbf{P})^{-1}\mathbf{x}_0 z + (z\mathbf{I} - \mathbf{P})^{-1}\mathbf{Q}\mathbf{U}(z),$$

involves $(z\mathbf{I} - \mathbf{P})^{-1}$ as a crucial term. Recalling Eq. (3-51), we define the solution matrix by

$$\mathbf{S}(k) = \mathbf{P}^k = \mathscr{Z}^{-1}(z\mathbf{I} - \mathbf{P})^{-1}z. \tag{3-53}$$

If all eigenvalues, that is the roots $z_1, \ldots, z_n$, of the characteristic equation

$$|z\mathbf{I} - \mathbf{P}| = 0$$

are distinct, then each element of the matrix $(z\mathbf{I} - \mathbf{P})^{-1}z$ is expandable into partial fractions:

$$\frac{k_1 z}{z - z_1} + \cdots + \frac{k_n z}{z - z_n},$$

and, knowing the inverse transform (see Chapter 1, Table 1-2)

$$\mathscr{Z}^{-1}\left[\frac{k_i z}{z - z_i}\right] = k_i z_i^k,$$

we can find the elements of the solution matrix (3-53).

3-6

The **P**-matrix can be diagonalized by a variable change $\mathbf{x}_k = \mathbf{T}\mathbf{x}_k^*$ if all eigenvalues are distinct. The modal matrix **T** for this purpose is given in terms of eigenvectors $\mathbf{v}^i$:

$$\mathbf{T} = [\mathbf{v}^1, \mathbf{v}^2, \ldots, \mathbf{v}^n],$$

where the eigenvectors are obtained from the equation

$$(z_i\mathbf{I} - \mathbf{P})\mathbf{v}^i = \mathbf{0}, \quad i = 1, \ldots, n.$$

The difference equation in the modal domain is

$$\mathbf{x}_{k+1}^* = \mathit{\Lambda}\mathbf{x}_k^* + \mathbf{Q}^*\mathbf{u}_k,$$

where

$$\mathit{\Lambda} = \mathbf{T}^{-1}\mathbf{P}\mathbf{T} = \begin{bmatrix} z_1 & & 0 \\ & \ddots & \\ 0 & & z_n \end{bmatrix}, \quad \mathbf{Q}^* = \mathbf{T}^{-1}\mathbf{Q}.$$

The solution matrix in the modal domain is also diagonal:

$$\mathbf{S}^*(k) = \begin{bmatrix} z_1^k & & 0 \\ & \ddots & \\ 0 & & z_n^k \end{bmatrix}. \tag{3-54}$$

The original matrix $\mathbf{S}(k)$ is related to $\mathbf{S}^*(k)$ by

$$\mathbf{S}(k) = \mathbf{T}\mathbf{S}^*(k)\mathbf{T}^{-1}. \tag{3-55}$$

Example 3-14 Let us determine the solution matrix of a second-order system represented by

$$\mathbf{P} = \begin{bmatrix} 0.4 & 0.2 \\ 0.3 & -0.1 \end{bmatrix}.$$

The eigenvalues are found to be $z_1 = -0.2$, $z_2 = 0.5$. It is then easy to fix the eigenvectors and choose the following modal matrix:

$$\mathbf{T} = \begin{bmatrix} 1 & 2 \\ -3 & 1 \end{bmatrix}, \quad \mathbf{T}^{-1} = \frac{1}{7}\begin{bmatrix} 1 & -2 \\ 3 & 1 \end{bmatrix}.$$

By Eq. (3-55) we find that

$$\mathbf{S}(k) = \begin{bmatrix} 1 & 2 \\ -3 & 1 \end{bmatrix} \begin{bmatrix} z_1^k & 0 \\ 0 & z_2^k \end{bmatrix} \frac{1}{7}\begin{bmatrix} 1 & -2 \\ 3 & 1 \end{bmatrix}$$

$$= \frac{1}{7}\begin{bmatrix} z_1^k + 6z_2^k & -2z_1^k + 2z_2^k \\ -3z_1^k + 3z_2^k & 6z_1^k + z_2^k \end{bmatrix}.$$

A faster way to obtain the same answer is to use Eq. (3-53). We have

$$(z\mathbf{I} - \mathbf{P})^{-1} = \frac{1}{(z + 0.2)(z - 0.5)}\begin{bmatrix} z + 0.1 & 0.2 \\ 0.3 & z - 0.4 \end{bmatrix},$$

and each element of $\mathbf{S}(k)$ can be determined by the inverse z-transformation applied on partial fractions. For instance, the upper left-hand element of $(z\mathbf{I} - \mathbf{P})^{-1}$ is

$$\frac{z + 0.1}{(z + 0.2)(z - 0.5)} = \frac{K_1}{z + 0.2} + \frac{K_2}{z - 0.5},$$

or, clearing the fractions,

$$z + 0.1 = (z - 0.5)K_1 + (z + 0.2)K_2.$$

Letting $z \to -0.2$, we find that $K_1 = \frac{1}{7}$. With $z \to 0.5$, K_2 is found to be $K_2 = \frac{6}{7}$ and thus

$$\mathscr{Z}^{-1} \frac{(z + 0.1)z}{(z + 0.2)(z - 0.5)} = \mathscr{Z}^{-1} \frac{\frac{1}{7}z}{z + 0.2} + \mathscr{Z}^{-1} \frac{\frac{6}{7}z}{z - 0.5}$$

$$= \frac{1}{7}(-0.2)^k + \frac{6}{7}(0.5)^k.$$

Note that the mode for the negative real eigenvalue oscillates. For example, $(-0.2)^k$ for $k = 0, 1, \ldots$ produces a sequence $1, -0.2, 0.04, -0.008, \ldots$

If a discrete-time system has conjugate complex eigenvalues, the modified canonical form Eq. (3–26) can be applied to avoid imaginary quantities in the modal-domain relations. The Jordan canonical form must be used in place of the diagonal matrix in some systems with high-order poles. For example, the Schwartz form for a second-order system with a double pole $z = z_1$ is

$$\Lambda = \mathbf{T}^{-1}\mathbf{P}\mathbf{T} = \begin{bmatrix} z_1 & 1 \\ 0 & z_1 \end{bmatrix}.$$

The solution matrix in the modal domain is then given by

$$\mathbf{S}^*(k) = \Lambda^k = \begin{bmatrix} z_1^k & kz_1^{k-1} \\ 0 & z_1^k \end{bmatrix}. \tag{3–56}$$

The kz_1^{k-1} is the second mode that corresponds to the $te^{p_1 t}$ in the continuous-time system with a double pole p_1 (see Jury [5] for the general case of the Jordan canonical form).

The rules developed in Section 3–4 for controllability and observability can also be applied to discrete-time systems. An nth-order discrete-time system, represented by a difference state equation for a scalar input u_k,

$$\mathbf{x}_{k+1} = \mathbf{P}\mathbf{x}_k + \mathbf{q}u_k, \tag{3–57}$$

is controllable if and only if

$$|\mathbf{q}, \mathbf{P}\mathbf{q}, \ldots, \mathbf{P}^{n-1}\mathbf{q}| \neq 0. \tag{3–58}$$

When the output equation of this system for a scalar response y_k is

$$y_k = \mathbf{c}\mathbf{x}_k, \tag{3–59}$$

the necessary and sufficient condition for the system to be observable is

$$|\mathbf{c}', \mathbf{P}'\mathbf{c}', \ldots, (\mathbf{P}')^{n-1}\mathbf{c}'| \neq 0 . \tag{3-60}$$

The reason for these conditions, which are identical in form to Eqs. (3–38) and (3–39), respectively, will be discussed in Chapter 11.

REFERENCES

1. Bellman, R., *Introduction to Matrix Analysis*, New York: McGraw-Hill, 1960.
2. Dorf, R. C., *Time-Domain Analysis and Design of Control Systems*, Reading, Mass: Addison-Wesley, 1965.
3. Newton, G. C., Gould, L. A., Kaiser, J. F., *Analytical Design of Linear Feedback Controls*, New York: Wiley, 1957.
4. Gupta, S. C., *Transform and State Variable Methods in Linear Systems*, New York: Wiley, 1966.
5. Jury, E. I., *Theory and Application of the z-Transform Method*, New York: J. Wiley, 1964.

PROBLEMS

3-1 Newton's equation of motion for a free inertial system is

$$m \frac{d^2x}{dt^2} + b \frac{dx}{dt} = 0 ,$$

where m is mass and b is a linear (or viscous) damping coefficient. Let displacement $x = x_1$ and velocity $dx_1/dt = x_2$, and determine the solution matrix $\mathbf{S}(t)$ by a direct computation of $e^{\mathbf{A}t}$.

3-2 Solve the preceding problem by means of

$$\mathbf{S}(t) = \mathscr{L}^{-1}(s\mathbf{I} - \mathbf{A})^{-1} .$$

3-3 Obtain the eigenvectors of the **A**-matrix for Problem 3-1 and diagonalize it. Sketch the phase plane trajectories.

3-4 The **A**-matrix of a second-order system that has distinct eigenvalues is given in terms of four real elements a, b, c, and d:

$$\mathbf{A} = \begin{bmatrix} a & b \\ c & d \end{bmatrix} .$$

The slope dx_2/dx_1 of a free trajectory, for use with the isocline method, is given by

$$\frac{dx_2}{dx_1} = \frac{c + dr}{a + br} ,$$

where $r = x_2/x_1$. Prove that the slopes of eigenvectors in the state plane are deter-

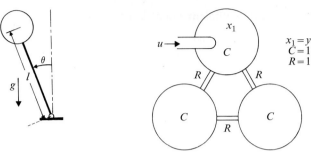

Fig. P3-5   Fig. P3-7

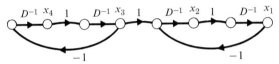

Fig. P3-9

mined by r which satisfies the condition

$$\frac{dx_2}{dx_1} = r .$$

3-5 The equation of motion of a frictionless inverted pendulum (shown in Fig. P3-5) is

$$\frac{d^2}{dt^2}\theta = \frac{g}{l}\sin\theta ,$$

where $\sin\theta \approx \theta$ when the angle measured from the upright position is small. Let $x_1 = \theta$, $d\theta/dt = x_2$, and determine the trajectory pattern of the linearized system by means of the isocline method.

3-6 Find the T-matrix for the A-matrix given, and convert the A-matrix into the canonical form

$$\mathbf{A} = \begin{bmatrix} -3 & 4 & 4 \\ 1 & -3 & -1 \\ -1 & 2 & 0 \end{bmatrix}.$$

3-7 Three identical tanks of $C = 1$ are connected into a delta form with unit resistors, as shown in Fig. P3-7 (this is the system given in Prob. 2-4). (a) Determine the modal matrix and the canonical form of the A-matrix. (b) For the input and output indicated in the figure, discuss controllability and observability in terms of $\mathbf{A}$, $\mathbf{b}$, $\mathbf{c}$ and also in the modal domain.

3-8 Determine the solution matrix $\mathbf{S}(t)$ of the modified canonical system

$$\frac{d}{dt}\mathbf{x}^* = \begin{bmatrix} -\sigma & \omega \\ -\omega & -\sigma \end{bmatrix} \mathbf{x}^* .$$

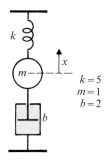

Fig. P3-10

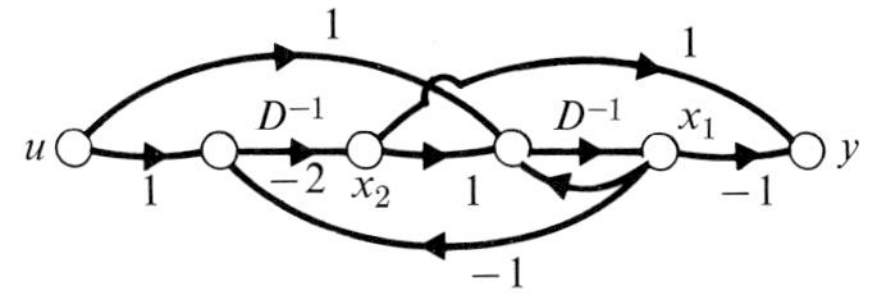

Fig. P3-11

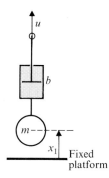

Fig. P3-12

3-9 A signal-flow diagram is shown in Fig. P3-9. Obtain (a) $(s\mathbf{I} - \mathbf{A})^{-1}$ and (b) the solution matrix $\mathbf{S}(t) = \mathscr{L}^{-1}(s\mathbf{I} - \mathbf{A})^{-1}$.

3-10 The displacement x of a free oscillatory system (Fig. P3-10) is described by

$$\frac{d^2x}{dt^2} + 2\frac{dx}{dt} + 5x = 0 \, .$$

Let $x = x_1$, $dx/dt = x_2$, and obtain the modal matrix $\mathbf{T}$ which converts the A-matrix into the modified canonical form.

3-11 The signal-flow diagram of a second-order system is given in Fig. P3-11. Determine (a) the canonical form $\mathbf{\Lambda}$, (b) the modal-domain signal-flow diagram, (c) discuss controllability and observability, and (d) obtain $G(s) = Y(s)/U(s)$.

3-12 A mass is pulled (or pushed) with a dashpot cylinder. The displacement of the mass is represented by the state variable x_1, and its velocity is x_2. The input u is a velocity forced at the top of the mechanism (Fig. P3-12), and the reaction force y is the output. The force transmitted through the dashpot is proportional to the relative velocity of piston and cylinder, and b is the proportionality constant. Is this system controllable and observable? Ignore the effects of gravity.

3-13 The A-matrix for a third-order system is

$$\mathbf{A} = \begin{bmatrix} -3 & 1 & 0 \\ -1 & 0 & -1 \\ -1 & 2 & -3 \end{bmatrix}.$$

For a scalar input, **b** is

$$\begin{bmatrix} 1 \\ 1 \\ 1 \end{bmatrix} \quad \text{or} \quad \begin{bmatrix} 0 \\ 1 \\ 1 \end{bmatrix} \quad \text{or} \quad \begin{bmatrix} 1 \\ 1 \\ 0 \end{bmatrix}.$$

For a scalar output, **c** is

$$[1 \quad -1 \quad 1] \quad \text{or} \quad [-1 \quad 1 \quad 0] \quad \text{or} \quad [0 \quad 1 \quad -1].$$

Discuss controllability and observability, first, without using the canonical form for the **A**-matrix, and then constructing the modal-domain signal-flow diagram for each **b** and **c**.

3-14 The empirical rule for the truncation of the matrix exponential computation given in Section 3-6,

$$\frac{1}{p!}(nqT)^p e^{nqT} = 0.001,$$

is based on a test of the double-oscillator response. If there exists significant truncation error, the amplitude would either build up or slowly decay. The number 0.001 was considered about right as a result of this test. Using a set of inequality conditions applied to the sum of remainder terms of the expanded matrix exponential, derive the left-hand-side expression of the rule.

3-15 A second-order sampled-data system has a double pole at -1, so that the **P**-matrix in the Schwartz form is

$$\mathbf{P} = \begin{bmatrix} -1 & 1 \\ 0 & -1 \end{bmatrix}.$$

Compute the response x_k, $k = 1, 2, \ldots$, for an initial state

$$\mathbf{x}_0 = \begin{bmatrix} 0 \\ 1 \end{bmatrix}.$$

3-16 The **A**-matrix of a second-order system is

$$\mathbf{A} = \begin{bmatrix} -\sigma & \omega \\ -\omega & -\sigma \end{bmatrix}, \quad \text{where} \quad \omega > 0.$$

Obtain the **P**-matrix of the system for a sampling period $T = \pi/(4\omega)$, and determine the general form for $\mathbf{P}^k$.

3-17 The digital controller shown in Fig. P3–17(a) is designed to cancel out the factors $(z + 1)$ and $(z - 0.9)$ of the control object. If such cancellation is really effective, the control system would reduce to the block diagram (b) of the figure, and the reference input to controlled-variable transfer function of the closed-loop system will be

$$G_{CL}(z) = \frac{1/(z-1)}{1 + 1/(z-1)} = z^{-1} \quad \text{or} \quad C(z) = z^{-1}R(z).$$

This means the output would follow the input with a delay of only one sampling period.

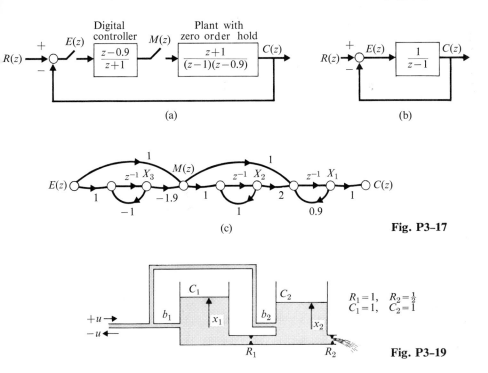

Fig. P3-17

Fig. P3-19

To investigate the scheme from the modern control viewpoint; (a) determine the modal-domain signal-flow diagram of the open-loop system [from $E(z)$ to $C(z)$ without the feedback], and (b) discuss the closed-loop system's controllability and observability. For your information, the signal-flow diagram of the original open-loop system is constructed in part (c) of the figure.

3-18 The fast and slow eigenvectors in Example 3-11 appear as straight lines in Fig. 3-13(a). Derive the equations for these lines and choose suitable initial-condition voltages for the integrators of the analog simulation that will yield a pure fast-eigenvector response and a pure slow-eigenvector response.

3-19 The system simulated in Example 3-11 had the first tank gravity-feeding the second tank. Find the system matrix, signal-flow diagram and analog computer diagram for the case where the bottoms of the two tanks are on the same level and connected by a conduit having the same resistance as in the first case. The second tank still drains out through the resistance R_2, as indicated in Fig. P3-19.

3-20 Sketch the phase portrait, similar to Fig. 3-13(a), for the system of Problem 3-19. Rework Problem 3-18 for this case.

3-21 Following Eq. (2-3), there is a discussion which concludes that the product (bc) is constant for the scalar case. Extend this discussion to the vector variable case, i.e.: "For r specified inputs $\mathbf{u}$ and m specified outputs $\mathbf{y}$, the matrix product $\mathbf{CB}$ is invariant, no matter what choice of the n state variables $\mathbf{x}$ is made." [*Hint*: Apply $\mathbf{x} = \mathbf{Tx}^*$.]

4

STABILITY OF LINEAR
LUMPED-PARAMETER SYSTEMS

In the previous chapter we turned our attention to obtaining the trajectory for a given system. We now change our focus and strive to deduce the behavior of the whole class of trajectories for a given system *without solving for them*. Such an approach is afforded by Lyapunov's second or direct method [1]. This chapter begins with a brief introduction to the Lyapunov concept of stability, and continues with applications of the method. It is the opinion of the authors that, even when considering linear and stationary systems, the importance of the Lyapunov method lies primarily in its point of view of system stability rather than in its application as a design tool. When dealing with nonlinear or time-varying systems, the application of Lyapunov's method becomes, at best, tedious and difficult. Nonetheless, the method is an important one in which there is a great deal of interest at the moment. We present it to the reader as a valuable way of viewing system stability and, in those cases where the system is described by linear, stationary, state equations, it is a convenient method for computing system stability.

It will be shown that Lyapunov's direct method is equally applicable to forced and to free systems. Suppose that a trajectory of a forced system is known. With a slight change in initial conditions we can get another trajectory which starts at a point near the start of the first trajectory. Lyapunov's direct method will enable us to determine whether the two trajectories converge (for a stable motion), or diverge (for an unstable one).

If the characteristic equation of a system, rather than its state equations, is available, a more convenient way to test system stability is by the Routh array [2] or the Hurwitz determinants [3]. Both of these techniques are used to determine whether any of the system eigenvalues (roots of the characteristic equation) lie in the right-half-plane. This is done without explicitly solving for the roots but by merely considering the coefficients of the characteristic equation. Recall that the modes of a linear, lumped, stationary system are represented by the eigenvalues and that the stability of a system depends on these modes. Intuitively it would appear that there is some connection between the Routh array and Lyapunov's direct method, and indeed the former has been arrived at by starting with the latter [4]. Here, the Routh method will be presented without proof. The convenience and importance of the Routh method as a design tool will be demonstrated.

The Nyquist stability theorem, which also applies to linear systems, will be discussed in the treatment of frequency-domain methods in Chapter 9.

As in previous chapters, consideration of the stability of discrete-time systems will be presented. Lyapunov's direct method will be applied to this problem. A convenient use of Lyapunov functions to evaluate quadratic performance criteria will also be discussed.

4-1 TRAJECTORY PATTERNS AND STABILITY

Our emphasis in this section will be on interpretation and application. The reader interested in either historical perspective or mathematically rigorous derivations and proofs is referred to the literature [4, 5, 6, 7].

Let us consider a dynamic system with a state vector $\mathbf{x}_a$ which satisfies the equation

$$\frac{d}{dt}\mathbf{x}_a(t) = \mathbf{F}(\mathbf{x}_a(t)) . \tag{4-1}$$

The system is said to be autonomous, or free and stationary. It may be linear or nonlinear. Let us assume that the system has an *equilibrium state* $\mathbf{x}_a = \mathbf{c}$. Then

$$\lim_{t \to \infty} \frac{d}{dt}\mathbf{x}_a(t) = \mathbf{F}(\mathbf{c}) = \mathbf{0} .$$

We now ask whether or not the system is stable relative to this equilibrium point. To investigate the stability, we perform a transformation of coordinates $\mathbf{x} = \mathbf{x}_a - \mathbf{c}$, and deal with the system

$$\frac{d}{dt}\mathbf{x} = \mathbf{F}(\mathbf{x}), \qquad \mathbf{F}(0) = \mathbf{0} . \tag{4-2}$$

If an autonomous system is linear, its equation is

$$\frac{d}{dt}\mathbf{x} = \mathbf{A}\mathbf{x} . \tag{4-3}$$

Since Eqs. (4-2) or (4-3) are considered to be representative of a physical phenomenon, the solution $\mathbf{x}(t)$ is assumed to be unique and continuous in all of its arguments (t and x_i, $i = 1, \ldots, n$). (For the mathematical conditions concerning the validity of these assumptions see [4] and [6].) If a system (4-2) is nonlinear, the region where the assumptions are valid may be limited to:

$$\|\mathbf{x}\| < R , \tag{4-4}$$

where R is some finite, real, positive number and

$$\|\mathbf{x}\| = (x_1^2 + \cdots + x_n^2)^{1/2}$$

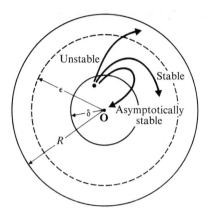

Fig. 4-1 Definition of stability.

is the distance of a point **x** from the origin of the state space. $\|\mathbf{x}\|$ is called the *norm* of **x**, and R defines the domain for which our conclusions about system stability will be valid. For linear systems, the restriction (4-4) is not necessary and the results of our stability analysis will be said to apply in the large (i.e., globally in the state space). For the more general case of (4-2), however, condition (4-4) means that we limit the stability analysis to the inside of a hyperspherical region of radius R in the state space.

In order to define the *stability* of the system (4-2) *in the sense of Lyapunov*, we consider two hyperspherical regions in the state space inside the hypersphere R of (4-4) such that (see Fig. 4-1)

$$\delta \leqslant \varepsilon < R.$$

The stability of the system (4-2) at an equilibrium state $\mathbf{x} = \mathbf{0}$ (the origin) is defined as follows:

It is *stable* if for every radius ε there exists a radius δ such that if a trajectory starts at a point $\mathbf{x}_0$ inside the hyperspherical region of radius δ (or on that hypersphere), then it will always remain in the hyperspherical region of radius ε (or on *that* hypersphere).

The systems 4 in Table 3-3 and 13 in Table 3-5 are all stable according to this definition. A stronger stability condition, whereby a perturbed motion returns to an equilibrium state at the origin, as in the systems 1 in Table 3-3, 6 and 9 in Table 3-4, and 12 in Table 3-5, is defined by the following:

An equilibrium state **0** is *asymptotically stable* if it is stable, and if every trajectory starting inside some hyperspherical region in the state space (δ in Fig. 4-1) converges to the origin as time increases indefinitely, $t \to \infty$. Note that asymptotic stability implies that the trajectory must always remain within the hyperspherical region of radius ε (i.e., the solution is bounded,

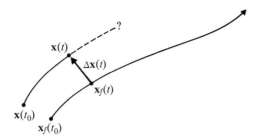

Fig. 4-2 Deviation of a motion $\mathbf{x}(t)$ from a fixed motion $\mathbf{x}_f(t)$.

which is the condition implied in plain stability). We also note that when a linear system described by Eq. (4–3) is asymptotically stable, it is also *asymptotically stable in the large* since (4–4) does not apply.

The systems 2, 3, and 5 in Table 3–3, 7, 8, and 11 in Table 3–4, and 14 in Table 3–5 are unstable. An equilibrium point (the origin **0** in our discussion) is said to be unstable whenever for some arbitrarily large ε inside R and *any* arbitrarily small δ, there is always a starting point $\mathbf{x}_0$ within the hyperspherical region of radius δ such that the trajectory from $\mathbf{x}_0$ passes beyond the boundary hypersphere of radius ε.

Generally, the question of stability concerns deviations about some fixed motion of a (free or forced) system instead of the stability about a fixed point (the origin) of a free system. We shall show that the stability problem of the general kind can be reduced to the simpler problem of the latter category. Consider a specific motion $\mathbf{x}_f(t)$ of a system

$$\frac{d}{dt}\mathbf{x}_f(t) = \mathbf{f}(\mathbf{x}_f, \mathbf{u}, t)$$

and other motions $\mathbf{x}(t)$ which start near $\mathbf{x}_f(t)$ at some time $t = t_0$, subject to the same forcing input $\mathbf{u}(t)$ (see Fig. 4–2). We ask the question: Will $\mathbf{x}(t)$ remain near $\mathbf{x}_f(t)$ as time proceeds, or will it diverge?

Our interest in this case is in the behavior of the deviation vector

$$\Delta \mathbf{x}(t) = \mathbf{x}(t) - \mathbf{x}_f(t).$$

The dynamic equation for $\Delta \mathbf{x}$ is

$$\frac{d}{dt}\Delta\mathbf{x}(t) = \frac{d}{dt}\mathbf{x}(t) - \frac{d}{dt}\mathbf{x}_f(t)$$

$$= \mathbf{f}(\mathbf{x}_f + \Delta\mathbf{x}, \mathbf{u}, t) - \mathbf{f}(\mathbf{x}_f, \mathbf{u}, t). \tag{4-5}$$

Since $\mathbf{x}_f(t)$ and $\mathbf{u}(t)$ are fixed functions of time, the right-hand side of this equation is a function of $\Delta\mathbf{x}(t)$ and t. If Eq. (3–14) applies to both $\mathbf{x}_f$ and $\mathbf{x}$ (that is, if the forced system is linear and stationary), then the convolution integral cancels when we take the difference $\mathbf{x} - \mathbf{x}_f$. Therefore,

$$\frac{d}{dt} \Delta \mathbf{x}(t) = \mathbf{g}(\Delta \mathbf{x}, t), \tag{4-6}$$

and from (4-5) we also have

$$\mathbf{g}(\mathbf{0}, t) = \mathbf{0} \quad \text{for all} \quad t.$$

Thus, writing F for g, we reduce the problem to one of investigating the stability of the equilibrium point $\mathbf{0}$ of a system

$$\frac{d}{dt} \mathbf{x}(t) = \mathbf{F}(\mathbf{x}, t), \qquad \mathbf{F}(\mathbf{0}, t) = \mathbf{0}. \tag{4-7}$$

The form of Eq. (4-7) is free (no forcing input), but the system may be time varying. (See [4] for definitions of stability in various cases of time-varying and/or nonlinear systems.)

In Lyapunov's direct method the stability properties discussed earlier in this section are tested by a scalar function $V(\mathbf{x})$ of the state. $V(\mathbf{x})$ is called the Lyapunov function when it satisfies certain conditions which we will define later. Until these conditions are shown to be satisfied, the test function $V(\mathbf{x})$ is only a *candidate* Lyapunov function, and no conclusions regarding system stability may yet be drawn.

Consideration of some simple systems enables us to apply our physical intuition to the construction of a test function. For instance, in some cases $V(\mathbf{x})$ may represent the energy level of an isolated system at the state $\mathbf{x}$. If the energy level $V(\mathbf{x})$ of a system continually decreases with time (that is, along a free trajectory), until it reaches its minimum value at the equilibrium state $\mathbf{x}_e = \mathbf{0}$, we can expect the system to be stable. Although candidate Lyapunov functions can seldom be identified with energy levels of a complex system, it is helpful to consider a simple example from the energy point of view. For a linear and undamped mass–spring system, the energy in the system is given by

$$E(\mathbf{x}) = c_1 x_1^2 + c_2 x_2^2,$$

where x_1 is the position of the mass, x_2 is its velocity, and c_1 and c_2 are parameter constants. The first term on the right-hand side is the potential energy in the system at any instant, and the second term is the kinetic energy. The state of the system is given by

$$\mathbf{x} = \begin{bmatrix} x_1 \\ x_2 \end{bmatrix}.$$

Trajectories of motion on the modal state-plane (or phase plane) will be circles, where increasing radii indicate higher energy levels at which the motion is started. Identifying $V(\mathbf{x}) = E(\mathbf{x}) = $ constant, we likewise obtain circular contours for the candidate Lyapunov function on the state plane.

In the undamped case the trajectory of motion will overlap a contour of constant $V(\mathbf{x})$, indicating the expected oscillatory response with constant amplitude. For a system with damping, the resulting trajectory will spiral into the origin of the state plane from any starting point $\mathbf{x}_0$. As the trajectory crosses the $V(\mathbf{x})$ contour lines, the value of $V(\mathbf{x})$ on the trajectory changes with time. In the case of this simple example, $V(\mathbf{x})$ decreases with time (due to energy dissipation by the damper) and the system is asymptotically stable.

Now let us generalize our thinking by considering nth-order systems for which $V(\mathbf{x})$ may *not* be identified with energy level. There is a geometric pattern in the state space which leads to another interpretation of the candidate Lyapunov function $V(\mathbf{x})$. Suppose a value of $V(\mathbf{x})$ represents the radius of a hypersphere or a hyperellipsoid (e.g., its major axes). We consider the state space filled with a set of such closed contours, all of different sizes, with center at the origin, and all nonintersecting. If a trajectory $\mathbf{x}(t)$ crosses all hyperspheres (or all hyperellipsoids) from outside to inside, the trajectory must converge to the origin. In turn, this means a monotonic decrease of the value of $V(\mathbf{x})$ along the trajectory.

The intuitive concepts contained in the last two sentences imply that for stable systems the candidate Lyapunov function $V(\mathbf{x})$ has the following properties:

1. $V(\mathbf{x})$ must be positive, becoming zero only at the origin, and
2. $V(\mathbf{x})$ evaluated along a trajectory must decrease as $\mathbf{x}$ proceeds on a trajectory; in other words, $dV(\mathbf{x}(t))/dt$ of a trajectory must be negative.

Before concluding this section with a statement of what conditions $V(\mathbf{x})$ must fulfill in order to be considered a Lyapunov function, we must first define some properties of scalar functions. A scalar function $V(\mathbf{x})$ is said to be *positive definite* when:

1. $V(\mathbf{0}) = 0$, and
2. $V(\mathbf{x}) > 0$ in some region of $\mathbf{x}$ outside the origin. Let us represent the region in state space by S. Then

$$V(\mathbf{x}) > 0, \quad \mathbf{x} \in S; \quad \mathbf{x} \neq \mathbf{0}.$$

3. $V(\mathbf{x})$ is continuous in S, and
4. $\partial V(\mathbf{x})/\partial x_i$, $i = 1, \ldots, n$ are also continuous.

If, in condition 2 above, $V(\mathbf{x}) \geqslant 0$ instead of > 0 (the stronger condition), then $V(\mathbf{x})$ is said to be *positive semidefinite*. The significance of this latter condition is that $V(\mathbf{x})$ may take on a zero value at points in the state space other than the origin. If the inequality sign is reversed, the definitions of *negative definite* and *negative semidefinite* follow immediately.

According to condition 4, $V(\mathbf{x})$ has partial derivatives. Therefore, it has a gradient vector:

$$\mathbf{grad}\ V(\mathbf{x}) = \begin{bmatrix} \dfrac{\partial V(\mathbf{x})}{\partial x_1} \\ \vdots \\ \dfrac{\partial V(\mathbf{x})}{\partial x_n} \end{bmatrix}. \tag{4-8}$$

The time derivative $dV(\mathbf{x})/dt$ along any trajectory of a system (4–2) is given by

$$\begin{aligned}\frac{d}{dt}(V(\mathbf{x})) &= \frac{\partial V(\mathbf{x})}{\partial x_1}\frac{dx_1}{dt} + \cdots + \frac{\partial V(\mathbf{x})}{\partial x_n}\frac{dx_n}{dt} \\ &= [\dot{x}_1, \dot{x}_2, \ldots, \dot{x}_n]\,[\mathbf{grad}\ V(\mathbf{x})] \\ &= \left(\frac{d\mathbf{x}}{dt}\right)' \cdot \mathbf{grad}\ V(\mathbf{x}) \\ &= (\mathbf{F}(\mathbf{x}))' \cdot \mathbf{grad}\ V(\mathbf{x}). \end{aligned} \tag{4-9}$$

Recall also that the gradient to the surface of $V(\mathbf{x})$ is perpendicular to the surface and pointing outward in the direction of increasing $V(\mathbf{x})$. If a trajectory $\mathbf{x}(t)$ for a stable system crosses this surface of $V(\mathbf{x})$, it must cross it from outside to inside with $V(\mathbf{x})$ values decreasing along the trajectory in the direction of the vector $d\mathbf{x}(t)/dt = \dot{\mathbf{x}}(t)$. Thus, the relation given by (4–9) expresses the projection of $\dot{\mathbf{x}}(t)$ on the gradient vector, and this projection must always be less than zero (or at worst equal to zero) for a system to be stable.

We may now state the conditions under which a candidate function becomes a Lyapunov function: If $V(\mathbf{x})$ is positive definite and $dV(\mathbf{x})/dt$ along a trajectory is negative definite or negative semidefinite (i.e., $\leqslant 0$) in a region S, then $V(\mathbf{x})$ is called a *Lyapunov function*. In the next section we shall use this definition to formulate the Lyapunov stability theorems, and in Section 4–3 we shall indicate methods of applying the theorems.

4-2 THE LYAPUNOV THEOREMS

Stability theorem If there exists a Lyapunov function $V(\mathbf{x})$ in some region S around the origin, then the origin is stable for all $\mathbf{x}_0$ contained in S.

Referring to the definition of stability in the sense of Lyapunov given in the preceding section and to Fig. 4–3, we can see the validity of the theorem in the following way:

1. Find a constant k such that the closed hypersurface defined by $V(\mathbf{x}) = k$ comes just inside the hypersphere of radius ε.

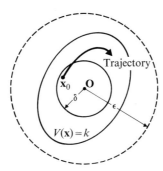

Fig. 4-3 Proof of stability via a Lyapunov function $V(\mathbf{x})$.

2. Select $\delta > 0$ such that the hypersphere of radius δ is contained within $V(\mathbf{x}) = k$, and

$$V(\mathbf{x}_0) < k \tag{4-10}$$

is satisfied for $\mathbf{x}_0$ in the hyperspherical region S. Since $V(\mathbf{x})$ is continuous and positive, and since $V(\mathbf{0}) = \mathbf{0}$, a δ that satisfies (4-10) exists.

3. $V(\mathbf{x})$ along a trajectory either continually decreases or reaches a constant value when $dV(\mathbf{x})/dt = 0$. Therefore, with (4-10) satisfied, the trajectory that starts at $\mathbf{x}_0$ can never reach the hypersurface $V(\mathbf{x}) = k$.

Therefore, the origin is stable. The spring–mass–dashpot system discussed earlier affords a graphical interpretation of the condition $\dot{V}(\mathbf{x}) = 0$. Suppose that at some instant t_a during motion along the trajectory of such a system, the dashpot is suddenly disconnected so that the system becomes conservative. The energy in the system, and thus $V(\mathbf{x})$ for that example, will remain constant at the value reached at t_a. Unless the dashpot is reconnected, motion in state space will be around the closed contour $V(\mathbf{x}) = k_a$, where k_a is the value of $V(\mathbf{x})$ at time t_a and $k_a < k$. By the definition given above, such a system would be considered to be stable in the sense of Lyapunov even though it might not be an acceptable engineering design. If $V(\mathbf{x}(t))$ is plotted against t, then the condition just discussed will occur whenever the slope of the plot reaches (and remains at) zero at a value of V greater than zero.

Asymptotic stability theorem If there exists a Lyapunov function $V(\mathbf{x})$ in some region S around the origin, and if in addition $-dV(\mathbf{x})/dt$ along a trajectory is positive definite in S, then the stability at the origin is asymptotic.

Note that this is a more restrictive definition of stability than the previous one in that the condition $\dot{V}(\mathbf{x}) = 0$ is not allowed. This means that a trajectory will not be allowed to "stall" on a closed hypersurface of $V(\mathbf{x})$ containing the origin, but will always be required to approach the origin with a monotonic decrease in V along the trajectory. Put more precisely, the asymptotic stability theorem states that since $dV(\mathbf{x})/dt < 0$ for all $\mathbf{x} \neq \mathbf{0}$, $V(\mathbf{x})$ along a trajectory tends steadily to zero, which means that the trajectory tends to the origin where $V(\mathbf{0}) = 0$.

If the stability theorems apply everywhere in the state space, then the limitations to a region S imposed on the statements may be deleted. An equilibrium state $\mathbf{0}$ is then said to be stable (or asymptotically stable) *in the large*. The term *global* stability is also used in this context. For linear systems, in which the properties do not depend on $\mathbf{x}$, no distinction between local and global stability is necessary.

It is important to note that failure of a candidate Lyapunov function to yield stability prediction does not denote instability (that is, the converse of the above stability theorems is not true). In such a case all that may be said is that the choice of candidate Lyapunov function is inconclusive, and that other candidate Lyapunov functions must be investigated in order to prove stability. As the theorems now stand, only stability may be proved, and not instability. However, there does exist an instability theorem [8] which states that if $V(\mathbf{x})$ is continuous and if $dV(\mathbf{x})/dt$ along a trajectory is negative definite, then:

1. the system is unstable in that finite region of the state space for which $V(\mathbf{x})$ is not positive semidefinite, and
2. the response of the system $\dot{\mathbf{x}} = \mathbf{F}(\mathbf{x})$ is unbounded as $t \to \infty$ if $V(\mathbf{x})$ is not globally positive semidefinite.

The converse of the instability theorem fortunately exists, namely, if for a negative definite $\dot{V}(\mathbf{x})$ a proper $V(\mathbf{x})$ can be obtained which does not satisfy the instability theorem conditions [that is, a positive semidefinite $V(\mathbf{x})$], then the conditions for stability are automatically met. This converse to the instability theorem could be a more powerful result than the stability theorems, since it is an easy task to pick a negative definite $\dot{V}(\mathbf{x})$. It is not always so easy, however, to find a proper $V(\mathbf{x})$ function for a given $\dot{V}(\mathbf{x})$. If the system is linear, the necessary integration procedure (by matrix methods to be explained) is straightforward. This latter approach based on the instability theorem will be used later in this chapter. The basic difference between the stability and instability theorems is that the former begins with a trial $V(\mathbf{x})$ and seeks to satisfy conditions applicable to $\dot{V}(\mathbf{x})$, whereas the latter reverses the procedure. For purposes of applying the instability theorem note that the signs may be reversed throughout without changing the meaning of the theorem.

The form of the Lyapunov function must be generalized to $V(\mathbf{x}, t)$ when we deal with a nonautonomous system (4-7), restated as follows:

$$\frac{d}{dt}\mathbf{x}(t) = \mathbf{F}(\mathbf{x}, t), \qquad \mathbf{F}(\mathbf{0}, t) = \mathbf{0} \qquad \text{for} \qquad t \geqslant 0.$$

4-2 The Lyapunov theorems

The positive definiteness of $V(\mathbf{x}, t)$ is defined by the following conditions:

1. $V(\mathbf{0}, t) = 0$ for $t \geq 0$,
2. continuous first partial derivatives of $V(\mathbf{x}, t)$ with respect to $\mathbf{x}$ and t exist,
3. there exists a continuous, nondecreasing scalar function $W(\|\mathbf{x}\|)$ such that $W(\mathbf{0}) = 0$ and $0 < W(\|\mathbf{x}\|) \leq V(\mathbf{x}, t)$ for $t \geq 0$, $\mathbf{x} \neq \mathbf{0}$.

The time derivative of $V(\mathbf{x}, t)$ taken along a trajectory that initiates at $t = 0$ is given by

$$\frac{d}{dt} V(\mathbf{x}, t) = \frac{\partial V}{\partial t} + (\mathbf{F}(\mathbf{x}, t))' \cdot \mathbf{grad}\, V(\mathbf{x}, t). \tag{4-11}$$

When $V(\mathbf{x}, t)$ is positive definite and

$$\frac{d}{dt} V(\mathbf{x}, t) \leq 0,$$

$V(\mathbf{x}, t)$ is said to be a Lyapunov function. There are a number of definitions for various kinds of stability of a nonautonomous system [4]; stability statements and proofs require more care than those for an autonomous system.

The Lyapunov theorems provide sufficient conditions for stability but they may not yield necessary conditions. Thus a particular Lyapunov function may result in a *strong* (i.e., more than necessary) condition for stability or asymptotic stability. We may think of the "sharpness" of a particular Lyapunov function as being a measure of how close we can come to the necessary condition. If we are using the Lyapunov theorems as a design tool to choose undetermined coefficients in the system matrix, for example, then the sharpness of the Lyapunov function chosen will determine how close to the actual stability limit we are designing. Although it is always possible to construct a "sharp" Lyapunov function which gives necessary and sufficient conditions of stability if a system is linear and stationary (see Section 4-3), this is not true if a system is nonlinear. There is no known general approach for constructing a "sharp" Lyapunov function for an arbitrary system. Hunting for a suitable Lyapunov function has been a popular undertaking among some applied mathematicians.

Example 4-1 We try

$$V(\mathbf{x}) = x_1^2 + x_2^2 \quad \text{(a circle)}$$

as a test function for a second-order linear stationary system for which [see Table 3-3, 1(a)]

$$\mathbf{A} = \begin{bmatrix} -2 & -1 \\ -2 & -3 \end{bmatrix}.$$

124 Stability of linear lumped systems

In Eq. (4-9),

$$\text{grad } V(\mathbf{x}) = \begin{bmatrix} 2x_1 \\ 2x_2 \end{bmatrix};$$

thus

$$\dot{V}(\mathbf{x}) = [-2x_1 - x_2, \ -2x_1 - 3x_2] \begin{bmatrix} 2x_1 \\ 2x_2 \end{bmatrix} = -4x_1^2 - 6x_1 x_2 - 6x_2^2$$

$$= -(2x_1 + (3/2)x_2)^2 - (15/4)x_2^2 \ .$$

Therefore $-\dot{V}(\mathbf{x})$ is positive definite. We conclude that the origin **0** of the system is asymptotically stable in the large.

If we apply the same $V(\mathbf{x})$ to a system described by

$$\mathbf{A} = \begin{bmatrix} 2 & -3 \\ 6 & -4 \end{bmatrix},$$

we have

$$\dot{V}(\mathbf{x}) = [2x_1 - 3x_2, \ 6x_1 - 4x_2] \begin{bmatrix} 2x_1 \\ 2x_2 \end{bmatrix} = 4x_1^2 + 6x_1 x_2 - 8x_2^2 \ ,$$

and $-\dot{V}(\mathbf{x})$ is not positive definite; for instance, $-4x_1^2 < 0$ when $x_2 = 0$. Thus, although this system is asymptotically stable at **0**, as shown in Table 3-5, 12(a), the Lyapunov stability theorem is not able to show this stability for the inconclusively chosen $V(\mathbf{x})$. Therefore, a careless converse of the stability theorems must be avoided.

As a clue toward constructing a positive definite function of **x**, suppose that $V(\mathbf{x})$ can be represented by a power series in x_i, $i = 1, \ldots, n$, in the vicinity of the origin. Then

$$V(\mathbf{x}) = V_k(\mathbf{x}) + V_{k+1}(\mathbf{x}) + \cdots ,$$

where k is a positive integer and $V_k(\mathbf{x})$ is a kth-order homogeneous polynomial in $x_1, \ldots, x_n$. For example, if $n = 2$,

$$V_1(\mathbf{x}) = a_1 x_1 + a_2 x_2 ,$$
$$V_2(\mathbf{x}) = a_{11} x_1^2 + 2a_{12} x_1 x_2 + a_{22} x_2^2 ,$$

with constant coefficients a_1, a_2, etc. Note that V_k for the lowest k dominates the behavior of $V = \sum V_i$ near the origin.

Introducing a set of parameters $r_1, \ldots, r_{n-1}$ such that

$$r_1 = x_1/x_n , \quad r_2 = x_2/x_n , \quad \ldots, \quad r_{n-1} = x_{n-1}/x_n ,$$

we obtain

$$V_k(\mathbf{x}) = x_n^k V_k(r_1, r_2, \ldots, r_{n-1}, 1) , \qquad (4\text{–}12)$$

where it is assumed that $r_1, \ldots, r_{n-1}$ have been so chosen that

$$V_k(r_1, \ldots, r_{n-1}, 1) \neq 0 \ .$$

If k is odd in (4–12), $(x_n)^k$ can be both positive and negative, blocking the possibility of $V_k > 0$ (positive definite). Therefore, a necessary condition for V_k to be positive definite is that k be even. We use $k = 2$ to construct a Lyapunov function for a linear stationary system in the next section.

4-3 STABILITY OF LINEAR, STATIONARY, LUMPED SYSTEMS

Let us consider a quadratic form (second-order homogeneous polynomial):

$$V(\mathbf{x}) = a_{11}x_1^2 + a_{22}x_2^2 + \cdots + a_{12}x_1x_2 + a_{13}x_1x_3 + \cdots$$
$$= \sum a_{ij}x_ix_j, \quad i,j = 1, \ldots, n$$
$$= [x_1\ x_2\ \ldots\ x_n] \begin{bmatrix} a_{11} & \cdots & a_{1n} \\ \vdots & & \vdots \\ a_{n1} & \cdots & a_{nn} \end{bmatrix} \begin{bmatrix} x_1 \\ x_2 \\ \vdots \\ x_n \end{bmatrix}$$
$$= \mathbf{x}'\mathbf{Y}\mathbf{x}, \tag{4-13}$$

where

$$\mathbf{Y} = \begin{bmatrix} a_{11} & \cdots & a_{1n} \\ \vdots & & \vdots \\ a_{n1} & \cdots & a_{nn} \end{bmatrix}. \tag{4-14}$$

Since

$$x_ix_j = x_jx_i,$$

we have no reason to consider the case $a_{ij} \neq a_{ji}$. Letting

$$a_{ij} = a_{ji} \tag{4-15}$$

in (4–13) and (4–14), we find that $\mathbf{Y}$ becomes symmetric, which simplifies the succeeding computations.

$V(\mathbf{x})$ in a quadratic form is positive definite when the coefficients of $\mathbf{Y}$ in (4–13) satisfy the Sylvester condition (4–16). If $n = 1$, then $\mathbf{x} = x_1$ and

$$V(\mathbf{x}) = a_{11}x_1^2.$$

This function is positive definite if $a_{11} > 0$. For $n = 2$,

$$V(\mathbf{x}) = a_{11}x_1^2 + 2a_{12}x_1x_2 + a_{22}x_2^2$$
$$= a_{11}(x_1 + (a_{12}/a_{11})x_2)^2 + (a_{22} - (a_{12}^2/a_{11})) \cdot x_2^2.$$

Thus the necessary and sufficient conditions for positive definiteness of $V(\mathbf{x})$ are:

$$a_{11} > 0 \quad \text{and} \quad \begin{vmatrix} a_{11} & a_{12} \\ a_{21} & a_{22} \end{vmatrix} > 0, \quad (\text{where } a_{12} = a_{21}).$$

Following similar steps for an nth-order system, we can prove [9] that the quadratic form is positive definite when all principal minors of $\mathbf{Y}$ are positive:

$$a_{11} > 0 \qquad \begin{vmatrix} a_{11} & a_{12} \\ a_{21} & a_{22} \end{vmatrix} > 0, \qquad \ldots, \qquad \begin{vmatrix} a_{11} & \cdots & a_{1n} \\ \vdots & & \vdots \\ a_{n1} & \cdots & a_{nn} \end{vmatrix} > 0. \tag{4-16}$$

According to analytic geometry, $V(\mathbf{x}) = $ constant describes a hyperellipsoid (in x-space) when $V(\mathbf{x})$ is a quadratic that satisfies the Sylvester inequalities (4–16). The axes of the hyperellipsoid can be determined by diagonalization of $\mathbf{Y}$. This diagonalization is simpler than the one discussed in Section 3–3, because $\mathbf{Y}$ is symmetric (see [9] for properties of symmetric matrices). The outline of the diagonalization process is as follows: We change coordinates from $\mathbf{x}$ to $\mathbf{x}^*$ by

$$\mathbf{x} = \mathbf{T}\mathbf{x}^*$$

so that $\mathbf{T}'\mathbf{Y}\mathbf{T}$ assumes a diagonal form. Since

$$V(\mathbf{x}) = \mathbf{x}'\mathbf{Y}\mathbf{x} = (\mathbf{T}\mathbf{x}^*)'\mathbf{Y}\mathbf{T}\mathbf{x}^* = (\mathbf{x}^*)'\mathbf{T}'\mathbf{Y}\mathbf{T}\mathbf{x}^*,$$

a canonical form for a hyperellipsoid

$$V(\mathbf{x}) = \sum_{i=1}^{n} \left(\frac{x_i^*}{r_i}\right)^2 = k \quad \text{(a constant)} \tag{4-17}$$

will follow when

$$\mathbf{T}'\mathbf{Y}\mathbf{T} = \begin{bmatrix} r_1^{-2} & & & 0 \\ & r_2^{-2} & & \\ & & \ddots & \\ 0 & & & r_n^{-2} \end{bmatrix}.$$

The $\mathbf{T}$ for the coordinate change will be a row matrix of orthogonal column vectors;

$$\mathbf{T} = [\mathbf{v}^1, \ldots, \mathbf{v}^n],$$

and these column vectors in turn are eigenvectors of $\mathbf{Y}$.

One must solve the characteristic polynomial

$$|\mathbf{Y} - p\mathbf{I}| = 0 \tag{4-18}$$

for eigenvalues (roots) $p = p_1, \ldots, p_n$. Then an eigenvector $\mathbf{v}^i$ is fixed (for real and distinct eigenvalues) by the conditions

$$\mathbf{Y}\mathbf{v}^i = p_i \mathbf{v}^i \qquad \text{or} \qquad (\mathbf{Y} - p_i\mathbf{I})\mathbf{v}^i = \mathbf{0}$$

and also by

$$\|\mathbf{v}^i\| = \left[\sum_{j=1}^{n} (v_j^i)^2\right]^{1/2} = 1,$$

where the second condition is needed to normalize all the eigenvectors. The magnitudes of the axes of the $V(\mathbf{x})$-hyperellipsoid in modal-state space are given by $r_i(k)^{1/2}$, where

$$r_i = (1/p_i)^{1/2}. \qquad (4\text{--}19)$$

If one of the eigenvalues of $\mathbf{Y}$ vanishes, $p_n = 0$ for example, we must have $\mathbf{Y}\mathbf{v}^n = \mathbf{0}$ for the eigenvector $\mathbf{v}^n$. This requires that $|\mathbf{Y}| = 0$ in order for a nontrivial solution $\mathbf{v}^n$ to exist. According to (4–19), $r_n = \infty$ when $p_n = 0$. Thus, with $|\mathbf{Y}| = 0$ [the last determinant in the Sylvester chain (4–16)] we get a cylinder where $\mathbf{v}^n$ becomes the axis of the cylinder.

Example 4-2 Given

$$\mathbf{Y} = \begin{bmatrix} 5 & -2 \\ -2 & 8 \end{bmatrix},$$

the roots of the characteristic equation $p^2 - 13p + 36 = 0$ are $p_1 = 9$, $p_2 = 4$. Therefore $r_1 = 1/3$, $r_2 = 1/2$. We find that

$$\mathbf{v}^1 = \begin{bmatrix} 1/\sqrt{5} \\ -2/\sqrt{5} \end{bmatrix}, \quad \mathbf{v}^2 = \begin{bmatrix} 2/\sqrt{5} \\ 1/\sqrt{5} \end{bmatrix}, \quad \mathbf{T} = \frac{1}{\sqrt{5}} \begin{bmatrix} 1 & 2 \\ -2 & 1 \end{bmatrix},$$

$$\mathbf{T}' = \frac{1}{\sqrt{5}} \begin{bmatrix} 1 & -2 \\ 2 & 1 \end{bmatrix},$$

and confirm that

$$\mathbf{T}'\mathbf{Y}\mathbf{T} = \begin{bmatrix} 9 & 0 \\ 0 & 4 \end{bmatrix}.$$

Thus, by Eq. (4–17)

$$V(\mathbf{x}) = V(\mathbf{x}^*) = 9x_1^{*2} + 4x_2^{*2} = k.$$

An ellipse for $k = 36$ is shown in Fig. 4-4. To check the equivalence of the two coordinate systems in Fig. 4-4, note that $\mathbf{x}^* = \mathbf{T}^{-1}\mathbf{x}$ so that $x_1^* = (x_1 - 2x_2)/\sqrt{5}$ and $x_2^* = (2x_1 + x_2)/\sqrt{5}$.

Consider now a linear stationary system

$$\frac{d}{dt}\mathbf{x} = \mathbf{A}\mathbf{x}.$$

In order to explore the asymptotic stability of the system at the origin, we try

$$V(\mathbf{x}) = \mathbf{x}'\mathbf{Y}\mathbf{x}.$$

It follows then that

$$\dot{V}(\mathbf{x}) = \dot{\mathbf{x}}'\mathbf{Y}\mathbf{x} + \mathbf{x}'\mathbf{Y}\dot{\mathbf{x}},$$

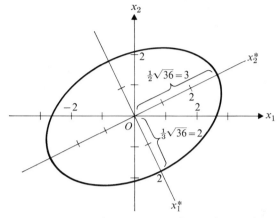

Fig. 4-4 An ellipse: $5x_1^2 - 4x_1x_2 + 8x_2^2 = 9x_1^{*2} + 4x_2^{*2} = 36$.

where $\dot{\mathbf{x}} = \mathbf{A}\mathbf{x}$ and $\dot{\mathbf{x}}' = (\mathbf{A}\mathbf{x})' = \mathbf{x}'\mathbf{A}'$, and thus

$$\dot{V}(\mathbf{x}) = \mathbf{x}'(\mathbf{A}'\mathbf{Y} + \mathbf{Y}\mathbf{A})\mathbf{x}.$$

Let us put

$$\dot{V}(\mathbf{x}) = -\mathbf{x}'\mathbf{W}\mathbf{x}.$$

Then

$$-\mathbf{W} = \mathbf{A}'\mathbf{Y} + \mathbf{Y}\mathbf{A}. \tag{4-20}$$

We now apply the *converse* of the instability theorem stated in the previous section. First we choose a convenient positive definite matrix $\mathbf{W}$ so that $\dot{V}(\mathbf{x})$ will be negative definite. Then, by solving (4–20), we find a matrix $\mathbf{Y}$. By (4–13), if $\mathbf{Y}$ is positive definite, then $V(\mathbf{x})$ is positive definite. This conclusion, however, violates the conditions of the instability theorem; thus the converse, or stability, is proved. The task, therefore, reduces to checking the matrix $\mathbf{Y}$ for positive definiteness by its satisfaction of the Sylvester inequalities.

The foregoing discussion may now be concisely summarized into an asymptotic stability theorem, for which the converse is also true.

Theorem The equilibrium state **0** of a linear stationary system $\dot{\mathbf{x}} = \mathbf{A}\mathbf{x}$ is asymptotically stable if *and only if*, given any symmetric positive definite matrix $\mathbf{W}$, there exists a symmetric positive definite matrix $\mathbf{Y}$ which is the unique solution of (4–20), and $V(\mathbf{x}) = \mathbf{x}'\mathbf{Y}\mathbf{x}$ is a Lyapunov function.

The statement implies that the theorem provides a necessary and sufficient condition for asymptotic stability. This theorem can be proved by using a diagonal or Jordan canonical form for the A-matrix (see Problem 4–3). Since $\mathbf{W}$ can be any symmetric positive definite matrix, it is usually convenient to use the identity matrix $\mathbf{I}$ for the purpose, and take the following steps:

1. For a given **A**, find **Y** by the equation

$$\mathbf{A'Y} + \mathbf{YA} = -\mathbf{I}. \tag{4-21}$$

If one cannot determine a unique symmetric matrix **Y** by (4–21), the system is not asymptotically stable at the origin. If, on the other hand, he finds such a **Y**, then he proceeds to the next step.

2. Test the positive definiteness of **Y** by the Sylvester chain (4–16). If **Y** satisfies (4–16), then **0** is asymptotically stable (in the large); if it does not, then the system is not asymptotically stable.

Example 4-3 Given

$$\mathbf{A} = \begin{bmatrix} -3 & -7 \\ 0 & -4 \end{bmatrix},$$

we make substitution in (4–21), and obtain

$$\begin{bmatrix} -3 & 0 \\ -7 & -4 \end{bmatrix} \begin{bmatrix} a & b \\ b & c \end{bmatrix} + \begin{bmatrix} a & b \\ b & c \end{bmatrix} \begin{bmatrix} -3 & -7 \\ 0 & -4 \end{bmatrix} = \begin{bmatrix} -1 & 0 \\ 0 & -1 \end{bmatrix}.$$

This matrix equation produces three scalar relations for three unknowns:

$$-3a = -0.5,$$
$$-3b - 7a - 4b = 0,$$
$$-7b - 4c = -0.5,$$

which yield

$$a = \tfrac{1}{6}, \quad b = -\tfrac{1}{6}, \quad c = \tfrac{5}{12}.$$

This set of values satisfies the Sylvester inequalities. Thus the system is asymp-

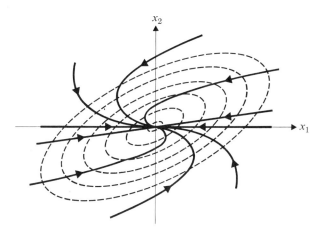

Fig. 4-5 Free trajectories of a system and contour lines of its Lyapunov function.

totically stable. By (4-13) the Lyapunov function is

$$V(\mathbf{x}) = \tfrac{1}{12}(2x_1^2 - 4x_1x_2 + 5x_2^2) = k \ .$$

Some trajectories of the system and a set of V-contours are shown in Fig. 4-5. Note that all trajectories cross all contours from outside to inside.

An interesting and potentially valuable application of Lyapunov functions is in the evaluation of performance indices which are quadratic in form. An integrated squared error is often used as a performance index of a control system. Consider, then, as an example,

$$J = \int_0^\infty e^2(t) \, dt \qquad (4\text{-}22)$$

or, more generally,

$$J = \sum_j^n \int_0^\infty w_j e_j^2(t) \, dt \ , \qquad (4\text{-}23)$$

where e_j is a component of the error and w_j is the weighting factor to be applied to that particular component. The quadratic Lyapunov function may be conveniently applied to evaluate an integral of the form of (4-23).* To simplify our initial consideration of the method, let us take a state variable $x_1(t)$ as the error of our interest. We then apply

$$\mathbf{W} = \begin{bmatrix} 1 & \cdot & \cdot & \cdot & 0 \\ 0 & 0 & & & 0 \\ \vdots & & & & \vdots \\ 0 & \cdot & \cdot & \cdot & 0 \end{bmatrix}$$

in the relation

$$\dot{V}(\mathbf{x}) = -\mathbf{x}'\mathbf{W}\mathbf{x}$$

and integrate:

$$\int_0^\infty \frac{dV}{dt} dt = -V(\mathbf{x}_0) + V(\mathbf{0}) = -\int_0^\infty x_1^2(t) \, dt \ .$$

The last integral is obtained through consideration of the previous equation. In evaluating the upper limit of integration to be $V(\mathbf{0})$ we are implicitly assuming that the system is asymptotically stable (that is, $\lim_{t \to \infty} \mathbf{x}(t) = \mathbf{0}$). Furthermore, since $V(\mathbf{0}) = 0$, we get

$$J = \int_0^\infty x_1^2(t) \, dt = V(\mathbf{x}_0) \ , \qquad (4\text{-}24)$$

---

* This method is far simpler than computation in the Fourier domain. The latter method is presented in [12].

where

$$V(\mathbf{x}_0) = \mathbf{x}_0' \mathbf{Y} \mathbf{x}_0 . \tag{4-25}$$

Therefore, J of Eq. (4-24) is found for a prescribed initial state by computing $\mathbf{Y}$ for a given $\mathbf{W}$ through the relation

$$\mathbf{A}' \mathbf{Y} + \mathbf{Y} \mathbf{A} = -\mathbf{W} . \tag{4-26}$$

Although $\mathbf{W}$, such as given above, is not positive definite (so that V is not a Lyapunov function), there is no problem in evaluating a meaningful performance index so long as the system has been previously proved to be asymptotically stable.

If

$$J = \sum_{j}^{n} \int_{0}^{\infty} w_j x_j^2(t) \, dt$$

is required, we follow the same approach as above, with $\mathbf{W}$ in the diagonal form:

$$\mathbf{W} = \begin{bmatrix} w_1 & & & 0 \\ & w_2 & & \\ & & \ddots & \\ 0 & & & w_n \end{bmatrix} . \tag{4-27}$$

For an index $J = \int_0^\infty y(t)^2 \, dt$ with $y = \mathbf{cx} = [c_1, \ldots, c_n]\mathbf{x}$, we compute

$$y^2 = \mathbf{x}' \mathbf{c}' \mathbf{c} \mathbf{x} ,$$

and thus obtain

$$\mathbf{c}'\mathbf{c} = \begin{bmatrix} c_1^2 & c_1 c_2 & c_1 c_3 & \cdots & c_1 c_n \\ c_2 c_1 & c_2^2 & c_2 c_3 & & c_2 c_n \\ \vdots & & & & \vdots \\ c_n c_1 & & \cdots & & c_n^2 \end{bmatrix} = \mathbf{W} . \tag{4-28}$$

We then substitute (4-28) in (4-26).

Example 4-4 Given

$$\mathbf{A} = \begin{bmatrix} -3 & -7 \\ 0 & -4 \end{bmatrix}, \quad \mathbf{x}_0 = \begin{bmatrix} 1 \\ 1 \end{bmatrix}, \quad y = x_1 + 2x_2 , \quad J = \int_0^\infty y^2 \, dt .$$

We have

$$\mathbf{W} = \begin{bmatrix} 1 \\ 2 \end{bmatrix} \begin{bmatrix} 1 & 2 \end{bmatrix} = \begin{bmatrix} 1 & 2 \\ 2 & 4 \end{bmatrix} .$$

Thus,

$$\begin{bmatrix} -3 & 0 \\ -7 & -4 \end{bmatrix} \begin{bmatrix} a & b \\ b & c \end{bmatrix} + \begin{bmatrix} a & b \\ b & c \end{bmatrix} \begin{bmatrix} -3 & -7 \\ 0 & -4 \end{bmatrix} = \begin{bmatrix} -1 & -2 \\ -2 & -4 \end{bmatrix}$$

or
$$-6a = -1, \quad -3b -7a -4b = -2, \quad -14b -8c = -4.$$

Therefore, $a = \frac{1}{6}$, $b = \frac{5}{42}$, $c = \frac{7}{24}$, and

$$J = V(\mathbf{x}_0) = a + 2b + c = \frac{1}{6} + \frac{5}{21} + \frac{7}{24} = \frac{117}{168}.$$

If a linear free system is time varying,

$$\frac{d}{dt}\mathbf{x}(t) = \mathbf{A}(t)\mathbf{x}(t), \qquad (4\text{--}29)$$

then we consider a quadratic test function

$$V(\mathbf{x}, t) = \mathbf{x}'\mathbf{Y}(t)\mathbf{x}, \qquad (4\text{--}30)$$

where the symmetric matrix $\mathbf{Y}(t)$ is a function of time. An additional term $d\mathbf{Y}/dt$ appears when we differentiate $V(\mathbf{x}, t)$ with respect to time:

$$\dot{V}(\mathbf{x}, t) = \mathbf{x}'\mathbf{A}'\mathbf{Y}(t)\mathbf{x} + \mathbf{x}'\mathbf{Y}(t)\mathbf{A}\mathbf{x} + \mathbf{x}'\dot{\mathbf{Y}}(t)\mathbf{x}.$$

Suppose that the condition

$$\dot{V}(\mathbf{x}) = -\mathbf{x}'\mathbf{W}\mathbf{x}$$

is to be satisfied. Then we get

$$\mathbf{A}(t)'\mathbf{Y}(t) + \mathbf{Y}(t)\mathbf{A}(t) + \dot{\mathbf{Y}}(t) = -\mathbf{W}, \qquad (4\text{--}31)$$

where $\mathbf{W}$ could also be a function of time. Equation (4–31) gives a unique relation between $\mathbf{Y}(t)$ and $\mathbf{W}(t)$ for a given initial condition $\mathbf{Y}(0)$. However, since it is equivalent to a set of inhomogeneous, linear, time-varying, first-order differential equations for all elements of $\mathbf{Y}(t)$, we encounter the difficulty of treating time-varying differential equations. No further approach that leads to sharp conditions of stability is yet known. In some cases, an intuitive selection of a constant positive definite matrix $\mathbf{Y}$ or $\mathbf{W}$ may lead to a less sharp but more practical statement of stability.

4–4 THE LYAPUNOV APPROACH TO DISCRETE-TIME SYSTEMS

Considerations of discrete-time systems generally parallel those of continuous-time systems with minor variations. The general equation (4–7) of a free dynamic system in the continuous-time domain must be replaced by a difference equation

$$\mathbf{x}_{k+1} = \mathbf{H}(\mathbf{x}_k, k), \qquad \mathbf{H}(\mathbf{0}, k) = \mathbf{0}. \qquad (4\text{--}32)$$

If the system is autonomous, then

$$\mathbf{x}_{k+1} = \mathbf{H}(\mathbf{x}_k), \qquad \mathbf{H}(\mathbf{0}) = \mathbf{0}. \qquad (4\text{--}33)$$

If it is linear and stationary, then

$$\mathbf{x}_{k+1} = \mathbf{P}\mathbf{x}_k. \tag{4-34}$$

The conditions for the positive definiteness of a test function $V(\mathbf{x})$ are the same as before. A variation to the approach appears when we compute the rate of change of $V(\mathbf{x})$ along a free trajectory. Instead of $dV(\mathbf{x})/dt$ we jump from point $\mathbf{x}_k$ to $\mathbf{x}_{k+1}$ and thus obtain

$$\frac{\Delta V(\mathbf{x})}{\Delta t} = \frac{V(\mathbf{x}_{k+1}) - V(\mathbf{x}_k)}{T}.$$

Except for the computation of a performance index, our interest is limited only to the (plus or minus) sign of $\Delta V(\mathbf{x})/\Delta t$. Therefore, we may let

$$T = 1$$

without losing generality in a stability investigation.

Consider a quadratic form:

$$V(\mathbf{x}_k) = \mathbf{x}_k' \mathbf{Y} \mathbf{x}_k, \tag{4-35}$$

where $\mathbf{Y}$ is a symmetric matrix. V is positive definite if $\mathbf{Y}$ satisfies the Sylvester condition (4-16).

We then have

$$\Delta V(\mathbf{x}_k) = V(\mathbf{x}_{k+1}) - V(\mathbf{x}_k) = \mathbf{x}_{k+1}' \mathbf{Y} \mathbf{x}_{k+1} - \mathbf{x}_k' \mathbf{Y} \mathbf{x}_k,$$

where

$$\mathbf{x}_{k+1} = \mathbf{P}\mathbf{x}_k \quad \text{and} \quad \mathbf{x}_{k+1}' = \mathbf{x}_k' \cdot \mathbf{P}'$$

for a linear stationary system.

Therefore,

$$\Delta V(\mathbf{x}_k) = \mathbf{x}_k'(\mathbf{P}'\mathbf{Y}\mathbf{P} - \mathbf{Y})\mathbf{x}_k$$

The system is asymptotically stable if

$$\Delta V(\mathbf{x}_k) = -\mathbf{x}_k' \mathbf{W} \mathbf{x}_k$$

for a positive definite matrix $\mathbf{W}$ or, in other words, if $\Delta V(\mathbf{x}_k)$ is a negative quantity.

We shall now state the necessary and sufficient condition for asymptotic stability of a free, linear, stationary, discrete-time system (without proof).

Theorem The equilibrium state $\mathbf{0}$ of a system $\mathbf{x}_{k+1} = \mathbf{P}\mathbf{x}_k$ is asymptotically stable if and only if, given any symmetric positive definite matrix $\mathbf{W}$, there exists a symmetric positive definite matrix **Y** which is the unique solution of

$$\mathbf{P'YP} - \mathbf{Y} = -\mathbf{W}, \tag{4-36}$$

and

$$V(\mathbf{x}) = \mathbf{x'Yx} \tag{4-37}$$

is a Lyapunov function.

Example 4-5 (See Section 2-5.) Given

$$\mathbf{P} = \begin{bmatrix} 0 & 1 & 0 \\ 0 & 0 & 1 \\ 0 & \frac{1}{4} & 0 \end{bmatrix},$$

we let

$$\mathbf{Y} = \begin{bmatrix} a & b & d \\ b & c & f \\ d & f & e \end{bmatrix}, \quad \mathbf{W} = \mathbf{I},$$

and substitute in (4-36):

$$\begin{bmatrix} 0 & 0 & 0 \\ 1 & 0 & \frac{1}{4} \\ 0 & 1 & 0 \end{bmatrix} \begin{bmatrix} a & b & d \\ b & c & f \\ d & f & e \end{bmatrix} \begin{bmatrix} 0 & 1 & 0 \\ 0 & 0 & 1 \\ 0 & \frac{1}{4} & 0 \end{bmatrix} - \begin{bmatrix} a & b & d \\ b & c & f \\ d & f & e \end{bmatrix}$$

$$= \begin{bmatrix} -1 & 0 & 0 \\ 0 & -1 & 0 \\ 0 & 0 & -1 \end{bmatrix}.$$

This matrix equation yields:

$$-a = -1,$$
$$-b = 0, \quad a + \tfrac{1}{2}d + \tfrac{1}{16}e - c = -1,$$
$$-d = 0, \quad b - \tfrac{3}{4}f = 0, \quad c - e = -1.$$

Therefore $a = 1$, $b = 0$, $c = \tfrac{11}{5}$, $d = 0$, $e = \tfrac{16}{5}$, $f = 0$, which satisfy the inequalities

$$a > 0, \quad \begin{vmatrix} a & b \\ b & c \end{vmatrix} > 0, \quad \begin{vmatrix} a & b & d \\ b & c & f \\ d & f & e \end{vmatrix} > 0.$$

The system is asymptotically stable.

It is interesting to consider the case where **P** is in a canonical form. If all eigenvalues are distinct and real (that is, $z_1, z_2, \ldots, z_n$), then

$$\mathbf{P} = \begin{bmatrix} z_1 & & & 0 \\ & z_2 & & \\ & & \ddots & \\ 0 & & & z_n \end{bmatrix}$$

and

$$\mathbf{P'YP} - \mathbf{Y} = -\mathbf{I}.$$

For this case the system equations are decoupled and we can expect to find a $\mathbf{Y}$ which is also diagonal, so that $\mathbf{YP} = \mathbf{PY}$ (matrix multiplication is commutative if both matrices are diagonal). We can then solve directly for $\mathbf{Y}$ and get

$$\mathbf{Y} = -(\mathbf{P}^2 - \mathbf{I})^{-1}$$

or

$$\mathbf{Y} = \begin{bmatrix} 1/(1-z_1^2) & 0 & 0 \cdots 0 \\ 0 & 1/(1-z_2^2) & 0 \cdots 0 \\ \vdots & & \ddots & \vdots \\ 0 & 0 & \cdots & 1/(1-z_n^2) \end{bmatrix},$$

which is indeed diagonal.

Therefore, the system is asymptotically stable if and only if

$$|z_i| < 1 \quad \text{for all } i, \quad i = 1, \ldots, n, \tag{4-38}$$

which is the familiar unit-circle stability criterion for sampled-data systems. Equation (4–38) generally holds; that is, it can be shown to be true when $\mathbf{P}$ is in the Jordan canonical form as well as for the case where some eigenvalues are complex conjugates [for the latter case a form similar to (3–26) is useful].

A performance index for a sampled-data control system,

$$J = \sum_{k=0}^{\infty} e_k^2, \tag{4-39}$$

or a general form,

$$J = \sum_{j=1}^{n} \sum_{k=0}^{\infty} w_j e_{jk}^2, \tag{4-40}$$

is given by a Lyapunov function

$$J = V(\mathbf{x}_0) = \mathbf{x}_0' \mathbf{Y} \mathbf{x}_0 \tag{4-41}$$

for a prescribed initial state $\mathbf{x}_0$ if the system is asymptotically stable (which must be the case if the performance index is to be meaningful). For this purpose we fix $\mathbf{Y}$ in (4–41) by the relation

$$\mathbf{P'YP} - \mathbf{Y} = -\mathbf{W}, \tag{4-42}$$

136 Stability of linear lumped systems

where

$$\mathbf{W} = \begin{bmatrix} w_1 & & & 0 \\ & w_2 & & \\ & & \ddots & \\ 0 & & & w_n \end{bmatrix}$$

for the general form (4–40). Note that particular diagonal elements of the **W**-matrix may be zero if the corresponding components of the error vector are of no interest.

We derive (4–41) in the following way: Let

$$\Delta V(\mathbf{x}_k) = V(\mathbf{x}_{k+1}) - V(\mathbf{x}_k) = -\mathbf{x}_k'\mathbf{W}\mathbf{x}_k = \sum_{j=1}^{n}(-w_j e_{jk}^2),$$

where $e_{jk} = x_{jk}$. Then

$$\sum_{k=0}^{\infty}\Delta V(\mathbf{x}_k) = V(\mathbf{x}_\infty) - V(\mathbf{x}_0) = -J,$$

where $V(\mathbf{x}_\infty) = 0$ when the system is asymptotically stable. Now

$$V(\mathbf{x}_0) = \mathbf{x}_0'\mathbf{Y}\mathbf{x}_0$$

if **Y** is determined by (4–42) for a prescribed **W**. We thus get Eq. (4–41).

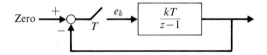

Fig. 4–6 Sampled-data control of an integrating process.

Example 4-6 We will find the performance index (4–39) of the sampled-data control system shown in Fig. 4–6. The difference equation is of first order:

$$e_{k+1} = (1 - KT)e_k,$$

that is,

$$P = (1 - KT).$$

Let $\mathbf{Y} = a$ (a scalar) in (4–42). Then

$$(1 - KT)a(1 - KT) - a = -1.$$

Solving the equation for a, we obtain

$$a = \frac{1}{2KT - K^2T^2}.$$

Hence

$$J = ae_0^2 = \frac{e_0^2}{2KT - K^2T^2}.$$

4-5 THE ROUTH TEST

The stability of an equilibrium point in the state space of a free, linear, stationary system depends solely on the roots (or eigenvalues, or zeros) of the characteristic equation,

$$|\mathbf{A} - p\mathbf{I}| = 0. \qquad (4\text{-}43)$$

The polynomial in p resulting from expansion of this determinant,

$$a_0 p^n + a_1 p^{n-1} + \cdots + a_{n-1} p + a_n = 0, \qquad (4\text{-}44)$$

is the common-denominator polynomial (the characteristic equation) of all transfer functions between an input vector $\mathbf{U}(s)$ and an output vector $\mathbf{Y}(s)$. In section 3-2 we observed that this same polynomial appeared in the Laplace-domain expression of a solution matrix [see Eq. (3-16)]. From the discussion that followed in Sections 3-3 and 3-4, it is obvious that the modes of a system's response are determined by the roots of Eq. (4-44).

We have seen that a linear, stationary, free system is asymptotically stable at an equilibrium state $\mathbf{x} = \mathbf{0}$ if and only if all roots of Eq. (4-44) are located in the left half of the complex p-plane. It is precisely this condition that the Lyapunov theorems are checking. However, in those cases where the characteristic equation is available or easily derived, there is an alternative to the Lyapunov approach which involves less work and is more straightforward.

Actually there are two such alternatives, both of which involve obtaining information about the asymptotic stability of a system from its characteristic polynomial without solving for the polynomial roots. One method was published by Routh in 1877 [2]; another, by Hurwitz [3], appeared in 1895. Both furnish much the same information and involve similar amounts of necessary computation. The Hurwitz theorem requires the expansion of high-order determinants (for high-order systems), whereas the Routh criterion does not. For this reason, and also because the Routh approach appears in recent control theory literature slightly more often than the Hurwitz approach, we will present only the Routh theorem. Since the proof of the Routh criterion is quite lengthy, involving rather complicated algebra, it will be presented without proof. For the same reason we shall not explore the relationship between the Routh and Lyapunov theorems.

The Routh theorem consists of two conditions. The first condition for asymptotic stability simply states that all coefficients of the characteristic polynomial must be nonzero and of the same sign. In other words, all coefficients must be positive:

$$a_i > 0, \qquad i = 0, 1, \ldots, n. \qquad (4\text{-}45)$$

Unfortunately, this inspection test is only a necessary condition and not a sufficient test for stability (i.e., a characteristic equation with all coefficients

138   Stability of linear lumped systems

nonzero and positive may still represent an unstable system). A heuristic proof of this first condition involves expanding, say, a third-order polynomial

$$(p + p_1)(p + p_2)(p + p_3) = 0$$

into

$$p^3 + (p_1 + p_2 + p_3)p^2 + (p_1p_2 + p_1p_3 + p_2p_3)p + p_1p_2p_3 = 0$$

and noting that a coefficient in the resulting polynomial will disappear or become negative only if one or more of the p_i's are negative (i.e., a root in the right half-plane, henceforth RHP) or zero. This argument could be duplicated for complex conjugate roots or higher-order polynomials.

The second portion of the Routh criterion supplies the sufficiency condition for asymptotic stability. An array of numbers in a triangular pattern is needed for this second portion of the test. The computation of the elements in the array consists of the following steps:

Step 1. Arrange the coefficients of the characteristic equation in an array;

$$\begin{array}{ccccccc} a_0 & & a_2 & & a_4 & \cdots a_n & 0 \cdots \\ & \searrow & & \searrow & & \nearrow & \\ & a_1 & & a_3 & & \cdots 0 & 0 \cdots \end{array} \qquad (4\text{-}46)$$

The array ends as shown if n is even. If n is odd, a_n falls in the second row. When the array ends at a_n, we consider all the rest of the elements to be zero.

Step 2. We now compute the numbers $b_1, b_2, \ldots$ of the third row:

$$\begin{array}{cccccc} a_0 & a_2 & a_4 & \cdots a_n & 0 & \\ a_1 & a_3 & & \cdots 0 & & \qquad (4\text{-}47) \\ b_1 & b_2 & b_3 \cdots b_m & & 0\,. & \end{array}$$

The rule for computing the b values obeys the following zig-zag pattern (see Fig. 4-7):

$$b_1 = a_2 - \left(\frac{a_0 a_3}{a_1}\right), \qquad b_2 = a_4 - \left(\frac{a_0 a_5}{a_1}\right), \qquad \text{and so on.}$$

The last element b_m in the array (4-47) comes one column to the left of a_n.

Step 3. Apply the same rule to the second and the third rows, and compute the numbers of the fourth row, $c_1, c_2, \ldots$

$$\begin{array}{llllll} \text{2nd row:} & a_1 & a_3 & a_5 \cdots a_{n-1} & 0 & \\ \text{3rd row:} & b_1 & b_2 & b_3 \cdots b_m & 0 & \qquad (4\text{-}48) \\ \text{4th row:} & c_1 & c_2 & c_3 \cdots c_{m-1} & 0 & 0\,. \end{array}$$

4-5 The Routh test

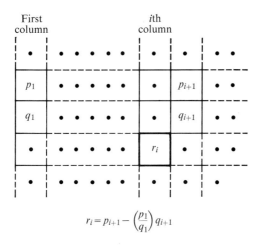

$$r_i = p_{i+1} - \left(\frac{p_1}{q_1}\right) q_{i+1}$$

Fig. 4-7 The zigzag pattern of the Routh array.

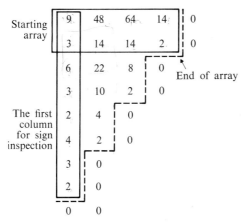

Fig. 4-8 Routh array for: $9p^7 + 3p^6 + 48p^5 + 14p^4 + 64p^3 + 14p^2 + 14p + 2 = 0$.

For instance,

$$c_1 = a_3 - \left(\frac{a_1 b_2}{b_1}\right), \quad c_2 = a_5 - \left(\frac{a_1 b_3}{b_1}\right).$$

This row ends one element sooner than the third row (unless n is odd).

Step 4 through the last step consist of computation in the same pattern. An array is completed when the $(n+1)$th row of the first column is filled. We will identify this last row as index "0", the first row as index "n", and all intermediate rows with a consecutively decreasing index. A sample calculation for a seventh-order polynomial is shown in Fig. 4-8.

140 Stability of linear lumped systems

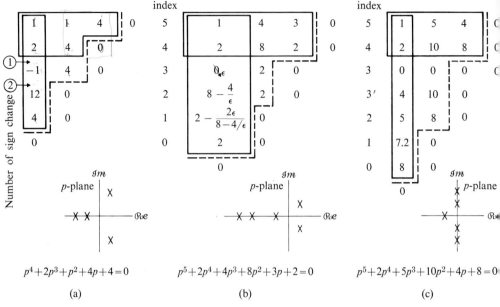

Fig. 4–9 Examples of the Routh test.

The second part of the theorem states: If all numerical values in the first column of the array are positive, then the system is asymptotically stable. If some negative numbers appear in the first column, then the number of unstable roots (i.e., in the RHP) is equal to the number of sign changes.

For example, the system treated in Fig. 4–8 is stable because all elements in the first column are positive. On the other hand, the system of Fig. 4–9(a) is unstable. Two sign changes occur, one from 2 to -1, and the other from -1 to 12. We conclude, in this case, that the total number of unstable roots is two. A possible pattern of root distribution of this system is also shown in the figure, although the theorem does not directly yield information about whether some roots (in particular, stable roots) are real or complex.

The rule regarding the number of unstable roots also applies to systems which fail the first test, as shown by the following example.

Example 4–7 Given

$$p^3 + 3p^2 + 9p - 13 = 0.$$

The numbers in the first column are found to be

$$1, 3, 13.33, -13.$$

Only the last one is negative. One sign change indicates one unstable root.

On occasion, the leading element in a row may be zero. The computation pattern then breaks down since this lead element appears in the denominator of all terms in the succeeding row. There are two ways of circumventing this problem. In actual practice, exactly zero seldom occurs since the coefficients of the characteristic polynomial are usually not known to the degree of exactitude necessary to produce precise zeros. However, for completeness and to provide understanding of those cases in which the lead element *approaches* zero, both methods will be briefly discussed.

The first method involves replacing the zero by a small positive quantity ε, and then proceeding with the calculations. The method is exemplified in Fig. 4-9(b). The zero appearing in the first column of row 3 is replaced by $+\varepsilon$. Note that the leading element of row 2 will now be negative and that of row 1 positive, indicating two unstable roots. It is interesting to observe that this same conclusion is arrived at by replacing the zero by $-\varepsilon$ instead of $+\varepsilon$. In any case, the appearance of a zero in the first column of an otherwise nonzero row (i.e., nonzero elements follow the leading zero) immediately demonstrates the existence of at least one unstable root.

The second method involves replacing p by $1/x$ in the characteristic polynomial and multiplying every term by x^n. The Routh array is constructed for the new polynomial in x, yielding the same results as found by the ε method. If this second method is applied to the example of Fig. 4-9(b), the terms in the left-hand column are found to be $+2$, $+3$, $+\frac{16}{3}$, $+\frac{13}{4}$, $-\frac{12}{39}$, and $+1$, indicating two unstable roots, as before.

In computing a Routh array, some labor can often be saved if we recall that we are not interested in the magnitude of the elements in the left-hand column, but only in their signs. Thus, in progressing through the array, we may multiply or divide any row by a positive constant before computing the succeeding row. Such simplifications of the computational schedule will not change the number of sign changes appearing in the left-hand column of the array.

Another possible complication that may arise in the computation of the Routh array is the appearance of a row of zeros. Again, this is largely an academic point for the reasons cited earlier in connection with a lead zero. Nevertheless, consideration of the technique necessary to bypass this situation offers additional insight into the significance of a row whose elements are all tending to zero. The meaning of a row of zeros in the Routh array is that the original characteristic polynomial contains equal and opposite roots, that is, roots radially placed with respect to the origin in the complex plane. Such roots may be real, imaginary, or complex. If these roots are of multiplicity m, there will appear m rows of zeros in the Routh array. Before proceeding further, we note that a single-zero-element row falls under the present situation and not the previous one.

The method of dealing with a row of zeros is best illustrated by an example (Fig. 4-9(c)). The row of index 3 is seen to be made up of all zeros. We form an auxiliary equation from the row preceding the row of zeros (row 4 in this example) as follows:

1. The highest-power term in the auxiliary equation will coincide with the index of the row now under inspection.
2. Succeeding terms in the auxiliary polynomial will decrease in power by twos.
3. The coefficients of this auxiliary equation will be the numbers appearing in the row under inspection.

Applying these rules to row 4 of Fig. 4–9(c), we get as the auxiliary equation:

$$2p^4 + 10p^2 + 8 = 0.$$

The auxiliary equation is now differentiated with respect to p, yielding (after division by two)

$$4p^3 + 10p = 0.$$

The coefficients of the differentiated auxiliary equation are now used to replace the row of zeros (row 3′ in the figure), and the computation proceeds in the regular fashion but with the row of zeros deleted. Examination of the resulting left-hand column reveals no sign changes and therefore no roots in the RHP. The appearance of a single row of zeros must thus indicate at least one pair of equal and opposite pure imaginary roots of multiplicity one. If a second row of zeros had appeared, indicating roots of multiplicity two on the imaginary axis, the procedure described above could be repeated by working on the row preceding the new row of zeros.

It is helpful to note that the roots of the auxiliary equation will also be roots of the original characteristic polynomial. In fact, the resulting roots will be precisely those equal and opposite roots of the characteristic equation which caused the row of zeros. For the auxiliary equation of the example just considered the roots are easily found to be $\pm j$ and $\pm 2j$. It is easily ascertained that these are indeed roots of the original characteristic polynomial. By division, it can be shown that the remaining root is -2, which does lie in the LHP.

Knowledge of the foregoing is often a convenient aid to the designer in avoiding a linear feedback control system which is stable in the sense of Lyapunov, but having roots of the characteristic equation on the imaginary axis which result in an oscillatory system response of constant amplitude. (For this reason the term "stability" usually means asymptotic stability.) Often the system design includes one or more adjustable parameters, which may be carried through the Routh array to determine their limiting ranges for either stability or marginal stability. There are numerous examples in the literature involving such use of the Routh array, and the reader is referred to Problems 4-14 and 4-18 for further clarification. The asymptotic *stability limit* is well defined by requiring every term in the left-hand column of the Routh array to be greater than zero.

Roots at the origin can be detected by inspection of the characteristic equation. For instance, if one root vanishes, a_n in (4-44) becomes zero and thus violates the first condition of the Routh theorem. The Routh array may be constructed for the remaining roots by dividing every term of the characteristic equation by p and then proceeding as before.

Throughout this section we have emphasized asymptotic stability rather than stability in the sense of Lyapunov (see Section 4-1) mainly because of the practical importance of asymptotic stability, but also because, in so doing, we have avoided the necessity of a detailed treatment of the stability of a system with multiple poles at the origin. Compare, for example, cases 10 and 7 in Table 3-4.

Before leaving the topic of the Routh criterion as applied to continuous-time systems, it is worth while to briefly discuss several transformations of the characteristic polynomial. If we let $p = -q$ and then construct the Routh array for the resulting polynomial, we find that the number of sign changes in the left-hand column indicates the number of roots in the LHP. This information is often helpful in identifying exactly how many roots fall on the imaginary axis; we can simply subtract the number of roots in the RHP *and* in the LHP from the order of the characteristic equation. Another transformation that can be valuable in the design process consists in letting $p = (q - a)$, where a is a positive constant. After computing the Routh array for the resulting polynomial in q, we find that the number of sign changes in the left-hand column indicates the number of roots which fall to the right of the vertical line $p = -a$. These roots in turn indicate the number of eigenvalues representing time constants greater than $1/a$. (See Problem 4-15.) As a final example of a potentially valuable transformation, consider letting $p = qe^{-j\theta}$ [10]. This substitution results in a rotation of the axes about the origin through an angle θ. The number of sign changes in the left-hand column of the Routh array for the resulting polynomial in q indicates the number of roots of the original polynomial that lie to the right of the radial lines at $\pm(90° + \theta)$. This latter number, of course, determines the minimum damping ratio of the system since $\theta = \sin^{-1}\zeta$. The reader is cautioned, however, that this particular transformation involves lengthy computations even for low-order systems.

The characteristic equation for a linear, stationary, sampled-data system was given in Section 3-6 as

$$|z\mathbf{I} - \mathbf{P}| = 0,$$

where $\mathbf{P}$ is the square matrix that appears in a free system's difference equation $\mathbf{x}_{k+1} = \mathbf{P}\mathbf{x}_k$. The determinant, $|z\mathbf{I} - \mathbf{P}|$, when expanded, takes a polynomial

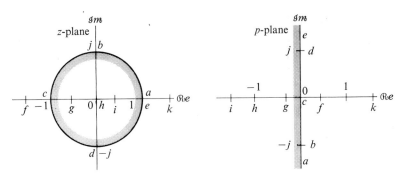

Fig. 4-10 Transformation $z = (p+1)/(p-1)$.

form:

$$F(z) = a_n + a_{n-1}z + a_{n-2}z^2 + \cdots + a_0 z^n, \qquad a_0 > 0. \qquad (4\text{-}49)$$

This polynomial is similar to (4-44), except that p in (4-44) is here replaced by z.

We have seen in the preceding section that a free, linear, stationary, sampled-data system is asymptotically stable if and only if all roots of $F(z) = 0$ lie *within the unit circle* centered at the origin of a complex plane.

If we transform the inside of the unit circle into the left half of another complex plane (labeled p), then the Routh theorem can be applied in the p-domain. A transformation

$$z = \frac{p+1}{p-1} \qquad (4\text{-}50)$$

satisfies this requirement, as shown in Fig. 4-10.

Example 4-8 We have

$$F(z) = a_2 + a_1 z + a_0 z^2 = a_2 + a_1 \frac{p+1}{p-1} + a_0 \left(\frac{p+1}{p-1}\right)^2 = 0.$$

It follows that

$$a_2(p-1)^2 + a_1(p^2-1) + a_0(p+1)^2 = 0$$

or

$$(a_0 + a_1 + a_2)p^2 + 2(a_0 - a_2)p + (a_0 - a_1 + a_2) = 0.$$

According to the Routh theorem, the three coefficients $(a_0 + a_1 + a_2)$, $(a_0 - a_2)$, and $(a_0 - a_1 + a_2)$ must have the same sign. For this second-order system, the necessary and sufficient conditions for asymptotic stability are

$$\frac{a_0 - a_2}{a_0 + a_1 + a_2} > 0 \quad \text{and} \quad \frac{a_0 - a_1 + a_2}{a_0 + a_1 + a_2} > 0.$$

A stability test for sampled-data systems does exist (the Jury theorem [11]) that does not require the transformation (4-50). The computations follow a tabular form somewhat similar to the Routh array.

REFERENCES

1. Lyapunov, M. A., *Dissertation*, Kharkov, USSR, 1892. (French Translation: "Problème général de la stabilité du mouvement," *Communications de la Société Mathématique de Kharkow*, 1893; also *Annales de la Faculté de Science*, Toulouse, 1907. Reprinted in *Annals of Mathematics Study*, No. 17, Princeton, N. J.: Princeton University Press, 1949.) Also *Stability of Motion*, translated by F. Abramovici and M. Shimshoni, New York: Academic Press, 1966.
2. Routh, E. J., *A Treatise on the Stability of a Given State of Motion*. London: Macmillan, 1877.
3. Hurwitz, A., "Ueber die Bedingungen unter welchen eine Gleichung nur Wurzeln mit negativen reelen Theilen besitzt." *Math. Ann.*, **46**, 1895, pp. 273–284.
4. Kalman, R. E., Bertram, J. E., "Control system analysis and design via the 'second method of Lyapunov.' 1. Continuous-time systems. 2. Discrete-time systems." *Trans. ASME, J. of Basic Eng.*, **80**, June 1960, pp. 371–400.
5. LaSalle, J., Lefschetz, S., *Stability by Lyapunov's Direct Method with Applications.* New York: Academic Press, 1961.
6. Lehnigk, S. H., *Stability Theorems for Linear Motions*. Englewood Cliffs, N. J.: Prentice Hall, 1966.
7. Alex, F. R., "A study of Russian feedback control theory," Part II, Vol. 2, "A survey of Russian control literature—Liapunov's theory of stability." *Convair Report ZU-044*, Flight Control Laboratory, Wright Air Development Division, Air Research and Development Command, U. S. Air Force, Wright-Patterson Air Force Base. Ohio, *AADD Technical Report 61-32*, Feb. 1961.
8. Gibson, J. E., *Nonlinear Automatic Control*. New York: McGraw-Hill, 1963, pp. 298–299.
9. Bellman, R., *Introduction to Matrix Analysis*. New York: McGraw-Hill, 1960.
10. Takahashi, T., *Mathematics of Automatic Control*. New York: Holt, Rinehart and Winston, 1966, pp. 321–323.
11. Jury, E. I., *Theory and Application of the z-Transform Method*. New York: Wiley, 1964.
12. Newton, G. C., et al., *Analytical Design of Linear Feedback Controls*. New York: Wiley, 1957.

PROBLEMS

4-1 Determine the axes and sketch the ellipse for

$$\mathbf{x}' \begin{bmatrix} 5 & 2 \\ 2 & 1 \end{bmatrix} \mathbf{x} = 1 \ .$$

4-2 Prove that all axes of a hyperellipsoid

$$V = \mathbf{x}'\mathbf{Y}\mathbf{x} = \text{constant}$$

are orthogonal to each other.

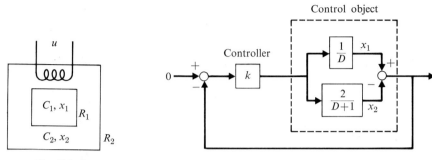

Fig. P4-4   Fig. P4-5

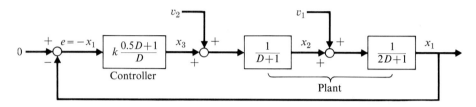

Fig. P4-6

4-3 Prove Lyapunov's theorem of asymptotic stability directly for the following second-order systems in the canonical form:

$$\frac{d\mathbf{x}^*}{dt} = \Lambda \mathbf{x}^*,$$

where

(a) $\Lambda = \begin{bmatrix} p_1 & 0 \\ 0 & p_2 \end{bmatrix}$, p_1 and p_2 are both real,

(b) $\Lambda = \begin{bmatrix} p & 1 \\ 0 & p \end{bmatrix}$,   where p is real,

(c) $\Lambda = \begin{bmatrix} \sigma & \omega \\ -\omega & \sigma \end{bmatrix}$, σ and ω both real and $\omega > 0$.

4-4 The thermal system shown in Fig. P4-4 is assumed to have two heat capacitances C_1 and C_2 with temperatures x_1 and x_2, respectively. The ambient temperature is zero. R_1 and R_2 are linear resistances to heat flow, so that the following heat balance will hold:

$$C_1 D x_1 = \frac{x_2 - x_1}{R_1}, \qquad C_2 D x_2 = \frac{x_1 - x_2}{R_1} - \frac{x_2}{R_2} + u,$$

where u is the rate of heat supply, which for stability analysis may be put equal to zero. Let $C_1 = 1$, $C_2 = 0.8$, $R_1 = 1$, $R_2 = 1/1.4$. (a) Prove that the system is asymptotically stable by means of an arbitrarily chosen test function

$$V = x_1^2 + x_2^2.$$

(b) Determine the Lyapunov function for $\mathbf{W} = \mathbf{I}$.

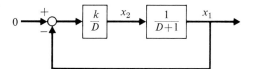

Fig. P4-8

4-5 A feedback control system is given in Fig. P4-5, where the control object is said to have reverse reaction because its unit-step input response (for zero initial state) first swings down and then goes up. Does a Lyapunov function exist when the controller gain is $k = 1$?

4-6 A feedback control system is shown in Fig. P4-6. Determine the integrated squared-error-performance index

$$J = \int_0^\infty x_1^2 \, dt ,$$

for the following two cases, and obtain the optimal value of gain k for which J becomes minimal:

Case (a) Disturbance $v_1 = 1$ for time $-\infty$ to zero, and 0 for $0 < t$; and v_2 is kept at zero for $-\infty < t < \infty$.

Case (b) $v_1 = 0$ for $-\infty < t < \infty$; and $v_2 = 1$ for $-\infty < t \leq 0$, zero for $0 < t$.

4-7 In the reverse-reaction-process control system shown in Fig. P4-5, the initial state of the process is zero [i.e., $x_1(0) = 0$ and $x_2(0) = 0$]. The reference input (labeled zero at the left of the block diagram) is changed to a unit step. Obtain the integrated squared-error index

$$J = \int_0^\infty e^2 \, dt ,$$

and determine the controller gain k that will minimize the index.

4-8 In the free feedback control system shown in Fig. P4-8, the performance index is defined to be

$$J = \int_0^\infty (x_1^2 + w x_2^2) \, dt ,$$

where w is a positive weighting factor. We normalize J:

$$J^* = \text{normalized } J = \frac{J}{\|\mathbf{x}_0\|} .$$

Determine the average value of J^* over the entire state space, and obtain the optimal value of gain k that will minimize the average J^*.

4-9 Introducing a third state variable x_3, obtain the **A**-matrix of a free third-order system that is equivalent to the forced system shown in Fig. P4-9. By means of this **A**-matrix and the Lyapunov approach, find the index

$$J = \int_0^\infty x_1^2 \, dt .$$

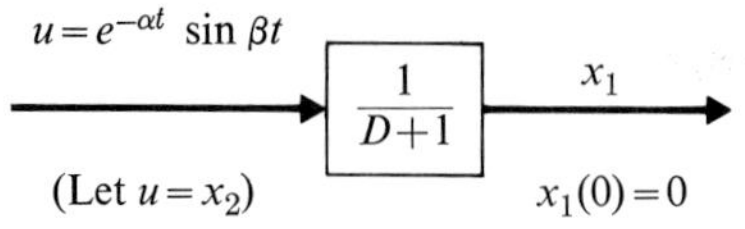

148   Stability of linear lumped systems

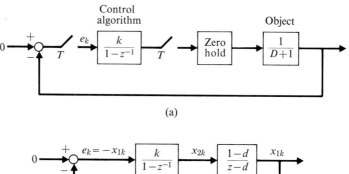

Fig. P4-9

Fig. P4-11

4-10 A free, second-order, discrete-time system is represented by

$$\mathbf{x}_{k+1} = \mathbf{P}\mathbf{x}_k, \qquad \mathbf{P} = \begin{bmatrix} \sigma & \omega \\ -\omega & \sigma \end{bmatrix},$$

where σ and ω are both real. Using the Lyapunov method, obtain the conditions on σ and ω for the system to be asymptotically stable.

4-11 A first-order object is under direct digital control via a zero hold, as shown in Fig. P4-11(a). The *DDC* algorithm is an I-action,* with gain k (the feedback control system is therefore very much like the continuous-time system given in Fig. P4-8). The z-domain block diagram is shown in Fig. P4-11(b) (see also Chapter 11). For an initial state $\mathbf{x}_0 = \begin{bmatrix}1\\1\end{bmatrix}$, find the squared-error summation

$$J = \sum_{k=0}^{\infty} x_{1k}^2$$

and determine the optimal value of k for which J becomes minimal. Assume that a sampling period is chosen such that $d = e^{-T} = \tfrac{1}{2}$. Is the control system oscillatory when k is adjusted to its optimal value?

4-12 By means of the Routh method, find the condition for gain k for the system of Fig. P4-11(b) to be asymptotically stable.

4-13 Prove that an instability caused by only one unstable (real) eigenvalue (i.e., an autocatalytic reaction we shall see in Chapter 5) can always be detected by the first condition of the Routh theorem.

---

*   See footnote next page.

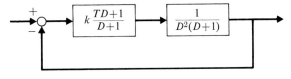

Fig. P4-14

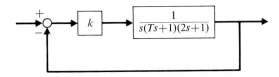

Fig. P4-18

4-14 $(TD + 1)$ in the control system of Fig. P4-14 acts as a PD^* controller, and contributes to the system's stability. Using the Routh array, determine the value of T at which gain k at the stability limit is maximized. (This may indicate an optimal value of T for minimum steady-state error.)

4-15 A system has a characteristic equation given by

$$s^3 + 7s^2 + 17s + k = 0,$$

where k is an alterable parameter. What is the allowable range for k so that the system response includes no terms having a time constant greater than $\frac{1}{2}$ second?

4-16 A sixth-order characteristic equation has the roots $s = \pm j, \pm 2, -1, -1$. Form the characteristic equation and corroborate the location of these roots by application of the Routh array. Use of the transformation $s = -q$ will be helpful as a final check.

4-17 Identify the equal and opposite roots, and their multiplicity, of the characteristic equation

$$s^6 - 2s^5 - 7s^4 + 16s^3 + 8s^2 - 32s + 16 = 0$$

by using the Routh array.

4-18 A feedback control system with alterable parameters k and T is shown in Fig. P4-18. By use of the Routh array, sketch the allowable design domain on a set of kT-axes.

4-19 The characteristic equation

$$s^4 + 2s^3 + s^2 + 2s + 3 = 0$$

yields a Routh array with a row having a zero as a leading element. Show that this zero may be replaced by *either* plus or minus ε (a small quantity) to give the same results. Comment on the system stability.

---

*   See the end of Chapter 5 for definition of integral (I) control, proportional (P) control, and derivative (D) control.

PART 2

FORMULATION OF DYNAMIC SYSTEMS

Chapter 5
Mathematical Modeling of Engineering Systems

Chapter 6
Bilaterally Coupled Systems

Chapter 7
Distributed-Parameter Systems

5
MATHEMATICAL MODELING OF ENGINEERING SYSTEMS

Our purpose in Chapters 5 through 7 is to show how the dynamic behavior of engineering systems and components is related to the mathematical constructs with which we describe these systems. This step, called *mathematical modeling*, is crucial in engineering design or analysis problems because many subsequent decisions are based on results derived from the mathematical model of the system. The nature of the mathematical model and the methods used in obtaining it are strongly influenced by the object of the investigation; for example, a system with electrical inputs and outputs might be analyzed as a thermal system if the effect of temperature on the main function of the system is our concern. Naturally, the complexity of the mathematical description is also dependent on the purposes of the investigation.

In the last three chapters we developed a mathematical structure consisting of integrators and static functions. This structure results in a finite-order, stationary, lumped-parameter, linear model that applies to a large class of engineering systems. In this chapter we introduce the generalized concepts of resistance, capacitance, and inductance to simplify the application of this mathematical model to many engineering systems. We then discuss nonlinear couplings of these systems by introducing reaction processes and nonideal sources. Lastly, we use the block diagram approach to analyze systems that can be used to realize typical control laws.

5-1 THE SYSTEMS CONCEPT

Our first step in proceeding from a physical system (or a concept of one) to its mathematical model is to isolate conceptually that part of the universe which interests us. The interesting portion we call the *system*, and the part of the universe that interacts in some way with the system we call its *environment*. A system and its environment are separated by an imaginary boundary. We will restrict ourselves to the consideration of engineering systems whose interaction with their environments is in the form of passage of matter and/or energy across the system boundary. It will be helpful in our analysis to define the system boundary in such a way that the points of interaction can be localized.

The definition of a system depends not only on the physical entity involved but also on the purpose of the investigation. For example, let us consider a thermal-power-generating station. If our interests centered on the relations between the power station and the community, that is, the amounts of fuel and cooling water used, the amount of electrical power generated, etc., we could define the entire power station as the system of interest, as shown in Fig. 5–1(a). On the other hand, we might be primarily interested in the speed regulation of the turbogenerator or in the thermal stresses in the turbine rotor. Then we would define a new system—one which would be characterized by the interactions relating directly to the problems of interest; perhaps the system shown in Fig. 5–1(b) is suitably defined for the study of the turbogenerator governor.

Our motive for defining a system in this manner is to enable us to study the system for purpose of design or analysis by making a model of it.

The model can be constructed at several levels of abstraction, depending on the needs and economics of the particular problem. On the most concrete level, a system can be studied by building and testing it full size and under actual operating conditions. This may well be the most economical way to design a simple and inexpensive item, but for something as complex and expensive as a turbogenerator, for example, the designer cannot experiment by building many such generators and choosing the best one. At an abstract

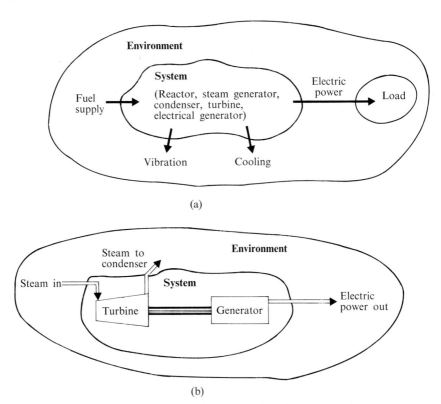

Fig. 5–1 System and environment, electric power station.

level, one can "build" a mathematical model, a set of equations or numerical procedures that describe the system. Using this model, the designer can try many possible candidates, sometimes thousands, to find the best design. A compromise between these two procedures is the use of "scale" models at a size that can be built and tested economically. In addition to the experimentation required, analytic insights into the system are required to relate the performance of the scale model to the expected performance of the full-sized version. In these chapters we will examine techniques for obtaining mathematical models of engineering systems.

Once we have selected a system and defined the interactions inside and across the interface, the mathematical model is derived by finding appropriate mathematical descriptions for the interacting quantities. The form these relations take depends on the purposes of our investigation, the nature of the system, and the mathematical techniques we would like to use. The scope of validity of the model depends on the assumptions we make in constructing the mathematical model.

In Part I (Chapters 2, 3, and 4) a mathematical structure was developed primarily for the analysis of linear, stationary, finite-order, lumped-parameter models. The generality and simplicity of this model make it very attractive, but we must first examine the assumptions we have to make about a particular system to make it amenable to analyses using these techniques. For example, is the system stationary? Virtually all systems experience some change in parameters with time, but if the changes are small or occur very slowly, a stationary model may be a good approximation to the real system. For a model to be finite order, properties like capacitance, inductance, and resistance must be "lumped" at discrete points in space. It is often the case that these properties are distributed in space (that is, they are expressed as capacitance or inductance per unit length), but we ask the question: Under what circumstances can they be considered to be lumped without seriously

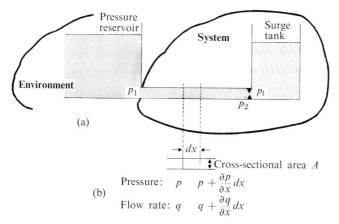

Fig. 5-2 Liquid-flow system.

reducing the accuracy of the model? The analysis of distributed-parameter systems is treated in Chapter 7, but let us consider a simple example here to illustrate the process of lumping.

Example 5-1 The system shown in Fig. 5-2(a) includes a pressure reservoir at constant pressure p_1, a conduit, an orifice, and a surge tank; the working fluid is a liquid. We make the assumption at the start that the resistance to flow in the conduit is negligible. To analyze the dynamics of the flow in the conduit, we must consider both the compressibility (capacitance) of the liquid and its inertia (inductance). In Fig. 5-2(b) we see a differential length of the conduit with the pressures and flows indicated at each end of the element. We write the momentum balance by equating the net force ($\sum F$) on the element to the rate of change of the momentum:

$$\sum F = A\left[p - \left(p + \frac{\partial p}{\partial x}dx\right)\right] = -A\frac{\partial p}{\partial x}dx = \rho A\, dx\left(\frac{\partial v}{\partial t}\right),$$

$$A \cdot \frac{\partial p}{\partial x} = -\rho A \frac{\partial}{\partial t}\left(\frac{q}{A}\right), \qquad \frac{\partial p}{\partial x} = -\frac{\rho}{A}\frac{\partial q}{\partial t}, \tag{5-1}$$

where ρ is the mass density of the fluid in the element. The continuity condition relates the net inflow of liquid to the rate of change of mass inside the element:

$$\rho q - \rho\left(q + \frac{\partial q}{\partial x}dx\right) = -\rho\frac{\partial q}{\partial x}dx = A\, dx\, \frac{\partial \rho}{\partial t}. \tag{5-2}$$

The rate of change of mass density is related to a change in pressure through the bulk modulus β,

$$\frac{dp}{\beta} = \frac{d\rho}{\rho},$$

so Eq. (5-2) becomes

$$\frac{\partial q}{\partial x} = -\frac{A}{\beta}\frac{\partial p}{\partial t}. \tag{5-3}$$

The partial differential equations (5-1) and (5-3) describe the variations of pressure and flow with time and distance. The inductance and capacitance *per unit length* are defined by Eqs. (5-1) and (5-3) as

$$I \equiv \frac{\rho}{A}, \qquad C \equiv \frac{A}{\beta}. \tag{5-4}$$

Note that the capacitance and inductance are coupled at the microscopic level and, at least at this point in the analysis, we are not justified in treating them as lumped on a macroscopic scale. By taking the Laplace transform with respect to time, we can reduce the partial differential equations to ordinary differential equations in x. Assuming zero initial conditions, we have

$$\frac{dP}{dx} = -IsQ, \tag{5-5a}$$

$$\frac{dQ}{dx} = -CsP, \tag{5-5b}$$

where $P = \mathscr{L}[p(t)]$ and $Q = \mathscr{L}[q(t)]$. Equations (5-5a) and (5-5b) can be solved together to yield a second-order equation in P,

$$\frac{d^2P}{dx^2} - ICs^2 P = 0 , \tag{5-6}$$

which has a solution in terms of exponentials:

$$P = a_1 e^{\sqrt{IC}sx} + a_2 e^{-\sqrt{IC}sx} , \tag{5-7}$$

where a_1 and a_2 are arbitrary constants. This solution can be substituted back into Eq. (5-5a) to yield Q,

$$Q = -\sqrt{C/I}(a_1 e^{\sqrt{IC}sx} - a_2 e^{-\sqrt{IC}sx}) . \tag{5-8}$$

Equations (5-7) and (5-8) give us the transform solutions for P and Q anywhere along the conduit. To return to the problem at hand (Fig. 5-2), we see that we are really only interested in the relation between the pressure and flow at the left end of the conduit and the pressure and flow at the right end. To find this, we successively substitute $x = 0$ and $x = l$ into Eqs. (5-7) and (5-8) and get the following system of equations:

$$P_1 = a_1 e^0 + a_2 e^0 , \tag{5-9a}$$
$$P_2 = a_1 e^{\sqrt{IC}sl} + a_2 e^{-\sqrt{IC}sl} , \tag{5-9b}$$
$$Q_1 = -\sqrt{C/I}(a_1 - a_2) , \tag{5-9c}$$
$$Q_2 = -\sqrt{C/I}(a_1 e^{\sqrt{IC}sl} - a_2 e^{-\sqrt{IC}sl}) . \tag{5-9d}$$

The constants a_1 and a_2 can be found by solving (5-9a) and (5-9c):

$$a_1 = \tfrac{1}{2}P_1 - \tfrac{1}{2}\sqrt{I/C}Q_1 , \qquad a_2 = \tfrac{1}{2}P_1 + \tfrac{1}{2}\sqrt{I/C}Q_1 .$$

a_1 and a_2 are substituted back into (5-9b) and (5-9d) to yield the desired result, P_2 and Q_2 as function of P_1 and Q_1:

$$P_2 = P_1 \cosh(\sqrt{IC}ls) - Q_1 \sqrt{I/C} \sinh(\sqrt{IC}ls) , \tag{5-10a}$$
$$Q_2 = -\sqrt{C/I}P_1 \sinh(\sqrt{IC}ls) + Q_1 \cosh(\sqrt{IC}ls) , \tag{5-10b}$$

where the exponential terms have been replaced by the corresponding hyperbolic functions. We can complete the analysis by writing the relations that describe the pressure in the tank p_t as a function of the pressure and flow just upstream of the orifice p_2 and q_2,

$$p_2 - p_t = Rq_2 \quad \text{and} \quad q_2 = C_t \frac{d}{dt} p_t ,$$

where R is the orifice resistance and C_t the tank capacitance. Referring again to Fig. 5-2, we see that the pressure p_1 is imposed on our system by its environment. It therefore makes good sense to solve the above equations for P_t as a function of P_1:

$$P_t(s) = \frac{P_1(s)}{(1 + RC_t s)\cosh(\sqrt{IC}ls) + C_t\sqrt{I/C}s \sinh(\sqrt{IC}ls)} . \tag{5-11}$$

The sinh and cosh can be expanded into power series to give an alternative form of Eq. (5-11):

$$P_t(s) = \frac{P_1(s)}{(1 + RC_t s)\left[1 + \frac{1}{2!}(\sqrt{IC}ls)^2 + \frac{1}{4!}(\sqrt{IC}ls)^4 + \cdots\right] + C_t \sqrt{\frac{I}{C}} s \left[\sqrt{IC}ls + \frac{1}{3!}(\sqrt{IC}ls)^3 + \cdots\right]}.$$

(5-12)

What do we learn from this example? First, we can see from Eq. (5-12) that the system's characteristic equation is infinite order and that it has, therefore, an infinite number of roots (eigenvalues). To convert it into the finite-order model we need if we are to apply linear-systems theory, the infinite series must be truncated into finite-order polynomials. But how will this affect the validity of the model? Application of the initial- and final-value theorems of Laplace transform theory answers this question, at least qualitatively. The initial-value theorem applied to Eq. (5-11) or (5-12) tells us that for a step input in p_1, $P_1(s) = 1/s$, $p_t(0^+)$ and all the derivatives of p_t for time 0^+ are zero; when the initial-value theorem is applied to a model using the truncated series, nonzero, finite-order derivatives are predicted. The behavior predicted from the approximate truncated series thus differs fundamentally from the behavior predicted by using the "exact" model for the time period immediately following the application of the disturbance. For times long after the application of the disturbance, as steady state is approached, the behavior predicted by applying the final-value theorem is the same for both the "exact" and the approximate models. Our final conclusion is that, if we are interested in knowing *in detail* the system's behavior immediately after it is disturbed, then we *cannot* use a finite-order model to describe this system. On the other hand, if these details are of no particular interest, a finite-order approximation will accurately predict the behavior as the system approaches steady state. These conclusions can also be reached by examining the frequency response of this system at high and low frequencies. See Chapter 9 for further details.

In arriving at a mathematical model for the liquid-flow system, we deduced the mathematical relations between the interacting signals directly from the physical description of the system. We will follow this practice throughout Chapter 5 in developing the techniques used in constructing mathematical models. An alternative to this procedure is to first transform the system into an equivalent system made up from a set of "ideal" elements. The set of ideal elements is chosen in such a way that a simple and direct procedure can be used to get the mathematical model of a system made up of these elements. This alternative approach will be treated in detail in Chapter 6.

In the preceding example, and as a general rule, we determine the causalities of a system ("who causes what") to derive a set of equations relating the "input" quantities to the "output" quantities. Either the standard vector equations,

$$\frac{d}{dt}\mathbf{x}(t) = \mathbf{A}\mathbf{x}(t) + \mathbf{B}\mathbf{u}(t), \qquad (5\text{--}13)$$

$$\mathbf{y}(t) = \mathbf{C}\mathbf{x}(t) + \mathbf{D}\mathbf{u}(t), \qquad (5\text{--}14)$$

or the scalar form of these equations in the Laplace domain,

$$Y_j(s) = G_{ij}(s)U_i(s), \qquad (5\text{--}15)$$

where $\mathbf{x}$ is the state vector, $\mathbf{u}$ the input vector, $\mathbf{y}$ the output vector, and G the transfer function (expressible as a ratio of polynomials in s), are used to express the input–output relationship (these equations are repeated from Chapters 2 and 3). Since we have already discussed at length the techniques for solving Eqs. (5–13) through (5–15) and determining system stability, our task in these chapters is reduced to:

1. deciding if linear-systems theory is applicable to the problem (as we had to decide in the above example), and
2. formulating these equations in terms of the system parameters.

In a great many instances *linear* proves to be the key word in answering the first question above. All engineering systems exhibit some sort of nonlinearities, so we are always faced with the necessity of finding suitable linear models. Fortunately, many components of engineering systems show a high degree of linearity so long as they are operated within some specified operating range. Linear-systems theory can be applied directly to systems made up of springs, masses, resistors, capacitors, inductors, uniform tanks, etc., with good agreement between theory and experiment over a wide operating range so long as none of the components are overloaded (tanks overflow, resistors overheat, capacitors arc, etc.). In other common problems, however, we find that nonlinear behavior is the rule, such as in fluid-flow systems with turbulence, or sliding friction. In these cases we must use much more caution in formulating linear models. For situations in which a single linear model cannot accurately describe a system, we often use a series of linear models covering overlapping ranges to get a "quasilinear" solution. This and other techniques for nonlinear systems are discussed in Chapter 12.

As was shown in Chapter 2, a nonlinear lumped-parameter system can be represented operationally on a signal-flow graph by using a set of linear dynamic operators, integrators, and nonlinear coefficient functions. Physically, the nonlinearity is expressed in the constitutive relation, or equation of state, for a component. For a spring, for example, the constitutive equation relates force to displacement through the linear equation, $F = kx$, where k, the spring constant, is a parameter that describes this particular spring, and the state of the spring at any time is determined from either the force

160   Mathematical modeling of engineering systems

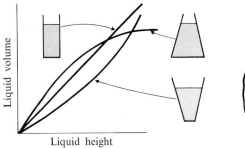
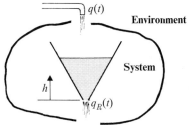

Fig. 5-3 Constitutive relations for liquid-storage tanks.

Fig. 5-4 Leaking tank system.

or the displacement. The constitutive equation for a liquid storage tank relates the pressure at the base of the tank (which is proportional to the height of liquid in the tank) to the volume of liquid stored. For a tank with straight sides, equal increments in the volume of liquid in the tank cause equal changes in the liquid height. But if the sides are not straight and the cross-sectional area varies with height, the constitutive relation is nonlinear and may appear as in Fig. 5-3. In the several example problems with tanks that were solved in earlier chapters (and in the first part of this chapter) it was always assumed that the tanks had linear constitutive relations and thus that the solutions were valid for large changes in liquid level, regardless of the particular initial state. As shown in Chapter 1, to formulate a linear model for a system with a strong nonlinearity, such as a tapered tank, we introduce the concept of an "operating point"—usually an equilibrium point or initial condition—and use a tangential approximation to the constitutive curve at the operating point to define a linearized relation. The transfer functions or system equations that are derived on the basis of this linearized relation are valid only so long as the state of the tank (liquid height) stays within the small region around the operating point for which the tangential approximation is an accurate representation of the true constitutive relation.

Example 5-2 Let us again consider the simple leaking-tank problem. However, this time the tank is tapered (Fig. 5-4). Assume that it has the shape of an inverted cone with its apex at the bottom of the tank, where the orifice is. We define the system to be the tank and orifice, which is "disturbed" by the environment by the inflow of liquid from the tap. Since the tank wall has a linear taper, the cross-sectional area is a function of the height:

$$A_x = \frac{\pi}{4} d^2, \quad d = K'h. \quad \therefore \quad A_x = \frac{\pi}{4}(K'h)^2 = Kh^2,$$

where A_x is the cross-sectional area, d the diameter, K' and K are constants. The constitutive relation for a tank, as we have noted, relates the liquid height to the

5-1 The systems concept 161

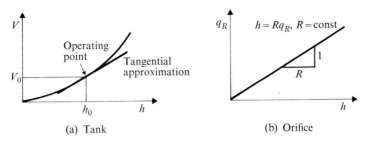

Fig. 5-5 Constitutive relation for the conical tank and the orifice.

volume V of liquid in the tank. It is easily derived from geometric considerations, that

$$V = \int_0^h A_x\, dh = \frac{K}{3} h^3 \; .$$

To define the problem more precisely, assume that the tank has been operating for a long time with a steady inflow q_0, and thus a steady liquid height $h_0\,(=Rq_0)$, and that we are interested in finding the effect of small changes in the inflow. Also assume that R for the orifice is a constant. We define a new set of variables, "perturbation" variables, which have their origin at the steady-state operating point,

$$\hat{h} = h - h_0\,, \qquad \hat{q} = q - q_0\,, \qquad \text{where} \qquad q_0 = \frac{h_0}{R}\; .$$

The rate of change of the liquid volume in the tank must at any time equal the *net* inflow of liquid. This statement of the continuity requirement leads to the differential equation

$$\frac{dV}{dt} = q(t) - q_R = q(t) - \frac{h}{R}\; . \tag{5-16}$$

The volume, however, is a nonlinear function of the height h, and before we can solve the differential equation (5-16) we have to find a linear approximation to the volume-vs.-height function for small changes of height around the operating point h_0. The Taylor series will give us a power-series expansion which we can then truncate into a linear relation:

$$V(h_0 + \hat{h}) = V(h_0) + \left.\frac{dV}{dh}\right|_{h_0} \hat{h} + \frac{1}{2!}\left.\frac{d^2 V}{dh^2}\right|_{h_0} \hat{h}^2 + \cdots,$$

$$V(h_0 + \hat{h}) \approx V(h_0) + \left.\frac{dV}{dh}\right|_{h_0} \hat{h} = \frac{K}{3} h_0^3 + K h_0^2 \hat{h}\; .$$

This is exactly the same result we would get by approximating the constitutive relation with a tangential straight line through the point (h_0, V_0) in Fig. 5-5(a). Substituting this expression into Eq. (5-16), we get

$$\frac{d}{dt}\left(\frac{K}{3} h_0^3 + K h_0^2 \hat{h}\right) = q_0 + \hat{q} - \frac{h_0 + \hat{h}}{R}\; ,$$

but

$$q_0 = \frac{h_0}{R} \quad \text{and} \quad \frac{d}{dt}\left(\frac{K}{3}h_0^3\right) = 0,$$

so

$$RKh_0^2 \frac{d\hat{h}}{dt} + \hat{h} = R\hat{q}.$$

This is a first-order equation with time constant:

$$T = RKh_0^2.$$

Note that the time constant is a function of the operating point h_0. In other words, the system will respond differently at different operating levels, even though the input is not changed.

5-2 RESISTANCE, CAPACITANCE, AND INDUCTANCE

It would be convenient if we could refer to components of dynamic systems generally, without having to identify the physical medium: electrical, fluid, mechanical, etc. Prior experience leads us to believe that this is possible, since we have seen, at least by example, that regardless of the medium the same class of differential equations applies to a great variety of dynamic systems. In this section we will begin the formalization of a general system of notation for dynamic systems by first defining generalized signal variables and then defining a set of elements which are classified according to the nature of their constitutive relations.

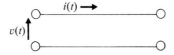

Fig. 5-6 Power transmission in electrical systems.

A pair of signal variables, $p(t)$ and $f(t)$, are the basis for this generalization. The values of p and f at any instant fix a state of energy interaction between two systems or parts of systems. For example, in an electrical system power is transmitted along a *pair* of wires characterized by instantaneous values of voltage and current (Fig. 5-6). We call $p(t)$ the potential, and it corresponds to voltage in electrical systems; $f(t)$ is called the flow and corresponds to current in electrical systems. The power transmitted instantaneously through the pair of wires is given by the product of voltage and current or potential times flow:

$$\text{Power} \equiv \mathsf{P}(t) = v(t) \cdot i(t) = p(t) \cdot f(t).$$

We will continue the analogy by extending the concepts of resistance, capacitance, and inductance to other types of systems.

It is easiest to define the generalized elements in terms of their constitutive relations in p and f, and then to determine how these definitions correspond to actual elements. Although the analogy is not limited to linear, lumped systems, the relations are presented in *linearized* form so that they apply to Eqs. (5-13) through (5-15).

A resistance is an energy-dissipating element—energy that leaves a system through a resistance is not recoverable to the system in the form in which it has left. A resistor is described by a static constitutive equation of the form

$$p(t) = R \cdot f(t). \tag{5-17}$$

It is called *static* because p and f have fixed proportionality, regardless of the past states of the resistor. We can multiply both sides of Eq. (5-17) by $f(t)$ to show that the power flow must always be in the same direction, independently of the direction of $f(t)$:

$$P = p(t) \cdot f(t) = R[f(t)]^2.$$

In the thermodynamic sense, a resistor is associated with a loss of availability.

Capacitance and inductance can be treated together since their constitutive relations are duals of each other. Because both are lossless energy-storage elements, they must have constitutive relations that express dynamics—in this case, the ability to retrieve energy stored in the past without loss.

The constitutive equation for a capacitor is:

$$p(t) = \frac{1}{C} \int_0^t f(\tau) \, d\tau; \tag{5-18}$$

and for an inductor:

$$f(t) = \frac{1}{I} \int_0^t p(\tau) \, d\tau, \tag{5-19}$$

where τ is a dummy variable of integration. These equations are more traditionally written in operational form as:

$$p(t) = \frac{1}{CD} f(t),$$

and

$$f(t) = \frac{1}{ID} p(t).$$

The definitions for generalized elements and signal variables are summarized in Table 5-1.

Application to Electrical Systems. Since our terminology and models have been drawn from electrical systems, the above generalizations fit readily when

Table 5-1 UNIFIED NOTATION FOR DYNAMIC SYSTEM ELEMENTS

	Potential $p(t)$	Flow $f(t)$	Resistance R	Capacitance C	Inductance I	Energy
Electrical	volt	ampere	ohm	farad	henry	watt·sec
Liquid	head (or pressure) m or kgf/m²	volume flow m³/sec	head/flow m/(m³/sec)	surface area m²	inertia $\dfrac{\text{kgf}/\text{m}^2}{\text{m}^3/\text{sec}^2}$	kgf·m
Gas	pressure kgf/m²	mass flow kgm/sec	$\dfrac{\text{pressure}}{\text{flow}} \dfrac{\text{kgf}/\text{m}^2}{\text{kgm}/\text{sec}}$	$\dfrac{\text{quantity}}{\text{pressure}} \dfrac{\text{kgm}}{\text{kgf}/\text{m}^2}$	inertia† $\dfrac{\text{kgf}/\text{m}^2}{\text{kgm}/\text{sec}^2}$	kgf·m
Thermal	temperature °C	heat flow kcal/sec	$\dfrac{\text{temperature}}{\text{heat flow}} \dfrac{\text{°C}}{\text{kcal}/\text{sec}}$	$\dfrac{\text{heat}}{\text{temperature}}$ kcal/°C	—	kcal
Mechanical*	velocity m/sec	force kgf	$\dfrac{\text{velocity}}{\text{force}} \dfrac{\text{m}/\text{sec}}{\text{kgf}}$	mass kgm	compliance m/kgf	kgf·m

* Replace velocity with angular velocity (radians/sec) and force with torque (kgf·m) for rotational mechanical systems.

† Since kg mass = (kg force)/(g m/sec²), the dimension reduces to m⁻¹. However, most of the dimensions in the Table are presented without reduction so that different units may be chosen for p and f, if desired.

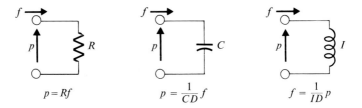

Fig. 5-7 Elements of electrical systems.

v is substituted for p and i for f (Fig. 5-7). We should note that it takes two wires to define each of these elements. In circuit diagrams one of the wires is often replaced by a ground circuit. A common practice is to put these elements in series or in shunt arrangements, forming two-terminal-pair (four-terminal) elements. The two possibilities are shown in Fig. 5-8, where $Z(D)$ is the series impedance, $Y(D)$ the shunt admittance, and D the time-

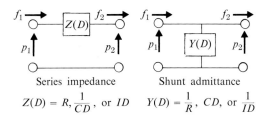

Fig. 5-8 Series impedance and shunt admittance.

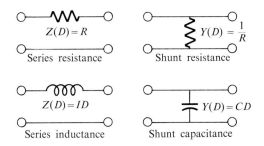

Fig. 5-9 Electrical elements in series and shunt.

differential operator d/dt. As we shall see in detail in the next chapter, using simple circuit rules we get the following matrix equations:

For the series impedance:

$$\begin{bmatrix} p_1 \\ f_1 \end{bmatrix} = \begin{bmatrix} 1 & Z(D) \\ 0 & 1 \end{bmatrix} \begin{bmatrix} p_2 \\ f_2 \end{bmatrix} \tag{5-20a}$$

and for the shunt admittance

$$\begin{bmatrix} p_1 \\ f_1 \end{bmatrix} = \begin{bmatrix} 1 & 0 \\ Y(D) & 1 \end{bmatrix} \begin{bmatrix} p_2 \\ f_2 \end{bmatrix}. \tag{5-20b}$$

Some specific examples are shown in Fig. 5-9.

Application to Liquid-flow Systems. A reasonable choice for p and f in liquid-flow systems is pressure (or head) and volume flow rate, respectively. Resistance to flow usually occurs because of wall friction, as in a long pipe, or because of restrictions such as orifices. Such resistances most often lead to constitutive relations which are nonlinear (usually quadratic), but we present linearized equations here. The orifice and wall friction both act as series resistances (Fig. 5-10a), where p is pressure and f volume rate of flow:

$$\begin{bmatrix} p_1 \\ f_1 \end{bmatrix} = \begin{bmatrix} 1 & R \\ 0 & 1 \end{bmatrix} \begin{bmatrix} p_2 \\ f_2 \end{bmatrix}. \tag{5-21}$$

166   Mathematical modeling of engineering systems

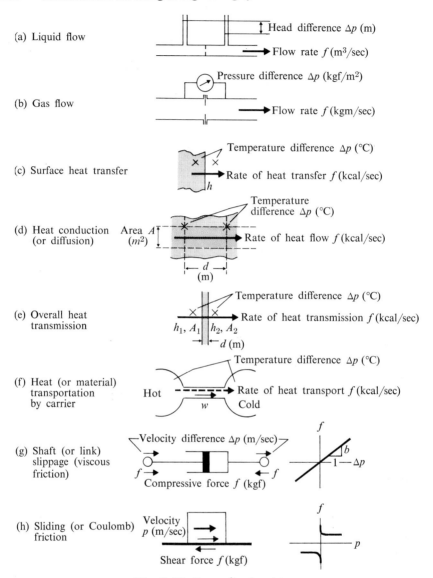

Fig. 5-10 Generalized resistance.

Capacitance, in an incompressible-liquid system, is represented by the open-surface tank which, because of its simplicity, has appeared in so many of our examples. It is assumed that any ripples or local pressure differences in the tank can be neglected. The tank, as shown in Fig. 5-11(a), is a shunt capacitance with pressure and flows related by

$$\begin{bmatrix} p_1 \\ f_1 \end{bmatrix} = \begin{bmatrix} 1 & 0 \\ CD & 1 \end{bmatrix} \begin{bmatrix} p_2 \\ f_2 \end{bmatrix}. \tag{5-22}$$

5-2 Resistance, capacitance, and inductance 167

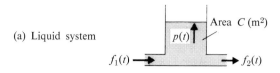

(a) Liquid system

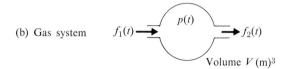

(b) Gas system

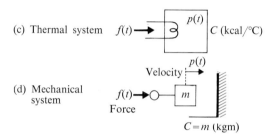

(c) Thermal system

(d) Mechanical system

Fig. 5-11 Generalized capacitance.

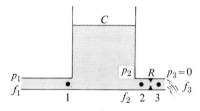

Fig. 5-12 Tank and orifice.

Example 5-3 Consider a tank with a downstream orifice which is open to atmosphere (Fig. 5-12). We wish to find a constitutive equation for this system in terms of p_1 and f_1. At first glance we do not know what sorts of components may be connected at point 1. Point 2 is just upstream of the orifice and downstream of the tank, and point 3 is downstream of the orifice. The pressure and flow at point 1 is related to the pressure and flow at point 2 by Eq. (5-22):

$$\begin{bmatrix} p_1 \\ f_1 \end{bmatrix} = \begin{bmatrix} 1 & 0 \\ CD & 1 \end{bmatrix} \begin{bmatrix} p_2 \\ f_2 \end{bmatrix};$$

and the state at point 2 is related to the state at point 3 by Eq. (5-21):

$$\begin{bmatrix} p_2 \\ f_2 \end{bmatrix} = \begin{bmatrix} 1 & R \\ 0 & 1 \end{bmatrix} \begin{bmatrix} p_3 \\ f_3 \end{bmatrix}.$$

We can combine these two equations to get an overall relation:

$$\begin{bmatrix} p_1 \\ f_1 \end{bmatrix} = \begin{bmatrix} 1 & 0 \\ CD & 1 \end{bmatrix} \begin{bmatrix} 1 & R \\ 0 & 1 \end{bmatrix} \begin{bmatrix} p_3 \\ f_3 \end{bmatrix} = \begin{bmatrix} 1 & R \\ CD & RCD+1 \end{bmatrix} \begin{bmatrix} p_3 \\ f_3 \end{bmatrix}.$$

But $p_3 = 0$, so

$$p_1 = Rf_3, \qquad f_1 = (RCD+1)f_3, \qquad p_1 = p_2,$$

or

$$p_2 = p_1 = \frac{R}{RCD+1} f_1,$$

which is the desired relation.

Series inductance in liquid systems corresponds to liquid inertia. For a compressible liquid, inductance and capacitance are interwoven on a microscopic scale (see the fluid-line sample problem in Section 5–1 and Chapter 7), but when the liquid can be considered incompressible, the inductance appears as a macroscopic element.

Application to Gas-flow Systems. As in liquid-flow systems, the pressure corresponds to the potential $p(t)$. Volume rate of flow, however, is poorly defined in a gas system because of the changes in density. A better choice for the flow variable $f(t)$ is the mass rate of flow. In analyzing gas-flow systems, however, great care must be taken in assessing the importance of thermodynamic effects, such as choking, because they can introduce strong nonlinearities into a system. The orifice is a series resistance which usually has a quadratic constitutive equation (at high velocities sonic effects begin to dominate the behavior) (Fig. 5–10b). For small perturbations in the flow, a linearized approximation can be used.

A tank (Fig. 5–11b) is a shunt capacitance in this analogy and thus must be described by

$$\begin{bmatrix} p_1 \\ f_1 \end{bmatrix} = \begin{bmatrix} 1 & 0 \\ CD & 1 \end{bmatrix} \begin{bmatrix} p_2 \\ f_2 \end{bmatrix}, \qquad (5-23)$$

where $p_1 = p_2 \, [=p \text{ in Fig. 5-11(b)}]$ is the gas pressure in the tank. The capacitance by definition relates the *net* mass flow into the tank to the increase in pressure. The rate of change of mass storage in the tank is

$$\frac{dm}{dt} = f_1 - f_2. \qquad (5-24)$$

If V = volume of the tank and v = specific volume of gas, then

$$m = \frac{V}{v}.$$

We can substitute this expression into the left-hand side of Eq. (5–24), differentiate with time, and then use the identity $dv/dt = (dv/dp)(dp/dt)$;

$$\frac{dm}{dt} = \frac{d(V/v)}{dt} = -\frac{V}{v^2}\frac{dv}{dt} = -\frac{V}{v^2}\frac{dv}{dp}\frac{dp}{dt} = f_1 - f_2. \qquad (5\text{-}25)$$

For a polytropic process $pv^n = \text{const}$ and therefore

$$v^n dp + npv^{n-1}\,dv = 0 \qquad \text{or} \qquad \frac{dv}{dp} = -\frac{v}{np}.$$

Substituting into Eq. (5-25), we obtain

$$-\frac{V}{v^2}\frac{dv}{dp}\frac{dp}{dt} = \left(-\frac{V}{v^2}\right)\left(-\frac{v}{np}\right)\frac{dp}{dt} = \frac{V}{nvp}\frac{dp}{dt} = f_1 - f_2.$$

But by Eq. (5-23),

$$f_1 - f_2 = C\frac{dp}{dt}.$$

We therefore see that

$$C = \frac{V}{nvp}.$$

p (absolute pressure) and v, however, are related by the equation of state for an ideal gas,

$$pv = R_g \Theta,$$

where R_g is the gas constant and Θ is its absolute temperature, so that

$$C = \frac{V}{nR_g\Theta}. \qquad (5\text{-}26)$$

If the process is isothermal, $n = 1$ and Θ is constant, so the capacitor is linear. Otherwise it is nonlinear because of variations in temperature.

Because of gas compressibility, series inductance (gas inertia) almost always appears coupled to shunt capacitance on a microscopic level.

Application to Thermal Systems. In thermal systems, temperature and heat flow are the signal variables which usually represent potential and flow. This choice leads to analogous thermal elements for resistance and capacitance, but no element corresponding to inductance. In addition, the product of p and f is not a power flow as it is in other types of systems. Variables that do have power as their product are temperature and entropy flow. No significant use seems to have been made of these variables as signal variables in transient analysis of thermal systems.

Resistance in heat transfer from a solid surface to a flowing fluid, or vice versa (Fig. 5-10c), is described by Newton's law of cooling, which states that the heat flow $f(t)$ across the surface is proportional to the contact area A and the temperature difference $\Delta p(t)$. The film coefficient h is the proportionality constant:

$$f(t) = hA\,\Delta p(t),$$

and therefore

$$R = \frac{1}{hA}.$$

A similar relation holds for heat conduction through a solid (Fig. 5–10d):

$$R = \frac{d}{kA},$$

where d is the thickness and k is the conductivity. This last relation also applies to a diffusion process. We are often interested in overall heat transfer from one fluid to another through a solid wall whose heat capacity can be neglected (Fig. 5–10e). The resistance then becomes a sum of two surface resistances and the resistance of the wall itself:

$$R = \frac{1}{UA} = \frac{1}{h_1 A_1} + \frac{d}{kA} + \frac{1}{h_2 A_2}, \tag{5-27}$$

where U is the overall heat transfer coefficient. There are systems where heat or material is transported by a carrier. For transport of heat (Fig. 5–10f),

$$w\Delta p = f,$$

where w is product of flow rate and specific heat of the carrier. The analogy for this case is subtle, but it holds (see Example 5–7):

$$R = \frac{1}{w}.$$

Heat storage elements form the capacitance of thermal systems (Fig. 5–11c). We assume that any local temperature differences in the element can be neglected, just as we neglected ripples and local turbulences in liquid storage tanks. For a heat storage element with uniform temperature throughout, the heat balance is

$$f(t) = C\frac{d}{dt}p(t),$$

where $p(t)$ is the temperature. From this we conclude that

C = heat capacitance
  = (mass of storage element) · (its specific heat).

5-2 Resistance, capacitance, and inductance

Application to Translational and Rotational Mechanical Systems. Force and velocity are widely accepted as useful signal variables in mechanical systems. In the cases we have considered so far, the choice of variables for potential and flow was straightforward. However, here it is possible to define *either* force or velocity as the potential, thus leading to two possible sets of analogs for mechanical systems. In this treatment, we will define the velocity as the potential p, and the force as the flow f:

$$\text{velocity} \to p(t), \qquad \text{force} \to f(t).$$

This definition will give us a set of analogous elements in which viscous or dry friction is resistance, the mass is a shunt capacitance to ground, and the spring is an inductance [1]. Although either analogy will lead ultimately to the same answer (they certainly had better!), in the opinion of the authors the velocity–potential (or velocity–voltage, to go all the way back to an electric-circuit model) analogy leads to a simpler and more logical formulation of a mathematical model for a mechanical system, especially if the system is a complex one. The usual textbook example for resistance in mechanical systems is the "dashpot" damper (Fig. 5–10g). In its linearized version the constitutive relation expresses a proportionality between force and velocity, $f = b(p_1 - p_2)$, where b is the damper coefficient. Remembering that force is the flow variable and velocity the potential variable, we see that the constitutive relation is

$$p_1 - p_2 = Rf, \tag{5-28}$$

where $R = 1/b$. Sliding friction (Coulomb friction) which is common in unlubricated mechanical systems also represents a resistance but is highly nonlinear, as shown in Fig. 5–10(h). The above discussion, and that to follow, on mechanical translational systems also applies to rotational systems if torque is substituted for force and angular velocity for translational velocity.

Newton's law of motion, written in terms of force F and velocity V is

$$F = m \frac{d}{dt} V,$$

where it is important to note that the velocity must be measured relative to an inertial frame (ground). We then have

$$\frac{1}{mD} f(t) = p(t), \tag{5-29}$$

which is the equation of a capacitance with $C = m$. Since the potential is always referred to a ground, it is a shunt capacitance to the ground (Fig. 5–11d).

The equation for a linear massless spring is usually given as $F = kx$, where k is the spring constant and x is the deflection. This equation is easily

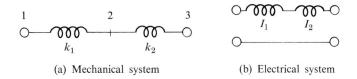

(a) Mechanical system (b) Electrical system

Fig. 5-13 Series inductances.

converted into

$$f(t) = \frac{k}{D} p(t), \qquad (5\text{-}30)$$

which is the constitutive equation for an inductance with $I = 1/k$.

One of the advantages of the force–flow analogy (also called the mobility analogy in the field of vibrations) is that the topological (series and parallel) identities are the same for mechanical systems as they are in electric circuits. This is demonstrated in the following example.

Example 5-4 Find an equivalent inductance for two springs in series, as shown in Fig. 5-13(a), and compare it to the equivalent inductance calculated for two electrical inductances in series (Fig. 5-13b).

The velocity of the right-hand end ③ relative to the left-hand end ① (p_{31}) is equal to the sum of the velocities across each of the springs:

$$p_{31} = p_{32} + p_{21}. \qquad (5\text{-}31)$$

For each of the springs the constitutive equation relates p to f by:

$$p_{21} = I_1 D f_1 \quad \text{and} \quad p_{32} = I_2 D f_2.$$

In addition, since the springs are considered to be massless, f_1 must equal f_2. We can then substitute the last relations into Eq. (5-31) to get

$$p_{31} = (I_1 + I_2) D f.$$

Hence the equivalent inductance I_{eq} (or the equivalent spring constant k_{eq}) is:

$$I_{eq} = I_1 + I_2 = \frac{1}{k_1} + \frac{1}{k_2} = \frac{1}{k_{eq}}.$$

Electric-circuit rules show that the same relation holds for the equivalent inductance of two inductances in series.

5-3 SOME EXAMPLES OF *RC*-SYSTEMS

In many systems of interest we find that there is only one type of dynamic element; that is, the system may contain capacitances or inductances but not both. In general, resistance is also present. In this section we will examine some examples of systems containing only resistances and capacitances.

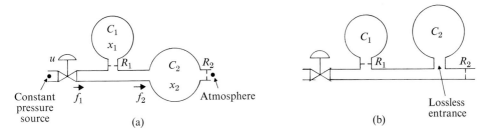

Fig. 5-14 Two-tank system of Example 5-5.

Since these systems can store only one type of energy, we expect that they will not be capable of oscillatory motion. In other words, we expect that linear *RC*-systems will have all negative real eigenvalues.

Each independent capacitor is represented dynamically by one integrator; thus, we will find that a system having n independent capacitors will have n state variables and therefore be nth order. In most of the example problems, close examination reveals that inductive effects are really present but are small enough to be ignored for the purposes of analysis.

Example 5-5 *Gas-pressure system* (Fig. 5-14). Note first that, for the purposes of formulating our mathematical model, the systems shown in Fig. 5-14(a) and 5-14(b) are identical since we are assuming that there are no local pressure gradients within the tank. In this problem we are interested in finding the response of the system due to small changes in the control valve position. We define the following incremental variables:

f_1: change in flow through the control valve,
f_2: change in flow in pipe between tanks 1 and 2,
x_1: change in pressure in tank 1,
x_2: change in pressure in tank 2,
u: change in valve position.

The flow deviation through the control valve f_1 is a function of u and x_2, since we are neglecting the effects of inertia in the connecting lines:

$$f_1 = k_1 u - k_2 x_2 , \qquad (5\text{-}32)$$

where k_1 and k_2 are constants, and second-order terms (that is, terms involving $u \cdot x_2$) have been neglected. In Example 5-3 we found the relation linking pressure and flow upstream of a tank which was vented to the atmosphere. Although we were concerned with a liquid-flow system in that case, we formulated the problem in terms of generalized resistance and capacitance, and thus the result can be applied directly to our present problem to give us the following relation between x_2 and f_2:

$$(R_2 C_2 D + 1)x_2 = R_2 f_2 . \qquad (5\text{-}33)$$

The flow into tank 1 is controlled by the orifice R_1:

$$(\text{flow into tank 1}) = f_1 - f_2 = \frac{x_2 - x_1}{R_1} , \qquad (5\text{-}34)$$

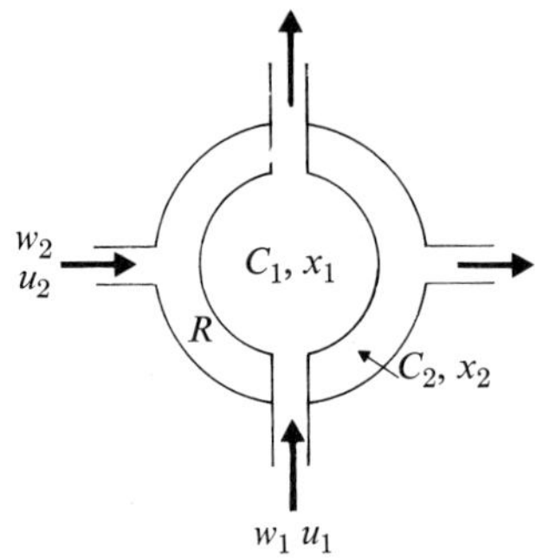

174 Mathematical modeling of engineering systems

Fig. 5-15 A heat exchanger with two lumped capacitances.

and the pressure build-up in tank 1 is governed by

$$x_1 = \frac{1}{C_1 D}(f_1 - f_2). \tag{5-35}$$

We combine Eqs. (5-32) through (5-35) to get the following pair of first-order differential equations:

$$C_1 \frac{d}{dt} x_1 = \frac{x_2 - x_1}{R_1}$$

and

$$C_2 \frac{d}{dt} x_2 = \left(k_1 u - k_2 x_2 - \frac{x_2 - x_1}{R_1}\right) - \frac{x_2}{R_2}.$$

The vector equation (5-13) can be obtained by simply rearranging terms in the last equation:

$$\mathbf{A} = \begin{bmatrix} -1/(R_1 C_1) & 1/(R_1 C_1) \\ 1/(R_1 C_2) & -1/(R_1 C_2) - 1/(R_2 C_2) - k_2/C_2 \end{bmatrix}.$$

$$\mathbf{B} = \begin{bmatrix} 0 \\ k_1/C_2 \end{bmatrix}. \tag{5-36}$$

If x_2 is the variable of interest, the output y is

$$y = \begin{bmatrix} 0 & 1 \end{bmatrix} \mathbf{x},$$

and by Eq. (2-21),

$$G(s) = \frac{k_1 R_2 (1 + R_1 C_1 s)}{R_1 C_1 R_2 C_2 s^2 + [R_1 C_1 (1 + k_2 R_2 + R_2/R_1) + R_2 C_2]s + (k_2 R_2 + 1)}. \tag{5-37}$$

The next example is a system with more than one input.

Example 5-6 *Heat exchanger system* (Fig. 5-15). By neglecting nonuniformities in temperature and assuming perfect mixing, we lump the heat capacitances into two: C_1 for the inside fluid, and C_2 for the jacket side. The resistance of the entire heating surface (see Eq. 5-27) is R. Let us ignore heat losses and let w_1 and w_2 be the products of flow rate and specific heat for each flow, with inlet temperatures u_1 and

u_2. Then the heat-balance equation becomes:

$$C_1 \frac{dx_1}{dt} = \frac{1}{R}(x_2 - x_1) + w_1(u_1 - x_1),$$
$$C_2 \frac{dx_2}{dt} = \frac{1}{R}(x_1 - x_2) + w_2(u_2 - x_2),$$
(5-38)

where the second terms on the right-hand side represent heat carried in by the flow. The equation is linear and stationary if in particular the flow rates w_1 and w_2 are constant. For this situation we obtain

$$\mathbf{A} = \begin{bmatrix} -1/(RC_1) - w_1/C_1 & 1/(RC_1) \\ 1/(RC_2) & -1/(RC_2) - w_2/C_2 \end{bmatrix}, \quad \mathbf{B} = \begin{bmatrix} w_1/C_1 & 0 \\ 0 & w_2/C_2 \end{bmatrix}.$$
(5-39)

If x_1 is the response, then

$$y = \begin{bmatrix} 1 & 0 \end{bmatrix}\mathbf{x}$$

and

$$Y(s) = G_1(s)U_1(s) + G_2(s)U_2(s),$$

where the transfer functions are given by

$$G_1(s) = \frac{w_1 R(1 + w_2 R + RC_2 s)}{\Delta}, \quad G_2(s) = \frac{w_2 R}{\Delta},$$

$$\Delta = C_1 C_2 R^2 s^2 + [(C_1 + C_2)R + (C_1 w_2 + C_2 w_1)R^2]s + (w_1 R + w_2 R + w_1 w_2 R^2).$$

If changes of limited magnitude in the flow rates w_1 and/or w_2 are the input, an approximate method of deriving linear, stationary, input-output relations can be applied. This will be shown in Example 5-9.

The coupling between two components is said to be *noninteracting*, or *unilateral*, if the state of one of the components (the "upstream" component) is not affected by the state of the other (the "downstream" component). The state of the downstream component, however, is affected by the state of the upstream component (if this were not true, the two components would be completely decoupled). In Example 5-5, the coupling was *bilateral*, not unilateral, because the pressure in the first tank affected the pressure in the second, and vice versa. This is immediately apparent since there are no zeros in the **A**-matrix. In the example to follow we will look at a system with a unilateral coupling. (Is the system of Example 3-11 unilateral or bilateral?)

Example 5-7 *Component concentration system* (Fig. 5-16a). A solution with a volumetric flow rate w and a concentration u (moles/volume) is passing through two mixing tanks of volumetric capacitances C_1 and C_2. If we assume complete mixing and neglect diffusion, the mole balance at each tank is given by

$$C_1 \frac{dx_1}{dt} = w(u - x_1), \quad C_2 \frac{dx_2}{dt} = w(x_1 - x_2), \quad (5\text{-}40)$$

176 Mathematical modeling of engineering systems

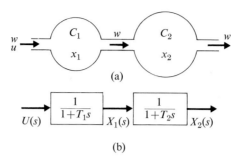

Fig. 5-16 A noninteracting concentration system.

where x_1 and x_2 are moles/volume concentrations in each tank, and therefore also in each outflow line. This relation corresponds to the vector equation

$$D\mathbf{x} = \begin{bmatrix} -w/C_1 & 0 \\ w/C_2 & -w/C_2 \end{bmatrix} \mathbf{x} + \begin{bmatrix} w/C_1 \\ 0 \end{bmatrix} u . \qquad (5\text{-}41)$$

If the carrier flow rate w is kept constant and the supply concentration u is an input, the action of the first tank on the second tank is unilateral. This means that the state of the second tank does not affect the first tank. Because of such simple causality, the block-diagram representation of Eq. (5-40), shown in Fig. 5-16(b), exactly fits the original structure. In the Laplace domain, we have

$$\frac{x_1(s)}{U(s)} = \frac{1}{1 + T_1 s}, \qquad \frac{x_2(s)}{U(s)} = \frac{1}{(1 + T_1 s)(1 + T_2 s)}, \qquad (5\text{-}42)$$

where $T_1 = C_1/w$, $T_2 = C_2/w$. This relation can be generalized to an arbitrary number of cascaded tanks. If, moreover, the capacitances of all tanks are the same,

$$C_i = \frac{C}{n}, \qquad i = 1, 2, \ldots, n ,$$

the final output is given by

$$x_n(s) = \frac{1}{[(T/n)s + 1]^n} U(s) , \qquad (5\text{-}43)$$

where $T = C/w$. A simplification due to the noninteracting (or cascade) coupling appears in the vector equation. The following **A**-matrix for the chain we just saw, with transposition (or defining state variables from right to left along the line), reduces to a special case of the Schwartz form (Eq. (3-28)):

$$\mathbf{A} = \begin{bmatrix} -n/T & 0 & \cdots & & 0 \\ n/T & -n/T & 0 & & 0 \\ 0 & n/T & & & 0 \\ \vdots & & & & \vdots \\ 0 & \cdots & 0 & n/T & -n/T \end{bmatrix}, \qquad \mathbf{B} = \begin{bmatrix} n/T \\ 0 \\ \vdots \\ 0 \end{bmatrix}. \qquad (5\text{-}44)$$

If we think of each of the tanks in the above example as a small portion of a continuous pipe, we can then use the relations derived to find the transfer function

5-3 Some examples of RC-systems

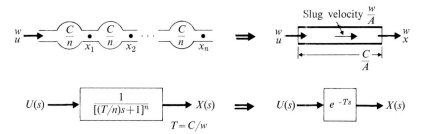

Fig. 5-17 Lumped-parameter system (left) *vs.* axially-distributed-parameter system (right).

for the uniform pipe, a distributed system, from its lumped-parameter microstructure. This is an alternative to the technique used in Example 5-1 in which the differential equation was solved directly in the Laplace domain.

Example 5-8 Let the number of "tanks" n increase indefinitely. Since the capacitance of each lump is defined in terms of the total capacitance, the individual capacitance decreases as the number of lumps increases. At the limiting condition $n \to \infty$ the following mathematical relation holds:

$$\lim_{n\to\infty} \frac{1}{[(T/n)s + 1]^n} = e^{-sT} . \tag{5-45}$$

Therefore, when a cascade tank system of total capacitance C reduces to a single pipeline of the same capacitance (Fig. 5-17), the end-to-end signal relation becomes a pure delay or dead time T given by a transfer function

$$G(s) = e^{-sT} . \tag{5-46}$$

We assume here a complete mixing in the radial direction and no mixing in the axial direction of the pipe. Note that C/A is the length of a pipe of a uniform cross section A, and w/A is the velocity of flow, so that $T = (C/A)/(w/A)$ becomes the time for a slug of fluid to pass through the pipe. This gives a physical interpretation of the delay operator e^{-sL} which was derived in Section 1-5. The vector equation (5-13) does not apply to this system because the number of state variables is not finite.

In Example 5-2 we analyzed a system in which a nonlinearity appeared because of the physical description of the system (in that case, the geometry). The following example is a mixing system in which nonlinearities or nonstationary terms appear because of flow-rate variations in the carrier fluid.

Example 5-9 *Dilution process* (Fig. 5-18). Let w_1 be the volumetric flow rate of the main flow, with concentration x_1 moles/volume. This concentration is diluted, at flow rate w_2, by a solvent with zero concentration. The mole balance at the mixing tank of a capacitance C becomes

$$C \frac{dx_2}{dt} = w_1 x_1 - (w_1 + w_2) x_2 , \tag{5-47}$$

where x_2 is the concentration of the outflow. The equation is linear but nonstation-

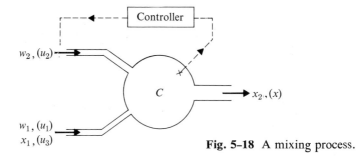

Fig. 5-18 A mixing process.

ary when one or both of the flow rates change with some known time pattern. The relation is nonlinear if, for instance, w_2 is changed by means of a feedback controller indicated in the figure.

A linear stationary equation can be derived for small deviations from an equilibrium state that corresponds to a normal operating condition. Let w_{1e}, w_{2e}, x_{1e}, and x_{2e} be a set of normal equilibrium values, and u_1, u_2, u_3, and x be the deviations such that

$$w_1 = w_{1e} + u_1, \qquad w_2 = w_{2e} + u_2,$$
$$x_1 = x_{1e} + u_3, \qquad x_2 = x_{2e} + x.$$

Substituting these relations in Eq. (5-47), and subtracting the equilibrium (static) relation, we have

$$w_{1e}x_{1e} - (w_{1e} + w_{2e})x_{2e} = 0,$$

which holds at the normal condition. Ignoring product terms of small deviations, we obtain the following equation:

$$C\frac{dx}{dt} = -(w_{1e} + w_{2e})x + (x_{1e} - x_{2e})u_1 - x_{2e}u_2 + w_{1e}u_3. \tag{5-48}$$

This equation is linear and stationary with respect to input variables u_1, u_2, u_3, and output (or state) variable x.

5-4 REACTION PROCESSES AND STABILITY

In the preceding section we dealt with mathematical modeling of some fluid, thermal, and concentration systems consisting of R and C elements and ideal sources. Ideal sources, such as pressure reservoirs, are assumed to function independently of the systems they feed, but in actual practice there is often some sort of coupling between the source and the system state. Reaction processes in particular offer many interesting examples of systems in which there is a coupling between a material system and a thermal system. In this section we will look into the analysis of systems with nonideal sources, particularly reaction processes.

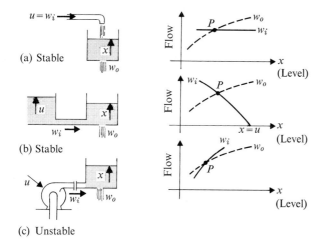

Fig. 5-19 Stability and instability of an equilibrium level for a constant input u.

The mathematical formulation of both chemical and nuclear reactions involves highly nonlinear forms. Depending on the aims of modeling, such forms can be programmed for analog or digital computer simulation [2]. However, since the nonlinearities are smooth, linearization is usually possible for small perturbations. Linearized differential equations derived for chemical and nuclear reactors in the following paragraphs fit directly in with the form of Eqs. (5-13) through (5-15), with all the advantages of the linear, lumped mathematical model.

The linearized equations of a reaction process can be used for the investigation of local stability, that is, stability for small fluctuations. If the system is linear, stability becomes independent of the magnitude of fluctuations. A qualitative explanation of stability is based on Fig. 5-19 for a liquid-level system. An equilibrium point, represented by P in the figure, is stable in (a) and (b), because the dependency of supply flow w_i and discharge flow w_o is such that plus or minus deviation in level from its equilibrium position diminishes with time. This is asymptotic stability as we saw in the last chapter. On the other hand, if a supply device has a characteristic such as shown in part (c) of the figure, the slightest deviation from P will build up, resulting in instability. Although the last case is highly fictitious for fluid and other machinery, it does happen in some reaction processes.

Figure 5-20 shows a static characteristic of a chemical reactor. Curve Q_2 represents the rate at which heat is removed. There are three intersections, P_1, P_2, and P_3. From what was discussed about the system of Fig. 5-19, it is now obvious that the equilibrium at P_1 is unstable, whereas those at P_2 and P_3 are (asymptotically) stable.

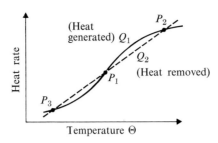

Fig. 5-20 Equilibrium states in a stirred-tank reactor.

As was discussed in Chapter 4, a linear system is unstable if it has characteristic root(s) in the right-hand half of the complex plane. If, for example, the neutron density y in a fission material is described by a dynamic relation

$$\frac{dy(t)}{dt} = ky(t) \tag{5-49}$$

with a positive constant k (which means power tends to rise indefinitely), the system is unstable. The free response of the system is an exponential rise:

$$y(t) = y(0)e^{kt}.$$

On the other hand, the processes given in the examples in the preceding section were all asymptotically stable. Moreover, it can be shown that all characteristic roots of these processes are real. With negative real roots, the free responses of the processes are expressed by superimposing exponential decays; they are nonoscillatory.

We now consider a continuous-flow stirred tank reactor [3] shown in Fig. 5-21. The state of the mixture when lumped in the tank of constant volume V is represented by the concentration y of the reactant, and the absolute temperature θ. Two equations are required to describe the system: a material-balance equation for y and a thermal balance for θ. The equations are coupled, however, because the reaction rate, which appears in the material-balance equation, is a function of the temperature, and the heat generation rate, which appears in the thermal balance, is a function of the conversion rate.

According to Arrhenius' law, the rate of a chemical reaction, which thus couples the two equations, is proportional to $V \cdot y$ and a factor which is given by

$$k_r = re^{-E/R\theta}, \tag{5-50}$$

where $E = $ energy of activation, $R = $ gas constant, and r is a coefficient, all constants for a given system. Thus the reaction rate constant k_r is a function of θ only.

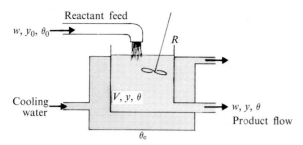

Fig. 5-21 Stirred-tank reactor.

The material-balance equation is,

$$V \frac{dy}{dt} = w(y_0 - y) - V k_r y, \tag{5-51}$$

where w is the volumetric flow rate of the feed of concentration y_0 (Fig. 5-21). Note that θ is hidden in k_r.

The heat-balance equation is

$$V c \rho \frac{d\theta}{dt} = Q_1 - Q_2,$$

where c and ρ are the specific heat and density, respectively, of the mixture. The rate of heat generated Q_1 is proportional to the conversion rate $V k_r y$, and is given by

$$Q_1 = H V k_r y,$$

where H, the heat of reaction, is positive for an exothermic reaction. This is a function of θ as shown in Fig. 5-20. Q_2 is the heat removed through the wall to the cooling water. An equation of the form of (5-38) applies for conduction from one liquid to another through a thin wall, as in a heat exchanger,

$$Q_2 = w c \rho (\theta - \theta_0) + \frac{\theta - \theta_c}{R},$$

where θ_0 is the inlet temperature of the reactant, θ_c is the mean temperature of the coolant, and R is the resistance of the cooling wall (Eq. 5-26). Combining these three equations, we get the second relation:

$$V c \rho \frac{d\theta}{dt} = w c \rho (\theta_0 - \theta) - \frac{1}{R}(\theta - \theta_c) + H V k_r y. \tag{5-52}$$

The reactor system of Fig. 5-21 is thus described by two nonlinear first-order differential equations (5-51) and (5-52). A third equation, for the heat balance of the cooling jacket, may be written in the form of Eq. (5–38) for the temperature θ_c. The system becomes third-order when this is done.

The coupling of the two equations through the reaction term $Vk_r y$ appears explicitly when the relation is linearized in the vicinity of an equilibrium state. Let us introduce small deviations u_1, u_2, u_3 in the inputs, and x_1, x_2 in the outputs such that

$$w = w_e + u_1, \qquad \theta_c = \theta_{ce} + u_2, \qquad y_0 = y_{0a} + u_3,$$
$$y = y_e + x_1, \qquad \theta = \theta_e + x_2, \qquad (5\text{–}53)$$

where a set of equilibrium values are designated by subscripts e. The inlet temperature of the reactant, θ_0, is assumed to remain constant.

We first linearize Eq. (5–50). Since we have

$$\frac{dk_r}{d\theta} = \frac{E}{R\theta^2} re^{-E/R\theta},$$

the deviation Δk_r from a steady-state value k_{re} is given by

$$\Delta k_r = k_{re} \frac{E}{R\theta^2} x_2. \qquad (5\text{–}54)$$

Substituting Eqs. (5–53) and (5–54) in Eqs. (5–51) and (5–52), and applying the same technique as was shown in Example 5-9, Section 5-3, we obtain the following linearized equations:

$$\frac{d\mathbf{x}}{dt} = \mathbf{A}\mathbf{x} + \mathbf{B}\mathbf{u}, \qquad (5\text{–}55)$$

where elements of **A** and **B** are:

$$a_{11} = -\frac{w_e}{V} - k_{re}, \qquad a_{12} = -k_{re} \frac{E}{R\theta_e^2} y_e,$$

$$a_{21} = \frac{Hk_{re}}{c\rho}, \qquad a_{22} = \frac{Hk_{re}}{c\rho} \frac{E}{R\theta_e^2} y_e - \frac{1}{RVc\rho} - \frac{w_e}{V},$$

$$b_{11} = \frac{y_{0e} - y_e}{V}, \qquad b_{12} = 0, \qquad b_{13} = \frac{w_e}{V},$$

$$b_{21} = \frac{\theta_0 - \theta_e}{V}, \qquad b_{22} = \frac{1}{RVc\rho}, \qquad b_{23} = 0.$$

Stability at the equilibrium state (Fig. 5–20) can be investigated by the methods of Chapter 4, or by determining the eigenvalues of **A**.

Our next topic in this section is an approximate formulation of the chain reaction process in a nuclear reactor [4]. As is well known, the capture of a neutron by an "atomic fuel" such as U^{235} causes fission, which releases free neutrons. These free neutrons, decelerated in a moderator, lead to the next fission. This is the chain reaction, and is characterized by a factor k_m such that

$$k_m = \frac{\text{number of neutrons in a generation}}{\text{number of neutrons in a preceding generation}}.$$

We have the dynamic equation for the number of neutrons y per unit volume:

$$\frac{dy}{dt} = \frac{k_m - 1}{l} y, \qquad (5\text{-}56)$$

where l is the mean lifetime of a neutron, in the order of 10^{-3} sec. This relation agrees with Eq. (5-49) when k_m is a constant greater than 1. Fortunately, the process in a reactor is much less explosive than Eq. (5-56) indicates, because a small proportion, about 0.75%, of the neutron emissions are delayed. Although over 99% of the fission neutrons are emitted at the instant of fission, the fission fragments carry along the remainder, $\beta \approx 0.0075$, and emit them in a radioactive-decay process a minute or so later. If we take into account a finite number m of groups of delayed-neutron emitters, then Eq. (5-56) becomes

$$\begin{aligned}\frac{dy}{dt} &= \frac{\delta k - \beta}{l} y + \sum_{i=1}^{m} d_i z_i, \\ \frac{dz_i}{dt} &= \frac{\beta_i}{l} y - d_i z_i, \qquad i = 1, \ldots, m,\end{aligned} \qquad (5\text{-}57)$$

where $\sum_{i=1}^{m} \beta_i = \beta \approx 0.0075$, $\delta k = k_m - 1$, $z =$ the number of fission fragments or delayed-neutron emitters per cm³, and $d_i =$ decay constant. If we

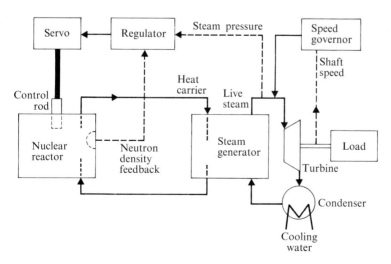

Fig. 5-22 Schematic diagram of a Rankine-cycle nuclear-power plant.

lump the m groups of delayed-neutron emitters into one, we may approximate Eq. (5–57) by writing

$$\frac{dy}{dt} = \frac{\delta k - \beta}{l} y + dz, \qquad \frac{dz}{dt} = \frac{\beta}{l} y - dz. \tag{5-58}$$

The power level of a nuclear reactor is regulated by control rods, which are neutron absorbers made of boron or cadmium (Fig. 5–22). Their position relative to the core determines δk. But δk is also affected by the system temperature. Usually the effect is negative and thus contributes to system stability.

To get an overall picture, let us assume, for instance, that

$$\delta k = \delta k_c - \alpha \theta,$$

where δk_c represents the effect of the control rods, and α, a positive proportionality constant, describes the effect of temperature. Then we have the following set of equations:

$$\frac{dy}{dt} = \frac{\delta k_c - \alpha \theta - \beta}{l} y + d \cdot z,$$

$$\frac{dz}{dt} = \frac{\beta}{l} y - d \cdot z, \tag{5-59}$$

$$C \frac{d\theta}{dt} = k_f y - \frac{1}{R} (\theta_c - \theta),$$

where the third equation is a heat balance of the fuel element. The system constants in the third equation are C = heat capacitance, R = thermal resistance for cooling heat flow, and k_f = proportionality constant for heat generation due to fission. Equation (5–59) can be considered a vector equation for a state vector,

$$\mathbf{x} = \begin{bmatrix} y \\ z \\ \theta \end{bmatrix},$$

with an input vector

$$\mathbf{u} = \begin{bmatrix} \delta k_c \\ \theta_c \end{bmatrix},$$

where the coolant temperature θ_c may be related to fuel temperature θ via a circulation loop (Fig. 5–22). Note that Eq. (5–59) is still nonlinear; small perturbations must be introduced to linearize the relation.

The time pattern of an *epidemic* is similar to that of a batch chemical reaction. A crude model of an epidemic can be derived on the basis of the Law of Mass Action, which states that the rate of growth in the number of

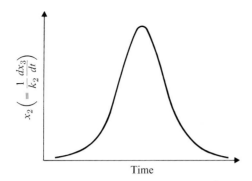

Fig. 5-23 Typical time-history of an epidemic.

patients is proportional to the number of contacts between susceptible people and patients [5]. Let

$x_1(t)$ = number of people susceptible to the disease,

$x_2(t)$ = number of patients not isolated,

$x_3(t)$ = number of people removed due to recovery (with immunity) plus isolation (to hospital) and death;

then the law stipulates that

$$\frac{dx_1}{dt} = -k_1 x_1 x_2, \qquad \text{where} \quad k_1 = \text{const}. \tag{5-60}$$

We further assume that the rate of removal dx_3/dt is proportional to x_2:

$$\frac{dx_3}{dt} = k_2 x_2, \qquad k_2 = \text{const}. \tag{5-61}$$

For a constant size of total population, $N = x_1 + x_2 + x_3$, $dx_2/dt = -dx_1/dt - dx_2/dt$; hence

$$\frac{dx_2}{dt} = k_1 x_1 x_2 - k_2 x_2. \tag{5-62}$$

Because of the constraint $N = $ constant, the system is second-order, and the state-plane trajectory in the $x_1 x_2$-plane can be constructed (Problem 5-13) and then used to obtain a typical pattern of an epidemic such as shown in Fig. 5-23. It is interesting to note that an epidemic grows only when the initial value of dx_2/dt is positive, that is

$$k_1 x_1(0) > k_2.$$

Since $x_1(0)$ is practically equal to N, the critical size of the population for an

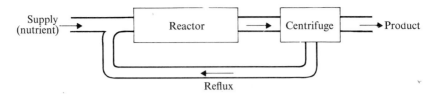

Fig. 5-24 Continuous fermentation process.

epidemic to grow is given by

$$N_c = \frac{k_2}{k_1}. \tag{5-63}$$

This model makes many simplifying assumptions but illustrates the basic nature of this type of system. An important factor that was not considered is the delay, or incubation period, between the time of contact and the actual appearance of symptoms. This would appear as a pure time delay in the system formulation and would make the finite state model invalid. A higher-order approximation could be used, however. In addition, in light of man's increased mobility, an isolated community is seldom found in modern society. The model may apply to such problems as insect control and animal diseases.

The *fermentation process* is another example where an autocatalytic reaction is involved [6]. A continuous fermentation process is illustrated in Fig. 5-24. The basic biodynamics involved in the continuous fermentator, however, are also found in a batch fermentation, on which we shall focus our attention in the following discussion. The reactor (fermentator) is assumed to have a complete mixing without any inflow or outflow. Let $x_1 =$ number of microorganisms per unit volume, $x_2 =$ concentration of nutrient. The rate of growth of the population of a microorganism species is considered to be proportional to its population, so we have the following relation:

$$\frac{dx_1}{dt} = kx_1, \tag{5-64}$$

where k is a reaction coefficient. The value of k depends on x_2, and the following approximate relation holds:

$$k = \frac{k_{\max} x_2}{K + x_2}, \tag{5-65}$$

where $k_{\max}$ and K are constants. It is also known that x_1 and x_2 are approximately related by a proportionality relation

$$-\frac{dx_1}{dx_2} = r = \text{positive const}. \tag{5-66}$$

Therefore the problem reduces to a first-order nonlinear differential equation in x_1.

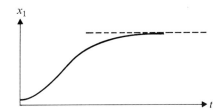

Fig. 5-25 Microorganism population in a batch fermentation process.

Since $x_1(0)$ is almost zero and $x_2(0)$ is far greater than K in Eq. (5-65), the growth in the first phase is logarithmic. This is followed by the second phase where the rate of growth levels off due to a limitation in available nutrient. The nutrient (x_2) approaches zero in the last phase so that

$$k \approx \frac{k_{\max}}{K} x_2,$$

$$x_1 \approx x_1 - x_{10} = r(x_{20} - x_2) \approx rx_{20},$$

and

$$\frac{dx_1}{dt} = -r\frac{dx_2}{dt} = \frac{k_{\max}}{K}(rx_{20})x_2.$$

x_2 in the last phase is thus an exponential decay toward zero. Hence, in the last phase, $x_1(t)$ is exponentially asymptotic to a fixed final value. The overall pattern $x_1(t)$ in the batch process consists of an exponential growth (instability), constant rate of growth, and finally, an asymptotic behavior. (See Fig. 5-25.)

5-5 CONTROLLERS AND CONTROL LAWS

The central element of typical feedback control systems is the "controller." This device has as its input a low-power-level signal, which is a measure of a process variable, and as its output a high-power-level signal, which is related to the input signal by some specified but usually adjustable input–output relation. Note that ideally, the controller is a strictly unilateral device—the conditions at the output end of the controller should in no way have any effect "backward" through the controller on the process. The output signal itself is, of course, used to manipulate the process in some way through an actuator. Because a high-power-gain device by itself can be very difficult to regulate, a common technique in building controllers is to use a passive feedback element around the high-gain device. This reduces the overall gain considerably, but the end result is an instrument that has sufficient power gain and is capable of precise regulation. This method is shown schematically in Fig. 5-26(a). The power unit has an external power supply (not shown)

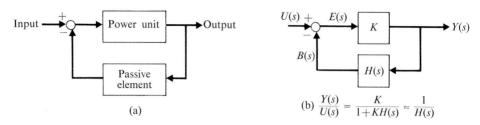

Fig. 5-26 Feedback structure of instruments and regulators.

to provide the power gain, and the input–output relation is controlled completely by the passive feedback.

To explain why this works, we shall assume the linear relation of Fig. 5-26(b). We have

$$Y(s) = KE(s), \qquad E(s) = U(s) - B(s), \qquad B(s) = H(s)Y(s),$$

and thus

$$Y(s) = \frac{K}{1 + KH(s)} \cdot U(s),$$

where K is the (high) gain of the amplifier. Therefore, an approximate input–output relationship is given by

$$Y(s) = \frac{1}{H(s)} U(s). \tag{5-67}$$

This is the property stated above.

It is conceptually possible for us to assign virtually any realizable function $H(s)$ to the feedback element and, by virtue of Eq. (5-67), determine the input–output relation for the controller. We will now investigate some of the most commonly used functions (control laws) and physical realizations of them. (Feedback control and the action of various control laws in closed-loop systems is discussed in Part 3.)

Let us first consider the case where $H(s)$ is static, so that the proportionality relation

$$y(t) = k_c u(t) \tag{5-68}$$

will be approximately maintained between input and output.

The baffle-nozzle pair shown in Fig. 5-27(a) is a power unit that has been used for over thirty years in pneumatic instruments and controllers. The nozzle back pressure is very sensitive to nozzle opening ε (Fig. 5-27b) when high-pressure air is supplied as shown. The high nonlinear gain characteristic of the power unit can be converted into a linear proportional gain by the feedback shown in Fig. 5-28(a). The booster in the figure is itself a

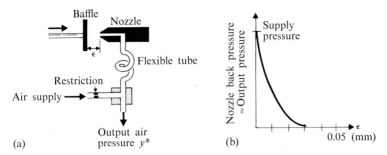

Fig. 5-27 Baffle-nozzle unit.

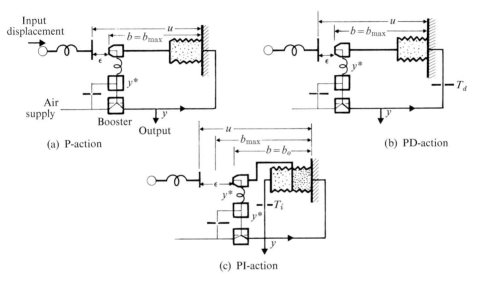

Fig. 5-28 Pneumatic controller configurations.

proportionality device of the type we are discussing, making the controller shown a two-stage instrument. By a feedback of its own (not shown in the figure), the booster establishes a negative proportionality relation between air pressure deviations y^* and y,

$$y = -k_b y^*,$$

where k_b is assumed to be 1 for the sake of brevity. The booster output pressure $y_t(t)$ is the output of the instrument where subscript t is used to imply total (or gage) pressure. This pressure also provides the feedback by moving the nozzle via a bellows mechanism. An input displacement [represented by $u(t)$] of the baffle position causes a proportional change in the output pressure $y_t(t)$ when $u(t)$ is very close to the extreme position b_{max} of the nozzle, which in turn corresponds to $y_t^* = 0$, $y_t = $ max. In other words, Eq. (5-68) holds in a range $b_{min} \leq b \leq b_{max}$. The range, measured on the indicator or recorder of the instrument, is called the proportional band. The constant k_c in Eq. (5-68), called the gain, is inversely proportional to the proportional band. The proportional band or the gain k_c can be adjusted by changing the gain of the feedback path, for instance, by inserting an adjustable-stroke linkage between the bellows and the nozzle in the system of Fig. 5-28(a).

Because of the high-gain characteristic of the pneumatic power unit (Fig. 5-27b), the nozzle opening ε is kept practically at zero when a system is operating in its proportional band. This is generally true for a feedback system (Fig. 5-26a) which has a high-gain power unit in its forward path. The system operates in such a way that the input $u(t)$ and feedback signal $b(t)$ balance:

$$u(t) - b(t) = \varepsilon(t) \approx 0. \tag{5-69}$$

Two displacements were balanced in the system we just saw, a displacement-balance type. The force-balance type is also common in pneumatic instruments and controllers [7]. Voltage or current balance is used in electrical and electronic devices.

In general, an error signal $e(t)$ is the input to a feedback controller. The signal is generated by taking the difference between a reference input (or a desired value) $r(t)$ and a (measured) value $c(t)$ of a variable to be controlled:

$$e(t) = r(t) - c(t). \tag{5-70}$$

The output of the controller, $m(t)$, becomes a controlling input to an actuator. Equation (5-68) applied to a controller yields the following control law:

$$m(t) = k_c e(t). \tag{5-71}$$

This is called a proportional control or P-control, with a gain k_c. It is a static action when all the dynamic effects involved in the controller mechanism are ignored. In other words, the output $m(t)$ of a P-action depends only on the magnitude of error $e(t)$ at the same instant.

By making a modification in the feedback component of a controller, it is possible to develop a control action which responds not only to the present magnitude of the error but also to its rate of change $de(t)/dt$. The idealized form of such an action is given by the control law,

$$m(t) = k_c(1 + T_d D)e(t). \tag{5-72}$$

This is proportional-plus-derivative control, or PD-action. We put a restriction [T_d in Fig. 5-28(b)] in the feedback path of a pneumatic controller to realize this action. This restriction causes a lag in pressure build-up in the feedback bellows and hence a lag in feedback motion b. Denoting by T_d the time constant of the lag, we modify Eq. (5–68) as follows,

$$T_d \frac{db}{dt} + b = \frac{y}{k_c}. \tag{5-73}$$

If a system is operating in its proportional band, the approximate linear relations

$$y^* = -\varepsilon K = -(u - b)K \quad \text{and} \quad y = -y^* = (u - b)K$$

hold, and thus Eq. (5–73) becomes

$$\frac{db(t)}{dt} = -\left(1 + \frac{K}{k_c}\right)\left(\frac{1}{T_d}\right)b(t) + \left(\frac{K}{k_c}\right)\left(\frac{1}{T_d}\right)u(t),$$
$$y(t) = -Kb(t) + Ku(t). \tag{5-74}$$

This is the state-variable formulation of the PD-action; replacing $b(t)$ by $x(t)$, we get the form of (5–13) and (5–14).

From Eq. (5–74) we derive the transfer function

$$G_c(s) = \frac{Y(s)}{U(s)} = \frac{K(T_d s + 1)}{T_d s + 1 + K/k_c}.$$

Since K is very large, we apply (5–67) to obtain the approximate form

$$G_c(s) = k_c(T_d s + 1), \tag{5-75}$$

the PD-action of (5–72).

The PI (proportional-plus-integral) control law is defined by

$$m(t) = k_c \left(1 + \frac{1}{T_i D}\right) e(t). \tag{5-76}$$

It is sometimes called proportional-plus-reset action, because an integral action "resets" the position of proportional band. Since the operator $1/D$ in Eq. (5–76) designates the operation of integration, we see from Eq. (5–76) that the manipulated variable $m(t)$ is equal to a constant proportion of the error plus the integrated past error.

The PI-control law can be approximately realized in a pneumatic controller by using a set of opposing feedback bellows, as shown in Fig. 5–28(c). One unit has a time constant T_i while the other has no lag. The block-diagram description for the operation of a PI-controller within its linear range (proportional band) (Fig. 5–29c) is constructed by combining elements of the block diagrams of the P and PD controllers (Fig. 5–29a and b). The feedback transfer function for the PI-controller is

$$H\{s\} = \frac{1}{k_c} - \frac{1/k_c}{1 + T_i s} = \frac{T_i s / k_c}{1 + T_i s},$$

and by the approximation (5–67) we get the PI-control law.

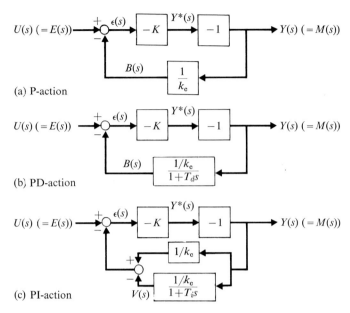

Fig. 5-29 Block diagrams of pneumatic controllers.

The following state-space formulation holds directly for this control law:

$$\left.\begin{array}{l}\dfrac{dx}{dt}=\dfrac{1}{T_i}u\\ y=k_c u+k_c x\end{array}\right\} \quad \text{or} \quad \left\{\begin{array}{l}\dfrac{dx}{dt}=\dfrac{k_c}{T_i}u,\\ y=x+k_c u.\end{array}\right. \tag{5-77}$$

If, on the other hand, we derive the state-space formulation from Fig. 5-29(c) with $V(s)$ as the state variable, a somewhat lengthy algebraic computation yields the following result:

$$\begin{aligned}\dfrac{dx}{dt}&=-\left(\dfrac{k_c/T_i}{K+k_c}\right)x+\left(\dfrac{Kk_c/T_i}{K+k_c}\right)u,\\ y&=\left(\dfrac{K}{K+k_c}\right)x+\left(\dfrac{Kk_c}{K+k_c}\right)u.\end{aligned} \tag{5-78}$$

When $K = \infty$, Eqs. (5-78) reduce to (5-77).

In the above derivations and transfer functions, we assumed that the controller was operating within its linear range. What happens, however, when the controller sees a very large error signal, such as during a start-up or at any other time a large change is made in the set point? For proportional controllers or PD-controllers there is some limitation on the output signal so that no matter how high the error is, the output $m(t)$ will reach a maximum.

5-5 Controllers and control laws

Table 5-2 SOME REALIZATIONS OF THE PID CONTROL LAW

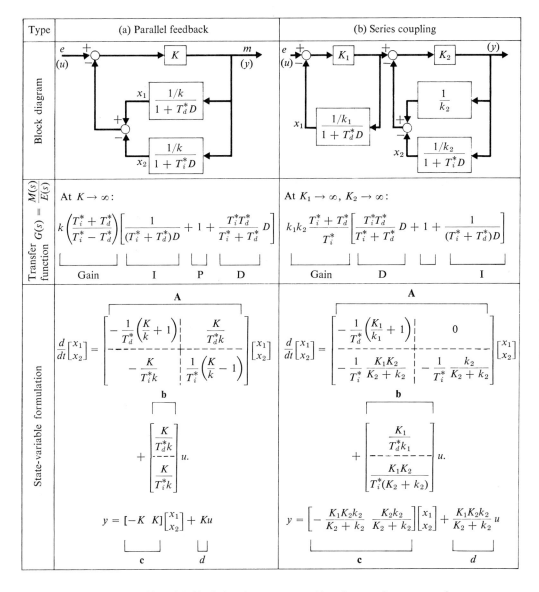

In a pneumatic controller this limit is often governed by the maximum supply pressure available to the controller. The output will thus *saturate* and stay at its maximum until the error signal comes within the linear range. The controller will revert to linear behavior as soon as the error comes within the proportional band. In a PI-controller, or any other controller with integral action, the saturation phenomenon also takes place but, in addition, we must contend with the phenomenon which is often called *reset windup*. This is easily understood if we try to visualize the response of the controller in the time following, for example, a large change in set point. As soon as it sees the large error the proportional part of the controller action responds by saturating. The integrator has no immediate response, but begins to integrate this large error signal, and after some time the integrator also saturates. As the signal comes within the proportional band, a proportional controller would again begin to act in a linear manner and reduce the output as the error decreases. In the PI-controller, however, since the error has not changed sign, the integral action continues to add to the already saturated value it is holding. The integrator cannot even begin to "discharge" until *after* the error signal has changed sign, thus holding the total controller output at or near its saturation value even though the process variable is very close to the new set point. This can cause a large (and undesirable) overshoot in the system response. One possible way to avoid reset windup is to design the controller with a device which will ensure that the integrator maintains a zero initial state until the error signal gets within the linear range. It is important to note that both saturation and reset windup are nonlinear phenomena which do not appear in the transfer function descriptions of the controllers.

PI-action and PD-action combined produce PID-action, or proportional-plus-integral-plus-derivative control. Two combinations are shown in Table 5–2, where (a) is realized by adding a restriction (T_d^*) on the right-hand feedback air line of Fig. 5–28(c), and (b) is realized by putting the systems of Fig. 5–28(b) and (c) in series.

Reset windup is minimized in the system of Tables 5–2(a) [8]. Simplified transfer functions and state-variable formulations are also included in Table 5–2.

Idealized PID-control laws are given in Table 5–3. The two forms are related by

$$k'_c = \frac{k_c}{r}, \qquad T'_i = \frac{T_i}{r}, \qquad T'_d = T_d r, \qquad (5\text{–}79)$$

where for physical realizability of the second canonical form r must be real and given by

$$r = \frac{2}{1 + \sqrt{1 - 4T_d/T_i}}.$$

Comparing the first canonical form with the transfer functions derived for the two structures in Table 5–2, we can express the parameters of the control law in terms of original system parameters. For example,

$$k_c = k_1 k_2 \frac{T_i^* + T_d^*}{T_i^*}$$

Table 5-3 PID CONTROL LAWS

	(a) First canonical form	(b) Second canonical form
Ideal control law	$k_c\left(1 + \dfrac{1}{T_i D} + T_d D\right)$	$k_c'(1 + T_d' D)\left(1 + \dfrac{1}{T_i' D}\right)$
Realizable control law — Transfer function	$\dfrac{k_c}{1+\tau D}\left(1 + \dfrac{1}{T_i D} + T_d D\right)$	$k_c'\dfrac{1 + T_d' D}{1 + \tau D}\left(1 + \dfrac{1}{T_i' D}\right)$
Realizable control law — Signal-flow diagram	(signal-flow diagram with nodes u, x_2, x_1, y and gains $1, -\frac{1}{\tau}, \frac{1}{D}, \frac{k_c}{\tau}\left(1 - \frac{T_d}{\tau}\right), k_c T_d/\tau, \frac{k_c}{T_i \tau}, \frac{1}{D}$)	(signal-flow diagram with nodes u, x_2, x_1, y and gains $1, -\frac{1}{\tau}, \frac{1}{T_i'}, \frac{1}{D}, \frac{k_c T_d'}{\tau}, \frac{1}{D}, k_c'\left(1 - \frac{T_d'}{\tau}\right)$)
Realizable control law — State-space model	$\dfrac{d}{dt}\begin{bmatrix}x_1\\x_2\end{bmatrix} = \underbrace{\begin{bmatrix}0 & 1 \\ \hline 0 & -\dfrac{1}{\tau}\end{bmatrix}}_{A}\begin{bmatrix}x_1\\x_2\end{bmatrix} + \underbrace{\begin{bmatrix}0\\1\end{bmatrix}}_{b} u$ $y = \underbrace{\left[\dfrac{k_c}{T_i \tau} \;\bigg|\; \dfrac{k_c}{\tau}\left(1 - \dfrac{T_d}{\tau}\right)\right]}_{c}\begin{bmatrix}x_1\\x_2\end{bmatrix} + \underbrace{\dfrac{k_c T_d}{\tau}}_{d} u$	$\dfrac{d}{dt}\begin{bmatrix}x_1\\x_2\end{bmatrix} = \underbrace{\begin{bmatrix}0 & \dfrac{1}{\tau T_i'} \\ \hline 0 & -\dfrac{1}{\tau}\end{bmatrix}}_{A}\begin{bmatrix}x_1\\x_2\end{bmatrix} + \underbrace{\begin{bmatrix}\dfrac{k_c' T_d'}{\tau T_i'} \\ k_c'\left(1 - \dfrac{T_d'}{\tau}\right)\end{bmatrix}}_{b} u$ $y = \underbrace{\begin{bmatrix}1 & 1\end{bmatrix}}_{c}\begin{bmatrix}x_1\\x_2\end{bmatrix} + \underbrace{\dfrac{k_c' T_d'}{\tau}}_{d} u$

for the series coupled structure. This shows an interaction among the adjustable parameters of the original system. For instance, a change in T_i^* affects not only T_i but also k_c and T_d. Such interactions are considered undesirable and can be minimized by modifying the structure of the instrument [9].

Since derivative action without a time lag is not realizable (see Chapters 6 and 9 for realizability condition), PID-control laws modified by a lag of (short) time constant τ are shown in the second row of Table 5-3 [10]. It is possible to construct state-space models for these modified transfer functions. The rule given in Fig. 2-8 was applied to derive the signal-flow diagrams and the vector equations that are listed in Table 5-3. When D-action is involved in a control law, K of the high-gain amplifier appears on the right-

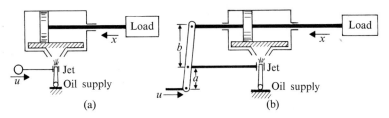

Fig. 5-30 Hydraulic regulator, I-action (a) and P-action (b).

hand side of the output equation. As a consequence, the right-hand side of Eq. (5-74), for example, forms a product of a large number (K) and a small number ($-b + u$). This computational difficulty (in the state-space formulation) may be avoided by applying the ideal control laws coupled with the dynamics of the object of control; τ in Table 5-3 may then be ignored relative to lags of the control object. The realization of P-, PD-, PI-, and PID-controllers with electronic components follows the basic form outlined in Fig. 5-29. Because the high-gain amplifier is basic to electronic controllers, as it also is to pneumatic controllers, the simplest mode to realize is P-action.

However, when a fluid power unit or electric motor is considered the basic component of a regulator, integral- (or I-) action becomes the basic mode, because the output is usually a mechanical motion whose velocity is proportional to the magnitude of the input. The I-control law is

$$\frac{dm(t)}{dt} = k_i e(t), \tag{5-80}$$

where $e(t)$ is the error input to the regulator (placed in a feedback control system), $m(t)$ is an output position (of a control valve, for example), and k_i (or $T_i = 1/k_i$) is a proportionality constant [11]. A jet-pipe type hydraulic motor is shown in Fig. 5-30(a). If the angle of the jet is deflected by an input displacement u, to the right for instance, oil injected from a nozzle enters the right port, pushes the piston to the left, and the oil on the left-hand side of the piston flows out of the left-hand port. A semiempirical relation,

$$\frac{dx}{dt} = k_i u,$$

holds over a limited range of operation. Integral action (5-80) is realized without feedback, although k_i is subject to some changes (due to viscosity change, for instance) because of the open-loop structure.

If the output displacement is fed back into the pilot side via a link (Fig. 5-30b), the I-action is converted into P-action with a lag:

$$\frac{X(s)}{U(s)} = \frac{k_c}{1 + Ts},$$

where $k_c = b/a$ and $T = (a + b)/(k_i a)$. This arrangement can be used either as a servomechanism or as a P-action regulator, depending on the application. The action can be made PI by gradually canceling the feedback action. It is possible to delete the jet pipe and deflect a free jet by applying static pressure from either side. Such a device is called a fluid jet amplifier, and is mostly used for logical operations (Chapter 15).

A digital computer can also be used as a controller (control systems using them are called direct digital control systems, DDC). The flexibility of programmed digital computers permits the control laws discussed here plus many more to be realized. The interested reader is referred to Chapter 11, Section 1.

In some instances we wish to identify the structure of a controller by observing it in action rather than analyzing its structure as we have been doing. In that case, the transfer function (or "black-box") approach is often more convenient than the state-variable approach to identify linear dynamic behavior when we do not know the internal structure. The human operator is an example. It has been generally accepted that the human operator acts as a PID-controller with a delay [12],

$$G(s) = ke^{-sL}\left[T_d s + 1 + \frac{1}{T_i s}\right],$$

when he controls a simple, linear dynamic object with a steering wheel or a stick with visual feedback. There exist personal differences in the quality of control. The quality also depends on the amount of practice. The dead time L is in the order of 0.15 to 0.2 sec. Since a human operator tends to optimize control quality for a given dynamic object, the values of parameters k, T_i, and T_d in the control law depend on the dynamics of his object. It has been observed that:

1. he changes gain k quite rapidly when the control object changes its gain,
2. the rate time T_d is larger for control objects with slower reactions (that is, longer time constants),
3. he strengthens integrating action by making T_i shorter for highly stable control objects and for slowly changing disturbing inputs.

Although the human operator adapts (tunes k, T_i, and T_d) to a given control object, there exists a certain range for the set of parameters k, t_i, and T_d for which he performs the job with least strain. For instance, extremely high or low gain (k), excessively long T_d and/or short T_i constitute a burden on the operator. It therefore follows that either a machine or its response display should be designed to best fit the human operator. For example, if a machine has a sluggish response, a response display with PD-action will release the man from his excessively high derivative action. A new predictive display scheme has thus been suggested [13] for submarine steering.

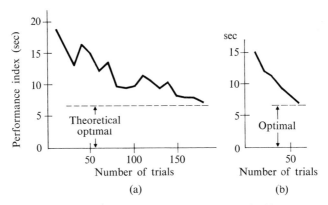

Fig. 5-31 Manual control of an inertial object.

A "preview control," where the operator gathers information in advance of an event, may also improve control quality [14].

To effectively combine manipulators provides a solution to a variety of multiple-input control problems, as in the steering stick of an aircraft and the shutter-speed/lens-opening control of a camera.

Another area of interest to control engineers is the learning and self-adaptation process of man. Time-optimal manual control of an inertial object was tested by Yoshimoto [15]. When the operator was not shown the pattern (or law) of optimal strategy beforehand, his learning process took 180 tests to approach the ultimate optimal, as shown in Fig. 5-31(a). Note that a wave pattern is superimposed on the decay. Such waves indicate the operator's change in strategy. The waves disappear when the operator is shown the optimal strategy (Fig. 5-31b). Manual control with visual feedback involves a wide range of complex bioengineering systems [16].

REFERENCES

1. Shearer, J. L., Murphy, A. T., Richardson, H. H., *Introduction to System Dynamics*. Reading, Mass.: Addison Wesley, 1967.
2. Rosenbrock, H. H., Storey, C., *Computational Techniques for Chemical Engineers*. Oxford: Pergamon Press, 1966.
3. Foss, A. S., "Chemical reaction-system dynamics." *Chem. Eng. Progr.*, *Symposium Series*, **55**, No. 25, 1959, pp. 40-51.
4. Schultz, M. A., *Control of Nuclear Reactors and Power Plants*. New York: McGraw-Hill, 1955.
5. Kendall, D. G., "Deterministic and stochastic epidemics in closed populations." *Proceedings of the Third Berkeley Symposium on Mathematical Statistics and Probability*, 1954, pp. 149-165.

6. Grieves, R. B., Pipes, W. O., Milbury, W. F., Wood, R. K., "Piston-flow model for continuous industrial fermentations." *J. Appl. Chem.*, November 1964, pp. 478–485.
7. Hind, E. C., Dollar, W., "Analysis of pneumatic controllers." *J. Basic Eng.*, *ASME-Trans.*, June 1966, pp. 287–294.
8. Clarridge, R. E., "An improved control system," *Trans. ASME*, **73**, 1951, pp. 297–305.
9. Young, A. J., *Process Control System Design*. London: Longmans, 1955.
10. *Terminology for Automatic Controls*, ASA C85, ASME, April 1961.
11. Blackburn, J. L., Reethof, G., Shearer, J. L., *Fluid Power Control*. New York: Wiley, 1960.
12. McRuer, D., Krendel, E., "The human operator as a servo-system element," *J. Franklin Inst.*, **267**, No. 5, 1959, pp. 381–403.
13. Kelley, C. R., "Development and testing effectiveness of the predictor instrument," *Tech. Report 252-60-1*, Office of Naval Res., Washington, D. C.
14. Bender, E. K., "Optimum linear preview control with application to vehicle suspension," *J. Basic Eng., Trans. ASME*, **90**, June 1968, pp. 213–221.
15. Yoshimoto, K., "Manual Control of a Multivariable System" M.S. Thesis, University of Tokyo, 1966.
16. Stark, L., *Neurological Control Systems*, New York: Plenum Press, 1968.

PROBLEMS

5-1 A water reservoir is a sphere of radius r (see Fig. P5-1).

(a) Obtain the constitutive relation between liquid height and volume, and sketch the curve.

(b) Define time t as the first state variable x_1, and h as the second, x_2. Determine the response curve in the state plane by means of the isocline method, for an initial state $h = 2r$ (full). Assume that the outflow orifice is linear.

5-2 A capillary tube is attached to an air reservoir (see Fig. P5-2). The inside diameter of the tube is 1 mm, and its length is 10 cm. The volume of the tank is 1000 cm³. Compute the resistance of the tube $R\,(\text{kgf/cm}^2)/(\text{kgm/sec})$, by means of

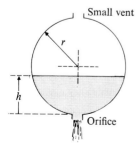

Fig. P5-1

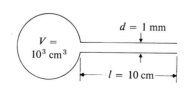

Fig. P5-2

Poiseuille's law:

$$\Delta p = \frac{32 \mu l v}{d^2},$$

where Δp = pressure drop (kgf/m^2), μ = coefficient of viscosity = 1.819×10^{-6} kgf · (sec/m^2) for 15°C air at atmospheric pressure, l = length of tube (m), d = inside diameter (m), v = mean velocity of flow (m/sec). Also compute the time constant of the tank–tube system, for air at 15°C and atmospheric pressure, for which the gas constant is $R_g = 29.3$ (kgf · m)/(kgm · °K), and density is $\rho = 1.23$ kgm/m^3. Assume that the system is isothermal.

5-3 According to the theory of elastic deformation, the following relation holds for a circular elastic pipe of outside diameter D_o and inside diameter D_i:

$$dA = \frac{A}{E(D_o/D_i - 1)} dp,$$

where E = Young's modulus of pipe material (kgf/m^2), dA = change in area A due to internal pressure change dp (kgf/m^2) (see Fig. P5-3). Determine C (capacitance) of the fluid system per unit length of line.

5-4 An adiabatic space of volume V (m^3) holds m (kgm) of an ideal gas. Prove that the following relation holds:

$$\begin{bmatrix} dp \\ d\theta \end{bmatrix} = \begin{bmatrix} \gamma P/m & J(\gamma - 1)/V \\ (\gamma - 1)\theta/m & J(\gamma - 1)/(mR_g) \end{bmatrix} \begin{bmatrix} dm \\ dq \end{bmatrix},$$

where dm and dq are, respectively, in kgm of gas and kcal of heat entering the system, and dp (kgf/m^2) and $d\theta$ (°C) are the resulting changes in pressure and temperature of the tank. Other symbols are: $J = 426.79$ kgf · m/kcal, P = absolute pressure (kgf/m^2), θ = absolute temperature (°K), γ = specific heat ratio of the gas, and R_g (kgf · m)/(kgm · °K) = gas constant.

5-5 One cm^3 of air is trapped by 1 cm^3 of water in a cylinder of 1 cm^2 cross section (see Fig. P5-5). Assume that the water moves like a piston in a frictionless cylinder. The cylinder holds a pressure sensor to measure the pressure in the air space. The average pressure of the air space is 100 kg/cm^2 abs. Assuming the system to be adi-

Fig. P5-3

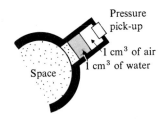

Fig. P5-5

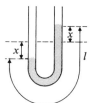

Fig. P5-6

abatic, determine the effect of the water (piston)–air (spring) system on the oscillograph record of the space pressure.

5-6 The total length of a mercury column (see Fig. P5-6) is l in a U-tube manometer of a uniform cross section. Friction is negligible. Find the period of natural oscillation.

5-7 Three pressure tanks forming a ring (see Fig. P5-7) are connected by resistors R_1, R_2, and R_3. Assume that $C_1 = C_2 = C_3 = 1$ and $R_1 = R_2 = R_3 = 1$.

(a) Obtain the A-matrix

(b) Determine eigenvectors and the canonical form of the A-matrix.

(c) One of the pressures x_1, x_2, or x_3 is measured. Is this system observable?

5-8 The three-tank system of Problem 5-7 is now connected to pressure reservoirs (or pressure sources of pressures u_1 and u_2 via resistors R_4 and R_5, as shown in Fig. P5-8). Assume that all resistances and capacitances are unity. (a) Determine $G(s) = X_3(s)/U_1(s)$. (b) Determine the initial values Dx_3 and D^2x_3 when $x_1 = 1$, $x_2 = 2$, $x_3 = 3$, and $u_1 = 10$ at $t = 0$ and when u_2 is kept at zero.

5-9 In some vehicle motion, systems dynamics becomes relevant due to the geometry of the motion. In automobile steering, for instance, the motion (or trajectory) for a constant steering angle u (radians) can be approximated by a circle of radius

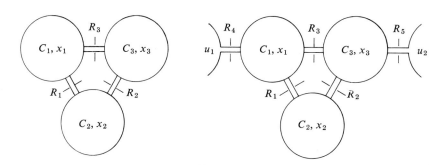

Fig. P5-7   Fig. P5-8

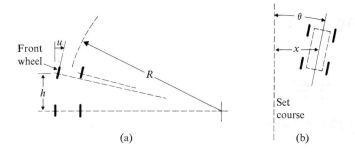

Fig. P5-9

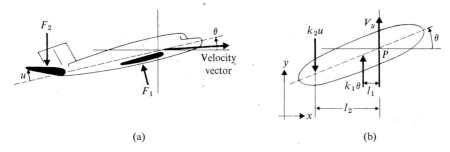

Fig. P5-10

R such that

$$R = \frac{h}{u},$$

where h is the wheel base (see Fig. P5-9a). Taking angle θ (radians) of an automobile's center line from a fixed straight-line course as x_1 [the first state variable; see Fig. P5-9(b)], and its course deflection as x_2, obtain the state-space formulation. If a P-control,

$$u = -k_c x_2, \quad k_c = \text{const},$$

is applied to the steering, what kind of motion would you expect? Assume the linear velocity v of the car to be constant.

5-10 Small deviations ΔF_1 and ΔF_2 of forces acting on an airplane (Fig. P5-10a) are assumed to be proportional to the pitch angle θ and angle of elevator u:

$$\Delta F_1 \approx k_1 \theta, \quad \Delta F_2 \approx k_2 u,$$

where a small difference between θ and the angle of attack is ignored. The problem is reduced to a free-body diagram in Fig. P5-10(b). Let m = mass of the rigid body, I = moment of inertia about center of gravity P, $bD\theta$ = aerodynamic damping, V_y = vertical velocity of P. Taking $V_y = x_1$, $\theta = x_2$, $D\theta = x_3$ as state variables, obtain the **A**- and **b**-matrices. Also show that

$$G(D) \triangleq \frac{V_y}{u} \approx k \frac{1 - T_2 D}{D(T_1 D + 1)},$$

where

$$k = \frac{k_2}{m} \frac{\Delta l}{l_1}, \quad T_1 = \frac{b}{k_1 l_1}, \quad T_2 = \frac{b}{k_1 \Delta l}, \quad \Delta l = l_2 - l_1,$$

and the effect of I is negligible. Sketch the pattern of unit step-input response of $G(D)$.

5-11 The forces acting on an isolated rocket system (Fig. P5-11) are: weight force W or mg (where g is the gravitational acceleration), drag f parallel to the velocity vector, lift h, thrust F, and an aerodynamic damping torque M about P_2 which is not

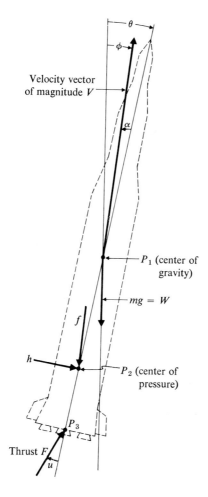

Fig. P5-11

shown in the figure. The distances between P_1, P_2, and P_3 are

$P_1P_2 = l_1$, $P_1P_3 = l_2$.

(a) Show that the moment equation about P_1 is

$$I\frac{d^2\theta}{dt^2} = -M - Fl_2 \sin u - fl_1 \sin \alpha - hl_1 \cos \alpha ,$$

and the force equation perpendicular to the velocity vector is

$$mV\frac{d\phi}{dt} = mg \sin \phi + h + F \sin (u + \alpha) .$$

I in the moment equation is the moment of inertia of the rocket about P_1, and M is the aerodynamic damping torque:

$$M = b\frac{d\theta}{dt}, \quad b = \text{const} .$$

204   Mathematical modeling of engineering systems

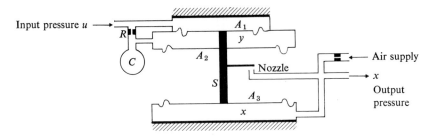

Fig.P5-15

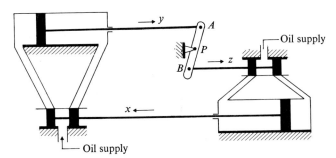

Fig. P5-16

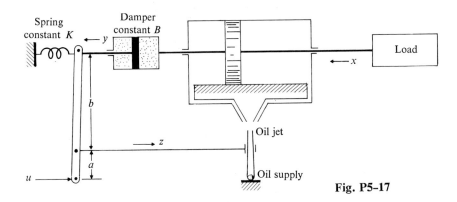

Fig. P5-17

(b) The third equation for a force balance in the direction of motion is not needed to formulate the system's behavior relative to the rocket's attitude behavior (change of angles). Assume that the drag force remains constant ($f =$ const), and the lift force is approximately proportional to the angle of attack α;

$h = k\alpha$, $k =$ lift coefficient $=$ const .

Define the state variables by

$$x_1 = \theta, \qquad x_2 = \frac{d\theta}{dt}, \qquad x_3 = \phi,$$

where all angles are measured in radians. Obtain the vector function $\mathbf{f}(\mathbf{x}, u)$ in the state equation $d\mathbf{x}/dt = \mathbf{f}(\mathbf{x}, u)$.

(c) Linearize the relation and obtain $\mathbf{A}$ and $\mathbf{b}$ in $d\mathbf{x}/dt = \mathbf{A}\mathbf{x} + \mathbf{b}u$.

5-12 In Eq. (5-58), δk is a step of magnitude 0.001, $l = 0.001$, $d = 0.1$, and $\beta = 0.0075$. Let $y(t) = y_0$ and $dz/dt = 0$ at $t = 0$, and find $y(t)$. Compare the response with $y(t)$ for $\beta = 0$.

5-13 Determine the state-plane trajectories of the epidemic process [Eqs. (5-60) and (5-62)] by means of the isocline method.

5-14 Derive the relation of Eq. (5-79), taking into account the condition that $T'_i > T'_d$, which is normally satisfied in PID-action.

5-15 A stack-type pneumatic instrument is shown in Fig. P5-15. Three diaphragm plates, with areas A_1 and $A_2 = A_3$, are rigidly coupled by a spindle S. The only time lag in the system is caused by C and R, in a form $y/u = 1/(TD + 1)$, $T = RC$, where y is the pressure between A_1 and A_2. Obtain the transfer function $G(D) = x/u$.

5-16 Two hydraulic servos are coupled via a link with a lever ratio $\overline{AP}/\overline{BP} = r$ (see Fig. P5-16). Equation (5-80) applies to each servo, so that

$$Dy = k_1 x, \quad Dx = k_2 z.$$

Discuss the pattern of the system's free motion.

5-17 Show that the hydraulic regulator of Fig. P5-17 has a PI-action with a lag.

6
BILATERALLY COUPLED SYSTEMS

This chapter presents a general approach to modeling interacting energetic systems based on identifying the system structure. The concept of an energy port—a point of energetic interaction between a system and its environment—is central to the approach. Formulations are made in terms of linear time-domain equations. Significant nonlinearities are mentioned and linearized for small perturbations.

The basic structural elements of the formulation are described by simple matrices. They are combined as building blocks to create various structures. The capability of digital computers to do logical manipulation can be exploited by writing a program to put simple blocks together systematically into a complex structure. Since the structural blocks very closely resemble the physical systems, it is possible to construct a computer simulation that uses the structural code of a system rather than a mathematical description as its input [1]. Although linearity is assumed in the matrix structures that appear in this chapter, the basic principle can be easily extended to nonlinear simulation.

This approach is particularly useful in the design or analysis of complex systems, since it gives structural as well as numerical information about the system. For simpler systems, especially those with strong nonlinearities, the approach of Chapter 5 may prove easier to apply.

Simple examples from fluid, electrical, and mechanical systems are given throughout the chapter, and a summary at the end relates the general formulation with the vector and scalar equations given in Chapter 5.

6-1 ENERGY PORTS AND ONE-PORT ELEMENTS

Our definition of a system, established in the last chapter, hinges on the identification of the energetic interactions between the system and its environment. After defining a system, we examined the mathematical relations among the external and internal variables to obtain a mathematical model for describing the system.

In this chapter, we will discuss a method of analysis in which we first attempt to deduce the structure of the system in terms of a set of ideal elements. Because ideal elements have simple mathematical descriptions, if we are successful in finding a structure composed of ideal elements which accurately represents the original system, we can obtain the mathematical model for the system with very little difficulty. The ideal elements are them-

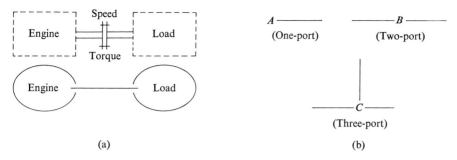

Fig. 6-1 Energy ports.

selves simple systems which interact with their environment through a limited number of *energy ports* [2]. An energy port is a point at which power can be transmitted between a system and its environment. Each energy port transmits a single type of power, and the power transmission is instantaneously described by a pair of signal variables. If several types of power are transmitted at the same point (for example, a rotating shaft with an electrical current through it), a separate energy port is designated for each of the power types. We will be concerned here with bilaterally coupled systems: systems whose structure contains mostly passive elements and ideal sources.

Systems or elements are coupled by interconnections of their energy ports. The interconnection is shown by a solid line, called a bond (Fig. 6-1a). A bond is really an ideal energy transmission element. It transmits energy from one point to another without any dissipation and with no delay time. Examples of components that can often be approximated with a bond are shafts, short pipelines, electrical wires, etc. In addition, the connection between two elements whose energy ports are directly connected, with no intervening "carrier," is also shown by a bond. A system (or element) is characterized by the number of energy ports it has, as a one-port, two-port, five-port, etc. (Fig. 6-1b). Since an energy port is a point at which other elements could be connected to the element, it is indicated with a "free bond."

The *portality* of an element (its number of ports) determines the complexity of systems that can be made by interconnecting elements of the same type. Two one-ports, for example, can only be connected together as shown in Fig. 6-2(a). Two-ports can be connected into a ring structure (Fig. 6-2b), or a chain terminated at each end by one ports (Fig. 6-2c). Systems constructed entirely out of one-and two-ports are thus severely limited in form. With the addition of three-ports, however, structures of any degree of complexity can be constructed; a five-port is shown in Fig. 6-2(d) made up of three three-ports. Since elements of any complexity can be constructed from three-ports, we will define a set of ideal elements that includes one-port,

208   Bilaterally coupled systems

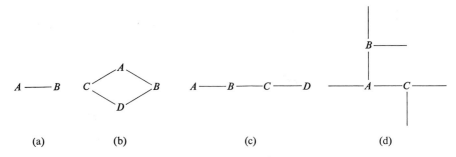

Fig. 6-2 Interconnection of one-, two-, and three-port elements.

two-port, and three-port elements. The graphical representation of a system obtained by using bonds and energy ports is called a *bond graph* [2], [13].

Let us begin with the simplest members of the family: the one-port elements. We define five basic one-ports: the resistor, capacitor, inductor, potential source, and flow source. Examples of electrical, fluid, and mechanical components which correspond to these basic elements are shown in Table 6-1. Thermal and concentration systems are not shown in this table because the energy-port approach does not usually yield any simplifications in the analysis of these systems. The methods presented in Chapter 5 can be used for them.

Each one-port element is characterized by an equation relating its port variables $p(t)$ and $f(t)$ to each other. The product of these port variables is the power transmitted at the port. In the mathematical sense, these simple equations can be written with either variable indicated as the input and the other as the output. In order for the models to make sense physically, however, we must examine each of the elements and determine the natural causality, i.e., which variable must be the input and which the output. Looking, for example, at the linear capacitor, we know from Chapter 5 (Eq. 5-18) that the port variables obey the relation

$$p(t) = \frac{1}{CD} f(t) . \tag{6-1}$$

If we were to try to impose a step change in voltage as an input to an electrical capacitor, we know from experience that an extremely high current and perhaps a spark would result. According to Eq. (6-1), the current should actually go to infinity. We thus conclude that the voltage, or potential, is *not* the natural input to a capacitor, since a source of infinite power would be necessary if Eq. (6-1) were to be satisfied. Yet, when we actually apply a voltage to a real capacitor we get very high but finite current, indicating that a small but significant resistance is acting to keep the current finite. This contradicts the structural model we had assumed since this model was a pure capacitor with no resistance; in this case it is not the

6-1 Energy ports and one-port elements

Table 6-1 ONE-PORT ELEMENTS

		Prototype elements			Ideal sources		Remarks
		$-R$	$-C$	$-I$	$p \rightarrow$	$f \rightarrow$	
Constitutive relation	Electrical systems	$p\,\substack{f\rightarrow\\ \gtrless R}$ or	$p\,\substack{f\rightarrow\\ \rightvert\leftvert C}$ or	$p\,\substack{f\rightarrow\\ \text{ooo}\, I}$ or	$p = 0$ if grounded	$f = 0$ if circuit is open	p is voltage of upper terminal relative to lower terminal; f is current, $+$ for left to right
	Fluid systems	$\substack{p\\ f\rightarrow\\ R}$	$\substack{C\\ f\uparrow\\ p}$ $\substack{\rightarrow p\\ Cp\\ f}$	$\substack{\leftarrow p\rightarrow\\ f\rightarrow I\\ \text{(slug of fluid)}}$	$\substack{\text{Area}\\ p\rightarrow\\ \\ p=0 \text{ for}\\ \text{open end}}$	$\substack{\text{Piston}\\ \rightarrow f\\ f=0 \text{ for}\\ \text{closed end}}$	p is pressure or head (relative to ambient), f is flow, $+$ for left to right
	Mechanical systems	$\substack{\leftarrow p\rightarrow\\ \rightarrow\\ f\quad R}$	$\substack{\leftarrow p\rightarrow\\ \boxed{C}\\ \text{Reference ground}}$	$\substack{\rightarrow p\\ f \text{ooo} I\\ f(\text{force})}$	velocity source (a cam) $p=0$ if fixed	applied force $f=0$ for free end	p is the velocity across an element, f is the force, both p and f positive for compression
Causal relation	Computing diagram for f input — Linear	$\boxed{R}\substack{p\\ f}$ $p = Rf$	$\boxed{\tfrac{1}{C}}\substack{p\\ q}$ $\boxed{\tfrac{1}{D}}\substack{f\\ q}$ $p = \tfrac{f}{CD} = \tfrac{q}{C}$	unnatural	$\boxed{\ }\substack{p}$ p		$p = Z(D)\cdot f$ $Z(D) = R$ or $\tfrac{1}{CD}$ (impedance)
	Computing diagram for f input — Nonlinear	$\boxed{H_R}\substack{p\\ f}$ $p = H_R(f)$	$\boxed{H_c}\substack{p\\ q}$ $\boxed{\tfrac{1}{D}}\substack{f\\ q}$ $\dot q = f$; $p = H_c(q)$		zero impedance	$\substack{\ \\ f}$	
	Computing diagram for p input — Linear	$\boxed{\tfrac{1}{R}}\substack{p\\ f}$ $f = \tfrac{1}{R} p$	unnatural	$\boxed{\tfrac{1}{D}}\substack{p\\ m}$ $\boxed{\tfrac{1}{I}}\substack{f\\ m}$ $f = \tfrac{p}{ID} = \tfrac{m}{I}$		$\substack{f\\ \ \\ f}$ f	$f = Y(D)\cdot p$ $Y(D) = \tfrac{1}{R}$ or $\tfrac{1}{ID}$ (admittance)
	Computing diagram for p input — Nonlinear	$\boxed{F_R}\substack{p\\ f}$ $f = F_R(p)$		$\boxed{\tfrac{1}{D}}\substack{p\\ m}$ $\boxed{F_I}\substack{f\\ m}$ $\dot m = p$, $f = F_I(m)$		zero admittance	
Energy E and Power $\dot E$	Linear	$\dot E_d =$ $Rf^2 = \tfrac{1}{R}p^2$	$E_p =$ $\tfrac{1}{2C}q^2 = \tfrac{C}{2}p^2$	$E_f = \tfrac{1}{2I}m^2 = \tfrac{I}{2}f^2$	$\dot E_s = pf$		$\dot E_d$ is rate of dissipation, E_p and E_f are stored energy, $\dot E_s$ is power supply
	Nonlinear	$\dot E_d =$ $fH_R(f) = pF_R(p)$	$E_p =$ $\int_0^q H_C(q)\,dq$	$E_f =$ $\int_0^m F_I(m)\,dm$			

equations that are wrong, but the model. Similar reasoning pertains to mechanical and fluid capacitances.

If the potential is not the natural input for a capacitor, the flow must be, and we can determine the state of the capacitor by observing the output quantity, the potential. Looking back into Chapters 2 and 3 for a moment, we recall that the state variables for a system could be chosen as the output of the system's integrators. Note in Eq. (6-1) above that the capacitor integrates the flow, and our conclusion that its state is determined by the potential agrees with what we discussed in another context in Chapters 2 and 3.

Similar arguments can be made with regard to the inductor to show that its natural causality is to have the potential as input and the flow as output. The state of an inductor can be determined from the flow. The constitutive relation for a resistor involves no dynamic action whatever, so it has no preferred natural causality. The arrows shown in the block diagrams of Table 6-1 indicate the natural causalities for the elements defined there.

The quantities q and m are defined as the output of the integrators in the computing diagrams for the capacitor and inductor (Table 6-1). The constitutive relation for these elements can then be defined by the scalar relations shown. This is particularly useful for nonlinear systems, since the state of an element can then be fixed directly by the output of the integrator, before the nonlinear block. For linear systems, it is equally convenient to use either q or p as the state variable for a capacitor and either m or f for an inductor.

The ideal sources are the two remaining one-ports: the potential source and the flow source. Their causality is obvious, and many of the physical examples shown Table 6-1 are quite familiar.

In connecting two one-ports into a system care must be taken to ascertain that the natural causalities of *both* elements are correct; the input into one must be the output from the other, and vice versa. Any model which violates natural causality is inadequate, for the reasons discussed above. Several examples of interesting systems that can be constructed solely from one-ports are shown below.

Example 6-1 *Resistor and capacitor* (Fig. 6-3a). The arrows above and below the bonds indicate causality. Since the resistor has no preferred causality, it takes whatever causality is necessary for compatibility with the capacitor. The equation for the resistor is

$$f = \frac{1}{R} p,$$

and for the capacitor

$$p = -\frac{1}{CD} f.$$

6-1   Energy ports and one-port elements   211

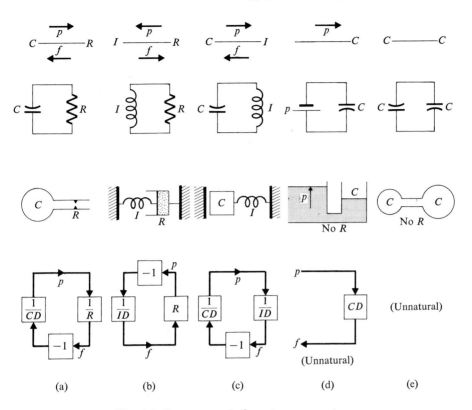

Fig. 6-3 Systems made from two one-ports.

The negative sign in this expression is necessary because the flow has been defined as positive toward C (Table 6-1), but it is away from C in Fig. 6-3(a). The resistor is static; therefore, the state of the system is determined by p, which fixes the state of the capacitor. Combining the two equations above to eliminate f, we get the differential equation,

$$RCDp + p = 0.$$

The free response of the system with an initial value p_0 is an exponential decay $e^{-t/T}p_0$, where the time constant T is RC.

Example 6-2 *Resistor and inductor* (Fig. 6-3b). In order for the resistor to be compatible with the inductor, we must give it the causality shown in Fig. 6-3(b), which is opposite to the causality of the resistor in the first example. The equations for the resistor and inductor are:

$$p = Rf \quad \text{and} \quad f = -\frac{1}{ID}p.$$

The state variable for the system is the flow in the inductor, and we get the following differential equation:

$$\frac{I}{R} Df + f = 0.$$

Example 6-3 *Inductor and capacitor* (Fig. 6-3c). Because the inductor and capacitor have opposite causality, they can be joined and still maintain natural causality. The equations for each of the elements are:

$$f = \frac{1}{ID} p \quad \text{and} \quad p = -\frac{1}{CD} f.$$

The state of the inductor is determined by the flow, and the state of the capacitor by the potential. We must therefore know both p and f to fix the state of the system. The system differential equation, in matrix form, is:

$$D \begin{bmatrix} p \\ f \end{bmatrix} = \begin{bmatrix} 0 & -1/C \\ 1/I & 0 \end{bmatrix} \begin{bmatrix} p \\ f \end{bmatrix}.$$

These, of course, may be solved for the differential equation of the oscillator:

$$CID^2 p + p = 0 \quad \text{or} \quad CID^2 f + f = 0.$$

A simple harmonic motion of period $2\pi \sqrt{CI}$ takes place as the free response because of energy surging between its two forms: the potential energy and the kinetic energy (see Table 6-1, last row).

Example 6-4 Some examples of systems with unnatural causality are shown in Fig. 6-3(d) and (e). The system of 6-3(d) was discussed earlier; it is the case of a voltage applied to a capacitor. It is obvious that a causal element cannot be combined with another of the same type, as in Fig. 6-3(e), but what physical significance does this fact have? Each of the capacitors, say, is a dynamic element that is represented mathematically by an integrator. By putting two capacitors (or inductors) together to form a system, we are implying that they are independent dynamic elements and thus *each one* can be set to an arbitrary initial state. If we tried to do so, however, infinite current would result because of the unbalanced voltage. A pair of the same type elements must be always lumped into one.

Fig. 6-4 Notation to show causality.

Systems with ideal power sources usually contain more than two elements and will be treated later. The above examples demonstrate that we need a simpler way to indicate both the sign convention and the causality on the bond graph. Since the notation to show causality by arrow requires us to put a lot of marks on each bond, an abbreviated notation using a single stroke mark is shown in Fig. 6-4. The sign we use in the constitutive equation for an element is determined by whether we define positive power flow

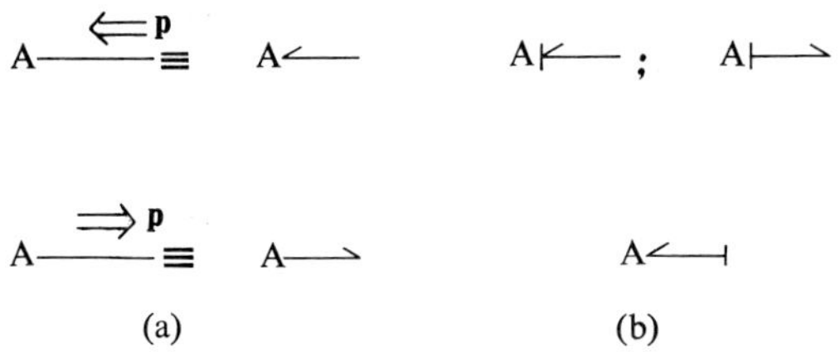

6-2   Ideal junction: Series impedance, shunt admittance   213

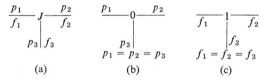

Fig. 6-5 (a) Definition of notation for power sign convention and (b) one-port element with power sign and causality shown.

to be into or out of the element. The constitutive relations given in Table 6-1 for the resistor, capacitor, and inductor, are for power defined as positive *into* the element. This can be shown on a bond graph with an arrow head (Fig. 6-5). The use of these conventions will be demonstrated as part of example problems later in this chapter.

6-2 IDEAL JUNCTION: SERIES IMPEDANCE, SHUNT ADMITTANCE

We will skip two-port elements for a moment to introduce ideal junctions. The prototype ideal junction is the three-port shown in Fig. 6-6(a).

$$\begin{array}{ccc} \dfrac{p_1}{f_1}\!-\!J\!-\!\dfrac{p_2}{f_2} & \dfrac{p_1}{}\!-\!0\!-\!\dfrac{p_2}{} & \dfrac{}{f_1}\!-\!1\!-\!\dfrac{}{f_2} \\ p_3\Big| f_3 & p_3\Big| & \Big| f_3 \\ & p_1=p_2=p_3 & f_1=f_2=f_3 \\ (a) & (b) & (c) \end{array}$$

Fig. 6-6   Ideal junctions.

It is lossless and in instantaneous power balance; that is, it neither stores nor produces energy. This fact can be expressed by the following equation:

$$p_1 f_1 + p_2 f_2 + p_3 f_3 = 0 \ . \tag{6-2}$$

The simplest junctions that satisfy this equation are those that have either all the *p*'s equal or all the *f*'s equal. H. M. Paynter called the first type a zero-junction [2]. Such a junction is drawn as shown in Fig. 6-6(b). It is described by the pair of equations:

$$p_1 = p_2 = p_3 \ , \tag{6-3a}$$

$$f_1 + f_2 + f_3 = 0 \ . \tag{6-3b}$$

The dual to this junction is the one-junction (Fig. 6-6c). It is described by:

$$p_1 + p_2 + p_3 = 0 \ , \tag{6-4a}$$

$$f_1 = f_2 = f_3 \ . \tag{6-4b}$$

Fig. 6-7 Electrical equivalents to ideal junctions.

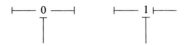

Fig. 6-8 Causality patterns for the zero- and one-junctions.

Note that the above equations (6-2 through 6-4) are written with the assumption that there is a symmetric power-sign convention; that is, the positive direction for power flow is either toward the junction on all its bonds or away from it on all bonds.

It is not usual to view electrical systems in terms of pure junction structures, but it is easy to verify, by application of Kirchhoff's rules, that the two junctions just defined correspond to the parallel and series electrical circuits shown in Fig. 6-7.

We can deduce the preferred causality for the zero- and one-junctions from Eqs. (6-3) and (6-4). Though these elements do not have any dynamic action, the zero-junction, for example, can have $p(t)$ input on only one bond if Eq. (6-3a) is to be satisfied. Therefore, regardless of its portality, a zero-junction must have p-input on one and only one bond, and f-input on the other bonds. On the other hand, in order to satisfy Eq. (6-4b), a one-junction must have f-input on one bond and p-input on all other bonds. These causality patterns are shown in bond diagram form in Fig. 6-8.

Consider the system shown in Fig. 6-9. Both the electrical circuit of Fig. 6-9(a) and the liquid-flow system of 6-9(b) are represented by the bond graph in Fig. 6-9(c), where P is an ideal p-source. The bond graph is shown with power signs and causality already indicated. Note that, without the resistor, the system reduces to the system of Fig. 6-3(d), which does not have compatible causality. The resistance, therefore, no matter how small, is significant. The state of the system is determined by the potential p_2, on

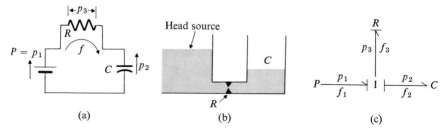

Fig. 6-9 Physical systems and bond graph for RC system with p source.

the capacitor, which we will take as the state variable for the system. The equations for each of the elements are:

one-junction: $p_1 - p_2 - p_3 = 0$
$f_1 = f_2 = f_3 = f$

resistor: $p_3 = Rf_3$

capacitor: $p_2 = \dfrac{1}{CD} f_2$

potential source: $p_1 = P$.

The signs in the one-junction equation are determined by inspection of the power-sign conventions on the bond graph. Terms corresponding to bonds with the arrow in the same direction (that is, either toward or away from the junction) appear with the same sign; the sign arrow points away from the junction for bonds 2 and 3, and therefore p_2 and p_3 appear with the same sign. We get a first-order differential equation by combining these equations:

$$RCDp_2 + p_2 = P.$$

Relating the p's and f's back to the electrical circuit, for example, we find that p_1, p_2, and p_3 correspond to the voltages shown, and f corresponds to the current.

Example 6-5 Figure 6-10 shows a parallel RIC-electrical circuit and the bond graph for that system. Causality and sign conventions are shown on the bond graph. The system state equation is derived from the following constitutive relations for the elements:

zero-junction: $p_1 = p_2 = p_3$,
$f_1 + f_2 + f_3 = 0$,

resistor: $f_1 = \dfrac{1}{R} p_1$,

capacitor: $p_2 = \dfrac{1}{CD} f_2$,

inductor: $f_3 = \dfrac{1}{ID} p_3$.

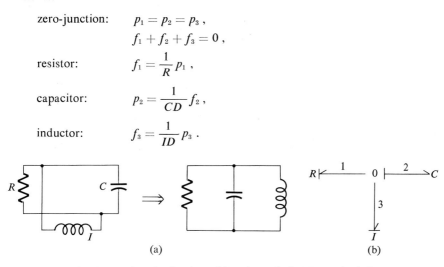

(a) (b)

Fig. 6-10 Electric circuit and bond graph for Example 6-5.

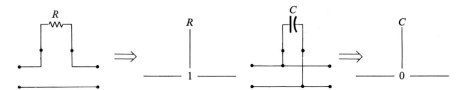

Fig. 6-11 Series resistance and shunt capacitance.

The system has two integrators; its state can be determined by the potential on the capacitor p_2 and the flow through the inductor f_3. The state equation in matrix form is:

$$D \begin{bmatrix} p_2 \\ f_3 \end{bmatrix} = \begin{bmatrix} -1/(RC) & -1/C \\ 1/I & 0 \end{bmatrix} \begin{bmatrix} p_2 \\ f_3 \end{bmatrix},$$

having the familiar characteristic equation

$$\left(CID^2 + \frac{I}{R}D + 1 \right) = 0$$

for either p_2 or f_3.

Depending on the parameter values, the system can have real, imaginary, or complex eigenvalues. It is easy to show that the eigenvalues will always have negative real parts.

Series impedance and shunt admittance (discussed in Chapter 5) are two-port elements that can be constructed by combining a three-port junction with a one-port element. For example, series resistance and shunt capacitance are shown in Fig. 6-11. The series resistance consists of a one-junction and a one-port resistor; the shunt capacitance is a three-port zero-junction combined with a one-port capacitor. Various examples of series and shunt systems are shown in Table 6-2. Note that there are no mechanical or fluid equivalents to the series capacitance. Power-sign conventions are indicated on the bond graphs, and the possible causalities for each element are shown by the block diagrams. The minus signs in the block diagrams appear as a consequence of the power-sign conventions that were chosen. There are a total of four possible causal combinations for a two-port element: in general, either variable at a port can become the input, the other becoming an output. For series impedance or shunt admittance elements, this is reduced to three possible combinations at most because of the causality restrictions imposed by the junction itself. As we see from Table 6-2(d), static two-ports (series or shunt resistance) have the maximum of three possible causal combinations. Dynamic two-ports have either one or two possible causal combinations.

The transfer (or transmission) matrix description of a series impedance, first discussed in Chapter 5, is a very convenient acausal representation. It is acausal because of the appearance of the potential and flow *at the same*

6–2 Ideal junction: Series impedance, shunt admittance

Table 6-2 PROTOTYPE TWO-PORT SYSTEMS

	Series impedance systems			Shunt admittance systems		
Bond diagrams	$R \atop \uparrow \atop \rightarrow 1 \rightarrow$	$C \atop \uparrow \atop \rightarrow 1 \rightarrow$	$I \atop \uparrow \atop \rightarrow 1 \rightarrow$	$R \atop \uparrow \atop \rightarrow 0 \rightarrow$	$C \atop \uparrow \atop \rightarrow 0 \rightarrow$	$I \atop \uparrow \atop \rightarrow 0 \rightarrow$
(a) Electrical systems	R (resistor)	C (capacitor)	I (inductor)	R	C	I
(b) Fluid systems	R (orifice or pipe friction)		I (fluid inertance)	R (leaky resistance)	C (liquid or gas tank)	I (inertance of leaking fluid)
(c) Mechanical systems	R (damper, $b = 1/R$)		I (spring, $k = 1/I$)	(viscous friction, $b = 1/R$)	$(\!-\!C\!-\!)$ (mass, $C=m$)	(spring, $k = 1/I$)
(d) Causality — Block diagram	$p_1 \xrightarrow{+} \overset{R}{\square} \xrightarrow{} p_2$, $f_1 \xleftarrow{} f_2$; $p_1 \xrightarrow{} \overset{1/R}{\square} \xrightarrow{} p_2$, $f_1 \xleftarrow{-} f_2$	$p_1 \xrightarrow{+} \overset{1}{CD} \xrightarrow{} p_2$, $f_1 \xrightarrow{} f_2$; $p_1 \xrightarrow{} \overset{1}{CD} \xrightarrow{} p_2$, $f_1 \xleftarrow{} f_2$	$p_1 \xrightarrow{+} \overset{1}{ID} \xrightarrow{} f_2$, $f_1 \xleftarrow{} f_2$	$p_1 \xrightarrow{} \overset{1/R}{\square} \xrightarrow{} p_2$, $f_1 \xrightarrow{+} f_2$; $p_1 \xrightarrow{} \overset{R}{\square} \xrightarrow{} p_2$, $f_1 \xrightarrow{1/R} f_2$	$p_1 \xrightarrow{} \overset{1}{CD} \xrightarrow{} p_2$, $f_1 \xrightarrow{+} f_2$; $p_1 \xrightarrow{} \overset{1}{CD} \xrightarrow{} p_2$, $f_1 \xrightarrow{} f_2$	$p_1 \xrightarrow{} \overset{1}{ID} \xrightarrow{} p_2$, $f_1 \xrightarrow{+} f_2$; $p_1 \xrightarrow{} \overset{1}{ID} \xrightarrow{} p_2$
(d) Causality — Bond graph	$\overset{R}{\underset{\downarrow}{\vdash 1 \vdash}}$; $\overset{R}{\underset{\uparrow}{\dashv 1 \vdash}}$	$\overset{C}{\underset{\downarrow}{\vdash 1 \vdash}}$; $\overset{C}{\underset{\uparrow}{\dashv 1 \vdash}}$	$\overset{I}{\underset{\downarrow}{\dashv 1 \vdash}}$	$\overset{R}{\underset{\downarrow}{\vdash 0 \vdash}}$; $\overset{R}{\underset{\uparrow}{\dashv 0 \vdash}}$	$\overset{C}{\underset{\downarrow}{\vdash 0 \dashv}}$	$\overset{I}{\underset{\downarrow}{\vdash 0 \vdash}}$; $\overset{I}{\underset{\uparrow}{\dashv 0 \vdash}}$
	$\vdash 1 \dashv$ is illegal because $f_1 = f_2$ must hold for 1-junction			$\dashv 0 \vdash$ is illegal because $p_1 = p_2$ must hold for 0-junction		
(e) State formulation	static: $f_1 = f_2 = \frac{1}{R}(p_1 - p_2)$	static: $q = C(p_1 - p_2)$ dynamic: $\dot{q} = f_1 = f_2$	static: $f_1 = f_2 = \frac{1}{I} m$ dynamic: $\dot{m} = p_1 - p_2$	static: $p_1 = p_2 = R(f_1 - f_2)$	static: $p_1 = p_2 = \frac{1}{C} q$ dynamic: $\dot{q} = f_1 - f_2$	static: $m = I(f_1 - f_2)$ dynamic: $\dot{m} = p_1 = p_2$
Transfer matrix	$\bullet\!-\!\boxed{Z}\!-\!\bullet$ $\overset{Z}{\underset{\downarrow}{-1-}}$	$\begin{bmatrix} p_1 \\ f_1 \end{bmatrix} = \begin{bmatrix} 1 & Z \\ 0 & 1 \end{bmatrix} \begin{bmatrix} p_2 \\ f_2 \end{bmatrix}$, $Z = R, \dfrac{1}{CD}, ID.$		$\boxed{Y}$ $\overset{Y}{\underset{\downarrow}{-0-}}$	$\begin{bmatrix} p_1 \\ f_1 \end{bmatrix} = \begin{bmatrix} 1 & 0 \\ Y & 1 \end{bmatrix} \begin{bmatrix} p_2 \\ f_2 \end{bmatrix}$, $Y = \dfrac{1}{R}, CD, \dfrac{1}{ID}$	

point on each side of the equation,

$$\begin{bmatrix} p_1 \\ f_1 \end{bmatrix} = \mathbf{M} \begin{bmatrix} p_2 \\ f_2 \end{bmatrix}, \tag{6-5}$$

where **M** is the transfer matrix. For a series impedance, **M** takes the form,

$$\mathbf{M} = \begin{bmatrix} 1 & Z \\ 0 & 1 \end{bmatrix}, \tag{6-6}$$

and for a shunt admittance,

$$\mathbf{M} = \begin{bmatrix} 1 & 0 \\ Y & 1 \end{bmatrix}. \tag{6-7}$$

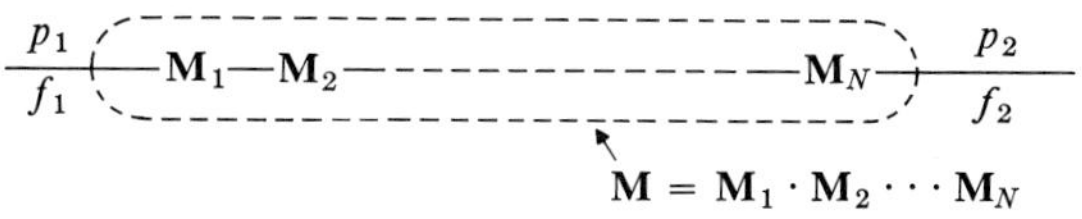

Fig. 6-12 Chain of cascaded two-ports.

The convenience of the transfer-matrix formulation comes in the analysis of systems made up of cascaded two-port elements. If a chain consists of two-port elements with transfer matrices $\mathbf{M}_1$, $\mathbf{M}_2$, etc. (Fig. 6-12), such that

$$\begin{bmatrix} p_1 \\ f_1 \end{bmatrix} = \mathbf{M}_1 \begin{bmatrix} p_a \\ f_a \end{bmatrix}, \quad \begin{bmatrix} p_a \\ f_a \end{bmatrix} = \mathbf{M}_2 \begin{bmatrix} p_b \\ f_b \end{bmatrix}, \ldots,$$

we find by successive substitution that:

$$\begin{bmatrix} p_1 \\ f_1 \end{bmatrix} = \mathbf{M}_1 \mathbf{M}_2 \begin{bmatrix} p_b \\ f_b \end{bmatrix} = \mathbf{M}_1 \mathbf{M}_2 \ldots \mathbf{M}_n \begin{bmatrix} p_2 \\ f_2 \end{bmatrix}$$

or

$$\mathbf{M} = \mathbf{M}_1 \mathbf{M}_2 \mathbf{M}_3 \ldots \mathbf{M}_n.$$

Example 6-6 *A spring-mass system* (Fig. 6-13). Referring to Table 6-2, we see that the mass is a shunt capacitance and the spring a series inductance, yielding the

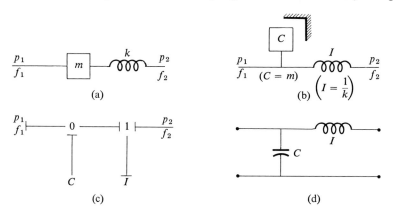

Fig. 6-13 Spring-mass system (a), redrawn (b) to show explicitly inertial reference frame for mass, its bond diagram (c), and equivalent electrical circuit (d).

6-2　Ideal junction: Series impedance, shunt admittance

following transfer matrices:

$$\begin{bmatrix} p_1 \\ f_1 \end{bmatrix} = \begin{bmatrix} 1 & 0 \\ CD & 1 \end{bmatrix} \begin{bmatrix} 1 & ID \\ 0 & 1 \end{bmatrix} \begin{bmatrix} p_2 \\ f_2 \end{bmatrix} = \begin{bmatrix} 1 & ID \\ CD & 1+ICD^2 \end{bmatrix} \begin{bmatrix} p_2 \\ f_2 \end{bmatrix}$$

$$= \begin{bmatrix} 1 & (1/k)D \\ mD & (m/k)D^2+1 \end{bmatrix} \begin{bmatrix} p_2 \\ f_2 \end{bmatrix}. \tag{6-8}$$

The potential p refers to velocity, and the flow f refers to force.

Example 6-7 *A simple model of an automobile suspension.* If we consider only the up-and-down mode of motion for an automobile, we may apply the spring-mass-dashpot model of Fig. 6-14. The velocity source represents the changing elevation of the road; the lower spring-dashpot the tire; m_1 the unsprung mass; the upper spring-dashpot the suspension system; and m_2 the mass of the car body and occupants.

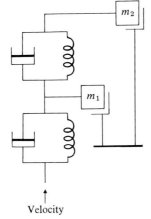

Fig. 6-14 Simple model for an automobile suspension system.

We can get the bond graph for this system by combining series impedance and shunt admittance elements. The spring and dashpot in parallel are represented by the more complex series impedance element shown in Fig. 6-15. Its transfer matrix is

$$\mathbf{M} = \begin{bmatrix} 1 & Z \\ 0 & 1 \end{bmatrix}, \quad \text{where} \quad \frac{1}{Z} = \frac{1}{R} + \frac{1}{ID}.$$

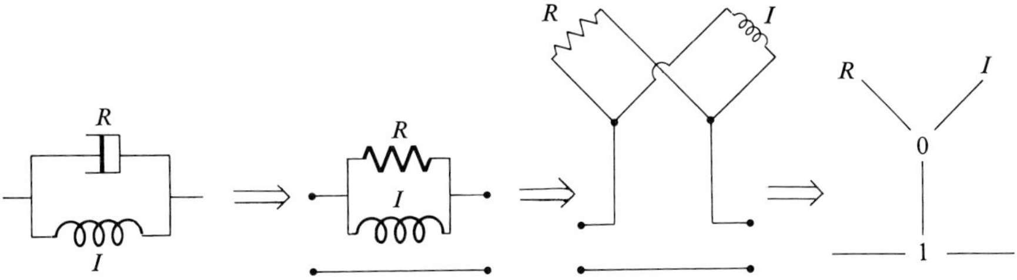

Fig. 6-15 Electrical circuit and bond graph equivalent for a parallel spring-damper system.

220 Bilaterally coupled systems

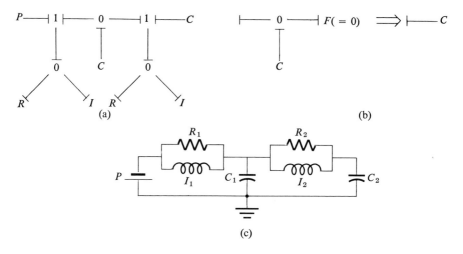

Fig. 6-16 Bond graph and electrical circuit for an automobile suspension system.

Proceeding from bottom-to-top (left-to-right on the bond graph), we get the bond graph of Fig. 6-16(a). The velocity source P is a potential source which terminates the chain at the left end. The series impedance just derived for the parallel spring-dashpot comes next, and then a shunt capacitance for m_1, etc. The termination at the right-hand end is a zero applied force which can be omitted. The shunt capacitance for m_2 then becomes a two-port zero-junction connected to the one-port capacitor. As shown in Fig. 6-16(b), the two-port zero-junction is equivalent simply to a bond. The equivalent electrical circuit is shown in Fig. 6-16(c).

When analogies for mechanical systems were introduced in Chapter 5, we saw that there were two commonly used analogies: the velocity–potential analogy that we have been using, and the force–potential analogy. A major advantage in using the velocity–potential analogy is the topological similarity that exists between the bond graph for the original system and the equivalent electrical circuit. For example, the spring and dashpot in parallel are connected by a zero-junction in the bond graph and appear as a parallel resistor-inductor in the electric circuit. The final differential equation must be the same regardless of the method used to derive it, but the intermediate structures, such as bond graphs and equivalent circuits, will differ. In the following example, we will illustrate these differences by again analyzing the automobile suspension system, only this time using the force–potential analogy.

Example 6-8 The easiest way to find the bond-graph structure for this system is first to establish the correspondence between the series impedance–shunt admittance elements and the various mechanical components. The following pairings are ap-

6-2 Ideal junction: Series impedance, shunt admittance 221

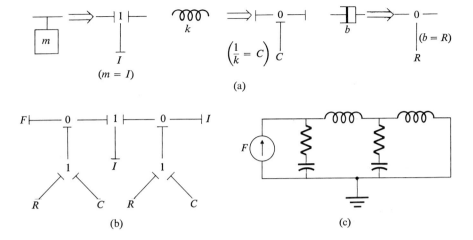

Fig. 6-17 Bond graph equivalents to mechanical components with force-potential analogy (a), an automobile suspension system (b), and equivalent circuit (c).

plicable:

force	potential
velocity	flow (current)
mass	inductance ($m = I$)
spring	capacitance ($1/k = C$)
dashpot	resistor ($b = R$).

The force on a mass (which is implicitly referred to ground) is the difference of the forces on the adjoining elements, and the velocity is the same at each end. Since force corresponds to potential in this analogy, Eq. (6-6) fits this description, and the mass is represented by a series impedance. Similar reasoning can be used to show that the series spring is represented by a shunt capacitance, and the series dashpot by a shunt resistance (Fig. 6-17a). The final bond graph and equivalent circuit are shown in Fig. 6-17(b) and (c).

The transfer (transmission) matrix method can be applied to derive a set of transfer functions for a chain of two-port systems with a known causality. Consider the following general acausal form:

$$\begin{bmatrix} p_1 \\ f_1 \end{bmatrix} = \begin{bmatrix} a & b \\ c & d \end{bmatrix} \begin{bmatrix} p_2 \\ f_2 \end{bmatrix}. \tag{6-9}$$

If u_1, y_1 are an input-output pair at the left end, and u_2, y_2 are a pair at the right end, four transfer functions will relate the two inputs and two outputs:

$$\begin{bmatrix} y_1 \\ y_2 \end{bmatrix} = \begin{bmatrix} G_{11}(D) & G_{12}(D) \\ G_{21}(D) & G_{22}(D) \end{bmatrix} \begin{bmatrix} u_1 \\ u_2 \end{bmatrix}. \tag{6-10}$$

222 Bilaterally coupled systems

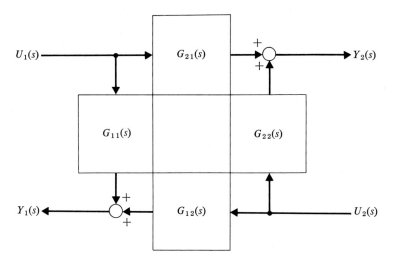

Fig. 6-18 End-to-end transfer functions of a two-port system.

Table 6-3 MATRIX TRANSFER FUNCTIONS $G(s)$

$$\begin{matrix} p_1 \rightarrow & \rightarrow p_2 \\ f_1 \leftarrow & \leftarrow f_2 \end{matrix} \qquad \begin{matrix} p_1 \rightarrow & \leftarrow p_2 \\ f_1 \leftarrow & \rightarrow f_2 \end{matrix}$$

$$\underset{y}{\begin{bmatrix} f_1 \\ p_2 \end{bmatrix}} = \underset{G}{\begin{bmatrix} c/a & \Delta/a \\ 1/a & -b/a \end{bmatrix}} \underset{u}{\begin{bmatrix} p_1 \\ f_2 \end{bmatrix}} \qquad \underset{y}{\begin{bmatrix} f_1 \\ f_2 \end{bmatrix}} = \underset{G}{\begin{bmatrix} d/b & -\Delta/b \\ 1/b & -a/b \end{bmatrix}} \underset{u}{\begin{bmatrix} p_1 \\ p_2 \end{bmatrix}}$$

$$\begin{matrix} p_1 \leftarrow & \rightarrow p_2 \\ f_1 \rightarrow & \leftarrow f_2 \end{matrix} \qquad \begin{matrix} p_1 \leftarrow & \leftarrow p_2 \\ f_1 \rightarrow & \rightarrow f_2 \end{matrix}$$

$$\underset{y}{\begin{bmatrix} p_1 \\ p_2 \end{bmatrix}} = \underset{G}{\begin{bmatrix} a/c & -\Delta/c \\ 1/c & -d/c \end{bmatrix}} \underset{u}{\begin{bmatrix} f_1 \\ f_2 \end{bmatrix}} \qquad \underset{y}{\begin{bmatrix} p_1 \\ f_2 \end{bmatrix}} = \underset{G}{\begin{bmatrix} b/d & \Delta/d \\ 1/d & -c/d \end{bmatrix}} \underset{u}{\begin{bmatrix} f_1 \\ p_2 \end{bmatrix}}$$

$$\Delta = \begin{vmatrix} a & b \\ c & d \end{vmatrix} = ad - bc$$

For instance, the prescribed causality

$$p_1 = u_1, \qquad f_1 = y_1, \qquad p_2 = u_2, \qquad f_2 = y_2$$

will yield the pair of equations

$$u_1 = au_2 + by_2, \qquad y_1 = cu_2 + dy_2.$$

Solving for the two responses, we see that

$$y_1 = \frac{d}{b} u_1 - \frac{\Delta}{b} u_2, \qquad y_2 = \frac{1}{b} u_1 - \frac{a}{b} u_2,$$

where

$$\Delta = ad - bc.$$

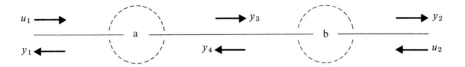

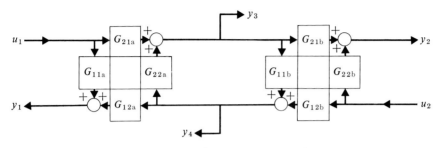

Fig. 6-19 Interchain outputs y_3 and y_4.

Comparing these equations with Eq. (6-10), we find that

$$G_{11} = \frac{d}{b}, \quad G_{12} = -\frac{\Delta}{b}, \quad G_{21} = \frac{1}{b}, \quad G_{22} = -\frac{a}{b}.$$

These transfer functions (Fig. 6-18) are listed in Table 6-3 for four causalities.

There are cases where the output of interest is not at a terminal, but inside a chain of two-port systems, such as y_3 and y_4 in Fig. 6-19. The input-output relation for these inside variables can be determined by applying Eq. (6-10) to systems a and b in the figure. The results are:

$$\begin{aligned} y_3 &= \frac{G_{21a}}{1 - G_{22a}G_{11b}} u_1 + \frac{G_{22a}G_{12b}}{1 - G_{22a}G_{11b}} u_2, \\ y_4 &= \frac{G_{21a}G_{11b}}{1 - G_{22a}G_{11b}} u_1 + \frac{G_{12b}}{1 - G_{22a}G_{11b}} u_2. \end{aligned} \quad (6\text{-}11)$$

The causal relations become simpler if a chain terminates with a one-port system (Fig. 6-20). The input-output relations of such a system can be found by combining the results of Tables 6-1 and 6-3. Transfer functions

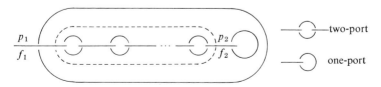

Fig. 6-20 A composite one-port system.

Table 6-4 INPUT-OUTPUT RELATIONS OF A COMPOSITE ONE-PORT SYSTEM

(a)	(b)
$\dfrac{f_1}{p_1} = \dfrac{c+dY}{a+bY} \quad \dfrac{p_2}{p_1} = \dfrac{1}{a+bY} \quad \dfrac{f_2}{p_1} = \dfrac{Y}{a+bY}$	$\dfrac{f_1}{p_1} = \dfrac{cZ+d}{aZ+b} \quad \dfrac{f_2}{p_1} = \dfrac{1}{aZ+b} \quad \dfrac{p_2}{p_1} = \dfrac{Z}{aZ+b}$
(c)	(d)
$\dfrac{p_1}{f_1} = \dfrac{a+bY}{c+dY} \quad \dfrac{p_2}{f_1} = \dfrac{1}{c+dY} \quad \dfrac{f_2}{f_1} = \dfrac{Y}{c+dY}$	$\dfrac{p_1}{f_1} = \dfrac{aZ+b}{cZ+d} \quad \dfrac{f_2}{f_1} = \dfrac{1}{cZ+d} \quad \dfrac{p_2}{f_1} = \dfrac{Z}{cZ+d}$

for all possible combinations are shown in Table 6–4. Note that p_1/f_1 and f_1/p_1 are the impedance and admittance of a composite one-port system, respectively.

Example 6-9 *The CR-Chain of Table* 6–5. The bond diagram at the top of the table applies to any of the systems listed in (a) through (d). The analogy also applies to a heat exchanger [(e) of the table], which is the system treated in Example 5–6, Section 5–3. The configuration of the heat exchanger is distorted to provide for two-ports with net heat flows or available heat energies. Note that pf is not an energy rate for the thermal units defined in Table 5-1. The authors feel that the method in the preceding chapter fits such thermal processes better than the present approach.

The signal flow diagram is constructed in (f) for a set of causalities chosen from Table 6–2(d). We suppress the static relations for the variables $f_1 = f_2$, $f_3 = f_4$, and $f_5 = f_6$, respectively, and then substitute the suppressed results into

$$\frac{dq_1}{dt} = f_2 - f_3, \qquad \frac{dq_2}{dt} = f_4 - f_5; \qquad y_1 = f_1, \qquad y_2 = f_6$$

to obtain the state vector equations

$$\frac{d\mathbf{x}}{dt} = \mathbf{Ax} + \mathbf{Bu}, \qquad \mathbf{y} = \mathbf{Cx} + \mathbf{Du},$$

where

$$\mathbf{x} = \begin{bmatrix} q_1 \\ q_2 \end{bmatrix}, \qquad \mathbf{y} = \begin{bmatrix} y_1 \\ y_2 \end{bmatrix}, \qquad \mathbf{u} = \begin{bmatrix} u_1 \\ u_2 \end{bmatrix},$$

$$\mathbf{A} = \begin{bmatrix} -(1/(C_1 R_1) + 1/(C_1 R_2)) & 1/(C_2 R_2) \\ 1/(C_1 R_2) & -(1/(C_2 R_2) + 1/(C_2 R_3)) \end{bmatrix}, \qquad \mathbf{B} = \begin{bmatrix} 1/R_1 & 0 \\ 0 & 1/R_3 \end{bmatrix}$$

$$\mathbf{C} = [-1/(C_1 R_1) \quad 1/(C_2 R_3)], \qquad \mathbf{D} = \begin{bmatrix} 1/R_1 & 0 \\ 0 & -1/R_3 \end{bmatrix}.$$

6-2 Ideal junction: Series impedance, shunt admittance

Table 6-5 A TWO-PORT CHAIN

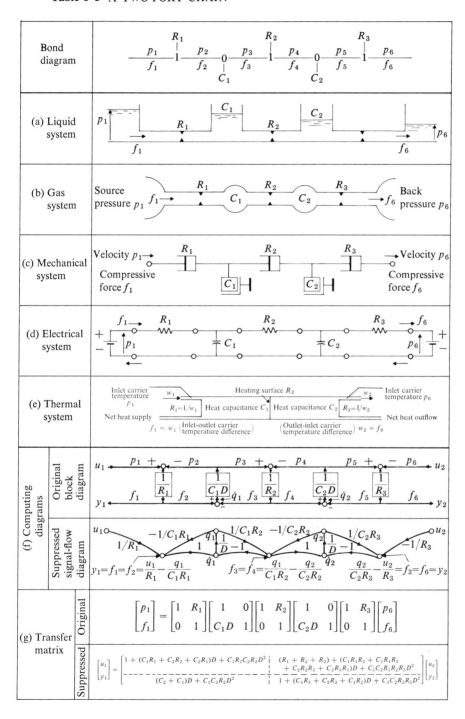

If a matrix or scalar transfer function is the result we desire, we write the transfer matrices and compute the product, as shown in Table 6-5(g). We obtain

$$b = (R_1 + R_2 + R_3) + (C_1 R_1 R_2 + C_1 R_1 R_3 + C_2 R_1 R_3 + C_2 R_2 R_3)D$$
$$+ C_1 C_2 R_1 R_2 R_3 D^2$$

in the overall transmission matrix

$$\begin{bmatrix} p_1 \\ f_1 \end{bmatrix} = \begin{bmatrix} a & b \\ c & d \end{bmatrix} \begin{bmatrix} p_6 \\ f_6 \end{bmatrix}$$

and, by Table 6-3, find, for instance,

$$\frac{f_6}{p_1} = \frac{y_2}{u_1} = G_{21}(D) = \frac{1}{b}.$$

6-3 TRANSMISSION LINES AND ITERATIVE CHAINS

In Section 5-1 we obtained a differential equation for the fluid transmission line in Example 5-1 directly from basic fluid mechanics. In many instances it is convenient to develop a *microstructure* for a transmission line in terms of series impedance and shunt admittance elements. The finite-length transmission line is then modeled by a chain of many cascaded microelements. The simplest case occurs when a line's microstructure contains only one kind of prototype two-port element, either a series impedance or a shunt admittance. In light of Eqs. (6-3) and (6-4), we see that it is possible to combine adjoining junctions of the same type into a single junction. We can thus define a general series impedance (and shunt capacitance) as shown in Fig. 6-21(a) and (b). If the microstructure of a transmission line contains only one type of two-port element, as in Fig. 6-21, it becomes very simple to compute the overall transmission matrix,

$$\prod_i \begin{bmatrix} 1 & Z_i \\ 0 & 1 \end{bmatrix} = \begin{bmatrix} 1 & \sum_i Z_i \\ 0 & 1 \end{bmatrix}, \quad \prod_j \begin{bmatrix} 1 & 0 \\ Y_j & 1 \end{bmatrix} = \begin{bmatrix} 1 & 0 \\ \sum_j Y_j & 1 \end{bmatrix}. \quad (6\text{-}12)$$

Therefore either Z or Y can be lumped, if they appear next to each other, by the rule

$$\sum_i Z_i = \sum_i \left(R_{iz} + \frac{1}{C_{iz}D} + I_{iz}D \right) = R_z + \frac{1}{C_z D} + I_z D \quad (6\text{-}13)$$

or

$$\sum_j Y_j = \sum_j \left(\frac{1}{R_{jy}} + C_{jy}D + \frac{1}{I_{jy}D} \right) = \frac{1}{R_y} + C_y D + \frac{1}{I_y D}, \quad (6\text{-}14)$$

where

$$R_z = \sum_i R_{iz}, \quad \frac{1}{C_z} = \sum_i \frac{1}{C_{iz}}, \quad I_z = \sum_i I_{zi}, \quad (6\text{-}15)$$

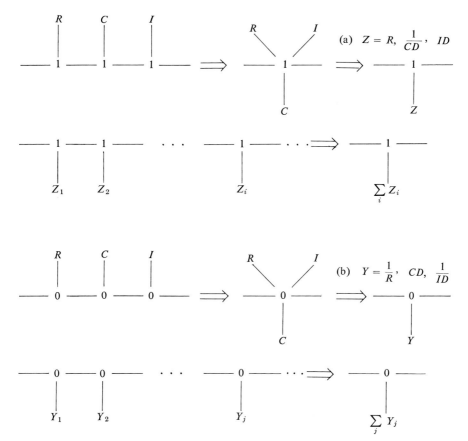

Fig. 6-21 Transmission-line microstructures with only one type of two-port.

and

$$\frac{1}{R_y} = \sum_j \frac{1}{R_{jy}}, \quad C_y = \sum_j C_{jy}, \quad \frac{1}{I_y} = \sum_j \frac{1}{I_{jy}} \tag{6-16}$$

for adjacent series elements and shunt elements, respectively. Parameters with i or j subscripts, C_{iz} or R_{jy}, refer to the property value of a single microelement; parameters without i or j subscripts refer to property values summed over the whole line, that is, they represent total resistance, total inductance, etc. If two elements belong to the same type, the matrix commutes; for instance,

$$\begin{bmatrix} 1 & Z_1 \\ 0 & 1 \end{bmatrix} \begin{bmatrix} 1 & Z_2 \\ 0 & 1 \end{bmatrix} = \begin{bmatrix} 1 & Z_2 \\ 0 & 1 \end{bmatrix} \begin{bmatrix} 1 & Z_1 \\ 0 & 1 \end{bmatrix},$$

which means that R_z and I_z in the fluid or solid system of Table 6-6(a) can

Table 6-6 COMPOSITE TWO-PORT STRUCTURES

		Z-system $\begin{bmatrix}p_1\\f_1\end{bmatrix}=\begin{bmatrix}1 & Z\\0 & 1\end{bmatrix}\begin{bmatrix}p_2\\f_2\end{bmatrix}$; —⊣⊢— $\frac{1}{Z}$			Y-system $\begin{bmatrix}p_1\\f_1\end{bmatrix}=\begin{bmatrix}1 & 0\\Y & 1\end{bmatrix}\begin{bmatrix}p_2\\f_2\end{bmatrix}$; —○— $\frac{1}{Y}$		
		$Z=R_z+\frac{1}{C_zD}+I_zD$	$Z=R_z+I_zD$ $(C_z=\infty)$	$Z=R_z$ $(C_z=\infty, I_z=0)$	$Y=\frac{1}{R_y}+\frac{1}{R_i+1/C_yD}+\frac{1}{I_yD}$	$Y=\frac{1}{R_y}+\frac{1}{R_i+1/C_yD}$ $(I_y=\infty)$	$Y=\frac{1}{R_y}$ $(I_y=\infty, C_y=0, R_i=0)$
(a) Examples							
(b) General block diagram		$f_1=f_2=\frac{1}{Z(D)}(p_1-p_2)$			$p_1=p_2=\frac{1}{Y(D)}(f_1-f_2)$		
(c) Detailed block diagram							
(d) State model	Output equation	$f_1=f_2=\frac{m}{I_z}$	$f_1=f_2=\frac{m}{I_z}$	$f_1=f_2=\frac{1}{R_z}(p_1-p_2)$	$p_1=p_2=\frac{R_y}{C_y(R_i+R_y)}q-\frac{R_iR_y}{I_y(R_i+R_y)}m+\frac{R_iR_y}{(R_i+R_y)}(f_1-f_2)$	$q=-\frac{1}{C_y(R_i+R_y)}q+\frac{R_y}{(R_i+R_y)}(f_1-f_2)$	$p_1=p_2=R_y(f_1-f_2)$
	State equation	$\dot{m}=-\frac{R_z}{I_z}m-\frac{1}{C_z}q+(p_1-p_2)$ $\dot{q}=\frac{1}{I_z}m$	$\dot{m}=-\frac{R_z}{I_z}m+(p_1-p_2)$ $q=0$	$m=0$ $q=0$	$\dot{q}=-\frac{1}{C_y(R_i+R_y)}q-\frac{R_y}{I_y(R_i+R_y)}m+\frac{R_y}{(R_i+R_y)}(f_1-f_2)$ $\dot{m}=\frac{R_y}{C_y(R_i+R_y)}q-\frac{R_iR_y}{I_y(R_i+R_y)}m+\frac{R_iR_y}{(R_i+R_y)}(f_1-f_2)$	$q=0$ $m=0$	$m=0$ $m=0$

be exchanged for each other without causing any change in the overall dynamics.

Some examples of series impedance and shunt admittance systems (which we shall call Z- and Y-systems, respectively) are listed in Table 6–6. Later in this section we shall use these as building blocks for a transmission line.

The input-output relations for natural causality are shown by block- and signal-flow diagrams in Table 6–6(b) and (c), respectively. The signal-flow diagrams, which were constructed by using Table 6–2(d) as a catalogue, reveal the internal or state variables q and m (the use of q and m as state variables facilitates nonlinear simulation; p and f could be used instead with linear systems). Writing an equation for each dq/dt, dm/dt and (scalar) output, we obtain the state models of Table 6–6(d). Nonlinear state models can be built in terms of the static nonlinear functions F and H given in Table 6–1.

Next we consider the coupling of a Z-system and a Y-system. Since the matrices for Z and Y do *not* commute, two possibilities arise, the ZY-pair and the YZ-pair:

$$\begin{bmatrix} 1 & Z \\ 0 & 1 \end{bmatrix} \begin{bmatrix} 1 & 0 \\ Y & 1 \end{bmatrix} = \begin{bmatrix} 1+YZ & Z \\ Y & 1 \end{bmatrix},$$

$$\begin{bmatrix} 1 & 0 \\ Y & 1 \end{bmatrix} \begin{bmatrix} 1 & Z \\ 0 & 1 \end{bmatrix} = \begin{bmatrix} 1 & Z \\ Y & YZ+1 \end{bmatrix}.$$

(6-17)

Various pairs are listed in Table 6–7, where Z and Y were taken from Table 6–6. A pair of signal-flow diagrams from Table 6–6(c), one for Z and the other for Y, when combined, produce the various configurations shown in Table 6–7(b), (c), and (d). Linearized equations for outputs and state variables are listed in Table 6–7.

Note in Table 6–7 the causality directions toward and away from a pair. In a ZY-pair the potential variable $p(t)$ is input at the left and output at the right, while the flow variable $f(t)$ takes the opposite causality. The directions are reversed in a YZ-pair. With these causalities we can construct the chains:

$$——|ZY——|ZY——|ZY—— \cdots$$

or

$$—— YZ|—— YZ|—— \cdots$$

For this reason, the pairs listed in Table 6–7 can serve as microelements in a lumped approximation of a transmission line.

A transmission line can be a pipe in a fluid system, a bar or a shaft in a mechanical system, a conductor for electricity, or a medium for heat or mass diffusion. An ideal limit of the transfer matrix for either power or

230   Bilaterally coupled systems

Table 6-7 *ZY* AND *YZ* PAIRS

	ZY-pair; $\quad\dashv 1\vdash\!\!\!\dashv 0\dashv$ $\quad\perp Z \quad \top Y$	YZ-pair; $\quad\vdash 0\dashv 1\vdash$ $\quad\top Y \quad \perp Z$
(a) Block diagram	$\begin{bmatrix}p_1\\f_1\end{bmatrix}=\begin{bmatrix}1&Z\\0&1\end{bmatrix}\begin{bmatrix}1&0\\Y&1\end{bmatrix}\begin{bmatrix}p_2\\f_2\end{bmatrix}=\begin{bmatrix}1+ZY&Z\\Y&1\end{bmatrix}\begin{bmatrix}p_2\\f_2\end{bmatrix}$ $p_2=\dfrac{1}{1+YZ}p_1-\dfrac{Z}{1+YZ}f_2$ $f_1=\dfrac{Y}{1+YZ}p_1+\dfrac{1}{1+YZ}f_2$	$\begin{bmatrix}p_1\\f_1\end{bmatrix}=\begin{bmatrix}1&0\\Y&1\end{bmatrix}\begin{bmatrix}1&Z\\0&1\end{bmatrix}\begin{bmatrix}p_2\\f_2\end{bmatrix}=\begin{bmatrix}1&Z\\Y&1+YZ\end{bmatrix}\begin{bmatrix}p_2\\f_2\end{bmatrix}$ $p_1=\dfrac{Z}{1+YZ}f_1+\dfrac{1}{1+YZ}p_2$ $f_2=\dfrac{1}{1+YZ}f_1-\dfrac{Y}{1+YZ}p_2$
	$Z=R_z+ID$ and $Y=\dfrac{1}{R_y}+CD$	
(b) Example 1	Output equations: $f_1=\dfrac{m}{I},\quad p_2=\dfrac{q}{C}$ State equations: $\dot q=-\dfrac{q}{CR_y}+\dfrac{m}{I}-f_2$ $\dot m=-\dfrac{q}{C}-\dfrac{R_z}{I}m+p_1$	Output equations: $p_1=\dfrac{q}{C},\quad f_2=\dfrac{m}{I}$ State equations: $\dot q=-\dfrac{q}{CR_y}-\dfrac{m}{I}+f_1$ $\dot m=\dfrac{q}{C}-\dfrac{R_z}{I}m-p_2$
	$Z=R_z$ and $Y=\dfrac{1}{R_y}+CD$	
(c) Example 2	Output equations: $f_1=-\dfrac{q}{CR_z}+\dfrac{p_1}{R_z}$ $p_2=\dfrac{q}{C}$ State equation: $\dot q=-\left(\dfrac{1}{CR_y}+\dfrac{1}{CR_z}\right)q+\dfrac{p_1}{R_z}-f_2$	Output equations: $p_1=\dfrac{q}{C}$ $f_2=\dfrac{q}{CR_z}-\dfrac{p_2}{R_z}$ State equation: $\dot q=-\dfrac{q}{CR_y}+f_1-\dfrac{p_2}{R_z}$
	$Z=1\Big/\left(\dfrac{1}{R}+\dfrac{1}{ID}\right)$ and $Y=CD$	
(d) Example 3	Output equations: $f_1=\dfrac{m}{I}-\dfrac{q}{RC}+\dfrac{p_1}{R}$ $p_2=\dfrac{1}{C}q$ State equations: $\dot m=-\dfrac{q}{C}+p_1$ $\dot q=\dfrac{m}{I}-\dfrac{q}{RC}-f_2$	Output equations: $p_1=\dfrac{q}{C}$ $f_2=\dfrac{m}{I}+\dfrac{q}{RC}-\dfrac{p_2}{R}$ State equations: $\dot m=\dfrac{q}{C}-p_2$ $\dot q=-\dfrac{m}{I}-\dfrac{q}{RC}+\dfrac{p_2}{R}$

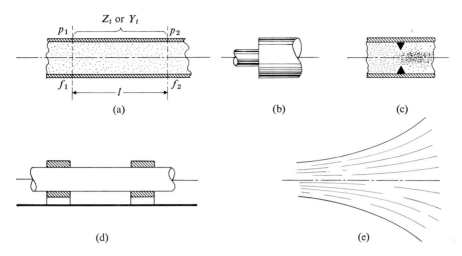

Fig. 6-22 Various transmission lines.

information transmission is given by

$$\mathbf{M} = \mathbf{I} \tag{6-18}$$

in the overall end-to-end relation. That is, power (or information) is transmitted without any loss or dynamic action. In general, power transmission is described by

$$\begin{bmatrix} p_1 \\ f_1 \end{bmatrix} = \mathbf{M} \begin{bmatrix} p_2 \\ f_2 \end{bmatrix}, \tag{6-19}$$

where p_1, f_1 are a pair of variables at the left (or source) end and p_2, f_2 are a pair at the right (or delivery) end. At the static limit, C and I drop away from $\mathbf{M}$, leaving only R, for losses in p and f. Under dynamic operating conditions, the combined effect of R, C, and I causes distortion in wave patterns.

A line may be uniform (Fig. 6-22a), piecewise uniform (b); or it may have some discrete effects (c and d), or a gradual change in parameters (e). If a section is uniform, the total impedance Z_t and admittance Y_t for some length l (Fig. 6-22a) are given by

$$Z_t = Z_0 l, \qquad Y_t = Y_0 l, \tag{6-20}$$

where Z_0 and Y_0 are the impedance and admittance per unit length. When a uniform line is cut into n equal sections,

$$\frac{Z_t}{n} = Z_0 \frac{l}{n} = Z^*, \qquad \frac{Y_t}{n} = Y_0 \frac{l}{n} = Y^*, \tag{6-21}$$

232 Bilaterally coupled systems

where Z^* and Y^* apply to each section (or microelement). The same subscripts and superscripts will also be used for R, C, and I in the sequel. A distinction between two different sets of R, C, and I was made by subscripts z and y in Tables (6–6) and (6–7), so that

$$R_{z\iota} = lR_{z0} = nR_z^*, \quad C_{z\iota} = \frac{C_{z0}}{l} = \frac{C_z^*}{n}, \quad I_{z\iota} = lI_{z0} = nI_z^* \quad (6\text{–}22)$$

for Z-systems and

$$R_{y\iota} = \frac{R_{y0}}{l} = \frac{R_y^*}{n}, \quad C_{y\iota} = lC_{y0} = nC_y^*, \quad I_{y\iota} = \frac{I_{y0}}{l} = \frac{I_y^*}{n} \quad (6\text{–}23)$$

for Y-systems. Since I rather than C is important in Z-systems and the converse applies for Y-systems, subscripts for I in Z-systems and C in Y-systems are sometimes omitted. The problem of actually determining the parameters in terms of equipment dimensions, configurations, material properties, and operating conditions spreads over various branches of engineering. Therefore, apart from the brief descriptions given in the following table, all details will be left to the literature in the respective fields.

		Fluid Systems	Solid Systems
Z-system:	R_z	pipe friction	slip, for example, at a friction coupling
	I_z (or simply I)	momentum effect of fluid	longitudinal or torsional elasticity
Y-system:	R_y	leakage	bearing and other external friction
	C_y (or simply C)	fluid compressibility and pipe elasticity	inertia effects

Some relations (particularly R_y and R_z) can be highly nonlinear. Therefore, the following linearized relations apply only to small perturbations. Moreover, the applicable frequency range of the following formulation depends on the degree of lumping. Nonlinear state-variable formulations can be derived by using the static nonlinear functions F and H in Table 6–2.

Let us consider, as an example, a uniform fluid line which is made up of ZY-cells (Fig. 6–23). Each cell represents pipe friction R_z^*, fluid and pipe elasticity C_y^* while leakage and fluid inertia are considered zero ($R_y^* = \infty$, $I^* = 0$). We can derive the state equation for a model with a finite number of cells. Using symbols $b = 1/R_z^*$ and $c = 1/C_y^*$ for brevity, we write the

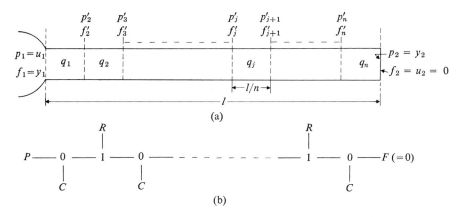

Fig. 6-23 *n*-lumping of a uniform fluid line.

equations from Table 6–7(c) for the first cell:

Inputs: $p_1 = u_1$ and f_2' from the second cell,

Outputs: $f_1 = -bcq_1 + bp_1 = -bcq_1 + bu_1$; $p_2' = cq_1$ to the second cell,

and

State: $\dfrac{dq_1}{dt} = -bcq_1 + bp_1 - f_2' = -bcq_1 + bu_1 - f_2'$,

where f_2' is given by an output equation of the second cell:

$$f_2' = -bcq_2 + bp_2' = -bcq_2 + bcq_1.$$

Thus,

$$\frac{dq_1}{dt} = -2bcq_1 + bcq_2 + bu_1.$$

Since $p_2' = cq_1$ is an input to the second cell, its state equation becomes

$$\frac{dq_2}{dt} = bcq_1 - 2bcq_2 + bcq_3,$$

and generally,

$$\frac{dq_j}{dt} = bcq_{j-1} - 2bcq_j + bcq_{j+1}$$

for $j = 2, \ldots, (n-1)$. A slight difference from the repetitive pattern appears in the last cell due to a right-end input $f_{n+1}' = f_2 = u_2$ (Fig. 6-23):

$$\frac{dq_n}{dt} = bcq_{n-1} - bcq_n - u_2.$$

234 Bilaterally coupled systems

Collecting all the state equations, we obtain the state vector equation:

$$\frac{d}{dt}\mathbf{x} = \mathbf{A}\mathbf{x} + \mathbf{B}\mathbf{u},$$

where

$$\mathbf{x} = \begin{bmatrix} q_1 \\ q_2 \\ \vdots \\ q_n \end{bmatrix}, \quad \mathbf{A} = \begin{bmatrix} -2bc & bc & 0 & 0 & \cdots & & & 0 \\ bc & -2bc & bc & 0 & \cdots & & & \vdots \\ 0 & bc & -2bc & bc & 0 & \cdots & & \\ \vdots & & \vdots & & \vdots & & & \\ & & & & 0 & bc & -2bc & bc \\ 0 & \cdots & & \cdots & 0 & 0 & bc & -bc \end{bmatrix},$$

$$\mathbf{B} = \begin{bmatrix} b & 0 \\ 0 & 0 \\ \vdots & \vdots \\ 0 & 0 \\ 0 & -1 \end{bmatrix}.$$

(6–24)

If the flow f_1 entering the system at the left end and pressure p_2 at the right end are considered outputs y_1 and y_2, respectively, they are given by

$$y_1 = -bcq_1 + bu_1, \quad y_2 = cq_n.$$

These equations can be rearranged in vector form:

$$\mathbf{y} = \mathbf{C}\mathbf{x} + \mathbf{D}\mathbf{u}.$$

A ZY-chain was chosen for the fluid line because its causality fits the sources at both ends: p-input at the left and f-input at the right (Fig. 6–23). Depending on the causality desired at the ends, different models must be used, as listed in Table 6–8(c).

We can use the transfer matrix if only a terminal relation of a linear uniform line is required. The overall transfer matrices are listed in Table 6–8(e). The results indicate that the difference among various models decreases as n is increased. We expect the differences to disappear completely at the limit $n \to \infty$. To see the pattern of the equation at the limit, let us consider the $C_y R_z$-system of Eq. (6–24) and Table 6–5. We have the following parameters for a differential section of length $\lim_{n \to \infty}(l/n) = dy$:

$$\frac{Z_t}{n} = R_{z0} \cdot \frac{l}{n} = R_{z0} \cdot dy,$$

$$\frac{Y_t}{n} = C_y D_0 \cdot \frac{l}{n} = C_{y0} \cdot dy.$$

Writing p, f, $p + (\partial p/\partial y)dy$, and $f + (\partial f/\partial y)dy$ for p_1, f_1, p_2, and f_2, respec-

Table 6-8 LUMPED MODELS FOR A UNIFORM LINE

		(a1)	(a2)	(a3)	(a4)
(a)	Terminal condition	$p_1 \rightarrow$ $\rightarrow p_2$; $f_1 \leftarrow$ $\leftarrow f_2$	$p_1 \leftarrow$ $\leftarrow p_2$; $f_1 \rightarrow$ $\rightarrow f_2$	$p_1 \rightarrow$ $\leftarrow p_2$; $f_1 \leftarrow$ $\rightarrow f_2$	$p_1 \leftarrow$ $\rightarrow p_2$; $f_1 \rightarrow$ $\leftarrow f_2$
(b)	Cell structure	Z^* series, Y^* shunt	Z^* series, Y^* shunt	$\tfrac{1}{2}Z^*$ — $\tfrac{1}{2}Z^*$ series, Y^* shunt	Z^* series, $\tfrac{1}{2}Y^*$ and $\tfrac{1}{2}Y^*$ shunts
(c)	Chain structure	(Z^*Y^*) --- (Z^*Y^*)	(Y^*Z^*) --- (Y^*Z^*)	$(\tfrac{1}{2}Z^*Y^*\tfrac{1}{2}Z^*)$ --- $\cdots$ $-(\tfrac{1}{2}Z^*Y^*\tfrac{1}{2}Z^*)$ or $(\tfrac{1}{2}Z^*Y^*)(Z^*Y^*)$ --- $\cdots$ $-(Z^*Y^*)(\tfrac{1}{2}Z^*)$	$(\tfrac{1}{2}Y^*Z^*\tfrac{1}{2}Y^*)$ --- $\cdots$ $-(\tfrac{1}{2}Y^*Z^*\tfrac{1}{2}Y^*)$ or $(\tfrac{1}{2}Y^*Z^*)(Y^*Z^*)$ --- $\cdots$ $-(Y^*Z^*)(\tfrac{1}{2}Y^*)$
(d) Examples	Fluid system	$f_2 = 0$; p_1 — Left: pressure source; Right: closed	$p_2 = 0$; f_1 — Left: flow source; Right: open	$p_2 = 0$; p_1 — Left: pressure source; Right: open	f_1; $f_2 = 0$ — Left: flow source; Right: closed
	Solid system	$p_1 \rightarrow$, $f_2 = 0$ — Left: displacement forced by velocity; Right: free	$p_2 = 0$, $f_1 \rightarrow$ — Left: force input; Right: fixed	$p_1 \rightarrow$, $p_2 = 0$ — Left: displacement forced by velocity; Right: fixed	$f_1 \rightarrow$, $f_2 = 0$ — Left: force input; Right: free
(e)	Transfer matrix	$\begin{bmatrix} 1+Y^*Z^* & Z^* \\ Y^* & 1 \end{bmatrix}^n$	$\begin{bmatrix} 1 & Z^* \\ Y^* & Y^*Z^*+1 \end{bmatrix}^n$	$\begin{bmatrix} 1+\tfrac{1}{2}Y^*Z^* & Z^*+\tfrac{1}{4}Y^*Z^{*2} \\ Y^* & \tfrac{1}{2}Y^*Z^*+1 \end{bmatrix}^n$	$\begin{bmatrix} 1+\tfrac{1}{2}Y^*Z^* & Z^* \\ Y^*+\tfrac{1}{4}Y^{*2}Z^* & \tfrac{1}{2}Y^*Z^*+1 \end{bmatrix}^n$

tively, we can express the relation for the differential section as:

$$\begin{bmatrix} p \\ f \end{bmatrix} = \begin{bmatrix} 1 & \dfrac{Z_t}{n} \\ \dfrac{Y_t}{n} & 1 \end{bmatrix} \begin{bmatrix} p_2 \\ f_2 \end{bmatrix} = \begin{bmatrix} 1 & R_{z0}dy \\ C_{y0}Ddy & 1 \end{bmatrix} \begin{bmatrix} p + \dfrac{\partial p}{\partial y}dy \\ f + \dfrac{\partial f}{\partial y}dy \end{bmatrix},$$

where all terms which have $1/n^2$ are neglected; hence the choice of cell structure becomes irrelevant. This matrix equation, ignoring factors with $(dy)^2$, yields the relation

$$\frac{\partial p}{\partial y} + R_{z0}f = 0, \qquad \frac{\partial f}{\partial y} + C_{y0}Dp = 0.$$

236 Bilaterally coupled systems

The symbol ∂ for partial differentiations is used because p and f depend on two independent variables, time t and line length y. Eliminating f from the last two equations, and writing $\partial/\partial t$ for D, we obtain

$$\frac{\partial^2 p}{\partial y^2} = T_0 \cdot \frac{\partial p}{\partial t}, \qquad (6\text{-}25)$$

where $T_0 = R_{z0}C_{y0}$. This is the well-known diffusion equation, in this case for a distributed-parameter system, which might, for example, be a one-dimensional heat conductor. Further discussion of distributed-parameter systems will be pursued in the next chapter.

6-4 TRANSDUCERS AND TRANSFORMERS

In the following, we will apply the viewpoint of two-port systems to systems and components that are sometimes called transformers, transducers, or energy converters. Some of these are shown in Fig. 6–24. A thermal-to-mechanical-energy converter (heat engine) could be added to the list, if we defined the port at the thermal side by its available energy, and expressed f_1 in entropy per unit time to give a flow of available energy by the product $p_1 f_1$. The direction of the main function is indicated in the figure by an arrow. We designate variables on the primary side and the secondary side by subscripts 1 and 2, respectively. (We avoid using the words input and output for these because confusion with the causality relation might arise.)

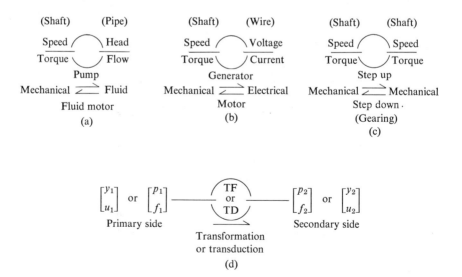

Fig. 6–24 Energy converters (a) and (b), transformer (c), and general configuration (d).

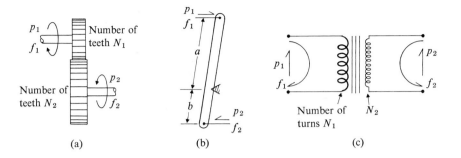

Fig. 6-25 Transformers.

Several examples of simple transformers are shown in Fig. 6-25. Looking, for example, at Fig. 6-25(b), the mechanical lever, if we assume that the lever is massless, we can balance the moments to get

$$f_1 a = f_2 b \quad \text{or} \quad f_1 = (b/a) f_2. \tag{6-26a}$$

Applying the continuity requirement, we find that

$$\frac{p_1}{a} = \frac{p_2}{b} \quad \text{(equal angular velocity)}$$

or

$$p_1 = \frac{a}{b} p_2. \tag{6-26b}$$

Equations (6-26a) and (6-26b) can be expressed by the following matrix relation:

$$\begin{bmatrix} p_1 \\ f_1 \end{bmatrix} = \begin{bmatrix} a/b & 0 \\ 0 & b/a \end{bmatrix} \begin{bmatrix} p_2 \\ f_2 \end{bmatrix}. \tag{6-27}$$

The diagonal matrix of Eq. (6-27) is the canonical form of the transfer matrix for ideal transformers,

$$\mathbf{M} = \begin{bmatrix} r & 0 \\ 0 & 1/r \end{bmatrix}, \tag{6-28}$$

where r is often called the *turns ratio*. The ideal electrical transformer [Fig. 6-25(c)] and the gear train [Fig. 6-25(a)] also satisfy equations of this form. Note that any element which has a transfer matrix of this form will always satisfy the condition that "power in" is instantaneously equal to "power out" since

$$p_1 f_1 = p_2 f_2.$$

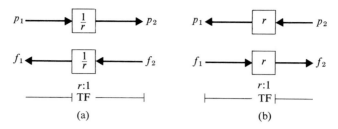

Fig. 6-26 Block diagram and causal bond graph for ideal transformers.

In the case of nonideal transformation, the effects of R, C, and I can be taken into account by inserting their matrices. If, for instance, the gears of Fig. 6-25(a) have moments of inertia J_1 and J_2 (which act as shunt capacitors), we write

$$\begin{bmatrix} p_1 \\ f_1 \end{bmatrix} = \begin{bmatrix} 1 & 0 \\ J_1 D & 1 \end{bmatrix} \begin{bmatrix} r & 0 \\ 0 & 1/r \end{bmatrix} \begin{bmatrix} 1 & 0 \\ J_2 D & 1 \end{bmatrix} \begin{bmatrix} p_2 \\ f_2 \end{bmatrix}$$

$$= \begin{bmatrix} r & 0 \\ \left(rJ_1 + \dfrac{J_2}{r}\right) D & \dfrac{1}{r} \end{bmatrix} \begin{bmatrix} p_2 \\ f_2 \end{bmatrix}. \tag{6-29}$$

These results show that an inertia of rJ_1 on the primary side is equivalent to an inertia of J_2/r on the secondary side (this property is often used intentionally for *impedance matching*). This concept can be extended to other shunt elements and, with slight modification, to series elements.

Since the ideal transformer is described by a static relation, any causality that is consistent with the algebraic equations is acceptable. From Eq. (6-27) or (6-28) we see that if p_1 is an input, then p_2 will be determined as an output. This leads to the two possible causalities shown in Fig. 6-26.

Transducers are two-port elements that convert energy from one medium to another. A simple example of an ideal fluid-to-mechanical transducer is shown in Fig. 6-27(a). Depending on the usual direction of energy transfer, this device can be either a fluid motor or a pump, that is, it is reversible. To find the defining equations for this ideal transducer, we write the force balance

$$p_1 A = f_2,$$

and the continuity requirement (the velocity of the piston face, p_2, must equal the fluid-flow rate divided by the piston area),

$$p_2 = \frac{f_1}{A}.$$

The second equation explains the locking action of a hydraulic actuator;

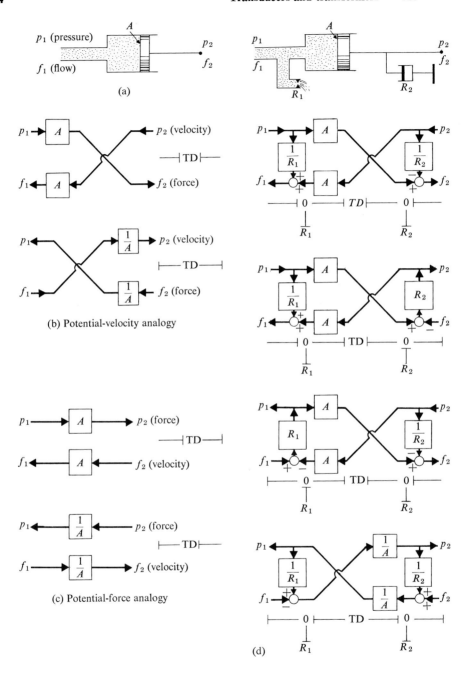

Fig. 6-27 Ideal fluid motor (a), its causal representations (b) and (c), fluid motor with leakage and friction (d).

240 Bilaterally coupled systems

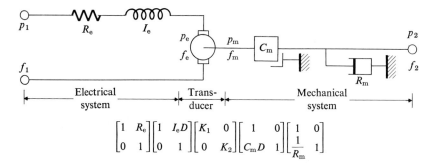

Fig. 6-28 DC-motor (with permanent magnet).

when $f_1 = 0$ (flow is stopped), the mechanical position is locked ($p_2 = 0$). Combining these equations, we obtain the governing relation in transfer matrix form,

$$\begin{bmatrix} p_1 \\ f_1 \end{bmatrix} = \begin{bmatrix} 0 & 1/A \\ A & 0 \end{bmatrix} \begin{bmatrix} p_2 \\ f_2 \end{bmatrix}. \tag{6-30}$$

Block diagrams and bond graphs showing the possible causalities for an ideal transducer are shown in Fig. 6-27(b). Note that the transfer matrix for this transducer has terms only in the off-diagonal positions. Whether the transfer matrix for an ideal transducer has only diagonal terms or only off-diagonal terms depends on the analogies being used on either side. If we had been using the potential-force analogy (instead of the potential-velocity analogy that we are using), the transfer matrix would have had only diagonal terms, and hence there would be no crossing in the block diagram (Fig. 6-27c).

Various secondary effects may be added to the ideal transduction matrices. For instance, a fluid motor with friction and leakage (Fig. 6-27d) is described by

$$\begin{bmatrix} p_1 \\ f_1 \end{bmatrix} = \begin{bmatrix} 1 & 0 \\ 1/R_{y1} & 1 \end{bmatrix} \begin{bmatrix} 0 & 1/A \\ A & 0 \end{bmatrix} \begin{bmatrix} 1 & 0 \\ 1/R_{y2} & 1 \end{bmatrix} \begin{bmatrix} p_2 \\ f_2 \end{bmatrix}. \tag{6-31}$$

Although the relation is still static, all four input-output combinations become possible, as shown in Fig. 6-27(d).

Another example of a transducer is the electromechanical transducer, shown in Fig. 6-28 as a *dc*-motor. The system is separated conceptually into an ideal transducer and associated dynamics and losses. In the ideal transduction process, the proportionality between p_e and p_m represents the counterelectromotive force produced by the rotor rotation p_m, and the relation between f_e and f_m describes the torque caused by the armature current f_e. Note that, in this case, the transfer matrix for the transduction process contains only diagonal terms.

6-4 Transducers and transformers

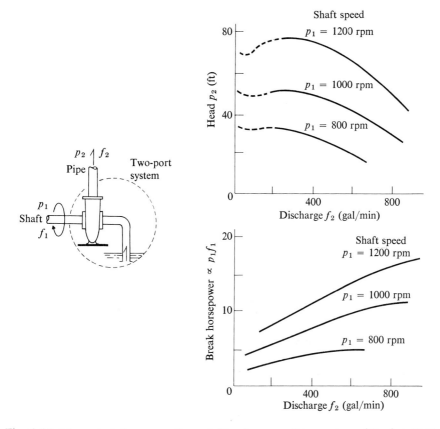

Fig. 6-29 Characteristic curves of a centrifugal pump. (By courtesy of Prof. A.D.K. Laird, University of California, Berkeley.)

The static characteristics of real transducers are generally nonlinear. The basic pattern of characteristic (or performance) curves depends on the basic transduction process, such as the piston or turbine type, as well as on complex miscellaneous effects such as friction (or eddy) losses and cavitation phenomena.

In general, the static behavior of a two-port system is represented by *two sets* of characteristic curves. Figure 6-29 is an example of experimentally obtained characteristic curves for a centrifugal pump. The choice of coordinates for the pair of charts depends upon the specific circumstances. For example, Fig. 6-29 belongs to a style typical in the field of fluid machinery. For characteristic curves of various machinery, see, for instance, Marks [3].

Taking a pair of responses y_1, y_2, one from each side, we can describe the nonlinear static relation by two functions F_1 and F_2 such that

$$y_1 = F_1(u_1, u_2), \qquad y_2 = F_2(u_1, u_2), \qquad (6\text{-}32)$$

or

$$\mathbf{y} = \mathbf{F}(\mathbf{u}) \qquad (6\text{-}33)$$

where

$$\mathbf{y} = \begin{bmatrix} y_1 \\ y_2 \end{bmatrix}, \quad \mathbf{u} = \begin{bmatrix} u_1 \\ u_2 \end{bmatrix}, \quad \mathbf{F}(\mathbf{u}) = \begin{bmatrix} F_1(\mathbf{u}) \\ F_2(\mathbf{u}) \end{bmatrix}.$$

The relation may be linearized for small perturbations in the variables (designated by Δ) as follows:

$$\Delta y_1 = k_{11}\Delta u_1 + k_{12}\Delta u_2, \qquad \Delta y_2 = k_{21}\Delta u_1 + k_{22}\Delta u_2, \qquad (6\text{-}34)$$

where

$$k_{11} = \left(\frac{\partial F_1}{\partial u_1}\right)_{u_2=\text{const}}, \qquad k_{12} = \left(\frac{\partial F_1}{\partial u_2}\right)_{u_1=\text{const}},$$

$$k_{21} = \left(\frac{\partial F_2}{\partial u_1}\right)_{u_2=\text{const}}, \qquad k_{22} = \left(\frac{\partial F_2}{\partial u_2}\right)_{u_1=\text{const}}.$$

The partial derivatives on the right-hand side of Eqs. (6–34) can be determined from a pair of charts (such as those in Fig. 6–29) at a prescribed operating point. It is sometimes advisable to replot the charts in a form for which the conditions $u_1 = $ const, $u_2 = $ const, clearly appear.

The transfer matrix for an ideal transformer is given by Eq. (6–28). We can define a second canonical transformation element which has an off-diagonal form for its transfer matrix,

$$\mathbf{M} = \begin{bmatrix} 0 & r \\ 1/r & 0 \end{bmatrix}. \qquad (6\text{-}35)$$

Like a transformer, this element, which is called a *gyrator*, is a two-port element in which both ports have the same medium of energy transfer; however, it crosses potential and flow acording to its governing relations

$$p_1 = rf_2 \qquad \text{and} \qquad f_1 = \frac{1}{r}p_2.$$

The bond-graph representation for a gyrator showing the possible causalities is shown in Fig. 6–30(a). The gyrator has several interesting properties when combined with other elements. For example, as shown in Fig. 6–30(b) through (d), two gyrators in cascade are equivalent to a transformer, a capacitor plus a gyrator is equivalent to an inductor, and a series impedance coupled at both ends through gyrators looks like a shunt admittance (these relations are easily verified by substitution of the proper transfer matrices). These results suggest that it is possible to define a minimal set of elements for the description of dynamic systems, since, for example, transformers

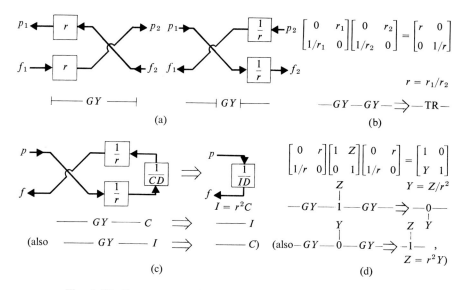

Fig. 6-30 Gyrator (a) and gyrational couplings (b) through (d).

can be constructed from gyrators, inductors from gyrators and capacitors, and one-junctions from gyrators and zero-junctions [13].

Perhaps the best-known example of a gyrator is the gyroscope or top. As is well known in the classical example of a top (Fig. 6-31), its spinning shaft OT moves towards OT' (*it precesses*) if the top is spinning clockwise, and a torque due to the force couple W (weight of top and reaction from support) works on the system. The angular velocity of precession (p) is given by

$$p = \frac{\text{torque}}{J_s \Omega_s}, \qquad (6\text{-}36)$$

where J_s is the moment of inertia of the wheel (or the top), and Ω_s is its spinning velocity. The equation stems from Newton's law of momentum

$$\frac{d\mathbf{H}}{dt} = \mathbf{T}, \qquad (6\text{-}37)$$

where $\mathbf{H}$ and $\mathbf{T}$ are vector expressions for the angular momentum $J_s\Omega_s$ and the external torque, repectively. Equation (6-37) is shown by a vector triangle in Fig. 6-31, from which Eq. (6-36) can be derived as follows:

$$\Delta\theta \approx \frac{|\mathbf{T}|\Delta t}{|\mathbf{H}|},$$

$$\therefore p = \lim_{\Delta t \to 0} \frac{\Delta\theta}{\Delta t} = \frac{\text{torque}}{J_s\Omega_s}.$$

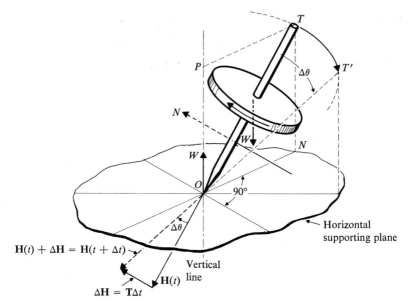

Fig. 6-31 Precession of a top.

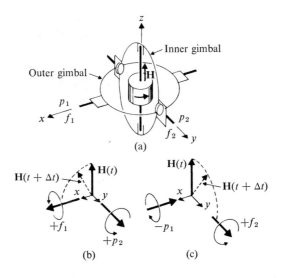

Fig. 6-32 Gyroscope as a two-port system.

Applying Eq. (6-36) to the gyroscope of Fig. 6-32, we relate the torque f_1 in the x-direction with the angular velocity of precession p_2 in the y-direction by $p_2 = f_1/(J_s \Omega_s)$ (see Fig. 6-32b). Similarly, f_2 in the y-direction and p_1 in the x-direction are related by $p_1 = -f_2/(J_s \Omega_s)$. The minus sign appears as a result of the vector relation shown in Fig. 6-32(c). Combining these two equations, we get the final result,

$$\begin{bmatrix} p_1 \\ f_1 \end{bmatrix} = \begin{bmatrix} 0 & -1/(J_s \Omega_s) \\ J_s \Omega_s & 0 \end{bmatrix} \begin{bmatrix} p_2 \\ f_2 \end{bmatrix}. \tag{6-38}$$

In a rate gyro, p_1 (angular rate of vehicle motion) is measured by f_2, restraining the motion of the inner gimbal (Fig. 6-32) by means of springs (thus $p_2 = 0$) and producing an angle proportional to f_2.

Many (but not all) of the transfer matrices discussed above have the interesting property,

$$|\mathbf{M}| = 1. \tag{6-39}$$

The prototype Z- and Y-matrices

$$\begin{bmatrix} 1 & Z \\ 0 & 1 \end{bmatrix}, \quad \begin{bmatrix} 1 & 0 \\ Y & 1 \end{bmatrix},$$

as well as the transduction and transformation matrices,

$$\begin{bmatrix} A & 0 \\ 0 & 1/A \end{bmatrix}, \quad \begin{bmatrix} r & 0 \\ 0 & 1/r \end{bmatrix},$$

and the gyro matrix (6-38), all satisfy Eq. (6-39). When these matrices are combined into one matrix, the determinant of the combined matrix is still equal to one. This can be proved by the following computation. Let

$$|\mathbf{M}_1| = \begin{vmatrix} a_1 & b_1 \\ c_1 & d_1 \end{vmatrix} = 1 \quad \text{and} \quad |\mathbf{M}_2| = \begin{vmatrix} a_2 & b_2 \\ c_2 & d_2 \end{vmatrix} = 1.$$

Then it follows that

$$\begin{aligned} |\mathbf{M}_1 \mathbf{M}_2| &= \begin{vmatrix} a_1 a_2 + b_1 c_2 & a_1 b_2 + b_1 d_2 \\ a_2 c_1 + c_2 d_1 & b_2 c_1 + d_1 d_2 \end{vmatrix} \\ &= a_1 a_2 d_1 d_2 + b_1 b_2 c_1 c_2 - a_1 b_2 c_2 d_1 - a_2 b_1 c_1 d_2 \\ &= (a_1 d_1 - b_1 c_1)(a_2 d_2 - b_2 c_2) = 1. \end{aligned}$$

When Eq. (6-39) holds, the relations in Table 6-3 are simplified because $\Delta = 1$.

6-5 MULTIPORT SYSTEMS

We have already considered the simplest of the multiport systems: the zero- and one-junctions. An interesting set of three-port systems can be constructed from zero- and one-junctions and one-ports, as shown in Table 6-9(a). The constitutive equations for these three-port "triangle" elements are most conveniently expressed in the form of impedance or admittance

246   Bilaterally coupled systems

Table 6-9 THREE-PORT STATIC SYSTEMS

	Shunt resistors	Series resistors
Bond diagrams	(a1)	(a2)
Nonelectrical systems	Solid system (b1)	Fluid system (b2)
Electrical systems	(c1)	(c2)
Equations	Ring voltage: p_1, p_2, p_3 Star current: $f_1 - f_2, f_2 - f_3, f_3 - f_1$. $p_1 = (f_1 - f_2)R_a + (f_1 - f_3)R_c$ $p_2 = (f_2 - f_3)R_b + (f_2 - f_1)R_a$ $p_3 = (f_3 - f_1)R_c + (f_3 - f_2)R_b$ $$\begin{bmatrix} p_1 \\ p_2 \\ p_3 \end{bmatrix} = \begin{bmatrix} R_a + R_c & -R_a & -R_c \\ -R_a & R_a + R_b & -R_b \\ -R_c & -R_b & R_b + R_c \end{bmatrix} \begin{bmatrix} f_1 \\ f_2 \\ f_3 \end{bmatrix}$$ **p = Zf** (d1)	Star current: f_1, f_2, f_3 Ring voltage: $p_1 - p_2, p_2 - p_3, p_3 - p_1$. $f_1 = (p_1 - p_2)/R_a + (p_1 - p_3)/R_c$ $f_2 = (p_2 - p_1)/R_a + (p_2 - p_3)/R_b$ $f_3 = (p_3 - p_2)/R_b + (p_3 - p_1)/R_c$. $$\begin{bmatrix} f_1 \\ f_2 \\ f_3 \end{bmatrix} = \begin{bmatrix} 1/R_a + 1/R_c & -1/R_a & -1/R_c \\ -1/R_a & 1/R_a + 1/R_b & -1/R_b \\ -1/R_c & -1/R_b & 1/R_b + 1/R_c \end{bmatrix} \begin{bmatrix} p_1 \\ p_2 \\ p_3 \end{bmatrix}$$ **f = Yp** (d2)

6-5 Multiport systems

relations,

$$\mathbf{p} = \mathbf{Z}\mathbf{f} \quad \text{or} \quad \mathbf{f} = \mathbf{Y}\mathbf{p}. \tag{6-40}$$

Both the **Z**- and **Y**-matrices are symmetric, as we see from their worked-out forms, shown in Table 6-9(d). The symmetry property can also be demonstrated from the bond graph (or one of the example systems): in either of the standard configurations, a change in p or f at, say, port 1 has the same effect at port 2 just as the same change at port 2 would have at the first port.

Moreover, both the column sums and the row sums of **Z** and **Y** vanish. The first property,

$$\text{column sum} = 0, \tag{6-41}$$

arises from the continuity conditions,

$$\text{sum of ring voltages} = [1\ 1\ 1]\mathbf{p} = 0, \text{ and thus } [1\ 1\ 1]\mathbf{Z} = \mathbf{0} \text{ in (d1)}$$

(which is the loop law of Kirchhoff), and

$$\text{sum of star currents} = [1\ 1\ 1]\mathbf{f} = 0, \text{ and thus } [1\ 1\ 1]\mathbf{Y} = \mathbf{0} \text{ in (d2)}$$

(node law of Kirchhoff). The second property,

$$\text{row sum} = 0, \tag{6-42}$$

is explained by the relativity condition. Since star currents are all expressed by current differences, a constant current f_0 can be added to $f_1, f_2,$ and f_3 without disturbing the relation $\mathbf{p} = \mathbf{Z}\mathbf{f}$; therefore,

$$\mathbf{p} = \mathbf{Z}\mathbf{f} = \mathbf{Z}\left(\mathbf{f} + \begin{bmatrix} 1 \\ 1 \\ 1 \end{bmatrix} f_0 \right)$$

or

$$\mathbf{Z} \begin{bmatrix} 1 \\ 1 \\ 1 \end{bmatrix} = \begin{bmatrix} 0 \\ 0 \\ 0 \end{bmatrix} \quad \text{for} \quad \text{(d1)};$$

and similarly,

$$\mathbf{Y} \begin{bmatrix} 1 \\ 1 \\ 1 \end{bmatrix} = \begin{bmatrix} 0 \\ 0 \\ 0 \end{bmatrix} \quad \text{for} \quad \text{(d2)}.$$

Both $|\mathbf{Z}|$ and $|\mathbf{Y}|$ are zero. Therefore we cannot solve $\mathbf{p} = \mathbf{Z}\mathbf{f}$ for $\mathbf{f}$, nor $\mathbf{f} = \mathbf{Y}\mathbf{p}$ for $\mathbf{p}$. This last and very interesting property is a consequence of the row (or column) sums being zero. We can demonstrate this from either mathematical or physical reasoning. If a square matrix has the property

248 Bilaterally coupled systems

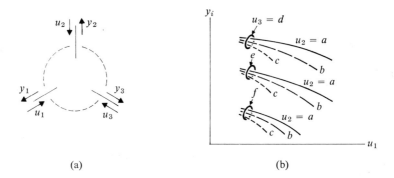

Fig. 6-33 Input-output and static characteristics of a three-port system (a, b, c, d, e, f are constant and $i = 1, 2,$ or 3).

that its row sums vanish, we can write it in the form

$$\begin{bmatrix} a_1 & a_2 & \cdots & a_{n-1} & -\sum_{i=1}^{n-1} a_i \\ b_1 & b_2 & \cdots & b_{n-1} & -\sum_{i=1}^{n-1} b_i \\ & & \cdots & & \end{bmatrix}.$$

Since the value of the determinant is not changed if we replace a column of the matrix with the sum of that column and any other column, the last column can be made to contain all zeroes by adding all the other columns to the last column. The matrix will then have a column of all zeroes and thus have a zero determinant. The fact that the determinant of **Z** is zero implies that the solution for **f** as a function of **p** is not unique. We saw above that the star currents satisfy the relativity condition, that is, the solution is unchanged if we add the same constant to all of the currents. Therefore, for any given **p** there are many **f**'s that will satisfy the equations in Table 6-9 (d1), and there can be no unique inverse for **Z**.

In general, the static characteristics of a three-port system are described by a three-dimensional vector equation

$$\mathbf{p} = \mathbf{H}_R(\mathbf{f}) \quad \text{or} \quad \mathbf{f} = \mathbf{F}_R(\mathbf{p}).$$

Taking responses y_1, y_2, y_3, one for each port (Fig. 6-33), we have

$$\mathbf{y} = \mathbf{F}(\mathbf{u}). \tag{6-43}$$

Equation (6-43) may be often linearized for small perturbations (designated by Δ) in the form

$$\Delta \mathbf{y} = \mathbf{K} \Delta \mathbf{u}, \tag{6-44}$$

where

$$\mathbf{K} = [k_{ij}], \quad k_{ij} = \left(\frac{\partial F_i}{\partial u_j}\right)_{u_k=\text{const}, u_m=\text{const}}$$

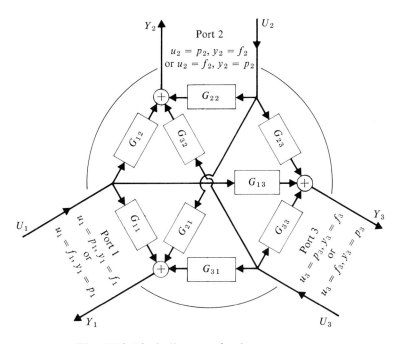

Fig. 6-34 Block diagram of a three-port system.

The static vector equation represents a set of three hyperplanes in a four-dimensional space. Hence a graphical description requires three charts where each chart is made of three sets of curves (Fig. 6-33).*

A two-port system has four ($= 2^2$) possible input-output combinations (Table 6-8), whereas a three-port system has eight ($= 2^3$). For each causality combination, a linear two-port system has four ($= 2^2$) transfer functions (Table 6-3); a three-port system has nine, that is a (3×3)-matrix transfer function $\mathbf{G}(s)$ in a vector input-output relation

$$\mathbf{Y}(s) = \mathbf{G}(s)\mathbf{U}(s).$$

The block diagram is shown in Fig. 6-34.

A valve, which is a three-port system, acts as a *modulator*, modulating the flow through it by the motion of a spindle. The flow, in this case, suffers a loss in available energy due to throttling (that is, entropy increase). A variable-resistance electrical system (Fig. 6-35) belongs to the same class of modulator.

An energy loss of this nature does not occur in other classes of modulators, such as the variable-ratio belt drive shown if Fig. 6-36. In the ideal

---

* Two planes in a three-dimensional space completely describe the static characteristic of a two-port system (see Fig. 6-29).

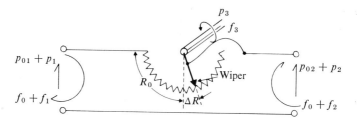

Fig. 6-35 Variable resistance system.

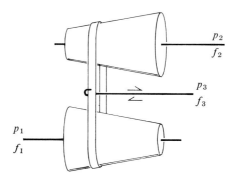

Fig. 6-36 Symbolic sketch of variable-ratio belt drive.

case, the drive ratio is varied by an input at port 3 and the device operates as a lossless, modulated transformer.

These modulation processes involve changes in parameters, for instance, a change ΔR in resistance (Fig. 6-35). A nonlinear relation caused by a parametric change can be linearized for small variations. For instance, consider the variable-resistance system: designating a set of normal values by subscript 0 (see Fig. 6-35), we have

$$\begin{bmatrix} p_{10} + p_1 \\ f_0 + f_1 \end{bmatrix} = \begin{bmatrix} 1 & R_0 + \Delta R \\ 0 & 1 \end{bmatrix} \begin{bmatrix} p_{20} + p_2 \\ f_0 + f_2 \end{bmatrix}.$$

A change ΔR is proportional to a deviation u of the shaft angle. Therefore,

$$\Delta R = \frac{kp_3}{D} = ku,$$

where k is a constant. Combining these equations and neglecting the products of two deviations, we obtain

$$\begin{bmatrix} p_1 \\ f_1 \end{bmatrix} = \begin{bmatrix} 1 & R_0 \\ 0 & 1 \end{bmatrix} \begin{bmatrix} p_2 \\ f_2 \end{bmatrix} + \begin{bmatrix} f_0 \\ 0 \end{bmatrix} ku. \qquad (6-45)$$

A similar linearization was shown for a valve in Section 1-2.

If we choose the transformer of Eq. (6–28) as an example of a lossless modulator, then a parametric change in r caused by u leads to the following relations:

$$r = r_0 + ku, \qquad \frac{1}{r} \approx \frac{1}{r_0} - \frac{1}{r_0^2} ku.$$

Therefore,

$$p_{10} + p_1 = r(p_{20} + p_2), \qquad f_{10} + f_1 = \frac{1}{r}(f_{20} + f_2),$$

$$p_{10} = r_0 p_{20}, \qquad f_{10} = \frac{1}{r_0} f_{20},$$

and

$$\begin{bmatrix} p_1 \\ f_1 \end{bmatrix} = \begin{bmatrix} r_0 & 0 \\ 0 & 1/r_0 \end{bmatrix} \begin{bmatrix} p_2 \\ f_2 \end{bmatrix} + \begin{bmatrix} p_{20} \\ -f_{20}/r_0^2 \end{bmatrix} ku. \tag{6-46}$$

The shaft (third port) of a variable resistor (Fig. 6–35) may have a friction effect only. To trace a causal relation in the system, let us assume viscous friction R_y^{-1} instead of Coulomb friction on the shaft:

$$f_3 = \frac{1}{R_y} p_3. \tag{6-47}$$

This equation, with Eq. (6–45), yields the block diagram of Fig. 6–37. The diagram shows that the electrical system (two-port) and the mechanical

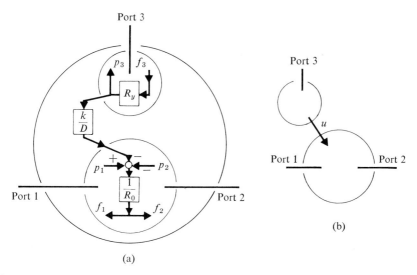

Fig. 6-37 Structure of a series-resistance modulator (a) and canonical structure of a modulator (b).

252   Bilaterally coupled systems

system (one-port) are *unilaterally* coupled by a signal u. The relation is symbolically restated in (b) of the figure. This may be considered a canonical structure for a modulator. The unilateral relations, which correspond to $G_{13} = 0$, $G_{23} = 0$ in the general block diagram (Fig. 6–34), mean that the state of the network has no influence on the state of the mechanical part of the resistor.

The state of a fluid may exert a reaction on the spindle of a valve. Since this reaction complicates the execution of a modulation process, a servomechanism (positioner) can be applied at the third port element to cut the loading effect. As we will see in the next paragraph, an ideal servo has a unilateral input. Taking this input as a modulating signal u, and combining the servo with the main system (Fig. 6–38), we see that the overall structure reduces to the canonical form of Fig. 6–37(b).

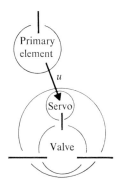
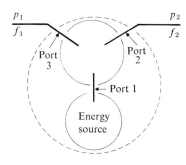

Fig. 6–38 Modulator with a servo.   **Fig. 6–39** Structure of an amplifier.

The *amplifier* is closely related to the modulator in structure. As shown in Fig. 6–39, the amplifier is obtained by connecting the left port of a modulator (port 1) to an energy source and putting the modulating port (port 3) at the primary side. This indicates that it is possible to look upon amplifiers as two-port systems instead of three. For instance, a transistor has three ports: base, collector, and emitter. However, only four variables are significant for the connection of Fig. 6–40(b). Two charts can relate four variables. An example of these charts is shown in Fig. 6–40(c).

In this example, the relation between $p_1 (= V_b)$ and $f_1 (= I_b)$ (see the upper chart) is not seriously affected by $p_2 (= V_c)$. In other words, the primary side (the base circuit) is practically decoupled from the secondary side (collector and output circuit), making the system a current amplifier. An approximation of a similar nature was already made on the baffle–nozzle unit in Fig. 5–27. In a fluid amplifier (Fig. 6–41a), a small pressure difference (u) between the control ports produces a sizable difference between

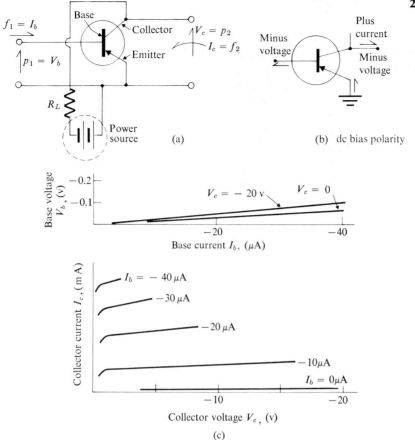

Fig. 6-40 A grounded-emitter transistor circuit and characteristic curves.

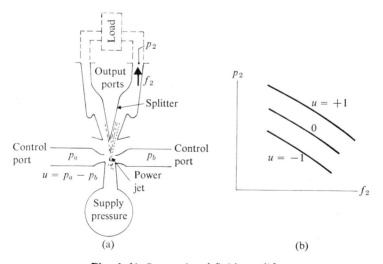

Fig. 6-41 Proportional fluid amplifier.

the flows through the two output ducts.* Since the pressure differential u acts unilaterally on the secondary (or output) side, the chart shown in Fig. 6–41(b), which gives information on output impedance and gain of amplification, is practically sufficient to describe the system's performance. Of course, far more detailed information is required to design a fluid jet amplifier. See, for example, the article by A. K. Simson [4].

The approximate decoupling of the secondary side from the primary side of an amplifier means a cut in a chain of two-port systems, where a bond is replaced by a directed signal u, as shown in Fig. 6–42. A necessary set of relations can be written (Fig. 6–42) without using a transfer matrix.

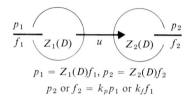

$$p_1 = Z_1(D)f_1, \; p_2 = Z_2(D)f_2$$
$$p_2 \text{ or } f_2 = k_p p_1 \text{ or } k_f f_1$$

Fig. 6–42 Reduced structure of an amplifier.

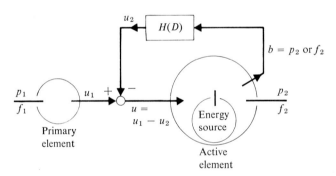

Fig. 6–43 Feedback amplifier.

Feedback from the secondary side is sometimes applied on the directed signal u. The loading effect of the feedback on the secondary side is usually negligible because the power level of the secondary side is much higher than the primary side. Therefore, the feedback action can be picked off as a signal b in Fig. 6–43, with negligible energy effect. This leads to the discussion presented in Section 5–5.

In the previous paragraphs, we saw some situations in which the couplings became unilateral. A similar viewpoint applies, in particular, to

---

* This unit is called a proportional fluid amplifier to distinguish it from mono- or bistable type units where the Coanda effect (tendency of a power jet to stick to the wall) is utilized. By changing the geometry, we avoid (or delay) the Coanda effect in a proportional fluid amplifier.

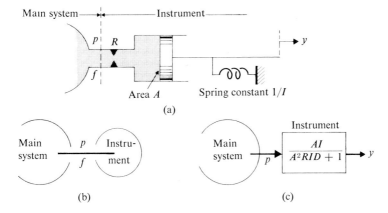

Fig. 6-44 Bilateral *vs.* unilateral measurement.

various *measuring problems*. For instance, if the outer gimbal of a gyroscope (Fig. 6-32) is fixed on a vehicle and the precession p_2 of the inner gimbal is restrained, measurement of the angular velocity p_1 of the vehicle reduces to measurement of the torque f_2. The system is called a *rate gyro*. Since the torque reaction of the instrument on the vehicle is negligible, the angular velocity measurement is unilateral. The opposite situation arises when a powerful gyroscope is installed on a vehicle as a direct stabilizer; the reaction torque acts as a stabilization, and the coupling is bilateral.

Let us consider, as a second example, the pressure indicator schematically shown in Fig. 6-44(a). Ignoring fluid compressibility, we write

$$\begin{bmatrix} p \\ f \end{bmatrix} = \begin{bmatrix} 1 & R \\ 0 & 1 \end{bmatrix} \begin{bmatrix} 0 & 1/A \\ A & 0 \end{bmatrix} \begin{bmatrix} 1 & 0 \\ 1/(ID) & 1 \end{bmatrix} \begin{bmatrix} Dy \\ 0 \end{bmatrix}.$$

Therefore, the instrument output (pressure indication) is given by

$$y = \left(\frac{AI}{A^2 RID + 1} \right) p.$$

and the input impedance of the instrument becomes

$$\frac{p}{f} = R + \frac{1}{A^2 ID} = Z(D).$$

Although $Z(D)$ thus determines the magnitude of the drain f, such flow may or may not affect the state of the main system. If it does, the coupling is bilateral (Fig. 6-44b); otherwise it is unilateral (Fig. 6-44c). Therefore, the value of $Z(D)$ relative to the main system determines the situation. Usually an impedance (mis)matching which yields a unilateral relation is preferred in this case.

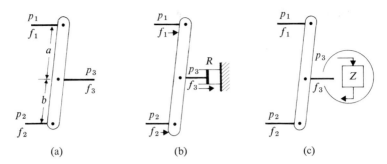

Fig. 6-45 Floating link.

Some gear trains and linkages have an interesting three-port structure. As discussed in Section 1-3, the following relation can be established by the link of Fig. 6-45(a) for the three linear motions indicated by arrows in the figure:

$$p_3 = \left(\frac{b}{a+b}\right)p_1 + \left(\frac{a}{a+b}\right)p_2.$$

This relation is used for addition or subtraction, such as shown in Fig. 5-30(b). If the arrows also indicate the direction of forces acting on the link, then the force and moment balance becomes

$$f_1 + f_2 + f_3 = 0, \qquad af_1 - bf_2 = 0.$$

The structure can be made into a two-port by closing any one of the three ports. For instance, if the variables of port 3 are related by an impedance Z [Fig. 6-45(b) and (c)], then

$$-p_3 = Zf_3.$$

Eliminating p_3 and f_3 from the above equations, we obtain a transfer matrix,

$$\begin{bmatrix} p_1 \\ f_1 \end{bmatrix} = \begin{bmatrix} -\dfrac{1}{r} & \dfrac{(1+r)^2}{r}Z \\ 0 & r \end{bmatrix} \begin{bmatrix} p_2 \\ f_2 \end{bmatrix}, \qquad (6\text{-}48)$$

where $r = b/a$. If the link were pivoted at the middle pin, $Z = 0$, and then the relation reduces to the transformer matrix, Eq. (6-27), where the minus sign due to the reversal of motion must be inserted.

A relation similar to Eq. (6-48) holds for differential gears. For instance, Fig. 6-46 shows the differential device of an automobile rear axle, where constraints to shafts are indicated by ball bearings. Since f_1 is uniquely related to f_2 (Eq. 6-48), no torque appears on one wheel if the other wheel gets off the ground. Similar gearings are used in servomechanisms for the purpose of addition or subtraction.

6-5   Multiport systems   257

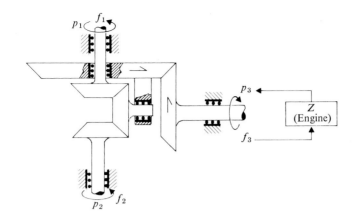

Fig. 6-46 Bevel-gear differential.

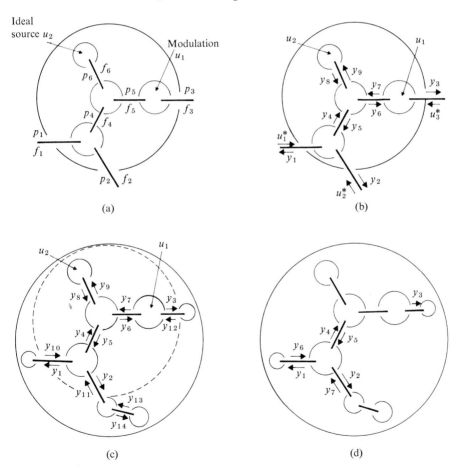

Fig. 6-47 Interaction of a system with its environment.

The effects of the R, C, and I of linkages may be combined with their main transfer matrix. Such treatment may yield an overall transfer characteristic where an interesting combination of displacement, inertia, and moment of inertia appear.

Let us close this chapter by summarizing briefly the approach to the dynamic-system representation presented here. Central to this approach is the idea of structural models based on bond graphs.

Consider, for example, the system bond graph given in Fig. 6–47(a), which has two unilateral inputs u_1, u_2, and three ports. If the system is linear, the variables may be related by

$$p_i = \sum_{j=1}^{3} P_{ij} f_j + \sum_{j=1}^{2} Q_{ij} u_j ,$$

$$f_k = \sum_{j=1}^{3} R_{kj} f_j + \sum_{j=1}^{2} S_{kj} u_j \qquad (6\text{–}49)$$

($i = 1, \ldots, 6$; $k = 4, 5, 6$), where P, Q, R, and S are expressed in terms of system constants and the derivative operator D. Causality is not considered. Therefore, P_{ij}, for example, may not be a realizable transfer function.

Suppose we consider the causality shown in Fig. 6–47(b). We then have a formulation in the form

$$y_i = \sum_{j=1}^{3} G_{ij}(D)^* u_j^* + \sum_{j=1}^{2} H_{ij}(D)^* u_j , \qquad i = 1, 2, \ldots, 9 , \qquad (6\text{–}50)$$

where $G_{ij}(D)^*$, $H_{ij}(D)^*$ are transfer functions; some of $H_{ij}(D)^*$ may agree with the corresponding Q_{ij} or S_{kj} in the preceding equation. The y's could be either p's or f's. If the total number of lumped C- and I-elements in the system is n^*, the state-space equation for an n^*-vector $\mathbf{x}^*$ becomes

$$D\mathbf{x}^* = \mathbf{A}^*\mathbf{x}^* + \mathbf{B}_1^*\mathbf{u}_1 + \mathbf{B}_2^*\mathbf{u}_2 ,$$

and $\qquad\qquad\qquad\qquad\qquad\qquad\qquad\qquad\qquad\qquad\qquad\qquad\qquad (6\text{–}51)$

$$\mathbf{y}^* = \mathbf{C}^*\mathbf{x}^* + \mathbf{D}_1^*\mathbf{u}_1 + \mathbf{D}_2^*\mathbf{u}_2 ,$$

where

$$\mathbf{u}_1 = \begin{bmatrix} u_1^* \\ u_2^* \\ u_3^* \end{bmatrix}, \qquad \mathbf{u}_2 = \begin{bmatrix} u_1 \\ u_2 \end{bmatrix}, \qquad \mathbf{y}^* = \begin{bmatrix} y_1 \\ \vdots \\ y_9 \end{bmatrix}.$$

Matrices $\mathbf{A}^*, \ldots, \mathbf{D}_2^*$ represent system constants.

Equations (6–50) and (6–51) are not complete in the sense that u_j^* may be related to the y_1, y_2, and y_3 through the environment. Although theoretically

$$D\mathbf{x}^* = \mathbf{A}^*\mathbf{x}^*$$

is the equation of a free system, it could be physically difficult to cut off some reaction u_j^*. If the boundary of the system is expanded so that the environment acts on the system unilaterally (Fig. 6–47c), we get

$$D\mathbf{x} = \mathbf{Ax} + \mathbf{Bu}, \qquad (6\text{–}52)$$

where $\mathbf{u} = \begin{bmatrix} u_1 \\ u_2 \end{bmatrix}$ becomes a unilaterally applicable vector input. If the total number of lumped C- and I-elements in the expanded system is n, then $\mathbf{x}$ must be an n-vector. The variables of our interest, in a vector form $\mathbf{y} = \begin{bmatrix} y_1 \\ \vdots \end{bmatrix}$, can be expressed by

$$\mathbf{y} = \mathbf{Cx} + \mathbf{Du}. \qquad (6\text{–}53)$$

As shown in Section 2–2, transfer functions $G_{ij}(D)$ in the scalar relation

$$y_i = \sum_j G_{ij} u_j \qquad (6\text{–}54)$$

can be derived from Eqs. (6–52) and (6–53). The system can be made free by letting

$$\mathbf{u} = \mathbf{0}.$$

The equation

$$D\mathbf{x} = \mathbf{Ax}$$

corresponds to the system of Fig. 6–47(d), which is completely isolated from its environment. A similar discussion holds for lumped-parameter non-linear systems.

REFERENCES

1. Rosenberg, R. C., "Computer-aided teaching of dynamic system behavior." *Ph. D. Thesis*, M. I. T., September 1965.
2. Paynter, H. M., *Analysis and Design of Engineering Systems*. Cambridge, Mass.: M. I. T. Press, 1961.
3. Marks, L. S., *Mechanical Engineers Handbook*. New York: McGraw-Hill, 1967.
4. Simson, A. K., "Gain characteristics of subsonic, pressure-controlled, proportional, fluid-jet amplifiers." *J. Basic Eng., Trans. ASME*, **88**, June 1966, pp. 295–305.
5. Raven, F. H., *Automatic Control Engineering*. New York: McGraw-Hill, 1961.
6. Lynch, W. A., Truxal, J. G., *Introductory System Analysis*. New York: McGraw-Hill, 1961.
7. Koenig, H. E., Blackwell, W. A., *Electromechanical Systems Theory*. New York: McGraw-Hill, 1961.

260   Bilaterally coupled systems

8. Shearer, J. L., Murphy, A. T., Richardson, H. H., *Introduction to Dynamic Systems*. Reading, Mass. Addison-Wesley, 1964.
9. Chorafas, D. N., *Systems and Simulation*. New York: Academic Press, 1965.
10. Koenig, H. E., et al., *Analysis of Discrete Physical Systems*. New York. McGraw-Hill, 1967.
11. Sagawa, K., "Analysis of the ventricular pumping capacity as a function of input and output pressure loads," Chapter 9 (pp. 141–149) in Saunders, W. B. (ed.), *Physical Basis of Circulatory Transport: Regulation and Exchange*. Reeve & Guyton, Philadelphia, Saunders, 1967.
12. Wood, J. E., "The venous system." *Scientific American*, 218, No. 1, January 1968, pp. 86–96.
13. Karnopp, D., Rosenberg, R. C., *Analysis and Simulation of Multiport Systems*. Cambridge, Mass.: MIT Press, 1968.
14. Cannon Jr., R. H., *Dynamics of Physical Systems*. New York: McGraw-Hill, 1967.

PROBLEMS

6-1 Determine the damper "resistance" R, for both $p = $ velocity and $p = $ force analogy, in terms of the net area A and fluid resistance R_f shown in Fig. P6-1. Assume that there is no leakage, no friction other than linear restriction R_f, and that the fluid is massless and incompressible. Compute the rate of energy dissipation.

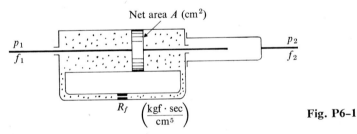

Fig. P6-1

6-2 According to the theory of elastic deformation, the following relation holds for a circular, thin-wall elastic pipe of outside diameter d_o and inside diameter d_i:

$$dA = \frac{A}{E(d_o/d_i - 1)} dp ,$$

where $E = $ Young's modulus of the pipe material, $dA = $ change in area due to internal pressure change dp (see Fig. P6-2a). (a) Obtain the capacitance of the fluid system per unit length of line due to this effect. (b) The left end of an elastic pipe of total length l is closed, and there is a leak through a linear resistor R at the right end (see Fig. P6-2b). Initial pressure of the fluid in the pipe is p_0. Determine the pressure transient $p(t)$.

6-3 A frictionless pipe with dimensions $A_1 = 4 \text{ cm}^2$ and $l = 200 \text{ cm}$ is connected to a water tank of cross-sectional area $A_2 = 200 \text{ cm}^2$ (see Fig. P6-3). Considering the

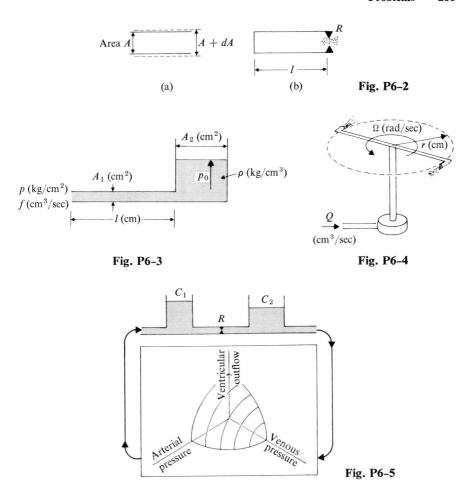

Fig. P6-2

Fig. P6-3

Fig. P6-4

Fig. P6-5

pipe as a "single lump," find the impedance $Z(D) = p/f$. Draw the block diagram and bond graph for the lumped system. Determine the pattern of level change for a step-input pressure change.

6-4 The water flow rate Q (cm³/sec) into a sprinkler (or water turbine) system is kept constant. The tangential jet velocity relative to the nozzle is v (cm/sec), and its radius is r (cm); see Fig. P6-4. Let ρ (kgm/cm³) = density of water, Ω (rad/sec) = angular velocity of rotor, and f (kgf·cm) = torque generated by the jet. Determine the torque–speed characteristic, and find the speed for maximum power.

6-5 The heart as a pump can be represented by a characteristic surface shown in Fig. P6-5. [11]. The ventricular pumping capacity is sensitive to both mean aortic pressure and arterial pressure, where the mean pressure approach holds for frequencies lower than about 1/5 cps. Assuming the lumped model for compliance of arteries and veins (with time constants of about 20 to 60 sec), and lumped resistance for capillaries, as shown in Fig. P6-5, sketch a causal block diagram of the overall system of circulation (see also [12]).

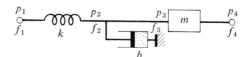

Fig. P6-6

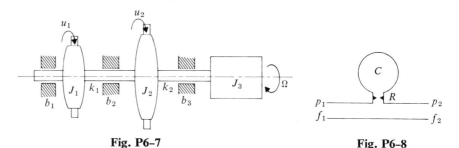

Fig. P6-7

Fig. P6-8

6-6 A spring–mass–damper system is shown in Fig. P6-6, where k = spring constant, b = damper constant (such that force = $b \cdot$ velocity) and m = mass. Replacing k, b, and m by I, R, and C for the potential–velocity analogy, and C, R, and I for the potential–force analogy, obtain (a) block diagram, (b) bond graph, and (c) transfer matrix representations.

6-7 A two-stage turbine rotor, shown in Fig. P6-7, has system constants

moments of inertia: $J_1 = J_2 = J_3 = 1$
linear resistances: $b_1 = b_2 = b_3 = 1$
spring constants (shaft elasticity): $k_1 = k_2 = 1$

Find transfer functions $G_1(D)$ and $G_2(D)$ which relate the torque at each stage, u_1 and u_2, to the angular velocity Ω of the load by an equation

$$\Omega = G_1(D)u_1 + G_2(D)u_2 .$$

Assume that the linear damping coefficients b_2 and b_3 may be halved and lumped at J_1, J_2, and J_3, respectively.

6-8 Obtain (a) the transfer matrix, (b) the block diagram and bond graph, and (c) the electrical-circuit analogy of the side-capacitance fluid system shown in Fig. P6-8.

6-9 Derive the input–output relation for the composite one-port system of Table 6-4(a).

6-10 A level h is measured by a float. Let m be the mass of the float, a its area, ρ the density of the liquid, and y the position of the float. (a) Relate y and h by a transfer function $G(D)$ when the area of the liquid surface is infinite (Fig. P6-10a). (b) Let A = area of liquid tank, f = rate of volumetric liquid supply (see Fig. P6-10b), and obtain the system block diagram. Assume $y = h$ for an equilibrium state of the float relative to the liquid level.

6-11 A hydraulic jack (fluid coupling or mechanical transformer) is shown in Fig. P6-11(a). Determine the **M**-matrix of the overall system. A surge tank (side capacitance) is inserted in the fluid circuit, and spring-mass effects are added on the output side of the system in Fig. P6-11(b). Obtain the new **M**-matrix. Is the second

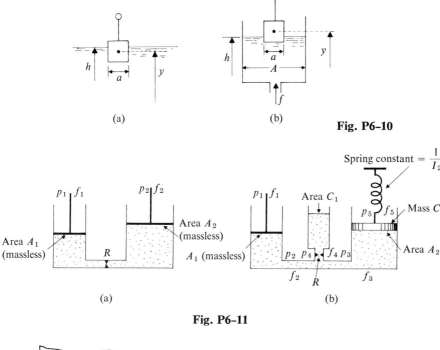

Fig. P6–10

Fig. P6–11

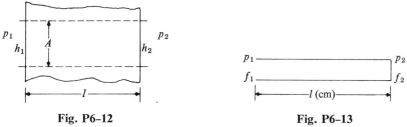

Fig. P6–12 **Fig. P6–13**

system observable when $A_1 = A_2 = 1$ and output piston speed is considered the response? To avoid gyration in transduction matrices, use the potential–force analogy all through this problem.

6-12 A heat-conducting wall is shown in Fig. P6–12. The system constants are: A (cm^2) = area, l (cm) = thickness, k (kcal/(cm·sec·°C)) = thermal conductivity, ρ (kgm/cm^3) = density of wall material, c (kcal/(kgm·°C)) = specific heat of the material, and h_1, h_2 (kcal/(cm^2·sec·°C)) = film coefficients on the surfaces. Let p_1, p_2 (°C) = fluid temperatures, p (°C) = wall temperature of a one-lump T-model. Find the transfer functions $G_1(D) = p/p_1$, $G_2(D) = p/p_2$. The result must be given in terms of the characteristic time $T_c = (l^2 \rho c)/k$ (sec) and the Biot numbers $Bi_1 = h_1 l/k$, $Bi_2 = h_2 l/k$.

6-13 A fluid of density ρ (kgm/cm^3) and bulk modulus β (kgf/cm^2) fills a uniform fluid line. The right end of the line is closed (see Fig. P6–13). Determine the transfer function $G(D) = p_2/p_1$ and sketch the pattern of a unit-step input response.

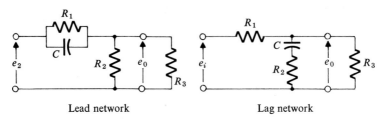

Lead network   Lag network

Fig. P6-14

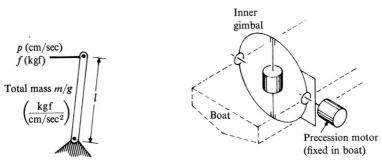

Fig. P6-15   **Fig. P6-16**

6-14 Determine the transfer function $G(D) = e_0/e_i$ of the networks shown in Fig. P6-14. These are lead and lag networks with a load represented by a resistor R_3. Also obtain the current which must flow through the left-hand terminal for a unit-step input voltage change when the initial capacitor charge is zero, and when there is no load (that is, $R_3 = \infty$).

6-15 Determine the impedance of a uniform rigid swing arm shown in Fig. P6-15. Assume that air resistance ("whipping loss") is negligible.

6-16 A very large gyro (antirolling gyro) is installed on a boat as shown in Fig. P6-16. Rolling angle θ is detected by a pilot gyro (not shown in the figure), amplified, and fed into a precession motor that tips the main gyro axis fore and aft according to the following relation:

$$p_2 = k\theta, \quad k = \text{const, and } p_2 \text{ is as defined in Fig. 6-32(c).}$$

Determine the transfer function $G(D) = p_1/u$, where $p_1 =$ angular velocity of rolling, and $u =$ rolling torque acting on the boat. Assume that $C_b =$ moment of inertia of the boat for rolling motion, and $C_p =$ moment of inertia of the boat for pitching $= \infty$. Sketch a block diagram of the system.

6-17 The plane flight shown in Fig. P6-17 is horizontal. The outer gimbal of an instrument gyro is fixed on the plane, while the inner gimbal is connected to the plane (wing in the figure) via a spring. The friction at the pin joint of the gimbals is represented by R. Angle y of the inner gimbal (not shown in the figure) is the output;

$$y = \frac{1}{D} p_2 \quad [\text{see Fig. 6-32(b) for } p_2].$$

What does y represent? Obtain the block diagram of the system.

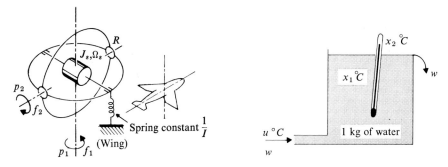

Fig. P6-17

Fig. P6-18

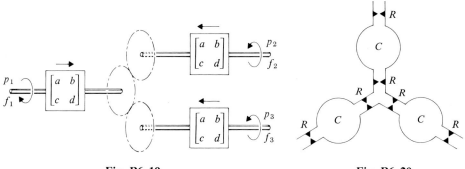

Fig. P6-19

Fig. P6-20

6-18 Temperature x_1 (°C) of an agitated tank [$C_1 = 1$ (kcal/°C)] is measured by a thermometer of a lumped capacitance $C_2 = 0.002$ (kcal/°C); see Fig. P6-18. The product of flow rate and specific heat of the water is $w = 0.1$ (kcal/sec·°C). The surface conductance of the thermometer is $Ah = 1/3600$ (kcal/sec·°C). There is no heat loss other than the carry-over of the outflow. Show that the temperature measurement is accomplished almost unilaterally.

6-19 Three identical shaft systems are coupled by an ideal gearing of 1 : 1 : 1 ratio, as sketched in Fig. P6-19. Assuming that $ad - bc = 1$, determine the matrix **Z** such that $\mathbf{p} = \mathbf{Zf}$.

6-20 Figure P6-20 shows a star mode of three identical pressure systems. Obtain **Z** for $\mathbf{p} = \mathbf{Zf}$.

7
DISTRIBUTED-PARAMETER SYSTEMS

It is sometimes necessary to consider variations in the state of a system with respect to time *and* distance. The practice of lumping, used extensively in Chapters 5 and 6, is equivalent to viewing a system as a set of discrete point functions, which can be described by ordinary differential equations. If, however, we need to know the continuous variation of variables with position in order to obtain an accurate description of a system, we must use a distributed-parameter model.

All systems considered in this chapter are linear, uniform, and one dimensional. The models describe the dynamic behavior of such systems and processes as fluid, mechanical, or electrical transmission lines, heat or mass diffusion and percolation processes, heat exchangers and distillation columns. The main variables of such systems and processes (velocity, flow, pressure, force, temperature, concentration, etc.) are functions of time t and position z along the line. We use the transfer matrix approach, which gives first-order vector differential equations with respect to the space coordinate, through most of the chapter. For many practically important problems, this approach leads to computational formulations which contain time-delay dynamic operators (as contrasted to the computational formulations for lumped systems, which contain integrators). Because digital computer simulation of time delays is straightforward, digital simulation techniques are effective for many distributed-parameter systems [4].

By using the transfer matrix approach we obtain descriptions of distributed-parameter systems as seen from the ends. Considering natural causalities at the ends, we can derive transfer functions for terminal input–output relations. However, since these transfer functions are irrational, it is generally difficult to use inverse transformation to get transient response. Frequency response, on the other hand, can be computed directly by replacing D or s in the transfer function with $j\omega$ (see Chapter 9). The classical method of solution, applicable only to simple problems, is demonstrated at the end of the chapter.

7-1 SOME CHARACTERISTICS AND EXAMPLES OF DISTRIBUTED-PARAMETER SYSTEMS

When is a system "distributed"? Or, more accurately, when should we use a distributed-parameter model to describe a system? We asked ourselves this question when trying to obtain a model for the liquid-filled line connecting the reservoir to the tank in Example 5-1. The answer, in that case, depended on the time period and the "fineness" of observation we were interested in (or alternatively, the frequency range, see Chapter 9). In Example 5-1 we were able to derive a finite-order transfer function for the line, but the response predicted by this transfer function would not be accurate in the time period immediately following a system disturbance. Since any physical system displays distributed characteristics if the input is varied rapidly enough, the sensitivity of interconnected systems to rapid changes and the precision we require in observing the system determine the type of model which must be used.

We generally designate an element as "distributed" if the state of the element depends on the value of a variable(s) as a function both of position within the element and of time. The mathematical model for such an element must be a partial differential equation, since there are two independent variables—position and time. We can eliminate differentiation with respect to time by taking either a Laplace or Fourier transform and, usually, derive a transcendental transfer function to describe the port-to-port relations of the element. If, however, it is possible to deduce a lumped microstructure for an element, we saw in Chapter 6 that a finite-order lumped structure can be constructed which will approximate a distributed element *to any desired degree of accuracy*. To achieve increased accuracy, more and smaller lumped-parameter elements must be added to the model. Under these circumstances it is obviously possible to use either a lumped- or distributed-parameter model for a particular system, and the choice must be based on the observation requirements and other factors such as computing efficiency. For example, if only one element in a system has distributed characteristics, it may be worthwhile to model it with even a relatively complex lumped-parameter model so that the whole system can be modeled as a finite-order lumped system. As we will see later in this chapter, it is generally easier to analyze a homogeneous system model which contains either all lumped or all distributed elements than it is to analyze a mixed model.

Example 7-1 The pure, unilateral, time-delay element has the transfer function:

$$G(s) = e^{-Ts}.$$

The time response, calculated from the inverse Laplace transform, is:

$$y(t) = u(t - T),$$

where $u(t)$ is the input function. The response to a ramp input $u(t) = kt$ is shown in Fig. 7-1(a). The transfer function for the time delay can be expressed in series form as:

$$G(s) = \frac{1}{1 + Ts + T^2 s^2/2 + T^3 s^3/3! + \cdots}.$$

Let us look at the first-order lag as an approximation to the actual system. Its

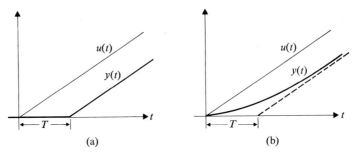

Fig. 7-1 Responses of a pure delay and a first-order lag to a ramp input.

transfer function $G(s) = 1/(1 + Ts)$ has the same first two terms as the original, and therefore represents an approximation by series truncation. With a ramp input $U(s) = k/s^2$, the system equation in the Laplace domain is:

$$Y(s) = \frac{1}{1 + Ts} \cdot \frac{k}{s^2}.$$

Taking the inverse transform, we have

$$y(t) = kT(e^{-t/T} - 1 + t/T).$$

This response is shown in Fig. 7-1(b). Depending on the circumstances, this might be a satisfactory model.

The shunt admittance and series impedance concepts can be applied to one-dimensional distributed systems to derive a vector partial differential equation which serves as a standard form for many problems. For a differential length of line dz, the potential and flow at the ends are related by:

$$\begin{bmatrix} p(z,t) \\ f(z,t) \end{bmatrix} = \begin{bmatrix} Y^*Z^*(dz)^2 + 1 & Z^*\,dz \\ Y^*\,dz & 1 \end{bmatrix} \begin{bmatrix} p(z,t) + \frac{\partial p}{\partial z}dz \\ f(z,t) + \frac{\partial f}{\partial z}dz \end{bmatrix}$$

$$= \begin{bmatrix} 1 & Z^*\,dz \\ Y^*\,dz & 1 \end{bmatrix} \begin{bmatrix} p(z,t) + \frac{\partial p}{\partial z}dz \\ f(z,t) + \frac{\partial f}{\partial z}dz \end{bmatrix}, \tag{7-1}$$

where the product $Y^*Z^*(dz)^2$ has been neglected. Note that since we are concerned with only a differential length of line, the distinction between the YZ-structure and the ZY-structure disappears. For a uniform line, Y and Z do not vary along the line, and we can interpret Y^* and Z^* as admittance or impedance *per unit length* of line (Fig. 7-2). In scalar form Eq. (7-1)

7-1 Characteristics of distributed-parameter systems

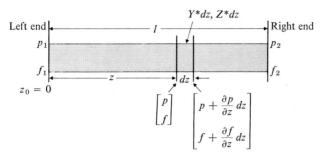

Fig. 7-2 Uniformly distributed YZ lines.

becomes

$$p(z, t) = p(z, t) + \frac{\partial p}{\partial z} dz + Z^* f(z, t) dz + Z^* \frac{\partial f}{\partial z} (dz)^2 ,$$

$$f(z, t) = Y^* p(z, t) dz + Y^* \frac{\partial p}{\partial z} (dz)^2 + f(z, t) + \frac{\partial f}{\partial z} dz .$$

Again, neglecting terms containing $(dz)^2$, we obtain:

$$\frac{\partial p(z, t)}{\partial z} = -Z^*(D) f(z, t) , \qquad \frac{\partial f(z, t)}{\partial z} = -Y^*(D) p(z, t) , \qquad (7\text{-}2)$$

or, in the matrix form which we shall use later,

$$\frac{\partial}{\partial z} \begin{bmatrix} p(z, t) \\ f(z, t) \end{bmatrix} = - \begin{bmatrix} 0 & Z^*(D) \\ Y^*(D) & 0 \end{bmatrix} \begin{bmatrix} p(z, t) \\ f(z, t) \end{bmatrix} . \qquad (7\text{-}3)$$

Eliminating one of the two variables, we can also write Eq. (7-2) as

$$\frac{\partial^2 p(z, t)}{\partial z^2} = Y^*(D) Z^*(D) p(z, t) , \qquad \frac{\partial^2 f(z, t)}{\partial z^2} = Y^*(D) Z^*(D) f(z, t) . \quad (7\text{-}4)$$

We have as yet made no assumptions about the nature of the functions $Y^*(D)$ and $Z^*(D)$. Equations (7-3) and (7-4) can be applied equally to the relatively simple systems shown in Tables 6-6 and 6-7 and, for example, to a uniform dispersive fluid line as done by Nichols [6] and Brown [7]. This fluid-line model, which includes boundary-layer effects, leads to complex expressions for Y and Z for which no simple microstructures exist.

Example 7-2 Heat conduction in a uniform rod can be modeled with a shunt capacitance-series resistance model (Fig. 7-3). Since we are dealing with a differential-sized microelement, we can use either of the structures shown in the figure. This structure is described by:

$$Z^* = R^* \quad \text{and} \quad Y^* = C^* D ,$$

Fig. 7-3 Models for heat conduction in a uniform rod.

where R^* is the heat resistance along the rod and C^* is the heat capacitance. In this analogy the potential is temperature θ, and the flow is heat flow Q. Substituting into Eq. (7-2), we get:

$$\frac{\partial \theta}{\partial z} = -R^*Q, \qquad \frac{\partial Q}{\partial z} = -C^* \frac{\partial \theta}{\partial t},$$

and from Eq. (7-4) we get the partial differential equation of heat flow:

$$\frac{\partial^2 \theta}{\partial z^2} = C^*R^* \frac{\partial \theta}{\partial t}.$$

We will now look at a class of one-dimensional systems in which power is transmitted in more than one mode. The uniform beam loaded under its own weight (or with a uniform applied load) is a typical example. We will consider two modes of power transmission: lateral deflection due to shear, and bending. To derive the differential equation, consider the force and moment balance in a dz-element of the beam (Fig. 7-4). The force balance in the vertical direction is

$$dS + \rho \frac{\partial^2 y}{\partial t^2} dz = 0$$

or

$$\frac{\partial S}{\partial z} = -\rho \frac{\partial^2 y}{\partial t^2}, \qquad (7\text{-}5)$$

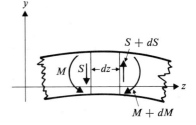

Fig. 7-4 Elastic beam.

where S is the shear force, ρ is the mass per unit length, y is lateral deflection, and $\rho(\partial^2 y/\partial t^2)\,dz$ is the inertia force of a dz-section.

The moment balance about any point on the right-hand face of the element is

$$dM - S\,dz = 0$$

or

$$\frac{\partial M}{\partial z} = S, \qquad (7\text{-}6)$$

where M is the bending moment. M in Eq. (7-6) is related to the curvature of the beam by the flexure equation

$$M = EJ\frac{\partial^2 y}{\partial z^2},\qquad(7\text{--}7)$$

where E is Young's modulus and J is area moment of inertia of the beam's cross section. Eliminating S and M from these relations, we obtain

$$EJ\frac{\partial^4 y}{\partial z^4} + \rho\frac{\partial^2 y}{\partial t^2} = 0.\qquad(7\text{--}8)$$

If we are interested in sinusoidal motion of the beam (see Chapter 9), the substitution $D \to j\omega$ yields

$$EJ\frac{\partial^4 y}{\partial z^4} - \omega^2 \rho y = 0.$$

This equation can be normalized to

$$\frac{\partial^4 y}{\partial \eta^4} - y = 0,\qquad(7\text{--}9)$$

where $\eta = bz$ = dimensionless distance, and $b = (\omega^2 \rho/EJ)^{1/4}$. This system will be discussed further in Section 7–3.

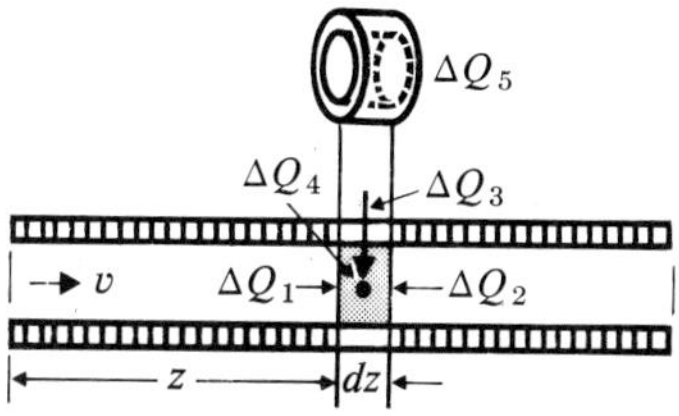

Fig. 7–5 Heat percolation process.

An important and interesting class of systems involves propagation of energy by a carrier flow. Figure 7–5 shows a coupled thermal and flow process where heat can be transferred from the fluid to the walls. Such a process would take place when, for example, the hot water is turned on in a shower and heat is transferred from the hot water to the pipes. A process of this nature, with either heat or material exchange, is called a percolation, and was first analyzed by P. Profos [16]. A heat balance for the liquid in the dz-section is

$$\Delta Q_4 = \Delta Q_1 - \Delta Q_2 + \Delta Q_3,$$

where

ΔQ_4 = heat stored in the fluid = $C^* \cdot dz \cdot (\partial \theta/\partial t) \cdot dt$,
C^* = heat capacitance of fluid per unit length,
θ = fluid temperature, assumed uniform in all transverse sections,
ΔQ_1 = heat carried in by fluid flow = $vC^*\theta\, dt$,
v = velocity of flow, assumed constant and uniform in a transverse section,
ΔQ_2 = heat carried over to the next section = $vC^*(\theta + (\partial\theta/\partial z)\,dz)\,dt$,
ΔQ_3 = heat transferred from the wall = $(dy/R^*)(\phi - \theta)\,dt$,
R^* = surface resistance per unit length,
ϕ = wall temperature.

Substituting these relations into the heat-balance equation, we obtain

$$\frac{\partial \theta}{\partial t} + v \frac{\partial \theta}{\partial z} = \frac{1}{R^*C^*}(\phi - \theta), \qquad (7\text{--}10)$$

where R^*C^*, which has the dimension of time, is called a time constant. Introducing the dimensionless time $\tau = t/L$ and dimensionless distance $\eta = z/l$, where $l = $ total length of the line, and $L = l/v = $ dead time, we can normalize Eq. (7–10) and obtain

$$\frac{\partial \theta}{\partial \tau} + \frac{\partial \theta}{\partial \eta} = a(\phi - \theta), \qquad (7\text{--}11)$$

where $a = l/(R^*C^*v) = L/(C^*R^*)$ is also dimensionless.

Metals are good heat conductors; hence the resistance of the pipe in the radial (or transverse) direction is usually negligible. The temperature gradient in the direction of fluid flow is generally not steep; heat conduction in that direction can be ignored. Assuming complete heat insulation outside the metal wall, we write the heat balance for the solid as follows:

$$\Delta Q_5 = -\Delta Q_3,$$

where

$\Delta Q_5 = $ heat stored in the solid per unit length
$\qquad = C_s^* \, dy \, (d\phi/dt) \, dt,$
$C_s^* = $ heat capacitance of solid per unit length.

After a rearrangement of terms, we obtain

$$\frac{\partial \phi}{\partial t} = \frac{1}{R^*C_s^*}(\theta - \phi) \qquad (7\text{--}12)$$

or, in the dimensionless form,

$$\frac{\partial \phi}{\partial \tau} = b(\theta - \phi), \qquad (7\text{--}13)$$

where $b = L/(C_s^*R^*)$ is dimensionless. We shall discuss this system in Section 7–5.

Equation (7–10) with slight modification in symbols applies to the heating coil of a mixing-type heat-exchanger shown in Fig. 7–6 [12]:

$$\frac{\partial \theta_h}{\partial t} + v_h \frac{\partial \theta_h}{\partial z} = \frac{1}{R_h^*C_h^*}(\phi - \theta_h), \qquad (7\text{--}14)$$

where the suffix h means "hot" (or tube-side) fluid. A term for the heat flow from the wall to the "cold" (or outside) fluid of temperature θ_c must

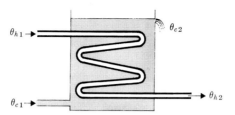

Fig. 7-6 Mixing-type heat exchanger.

be added to the right-hand side of Eq. (7–12). Therefore,

$$\frac{\partial \phi}{\partial t} = \frac{1}{R_h^* C_s^*}(\theta_h - \phi) + \frac{1}{R_c^* C_s^*}(\theta_c - \phi), \tag{7-15}$$

where c is the suffix for the cold side. Finally, Eq. (5–38) holds for the mixed fluid of temperature θ_{c2};

$$C_c \frac{d\theta_{c2}}{dt} = \frac{l}{R_c^*}(\phi_{av} - \theta_{c2}) + w_c(\theta_{c1} - \theta_{c2}), \tag{7-16}$$

where C_c is total heat capacitance of the mixed fluid, w_c is product of specific heat and flow rate of the fluid, and l is net total length of the heating tube. The average temperature of the tube wall is

$$\phi_{av} = \frac{1}{l}\int_0^l \phi(z)\,dz. \tag{7-17}$$

If heat capacitance of the wall C_s^* is negligible relative to heat capacitances of the fluids, Eq. (7–15) becomes static,

$$\phi(z) = \frac{R_c^* \theta_h + R_h^* \theta_c}{R^*}, \tag{7-18}$$

where $R^* = R_h^* + R_c^*$ is the overall resistance across the heating surface (per unit length) (Eq. 5–27) and

$$\begin{aligned}
\frac{\partial \theta_h}{\partial t} + v_h \frac{\partial \theta_h}{\partial z} &= \frac{1}{C_h^* R^*}(\theta_c - \theta_h), \\
C_c \frac{d\theta_{c2}}{dt} &= \frac{1}{R^*}\int_0^l \theta_h\,dz - \left(\frac{l}{R^*}\right)\theta_{c2} + w_c(\theta_{c1} - \theta_{c2}).
\end{aligned} \tag{7-19}$$

Introducing the dimensionless distance $\eta = z/l$ and dimensionless time $\tau = t/L$, where $L = l/v_h$, we may normalize Eq. (7–19) as follows:

$$\begin{aligned}
\frac{\partial \theta_h}{\partial \tau} + \frac{\partial \theta_h}{\partial \eta} &= a_h(\theta_{c2} - \theta_h), \\
\frac{d\theta_{c2}}{d\tau} &= a_c\left(\int_0^1 \theta_h(\eta)\,d\eta - \theta_{c2}\right) + \frac{1}{r}(\theta_{c1} - \theta_{c2}),
\end{aligned} \tag{7-20}$$

Distributed-parameter systems

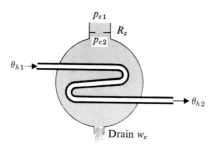

Fig. 7-7 Vapor condenser.

where $a_h = l/(v_h C_h^* R^*) = L/(C_h^* R^*)$, $a_c = l^2/(R^* C_c v_h) = lL/(R^* C_c)$, and $r = C_c/(L w_c)$.

A set of equations similar to Eqs. (7-19) approximately describe a vapor condenser (Fig. 7-7) where the tube flow is a cooling medium. The equation for the tube flow in (7-19) needs no change, but the equation for the shell-side fluid must be modified to account for the condensation. The rate of heat absorption by the coolant is

$$Q = \left(\frac{l}{R^*}\right) \theta_{c2} - \frac{1}{R^*} \int_0^l \theta_h(z) \, dz \,.$$

Therefore the rate of condensation is given by

$$w_c = Q/H \,,$$

where H is the latent heat of the condensing vapor. We can look upon the vapor side as a pressure system with negligible heat capacitance, and write the material balance:

$$\frac{p_{c1} - p_{c2}}{R_z} - w_c = C_v \frac{dp_{c2}}{dt} \,,$$

where R_z is lumped resistance as shown in Fig. 7-7, and C_v is the capacitance for a pressure system (Eq. 5-26). Combining these three equations, and approximating the saturation curve by a linear relation $\theta_{c2} = k_c p_{c2}$, where k_c is a constant to be determined for a given vapor, we obtain

$$C_v \frac{dp_{c2}}{dt} = \frac{1}{R^* H} \int_0^l \theta_h \, dz - \frac{k_c}{R^* H} + \left(\frac{1}{R_z}\right) p_{c2} + \frac{p_{c1}}{R_z} \,. \qquad (7\text{-}21)$$

This equation has a form very similar to Eq. (7-16).

The process of heating material on a conveyor (Fig. 7-8) is also a percolation process to which Eq. (7-10) applies:

$$\frac{\partial \theta}{\partial t} + v \frac{\partial \theta}{\partial z} = \frac{1}{R^* C^*} (u - \theta) \,, \qquad (7\text{-}22)$$

where v = conveyor speed, θ = temperature of heated layer on conveyor

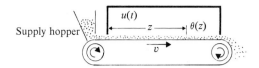

Fig. 7-8 Conveyor system.

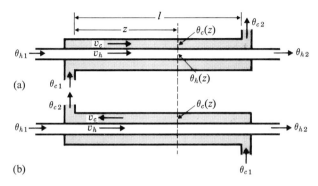

Fig. 7-9 Parallel-flow (a) and counterflow (b) heat exchangers.

(assumed thin) and $u =$ uniform temperature in oven unilaterally imposed on the system. If the solid layer (or tube wall in heat percolation) is thick, and lumping in the transverse direction becomes impossible, we can apply Eq. (6–25) with proper boundary conditions on both sides.

It is interesting to note that a new coordinate,

$$\zeta = t + z/v, \tag{7-23}$$

will combine the left-hand terms of Eq. (7–22) into a single term:

$$\frac{d\theta}{d\zeta} = \frac{\partial \theta}{\partial t} \cdot \frac{\partial t}{\partial \zeta} + \frac{\partial \theta}{\partial z} \cdot \frac{\partial z}{\partial \zeta} = \frac{\partial \theta}{\partial t} + v \frac{\partial \theta}{\partial z}. \tag{7-24}$$

$d/d\zeta$ is called the *hydrodynamic time derivative*. It is the rate of property change of a particle on the conveyor (that is, the change as viewed by an observer moving with a particular particle). A similar but somewhat complicated formulation holds for a drying process via a conveyor [10].

We can derive equations similar to (7–10) for the heating and cooling flows in tubular heat exchangers. For a parallel-flow heat exchanger (Fig. 7–9a) with negligible wall capacitance, let $R^* =$ overall tube resistance per unit length, $C^* =$ heat capacitance of a fluid per unit length of tube, and $v =$ fluid velocity. We now write the pair of equations:

$$\begin{aligned}\frac{\partial \theta_h}{\partial t} + v_h \frac{\partial \theta_h}{\partial z} &= \frac{1}{R^* C_h^*}(\theta_c - \theta_h), \\ \frac{\partial \theta_c}{\partial t} + v_c \frac{\partial \theta_c}{\partial z} &= \frac{1}{R^* C_c^*}(\theta_h - \theta_c),\end{aligned} \tag{7-25}$$

where the subscripts h and c designate tube side and shell side, respectively. The direction of the shell-side flow is reversed in a counterflow heat exchanger (Fig. 7-9b). As a consequence, we must change the sign of v_c in Eqs. (7-25) for counterflow heat exchangers:

$$\frac{\partial \theta_h}{\partial t} + v_h \frac{\partial \theta_h}{\partial z} = \frac{1}{R^* C_h^*} (\theta_c - \theta_h),$$

$$\frac{\partial \theta_c}{\partial t} - v_c \frac{\partial \theta_c}{\partial z} = \frac{1}{R^* C_c^*} (\theta_h - \theta_c).$$
(7-26)

These equations may be normalized as follows:

$$\frac{\partial \theta_h}{\partial \tau} + \frac{\partial \theta_h}{\partial \eta} = a_h (\theta_c - \theta_h),$$

$$r \frac{\partial \theta_c}{\partial \tau} \pm \frac{\partial \theta_c}{\partial \eta} = a_c (\theta_h - \theta_c),$$
(7-27)

where $\tau = t/L_h$, $\eta = z/l$, $L_h = l/v_h$, $L_c = l/v_c$, $r = L_c/L_h$, $a_h = L_h/(R^* C_h^*)$, $a_c = L_c/(R^* C_c^*)$; the minus in $\pm$ is for counterflow, and plus is for parallel-flow. If the heat-storage effect of the structure (tube and/or shell) is not negligible, we modify Eq. (7-25) or (7-26) to the form of Eqs. (7-14) and (7-15).

The distillation of a binary mixture in a column [1] yields a formulation which is similar to Eq. (7-26). Vapor and liquid streams with molar flow rates w_v and w_l, respectively, flow against the current in a column (with a sufficiently large number of plates). Material exchange through an interface takes place at a rate

$$f(\theta_l, \theta_v) \approx a\theta_l + b\theta_v + c,$$

where θ_l and θ_v are the concentrations of volatile components in the liquid and vapor, respectively, and a, b, c are approximately constant. The material balance yields the following relations:

$$C_l^* \frac{\partial \theta_l}{\partial t} = w_l \frac{\partial \theta_l}{\partial z} - f(\theta_l, \theta_v),$$

$$C_v^* \frac{\partial \theta_v}{\partial t} = -w_v \frac{\partial \theta_v}{\partial z} + f(\theta_l, \theta_v)$$
(7-28)

for the liquid and vapor, respectively, where the distance z is measured upward. C_l^* and C_v^* in the equations are the molar storages of liquid and vapor per unit length, respectively.

Highway traffic flow is quite different from fluid flow. A basic difference stems from the fact that a driver tends to follow the car ahead of him and, as a consequence, is governed by an oriented signal. According to the "car following" theory [17], the acceleration $Df_{n+1}(t + L)$ of the $(n + 1)$th car at time $t + L$ is approximately proportional to the difference between his speed at time t, $f_{n+1}(t)$, and the speed of the car ahead of him, $f_n(t)$. Thus,

$$Df_{n+1}(t + L) = k(f_n(t) - f_{n+1}(t)), \quad (7\text{-}29)$$

where k in the equation is the response coefficient of the $(n + 1)$th driver, and L is the delay in his response. Equation (7-29) or a more elaborate law of driving, programmed on a digital computer with proper bounds on acceleration and speed (which is always positive), with the probabilistic distribution of k and L, can simulate a platoon of cars on a highway (without any lane change) and yield such information as wave propagation, collision possibility, etc. A qualitative demonstration of the wave propagation phenomenon is possible if we use an extremely simplified distributed-parameter model:

$$Df(t) = kh \frac{\partial f}{\partial z}, \quad (7\text{-}30)$$

where f is a small perturbation in car speed, h the normal distance from the front bumper to the front bumper of the next car, and where L in Eq. (7-29) is ignored. The derivative on the right-hand side of Eq. (7-30) is obtained by noting that

$$\frac{f_n(t) - f_{n+1}(t)}{h} \approx \frac{\partial f}{\partial z}.$$

Since D on the left-hand side of Eq. (7-30) is the hydrodynamical time derivative [$d/d\zeta$ in Eq. (7-24)], the equation can be rewritten in the following form:

$$\frac{\partial f}{\partial t} + v \frac{\partial f}{\partial z} = kh \frac{\partial f}{\partial z}$$

or

$$\frac{\partial f}{\partial t} + (v - kh) \frac{\partial f}{\partial z} = 0, \quad (7\text{-}31)$$

in which v is the mean speed of the platoon. This is the equation of a delay-line†, and kh is the speed of a wave traveling backward. If k is 0.5 sec^{-1} and $h = 20$ ft, the velocity of the wave relative to the platoon is 6.82 mph.

---

† The equation $\partial x/\partial t + v \partial x/\partial z = 0$ describes a signal x moving in a line with velocity v. Solving $Dx + v\, dx/dz = 0$ for x, we obtain

$$x(z, t) = e^{-(z/v)D} x(0, t),$$

which gives the transfer function (5-46) for a dead time $T = z/v$.

7-2 UNIFORM, TWO-PORT TRANSMISSION LINE

The uniform transmission line is of great interest in many fields. Electrical transmission lines, fluid-filled pipes and tubes, and elastic shafts are all examples of uniform transmission lines. Equation (7–3) provides a general mathematical basis for the analysis of uniform transmission lines. Restated in vector form, Eq. (7–3) is

$$\frac{\partial \mathbf{q}(z, t)}{\partial z} = -\mathbf{R}\mathbf{q}(z, t), \qquad (7\text{--}32)$$

where

$$\mathbf{q}(z, t) = \begin{bmatrix} p(z, t) \\ f(z, t) \end{bmatrix}, \qquad \mathbf{R} = \begin{bmatrix} 0 & Z^*(D) \\ Y^*(D) & 0 \end{bmatrix}.$$

This equation has the solution (see Chapter 3):

$$\mathbf{q}(z, t) = e^{-\mathbf{R}(z-z_0)} \mathbf{q}(z_0, t)$$

or $\qquad (7\text{--}33)$

$$\mathbf{q}(z_0, t) = e^{\mathbf{R}(z-z_0)} \mathbf{q}(z, t).$$

Since **R** has a simple form, its products are also simple;

$$\mathbf{R}^2 = \begin{bmatrix} Y^*Z^* & 0 \\ 0 & Y^*Z^* \end{bmatrix}, \qquad \mathbf{R}^4 = \begin{bmatrix} (Y^*Z^*)^2 & 0 \\ 0 & (Y^*Z^*)^2 \end{bmatrix}, \quad \ldots,$$

$$\mathbf{R}^3 = \begin{bmatrix} 0 & Z^*(Y^*Z^*) \\ Y^*(Z^*Y^*) & 0 \end{bmatrix},$$

$$\mathbf{R}^5 = \begin{bmatrix} 0 & Z^*(Y^*Z^*)^2 \\ Y^*(Z^*Y^*)^2 & 0 \end{bmatrix}, \quad \ldots,$$

and, by using the series expansion for the matrix exponential (Eq. 3–1), we obtain

$$e^{\mathbf{R}(z-z_0)} = \begin{bmatrix} \cosh \Gamma^*(z-z_0) & Z_c \sinh \Gamma^*(z-z_0) \\ Z_c^{-1} \sinh \Gamma^*(z-z_0) & \cosh \Gamma^*(z-z_0) \end{bmatrix}, \qquad (7\text{--}34)$$

where

$$\Gamma^* = \sqrt{Y^*Z^*}, \qquad Z_c = \sqrt{Z^*/Y^*}.$$

Γ is called the propagation operator, and Z_c the characteristic impedance. We can easily verify that the determinant of this matrix is one (see Eq. 6–39). Therefore,

$$e^{-\mathbf{R}(z-z_0)} = \begin{bmatrix} \cosh \Gamma^*(z-z_0) & -Z_c \sinh \Gamma^*(z-z_0) \\ -Z_c^{-1} \sinh \Gamma^*(z-z_0) & \cosh \Gamma^*(z-z_0) \end{bmatrix}. \qquad (7\text{--}35)$$

7-2 Uniform, two-port transmission line

An acausal end-to-end relation can be obtained by letting

$$z - z_0 = l, \qquad \Gamma = \Gamma * l = \sqrt{Y_t Z_t}$$

in Eqs. (7–34) and (7–35), where l is the total length of the line and Y_t, Z_t are the total shunt admittance and series impedance, respectively. The right-end pair $\begin{bmatrix} p_2 \\ f_2 \end{bmatrix}$ in Fig. 7-2, is given in terms of the left-end pair $\begin{bmatrix} p_1 \\ f_1 \end{bmatrix}$, and vice versa, by the transfer matrices

$$\begin{bmatrix} p_2 \\ f_2 \end{bmatrix} = \begin{bmatrix} \cosh \Gamma & -Z_c \sinh \Gamma \\ -Z_c^{-1} \sinh \Gamma & \cosh \Gamma \end{bmatrix} \begin{bmatrix} p_1 \\ f_1 \end{bmatrix}$$

and (7–36)

$$\begin{bmatrix} p_1 \\ f_1 \end{bmatrix} = \begin{bmatrix} \cosh \Gamma & Z_c \sinh \Gamma \\ Z_c^{-1} \sinh \Gamma & \cosh \Gamma \end{bmatrix} \begin{bmatrix} p_2 \\ f_2 \end{bmatrix}.$$

The characteristic equation of (7–32) is

$$|s\mathbf{I} - \mathbf{R}| = 0,$$

and we find that the eigenvalues are:

$$s_1, s_2 = \pm \Gamma *.$$

We can diagonalize the relation (Section 3–3) by changing variables from $\mathbf{q}$ to $\mathbf{q}^*$:

$$\mathbf{q} = \mathbf{T}\mathbf{q}^*.$$

We then obtain the diagonal relation

$$\frac{\partial \mathbf{q}^*(z, t)}{\partial z} = \begin{bmatrix} -\Gamma * & 0 \\ 0 & \Gamma * \end{bmatrix} \mathbf{q}^*(z, t),$$

which can be integrated with respect to z, yielding

$$\mathbf{q}^*(z, t) = \begin{bmatrix} e^{-\Gamma *(z-z_0)} & 0 \\ 0 & e^{\Gamma *(z-z_0)} \end{bmatrix} \mathbf{q}^*(z_0, t). \qquad (7\text{--}37)$$

The end-to-end relation becomes

$$\mathbf{q}^*(z_0, t) = \begin{bmatrix} e^{\Gamma} & 0 \\ 0 & e^{-\Gamma} \end{bmatrix} \mathbf{q}^*(l, t), \qquad (7\text{--}38)$$

where $z_0 = 0$.

To determine $\mathbf{T} = [\mathbf{v}^1 \ \mathbf{v}^2]$, we compute

$$\mathbf{R}\mathbf{v}^1 = \Gamma * \mathbf{v}^1, \qquad \mathbf{R}\mathbf{v}^2 = -\Gamma * \mathbf{v}^2,$$

and letting $v_1^1 = v_1^2 = \sqrt{Z_c/2}$, we get what will prove to be a convenient form of **T**,

$$\mathbf{T} = \sqrt{\frac{1}{2}} \begin{bmatrix} \sqrt{Z_c} & \sqrt{Z_c} \\ 1/\sqrt{Z_c} & -1/\sqrt{Z_c} \end{bmatrix}. \tag{7-39}$$

The relation $\mathbf{q} = \mathbf{T}\mathbf{q}^*$, applied on both ends,

$$\mathbf{q}(z_0, t) = \begin{bmatrix} p_1(t) \\ f_1(t) \end{bmatrix}, \quad \mathbf{q}^*(z_0, t) = \begin{bmatrix} \theta_1(t) \\ \phi_1(t) \end{bmatrix} \text{ at the left end}$$

and

$$\mathbf{q}(l, t) = \begin{bmatrix} p_2(t) \\ f_2(t) \end{bmatrix}, \quad \mathbf{q}^*(l, t) = \begin{bmatrix} \theta_2(t) \\ \phi_2(t) \end{bmatrix} \text{ at the right end,}$$

gives the following static relations:

$$p_1 = \sqrt{\frac{Z_c}{2}} (\theta_1 + \phi_1), \quad p_2 = \sqrt{\frac{Z_c}{2}} (\theta_2 + \phi_2)$$

and $\hspace{6cm}$ (7-40)

$$f_1 = \frac{1}{\sqrt{2}\sqrt{Z_c}} (\theta_1 - \phi_1), \quad f_2 = \frac{1}{\sqrt{2}\sqrt{Z_c}} (\theta_2 - \phi_2).$$

The variables θ and ϕ are called wave-scattering variables [22] and are also related to the characteristic variables used in the solution of initial-value partial differential equations [23] when the system parameters Z_c and Γ are such that Eq. (7–32) is a hyperbolic differential equation. It is interesting to note that, while power flow in the potential-flow domain is expressed as a product, in the wave-scattering domain, power is a difference of squares:

$$P = p \cdot f = \sqrt{\frac{Z_c}{2}} (\theta + \phi) \cdot \frac{1}{\sqrt{2}\sqrt{Z_c}} (\theta - \phi) = \frac{\theta^2}{2} - \frac{\phi^2}{2}.$$

A block diagram for the port-to-port relations of a two-port transmission line is shown in Fig. 7–10. We shall see later that the causality chosen for θ and ϕ is natural; since the potential-flow variables are related to the wave-scattering variables by Eq. (7–40), which is static, all causalities shown in Fig. 7–10 for p and f are possible. Note that the block diagram demonstrates the symmetry of the uniform transmission line in the sense that if the system is reversed, end-for-end, the behavior is not changed.

The four transfer functions, G_{11}, G_{12}, G_{21}, and G_{22}, of Fig. 6–18 refer to a general causal two-port and are an alternative to the block diagram of Fig. 7–10. One such set is used to describe a two-port having a given causality; using the results of Table 6–3, we can find the transfer functions for a uniform transmission line with any desired causality by substituting

$$a = d = \cosh \Gamma, \quad b = Z_c \sinh \Gamma, \quad c = Z_c^{-1} \sinh \Gamma,$$
$$\text{and} \quad \Delta = 1.$$

7-2   Uniform, two-port transmission line

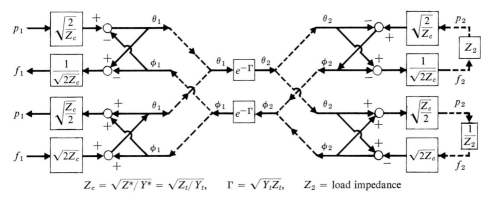

$Z_c = \sqrt{Z^*/Y^*} = \sqrt{Z_t/Y_t}, \quad \Gamma = \sqrt{Y_t Z_t}, \quad Z_2 = \text{load impedance}$

Fig. 7-10 Block diagram for a uniform, two-port transmission line.

If one end of a line is coupled to a one-port system, the transfer function of the overall system as seen at the other end can be determined. For instance, if the right end of a line is connected to a load of impedance Z_2, and p_1 at the left end is an input (Fig. 7-10, dashed lines), we can obtain the transfer function of the overall system as seen from the left by eliminating p_2 and f_2 from Eq. (7-36) by the loading condition $p_2 = Z_2 f_2$,

$$G(D) = \frac{f_1}{p_1} = \frac{Z_2 Z_c^{-1} \sinh \Gamma + \cosh \Gamma}{Z_2 \cosh \Gamma + Z_c \sinh \Gamma}. \tag{7-41}$$

The scattering variables θ_1 and ϕ_1 can be expressed in terms of the input when the overall system transfer function is known. For the case we just discussed, $\mathbf{T}^{-1}$ is found from Eq. (7-39) or (7-40),

$$T^{-1} = \frac{1}{\sqrt{2}} \begin{bmatrix} 1/\sqrt{Z_c} & \sqrt{Z_c} \\ 1/\sqrt{Z_c} & -\sqrt{Z_c} \end{bmatrix}, \tag{7-42}$$

and thus

$$\theta_1(t) = \frac{1}{\sqrt{2}} \left(\frac{p_1}{\sqrt{Z_c}} + \sqrt{Z_c} f_1 \right) = \frac{p_1}{\sqrt{2}} \left[\frac{1}{\sqrt{Z_c}} + \sqrt{Z_c} G(D) \right]$$

or

$$\theta_1(t) = \frac{1}{\sqrt{2Z_c}} [1 + Z_c G(D)] p_1(t) ;$$

and similarly,

$$\phi_1(t) = \frac{1}{\sqrt{2Z_c}} [1 - Z_c G(D)] p_1(t). \tag{7-43}$$

The distributed signal $q(z, t)$ can be expressed in terms of the input $p_1(t)$ by

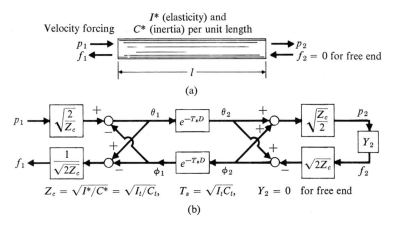

Fig. 7-11 Ideal shaft with free right end (a) and block diagram (b).

using the results of Eqs. (7–33), (7–35) and the equation relating p_1 to f_1, (7–41):

$$\mathbf{q}(z, t) = \begin{bmatrix} p(z, t) \\ f(z, t) \end{bmatrix} = \begin{bmatrix} \cosh \Gamma^* z & -Z_c \sinh \Gamma^* z \\ -(1/Z_c) \sinh \Gamma^* z & \cosh \Gamma^* z \end{bmatrix} \begin{bmatrix} p_1(t) \\ f_1(t) \end{bmatrix},$$

where z_0 has been taken equal to zero. However, $f_1(t) = G(D)p_1(t)$, so

$$p(z, t) = \left(\cosh \Gamma^* z - G(D) Z_c \sinh \Gamma^* z\right) p_1(t),$$
$$f(z, t) = \left(G(D) \cosh \Gamma^* z - Z_c^{-1} \sinh \Gamma^* z\right) p_1(t).\tag{7-44}$$

Example 7-3 *A frictionless elastic shaft* (Fig. 7–11).

$$Z^*(D) = I^* D, \qquad Y^*(D) = C^* D,$$

where I^* = elasticity per unit length, and C^* = inertia per unit length. Letting $I_t = I^* l$, $C_t = C^* l$, where l = total length of a length of a line, we find that

$$\Gamma = T_s D,\tag{7-45}$$

where $T_s = l/c$ = time for a sonic wave to travel from end to end, $c = (I^* C^*)^{-1/2}$ = sonic velocity. Also we find that $Z_c = \sqrt{I_t/C_t}$. Substituting Eq. (7–45) into Eq. (7–36), we find that

$$\begin{bmatrix} p_1 \\ f_1 \end{bmatrix} = \begin{bmatrix} \cosh T_s D & Z_c \sinh T_s D \\ Z_c^{-1} \sinh T_s D & \cosh T_s D \end{bmatrix} \begin{bmatrix} p_2 \\ f_2 \end{bmatrix}.\tag{7-46}$$

The block diagram (Fig. 7–11b) is constructed in the same manner as the block diagram of Fig. 7–10, but the causality specified for the shaft in Fig. 7–11(a) is applied. Physically, this causality corresponds to an angular velocity input at the left-end of the shaft (p_1) and a free right end (f_2 = torque = 0). The end condition is represented on the block diagram by a zero shunt admittance Y_2.

Consider the response of this system to a unit-step input in angular velocity. The solution for $p_2(t)$ and $f_1(t)$ (i.e., the angular velocity at the right end and the

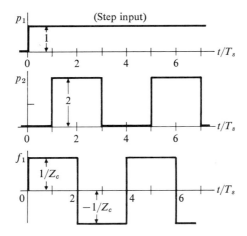

Fig. 7-12 Response of an ideal shaft.

torque at the left end, respectively) can be obtained by direct application of the block diagram. The wave-scattering variables θ and ϕ act as noninteracting traveling waves in the shaft; we see from the block diagram that a signal introduced at the left end θ_1, propagates undistributed down the line and appears at the right end T_s time units later as the signal θ_2. Since Z_c is a real number, the interaction at the right end is static and the reflected wave appears immediately as the signal ϕ_2, representing a wave traveling to the left. Furthermore, because f_2 is always zero, the reflected wave is equal to the incident wave. At the left end, this wave, reflected negatively (that is, with a reflection coefficient of -1), and the additive factor resulting from the input combine to form a new wave traveling to the right. This method is equivalent to delaying and then superposing the input function, each successive superposition delayed by the proper time and multiplied by a coefficient representing the reflections. This can be done for any input function but is most easily visualized for a step. The input and responses are shown in Fig. 7-12. At the right end, the response remains at zero until time $t/T_s = 1$ when the first wave arrives. At that time we have (because of the $+1$ reflection coefficient)

$$\phi_2 = \theta_2 = \sqrt{\frac{2}{Z_c}}\,; \qquad p_2 = \sqrt{\frac{Z_c}{2}}\,(\theta_2 + \phi_2) = 2\,.$$

At the left end, the torque f_1, attains a value of $1/Z_c$ as soon as the input is applied. It remains at that value until time $t/T_s = 2$ when the wave reflected from the right end arrives. It then jumps to its new value of $-1/Z_c$. This process is repeated to yield the response curve for as long as desired (fortunately, in this case, a repetitive pattern appears).

An alternative formulation to the same solution can be obtained from Eq. (7-46); with f_2 set equal to zero,

$$p_2 = \frac{1}{\cosh T_s D}\, p_1(t)\,, \qquad f_1 = \left(\frac{1}{Z_c} \sinh TD\right) p_2 = \left(\frac{1}{Z_c} \tanh T_s D\right) p_1(t)\,.$$

The operators $\cosh T_s D$ and $\tanh T_s D$ can be expanded in terms of the delay operator $e^{-(T_s D)}$,

$$\tanh T_s D = \frac{1 - e^{-2T_s D}}{1 + e^{-2T_s D}} = 1 - 2e^{-2T_s D} + 2e^{-4T_s D} - \cdots,$$

$$\begin{aligned} 1/\cosh T_s D &= 2e^{-T_s D}/(1 + e^{-2T_s D}) \\ &= 2(e^{-T_s D} - e^{-3T_s D} + e^{-5T_s D} - \cdots). \end{aligned}$$

Application of these operators is equivalent to the superposition process described above. The first method of obtaining the solution is often more convenient for computer solution [4], while the latter method is often easier for hand solution.

For cases such as the above example, in which we obtain a wavelike solution, we see that the wave scattering variables θ and ϕ represent the right- and left-traveling waves in the line. When the right-traveling wave is incident at the right end of the line, it is reflected to form the new left-traveling wave. The reflection coefficient can be found directly from the block diagram (Fig. 7–11b),

$$\phi_2 = \theta_2 - \sqrt{2Z_c}\, Y_2 \sqrt{\frac{Z_c}{2}}\,(\phi_2 + \theta_2)$$

or

$$\frac{\phi_2}{\theta_2} = \frac{1 - Z_c Y_2}{1 + Z_c Y_2} = \frac{Z_2 - Z_c}{Z_2 + Z_c}, \qquad (7\text{--}47)$$

where $Z_2 = 1/Y_2$. A reflection-free condition, that is, no left-traveling wave at any time, $\phi = 0$, occurs when $Z_2 = Z_c$. This would be the case, for example, if the line were terminated with a resistor $R_2 = Z_c$, or a semi-infinite line having the same characteristic impedance Z_c.

Example 7-4 A fluid line has momentum I^*, compressibility effect C^*, and line friction R^*, so that

$$Z^* = I^* D + R^*, \qquad Y^* = C^* D.$$

We then have

$$\varGamma = (Y^* Z^*)^{1/2} l = \sqrt{C^* I^*}\, \sqrt{1 + R^*/(I^* D)} \cdot lD,$$

and

$$Z_c = (Z^*/Y^*)^{1/2} = \sqrt{I^*/C^*}\, \sqrt{1 + R^*/(I^* D)}.$$

The approximation

$$\sqrt{1 + R^*/(I^* D)} \approx 1 + \frac{1}{2}\frac{R^*}{I^* D}$$

holds if R^* is sufficiently small, and therefore

$$\varGamma \approx T_s(a + D), \qquad Z_c \approx Z_i(1 + a/D), \qquad (7\text{--}48)$$

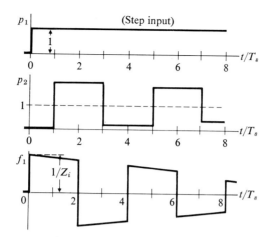

Fig. 7-13 Response of an *RIC*-line.

where

$$T_s = (\sqrt{C^*I^*})l, \qquad Z_i = \sqrt{I^*/C^*}, \qquad \text{and} \qquad a = \frac{1}{2}\frac{R^*}{I^*}.$$

Using the same terminal causality as in Fig. 7-11 where $f_2 = 0$ in a fluid line means that the right end is closed, we find that Eq. (7-36) with (7-48) yields the following response:

$$f_1(t) = \left[\frac{1}{Z_i(1 + a/D)}\tanh T_s(a + D)\right]p_1(t) \approx \left[\frac{1}{Z_i}(1 - a/D)\tanh T_s(a + D)\right]p_1(t),$$
(7-49)

$$p_2(t) = [\cosh T_s(a + D)]^{-1}p_1(t).$$

Making use of the expansion,

$$\tanh T_s(a + D) = 1 - 2e^{-2T_s a}e^{-2T_s D} + 2e^{-4T_s a}e^{-4T_s D} - \cdots,$$

we can deduce two parts f_{1a} and f_{1b} of the time-domain pattern $f_1 = f_{1a} - f_{1b}$:

$$f_{1a} = \frac{1}{Z_i}(1 - 2e^{-2T_s a}e^{-2T_s D} + \cdots)p_1(t),$$
(7-50)
$$f_{1b} = a\int_0^t f_{1a}\,dt,$$

where $e^{-2T_s D}$, etc., are delay operators as before, and $e^{-2T_s a}$, etc., are amplitude attenuations. If $p_1(t)$ is a unit step, we can estimate the response pattern by Eq. (7-50) as shown in Fig. 7-13. The response can also be calculated directly from the block diagram, as in Example 7-3; however, the situation is a little more complicated in this case because the transformation operators from wave-scattering variables to potential-flow variables include dynamic operators.

Example 7-5 Heat conduction of a semi-infinite solid. The equation of diffusion was derived in Example 7-2. Let

$$Z^* = R^*, \qquad Y^* = C^*D, \qquad l = \infty.$$

286 Distributed-parameter systems

The system constants combine into two:

$$\Gamma^* = \sqrt{T^*D} \quad \text{and} \quad Z_c = \sqrt{R^*/(C^*D)},$$

where $T^* = R^*C^*$. Since there is no reflection wave when $l = \infty$ in a stable passive system, $\theta_2 = 0$ in Fig. 7-10. If $p_1(t)$ (surface temperature) is an input and $p(z, t)$ (temperature at depth z) is an output, then

$$G(D) = p(z, t)/p_1(t) = e^{-\Gamma * z}.$$

Although the form of the transfer function is compact, it is irrational, and therefore transient response computation is not easy [15].

7-3 A GENERALIZATION OF THE TRANSFER MATRIX APPROACH[†]

The class of systems we discussed in the preceding section was characterized by a 2×2 matrix **R** with zero diagonal elements. Systems represented by such matrices are said to be *symmetric*, for they exhibit the same behavior when turned around 180 degrees (see Fig. 7–10). This concept can be generalized to systems for which **q**(z, t) is a 2n-vector such that

$$\mathbf{q}(z, t) = \begin{bmatrix} \mathbf{p}(z, t) \\ \mathbf{f}(z, t) \end{bmatrix}, \quad \mathbf{p}(z, t) = \begin{bmatrix} p_1(z, t) \\ \vdots \\ p_n(z, t) \end{bmatrix},$$

$$\mathbf{f}(z, t) = \begin{bmatrix} f_1(z, t) \\ \vdots \\ f_n(z, t) \end{bmatrix}. \tag{7-51}$$

The vector equation for a symmetric system is

$$\frac{\partial}{\partial z}\begin{bmatrix} \mathbf{p} \\ \mathbf{f} \end{bmatrix} = -\begin{bmatrix} 0 & \mathbf{Z}^* \\ \mathbf{Y}^* & 0 \end{bmatrix}\begin{bmatrix} \mathbf{p} \\ \mathbf{f} \end{bmatrix}. \tag{7-52}$$

This has the solution

$$\begin{bmatrix} \mathbf{p}(z_0) \\ \mathbf{f}(z_0) \end{bmatrix} = e^{\mathbf{R}(z-z_0)} \begin{bmatrix} \mathbf{p}(z) \\ \mathbf{f}(z) \end{bmatrix}, \tag{7-53}$$

where

$$\mathbf{R} = \begin{bmatrix} 0 & \mathbf{Z}^* \\ \mathbf{Y}^* & 0 \end{bmatrix}$$

[†] This section is based mostly on F. T. Brown's paper [3].

is a $(2n \times 2n)$-matrix such that

$$e^{\mathbf{R}(z-z_0)} = \begin{bmatrix} \cosh\sqrt{\mathbf{Y}^*\mathbf{Z}^*}(z-z_0) & \mathbf{Z}^*(\sqrt{\mathbf{Y}^*\mathbf{Z}^*})^{-1}\sinh\sqrt{\mathbf{Y}^*\mathbf{Z}^*}(z-z_0) \\ \mathbf{Y}^*(\sqrt{\mathbf{Z}^*\mathbf{Y}^*})^{-1}\sinh\sqrt{\mathbf{Y}^*\mathbf{Z}^*}(z-z_0) & \cosh\sqrt{\mathbf{Y}^*\mathbf{Z}^*}(z-z_0) \end{bmatrix}.$$
(7-54)

The variables p and f were defined in Section 5-2 as generalized potential and flow, respectively. We note the following properties of the pair:

1. The product $p \cdot f$ represents power.
2. The sign of one of the variables does not depend on the direction of a line; it may be said to be *symmetric* for this reason.
3. The sign of the other depends on the direction from which one sees the system; it is *asymmetric*.

Replacing the concept of potential-vs.-flow variables (which carries a direct analogy to electrical networks with it) by a symmetric-vs.-asymmetric pair of variables, F. T. Brown expands the scope of application of the transfer matrix approach. To demonstrate the efficiency of his approach, we shall treat the bending-beam problem. The basic formulation of the system was presented in Section 7-1 (Fig. 7-4). We choose the lateral deflection as the first symmetric variable:

$$y = p_1.$$

This variable is associated with shear force S, which is asymetric. We take a normalized shear force as f_1:

$$\frac{Sb}{\rho\omega^2} = f_1,$$

where the normalization stems from Eq. (7-9). The product $p_1 f_1$ represents the first kind of power flow. With $D = \partial/\partial t \to j\omega$, Eq. (7-5) yields the relation we need for the new formulation:

$$\frac{\partial f_1}{\partial \eta} = p_1, \qquad (7\text{-}55)$$

where $\eta = bz$ as we saw in Eq. (7-9).

The moment M is the second symmetric variable, which is associated with the slope of the beam $\partial y/\partial z$. We normalize M and define p_2 by

$$\frac{M}{b^2 EJ} = p_2,$$

while the normalized beam deflection $\partial y/\partial \eta$ is f_2. Since $p_1 = y$, we have

$$\frac{\partial p_1}{\partial \eta} = f_2, \tag{7-56}$$

and the flexure equation (7-7) becomes

$$\frac{\partial f_2}{\partial \eta} = p_2. \tag{7-57}$$

The last equation is given by normalizing the moment balance Eq. (7-6)

$$\frac{\partial p_2}{\partial \eta} = f_1. \tag{7-58}$$

Collecting Eqs. (7-55) through (7-58), we obtain the following vector equation which has the form of Eq. (7-52);

$$\frac{\partial}{\partial \eta} \begin{bmatrix} p_1 \\ p_2 \\ f_1 \\ f_2 \end{bmatrix} = \begin{bmatrix} 0 & 0 & 0 & 1 \\ 0 & 0 & 1 & 0 \\ 1 & 0 & 0 & 0 \\ 0 & 1 & 0 & 0 \end{bmatrix} \begin{bmatrix} p_1 \\ p_2 \\ f_1 \\ f_2 \end{bmatrix}. \tag{7-59}$$

The characteristic equation $|s\mathbf{I} - \mathbf{R}| = 0$ is

$$s^4 - 1 = 0,$$

and the characteristic roots are

$$s_1 = 1, \qquad s_2 = -1, \qquad s_3 = j, \qquad s_4 = -j.$$

Therefore, the set of modes

$$e^\eta, \quad e^{-\eta}, \quad e^{j\eta}, \quad \text{and} \quad e^{-j\eta},$$

or

$$\cosh \eta, \quad \sinh \eta, \quad \cos \eta, \quad \text{and} \quad \sin \eta,$$

is expected to appear in the solution. We obtain the following solution by means of the diagonalization method or by computing the matrices on the right-hand side of Eq. (7-54):

$$\begin{bmatrix} p_1(0) \\ p_2(0) \\ f_1(0) \\ f_2(0) \end{bmatrix} = \frac{1}{2} \begin{bmatrix} m_1 & m_2 & m_4 & m_3 \\ m_2 & m_1 & m_3 & m_4 \\ m_3 & m_4 & m_1 & m_2 \\ m_4 & m_3 & m_2 & m_1 \end{bmatrix} \begin{bmatrix} p_1(\eta) \\ p_2(\eta) \\ f_1(\eta) \\ f_2(\eta) \end{bmatrix}, \tag{7-60}$$

where

$$\begin{aligned} m_1 &= \cosh \eta + \cos \eta, & m_2 &= \cosh \eta - \cos \eta, \\ m_3 &= -\sinh \eta + \sin \eta, & m_4 &= -\sinh \eta - \sin \eta. \end{aligned} \tag{7-61}$$

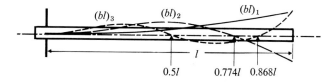

Fig. 7-14 The first three modes of vibration of a cantilever beam.

Equation (7-60), with boundary conditions at both ends, reduces to the solutions known in the field of elastic vibration. If, for instance, the uniform beam is a cantilever, the boundary conditions for a rigidly fixed left end are

$$p_1(0) = 0, \qquad f_2(0) = 0, \tag{7-62}$$

and, for the free right end,

$$p_2(bl) = 0, \qquad f_1(bl) = 0. \tag{7-63}$$

Letting $\eta = bl$ in Eq. (7-60), and applying Eqs. (7-62) and (7-63), the equations for the first and fourth row become

$$m_1 p_1(bl) + m_3 f_2(bl) = 0, \qquad m_4 p_1(bl) + m_1 f_2(bl) = 0.$$

Therefore, for a nontrivial solution to exist, we require that

$$\begin{vmatrix} m_1 & m_3 \\ m_4 & m_1 \end{vmatrix} = 0 \qquad \text{or} \qquad m_1^2 = m_3 m_4.$$

With Eq. (7-61), this condition reduces to the following:

$$1 + \cosh bl \cos bl = 0. \tag{7-64}$$

This equation has the roots

$$(bl)_1 = 1.875, \qquad (bl)_2 = 4.694, \qquad (bl)_3 = 7.855, \quad \text{etc.} \tag{7-65}$$

These roots give the modes of lateral vibration shown in Fig. 7-14.

The diagonal elements in matrix **R** of Eq. (7-3) or (7-52) become nonzero when, for instance, a transport motion is superimposed on the transient motion. If there is a bias velocity of fluid comparable to the sonic velocity, **R** for the fluid line discussed in Section 7-2 will take the form

$$\mathbf{R} = \begin{bmatrix} W & Z^* \\ Y^* & V \end{bmatrix} \tag{7-66}$$

(where the terms W and V depend directly on the bias velocity) for which the solution (7-36) does not apply any more. If the bias velocity is directed to the right, it will speed up the waves traveling to the right and slow down the other waves: the system will thus lose the symmetric behavior we saw in Fig. 7-10.

In Section 7-1 we saw a set of processes in which fluid flow acts as a heat or material carrier. Since f (or $\mathbf{f}$) is externally fixed in these processes, it becomes convenient and natural to look upon f as a system parameter rather than as a variable, and delete it from $\mathbf{q}$. We thus write an equation only for p or $\mathbf{p}$:

$$\frac{\partial \mathbf{p}(z, t)}{\partial z} = -\mathbf{W}\mathbf{p}(z, t). \tag{7-67}$$

If $\mathbf{W}$ is 2×2, then

$$\frac{\partial}{\partial z}\begin{bmatrix} p_1 \\ p_2 \end{bmatrix} = -\begin{bmatrix} w_{11} & w_{12} \\ w_{21} & w_{22} \end{bmatrix}\begin{bmatrix} p_1 \\ p_2 \end{bmatrix}. \tag{7-68}$$

In the following discussion we shall solve Eq. (7-68). A two-port non-symmetric system governed by $\mathbf{R}$ of Eq. (7-66) also has the same type of solution. The characteristic equation of (7-68) is

$$\begin{vmatrix} s + w_{11} & w_{12} \\ w_{21} & s + w_{22} \end{vmatrix} = 0,$$

or

$$s^2 + (w_{11} + w_{22})s + (w_{11}w_{22} - w_{12}w_{21}) = 0, \tag{7-69}$$

which has the roots

$$s_1, s_2 = -\tfrac{1}{2}(w_{11} + w_{22}) \pm \sqrt{\tfrac{1}{2}(w_{11} - w_{22})^2 + w_{12}w_{21}}. \tag{7-70}$$

To solve Eq. (7-68), we diagonalize the matrix by transforming from $\mathbf{p}$ to $\mathbf{p}^*$,

$$\mathbf{p} = \mathbf{T}\mathbf{p}^*,$$

in such a way that

$$\frac{\partial}{\partial z}\begin{bmatrix} p_1^* \\ p_2^* \end{bmatrix} = \begin{bmatrix} s_1 & 0 \\ 0 & s_2 \end{bmatrix}\begin{bmatrix} p_1^* \\ p_2^* \end{bmatrix}. \tag{7-71}$$

As shown in Section 3-3, $\mathbf{T} = (\mathbf{v}^1 \ \mathbf{v}^2)$ must satisfy the following conditions:

$$-\mathbf{W}\mathbf{v}^1 = s_1\mathbf{v}^1, \qquad -\mathbf{W}\mathbf{v}^2 = s_2\mathbf{v}^2. \tag{7-72}$$

Arbitrarily choosing $v_1^1 = v_1^2 = 1$, and letting

$$\frac{w_{11} + s_1}{w_{12}} = -\gamma_1, \qquad \frac{w_{11} + s_2}{w_{12}} = -\gamma_2,$$

we obtain

$$\mathbf{T} = \begin{bmatrix} 1 & 1 \\ \gamma_1 & \gamma_2 \end{bmatrix}, \qquad \mathbf{T}^{-1} = \frac{1}{\gamma_1 - \gamma_2}\begin{bmatrix} -\gamma_2 & 1 \\ \gamma_1 & -1 \end{bmatrix}. \tag{7-73}$$

We now have

$$\mathbf{p}^*(z) = \begin{bmatrix} e^{s_1 z} & 0 \\ 0 & e^{s_2 z} \end{bmatrix} \mathbf{p}^*(0)$$

and

$$\mathbf{p}^*(0) = \begin{bmatrix} e^{-s_1 z} & 0 \\ 0 & e^{-s_2 z} \end{bmatrix} \mathbf{p}^*(z). \tag{7-74}$$

Therefore

$$\mathbf{p}(0) = \mathbf{T} \begin{bmatrix} e^{-s_1 z} & 0 \\ 0 & e^{-s_2 z} \end{bmatrix} \mathbf{T}^{-1} \mathbf{p}(z)$$

$$= \frac{1}{\gamma_1 - \gamma_2} \begin{bmatrix} -\gamma_2 e^{-s_1 z} + \gamma_1 e^{-s_2 z} & -e^{-s_1 z} - e^{-s_2 z} \\ \gamma_1 \gamma_2 (-e^{-s_1 z} + e^{-s_2 z}) & \gamma_1 e^{-s_1 z} - \gamma_2 e^{-s_2 z} \end{bmatrix} \mathbf{p}(z). \tag{7-75}$$

We shall apply this result to various process systems in the next section.

7-4 PROCESSES WITH CARRIER FLOW

In this section we consider the heat exchangers of Section 7-1, where $\mathbf{p}$ consists of fluid temperatures. We take the parallel-flow heat exchanger as the first example. The normalized equation (7-27), rearranged, takes the canonical form:

$$\frac{\partial}{\partial \eta} \begin{bmatrix} \theta_h \\ \theta_c \end{bmatrix} = \begin{bmatrix} -(a_h + D) & a_h \\ a_c & -(a_c + rD) \end{bmatrix} \begin{bmatrix} \theta_h \\ \theta_c \end{bmatrix}. \tag{7-76}$$

Comparing Eqs. (7-76) and (7-68), we find that

$$\begin{aligned} w_{11} &= a_h + D, & w_{12} &= -a_h, \\ w_{21} &= -a_c, & w_{22} &= a_c + rD, \end{aligned} \tag{7-77}$$

where D represents the time derivative in the nondimensional time domain $\partial/\partial \tau$. We also find that

$$\gamma_1 = \frac{a_h + D + s_1}{a_h}, \qquad \gamma_2 = \frac{a_h + D + s_2}{a_h}.$$

Collecting terms in the solution (Eq. 7-75), we write

$$\begin{aligned} \theta_h(\eta, \tau) &= \alpha(\tau) e^{s_1 \eta} + \beta(\tau) e^{s_2 \eta}, \\ \theta_c(\eta, \tau) &= \gamma_1 \alpha(\tau) e^{s_1 \eta} + \gamma_2 \beta(\tau) e^{s_2 \eta}, \end{aligned} \tag{7-78}$$

where

$$\alpha(\tau) = \frac{\gamma_2 \theta_{h1} - \theta_{c1}}{\gamma_2 - \gamma_1}, \qquad \beta(\tau) = \frac{-\gamma_1 \theta_{h1} + \theta_{c1}}{\gamma_2 - \gamma_1}. \tag{7-79}$$

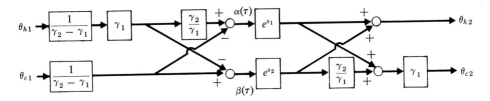

Fig. 7-15 Block diagram of a parallel-flow heat exchanger.

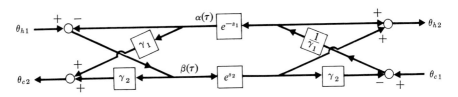

Fig. 7-16 Block diagram of a counterflow heat exchanger.

These two time functions appear as a pair of traveling waves in the block diagram (Fig. 7–15). Terminal relations are given by four transfer functions in the following form:

$$\begin{bmatrix} \theta_{h2} \\ \theta_{c2} \end{bmatrix} = \begin{bmatrix} G_{hh}(D) & G_{ch}(D) \\ G_{hc}(D) & G_{cc}(D) \end{bmatrix} \begin{bmatrix} \theta_{h1} \\ \theta_{c1} \end{bmatrix}; \qquad (7\text{–}80)$$

they are found to be

$$\begin{aligned}
G_{hh} &= (e^{-\lambda_1}/\mu)(\mu \cosh \mu - \lambda_2 \sinh \mu), \\
G_{ch} &= a_h(e^{-\lambda_1}/\mu) \sinh \mu, \\
G_{hc} &= a_c(e^{-\lambda_1}/\mu) \sinh \mu, \\
G_{cc} &= (e^{-\lambda_1}/\mu)(\lambda_2 \sinh \mu + \mu \cosh \mu),
\end{aligned} \qquad (7\text{–}81)$$

where

$$\lambda_1 = \tfrac{1}{2}(a_h + a_c + (1+r)D), \qquad \lambda_2 = \tfrac{1}{2}(a_h - a_c + (1-r)D),$$
$$\mu = \tfrac{1}{2}\sqrt{(a_h + a_c)^2 + 2(a_h - a_c)(1-r)D + (1-r)^2 D^2}.$$

λ_1 and μ are related to the eigenvalues by

$$s_1 = -\lambda_1 + \mu, \qquad s_2 = -\lambda_1 - \mu.$$

The relation for a parallel-flow heat exchanger simplifies when the delay ratio of the two flows is one: $r = L_c/L_h = 1$. For this special case

$$\mu = \tfrac{1}{2}(a_h + a_c)$$

and
$$\lambda_1 = \tfrac{1}{2}(a_h + a_c) + D, \qquad \lambda_2 = \tfrac{1}{2}(a_h - a_c).$$

Therefore all transfer functions become proportional to a single delay e^{-D}.

We next consider a counterflow heat exchanger. Equation (7–27), in the canonical form, becomes

$$\frac{\partial}{\partial \eta}\begin{bmatrix}\theta_h \\ \theta_c\end{bmatrix} = \begin{bmatrix}-(a_h + D) & a_h \\ -a_c & (a_c + rD)\end{bmatrix}\begin{bmatrix}\theta_h \\ \theta_c\end{bmatrix} \qquad (7\text{–}82)$$

or

$$w_{11} = a_h + D, \qquad w_{12} = -a_h,$$
$$w_{21} = a_c, \qquad w_{22} = -(a_c + rD).$$

The solution Eq. (7–78) applies, but the eigenvalues s_1 and s_2 are different from the parallel-flow case. They are given by

$$s_1 = -\lambda_2 + \mu, \qquad s_2 = -\lambda_2 - \mu,$$

where λ_2 is the same as before, but

$$\mu = \tfrac{1}{2}\sqrt{(a_h - a_c)^2 + 2(a_h + a_c)(1 + r)D + (1 + r)^2 D^2}.$$

$\alpha(\tau)$ and $\beta(\tau)$ in Eq. (7–78) must be fixed by the two-point boundary values;

$$\theta_h = \theta_{h1} = \alpha + \beta \quad \text{at} \quad \eta = 0 \quad \text{and}$$
$$\theta_c = \theta_{c1} = \gamma_1 \alpha e^{s_1} + \gamma_2 \beta e^{s_2} \quad \text{at} \quad \eta = 1.$$

We find:

$$\alpha(\tau) = \frac{-\lambda_2 e^{s_2} \theta_{h1} + \theta_{c1}}{\lambda_1 e^{s_1} - \gamma_2 e^{s_2}}, \qquad \beta(\tau) = \frac{\gamma_1 e^{s_1} \theta_{h1} - \theta_{c1}}{\gamma_1 e^{s_1} - \gamma_2 e^{s_2}}.$$

The block diagram of Fig. 7–16 can be constructed from this result. The four transfer functions in Eq. (7–81) are given by:

$$\begin{aligned}G_{hh} &= \mu e^{-\lambda_2}/H, & G_{ch} &= a_h \sinh \mu / H, \\ G_{hc} &= a_c \sinh \mu / H, & G_{cc} &= \mu e^{\lambda_2}/H,\end{aligned} \qquad (7\text{–}83)$$

where

$$H = \mu \cosh \mu + \lambda_1 \sinh \mu.$$

The results simplify when the relation is symmetric,

$$a_h = a_c = a, \quad r = 1.$$

We then have

$$\lambda_1 = a + D, \qquad \lambda_2 = 0, \qquad \mu = \sqrt{2aD + D^2},$$

which means in Fig. 7–16 that

$$e^{-s_1} = e^{s_2} = e^{-\sqrt{2aD+D^2}},$$

$$\gamma_1 = \frac{a + D + \sqrt{2aD + D^2}}{a}, \qquad \gamma_2 = \frac{a + D - \sqrt{2aD + D^2}}{a}.$$

The transfer functions G_{hc} and G_{ch} for this case become:

$$G_{hc} = G_{ch} = \frac{a}{(a + D) + \sqrt{2aD + D^2}\coth\sqrt{2aD + D^2}}. \qquad (7\text{-}84)$$

A direct computation of time response is not easy even for this special case [3 and 18]. We can derive a single time-constant model for G_{hc} and G_{ch} of Eq. (7-84) by making use of the initial- and final-value concepts in the Laplace domain to match the behavior at high and low frequencies, which correspond to $t \to 0^+$ and $t \to \infty$, respectively, in the transient response. As $D \to 0$, we have

$$x \coth x \bigg|_{x \to 0} = \frac{x(e^x + e^{-x})}{e^x - e^{-x}} \bigg|_{x \to 0} = 1\,;$$

thus

$$G_{hc} = G_{ch} \approx \frac{a}{a + 1 + D} = \frac{k_1}{1 + T_1 D},$$

$$k_1 = \frac{a}{a + 1}, \qquad T_1 = \frac{1}{a + 1}.$$

When $D \to \infty$, on the other hand,

$$\sqrt{D^2 + 2aD} = D\sqrt{1 + 2a/D} \approx D(1 + a/D) = D + a$$

and

$$G_{hc} = G_{ch} \approx \frac{a/2}{a + D} = \frac{k_2}{1 + T_2 D}, \qquad k_2 = \tfrac{1}{2}, \qquad T_2 = 1/a\,.$$

A compromise single time-constant model $k/(1 + TD)$ must have gain $k = k_1$ at $D \to 0$ and high-frequency behavior $k/(TD) = k_2/(T_2 D)$ at $D \to \infty$. Therefore

$$k = k_1 = \frac{a}{a + 1} \qquad \text{and} \qquad \frac{k_1 T_2}{k_2} = \frac{2}{a + 1}$$

or

$$G_{hc} = G_{ch} \approx \frac{a}{(a + 1) + 2D}. \qquad (7\text{-}85)$$

Equation (7-85) is a first approximation of a symmetric counterflow heat exchanger with negligible heat-storage effects on the tube walls.

The tube and shell heat capacitances may be taken into consideration by adding heat-balance equations for these in the first formulation [12]. Further complication in formulation arises when the tube makes more than one pass [19].

The last case we treat in this section is a mixed system, the heat exchanger of Fig. 7-6. It is a hybrid system of lumped- and distributed-parameter elements because it has a tube and a stirred (lumped) capacity. A pair of normalized equations for the system was given by Eq. (7-20), restated:

$$\frac{\partial \theta_h}{\partial \tau} + \frac{\partial \theta_h}{\partial \eta} = a_h(\theta_{c2} - \theta_h), \tag{7-86}$$

$$\frac{d\theta_{c2}}{d\tau} = a_c \left(\int_0^1 \theta_h(\eta)\, d\eta - \theta_{c2} \right) + \frac{1}{r}(\theta_{c1} - \theta_{c2}). \tag{7-87}$$

Due to the hybrid nature of the system, the transfer-matrix method does not apply. We therefore take a step-by-step approach. As a first step, we integrate Eq. (7-86) with respect to η and obtain

$$\theta_h(\eta) = \theta_{h1} e^{-(D+a_h)\eta} + \theta_{c2} \frac{a_h}{D + a_h}(1 - e^{-(D+a_h)\eta}), \tag{7-88}$$

where D is the operator $\partial/\partial\tau$. We then compute the average temperature:

$$\int_0^1 \theta_h(\eta)\, d\eta = \frac{1 - e^{-(D+a_h)}}{D + a_h} \theta_{h1} + \left(1 - \frac{1 - e^{-(D+a_h)}}{D + a_h}\right) \frac{a_h}{D + a_h} \theta_{c2}. \tag{7-89}$$

Substituting Eq. (7-89) into (7-87), we obtain

$$D\theta_{c2} = \frac{1}{r}\theta_{c1} + \frac{a_c}{D + a_h}(1 - e^{-(D+a_h)})\theta_{h1}$$

$$- \left[\frac{a_c D}{D + a_h} + \frac{1}{r} + \frac{a_h a_c(1 - e^{-(D+a_h)})}{(D + a_h)^2}\right] \theta_{c2}. \tag{7-90}$$

The four transfer functions in the matrix relation

$$\begin{bmatrix} \theta_{h2} \\ \theta_{c2} \end{bmatrix} = \begin{bmatrix} G_{hh} & G_{ch} \\ G_{hc} & G_{cc} \end{bmatrix} \begin{bmatrix} \theta_{h1} \\ \theta_{c1} \end{bmatrix}$$

can be determined from Eqs. (7-88) and (7-90) in the following form:

$$G_{hh} = e^{-(D+a_h)} + \frac{a_c a_h (1 - e^{-(D+a_h)})^2}{H},$$

$$G_{ch} = \frac{(a_h/r)(D + a_h)(1 - e^{-(D+a_h)})}{H}, \tag{7-91}$$

$$G_{hc} = \frac{a_c(D + a_h)(1 - e^{-(D+a_h)})}{H}, \quad G_{cc} = \frac{(1/r)(D + a_h)^2}{H},$$

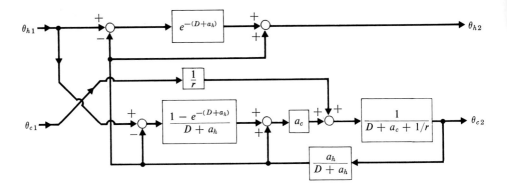

Fig. 7-17 Block diagram for a mixing-type heat exchanger.

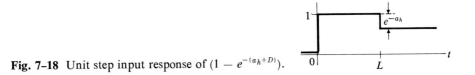

Fig. 7-18 Unit step input response of $(1 - e^{-(a_h+D)})$.

where

$$H = [D + (1/r)(D + a_h)^2 + a_c D(D + a_h) + (a_h a_c)(1 - e^{-(D+a_h)})].$$

The hybrid characteristic of the system is clearly seen in the block diagram (Fig. 7-17) in which delay with attenuation, $e^{-(D+a_h)}$, is combined with lumped lags. The factor $e^{-(D+a_h)}$ which repeatedly appears in the relation comes from a slug passing through the tube. The unit-step input response of $1 - e^{-(a+D)}$ has the simple pattern shown in Fig. 7-18. Since the transfer functions are rational except for the pure delay e^{-D}, there is no difficulty in writing a digital-computer simulation program or simulating the system on a hybrid computer.

7-5 PERCOLATION PROCESSES AND SOME REMARKS ON DISTRIBUTED-PARAMETER SYSTEMS

Most of the partial differential equations that appeared in the preceding sections are represented by the following second-order form:

$$\alpha \frac{\partial^2 y}{\partial t^2} + \beta \frac{\partial^2 y}{\partial t \partial z} + \gamma \frac{\partial^2 y}{\partial z^2} = \delta, \tag{7-92}$$

where α, β, γ, and δ are functions of t, z, y, and the first derivatives $\partial y/\partial t$, $\partial y/\partial z$, but not of the second derivatives.

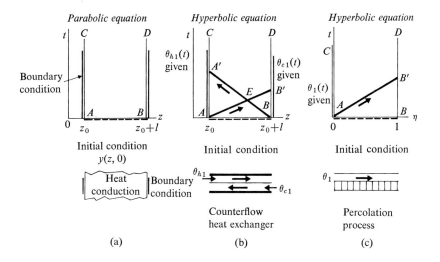

Fig. 7-19 Initial and boundary conditions.

Equation (7–92) is said to be *elliptic* when $\beta^2 < 4\alpha\gamma$. The solution $y(z, t)$ is then determined for a field enclosed by a closed contour in the zt-plane when boundary values are specified on the contour. Equation (7–92) becomes *parabolic* when $\beta^2 = 4\alpha\gamma$. The diffusion equation is parabolic because $\alpha = 0$, $\beta = 0$, $\gamma = 1$, $\delta = T\,\partial y/\partial t$ in Eq. (7–92) and thus satisfy the condition $\beta^2 = 4\alpha\gamma$. In general, the solution $y(z, t)$ of a parabolic equation is determined in the semi-infinite range $CABD$ in Fig. (7–19a), where the initial condition $y(z, 0)$ for $z_0 \leqslant z \leqslant z_0 + l$ (AB in the figure) and boundary conditions on edges AC and BD must be specified.

Equation (7–92) is *hyperbolic* when $\beta^2 > 4\alpha\gamma$. For instance, with $\beta = 1$, $\alpha = 0$, $\gamma = 0$, and $\delta = 0$ Eq. (7–92) reduces to

$$\partial^2 y/\partial t^2 = \partial^2 y/\partial z^2,$$

which is the wave equation we discussed in Example 7–3. An initial condition along the edge AB in Fig. 7–19(b) affects the solution $y(z, t)$ of a hyperbolic equation in the semi-infinite range bounded by AC and BD. The boundary conditions along these borders, however, affect only the areas $CAB'D$ and $CA'BD$. The solution within AEB is determined by initial condition alone. The lines AB', $A'B$ in the figure are called characteristics. These correspond, for example, to the propagation of the first wave fronts of hot and cold fluids in a counterflow heat exchanger.

An exact solution $y(z, t)$ of Eq. (7–92) for a prescribed set of initial and boundary conditions is usually complex. To demonstrate the complexity, we shall give a set of exact solutions for a percolation process.

298 Distributed-parameter systems

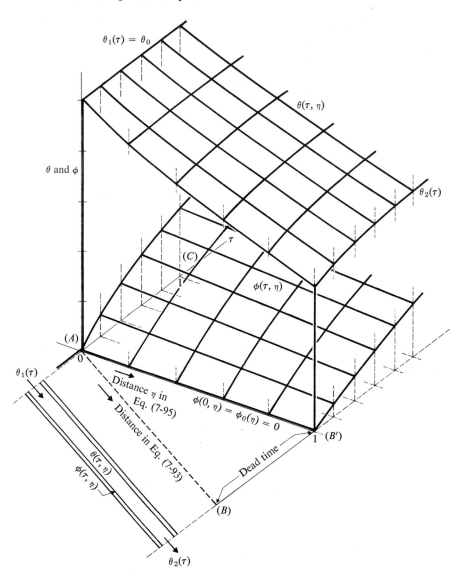

Fig. 7-20 Time-space-temperature pattern of a percolation process.

The process was described by Eqs. (7–11) and (7–13), restated:

$$\frac{\partial \theta}{\partial \tau} + \frac{\partial \theta}{\partial \eta} = a(\phi - \theta), \tag{7-93}$$

$$\frac{\partial \phi}{\partial \tau} = b(\theta - \phi). \tag{7-94}$$

We simplify Eq. (7–93) by taking the hydrodynamical derivative (Eq. 7–24). The equation then reduces to

$$\frac{\partial \theta}{\partial \eta} = a(\phi - \theta). \tag{7-95}$$

In this new coordinate system, η represents the distance a slug has traveled since entering the line, and τ represents the time at which it entered. $\tau = 0$ thus corresponds to the line AB' in Fig. 7–19(c) (see also Fig. 7–20). Eliminating one variable at a time from Eqs. (7–94) and (7–95), we obtain the following hyperbolic equations:

$$\frac{\partial^2 \theta}{\partial \eta \partial \tau} + a \frac{\partial \theta}{\partial \tau} + b \frac{\partial \theta}{\partial \eta} = 0, \qquad \frac{\partial^2 \phi}{\partial \eta \partial \tau} + a \frac{\partial \phi}{\partial \tau} + b \frac{\partial \phi}{\partial \eta} = 0. \tag{7-96}$$

Let the initial conditions along line AB' of Fig. 7–19(c) be

$$\phi(0, \eta) = \phi_0(\eta),$$

and, by Eq. (1–18) applied to Eq. (7–95),

$$\theta(0, \eta) = \left\{ \theta_1(0) + \int_0^\eta e^{av} a \phi_0(v) \, dv \right\} e^{-a\eta}.$$

The boundary conditions along line AC of Fig. 7–19(c) are:

$$\theta(\tau, 0) = \theta_1(\tau) = \text{inlet temperature of fluid:}$$

and solving Eq. (7–94) by Eq. (1–18), we have

$$\phi(\tau, 0) = \left[\phi_0(0) + \int_0^\tau e^{bv} \theta_1(v) \, dv \right] e^{-b\tau}.$$

The exact solution of Eq. (7–96) for the sets of initial and boundary conditions is known to be [14] as follows:

$$\theta(\tau, \eta) = \theta_1(\tau) e^{-a\eta} + e^{-a\eta} \int_0^{ab\tau\eta} \frac{I_1(2\sqrt{w})}{\sqrt{w}} e^{-w/a\eta} \theta_1\left(\tau - \frac{w}{ab\eta}\right) dw$$

$$+ \frac{e^{-b\tau}}{b\tau} \int_0^{ab\tau\eta} I_0(2\sqrt{w}) e^{-w/b\tau} \phi_0\left(\eta - \frac{w}{ab\tau}\right) dw, \tag{7-97}$$

$$\phi(\tau, \eta) = \phi_1(\eta) e^{-b\tau} + e^{-b\tau} \int_0^{ab\eta\tau} \frac{I_1(2\sqrt{w})}{\sqrt{w}} e^{-w/b\tau} \phi_0\left(\eta - \frac{w}{ab\tau}\right) dw$$

$$+ \frac{e^{-a\eta}}{a\eta} \int_0^{ab\tau\eta} I_0(2\sqrt{w}) e^{-w/a\eta} \theta_1\left(\tau - \frac{w}{ab\eta}\right) dw, \tag{7-98}$$

where I_0, I_1 are modified Bessel functions of the first kind, zero and first order, respectively. If the initial wall temperature is zero everywhere,

$\phi_0(\eta) = 0$, $0 \leq \eta \leq 1$, and the fluid inlet temperature is a step of magnitude θ_0, $\theta_1(\tau) = \theta_0$, $0 \leq \tau$, the solution reduces to:

$$\theta(\tau, \eta) = \theta_0 e^{-a\eta} \left[1 + \int_0^{ab\tau\eta} \frac{I_1(2\sqrt{w})}{\sqrt{w}} e^{-w/a\eta} \, dw \right],$$

$$\phi(\tau, \eta) = \theta_0 \frac{e^{-a\eta}}{a\eta} \left[\int_0^{ab\tau\eta} I_0(2\sqrt{w}) e^{-w/a\eta} \, dw \right],$$

(7-99)

Equation (7-99) represents a pair of surfaces as shown in Fig. 7-20. An expanded form for the step input response $\theta_2(\tau) = \theta(\tau, 1)$ can be derived from this result by letting $\eta = 1$ and writing the series for $I_1(2\sqrt{w})$:

$$\theta_2(\tau)/\theta_0 = e^{-a} + e^{-a} \int_0^{ab\tau} \frac{I_1(2\sqrt{w})}{\sqrt{w}} e^{-w/a} dw$$

$$= e^{-a} + e^{-a} \sum_{i=0}^{\infty} \int_0^{ab\tau} \frac{w^i e^{-w/a}}{i!(i+1)!} dw$$

$$= e^{-a} + e^{-a} \sum_{i=0}^{\infty} \frac{a^{i+1}}{(i+1)!} \left[1 - e^{-b\tau} \left(1 + b\tau + \frac{(b\tau)^2}{2!} + \cdots \right. \right.$$

$$\left. \left. + \frac{(b\tau)^i}{i!} \right) \right]. \quad (7\text{-}100)$$

We have seen four general methods used in the analysis of distributed-parameter systems. The first is the lumped approximation which we will discuss further below. The exact-solution approach just shown is the second; this belongs to the classic mode-superposition method, often used in the theory of heat conduction, elastic vibration, etc (see Section 10-4). In some cases, it is possible to trace only individual waves, as shown in Fig. 7-10. This is the third approach. The transfer matrix approach belongs to the last category. In this chapter we dealt with vector differential equations having first derivatives in position z, while time derivatives were replaced with the operator D. It is also possible to replace the position derivatives $(\partial/\partial z$, etc.) by an operator E (that is, take the position Laplace transform of all relations) and convert the partial differential equations to a first-order vector differential equation of time. Equations (7-93) and (7-94) for a percolation process, for instance, may be written in the form

$$\frac{d}{d\tau} \mathbf{x}(\tau, \eta) = \mathbf{A} \mathbf{x}(\tau, \eta), \quad (7\text{-}101)$$

where

$$\mathbf{x} = \begin{bmatrix} \theta \\ \phi \end{bmatrix} = \text{state vector}, \quad \text{and} \quad \mathbf{A} = \begin{bmatrix} -(a+E) & a \\ b & -b \end{bmatrix}.$$

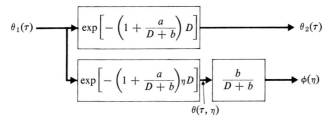

Fig. 7-21 Block diagram of a percolation process.

For this system (and many others), however, the transfer matrix method seems to be more convenient. For the percolation process, the formulation with position derivatives becomes scalar. Letting $\partial/\partial\tau \to D$, we find that Eq. (7-94) yields

$$\phi = \frac{b}{b+D}\theta,$$

and, substituting this into Eq. (7-93),

$$\frac{d\theta}{d\eta} = -\left(1 + \frac{a}{D+b}\right)D\theta. \tag{7-102}$$

Integrating Eq. (7-102) with respect to η, we get

$$\theta(\tau, \eta) = e^{-[1+(a/(D+b))]D\eta}\theta_1(\tau). \tag{7-103}$$

The outgoing fluid temperature is given by

$$\theta_2(\tau) = e^{-[1+(a/(D+b))]D}\theta_1(\tau) \tag{7-104}$$

or by the normalized transfer function

$$G(D) = e^{-D}e^{-aD/(D+b)}, \tag{7-105}$$

where e^{-D} represents the normalized transportation lag (dead time), and the second term on the right-hand side shows dispersive behavior. The block diagram of Fig. 7-21 can be constructed from these relations. The following approximation is allowed if a is small:

$$e^{-aD/(D+b)} \approx 1 - \frac{aD}{D+b} = (1-a) + \frac{ab}{D+b}, \tag{7-106}$$

which allows us to approximate a unit-step input response by the simple response shown in Fig. 7-22. It is also possible to derive the exact solution Eq. (7-100) by a series expansion of the exponential transfer function $e^{ab/(D+b)}$:

$$e^{-aD/(D+b)} = e^{-a}e^{ab/(D+b)} = e^{-a}\left(1 + \frac{ab}{D+b} + \frac{1}{2!}\frac{(ab)^2}{(D+b)^2} + \cdots\right).$$

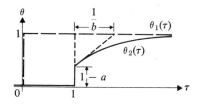

Fig. 7-22 Approximate unit-step input response of a percolation process.

Since the transportation delay e^{-D} in Eq. (7–105) was taken into account when we took a characteristic AB' of Fig. 7–19(c) and wrote Eq. (7–95), the outlet temperature with a time shift BB' of Fig. 7–19(c) will be given by the following inverse Laplace transformation:

$$\theta_2(\tau) = \mathscr{L}^{-1}\left[\frac{\theta_0}{s}e^{-a}e^{ab/(s+b)}\right]$$

$$= \theta_0 e^{-a}\mathscr{L}^{-1}\left[\frac{1}{s}\left(1 + \frac{ab}{s+b} + \frac{1}{2!}\frac{(ab)^2}{(s+b)^2} + \cdots\right)\right], \quad (7\text{--}107)$$

where D in the preceding relations is replaced by the Laplace operator s and θ_0/s is a step of magnitude θ_0. The Laplace transform pairs,

$$\mathscr{L}^{-1}\left[\frac{1}{s}\right] = 1,$$

$$\mathscr{L}^{-1}\left[\left(\frac{1}{s}\right)\frac{ab}{s+b}\right] = a(1 - e^{-b\tau}),$$

$$\mathscr{L}^{-1}\left[\left(\frac{1}{s}\right)\frac{(ab)^2}{(s+b)^2}\right] = a^2[1 - e^{-b\tau}(1 + b\tau)],$$

$$\vdots$$

applied in Eq. (7–107) give Eq. (7–100).

The lumped approximation to a distributed-parameter system looks straightforward in its formulation but there is a pitfall, as pointed out by Rosenbrock and Storey [1]. To demonstrate the pitfall let us derive a difference equation from a partial differential equation, using the example of the percolation process. Replacing the distributed-parameter model of Fig. 7–5 by the lumped approximation of Fig. 7–23(a), we designate by θ^0, $\theta^1, \ldots, \phi^0, \phi^1 \ldots$, the temperatures at points $\eta = 0, \Delta\eta, 2\Delta\eta, \ldots$ The position derivative $\partial/\partial\eta$ of Eq. (7–93) is now approximated by the difference relation

$$\frac{\partial\theta(\tau)}{\partial\eta} \approx \frac{\theta^{k+1}(\tau) - \theta^k(\tau)}{\Delta\eta} \quad (7\text{--}108)$$

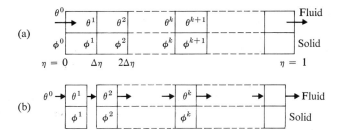

Fig. 7-23 Lumped models of a distributed-parameter system.

for the $(k+1)$th section, $k = 0, 1, \ldots$ We also replace the time derivative $\partial/\partial \tau$ in Eq. (7–93) by the difference form

$$\frac{\partial \theta(\tau)}{\partial \tau} \approx \frac{\theta(\tau + \Delta\tau) - \theta(\tau)}{\Delta\tau}. \tag{7-109}$$

With approximations (7–108) and (7–109), the original equation (7–93) becomes a difference equation:

$$\frac{\theta^k(\tau + \Delta\tau) + \theta^{k+1}(\tau + \Delta\tau) - \theta^k(\tau) - \theta^{k+1}(\tau)}{2\Delta\tau} + \frac{\theta^{k+1}(\tau) - \theta^k(\tau)}{\Delta\eta}$$

$$= a \left[\frac{\phi^k(\tau) + \phi^{k+1}(\tau)}{2} - \frac{\theta^k(\tau) + \theta^{k+1}(\tau)}{2} \right]. \tag{7-110}$$

With a zero initial condition,

$$\theta^k(0) = 0, \qquad \phi^k(0) = 0 \qquad \text{for all } k,$$

and a unit-step input as a boundary condition,

$$\theta^0(N\Delta\tau) = 1, \qquad N = 1, 2, \ldots,$$

Eq. (7–110) at $\tau = 0$ gives

$$\frac{\theta^k(\Delta\tau) + \theta^{k+1}(\Delta\tau)}{2} = 0.$$

Therefore

$$\begin{aligned}
1 + \theta^1(\Delta\tau) &= 0 \quad \text{or} \quad \theta^1(\Delta\tau) = -1 \quad \text{for} \quad k = 0, \\
\theta^1(\Delta\tau) + \theta^2(\Delta\tau) &= 0 \quad \text{or} \quad \theta^2(\Delta\tau) = +1 \quad \text{for} \quad k = 1, \\
\theta^2(\Delta\tau) + \theta^3(\Delta\tau) &= 0 \quad \text{or} \quad \theta^3(\Delta\tau) = -1 \quad \text{for} \quad k = 2, \quad \text{etc.}
\end{aligned} \tag{7-111}$$

Equations (7–111) are obviously wrong! From a mathematical point of view, the solution of the difference equation based upon the relations (7–108) and (7-109) does not tend to the exact solution of the original partial differential equation as $\Delta\tau$ and $\Delta\eta$ tend to zero. Such investigations are difficult; see, for instance, the complicated form of Eq. (7-99). Therefore it is far better to take the physical point of view to avoid just such a false step.

The above difficulty can be avoided if we set up difference equations for a physically meaningful model. The partial differential equations for a percolation process were derived in Section 7-1 by using a physical model which is similar to Example 5-6, Section 5-3. We therefore construct the discrete physical model of Fig. 7-23(b). In this model fluid passes through a series of agitated tanks where each tank has a lumped heat capacitance. The heat balance equation, similar to Eq. (5-38), for the fluid in the kth tank is

$$C^* \Delta z \frac{d\theta^k}{dt} = C^* v(\theta^{k-1} - \theta^k) + \frac{\Delta z}{R^*}(\phi^k - \theta^k),$$

which, with normalization by the relations

$$\tau = t/L, \quad \eta = z/l, \quad \Delta\eta = \Delta z/l, \quad L = l/v, \quad \text{and} \quad a = L/(R^* C^*),$$

becomes

$$\frac{d\theta^k}{d\tau} + \frac{\theta^k - \theta^{k-1}}{\Delta\eta} = a(\phi^k - \theta^k). \tag{7-112}$$

A similar heat balance on the kth lumped solid yields

$$\frac{d\phi^k}{d\tau} = b(\theta^k - \phi^k). \tag{7-113}$$

These equations predict the correct answer when the response is computed and under the limiting process, $\Delta\eta \to 0$ (we shall omit the justification). Moreover, a lumped model in this form may sometimes represent a real percolation process better than the partial differential equation, because a physical system always has some longitudinal diffusion [21].

In general, distributed-parameter systems may have two kinds of inputs and outputs. These are:

$\mathbf{u}_e(t)$ = end (or boundary) input,
$\mathbf{y}_e(t)$ = end (or boundary, or terminal) output,
$\mathbf{u}_d(t, \mathbf{z})$ = distributed input,
$\mathbf{y}_d(t, \mathbf{z})$ = distributed response (Fig. 7-24),

where z is a space coordinate. Usually a distributed controlling input

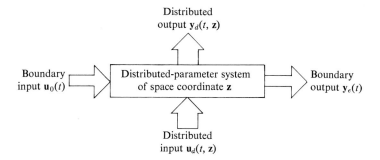

Fig. 7-24 Input and output of a distributed-parameter system.

(which is a member of u_d) and a distributed controlled variable (which is a member of y_d) must be discretized for realization by on-line computing control. To show the relations between these signals in a percolation process, we introduce a distributed input $u(t, z)$ as the ambient temperature along a pipe, and write an equation similar to Eq. (7–15):

$$\frac{\partial \phi}{\partial t} = \frac{1}{R^* C_s^*}(\theta - \phi) + \frac{1}{R_0^* C_s^*}(u - \phi)$$

or, in a normalized form,

$$\frac{\partial \phi}{\partial \tau} = b(\theta - \phi) + b_0(u - \phi), \qquad (7\text{--}114)$$

where $b_0 = L/(R_0^* C_s^*)$ and R_0^* = outside surface resistance per unit length. Other symbols were defined in Section 7–1. Solving Eq. (7–114) for $\phi(D)$ in operational form, we have

$$\phi = \frac{b\theta + b_0 u}{D + b_1}, \qquad (7\text{--}115)$$

where $b_1 = b + b_0$. Substituting Eq. (7–115) in Eq. (7–93), we obtain

$$\frac{d\theta}{d\eta} = -H(D)\theta + B(D)u, \qquad (7\text{--}116)$$

where

$$H(D) = \frac{a(D + b_0)}{D + b_1} + D, \qquad B(D) = \frac{ab_0}{D + b_1}.$$

Integrating Eq. (7–116) with respect to η, we find the distributed output:

$$\theta(\tau, \eta) = e^{-H(D)\eta}\theta_1(\tau) + e^{-H(D)}\int_0^\eta e^{H(D)v} B(D)u(\tau, v)\, dv. \qquad (7\text{--}117)$$

306 Distributed-parameter systems

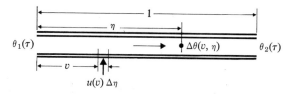

Fig. 7-25 A percolation process with distributed input.

$\theta(\tau, \eta)$ becomes an end output when $\eta = 1$; thus

$$\theta_2(\tau) = e^{-H(D)}\theta_1(\tau) + e^{-H(D)} \int_0^1 e^{H(D)v} B(D)u(\tau, v)\, dv. \qquad (7\text{-}118)$$

Defining four transfer functions,

$G_{ed}(D, \eta) = e^{-H(D)\eta}$ for end input to distributed output,

$G_{dd}(D, \eta, v) = \lim\limits_{\Delta\eta \to 0} \dfrac{\Delta\theta(\eta)}{u(v)\Delta\eta} = B(D)e^{-H(D)(\eta-v)}$ for distributed output at η due to distributed input at v (see Fig. 7-25),

$G_{ee}(D) = e^{-H(D)}$ for end input to end output, and

$G_{de}(v) = B(D)e^{-H(D)(1-v)}$ for distributed input at v to end output, derived by the relation

$$\Delta\theta_2(\tau) = e^{-H(D)} \int_v^{v+\Delta\eta} e^{H(D)w} B(D)u(\tau, w)\, dw \approx e^{-H(D)(1-v)} B(D)u(\tau, v)\Delta\eta,$$

both distributed output and end output are given in terms of distributed input, end input, and transfer functions:

$$\theta(\tau, \eta) = G_{ed}(D, \eta)\theta_1(\tau) + \int_0^\eta G_{dd}(\eta, v)u(\tau, v)\, dv \qquad (7\text{-}119)$$

and

$$\theta_2(\tau) = G_{ee}(D)\theta_1(\tau) + \int_0^1 G_{de}(v)u(\tau, v)\, dv. \qquad (7\text{-}120)$$

However, as noted above, discrete models rather than distributed-parameter models are more meaningful for the design of multivariable direct digital control systems.

REFERENCES

1. Rosenbrock, H. H., Storey, C., *Computational Techniques for Chemical Engineers.* Oxford: Pergamon Press, 1966, Chapter 7.
2. Magnusson, P. C., *Transmission Lines and Wave Propagation.* Boston: Allyn and Bacon, 1965.

3. Brown, F. T., "A unified approach to the analysis of uniform one-dimensional distributed systems." *ASME Paper 66-WA/AUT-20*, preprint for the ASME 1966 Winter Annual Meeting.
4. Auslander, D. M., *Analysis of Networks of Wavelike Transmission Elements*, Doctoral Thesis, M.I.T., August, 1966.
5. Blodgett, R. E., King, R. E., "Hydraulic system simulation using time-delay elements," *J. Basic Eng., ASME Trans.*, **89**, June 1967, pp. 315–320.
6. Nichols, N. B., "The linear properties of pneumatic transmission lines." *ISA Trans.*, **1**, No. 1, January 1962.
7. Brown, F. T., "The transient response of fluid lines." *J. Basic Eng., ASME Trans.*, **84**, 1962, p. 547.
8. Oldenburger, R., Goodson, R. E., "Simplification of hydraulic line dynamics by use of infinite products." *J. Basic Eng., ASME Trans.*, **86**, 1964, pp. 1–10.
9. McCausland, I., "Computation of time-optimal inputs for linear systems." *1965 Tokyo-IFAC* preprint volume, pp. III-37–45.
10. Yeh, H. H., Tou, J., "Design of optimal control for a class of distributed-parameter systems." *1966-JACC* preprint volume, pp. 684–693.
11. Wang, P. K. C., Tung, F., "Optimal control of distributed-parameter systems," *J. of Basic Eng., ASME Trans.*, **86**, 1964, pp. 67–69.
12. Takahashi, Y., "Transfer function analysis of heat exchangers," *Automatic and Manual Control*. London: Butterworths, 1952, pp. 235–248.
13. Thal-Larsen, H., "Dynamics of heat exchangers and their models." *J. Basic Eng., ASME Trans.*, **82**, 1960, pp. 489–504.
14. Nusselt, W., "Kreuzstrom Wärmeaustausch." *VDI-Z*, **55**, No. 48, 1911, pp. 2021–2024.
15. Carslaw, H. S., Jaeger, J. C., *Conduction of Heat in Solids*. London: Clarendon Press, 1947.
16. Profos, P., *Die Behandlung von Regelproblemen vermittels des Frequenzganges des Regelkreises und ihre Anwendung auf die Temperaturregelung durchströmter Rohrsysteme*, Doctoral Thesis, Eidgenössischen Technischen Hochschule, Zürich, 1943.
17. Chandler, R. E., et al., "Traffic dynamics: studies in car-following." *Oper. Research*, **6**, 1958, pp. 165–184.
18. Paynter, H. M., Takahashi, Y., "A new method of evaluating dynamic response of counterflow and parallel-flow heat exchangers." *ASME Trans.*, **78**, May 1956, pp. 749–758.
19. Masubuchi, M., "Dynamic response and control of multipass heat exchangers." *J. of Basic Eng., ASME Trans.*, **82**, 1960, pp. 51–65.
20. Churchill, R. V., *Operational Mathematics* (2nd ed.). New York: McGraw-Hill, 1958.
21. Wagner, R., "Dynamic study of temperature transducers by use of an optical method." *J. of Basic Eng., ASME Trans.*, **89**, June 1967, pp. 287–294.

22. Paynter, H. M., *Analysis and Design of Engineering Systems*. Cambridge, Mass.: M.I.T. Press, 1961.

23. Shapiro, A., *The Dynamics and Thermodynamics of Compressible Fluid Flow*. New York: Ronald Press, 1953.

PROBLEMS

7-1 A uniform fluid line (an artery, for instance) is exposed to zero back pressure at the right-hand end. The line has total resistance R_t and total capacitance (due to compliance) C_t. Using a one-lump T-model (see Fig. P7-1a), a two-lump T-model (Fig. P7-1), and a distributed-parameter model, obtain k, T_1, and T_2 in the first-order approximation of the line impedance $Z(D) = p_1/f_1 = k(1 + T_1 D)/(1 + T_2 D)$.

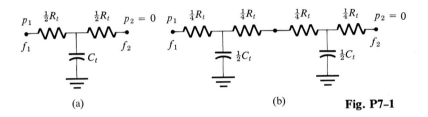

Fig. P7-1

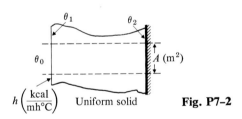

Fig. P7-2

7-2 The right-hand side of a uniform, heat-conducting solid is completely heat insulated, and the left-hand surface [of area A (m²)] is exposed to an ambient temperature θ_0 (°C) (see Fig. P7-2). Heat conduction in the solid is one-dimensional, and the thermal conductivity of the solid is k (kcal/m·hr·°C). Thermal diffusivity α (m²/hr) of the solid is defined by

$$\alpha = \frac{k}{wc},$$

where w is the specific weight of the solid (kg/m³) and c is its specific heat (kcal/kg·°C). The left-hand surface heat transfer (Newton's law of cooling) is characterized by heat-transfer coefficient h (kcal/m²·hr·°C), which in turn is expressed by a dimensionless number Bi, called the Biot number. For a solid of thickness l (m),

$$Bi = \frac{hl}{k}.$$

Obtain the transfer function that relates insulated surface temperature θ_2 (°C) to the fluid temperature θ_0:

$$G(D) = \frac{\theta_2}{\theta_0}.$$

This result must be expressed in terms of α, Bi, and l. Also obtain a second-order approximation of the transfer function. Determine a characteristic time that will completely normalize the end results.

7-3 For an IC-line of 20°C air at sea level, determine a characteristic length l_e for which the characteristic time $T_s = l\sqrt{I^*C^*} = \sqrt{I_tC_t}$ is one.

7-4 The right end of an ideal uniform lossless fluid line (an IC-line) is closed, and flow into the left end is the input. Determine the eigenvalues of the distributed-parameter system and compare the results with those for the single-π model. Also discuss the single T model.

7-5 In an ideal IC-line, let $T_s = \sqrt{IC}$ and obtain an analytical solution for the following conditions:

Initial condition: $p(z, 0) = 0$, $f(z, 0) = f_0 = $ const for all z;
Boundary condition: $p(0, t) = p_1(t) = $ a given function of time at $z = 0$, and $\lim_{z \to \infty} p(z, t) = 0$ (i.e., the line is semi-infinite).

The required solution is $p(z, t)$ for $0 \leq z < \infty$, $0 \leq t$.

7-6 For an ICR-line of total length l, define an energy function

$V = $ total energy of the line.

Obtain dV/dt and prove that the system is asymptotically stable if and only if $R > 0$.

7-7 An elastic fluid ($C > 0$, $I > 0$, $R = 0$) in a uniform line has a left-to-right velocity v. It is subsonic; i.e., $M < 1$ where $M = v/c$ (Mach number) and $c = $ sonic velocity. Determine R in the matrix equation

$$\frac{\partial}{\partial z}\begin{bmatrix} p \\ f \end{bmatrix} = -\mathbf{R}\begin{bmatrix} p \\ f \end{bmatrix}.$$

7-8 Diagonalize the **R**-matrix obtained in Problem 7-7 and construct a block diagram in the diagonalized domain. Assume the flow speed to be subsonic.

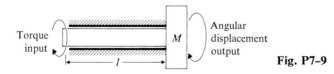

Fig. P7-9

7-9 A load is connected to an elastic shaft of total length l (see Fig. P7-9). The shaft is characterized by uniformly distributed parameters defined per unit length of the shaft: b_i (damping coefficient) for internal friction, b_e (damping coefficient) for external friction, k for "spring constant" (elasticity), and m for mass (polar moment of inertia of a rotating system). Mass (or polar moment of inertia) of the right-hand load is M. Input to the system is torque applied at the left end, and

310 Distributed-parameter systems

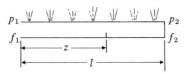

Fig. P7-10

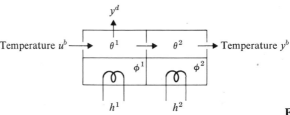

Fig. P7-11

output is displacement of the load. Determine the transfer function of the system by using both the velocity–potential analogy and the force–potential analogy.

7-10 The right end of a uniformly perforated hose is closed (see Fig. P7-10). The momentum effect of water flowing through the hose is negligible. System parameters per unit length of the line are R_z^* for line resistance, R_y^* for shunt resistance (causing leakage), and C^* for hose compliance (and fluid elasticity which is negligible). Determine the impedance of the system. Also obtain a transfer function that will relate leakage flowrate (per unit length) at location z to supply pressure p_1.

7-11 Show that a two-lump model of a percolation process (see Fig. P7-11) is described by

$$\frac{d}{dt}\begin{bmatrix}\theta^1\\\phi^1\\\theta^2\\\phi^2\end{bmatrix} = \begin{bmatrix}-(2+a) & a & 0 & 0\\b & -b & 0 & 0\\2 & 0 & -(2+a) & a\\0 & 0 & b & -b\end{bmatrix}\begin{bmatrix}\theta^1\\\phi^1\\\theta^2\\\phi^2\end{bmatrix} + \begin{bmatrix}2 & 0 & 0\\0 & 1 & 0\\0 & 0 & 0\\0 & 0 & 1\end{bmatrix}\begin{bmatrix}u^b\\u^{1d}\\u^{2d}\end{bmatrix},$$

$$\begin{bmatrix}y^b\\y^d\end{bmatrix} = \begin{bmatrix}0 & 0 & 1 & 0\\1 & 0 & 0 & 0\end{bmatrix}\begin{bmatrix}\theta^1\\\phi^1\\\theta^2\\\phi^2\end{bmatrix},$$

where θ^1, θ^2, ϕ^1, and ϕ^2 are lumped fluid and solid temperatures as shown in the figure, u^b is incoming fluid temperature (boundary input), y^b is outgoing fluid temperature (boundary output), and $\theta^1 = y^d$ is the fluid temperature in the first lump taken as a distributed output. Distributed heat input to the solid is represented by the heat input rate per unit length of line for each lump h_1 and h_2, from which a distributed input that has the dimension of temperature is defined by $(L/C_s^*)h^k = u^{kd}$, $k = 1$ and 2. See Eqs. (7-10) through (7-13) for the meaning of L, C_s^*, a, b, and τ. Generalize the vector equations to an N-lump model.

7-12 The lumped model of a percolation process where time is also discretized is more appropriate than the model of Problem 7-11 for simulating the "piston flow"

Problems 311

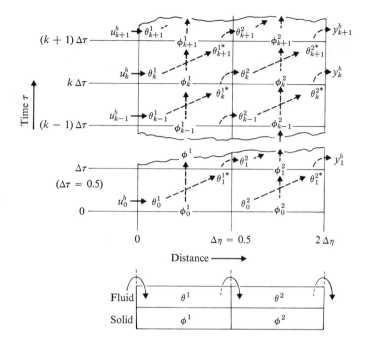

Fig. P7-12

characteristic of the fluid. The scheme of time and space discretization is shown in Fig. P7-12 for a two-lump model. Time quantization size $\varDelta\tau$ must be equal to the size (length of tube) of the space lump $\varDelta\eta$ in the normalized time and space coordinate variables τ and η, so that the characteristic line [AB' in Fig. 7-19(c)] will form a diagonal in all cubes of the discretization map. At the kth instant ($\tau = k\varDelta\tau$) solid temperatures are ϕ_k^1 and ϕ_k^2, which, $\varDelta\tau$ time units later, take on the values ϕ_{k+1}^1, ϕ_{k+1}^2 respectively. The temperature change is due to batchwise heat transfer between the solid and fluid in each lump. We consider the left-hand lumped space of fluid to be filled by supply flow of temperature u_{k-1}^b at time $(k - 1)\varDelta\tau$. Fluid temperature θ^1 of this space changes to $(\theta_k^1)^*$ at time $k\varDelta\tau$, at which instant this fluid is instantaneously moved (transferred) into the right-hand space so that the fluid temperature of the right-hand space θ_k^2 is equal to $(\theta_k^1)^*$. In the fluid space, θ_k^2 changes to $(\theta_{k+1}^2)^*$ in $\varDelta\tau$ time units, and this fluid will be instantaneously discharged at time point $(k + 1)\varDelta\tau$. Although there is no third lump in the model, the temperature of the outgoing fluid may be expressed by θ^3, which at the same time is the boundary output. Hence $(\theta_{k+1}^2)^* = \theta_{k+1}^3 = y_{k+1}^b$. Show that the following difference equation holds:

$$\begin{bmatrix} \phi_{k+1}^1 \\ \theta_{k+1}^2 \\ \phi_{k+1}^2 \\ \theta_{k+1}^3 \end{bmatrix} = \begin{bmatrix} p_{22} & 0 & 0 & 0 \\ p_{12} & 0 & 0 & 0 \\ 0 & p_{21} & p_{22} & 0 \\ 0 & p_{11} & p_{12} & 0 \end{bmatrix} \begin{bmatrix} \phi_k^1 \\ \theta_k^2 \\ \phi_k^2 \\ \theta_k^3 \end{bmatrix} + \begin{bmatrix} p_{21} \\ p_{11} \\ 0 \\ 0 \end{bmatrix} \cdot u_k^b, \quad y_k = [0\ 0\ 0\ 1] \begin{bmatrix} \phi_k^1 \\ \theta_k^2 \\ \phi_k^2 \\ \theta_k^3 \end{bmatrix},$$

312   Distributed-parameter systems

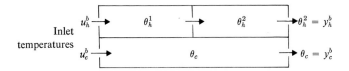

Fig. P7-13

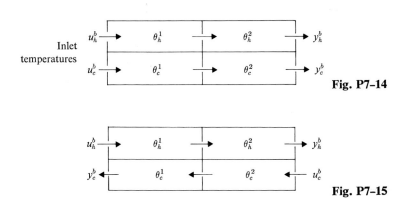

Fig. P7-14

Fig. P7-15

where

$$p_{11} = (b + ad)/(a + b), \quad p_{12} = a(1 - d)/(a + b),$$
$$p_{21} = b(1 - d)/(a + b), \quad p_{22} = (a + bd)/(a + b),$$
$$d = e^{-(a+b)/N}, \quad N = \text{number of lumps},$$

and a and b are the same parameters as in Problem 7–11.

7-13 Figure P7-13 shows a lumped model of the hybrid heat exchanger of Fig. 7-6. The model for the tube flow is the continuous-flow type given in Problem 7-11. Using dimensionless system parameters that appear in Eq. (7-20), obtain a vector differential equation for the lumped model. Is this lumped system observable for distributed output $y^d = \theta_h^1$?

7-14 Using a continuous-flow type two-lump approximation (same as the tube flow in Problem 7-11) on both hot and cold flows, obtain the vector differential equation for the parallel-flow heat exchanger shown in Fig. P7-14. Is this system observable for distributed output $y^{d1} = \theta_h^1$?

7-15 Obtain the vector differential equation of a counterflow heat exchanger (see Fig. P7-15) approximated by a two-lump flow-type model similar to that of Problem 7-14. Determine eigenvalues for $a_h = a_c = 1, r = 1$, and compare the result with the approximation given by Eq. (7-85).

7-16 A cooler system is shown in Fig. P7-16, which is the "one-side-mixed" or hybrid type. The mixed fluid temperature θ_i is the forced input; it is imposed on the system unilaterally, such as in the case of a vapor condenser where vapor pressure is the input. The inlet temperature of cooling water θ_{c1} is kept at zero. Solid wall capacitances are negligible. Slug velocity of the cooling water in the tube is

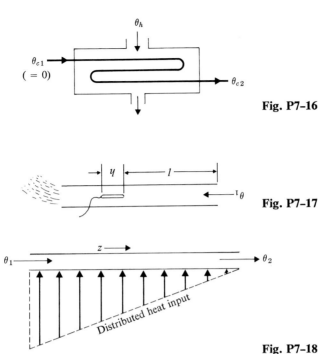

Fig. P7-16

Fig. P7-17

Fig. P7-18

$v = 1$ m/sec, and total effective tube length is $l = 1$ m. Let

$$a = \frac{\text{area of heating surface} \times \text{overall heat-transfer coefficient}}{\text{total heat capacitance of water in the tube}}.$$

(a) Obtain the transfer function for the cooling-water outlet temperature,

$$G(D) = \frac{\theta_{c2}}{\theta_h},$$

and (b) determine the unit-step input response of θ_{c2} for zero initial state.

7-17 The inlet temperature $\theta_1(t)$ of a piston flow of velocity v through an insulated uniform pipe is considered to be an input. Temperature of the fluid at location l is measured by a sensing bulb of axial length h (see Fig. P7-17). The sensor is assumed to be ideal, i.e., it has no heat capacitance and its gain is one. Let $\theta_m(t)$ be the sensor output and obtain its transfer function $G(D) = \theta_m/\theta_1$ in terms of $h/v = \tau$ and $l/v = L$.

7-18 The length of a pipe, fluid velocity through the pipe, and the capacitance of the fluid contained in the pipe are all 1. Distributed heat input along the line (see Fig. P7-18) is

$$u_d = (1 - z)u(t).$$

Heat capacitance of the pipe wall is negligible, and there is no heat loss. Determine

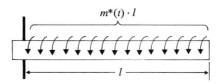

Fig. P7-19

the transfer function that will relate the moduling signal $u(t)$ of heat input and output fluid temperature θ_2. Sketch the step-input response pattern for zero initial state.

7-19 A uniform Y^*Z^* shaft of length l is subjected to a uniformly distributed torque input of magnitude $m^*(t)$ per unit length. The left end of the shaft is fixed, and the right end is free (see Fig. P7-19). Obtain the transfer function

$$G(D) = \text{right end velocity/torque magnitude } m^*(t).$$

Try both velocity-potential and force-potential analogies.

7-20 The transmission of a signal $\theta(t)$ through a pipe is described by

$$\frac{\partial \theta}{\partial t} + v \frac{\partial \theta}{\partial z} = 0,$$

where v is slug velocity of the carrier, $\theta = \theta_1(t)$ at the left end $(z = 0)$, and $\theta = \theta_2(t)$ at the right end $(z = l)$. Pressure of the carrier medium at the end is

$$p = p_1 = \text{const} \quad \text{at} \quad z = 0 \quad \text{and} \quad p = p_2 = 0 \quad \text{at} \quad z = l.$$

The line has uniformly distributed series resistance R_z^* and shunt resistance R_y^* (the carrier is leaking) per unit length of the line for the carrier flow, where the two resistances are related to each other by

$$\sqrt{R_z^*/R_y^*}\, l = 1.$$

Determine the transfer function

$$G(D) = \theta_2/\theta_1.$$

PART 3

CONTROL OF LINEAR SYSTEMS

Chapter 8
Scalar Input–Output Linear Systems and Feedback Control

Chapter 9
Frequency Response

Chapter 10
Multivariable Systems

Chapter 11
Linear Digital Control

8

SCALAR INPUT-OUTPUT
LINEAR SYSTEMS AND FEEDBACK CONTROL

Time-domain analysis and design of linear, stationary, continuous-in-time, single-loop feedback control systems are the main subjects of discussion in this chapter. We use conventional or classical control theory throughout this chapter, as it is most appropriate to the subject matter. Section 1 presents, in terms of the poles and zeros in the s-domain, the basic correlation between a system without control, called an open-loop system, and the system under feedback control, a closed-loop system. The response of a typical feedback control system is usually characterized by conspicuous damped oscillation. This oscillation plays an important role in time-domain specifications commonly accepted as design objectives for various control systems. The dominant mode of oscillation and its time-domain pattern are discussed in Section 2. The root-locus technique is introduced in Section 3. From the root-locus viewpoint a control system's eigenvalues (closed-loop poles) are functions of a variable system parameter. The method often serves as a useful guide in control system design. Performance indices, optimum tuning of controller adjustments, and design of compensation components are the main topics discussed in Section 4.

8-1 STRUCTURE OF A FEEDBACK CONTROL SYSTEM

Figure 8-1 shows a common structure of single variable, single-loop feedback control systems. Such configurations are widely used in various kinds of engineering systems and subsystems, with the purpose of keeping a plant (controlled) variable $C(s)$ close to its desired (reference) value $R(s)$. The variable s indicates that all signals are considered in the Laplace domain. The scalar nature of the relationship (see below) between the input $R(s)$ and the output $C(s)$ permits the efficient application of classical control theory. Although it will work, a vector approach is not necessary in this case.

In Fig. 8-1, a feedback controller represented by the transfer function, $G_c(s)$, with an actuating error input $E(s)$, produces a manipulated variable $M(s)$. The object of control, the plant, is represented by a transfer function $G_p(s)$ and a set of transfer functions $N_j(s)$, $j = 1, \ldots, u$. $G_p(s)$ acts on the manipulated variable $M(s)$, and $N_j(s)$ on the disturbance inputs (load upsets, etc.) $V_j(s), j = 1, \ldots, u$. All transfer functions are arrived at in the classical manner, by taking the Laplace transform of the output–input ratio for the

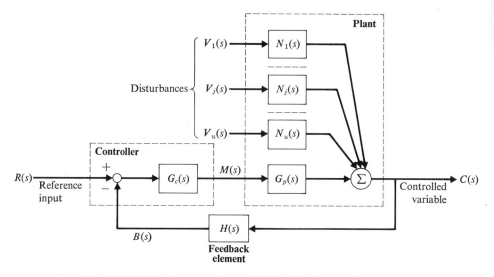

Fig. 8-1 Block diagram of a typical feedback control system.

element in question with zero initial conditions assumed. The feedback path may or may not have a feedback element $H(s)$. The feedback signal is $B(s)$. The difference between $B(s)$ and the reference input $R(s)$, $E(s) = R(s) - B(s)$, is fed into the controller to produce $M(s)$. This manipulated variable controls the plant in such a way that the disturbing effects of $V_j(s)$ on $C(s)$ are minimized and $C(s)$ is held in close correspondence with its desired value. If a system has unity feedback, $H(s) = 1$, then $B(s) = C(s)$ and

$$E(s) = R(s) - C(s).$$

$E(s)$ in this case is the error of the feedback control system. Otherwise a distinction between the error $E^*(s) = R(s) - C(s)$ and the controller input signal $E(s) = R(s) - B(s)$ must be made; the latter is called an actuating error or signal. In the following we most often consider unity-feedback systems.

Since the s-domain relations shown in Fig. 8-1 are algebraic, we can easily determine various responses in terms of inputs $R(s)$ and $V_j(s)$:

$$C(s) = \frac{G_c(s)G_p(s)}{1 + G(s)} R(s) + \sum_{j=1}^{u} \frac{N_j(s)}{1 + G(s)} V_j(s), \qquad (8\text{-}1)$$

$$E(s) = \frac{1}{1 + G(s)} R(s) - \sum_{j=1}^{u} \frac{H(s)N_j(s)}{1 + G(s)} V_j(s). \qquad (8\text{-}2)$$

In Eqs. (8-1) and (8-2),

$$G(s) = G_c(s)G_p(s)H(s) \qquad (8\text{-}3)$$

is called the open-loop transfer function. [The term "open loop" is used both for the quantity defined by Eq. (8–3) and to describe the response of a plant that is *not* under feedback control.] According to these relations, an ideal unity-feedback control, or one having

$$H(s) = 1, \quad C(s) = R(s), \quad E(s) = 0 \quad \text{for all } R(s) \text{ and } V_j(s),$$

can be reached if the system remains stable when the controller gain is made infinitely high:

$$|G_c(s)| = k_c \to \infty.$$

Note that the 1 is negligible in the denominators of Eqs. (8–1) and (8–2) at this limit. Since almost all engineering systems become unstable before reaching the ideal limit, designers must make an optimal compromise between control accuracy and system stability. Equations (8–1) and (8–2) have a common denominator $1 + G(s)$. The roots of the denominator equation,

$$1 + G(s) = 0, \tag{8-4}$$

are the eigenvalues of the system. They are called the closed-loop poles. Since a system's stability depends on its eigenvalues (see Chapter 4), the characteristic equation (8–4) plays a crucial role in a feedback control system's design and analysis.

Let us consider a linear, lumped-parameter system described by an open-loop transfer function in the general form,

$$G(s) = G_c(s)G_p(s) = \frac{b_0 + b_1 s + \cdots + b_m s^m}{a_0 + a_1 s + \cdots + a_n s^n}, \tag{8-5}$$

where unity feedback is assumed, $H(s) = 1$. For physically realizable transfer functions, the order of the denominator polynomial must be equal to or greater than that of the numerator polynomial (Section 6–1),

$$m \leqslant n. \tag{8-6}$$

Writing the polynomials of (8–5) in a factored form, we obtain the second canonical form,

$$G(s) = K \frac{\prod (s - z_j)}{\prod (s - p_i)}, \tag{8-7}$$

where K is the open-loop gain [equal to b_m/a_n from Eq. (8–5)], and z_j and p_i, $j = 1, \ldots, m$, $i = 1, \ldots, n$, are called the *open-loop* zeros and poles, respectively. Since all coefficients of the polynomials in Eq. (8–5) are real, poles and/or zeros which are complex must appear in conjugate pairs:

$$p_i \quad \text{or} \quad z_j = -\alpha \pm j\beta, \tag{8-8}$$

where α and β are both real, and $\beta > 0$. It is sometimes convenient to write

(8-7) in the following canonical form where all numbers are real,

$$G(s) = K \frac{\prod (s - z_j) \prod [(s + \alpha_j)^2 + \beta_j^2]}{\prod (s - p_i) \prod [(s + \alpha_i)^2 + \beta_i^2]}, \qquad (8\text{-}9)$$

and where z_j and p_i in Eq. (8-9) are real zeros and poles, respectively.

The *closed-loop poles* $s = P_1, P_2, \ldots, P_n$ are the roots of Eq. (8-4). Applying (8-5), we find that the closed-loop system's characteristic equation becomes

$$(a_0 + a_1 s + a_2 s^2 + \cdots + a_n s^n) + (b_0 + b_1 s + \cdots + b_m s^m)$$
$$= \prod_1^n (s - P_i) = 0. \qquad (8\text{-}10)$$

We write the closed-loop response $C(s)$ of Eq. (8-1) in the following form:

$$C(s) = G_{LR}(s) R(s) + \sum_{j=1}^{u} G_{LV_j} V_j(s), \qquad (8\text{-}11)$$

where the *closed-loop transfer functions* are defined by

$$G_{LR}(s) = \frac{G(s)}{1 + G(s)}, \qquad G_{LV_j}(s) = \frac{N_j(s)}{1 + G(s)}, \qquad (8\text{-}12)$$

and unity feedback, $H(s) = 1$, is assumed. By Eqs. (8-7) and (8-10) the closed-loop transfer function for reference input $G_{LR}(s)$ can be written in a factored form:

$$G_{LR}(s) = \frac{K \prod_1^m (s - z_j)}{\prod_1^n (s - p_i) + K \prod_1^m (s - z_j)} = \frac{K \prod_1^m (s - z_j)}{\prod_1^n (s - P_i)}, \qquad (8\text{-}13)$$

where n is assumed to be greater than m, which is a stronger condition than (8-6) but is generally satisfied in real systems. Letting

$$P_i = -\sigma_i \pm j\omega_i \qquad (8\text{-}14)$$

for conjugate complex closed-loop poles, where σ_i and ω_i are real and $\omega_i > 0$, we obtain an alternative form of (8-13):

$$G_{LR}(s) = \frac{B(s)}{\prod (s - P_i) \prod [(s + \sigma_i)^2 + \omega_i^2]}, \qquad (8\text{-}15)$$

where $B(s)$ is the numerator polynomial in s, and P_i are the real closed-loop poles.

A linear, lumped-parameter, stationary feedback control system is asymptotically stable if and only if all real poles and all real parts of conjugate complex poles are negative:

$$P_i = -\lambda_i, \qquad \sigma_i > 0, \qquad (8\text{-}16)$$

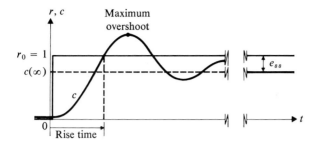

Fig. 8-2 A typical unit-step input response of a control system.

where $\lambda_i > 0$. According to Eqs. (8-1), (8-2), and (8-15), the following modes of motion will appear in a stable control system's response:

$$e^{-\lambda_i t}, \quad e^{-\sigma_i t} \sin \omega_i t, \quad \text{and} \quad e^{-\sigma_i t} \cos \omega_i t, \tag{8-17}$$

where all closed-loop poles are assumed to be distinct. All modes of (8-17) decay exponentially with time, where the decay rate is proportional to λ_i or σ_i. The time constants T_i associated with real poles are identically equal to $1/\lambda_i$. For complex conjugate poles one may think of the time constant of the envelope of the oscillatory response as equal to $1/\sigma_i$. As we shall see in the next section, a control system quite often has an oscillatory mode $e^{-\sigma_1 t} \sin \omega_1 t$ and/or $e^{-\sigma_1 t} \cos \omega_1 t$ whose decay is far slower than the remaining modes. In other words, the system has a pair of predominating conjugate complex poles,

$$P_{d1}, P_{d2} = -\sigma_1 \pm j\omega_1, \tag{8-18}$$

where σ_1 is considerably *smaller* than all real poles and the real parts of all other complex conjugate poles (that is, the time constant of the envelope of the oscillatory response is greater than all the remaining system time constants). Such a pair of roots of the characteristic equation are called *dominant* poles of the system.

8-2 RESPONSE OF FEEDBACK CONTROL SYSTEMS

Figure 8-2 shows the response, due to a unit-step reference input, of a typical control system starting from a zero initial state. By inverse Laplace transformation, we have

$$c(t) = \mathscr{L}^{-1}[C(s)] = \mathscr{L}^{-1}[R(s)G_{LR}(s)] = \mathscr{L}^{-1}\left[\frac{1}{s} G_{LR}(s)\right]. \tag{8-19}$$

An expansion of $C(s)$ into partial fractions (Section 2-4) yields

$$C(s) = \frac{C_0}{s} + \frac{C_1 s + C_2}{(s + \sigma_1)^2 + \omega_1^2} + \left\{\begin{matrix}\text{terms for the rest of} \\ \text{the closed-loop poles}\end{matrix}\right\}, \tag{8-20}$$

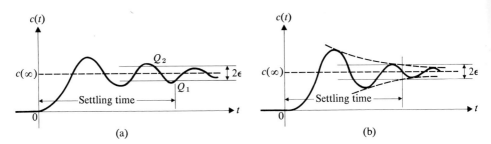

Fig. 8-3 Settling time.

where C_1, C_2 = coefficients of the expansion for the dominant poles. The step-input response pattern is dominated by the steady-state component, $c(\infty)$, introduced by the step input, and a damped oscillation of the dominant mode. All other modes usually vanish shortly after the step-input application, and would appear as ripples and distortions of the $c(t)$-curve in the vicinity of the origin of Fig. 8-2. Note carefully that the nature of the response $c(t)$ to a particular input $r(t)$ depends solely on the roots of the characteristic equation [Eq. (8-4), that is, the poles of Eq. (8-15)].

The response of Fig. 8-2 is characterized in terms of such quantities as the rise time, the percentage overshoot, the settling time, and the final value of error. Some or all of these are currently used as time-domain specifications [3], but various authors define them somewhat differently. For this reason we shall give physical explanations of each quantity rather than exact definitions. The rate of the initial rise of a step-input response is represented by the rise time. It may be the time when $c(t)$ first equals the magnitude of input $r_0 = 1$, as shown in Fig. 8-2, or the time for the rise of $c(t)$ across a prescribed range such as from $0.1\,r_0$ to $0.9\,r_0$. The percentage overshoot PO can be defined by

$$PO = 100\frac{[\text{maximum value of } c(t)] - [\text{final value}, c(\infty)]}{c(\infty)}.$$

The settling time is defined as the time at which the response attains and, from that point on, remains within $\pm\varepsilon$ of its final value $c(\infty)$, where 5% is a typical value for ε. With this definition, the settling time may jump from Q_1 to Q_2 if the response pattern undergoes a very small change, Fig. 8-3(a). To avoid a discontinuity, the settling time is sometimes defined for a smooth envelope of the decaying oscillation, as shown in Fig. 8-3(b).

The final value of error, defined as

$$e_{ss} = \lim_{t \to \infty}(r(t) - c(t)) = r_0 - c(\infty),$$

is a measure of static accuracy (Fig. 8-2). Exact analytical expressions for the rise time, the percentage overshoot, and the settling time become prohibitively complicated for systems of order higher than two (see, for instance, [4] and [5] for the analysis of second-order system response). However, a general discussion of the steady-state error (including the final value of error e_{ss} for a step reference input) is possible with an application of the final-value theorem:

$$\lim_{t \to \infty} e(t) = \lim_{s \to 0} sE(s) . \tag{8-21}$$

To determine the steady-state error or $\lim_{t \to \infty} e(t) = e(\infty)$ more generally, we use a different canonical form:

$$G(s) = k \frac{\prod (1 + T_j s) \prod [1 + 2\zeta_j(T_j s) + (T_j s)^2]}{s^N \prod (1 + T_i s) \prod [1 + 2\zeta_i(T_i s) + (T_i s)^2]} \tag{8-22}$$

for the open-loop transfer function of a system with unity feedback. A similar form with k replaced by k_n can be used for the plant transfer function $N(s)$ for disturbance input $V(s)$. T_i and T_j in Eq. (8–22) are time constants, and the quadratic factors with damping ratios $|\zeta| < 1$ represent conjugate complex poles or zeros. We also assume the same form for inputs $R(s)$ and $V(s)$, where k in Eq. (8–22) is replaced by k_r and k_v, respectively, and the order of the denominator (in s) is assumed greater than that of the numerator so that the amplitude of inputs is kept finite. The canonical form of (8–22) is called *type* 0 when $N = 0$, *type* 1 when $N = 1$ and *type* 2 when $N = 2$.

Before applying Eq. (8–21) to a discussion of steady-state error, we digress for a moment to compare Eqs. (8–9) and (8–22). The canonical form of Eq. (8–9) is used for the root-locus approach, to be discussed in the next section, and the canonical form of Eq. (8–22) is used for the frequency-response approach, which will be discussed in the next chapter. It is desirable to be able to interpret the results found by one approach in terms of the other. For such purpose the following easily derived relations are listed [in all cases the variables from Eq. (8–22), for frequency-response calculations, appear on the left]:

$N =$ the number of poles at the origin ($p_i = 0$),
$T_j = -1/z_j$ for the m' real zeros,
$T_i = -1/p_i$ for the n' real poles,
$T_j^2 = 1/(\alpha_j^2 + \beta_j^2) = 1/\omega_j^2$ for the $m - m'$
  complex conjugate zeros, $\tag{8-23}$
$T_i^2 = 1/(\alpha_i^2 + \beta_i^2) = 1/\omega_i^2$ for the $n - n'$ complex conjugate poles,
$k = K(\prod_1^m T_j / \prod_1^n T_i)$
$\zeta_j = \alpha_j/\sqrt{\alpha_j^2 + \beta_j^2}$ for the $m - m'$ complex conjugate zeros,
$\zeta_i = \alpha_i/\sqrt{\alpha_i^2 + \beta_i^2}$ for the $n - n'$ complex conjugate poles.

The denominator of the last expression is the undamped natural frequency ω_{n_i} of the oscillatory response. The damped actual frequency of the response is given by the imaginary part of the complex conjugate pole:

$$\beta_i = \omega_{n_i}\sqrt{1 - \zeta_i^2}. \tag{8-24a}$$

The real part of the complex conjugate pole is given by

$$\alpha_i = \zeta_i \omega_{n_i}. \tag{8-24b}$$

In the next section this will all be interpreted in terms of pole locations in the s-plane. We now return to a discussion of steady-state error by considering Eq. (8–21).

A step-input signal is a type 1 input. Generally, a type 1 signal has a steady-state value when all T's and ζ's of the signal are positive:

$$\lim_{t \to \infty} r(t) = r(\infty) = k_r, \qquad \lim_{t \to \infty} v(t) = v(\infty) = k_v. \tag{8-25}$$

A ramp input $r(t)$ or $v(t) = kt$ with $R(s)$ or $V(s) = k/s^2$ is a type 2 input, where k represents the slope of the ramp.

A feedback control system with a type 0 open-loop transfer function (and unity feedback) produces a steady-state error for a type 1 reference input. By Eqs. (8–2), (8–21), and (8–22), the error is found to be

$$e(\infty) = \lim_{s \to 0} [(1 + G(s))^{-1} s R(s)] = \frac{r(\infty)}{1 + k}. \tag{8-26}$$

The steady-state error of a control system with a type 0 open-loop transfer function subject to a type 1 reference input is called an *offset*. If $N(s)$ is also type 0, an offset due to disturbance is given by:

$$e(\infty) = \lim_{s \to 0} [(-N(s))(1 + G(s))^{-1} s V(s)] = \frac{-k_n v(\infty)}{1 + k}. \tag{8-27}$$

This is the case of a type 0 object with a *P*-control under a sustained disturbance of magnitude $v(\infty)$. The steady-state errors for type 1 inputs, both (8–26) and (8–27), vanish for an *I*-action control law (open-loop transfer function for a type 0 object is made type 1). The discussion of steady-state errors, given above only for practically important cases, can be extended to various combinations of input signal types and transfer function types [1, 2, 3, 6].

A major aim in system design is usually to make the system as insensitive as possible to changes in its parameters (such changes often occur because of aging, wear, manufacturing tolerances, etc). Systems designed with feedback configurations are preferred because of their effectiveness in reducing the sensitivity of the system's input–output relation to parameter changes.

8-2 Response of feedback control systems

Let us consider a control system's response to a reference input $R(s)$:

$$C(s) = \frac{G_f(s)}{1 + G_f(s)H(s)} R(s),$$

where $G_f(s) = G_c(s)G_p(s)$ is the forward-loop transfer function (Fig. 8-1). Suppose that a deviation ΔC appears for a change ΔG_f so that

$$C(s) + \Delta C = \frac{G_f(s) + \Delta G_f}{1 + G_f(s)H(s) + \Delta G_f H(s)} R(s).$$

It follows then, that

$$\Delta C = \left(\frac{G_f + \Delta G_f}{1 + G_f H + \Delta G_f H} - \frac{G_f}{1 + G_f H} \right) R(s)$$

$$= \left(\frac{(G_f + \Delta G_f)/G_f}{(1 + G_f H + \Delta G_f H)/(1 + G_f H)} - 1 \right) C(s)$$

or

$$\frac{\Delta C}{C} = \frac{(\Delta G_f)/G_f}{1 + G_f H + \Delta G_f H}. \tag{8-28}$$

If, for instance, G_f is a pure gain k_f and $H = 1$, then (8-28) gives a time-domain relation

$$\frac{\Delta c}{c} = \frac{\Delta k_f/k_f}{1 + k_f + \Delta k_f}. \tag{8-29}$$

When $k_f = 20$, a gain change $\Delta k_f = 2$ will produce a 10% change of $\Delta c/c$ in an open-loop system, whereas it is reduced to $(\frac{10}{23})\%$ by feedback.

As noted in Section 5-5, however, feedback is not effective in making $\Delta C/C$ insensitive to a feedback-gain change $\Delta H/H$. The output error ratio is expressed by [7]

$$\frac{\Delta C}{C} = \frac{-\Delta H/H}{1 + (G_f H)^{-1} + (\Delta H/H)}, \tag{8-30}$$

and thus

$$\frac{\Delta C}{C} \approx -\frac{\Delta H}{H}$$

if $|G_f H| \gg 1$ and $|\Delta H| \ll |H|$.

Example 8-1 *Servomechanism.* Let us drive the motor we discussed in Fig. 6-28 by a power amplifier that produces a voltage proportional to its input voltage. The input voltage is made proportional to an error in output position by the wiper mechanism shown in Fig. 8-4(a). The supply voltage p_1 in Fig. 6-28 now becomes the manipulated variable m, and output speed p_2 in the figure is equal to dc/dt or Dc when the

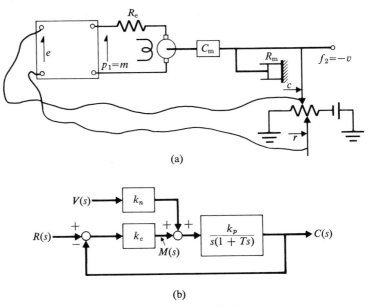

Fig. 8-4 A servomechanism.

output position is chosen as the controlled variable c. Assume there is a disturbing torque v such that $v = -f_2$, where f_2 is the right-hand torque shown in Fig. 6-28. If the inductance I_e of the motor circuit is negligible, the transfer matrix relation shown in Fig. 6-28 becomes

$$\begin{bmatrix} m \\ f_1 \end{bmatrix} = \begin{bmatrix} 1 & R_e \\ 0 & 1 \end{bmatrix} \begin{bmatrix} k_1 & 0 \\ 0 & k_2 \end{bmatrix} \begin{bmatrix} 1 & 0 \\ C_m D & 1 \end{bmatrix} \begin{bmatrix} 1 & 0 \\ R_m^{-1} & 1 \end{bmatrix} \begin{bmatrix} Dc \\ -v \end{bmatrix}.$$

Therefore,

$$m = -k_2 R_e v + [(k_2 R_e R_m^{-1} + k_1) + k_2 R_e C_m D] Dc . \tag{8-31}$$

Let us assume that the output impedance of the amplifier is sufficiently low so that the output voltage m is not affected by the load current f_1, and that m is proportional to position error $e = r - c$, with a proportionality constant k_c, that is

$$m = k_c e . \tag{8-32}$$

From Eqs. (8-31) and (8-32) we draw the block diagram of Fig. 8-4(b) where original system parameters are represented by k_p, k_n, and T such that

$$k_p = \frac{1}{k_2 R_e R_m^{-1} + k_1}, \quad k_n = k_2 R_e, \quad T = \frac{k_2 R_e C_m}{k_2 R_e R_m^{-1} + k_1}.$$

The open-loop transfer function of this system is

$$G(s) = \frac{k}{s(Ts + 1)},$$

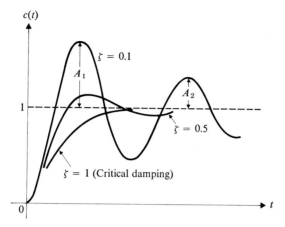

Fig. 8-5 Unit-step reference input response of a second-order servomechanism.

where

$$k = k_c k_p.$$

The characteristic equation of the servomechanism is

$$Ts^2 + s + k = 0, \qquad (8\text{--}33)$$

which has two closed-loop poles as its roots. The poles are both negative and real when $k < 1/(4T)$, while they become conjugate complex if $k > 1/(4T)$. A double pole appears at the borderline condition $k = 1/(4T)$, which is a condition sometimes called critical damping. A second-order system is said to be underdamped when it is oscillatory. We write Eq. (8-33) in the following canonical form when a system is underdamped:

$$s^2 + 2\zeta\omega_n s + \omega_n^2 = 0, \qquad \zeta < 1, \qquad (8\text{--}34)$$

where $\omega_n = \sqrt{k/T}$ and $\zeta = 1/(2\sqrt{kT})$ in the system of Fig. 8-4(b). The two roots of (8-34) are

$$s_1, s_2 = -\zeta\omega_n \pm j\sqrt{1-\zeta^2}\cdot\omega_n \qquad (8\text{--}35)$$

[compare to Eqs. (8-23) and (8-24)].

The period of oscillation is then given by

$$P = \frac{2\pi}{\sqrt{1-\zeta^2}\cdot\omega_n}, \qquad (8\text{--}36)$$

and the amplitude ratio per period A_2/A_1, in Fig. 8-5 is

$$r = \exp(-\zeta\omega_n P) = \exp(-2\pi\zeta/\sqrt{1-\zeta^2}). \qquad (8\text{--}37)$$

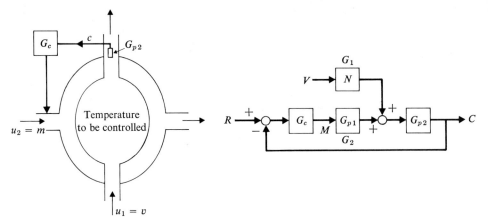

Fig. 8-6 Temperature control of a heat-exchanger system.

An exact analytical expression of the initially quiescent, underdamped system's response for a unit-step input (Fig. 8-5) is given by:

$$c(t) = \mathscr{L}^{-1}[C(s)]$$
$$= \mathscr{L}^{-1}\left[\frac{k}{Ts^2 + s + k} R(s)\right] = \mathscr{L}^{-1}\left[\frac{\omega_n^2}{s(s^2 + 2\zeta\omega_n s + \omega_n^2)}\right]$$
$$= 1 - \exp(-\zeta\omega_n t)\cos\sqrt{1-\zeta^2}\,\omega_n t$$
$$\quad - \frac{\zeta}{\sqrt{1-\zeta^2}} \exp(-\zeta\omega_n t)\sin\sqrt{1-\zeta^2}\,\omega_n t. \tag{8-38}$$

At $t \to \infty$, $c(\infty) = 1$ and there is no steady-state error. This is because the object is type 1. However, since the control law (that is, the amplifier transfer function k_c) is a P-action, an offset $+(k_n/k_c)v(\infty)$ appears for a disturbing torque of a sustained magnitude $v(\infty)$.

Example 8-2 *Temperature control of a heat exchanger system.* We consider the temperature control of the heat exchanger system already modeled in Example 5-6, Section 5-3. A controller G_c is introduced into the process of Fig. 5-15 (see Fig. 8-6a). A sensing bulb measures the temperature, and we assume a unilateral measurement (see Section 6-5) with a single lag for the bulb:

$$G_{p2} = \frac{1}{1 + T_3 s}. \tag{8-39}$$

The input u_2 in Fig. 5-15 now becomes the manipulated variable, and u_1 is considered a disturbance v. The transfer functions that relate these inputs and the temperature to be controlled were given by G_1 and G_2 in Section 5-3. Writing N for G_1 and G_{p1} for G_2, respectively, we obtain the block diagram of Fig. 8-6(b). Since G_2 (which is now

G_{p1}) obtained in Example 5-6, Section 5-3, can be expressed in the canonical form

$$G_{p1} = \frac{k_p}{(1 + T_1 s)(1 + T_2 s)},$$

the open-loop transfer function of the control system with a P-action, $G_c = k_c$, is found to be:

$$G(s) = G_c G_{p1} G_{p2} = \frac{k}{(1 + T_1 s)(1 + T_2 s)(1 + T_3 s)}, \qquad (8\text{-}40)$$

where

$$k = k_c k_p .$$

The characteristic equation of this system is third-order. The work involved in finding an exact response solution was already tedious in the preceding example, which was for a second-order system; mounting complexity makes such an approach impractical for a third-order system. A stability map or nomogram [8] found by the Routh test may serve as a design guide in some cases. In general the root-locus (next section) and the frequency-response method (next chapter) are more efficient methods of dealing with systems of order higher than two.

A time-domain solution for an open- or closed-loop response is also available in the form of the convolution integral (Eq. 3–14). By extending the limits of the convolution (which was shown in Fig. 3–1) over the entire span of time, we obtain the following expressions:

$$y(t) = \int_{-\infty}^{+\infty} g(t - \tau) u(\tau) \, d\tau = \int_{-\infty}^{+\infty} g(\eta) u(t - \eta) \, d\eta , \qquad (8\text{-}41)$$

where $u(t)$ = input, $y(t)$ = response, and τ or $\eta = t - \tau$ are dummy variables. A linear system, open or closed loop, is represented by its unit-impulse input response $g(t) = \mathscr{L}^{-1}[G(s)]$ which appears in the convolution. For numerical computation of the response for a given system and a prescribed input, it is generally advisable to rewrite the convolution in a summation form (like Eq. 1–26) in the discrete-time domain. Of course, if $u(t)$ and $g(t)$ are simple analytical expressions, direct integration of Eq. (8–41) is possible. If, on the other hand, both input $u(t)$ and response $y(t)$ in Eq. (8–41) are known and $g(t)$ is an unknown, then a deconvolution is required to find the unknown and thus *identify* the process. Although the computational difficulty of the deconvolution may be reduced in the discrete-time domain, another difficulty lies in the required length of observation time. Suppose a record of $u(t)$ and $y(t)$ starts at $t = 0$. Then one must face the problem of what to do about a response component that appears for $t > 0$, due to an input at $t < 0$ or, equivalently, due to a nonzero initial state [31]. This difficulty can be avoided if state vector measurement is possible, as we shall see in Chapter 10.

8-3 THE ROOT-LOCUS METHOD

There is a close relation between the eigenvalues (closed-loop poles) of a control system and the quality of control. Therefore, a useful design pro-

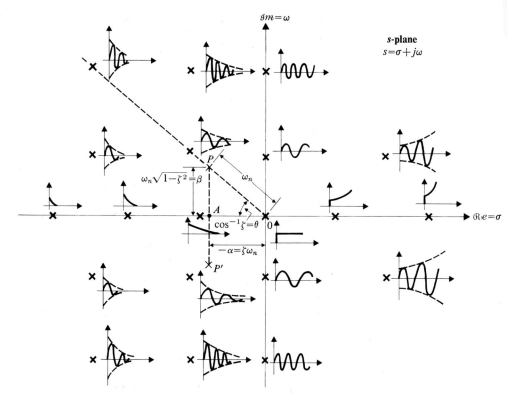

Fig. 8-7 Significance of root location in the *s*-plane.

cedure would be one in which some unknown system parameters are fixed in such a way that the pole distribution of the system assures favorable control quality. The simplest design situation occurs when only one parameter of a control system is unknown. If this parameter is considered an independent variable, all poles of the system become dependent variables of the unknown parameter. Therefore, a set of loci appears in the complex plane (called the *s*-plane) when the roots move in the plane as functions of this parameter. Such a set of loci is called the *root locus*. Once a root locus is determined in the *s*-plane, the unknown parameter value required for a particular control system quality can be chosen from the locus. With practice and insight gained from experience, in addition to fixing a main variable parameter, we can also use the method as a guide to proper selection of secondary control system parameters.

Before developing methods of plotting the root locus as a function of a main system parameter, let us first examine the significance of the location in the *s*-plane of eigenvalues of the system's characteristic equation (Eq. 8-4). The *s*-plane, having a real or σ-axis and an imaginary or ω-axis, consists of the infinite number of possible locations of the eigenvalues $s = \sigma + j\omega$ of the characteristic equation. Consider for the moment a characteristic equation having a single real pole or a single pair of complex conjugate poles, and subjected to an impulse input. We ask how the impulse response changes as the pole location is varied throughout the s-plane. Figure 8–7 indicates the nature of the impulse response associated with various pole locations. Each complex conjugate *pair* of poles results in the single oscillatory impulse response drawn twice on the figure. If the pole in question appears in the characteristic equation with a multiplicity n, then the responses must be multiplied by $(t^n/n!)$.

In particular, let us consider a pair of complex conjugate poles at points P and P' in Fig. 8–7, having a real part $-\alpha = -\zeta\omega_n$, and an imaginary part $\pm\beta = \omega_n\sqrt{1-\zeta^2}$ [see Eqs. (8–23) and (8–24)]. The hypotenuse OP of the triangle OAP is found to be ω_n, which is the undamped natural frequency of the response. The angle AOP is given by

$$\theta = -\cos^{-1}\zeta. \tag{8-42}$$

Several general observations may now be made with reference to Fig. 8–7:

1. All poles lying in the left-half-plane, of whatever multiplicity, lead to responses which decay with time. The farther into the left-half-plane they lie, the faster the responses decay. The time constant of the response, or the envelope of the response, is given by -1 over the real part of the pole. Thus, all poles lying along a particular vertical line have the same time constant.

2. All poles of multiplicity one lying on the imaginary axis lead to constant-amplitude responses (that is, limited stability). For multiplicity greater than one, the response grows with time and is unstable.

3. All poles in the right-half-plane lead to unbounded responses and thus indicate unstable systems.

4. All poles lying along the same horizontal line have the same damped natural frequency of oscillation, $\beta = \omega_n\sqrt{1-\zeta^2}$, appearing in the response. The farther from the real axis the poles fall, the higher the frequencies of the response.

5. All poles lying along the same radial line through the origin have the same damping ratio ζ. All such poles will thus indicate a constant amplitude ratio between successive peaks of the oscillatory response. Alternatively, all such poles will result in the same number of total oscillations before the response dies out.

6. For a high-order system which is linear, superposition may be applied, and the impulse response is then the sum of the various individual responses indicated on Fig. 8–7. To be able to say that a system has a predominant second-order system response, therefore, requires that

two of the system poles be complex conjugate lying relatively close to the imaginary axis, while all the remaining system poles must lie well into the left-half-plane. The farther the separation, the better the approximation of the system as being represented by a predominant second-order system response.

When designing to meet minimum specifications in the time domain, certain allowable regions in the s-domain are therefore clearly delineated. It would be advantageous to be able to predict the migration (root locus) of the eigenvalues of the characteristic equation as a main parameter is varied, so that a parameter value may be chosen which satisfies the time-domain (and, accordingly, s-domain) specifications. A graphical method of constructing the root-locus plot was devised by W. R. Evans [9], and is presented below.

Consider an open-loop transfer function in the canonical form of Eq. (8–7) or Eq. (8–9), where gain K is the required unknown. The characteristic equation of the closed-loop system takes the form

$$1 + KG_0(s) = 0,$$

or

$$G_0(s) = -\frac{1}{K}, \qquad K \geqslant 0, \tag{8-43}$$

where

$$G_0(s) = \frac{\prod_{j=1}^{m}(s - z_j)}{\prod_{i=1}^{n}(s - p_i)}, \qquad n \geqslant m. \tag{8-44}$$

In general, $G_0(s)$ is a complex number, or a phasor in the complex plane. Therefore, Eq. (8–43) requires a phasor $G_0(s)$ to satisfy two conditions, one for angle, the other for magnitude:

Angle: $\angle G_0(s) = -180° \pm 360°N$, $N =$ an integer, (8-45)

Length: $|G_0(s)| = \dfrac{1}{K}$. (8-46)

The general method of finding a root of Eq. (8–43) consists in the following three steps:

1. Mark, in the s-plane, the given set of open-loop poles and zeros (the numerator and the denominator must be factored in the canonical form; if unfactored, apply $s = -\alpha + j\beta$ and use synthetic division [10]) by ×'s and ○'s, respectively (Fig. 8–8a). Note that the number of open-loop zeros is equal to or less than the number of open-loop poles for a system that is physically realizable, $m \leqslant n$.

2. Find, by trial examination of points in the s-plane, a point P which satisfies the angle condition (Eq. 8–45). As shown below, each trial

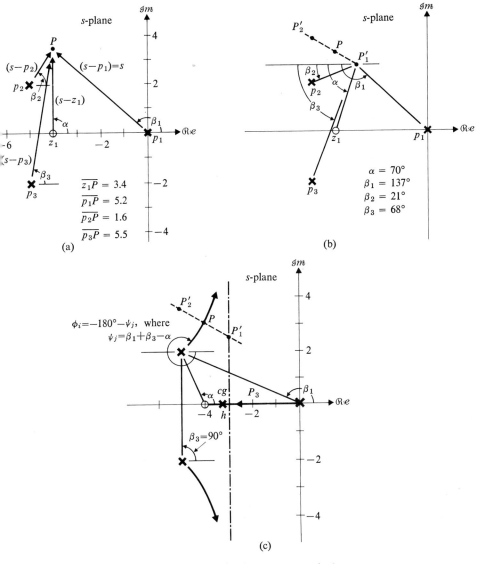

Fig. 8-8 Principles of root-locus plotting: an example for
$G_0(s) = (s + 4)/[s\{(s + 5)^2 + 4\}]$.

is a process of adding (and subtracting) angles measured in the s-plane. This can be done with a "Spirule" (available from The Spirule Company, Whittier, California) or a protractor.

3. For the point P thus found, compute K by Eq. (8–46). As shown below, the computation is a multiplication (and division) of lengths that appear in the s-plane. Note that Eq. (8–45) is the plotting equation, and Eq.

(8-46) is the scaling equation which is used after the plot has been made.

By repeated application of the method, a root-locus plot can be made in the s-plane. Usually a good guess for trials, proper choice of a line of scan [$P_1'P_2'$ in Fig. 8-8(c)], and application of guide rules given below reduce the number of trials to a very few, only two or three in many cases.

The trial process for finding a closed-loop pole will be shown for a system of an open-loop transfer function

$$KG_0(s) = K \frac{(s - z_1)}{(s - p_1)(s - p_2)(s - p_3)} . \tag{8-47}$$

If P in Fig. 8-8(a) is a closed-loop pole, it must satisfy the angle condition (8-45), which, for the canonical form of $G_0(s)$ given by Eq. (8-44), gives the general conditions

$$\angle G_0(s) = \sum_{j=1}^{m} \angle (s - z_j) - \sum_{i=1}^{n} \angle (s - p_i)$$
$$= -180° \pm N \cdot 360° . \tag{8-48}$$

Therefore, for our example we must have

$$\alpha - (\beta_1 + \beta_2 + \beta_3) = -180° \pm N \cdot 360° ,$$

where angles α and β_1 through β_3 are measured as shown in Fig. 8-8(a). Let us suppose, for instance, that we put the center of a protractor at a trial point P_1' in Fig. 8-8(b) and find an angle:

$$70 - (137 + 21 + 68) = -156° .$$

If an angle $-208°$ is found at another trial point P_2', then point P where the angle is $-180°$ can be found by interpolation. Repeating the procedure, the root locus shown in part (c) of the figure can be plotted.

Noting that the component phasors $(s - z_1), (s - p_1), \ldots$ appear as shown in Fig. 8-8(a), we can determine the value K at P by measuring the lengths of these phasors. From Eqs. (8-44) and (8-46) we have in general

$$K = \frac{\prod_{i=1}^{n} |(s - p_i)|}{\prod_{j=1}^{m} |(s - z_j)|} . \tag{8-49}$$

Therefore, for point P in Fig. 8-8(a), we compute:

$$K = \frac{(5.2)(1.6)(5.5)}{3.4} = 13.5 .$$

For a root locus to be complete, the value of K must be indicated along each branch at appropriate intervals. Such gain scaling will be omitted, however, in all that follows.

Root-locus construction follows nine basic rules. (For derivations of the rules, see Chapter 21 of [32].) These rules are:

1. The locus is symmetric with respect to the real (horizontal) axis.
2. If $G(s)$ has n poles, the locus consists of n branches. Some branches may cross each other, but they never overlap for a finite range of K.
3. All branches start at open-loop poles with $K = 0$, and end at open-loop zeros with K tending to infinity. If $n > m$ in Eq. (8-44), as is usually the case with engineering systems, $n - m$ branches go to infinity in the direction of the asymptotes of Rule 4.
4. The asymptotes are straight lines with directions given by
$$\pm 180°q/(n - m)$$
from the real axis, where q sequentially takes on the values 1, 3, 5, ... until all $n - m$ asymptotes have been determined.
5. The hub of the asymptotes, when $n - m \geq 2$, is at
$$h = \frac{\sum_{i=1}^{n} p_i - \sum_{j=1}^{m} z_j}{n - m}.$$
6. If $n - m \geq 2$, the center of gravity of the closed-loop poles, defined by
$$\text{c.g.} = \sum_{i=1}^{n} \frac{P_i}{n},$$
is independent of K, that is, remains fixed in the s-plane as K varies.
7. The root locus coincides with those portions of the real axis which have an odd number of open-loop poles plus zeros lying to the right.
8. Contribution to the total phase of $G_0(s)$ due to open-loop poles and zeros to the right of the point of departure of the locus from the real axis are balanced by the total phase of those to the left. If all open-loop poles and zeros are real, the coordinate b of the point of departure can be fixed by solving the following equation for b by trial:
$$\sum_{i=1}^{n} \frac{1}{b - p_i} = \sum_{j=1}^{m} \frac{1}{b - z_j}.$$
9. The angles θ_i of departure of a branch from a complex open-loop pole p_i (see Fig. 8-8c) is given by
$$\theta_i = -180° - \psi_j, \quad i \neq j,$$
where ψ_j is the angle at p_i contributed by other open-loop poles (measured positively) and zeros (measured negatively). The angle of approach to an open-loop complex zero may be similarly calculated.

Example 8-3 As an example, let

$$G_0(s) = \frac{(s+4)}{s[(s+5)^2 + 4]}.$$

The locus (Fig. 8-8c) consists of three branches (Rule 2), one of which ends at $z_1 = -4$ (Rule 3). Two branches are asymptotic to a straight line perpendicular to the real axis (Rule 4) at $h = -3$ (Rule 5). Since $n - m = 2$, the center of gravity of the closed-loop poles is invariant. It is located at $[(-5 + 2j) + (-5 - 2j) + 0]/3 = -3.33$ (Rule 6). The segment of the real axis between the origin and -4 is a branch (Rule 7). The angle of departure in the figure is computed by Rule 9.

The use of the root-locus method is not limited to the case where the gain K in Eq. (8-7) changes from 0 to infinity. The method is applicable to variations of any system parameter. For instance, the characteristic equation of a servomechanism (Eq. 8-33) can be rearranged as

$$1 + \frac{1}{T}\frac{s+k}{s^2} = 0, \tag{8-50}$$

and we can find a root locus for a span of $1/T$ from zero to plus infinity. For a sensitivity investigation of this kind it is generally advisable to reduce the closed-loop characteristic equation $1 + G(s) = 0$ (or Eq. (4-43)) to the canonical form

$$1 + f(y)\frac{\prod_{j=1}^{m'}(s - z'_j)}{\prod_{i=1}^{n}(s - p'_i)} = 0, \tag{8-51}$$

where $m' \leqslant n$. $f(y)$ in this expression is a function of the variable parameter y; for instance, $f(y) = 1/T$, $y = T$ in Eq. (8-50). If $f(y)$ in Eq. (8-51) or K in $1 + KG_0(s) = 0$ is negative (that is, a positive feedback system), the angle condition (8-45) must be revised to

$$\angle G_0(s) = \pm 360°N, \qquad N = 0, 1, 2, \ldots \tag{8-52}$$

Consequently, Eq. (8-48) and guide rules 4, 7, and 9 must be revised accordingly.

For comparison, the root-locus patterns of first-, second-, third-, and some fourth-order systems are listed in Table 8-1. Reference [11] advances simple drawing rules for root-locus branches of systems up to fourth order, while [12] gives analytical expressions for some of the loci shown in the table. An inspection of loci in Table 8-1 may help in the understanding of the guide rules. The ensemble serves to demonstrate the correlation between the trends of root loci and the configurations of open-loop poles and zeros. For instance, it is possible to observe how an open-loop zero to the left pulls branches away from the imaginary axis. This is the reason why derivative action stabilizes a system. Some patterns also show where a dominant branch (the branch on which a dominant pole is located) comes from, and how it is affected by open-loop poles and zeros. A control engineer develops a sensitivity to these characteristics of his system as he uses the root-locus approach. It is this sensitivity that guides him to a better design. Table 8–1 is meant to aid the development of this intuitive feeling regarding pole-zero array configurations. The table is not intended to be a catalog of root loci. Such a catalog is meaningless, because the variety of systems is almost limitless and the order of systems of practical interest is generally higher than those in the table. Furthermore, a designer's attention is usually focused on some specific aspects of the root locus and, in particular, on the dominant branch that runs closest to the imaginary axis, rather than on the general pattern of loci.

Everywhere along the root locus, s takes on values such that when substituted into the characteristic equation (Eq. 8–43), the result is the *real* number $-1/K$. In particular, we find that, as a branch of the root locus crosses from the left into the right-half-plane, $s = j\omega$. This, of course, constitutes the stability limit and is intimately connected to the frequency-response approach discussed in the next chapter. When this value of s is substituted into Eq. (8–43), we get the following two conditions for an oscillatory stability limit:

$$\mathscr{I}m\, G_0(j\omega_u) = 0, \qquad \mathscr{R}e\, G_0(j\omega_u) = -\frac{1}{K_u}. \qquad (8\text{–}53)$$

Two unknowns, a stability limit gain $K = K_u$, and a period of cycling $P_u = 2\pi/\omega_u$, are determined by Eq. (8–53). Since the real and imaginary parts have been considered separately, the order of the equations to be solved is one-half the original system order, and trial methods are sometimes unnecessary.

Example 8–4 Let us determine k_u of Eq. (8–40) for the thermal system of Example 2 when $T_1 = 2$, $T_2 = 1$, and $T_3 = \frac{1}{2}$. Rearranging Eq. (8–40) in the canonical form of Eq. (8–43) and substituting $s = j\omega$, we have:

$$\frac{1}{(j\omega + \tfrac{1}{2})(j\omega + 1)(j\omega + 2)} = -\frac{1}{K_u},$$

where, due to the choice of time constants, $k_u = K_u$ in this case (see Eq. 8–23). The root locus of Eq. (8–40) will appear as item 9 in Table 8–1, and our approach now is to find the frequency along the imaginary axis at which the root locus crosses it and the gain at which the crossing occurs. To do so, we apply Eq. (8–53):

$$\mathscr{I}m\left(\frac{1}{(-3.5\omega^2 + 1) + j\omega(3.5 - \omega^2)}\right) = 0$$

or $\omega = \sqrt{3.5} = 1.87$ and $P_u = 2\pi/1.87 = 3.4$, and

$$\mathscr{R}e\left(\frac{1}{(-3.5\omega^2 + 1)}\right) = -\frac{1}{K_u} \qquad \text{or} \qquad K_u = k_u = 11.25.$$

Table 8-1 TYPICAL PATTERNS OF ROOT LOCI

Number	$G_0(s)$	Root locus	Number	$G_0(s)$	Root locus
1	$\dfrac{1}{s - p_1}$		8	same as 5	
2	$\dfrac{s - z_1}{s - p_1}$		9	$\dfrac{1}{(s - p_1)(s - p_2)(s - p_3)}$	
3	$\dfrac{1}{(s - p_1)(s - p_2)}$		10	$\dfrac{1}{s(s - p_2)(s - p_3)}$	
4	$\dfrac{1}{(s - p_1)(s - p_2)}$ p_1, p_2 complex		11	$\dfrac{1}{(s - p)^3}$	
5	$\dfrac{(s - z_1)}{(s - p_1)(s - p_2)}$ $z_1 < p_1, p_2$		12	$\dfrac{1}{(s - p_1)(s - p_2)(s - p_3)}$ p_2, p_3 complex	
6	same as 5 $p_2 < z_1 < p_1$		13	same as 12	
7	same as 5 $p_1, p_2 < z_1$		14	same as 12	

8–3 The root-locus method

Number	$G_0(s)$	Root locus
15	$\dfrac{(s-z_1)}{(s-p_1)(s-p_2)(s-p_3)}$	
16	same as 15	
17	same as 15	
18	same as 15	
19	$\dfrac{(s-z_1)(s-z_2)}{(s-p_1)(s-p_2)(s-p_3)}$	
20	same as 19	
21	$\dfrac{(s-z_1)(s-z_2)}{(s-p)^3}$ $z_2 < z_1 < p$	

Number	$G_0(s)$	Root locus
22	$\dfrac{(s-z_1)(s-z_2)}{s(s-p_2)(s-p_3)}$ $z_2 < p_3 < z_1 < p_2 < 0$	
23	same as 22	
24	same as 22	
25	$\dfrac{1}{(s-p_1)(s-p_2)(s-p_3)(s-p_4)}$ all poles real	
26	same as 25 2 poles real 2 poles complex	
27	same as 26	
28	same as 25 all poles complex	

340 Scalar input–output linear systems and feedback control

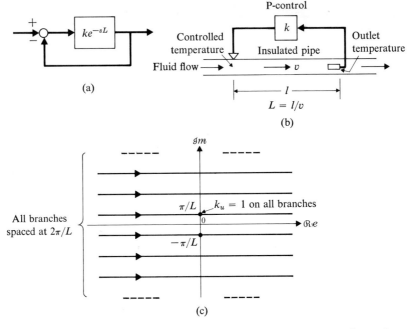

Fig. 8–9 Control loop of a pure delay (a) and an example (b). Part (c) shows the root locus of the system.

By letting

$$s = -\sigma + j\omega \tag{8-54}$$

in Eq. (8–43), we get [12]

$$\mathscr{I}m\,[G_0(-\sigma + j\omega)] = 0, \qquad \mathscr{R}e\,[G_0(-\sigma + j\omega)] = -\frac{1}{K}. \tag{8-55}$$

Although these are similar to the stability limit conditions of Eq. (8–53), they represent a shift of the vertical axis into the left-half-plane by σ units. As such, Eq. (8–55) permits us to analytically find the exact location of roots along the branches of the locus. Specifically, it is sometimes possible to find either a dominant pair of closed-loop poles of a given system, or a value of gain K that produces a prescribed decay pattern of the dominant oscillation. This general approach is valid for either lumped- or distributed-parameter systems, as we shall see in the next two examples.

Since the number of branches of a root locus equals the order of the system, the root locus of a distributed-parameter system will consist of an infinite number of branches. This makes the graphical method of plotting impossible. Root loci of systems with a dead time and some other simple distributed-parameter systems are known [13, 14, 15]. In the following we shall show the root-locus pattern for the simplest case: a control system of a pure delay process (Fig. 8-9). The characteristic equation of the control system is:

$$1 + ke^{-sL} = 0, \tag{8-56}$$

where L = dead time. Substituting from Eq. (8-54), Eq. (8-56) becomes

$$1 + ke^{\sigma L}(\cos \omega L - j \sin \omega L) = 0 .$$

Therefore, two relations,

$$ke^{\sigma L} \sin \omega L = 0, \tag{8-57}$$

and

$$ke^{\sigma L} \cos \omega L = -1, \tag{8-58}$$

must be satisfied by σ and ω simultaneously. For a positive value of k we must therefore have:

$$\omega L = \pm \pi, \pm 3\pi, \ldots \tag{8-59}$$

and

$$k = e^{-\sigma L} . \tag{8-60}$$

By Eq. (8-59), the imaginary part ω of all closed-loop poles is independent of k. Thus all branches of the root locus are parallel to the real axis (Fig. 8-9c). Since the real part σ vanishes at the stability limit, we find the stability limit gain by (8-60) as

$$k_u = 1 . \tag{8-61}$$

The closed-loop impulse response of the system will consist of a spectrum of decaying oscillatory responses, with the decay rate of each component depending on the system gain k.

Example 8-5 *Stability limit of a lag and delay system* (Fig. 8-10(a) and [16]). By Eq. (8-53) we have

$$\sin \omega_u L = -T\omega_u \cos \omega_u L, \quad k_u \cos \omega_u L = -1 .$$

Thus,

$$\tan \omega_u L = -\omega_u T, \quad k_u = -1/\cos \omega_u L ,$$

and we can determine ω_u by the intersection of a tangent curve and a straight line, getting point P_1 in Fig. 8-10(b). The first intersection, P_1 in the figure, corresponds to the imaginary-axis crossover of the dominant branch in the root-locus plot. Gains at other intersections (P_2, etc.) are higher than k_u and hence make the system unstable [16].

Example 8-6 *Quarter-decay condition of a delayed integrator system* (Fig. 8-11). The characteristic equation of this system can be normalized by the substitutions $\kappa = kL$

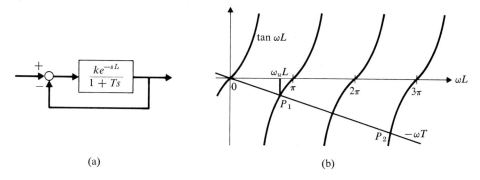

Fig. 8-10 Control loop with a delay and a lag (a) and graphical determination of its eigenvalue.

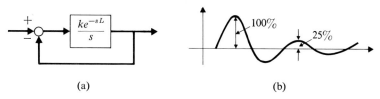

Fig. 8-11 Control loop for a delayed integrator (a) and a quarter decay of a dominant mode.

and $\lambda = sL$, which yield

$$\lambda + \kappa e^{-\lambda} = 0 . \tag{8-62}$$

Letting

$$\lambda = sL = (-\sigma + j\omega)L = -a + jb , \qquad a = \sigma L , \qquad b = \omega L ,$$

we find that Eq. (8-55) becomes

$$a = \kappa e^a \cos b , \qquad b = \kappa e^a \sin b ,$$

and thus

$$\frac{b}{a} = \tan b .$$

By Eq. (8-37) "quarter damping" is defined to be

$$r = \tfrac{1}{4} = e^{-aP} , \qquad \text{where} \quad P = \text{period of oscillation} = 2\pi/b .$$

Hence

$$\frac{a}{b} = \frac{\log_e 4}{2\pi} = 0.22 . \tag{8-63}$$

We now have the condition

$$\frac{b}{a} = \frac{1}{0.22} = \tan b ;$$

therefore
$$b \approx 1.36, \quad a = 0.22b \approx 0.3$$
and
$$\kappa = kL = \frac{ae^{-a}}{\cos b} \approx 1.0. \tag{8-64}$$

This result, which is surprisingly simple, will serve as the basis of the Ziegler-Nichols formulas, which are given in the next section.

8-4 QUALITY OF CONTROL AND ITS IMPROVEMENT BY COMPENSATION

Finding optimum adjustments of a controller for a given process is one of the most practical problems control and instrument engineers face in the process industry. The rule of thumb of Ziegler and Nichols [17] is by far the simplest procedure devised for such purposes. There are two methods: one is based on the step-input response pattern of a process, and the other utilizes information obtained at the stability limit of a process under P-control.

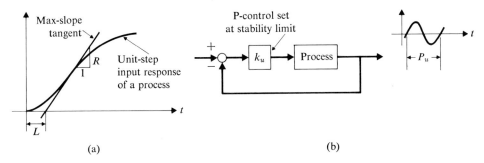

Fig. 8-12 Plant parameters for the Ziegler-Nichols method.

The process dynamics crucial to control behavior are reduced to only two parameters, R and L, in the first method. R is the maximum slope of a tangent drawn to the unit-step input response of the process to be controlled, and L is the time at which the tangent intersects the time axis (Fig. 8-12a). Ziegler and Nichols recommend the following controller adjustments as optimum:

$$k_c = \frac{1}{RL} \quad \text{for P-control,}$$

$$k_c = \frac{0.9}{RL}, \quad T_i = 3.3L \quad \text{for PI-control,} \tag{8-65}$$

$$k_c = \frac{1.2}{RL}, \quad T_i = 2L, \quad T_d = 0.5L \quad \text{for PID-control.}$$

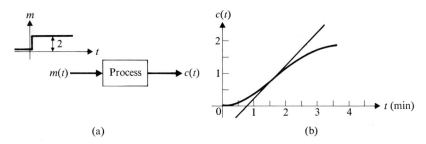

Fig. 8-13 Step-input response of a process.

In Eqs. (8-65), k_c, T_i, and T_d are adjustments of gain, integral action, and derivative action, respectively, that appear in the following control law:

$$G_c(s) = k_c \left(1 + \frac{1}{T_i s} + T_d s\right) = \text{controller transfer function.}$$

In the second method the optimal adjustments are given in terms of the P-control stability limit gain k_u and period of cycling P_u (Fig. 8-12b):

$k_c = 0.5 k_u$ for P-control,

$k_c = 0.45 k_u$, $T_i = 0.83 P_u$ for PI-control,   (8-66)

$k_c = 0.6 k_u$, $T_i = 0.5 P_u$, $T_d = 0.125 P_u$ for PID-control.

The rules were developed from Ziegler's experiments on various processes and Nichols' analysis. The criterion for optimality is the minimization of the integrated absolute error,

$$IAE = \int_0^\infty |e(t)|\, dt,$$

evaluated for a free unit-step response which starts from an equilibrium state. They observed that a dominant mode of oscillation, closely approximating a quarter-decay response, appeared in most of the optimal responses. Example 8-6 of the preceding section confirms this observation. The system of Fig. 8-11(a) can be interpreted as a P-control of a process approximated by R and L. We saw that $\kappa \approx 1$ for a quarter-decay response, where $\kappa = k_c R L$; hence $k_c = 1/RL$. Note that $k_c R$ replaces the k factor of Example 8-6, where R is the reciprocal of the time constant of the predominant closed-loop step response and k_c is the controller proportional gain.

Example 8-7 *The first method* (Eq. 8-65). The equilibrium state of a process was changed by a step test signal of magnitude 2 psi applied to a pneumatic control valve (Fig. 8-13a). The slope of the tangent at the inflection point P of the recorded response (Fig. 8-13b) was found to be 1.1 in./min, and $L = 0.8$ min. Assuming a static linear relation between m and c, we find that $R = 1.1/2 = 0.55$ in./min · psi, and

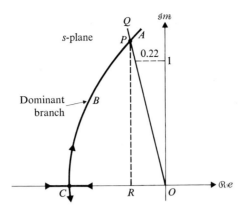

Fig. 8-14 A quarter decay (P) and critical damping (C) of a closed-loop pole.

thus $RL = 0.44$ in./psi. The optimal adjustments of a PID-controller for this process are: $k_c = 1.2/0.44 = 2.7$ psi/in., $T_i = 1.6$ min, and $T_d = 0.4$ min.

Example 8-8 *The second method* (Eq. 8-66). In Example 8-4, we found, for the heat exchanger considered there, $k_u = 11.25$ and $P_u = 3.4$ min. Referring back to Example 8-2 for the same system, we recall that $k = k_c k_p$, where k_c is the controller gain and k_p is the process gain. Thus, we have $(k_c)_u = 11.25/k_p$. If PI-control is desired for the process, its optimal adjustments are given by Eqs. (8-66) as:

$$k_c = 0.45(k_c)_u = \frac{5}{k_p} \quad \text{and} \quad T_i = 0.83 P_u = 2.8 \text{ min}.$$

Keeping in mind the rule of thumb that a quarter-decay response is often a good compromise for quick response and adequate stability, we can apply the root-locus method to a control system's design in the following manner: As given by Eq. (8-63), a complex pole for a quarter decay lies on the line OQ (Fig. 8-14) drawn from the origin of the s-plane with a slope of $1:0.22$. If AB in the figure is the dominant branch of the control system, we choose P as a (dominant) closed-loop pole, and fix the gain K (or other system constant) accordingly. Since the distance OP is proportional to the magnitude of the real part OR of the pole, it is advisable to make OP as large as possible by properly relocating the open-loop poles and zeros. The design principle is not limited to quarter decay. If, for instance, a dead-beat dominant mode is desired, we choose C in the figure as a real double pole, and try to make the distance OC as large as possible by reshaping the locus. For other decay rates, Eq. (8-42) may be used. The compensation network discussed below is one means of reshaping the locus to yield the desired closed-loop poles.

For transfer functions of the form

$$G(s) = \frac{b_0}{a_0 + a_1 s + a_2 s^2 + \cdots + a_n s^n}, \tag{8-67}$$

Gustafson [18] has shown the meaningfulness of the following second-order approximations:

$$G_{a1}(s) = \frac{b_0}{a_0 + a_1 s + a_2 s^2}, \qquad (8\text{-}68)$$

$$G_{a2}(s) = \frac{b_0}{r_{n+1,1} + r_{n,1} s + r_{n-1,1} s^2}, \qquad (8\text{-}69)$$

where the denominator coefficients in (8-69) are given by the last three elements of the Routh array:

$$
\begin{array}{cccc}
a_n & a_{n-2} & \cdots & a_2 \quad a_0 \\
a_{n-1} & a_{n-3} & \cdots & a_1 \\
\vdots & & & \\
r_{n-2,1} & r_{n-2,2} & & \\
r_{n-1,1} & r_{n-1,2} & & \\
r_{n,1} & & & \\
r_{n+1,1} & & &
\end{array}
$$

The first form (Eq. 8-68) has its first three time moments

$$m_0 = \int_0^\infty g_{a1}(t)\, dt, \qquad m_1 = \int_0^\infty g_{a1}(t) t\, dt, \qquad m_2 = \int_0^\infty g_{a1}(t) t^2\, dt,$$

where $g_{a1}(t) = \mathscr{L}^{-1}[G_{a1}(s)]$, equal to those of the original function. The integral of the squared impulse response of the second form (Eq. 8-69) is equal to that of the original system. Since these quantities include crucial response characteristics in them, the pair of Eqs. (8-68) and (8-69) is an approximation of the high-order transfer function. When a closed-loop transfer function $G_{LR}(s)$ (Eq. 8-12) is considered as the original function (8-67), the overshoot of the step-input response (in the pattern of Fig. 8-2) is generally bounded by the overshoots of the G_{a1} and G_{a2} approximations. This provides an approach to control system design by focusing the attention on an approximate second-order response.

Refinements of optimum-setting formulas [19 through 22] in general make the results restrictive to specific plant dynamics and other limiting conditions. A theoretical approach to the problem of parameter optimization [23, 24, 25] requires the definition of a performance index on which the optimization is performed. Some typical forms of the index are the following [3, 25]:

$$\text{IAE (integrated absolute value of error)} = \int_0^\infty |e(t)|\, dt, \qquad (8\text{-}70)$$

$$\text{ISE (integral of squared error)} = \int_0^\infty e^2(t)\, dt, \qquad (8\text{-}71)$$

ITAE (integral of time multiplied by absolute value of error)

$$= \int_0^\infty t\,|e(t)|\,dt, \qquad (8\text{-}72)$$

ITSE (integral of time multiplied by squared error)

$$= \int_0^\infty t e^2(t)\,dt, \quad \text{etc.} \qquad (8\text{-}73)$$

Once an index is selected, a hill-climbing technique on a digital computer could give the values of optimal parameters. In general, the results will apply only to the system or limited class of systems and their operating conditions for which the computer program was written. The choice of index in a machine computation is not limited to those given by Eqs. (8-70) through (8-73); magnitude of maximum overshoot, constraint on range of controlling variable, parameter sensitivity [26], and other items of interest could be added to an error criterion. On the other hand, the choice of an index for an analytical approach is extremely limited. The following example will show how the choice of an optimum parameter value depends on operating conditions.

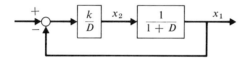

Fig. 8-15 State variables of a control system.

Example 8-9 *Gain of a second-order control system for minimal ISE.* For a control system similar to the servomechanism of Fig. 8-4, we introduce a state variable formulation as shown in Fig. 8-15:

$$\frac{d}{dt}\begin{bmatrix} x_1 \\ x_2 \end{bmatrix} = \begin{bmatrix} -1 & 1 \\ -k & 0 \end{bmatrix} \begin{bmatrix} x_1 \\ x_2 \end{bmatrix}, \qquad (8\text{-}74)$$

and define a performance index by

$$J = \int_0^\infty (w x_2^2 + x_1^2)\,dt, \qquad (8\text{-}75)$$

where w is a weighting factor. By Eq. (4-26), Section 4-3, we can compute J by using a quadratic Lyapunov function. Letting

$$\mathbf{A} = \begin{bmatrix} -1 & 1 \\ -k & 0 \end{bmatrix}, \quad \mathbf{Y} = \begin{bmatrix} a & b \\ b & c \end{bmatrix}, \quad \mathbf{W} = \begin{bmatrix} 1 & 0 \\ 0 & w \end{bmatrix},$$

we fix a, b, and c of $\mathbf{Y}$ by $\mathbf{A}'\mathbf{Y} + \mathbf{Y}\mathbf{A} = -\mathbf{W}$:

$$a = \frac{1}{2} + \frac{wk}{2}, \quad b = -\frac{w}{2}, \quad c = \frac{1}{2k} + \frac{w}{2} + \frac{w}{2k}. \qquad (8\text{-}76)$$

The index J is expressed in terms of the elements of $\mathbf{Y}$ as

$$J = \mathbf{x}'\mathbf{Y}\mathbf{x}_0 = ax_1(0)^2 + 2bx_1(0)x_2(0) + cx_2(0)^2 \ . \tag{8-77}$$

Substituting (8-76) into (8-77), and letting $\partial J/\partial k = 0$, we find the optimal gain for a minimal ISE to be

$$k = \frac{x_2(0)}{x_1(0)} \sqrt{\frac{1+w}{w}} \ . \tag{8-78}$$

The optimal gain thus depends on the initial state of the system.

It has been pointed out that modifying the shape of a root locus is usually effective in improving a control system's performance. A compensating element which favorably alters a root-locus configuration is sometimes added to a control loop, either as a series compensation (in series with forward-path elements) or as a feedback compensation (placed in an inner feedback path around some element) [27]. The most commonly used elements are combinations of lead and lag actions, in the form

$$\frac{1 + T_1 s}{1 + T_2 s},$$

where lead and lag refer to phase shifts in frequency response considerations (see the next chapter). Feedforward is also effective in improving control quality [28]. By feedforward compensation, a load upset (or a disturbance) is measured and fed into a controller to initiate a corrective action without waiting for a process upset and a resulting error signal. Cascade control is thus effective in reducing the magnitude of process disturbances [29]. The transient overshoot of a start-up process (Fig. 8-2) can be improved by a compensating network that involves a dead time ("Posicast" in [14]; see also [30]).

Compensation in some cases involves cancellation of a pole by a zero, but caution is often needed in treating such problems. We shall consider a simple problem of this kind in the following example:

Example 8-10 *PI-control of a first-order plant.* We add P-action to the I-control of Fig. 8-15 and obtain a PI-control system (Fig. 8-16a). The added P-action introduces a zero, $s = -1/T_i$, in the open-loop transfer function $G(s) = k(1 + T_i s)/s(1 + Ts)$, and this contributes to the stability of the closed-loop system. The stabilizing effect is similar to a lead network.

If the zero for stabilization exactly coincides with a system's pole, $T = T_i$ in our case, our second-order system appears to become first-order (Fig. 8-16b). To investigate the system's mode in this condition, we draw a signal-flow diagram (Fig. 8-16c) and write an open-loop system state-space equation:

$$\frac{d}{dt}\begin{bmatrix} x_1 \\ x_2 \end{bmatrix} = \begin{bmatrix} -1 & 1 \\ 0 & 0 \end{bmatrix} \begin{bmatrix} x_1 \\ x_2 \end{bmatrix} + \begin{bmatrix} k \\ k \end{bmatrix} u \ , \tag{8-79}$$

where, for brevity, T and T_i are both assumed to be unity. Changing the variable

8-4 Quality of control and its improvement by compensation

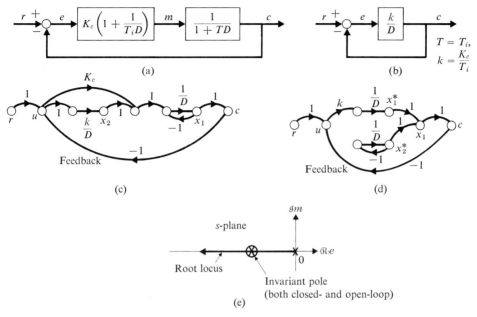

Fig. 8-16 A PI control system with a pole cancellation.

from $\mathbf{x}$ to $\mathbf{x}^*$ by $\mathbf{x} = \mathbf{T}\mathbf{x}^*$, where

$$\mathbf{T} = \begin{bmatrix} 1 & 1 \\ 1 & 0 \end{bmatrix}$$

(see Section 3-3 for the derivation of $\mathbf{T}$), we obtain the diagonal form,

$$\frac{d}{dt}\begin{bmatrix} x_1^* \\ x_2^* \end{bmatrix} = \begin{bmatrix} 0 & 0 \\ 0 & -1 \end{bmatrix}\begin{bmatrix} x_1^* \\ x_2^* \end{bmatrix} + \begin{bmatrix} k \\ 0 \end{bmatrix} u . \tag{8-80}$$

According to the signal-flow diagram for the new state vector $\mathbf{x}^*$ (Fig. 8-16d), x_2^* is not controllable (Section 3-4). In addition to the closed-loop pole, $-k$, of the reduced system (Fig. 8-16b), there still exists another pole $s = -1/T = -1$. Consequently the system's response has two modes, e^{-t} and e^{-kt} when $k \neq 1$, and the two real poles are distinct. We conclude:

(a) The pole cancellation produced an uncontrollable mode, but it was harmless because the mode was stable,

(b) If the control system's characteristic equation were written in the conventional form,

$$1 + \frac{k(1+Ts)}{s(1+Ts)} = 0 ,$$

we must not strike off $(1 + Ts)$ from the numerator and denominator of $G(s)$. Leaving $(1 + Ts)$ as it is, a rearrangement gives $(1 + Ts)(s + k) = 0$ and thus the two poles.

(c) The zero introduced by the compensation does contribute to the system's stability, by changing the root-locus pattern from Number 3 to Number 5 of Table 8-1. Further, the locus is reduced to the negative real axis (Fig. 8-16e) at the limit $T = T_i$, where the open-loop pole $s = -1/T$, on which an open-loop zero overlaps, becomes an invariant closed-loop pole.

In general, it can be shown that a closed-loop system is uncontrollable (or unobservable) if and only if its open-loop system is uncontrollable (or unobservable). (See Problem 8-17.)

REFERENCES

1. Truxal, J. G., (ed.), *Control Engineers Handbook*. New York: McGraw-Hill, 1958.
2. Truxal, J. G., *Automatic Control System Synthesis*. New York: McGraw-Hill, 1955.
3. Gibson, J. E., et al., "A set of standard specifications for linear automatic control systems." *Application and Ind., AIEE Trans.*, May 1961.
4. Langill, A. W., Jr., *Automatic Control Systems Engineering*, Vol. 1. Englewood Cliffs, N. J.: Prentice Hall, 1965.
5. Jones, G. A., Vahouny, R. E., "Finding transient descriptions graphically." *Control Eng.*, **12**, No. 10, October 1965, p. 97; No. 11, November 1965, p. 99.
6. Kobylarz, T. J., "Finding error coefficients from poles and zeros." *Control Eng.*, **11**, No. 3, March 1964, pp. 89-90.
7. Carter, W. C., "Accuracy and control system gain." *Control Eng.*, **10**, No. 3, March 1963, pp. 102-104.
8. Raymond, W. J., Harrison, H. L., "Check stability of third-order systems." *Control Eng.*, **12**, No. 8, August 1965, pp. 94-95.
9. Evans, W. R., *Control System Dynamics*. New York: McGraw-Hill, 1953.
10. Klock, H. F., "To plot root locus: use synthetic division." *Control Eng.*, **9**, No. 4, April 1962, pp. 115-117.
11. Bjorkstam, A. L., "Simplifying root-locus plotting," I and II. *Control Eng.*, **13**, No. 3, March 1966, pp. 99-100, No. 4, April 1966, pp. 95-96.
12. Wojcik, C. K., "Analytical representation of the root locus," *J. Basic Eng., ASME Trans.*, **86**, D, March 1964, pp. 37-50.
13. Jawor, T., "Using the root locus," I and II. *Control Eng.*, **6**, No. 10, October 1959, pp. 96-102; No. 11, November 1959, pp. 119-122.
14. Smith, O. J. M., *Feedback Control Systems*. New York: McGraw-Hill, 1958.
15. Tyner, M., "Computing time response of a deadtime process." *Control Eng.*, **11**, No. 4, April 1964, pp. 79-81.
16. Oldenbourg, R. C., Sartorius, H., *Dynamic selbsttätiger Regelungen*, Bd. 1 (2nd ed.). Munich: Oldenbourg, 1951. (English Translation of the first edition: Mason, H. L., *The Dynamics of Automatic Controls*, ASME, 1948.)

17. Ziegler, J. G., Nichols, N. B., "Optimum settings for automatic controllers." *ASME Trans.*, **64**, No. 8, 1942, p. 759; "Process lags in automatic control circuits." *ASME Trans.*, **65**, No. 5, 1943, p. 433; "Optimum settings for controllers." *ISEJ.*, June 1964, pp. 731-734.
18. Gustafson, R. D., "A paper and pencil control system design." *J. Basic Eng., ASME Trans.*, **88**, June 1966, pp. 329-336.
19. Willis, D. M., "Tuning maps for three-mode controllers." *Control Eng.*, April 1962, pp. 104-108; "A guide to controller tuning." *Control Eng.*, **9**, No. 8, August 1962, pp. 93-95.
20. Van der Grinten, P. M. E. M., "Finding optimum controller settings." *Control Eng.*, **10**, No. 12, December 1963, pp. 51-56.
21. Haalman, A., "Adjusting controllers for a deadtime process." *Control Eng.*, **12**, No. 7, July 1965, pp. 71-73.
22. Law, V. J., Weaver, R. E. C., "Systematic determination of optimal controller parameters." *1964 JACC Preprint*, pp. 76-86.
23. Newton, G. C., Gould, L. A., Kaiser, J. F., *Analytical Design of Linear Feedback Control*. New York: Wiley, 1957.
24. Bach, R. E., Jr., "A practical approach to control system optimization." *IFAC Tokyo Symposium Preprint*, 1965, pp. III-46-III-56.
25. Puri, N. N., Weygandt, C. N., "Calculation of quadratic moments of high-order linear systems via Routh canonical transformation." *1964 JACC Preprint*, pp. 70-75.
26. Hoffman, C. H., "How to check linear systems stability—v. parameter sensitivity." *Control Eng.*, **13**, No. 3, March 1966, pp. 81-87.
27. McCamey, R. E., "Design of some active compensators of feedback controls." *1962 JACC Preprint, 18-1*, pp. 1-8.
28. Bernard, J. W., Lefkowitz, I., "An approach to optimizing control based on a generalized dynamic model." *1962 JACC Preprint, 3-2*, pp. 1-9.
29. Webb, P. U., "Reducing process disturbances with cascade control." *Control Eng.*, **8**, No. 8, August 1961, pp. 73-76.
30. Fairfield, R. L., "Designing a deadbeat compensating network." *Control Eng.*, **13**, No. 8, August 1966, pp. 75-77.
31. Mishkin, E., *et al.*, *Adaptive Control Systems*. New York: McGraw-Hill, 1961.
32. Cannon, R. H., Jr., *Dynamics of Physical Systems*. New York: McGraw-Hill, 1967.

PROBLEMS

8-1 By the regulation of a volume of air space, a buoyant force f_b acting on a submarine is kept proportional to the submarine depth y;

$$f_b = ky, \quad k = \text{proportionality constant}.$$

Let m be the mass of the boat. Discuss the pattern of its depth control and suggest

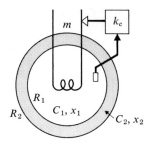

Fig. P8-2

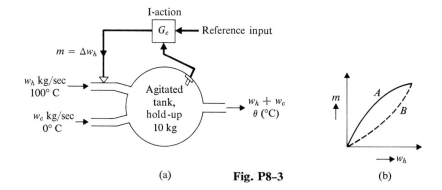

(a)   Fig. P8-3   (b)

modifications if necessary. Assume that the boat is a mass point, and ignore friction for vertical motion.

8-2 Heat capacitances of a furnace system are lumped into C_1 and C_2, with respective temperatures x_1 and x_2, as shown in Fig. P8-2. R_1 and R_2 are resistances, and the ambient temperature is zero. Determine $G_1(D) = x_1/m$ and $G_2(D) = x_2/m$, where m is a heat input. If m is controlled by a P-action, show that the response never oscillates. Is this last statement true in a real situation?

8-3 In a mixing heater shown in Fig. P8-3(a), the flowrate of hot water w_h is

$$w_h = w_{h0} + \Delta w_h ,$$

where w_{h0} is the normal flowrate such that, at a steady state when $\theta = 50°C$,

$w_{h0} = $ flowrate of cold water w_c .

Let $\Delta\theta = \theta - 50$ (temperature deviation) and find the process transfer function

$$G_p(D) = \frac{\Delta\theta}{\Delta w_h}$$

for two load conditions $w_c = 1$ and $w_c = 5$. Of the two types of available valve characteristics, A and B in Fig. P8-3(b), which will better adapt (or compensate) for a change of process gain due to the load?

8-4 A linear liquid-level system shown in Fig. P8-4 is the control object. Determine: (a) the **A**-matrix of the system with proportional control (gain k_c), (b) the

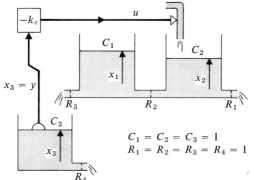

$$C_1 = C_2 = C_3 = 1$$
$$R_1 = R_2 = R_3 = R_4 = 1$$

Fig. P8-4

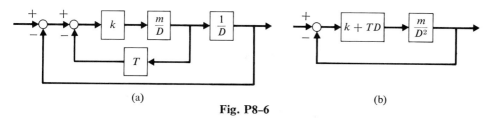

(a)　　　　　　　　　　　　　　(b)

Fig. P8-6

range of k_c for the system to be asymptotically stable, and (c) the scalar differential equation of the control system for output y, and a set of initial conditions on y that corresponds to initial state $\mathbf{x}_0$.

8-5 Prove that the complex branch of the root loci in Table 8-1, Numbers 5 and 8, are circles. Also prove that "the center of gravity" of the closed-loop poles is invariant if $(n - m) \geqslant 2$ (root-locus rule number 6, section 8-3).

8-6 An attitude control of a space vehicle (or any mass point) with velocity feedback is shown in Fig. P8-6(a). Let $km = K$ and find the root locus for K and T, respectively. Is this control system equivalent to the PD-control (Fig. P8-6b)?

8-7 An open-loop transfer function is given by

$$G(s) = \frac{k}{(s^2 + 6s - 16)(0.1s + 1)}.$$

Determine: (a) the stability condition of the control system, (b) the center of gravity of the poles, (c) the breakaway point and gain k at this point, (d) the intersection of the loci branches with the imaginary axis; and sketch the root locus.

8-8 Consider I-control of a damped oscillatory object. The open-loop transfer function is given by

$$G(s) = \frac{k}{s(ms^2 + bs + 1)}.$$

Let $k = 0.16$, $b = 1$ and determine the root locus for m.

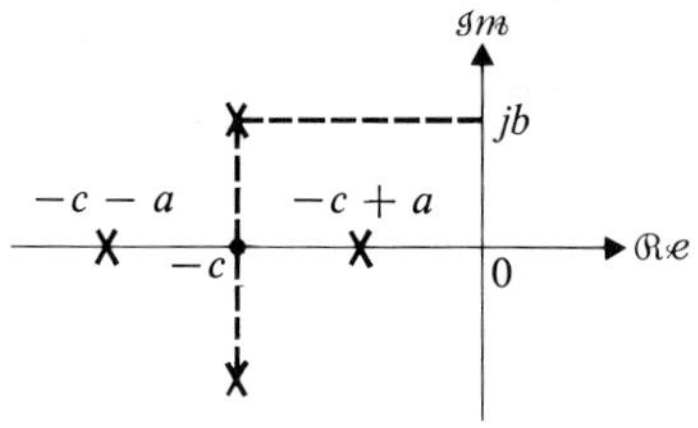

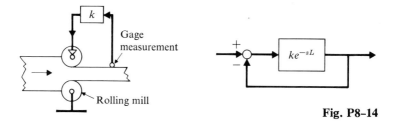

Fig. P8-11

Fig. P8-14

8-9 The open-loop transfer function of a control system is

$$G(s) = K \frac{(s + 0.2)}{s^2(s + 1)^2}.$$

It is desired to maximize the damping (that is, minimize the amplitude ratio per cycle) of the dominant closed-loop system oscillation. Obtain K for this condition.

8-10 A PID-control law

$$G_c(s) = k_c \left(1 + \frac{1}{T_i s} + T_d s\right)$$

is applied to a process

$$G_p(s) = \frac{k_p}{s(Ts + 1)^3}.$$

Time constants are $T_i = 4$, $T_d = 1$, and $T = 1$. Determine the center of gravity of poles, asymptotes, and breakaway points. Find the root locus for $K = k_c k_p$.

8-11 Discuss the root-locus pattern of a 4th-order system which has a symmetric distribution of open-loop poles, as shown in Fig. P8-11.

8-12 An open-loop transfer function is given by

$$G(s) = \frac{k}{s(Ts + 1)^4}.$$

Obtain the center of gravity of poles, asymptotes, breakaway points, intersection of branches with the imaginary axis, and sketch the overall pattern of the root locus.

8-13 A free system is described by the vector equation

$$\frac{d}{dt}\begin{bmatrix} x_1 \\ x_2 \\ x_3 \end{bmatrix} = \begin{bmatrix} -1 & 1 & 0 \\ 0 & -0.2 & k_1 \\ 1 - k_2 & -1 & 0 \end{bmatrix} \begin{bmatrix} x_1 \\ x_2 \\ x_3 \end{bmatrix}.$$

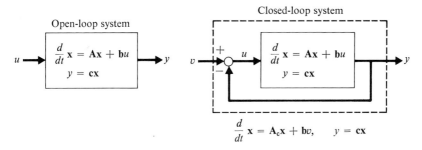

Fig. P8-17

Obtain the root locus for the following cases: (a) k_1 is variable, and $k_2 = 2$, (b) $k_1 = 0.2$ and k_2 is variable.

8-14 A pure dead-time process (gage control, for instance) is controlled by P-action of gain k (see Fig. P8-14). Determine the time-domain response for a unit-step reference input. Assume zero initial state. Also solve the problem using a difference equation, taking the discrete-time interval set equal to the dead time L.

8-15 The following transfer functions are used to approximate dead time on an analog computer:

$$G_1(s) = \frac{1 - \tfrac{1}{2}Ls}{1 + \tfrac{1}{2}Ls}, \qquad G_2(s) = \frac{(Ls)^2 - 6Ls + 12}{(Ls)^2 + 6Ls + 12}.$$

(a) Obtain the step-input response (for zero initial state) of G_1 and G_2.

(b) Determine the root loci of $G_1(s)$ and $G_2(s)$ with proportional control.

8-16 An open-loop transfer function of a process system is known to be

$$G(s) = \frac{ke^{-0.5s}}{(2s + 1)(s + 1)}.$$

Determine the optimum value of k via the two methods of Ziegler and Nichols.

8-17 An open-loop system is given by

$$\frac{dx}{dt} = Ax + bu, \qquad y = cx.$$

The loop is closed by letting $u = v - y$, where $v =$ input to the closed-loop system (see Fig. P8-17). For a third-order system show that

$$|b, Ab, A^2b| = |b, A_c b, A_c^2 b| \qquad \text{and} \qquad \begin{vmatrix} c \\ cA \\ cA^2 \end{vmatrix} = \begin{vmatrix} c \\ cA_c \\ cA_c^2 \end{vmatrix},$$

where $A_c = A - bc$. The result may be generalized to nth-order systems. It implies that a closed-loop system is controllable (or uncontrollable, observable, or unobservable) if and only if its open-loop system is controllable (or uncontrollable, observable, or unobservable), and vice versa.

9

FREQUENCY RESPONSE

This chapter deals with the sinusoidal steady state, or the frequency response, of a system. If the input to a linear and stationary system is a sinusoid, after all transients have died out, the output will also be a sinusoid of the same frequency. The system can be completely described by the output–input amplitude ratio and phase difference over the entire range of forcing frequency from zero to infinity. In Section 1 two formats for plotting frequency-response data are presented: the polar or Nyquist plot and the semilog or Bode plot. Examples of both lumped- and distributed-parameter linear systems are considered. Sections 2 through 4 deal mostly with the frequency-response approach to feedback control system design. The Nyquist Stability Theorem, derived and applied in Section 2, is the basis of this design approach. The major advantage of the frequency-response approach lies in the fact that it easily lends itself to the use of open-loop frequency-response data in the design of closed-loop systems. The design method is discussed in Section 3. In Section 4 the open-loop frequency response of a system is explicitly related to its closed-loop frequency response via the Nichols diagram. This section also presents the relation between the frequency and time domain by a Fourier-series approximation of a time signal. Included in the final section is a brief view of frequency-response results in the state space.

Since feedback control-system design and analysis by the frequency-response method constitutes the core of conventional or classical control theory, many books (see, for example, [1] and [2] and the majority of books on feedback control published in the 1950's and early 1960's) and papers are available on the subject. Our approach in this chapter will be to present a review and summary of the high points of the frequency-response approach. The content is therefore limited to theoretical and practical fundamentals, without consideration of the many important details covered in the earlier literature.

9-1 SINUSOIDAL INPUT AND OUTPUT

The sinusoidal steady state is realized when all transients have died out. Alternatively, it may be realized if a linear system is set at a proper initial state $\mathbf{x}(0)$ when the sinusoidal forcing begins. A spring–mass–damper system, for instance, must have a specific set of initial values,

$y(0)$ for spring deflection and $\dfrac{dy(0)}{dt}$ for mass velocity,

9-1 Sinusoidal input and output

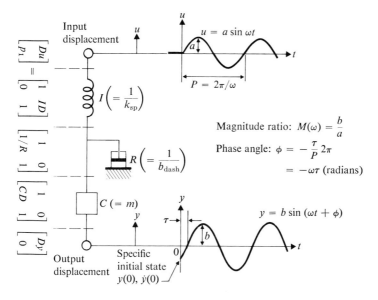

Fig. 9-1 Sinusoidal steady-state input-output of a damped oscillator.

for the sinusoidal steady state shown in Fig. 9-1 to take place with no preliminary transient. In general, the steady-state sinusoidal output–input amplitude ratio and output phase (with respect to the input) can be determined by substituting $j\omega$ for s in the system's transfer function $G(s)$. The proof of this important and basic concept follows.

For a sinusoidal input,

$$u(t) = a \sin \omega t,$$

and

$$U(s) = \frac{a\omega}{s^2 + \omega^2}. \tag{9-1}$$

By the definition of the transfer function $G(s)$ we get

$$Y(s) = G(s)U(s) = \frac{B(s)a\omega}{A(s)(s^2 + \omega^2)}, \tag{9-2}$$

where $G(s) = B(s)/A(s)$, and the roots of $A(s)$ are the n open-loop poles p_i. Expanding Eq. (9-2) into partial fractions, we have

$$Y(s) = \frac{K_0}{s - j\omega} + \frac{K_0^*}{s + j\omega} + \frac{K_1}{s - p_1} + \frac{K_2}{s - p_2} + \cdots \tag{9-3}$$

where K_0 and K_0^* are complex conjugates. They are evaluated in standard

358 Frequency response

fashion, *by setting s equal to jω*, as

$$K_0 = \left[\frac{G(s)a\omega}{s+j\omega}\right]_{s=j\omega} = \frac{G(j\omega)a\omega}{2j\omega} = \frac{aG(j\omega)}{2j} \quad (9\text{-}4a)$$

and

$$K_0^* = \frac{aG(-j\omega)}{-2j}. \quad (9\text{-}4b)$$

For a stable system, the steady-state output $y(t)_{ss}$ will be made up of the inverse transform of only the first two terms on the right-hand side of Eq. (9–3). Since all the p_i are negative or have negative real parts, terms involving them in Eq. (9–3), upon inverse transformation, will lead to transient terms in the time domain that die out as time progresses. We thus get

$$y(t)_{ss} = \mathscr{L}^{-1}\left[\frac{K_0}{s-j\omega} + \frac{K_0^*}{s+j\omega}\right],$$

and by substitution of Eqs. (9–4a) and (9–4b), we get, after inversion,

$$y(t)_{ss} = a\left[\frac{G(j\omega)e^{+j\omega t}}{2j} - \frac{G(-j\omega)e^{-j\omega t}}{2j}\right]. \quad (9\text{-}5)$$

Now, $G(j\omega)$ is generally a complex number having a magnitude and phase, and may be written as

$$G(j\omega) = M(\omega)\underline{/\phi(\omega)} = M(\omega)e^{+j\phi(\omega)} \quad (9\text{-}6a)$$

and

$$G(-j\omega) = M(\omega)\underline{/-\phi(\omega)} = M(\omega)e^{-j\phi(\omega)}. \quad (9\text{-}6b)$$

In Eq. (9–6)

$$M(\omega) = |G(j\omega)| = \sqrt{(\mathscr{R}e\, G(j\omega))^2 + (\mathscr{I}m\, G(j\omega))^2} \quad (9\text{-}7a)$$

and

$$\phi(\omega) = \underline{/G(j\omega)} = \tan^{-1}\left[\frac{\mathscr{I}m\, G(j\omega)}{\mathscr{R}e\, G(j\omega)}\right], \quad (9\text{-}7b)$$

where "$\mathscr{R}e$" means take the real part and "$\mathscr{I}m$" means the imaginary part. Substituting Eqs. (9–6a) and (9–6b) into Eq. (9–5) and rearranging, we get

$$y(t)_{ss} = aM(\omega)\left[\frac{e^{+j[\omega t+\phi(\omega)]} - e^{-j[\omega t+\phi(\omega)]}}{2j}\right].$$

Since

$$\sin(\beta) = \frac{e^{+j\beta} - e^{-j\beta}}{2j},$$

we arrive at

$$y(t)_{ss} = aM(\omega) \sin[\omega t + \phi(\omega)]. \tag{9-8}$$

However, in Fig. 9-1 the steady-state sinusoidal output may be represented as

$$y(t)_{ss} = b \sin[\omega t + \phi(\omega)]. \tag{9-9}$$

Comparing Eqs. (9-8) and (9-9), we see that

$$M(\omega) = |G(j\omega)| = \frac{b}{a}.$$

In other words, the sinusoidal steady-state output-over-input amplitude ratio is obtained by substituting $j\omega$ for s in the transfer function $G(s)$ and taking the amplitude (Eq. 9-7a) of the resulting complex number. The number thus calculated is called the *magnitude ratio* or *gain* and is clearly a function of frequency. Since the order of the numerator of $G(s)$ is usually lower than the order of the denominator, we can expect the gain to become very small as the frequency becomes very high. To obtain the *phase* between the input and output sinusoids in steady state, the procedure is to substitute $j\omega$ for s in $G(s)$ and take the argument (Eq. 9-7b) of the resulting complex number. For negative values of $\phi(\omega)$, the output lags the input as shown in the example of Fig. 9-1. Since this phase shift is a function of frequency, we may ask what the asymptotic limit of the phase shift is at high frequencies. If n and m are the orders of the denominator and numerator polynomials of $G(s)$, respectively, then it is easily deduced that the phase shift at high frequencies is given by $(n - m)(\pi/2)$. This, however, is only true for normal transfer functions that are called minimum phase (see below).

Considering the fact that $G(j\omega)$ is a complex number, we may plot it as a phasor function of ω on a complex plane having real and imaginary axes. We shall call this plane the $G(j\omega)$-plane, and the resulting curve is called the polar or Nyquist plot. The phasor $G(j\omega)$ has a length given by Eq. (9-7a) and a phase angle given by Eq. (9-7b).

Example 9-1 Draw the Nyquist plot for the system of Fig. 9-1, for a damping ratio of 0.40. We first write the transfer function in the canonical form:

$$G(s) = \frac{Y(s)}{U(s)} = \frac{1}{1 + (I/R)s + CIs^2} = \frac{1}{1 + 2\zeta Ts + (Ts)^2}, \tag{9-10}$$

where

$$CI = T^2, \quad \frac{I}{R} = 2\zeta T, \quad T = \text{time constant}, \quad \zeta = \text{damping ratio}.$$

To enable us to draw one universal plot for all time constants, we normalize the

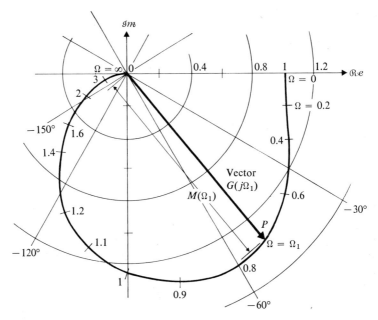

Fig. 9-2 Polar plot for $G(j\Omega) = 1/\{1 + 2\zeta j\Omega + (j\Omega)^2\}$, with $\zeta = 0.4$, $\Omega = \omega T$.

frequency to get

$$G(j\omega) = \frac{1}{1 + 2\zeta j\Omega + (j\Omega)^2}, \tag{9-11}$$

where

$$\Omega = \omega T = \text{normalized frequency.} \tag{9-12}$$

Equations (9-7a) and (9-7b), with Ω substituted for ω, may now be used to compute the magnitude and phase of $G(j\Omega)$ for various values of Ω, all computations being at the required damping ratio of 0.40. The phasor OP in Fig. 9-2 shows $G(j\Omega)$ at a single value of the normalized frequency Ω_1. It is now possible to completely describe a system $G(s)$ by a set of data $M(\Omega)$ and $\phi(\Omega)$ taken over the entire frequency range $0 \leq \Omega \leq \infty$. The term *frequency response* refers to such a set of data. The frequency response of the system is therefore given by the locus of point P over the semi-infinite frequency range. This locus, shown in Fig. 9-2, is the polar plot or the Nyquist plot referred to in the previous paragraph.

The Nyquist plot has been invaluable in the development of stability concepts useful in design procedure, as will be apparent in the following section. It is, however, a tedious job to obtain the Nyquist plot for a complex system. Further, frequency values, which play an important role in system design, appear as a labeled parameter in a nonlinear fashion along the Nyquist plot. To overcome both of these disadvantages an alternative form of presenting frequency-response data, the Bode diagram, will now be introduced.

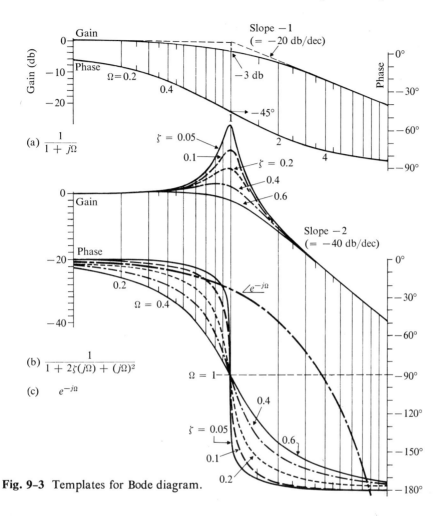

Fig. 9-3 Templates for Bode diagram.

The same magnitude and phase information found on a Nyquist plot is also presented by a Bode diagram. The latter consists of two plots: one of the phase angle $\phi(\omega)$, the other of the logarithm of the magnitude ratio $M(\omega)$, both plotted against ω, on a log scale. Plotting the *logarithm* of $M(\omega)$ greatly simplifies the plotting procedure since, as will be shown shortly, straight-line asymptotes may be used to advantage. The log magnitude ratio is commonly expressed in decibels, where the decibel (db) quantity of a number N is given by

$$N(\text{db}) = 20 \log_{10} N. \tag{9-13}$$

Most transfer functions are made up of combinations of the factors

$$(1 + j\Omega)^{\pm 1} \quad \text{and} \quad [1 + 2\zeta j\Omega + (j\Omega)^2]^{\pm 1}, \tag{9-14}$$

where the nondimensionalized frequency is defined by Eq. (9–12). Bode diagram templates for the -1 superscript case of Eq. (9–14) are shown in Fig. 9–3. For the $+1$ superscript case (that is, factors in the numerator of the transfer function) the signs on the ordinate scales of Fig. 9–3 change to plus. Alternatively, the plots for the $+1$ case are the mirror image about the 0-db and 0° lines of Fig. 9–3. With this set of templates we can construct the Bode diagram of a transfer function in the general form

$$G(s) = k \frac{\prod (1 + T_j s) \prod [1 + 2\zeta_l T_l s + (T_l s)^2]}{s^N \prod (1 + T_i s) \prod [1 + 2\zeta_m T_m s + (T_m s)^2]}, \quad k > 0. \tag{9-15}$$

This function has a product form in terms of building blocks $C_i(s)$:

$$G(s) = \prod C_i(s).$$

For the product form, the log magnitude and phase angle in the Bode diagram are both expressed by a summation:

$$\log |G(j\omega)| = \sum_i \log |C_i(j\omega)|, \quad \angle G(j\omega) = \sum_i \angle C_i(j\omega).$$

Therefore the Bode diagram plotting of Eq. (9–15) reduces to a graphical superposition of template patterns, except for k/s^N, which has the simple behavior

$$\left| \frac{k}{(j\omega)^N} \right| = \frac{k}{\omega^N}, \quad \angle \frac{k}{(j\omega)^N} = -90N \text{ deg}.$$

The last expression introduces a constant phase shift and a log magnitude plot with a slope of $-20N$ db/dec (decade) and a 0-db intercept at frequency $\omega = (k)^{1/N}$. Each building block $C_i(s)$ has a corner frequency ω_i defined as $\omega_i = 1/T_i$. For example, the corner frequencies of $C_m = 1/[1 + 2\zeta_m T_m s + (T_m s)^2]$ and $C_i = 1/[1 + T_i s]$ are $1/T_m$ and $1/T_i$ respectively. The general plotting procedure is best illustrated by an example.

Example 9-2 Let us consider

$$G(s) = \frac{\frac{1}{2}(1 + 0.5s)}{s(1 + 0.8s + s^2)}, \tag{9-16}$$

and follow the steps of the Bode diagram plotting.

Step 1. Letting $C_1 = 1/(2s)$, $C_2 = 1/(1 + 0.8s + s^2)$, $C_3 = 1 + 0.5s$, for each component we find the corner frequency where its normalized frequency Ω_i becomes 1. In our example, these values are 0.5, 1, and 2 for C_1, C_2, and C_3, respectively. In general, the horizontal location of a template or the position of the k/ω^N line is fixed by these frequencies. The frequencies are marked by arrows in Fig. 9–4.

Step 2. We draw an approximate log magnitude diagram by segments of straight lines. In our example, $\log |C_1(j\omega)|$ is a straight line of slope -20 db/dec of frequency

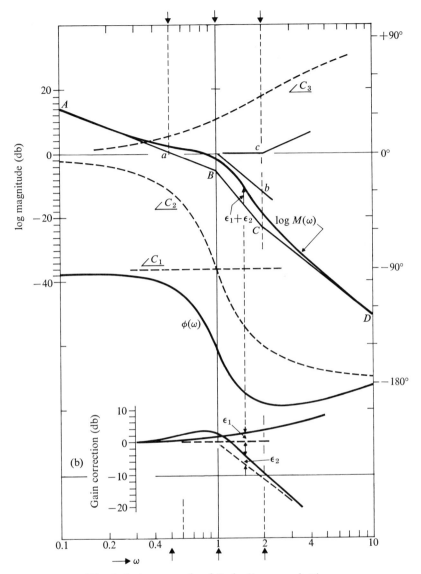

Fig. 9-4 An example of Bode-diagram plotting.

(also called a -1 slope), crossing the 0-db level at the marked frequency 0.5 (*a* in Fig. 9-4). The log magnitude diagram of $C_2(j\omega)$ is approximated by two asymptotes: $\lim_{\omega \to 0} |C_2(j\omega)| = 1$, which yields the 0-db line (zero slope) for $\omega \ll 1$ (low frequency); and $\lim_{\omega \to \infty} |C_2(j\omega)| = 1/\omega^2$, which yields a -2-slope line for $\omega \gg 1$ (*b* in the figure). Similarly, the asymptote of $C_3(j\omega)$ is the 0-db line for $\omega \ll 2$ and $+1$-slope line for $\omega \gg 2$ (*c* in the figure). Superimposing all of these curves, we obtain the approximate plot *ABCD*. The slope of this graph changes at *B* (from -1 to -3) and at *C* (from -3 to -2). The frequency at which line segments of different slopes meet is the *corner frequency* defined above. The corner frequencies of our system ($\omega = 1$ and 2) were already marked in step 1.

Step 3. Applying the templates for C_2 and C_3, we find the gain corrections, as indicated by ε_1 and ε_2, in Fig. 9-4(b). Note that the template must be applied in an upside-down position for the numerator factor C_3. Making the corrections mostly in the vicinity of each corner frequency, we obtain the log magnitude diagram labeled log $M(\omega)$ in the figure. With only 2 or 3 corrections, a fairly accurate plot may be obtained between the asymptotes.

Step 4. Phase angle plotting. We plot $\angle C_2(j\omega)$ and $\angle C_3(j\omega)$ by template (where the angle template for $\angle C_3$ must be turned upside down, as noted before). Adding these angles and $\angle C_1 = -90°$, we obtain the phase diagram $\phi(\omega)$. For details of the plotting technique, see, for instance, [2].

With experience, the actual superposition outlined in Example 9-2 becomes unnecessary and the Bode diagram may be drawn directly by considering the corner frequencies sequentially. Thus, for the transfer function of Eq. (9-16), since $N = 1$, we would start by drawing an initial slope line of -20 db/dec, which crosses the 0-db line at $\omega = 0.50$. At the first corner frequency of $\omega = 1$ there occurs a *double* break in slope (due to the quadratic), *downward* (since the quadratic is in the denominator) to a new slope of -60 db/dec. The slopes of the asymptotes will always be ($+$ or $-$) multiples of 20 db/dec, changing at a corner frequency to a new multiple. The multiplicity of the *change* in slope will correspond to the order of the corner frequency factor; hence the "double break" in slope at $\omega = 1$ above. Finally, there is a single upward break at the second corner frequency of $\omega = 2$, due to the single time-constant factor in the numerator, to a slope of -40 db/dec. The resulting straight line approximation to the log magnitude curve, *ABCD* in Fig. 9-4, may be corrected, as before, by using the templates in the vicinity of the corner frequencies to yield the accurate $M(\omega)$ plot. The phase shift diagram may be plotted by superposition, as in Example 9-2, or it may be calculated directly from Eq. (9-16) with attention focused on the vicinity of the corner frequencies and asymptotic behavior utilized elsewhere. In applying this technique of plotting the Bode diagram, it often happens that the first corner frequency is lower than the 0-db intercept of the initial slope line. In such cases the procedure is to draw the initial slope line *extended* to the 0-db intercept and then back up along this slope line to the first corner frequency, proceeding as before. In the first stages of a control-system design the straight-line approximation to the log magnitude plot is often sufficient for preliminary design purposes. In the final analysis, however, the curve should be corrected by use of the templates. When corner frequencies are close together, care must be exercised in the proper ($+$ or $-$) superposition of corrections.

The graphical method presented in the preceding paragraphs applies when the transfer function is in factored form. A graphical method for esti-

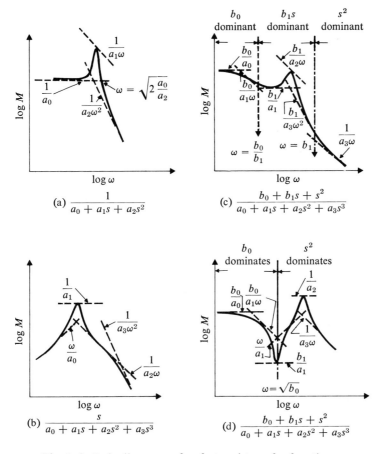

Fig. 9-5 Bode diagrams of unfactored transfer functions.

mating the log magnitude diagram from transfer functions in their unfactored polynomial form, $G(s) = B(s)/A(s)$, was developed by Ausman [3]. As an example, consider the second-order form

$$G(s) = \frac{1}{a_0 + a_1 s + a_2 s^2}.$$

According to the preceding discussion, asymptotes $1/a_0$ and $1/(a_2\omega^2)$ apply at low and high frequencies, respectively. Ausman points out that the middle term $1/(a_1\omega)$ predominates over an intermediate range of frequencies, as shown in Fig. 9-5(a). A third-order system with a resonance peak (that is, a pair of conjugate complex poles) is treated in part (b) of the figure. A stable third-order system must satisfy the Routh condition,

$$a_1 a_2 > a_3 a_0.$$

This condition can be interpreted as

$$\frac{1}{a_1} < \frac{1}{a_3 \omega_1^2} \quad \text{at} \quad \omega_1^2 = \frac{a_0}{a_2},$$

Therefore the $1/(a_3\omega^2)$ line is located higher than the $1/a_1$ line at frequency $\omega_1 = \sqrt{a_0/a_2}$ where the two lines ω/a_0 and $1/(a_2\omega)$ intersect.

Figure 9–5(c) and (d) represent third-order systems with second-order numerators. According to Ausman, a number of fourth- or higher-order systems can be treated in the same manner as outlined by the examples in the figure.

For linear systems described by the ratio of finite polynomials that have no poles or zeros in the right half s-plane, a direct relationship exists between the magnitude ratio diagram of the entire frequency range ($0 \leq \omega \leq \infty$) and the phase angle. The mathematical relationships between the two were formulated by Bode [4]. The first of the two Bode theorems states, in essence, that if the magnitude ratio of a system to be designed is arbitrarily fixed over the entire frequency range, then the phase angle is also fixed and may not be arbitrarily chosen. The second Bode theorem states that if the magnitude ratio is fixed over one frequency range, and the phase angle is fixed over the remainder of the frequency range, then the system is completely specified. From these Bode theorems it may be shown that, for a system to be stable, the log magnitude plot on the Bode diagram must have a slope ≤ -20 db/dec in the vicinity (from about $+12$ to -12 db) of the 0-db line. A system is said to be *minimal phase* when its $M(\omega)$ and $\phi(\omega)$ are related by the Bode theorems over the entire frequency range. $G(s)$ of Eq. (9–15) is minimal phase when all time constants and damping factors are nonnegative. For convenience in further discussion, let us write the transfer function in a different form:

$$G_0(s) = \frac{\prod (1 + b_j s) \prod [(1 + \beta_l s)^2 + \delta_l^2]}{\prod (1 + a_i s) \prod [(1 + \alpha_k s)^2 + \gamma_k^2]}. \tag{9-17}$$

This is a minimal phase system if all parameters (real numbers) a_i, b_j, etc., are positive. There exist transfer functions $G_m(s)$, $m = 1, 2, \ldots$ such that when $G(s) = G_0(s) G_m(s)$

$$|G(j\omega)| = |G_0(j\omega)|$$

but

$$\angle G(j\omega) \neq \angle G_0(j\omega).$$

The index m refers to the number of nonminimal phase factors in $G(s)$. The modifying factors $G_m(s)$ are given by

$$G_m(s) = \frac{1 - cs}{1 + cs}, \quad c > 0, \tag{9-18}$$

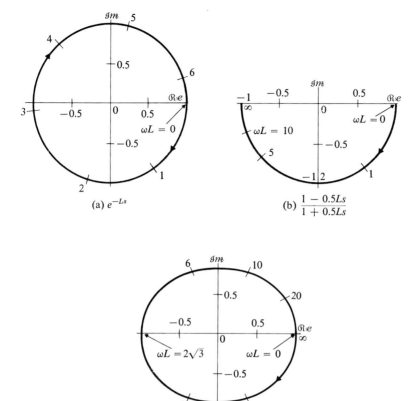

Fig. 9-6 Dead time and its approximations.

where c can be a_i, b_j, α_k, and/or β_l in (9–17). For example, $(1 + b_j s)$ in $G_0(s)$ is replaced by $(1 - b_j s)$ in $G(s)$, $b_j > 0$, or $(1 + \alpha_k s)$ is replaced by $(1 - a_i s)$, $a_i > 0$, and so on. $G_m(s)$ is a pure phase-shift element, because $|G_m(j\omega)| = 1$ but $\angle G_m(j\omega)$ varies from 0 to $-180°$ as ω is increased from zero to infinity, as shown in Fig. 9–6(b). This is why $G_m(s)$ introduces only phase change. All systems $G(s)$ that differ from $G_0(s)$ by one or more factors $G_m(s)$ are said to be *nonminimal phase*. Unstable systems (that is, systems with some positive poles) or those with positive zero(s) are nonminimal phase.

The templates (a) and (b) in Fig. 9–3 must be applied with caution for the phase plotting of nonminimal phase systems. If a factor $(1 - j\Omega)$ or $\left(1 - 2\zeta j\Omega + (j\Omega)^2\right)$ appears anywhere in a transfer function that is being plotted, the angle template for $(1 + j\Omega)$ or $\left(1 + 2\zeta j\Omega + (j\Omega)^2\right)$ must be turned over from its normal position to produce an upside-down phase plot.

Table 9-1 FREQUENCY RESPONSE OF SOME RATIONAL TRANSFER FUNCTIONS

Transfer function	Nyquist diagram	Bode diagram $\log M(\omega)$	Bode diagram $\phi(\omega)$	Transfer function	Nyquist diagram	Bode diagram $\log M(\omega)$	Bode diagram $\phi(\omega)$
(a) A Scalar 1		0 db	0°	(h) PI-action $1 + \dfrac{1}{T_i s}$			
(b) Integrator $\dfrac{1}{s}$		−1 slope, 0 db	−90°	(i) (Ideal) PID-action $1 + \dfrac{1}{T_i s} + T_d s$			+90°, 0°, −90°
(c) Single lag $\dfrac{1}{s+1}$		0 db, $\omega = 1$	0°, −90°	(j) Inertial system $\dfrac{1}{s^2}$		−2	0°, −180°
(d) Autocatalytic reaction $\dfrac{1}{s-1}$ *		0 db	0°, −90°, −180°	(k) Integrator with lag $\dfrac{1}{s(s+1)}$		−1, −2	0°, −90°, −180°
(e) Derivative with lag $\dfrac{s}{s+1}$		0 db, +1	0°, +90°	(l) Double lag $\dfrac{1}{(s+1)(Ts+1)}$ $0 < T < 1$		−1, −2	0°, −180°
(f) Phase lead $\dfrac{Ts+1}{s+1}$, $T > 1$		+1, 0 db	0°	(m) Damped oscillator $\dfrac{1}{s^2 + 2\zeta s + 1}$ $1/\sqrt{2} > \zeta$		−2	0°, −180°
(g) Phase lag $\dfrac{Ts+1}{s+1}$, $T < 1$		0 db, −1	0°	(n) Reverse reaction $\dfrac{k}{Ts+1} - \dfrac{1}{s+1}$ * $1 < k < T$		−1, −2, 0 db	0°, −270°

*nonminimum phase system

9–1　Sinusoidal input and output　369

Table 9-2 RESPONSE PATTERNS OF SYSTEMS WITH A DEAD TIME

	Series		Series-parallel	Parallel	
System	(a) $U \to \boxed{e^{-Ls}} \to \boxed{\dfrac{k}{s}} \to Y$	(b) $U \to \boxed{e^{-Ls}} \to \boxed{\dfrac{k}{Ts+1}} \to Y$	(c) $U \to \boxed{\dfrac{k}{s}} \to $ [1 and e^{-Ls} in parallel, subtracted] $\to Y$	(d) $U \to$ [1 and e^{-Ls} in parallel, subtracted] $\to Y$	(e) $U \to$ [$\dfrac{k_1}{Ts+1}$ and $k_2 e^{-Ls}$ in parallel, summed] $\to Y$
Step-input response	ramp after delay	exponential rise after delay	ramp then leveling at L	pulse of width L	rise then level at L
Nyquist diagram	inward spiral	inward spiral to origin	spiral to small loop	circle	spiral with loops

The frequency-response diagrams of some lumped-parameter systems are listed in Table 9-1. The arrow on the Nyquist diagrams of the Table indicates the direction of increasing frequency. The magnitude responses in the Bode diagrams are shown by their asymptotes. The name "reverse reaction" of the last system in the table stems from the system's step-input response pattern where a negative dip appears.

A delay (or dead-time) process

$$G(s) = e^{-Ls} \tag{9-19}$$

produces only phase shift, without any amplitude distortion in signal wave form. Its frequency response is

$$|G(j\omega)| = 1, \qquad \angle G(j\omega) = -\omega L \text{ (rad)} = -57.30\omega L \text{ (deg)}. \tag{9-20}$$

This is nonminimal phase. A template can be made for phase-angle plotting in the Bode diagram. It is shown in Fig. 9-3, where the normalized frequency is $\Omega = \omega L$. The Nyquist diagram of the dead-time transfer function is a unit circle (Fig. 9-6a). Some approximations of the delay process are shown in Fig. 9-6(b) and (c). They are used mostly for the purpose of analog computer simulation. Interesting response patterns appear when a dead time is coupled with rational transfer functions. Some of these patterns are shown in Table 9-2, where Number 3 is the zero-order holder of digital control systems. The zero-order holder and system 5 of the table already suggest the complexity of frequency-response patterns we face in various distributed-parameter systems.

For the majority of distributed-parameter systems it is no longer possible to approximate their magnitude response by segments of straight lines, nor to estimate their phase angle from the slope of the magnitude diagram, because they are nonminimal-phase systems. A ripple in the magnitude and/or phase angle diagrams may appear in some distributed-parameter systems. Referring back to Example 5-1, we note that a distributed-parameter system will have an infinite number of corner frequencies as we proceed to the right along the Bode diagram abscissa. We may represent the system in lumped fashion only when these higher-frequency breaks on the Bode diagram fall well to the right of signal frequencies that must be transmitted by the system. If this is not the case, then the lumped model arrived at in Example 5-1 is not sufficient, and the techniques of Chapter 7 must be applied.

Although the frequency response of distributed-parameter systems is in general not simple, it often provides the only approach to engineering problems of identification and stability analysis. This is so because the inverse Laplace transformation of an irrational function of s is generally difficult, and the Routh–Hurwitz stability criteria (Chapter 4) do not apply to characteristic equations of an infinite order. In Chapter 7 we saw the appearance of an exponential transfer function in the general form

$$G(s) = e^{F(s)} \tag{9-21}$$

9-1 Sinusoidal input and output

as a crucial dynamic element for a variety of distributed-parameter systems (for example, see Figs. 7-10, 7-15, 7-16, 7-20). We can compute the frequency response of (9-21) by the following expressions;

$$|G(j\omega)| = \exp[\mathscr{R}e\, F(j\omega)], \quad \angle G(j\omega) = \mathscr{I}m\,[F(j\omega)]\,(\text{rad}). \quad (9\text{-}22)$$

The function $F(s)$ itself is sometimes irrational. For example, $G(s)$ of a diffusion process was given in Section 7-2 and restated as

$$G(s) = e^{-\sqrt{T^*s}z}.$$

Letting $s = j\omega$ and introducing a dimensionless frequency $\Omega = z^2\omega T^*$, we obtain

$$G(j\Omega) = e^{-\sqrt{j\Omega}}.$$

We next apply the relation

$$\sqrt{j\Omega} = \pm(1+j)\sqrt{\Omega/2}, \quad (9\text{-}23)$$

and obtain the frequency response of the diffusion process:

$$|G(j\Omega)| = e^{-\sqrt{\Omega/2}}, \quad \angle G(j\Omega) = -\sqrt{\Omega/2}\,(\text{rad}). \quad (9\text{-}24)$$

The minus sign of (9-23) would give a magnitude amplification and phase advance that violate causality and thus is rejected.

If $F(s)$ has a square-root form,

$$F(s)_{s=j\omega} = F(j\omega) = \sqrt{a(\omega) + jb(\omega)}, \quad a(\omega) \text{ and } b(\omega) \text{ both real,}$$

we apply the well-known relation,

$$\sqrt{a + jb} = \pm \frac{1}{\sqrt{2}}\left(\sqrt{\sqrt{a^2+b^2}+a} + j\sqrt{\sqrt{a^2+b^2}-a}\right), \quad (9\text{-}25)$$

where we discard one of the two signs. A further generalization of (9-25) is given by DeMoivre's theorem. Writing the polar expression for a complex variable,

$$a + jb = re^{j\theta},$$

we find that the general relation is given by the following expressions:

$$(a+jb)^N = r^N(\cos N\theta + j\sin N\theta),$$
$$(a+jb)^{1/N} = r^{1/N}\left(\cos\left[\frac{\theta+2k\pi}{N}\right] + j\sin\left[\frac{\theta+2k\pi}{N}\right]\right), \quad (9\text{-}26)$$

where N is an integer, $k = 0, 1, 2, \ldots, N-1$, some of which must be rejected.

The frequency-domain representation of a linear system's dynamics (mostly in the format of the Bode diagram) is useful for various engineering problems: in particular, (a) to identify an unknown plant transfer function from frequency test data [5, 6, 7], (b) to derive approximate models of complex systems in the frequency domain [8], and (c) to confirm a theoretical result by frequency-response test. On this last point, for instance, Nichols' theory on linear pneumatic line dynamics [7, Chapter 7] was favorably confirmed by Karam, who reports experimentally determined magnitude plots of pneumatic lines, tested up to 1,000 cps [7]. As noted in the preceding paragraph, the frequency response approach is especially preferable in theoretical or experimental investigations of distributed-parameter systems [9, 10, 11]. In chapter 12 we shall see that frequency response techniques are useful in the analysis of nonlinear systems.

9-2 THE NYQUIST STABILITY THEOREM

Before presenting a derivation of the Nyquist stability criterion [12], some preliminary comments are in order. Consider the function $F(s) = 1/(s - p_1)$. If we restrict s to values along the imaginary axis of the s-plane ($s = j\omega$, $-\infty < \omega < +\infty$), then the denominator of $F(s)$ may be represented by the phasor $\overline{AQ}$ in Fig. 9-7 and $F(s)$ itself by $1/\overline{AQ}$. As ω varies from $-\infty$ to $+\infty$, the point Q moves up the imaginary axis, and the phasor $\overline{AQ}$ changes phase by $+\pi$ radians. The phasor $1/\overline{AQ}$ and therefore $F(s)$ changes phase by $-\pi$ radians. *If the point A lies in the right half-plane* (RHP), *then the signs of the above phase changes are reversed.* Similarly, the function $F(s) = (s - z_1)$ may be represented by the phasor $\overline{BQ}$. As Q sweeps the ω-axis the function $F(s)$, or the phasor $\overline{BQ}$, changes phase by $+\pi$ radians if B is in the left half-plane (LHP), and by $-\pi$ radians if point B is in the RHP.

The interpretation of this phase change is better understood if we examine the plot of the $F(s)$-function in the complex $F(s)$-plane for the stated values of s in the s-plane. The resulting plot is called the conformal map of the contour in the s-plane (the imaginary axis here) onto the $F(s)$-plane through the mapping function $F(s)$. The conformal maps for the two elementary functions considered above are shown in Fig. 9-8 for the case where A and B are in the LHP. The phase changes in the function as s sweeps the imaginary axis, *with respect to the origin of the $F(s)$-plane*, are clearly shown by the enlargement of the plot in the vicinity of the origin. Thus, for $F(s) = 1/(s - p_1)$, the phasor $F(j\omega)$ (shown at $\omega = \omega_1$) points straight up at $\omega = -\infty$ and straight down at $\omega = +\infty$. The phase change in $F(s)$ in this case is $-\pi$ radians. For $F(s) = (s - z_1)$, the phasor $F(j\omega)$ starts out pointing downward at $\omega = -\infty$ and ends up pointing upward at $\omega = +\infty$. The phase change in $F(s)$ here is therefore $+\pi$ radians. If A and B lie in the RHP, the arrowheads on the two plots reverse directions and the limit values of ω change positions. The phase changes in $F(j\omega)$ just described for both cases also change signs when A and B lie in the RHP of Fig. 9-7.

9–2 The Nyquist stability theorem

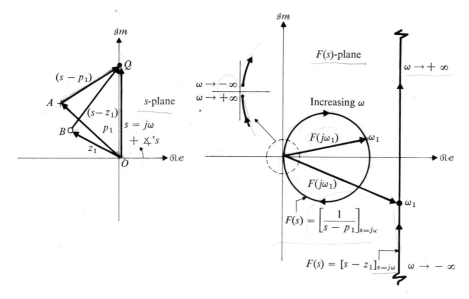

Fig. 9-7 Phasor representations in the s-plane.

Fig. 9-8 Conformal maps of the imaginary axis on the s-plane for two elementary functions.

It should be clear that the phase change of more complex functions, as point Q sweeps the imaginary axis on the s-plane, may be determined by superposition of these simpler cases. Thus the net phase change of $F(s) = (s - z_1)/(s - p_1)$ is zero when both the zero and the pole lie in the LHP. When the pole is in the LHP and the zero is in the RHP, the net change in phase of $F(s)$ as Q sweeps the imaginary axis is -2π radians. If the pole and zero are reversed, the net phase change is $+2\pi$ radians. We will employ this point of view in our derivation below.

At the beginning of chapter 8 we discussed the characteristic equation of a closed-loop system. We repeat some of that presentation in order to emphasize several points of importance for the derivation that follows. The characteristic equation of a closed-loop feedback control system is:

$$F(s) = 1 + G(s)H(s) = 1 + \frac{B(s)}{A(s)} = \frac{A(s) + B(s)}{A(s)} = 0, \quad (9\text{-}27)$$

where

$$G(s)H(s) = \frac{B(s)}{A(s)} = \frac{\prod_{j=1}^{m}(s - z_j)}{\prod_{i=1}^{n}(s - p_i)}, \quad (9\text{-}28)$$

is the open-loop transfer function with m open-loop zeros (the z_j) and n open-loop poles (the p_i). As usual, $m \leq n$ (due to causality restrictions) and the z_j and p_i may be real or complex. Noting in the last form of Eq. (9-27) that the order of numerator and denominator are the same, we may rewrite Eq.

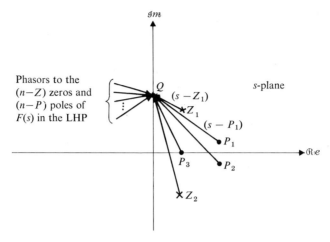

Fig. 9-9 Phasor representation of $F(s)$ when $s = j\omega$. (Two zeros and three poles in RHP.)

(9-27) in the factored form:

$$F(s) = \frac{\prod_{j=1}^{n}(s - Z_j)}{\prod_{i=1}^{n}(s - p_i)}, \tag{9-29}$$

where the Z_j are the n zeros of the characteristic equation. These Z's are of course the poles of the closed-loop transfer function, and their values determine the nature of the closed-loop response (see Fig. 8-7). For a stable system, all the Z_j must lie in the LHP. The Nyquist stability criterion determines this fact by using frequency-domain techniques applied to the open-loop transfer function.

We now proceed to the derivation.

Let $F(s)$ have Z zeros and P poles in the RHP. There are, thus, by examination of Eq. (9-29), $(n - Z)$ zeros and $(n - P)$ poles in the LHP. The situation is shown in Fig. 9-9 for $Z = 2$ and $P = 3$.

We now let the point Q sweep the imaginary axis (from $-\infty$ to $+\infty$) on the s-plane and tabulate the net phase change in $F(s)$ on the $F(s)$-plane:

Positive change in $\angle F(s)$ due to the $(n - Z)$ zeros in LHP $= (n - Z)\pi$
Positive change in $\angle F(s)$ due to the P poles in RHP $= P\pi$
Negative change in $\angle F(s)$ due to the $(n - P)$ poles in LHP $= -(n - P)\pi$
Negative change in $\angle F(s)$ due to the Z zeros in RHP $= -Z\pi$

Net positive change in phase of $F(s) = (P - Z)(2\pi)$

If we divide both sides of the last equation by 2π, we get on the left the number of *counterclockwise* revolutions (called N) of the function $F(s)$ with respect

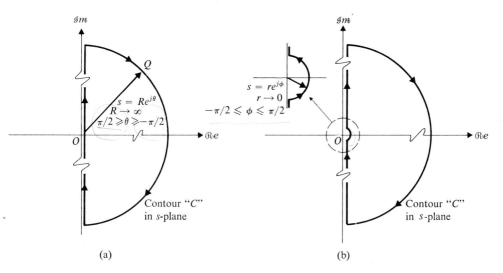

Fig. 9-10 Contour C in the s-plane: (a) no indentations, and (b) one indentation at the origin.

to the origin of the $F(s)$-plane. Thus,

$$N = (P - Z). \tag{9-30}$$

Equation (9-30) is the statement of the Nyquist stability criterion. Some texts, defining the net *negative* change in the phase of $F(s)$, present it as $N = Z - P$. In that case, N is counted as the number of *clockwise* revolutions of the plot of $F(s)$ with respect to the origin. Before proceeding to an interpretation and the use of Eq. (9-30), several additional comments are necessary to clarify and simplify the application of the criterion.

The contour C in the s-plane, consisting of the imaginary axis so far, can be closed by drawing an infinite semicircle in the RHP from $\omega = +\infty$ to $\omega = -\infty$ with center at the origin.* The closed contour C then appears as in Fig. 9-10(a). As the moving point Q sweeps out the infinite semicircle, the phase of $F(s)$ is unchanged since the order of the numerator and denominator polynomials of $F(s)$ are the same and *all* phasors from poles and zeros rotate in the same direction. This merely states that all poles and zeros lie to the left of the infinite semicircle of Fig. 9-10(a).

If any poles of $F(s)$ lie on the imaginary axis, the derivation must be modified slightly to account for discontinuities in the mapping procedure. The modification consists in indenting the contour C around the pole in

---

* Strictly speaking, this closure of the C-contour in the s-plane by use of the infinite semicircle is not necessary for the Nyquist stability criterion proof presented here. It is included to maintain the one-to-one mapping of a *closed* contour in the s-plane into a *closed* contour in the $F(s)$-plane, and to delineate clearly the entire right-hand half of the s-plane.

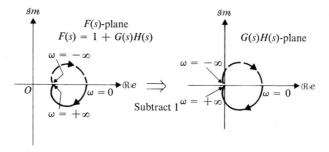

Fig. 9-11 Transformation from the $F(s)$- to the $G(s)H(s)$-plane.

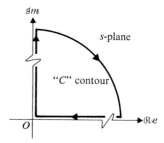

Fig. 9-12 Contour C for plotting open-loop frequency response.

question. Figure 9–10(b) shows the contour C needed when $F(s)$ has a pole at the origin. The proof is unchanged whether the indentation is to the right or to the left. However, when indenting to the left, the pole must be counted as one of the P poles lying in the RHP. What maintains the validity of the proof is that N changes by $+1$ for each indentation to the left. This can be corroborated by actually plotting $F(s)$ for each of the two cases (see Problems 9–11 and 9–12).

Instead of plotting $F(s) = 1 + G(s)H(s)$, we may consider the plot of $G(s)H(s)$ by subtracting one from $F(s)$ and counting phase change or revolutions with respect to the -1 point in the $G(s)H(s)$-plane. A typical transformation of this type appears in Fig. 9–11.

We may further simplify the use of Eq. (9–30) by making use of symmetry about the real axis in the $G(s)H(s)$-plane. Note that

$$G(j\omega)H(j\omega) = [G(-j\omega)H(-j\omega)]^*,$$

where the $*$ denotes conjugation. Thus we need only plot $G(s)H(s)$ for positive values of ω, using the contour C in the s-plane as shown in Fig. 9–12 (with indentations added as necessary). For such a contour we will get, in the $F(s)$-plane, half as many revolutions as before, so we must double the count of N determined from $G(j\omega)H(j\omega)$ obtained for $0 < \omega < +\infty$. This latter plot is the open-loop frequency response discussed in the previous section.

9-2 The Nyquist stability theorem

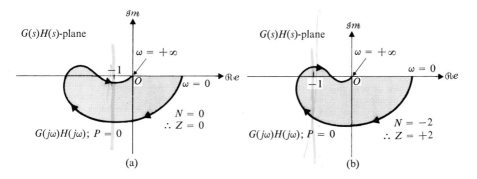

Fig. 9-13 Application of the modified Nyquist stability criterion to (a) a stable system, and (b) an unstable system.

Note, however, an important distinction: here s is set equal to $j\omega$ to delineate the left and right half-planes, whereas in the previous section s was set equal to $j\omega$ to compute the frequency response. That the calculation scheme is the same for both cases is a fortunate convenience.

Recall, in the equation $N = P - Z$, that P is the number of open-loop poles in the RHP and Z is the number of closed-loop poles in the RHP. For a system to be stable, Z must equal zero. Thus, N must equal P, where P is usually determined from a Routh array test on $A(s)$ in Eq. (9-28). In most cases of engineering interest $P = 0$ (which means that the system is open-loop stable). Then, for the system to be stable, N must equal zero. This requirement is referred to as the "modified Nyquist stability criterion." Now N is the number of counterclockwise revolutions of $G(s)H(s)$ with respect to the -1 point in the $G(s)H(s)$-plane as s sweeps the contour C in the s-plane. A convenient method for easily determining N is to imagine a straight line which crosses the $G(j\omega)H(j\omega)$-plot, extending from the -1 point to anywhere on the infinite circle. N will then be the net number of times that $G(j\omega)H(j\omega)$ crosses this straight line, for increasing ω, with $+$ and $-$ signs being assigned to right-to-left and left-to-right crossings, respectively. Another means of counting N is to imagine a phasor from the -1 point to the $G(j\omega)H(j\omega)$-plot, and then count the counterclockwise revolutions of this phasor as ω varies from 0 to ∞. (Remember to double the resulting count to get N!)

Another way of saying that N must equal zero (for the case when $P = 0$) is to require that the -1 point in the $G(s)H(s)$-plane not be encircled by the $G(j\omega)H(j\omega)$-plot. To check encirclement, imagine the $G(s)H(s)$-plane shaded to the right (looking in the direction of increasing ω) of the $G(j\omega)H(j\omega)$-plot. For N to be zero, the -1 point must not fall within this shaded area. The two possible situations are depicted in Fig. 9-13 for a typical $G(j\omega)H(j\omega)$-plot. While this same qualitative information regarding stability may be determined by application of the Routh array to $[A(s) + B(s)]$ in Eq. (9-27), such an approach would lead to a dead end in that it would afford us no insight

Table 9-3 STABILITY TEST VIA THE NYQUIST THEOREM

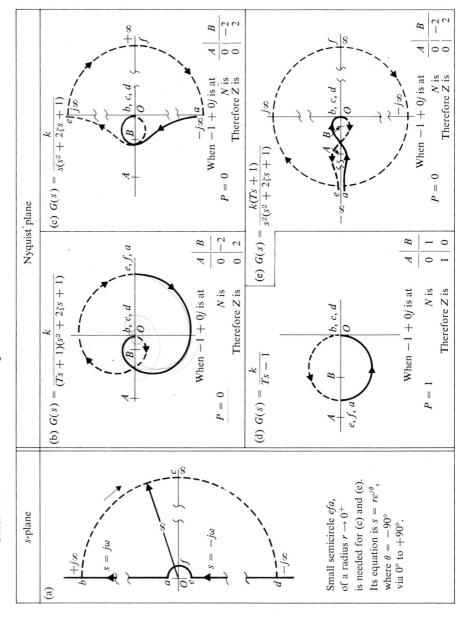

as to any means of system stabilization. We shall see in subsequent sections that application of the Nyquist stability criterion will enable us to determine suitable compensating networks for stabilizing an otherwise unstable system. In fact, application of the Nyquist stability theorem is not limited to feedback control systems, and it is one of the most important stability theorems in linear systems input–output analysis ([6] of Chapter 4). We must keep in mind, however, that application of the Nyquist stability theorem will not account for uncontrollable and/or unobservable modes that might arise, for example, in pole-zero cancellation.

If $G(s)$ has no pole at the origin ($s = 0$), $G(0)$ becomes a real number, and thus the entire Nyquist diagram is completed, as shown in parts (b) and (d) of Table 9–3. (For brevity, $H(s)$ is taken as unity here.) If, on the other hand, $G(s)$ has a pole at $s = 0$, the singular point is bypassed, as explained earlier, by a small semicircle in the s-plane (*efa* in Table 9–3a). The equation of the circle in polar coordinates is

$$s = re^{j\theta},$$

where r is positive and very small. We now evaluate $G(s)$ as $s \to 0$ by the following expression:

$$\lim_{s \to 0} G(s) = \lim_{s \to 0} \left(\frac{k}{s}\right) = \left(\frac{k}{r}\right) e^{-j\theta}, \qquad (9\text{--}31)$$

where $k = \lim_{s \to 0} sG(s)$ is the gain constant in the canonical form (9–15) where $N = 1$ is the system type. When θ in Eq. (9–31) is changed from $-90°$ via $0°$ to $+90°$, a large semicircle of radius $k/r \to \infty$ appears in the Nyquist plane, as shown by *efa* on the Nyquist plot in Table 9–3(c). If $G(s)$ has a double pole at $s = 0$, as in (e) of the table, we have

$$\lim_{s \to 0} G(s) = \left(\frac{k}{r^2}\right) e^{-2j\theta}.$$

As a consequence, a large complete circle appears in the Nyquist plane. A singularity at $s = 0$ of an order higher than two, or singularities on the imaginary axis of the s-plane (due to pure imaginary open-loop poles of an undamped oscillator) are all treated in exactly the same way (see Problem 9–11).

We are now ready to apply the Nyquist stability theorem to systems (b) through (e) in Table 9–3. If the gain k of these systems is small, the -1 point on the negative real axis of the Nyquist plane may be located at A in these figures. On the other hand, the location of the -1 point relative to the Nyquist diagram may move to B if k is increased. The results for these two relative locations are shown at the bottom of each space in the table, where the number of counterclockwise rotations N of $G(s)$ is counted about A and B, respectively. According to the tabulated results, systems (b), (c),

Table 9-4 FREQUENCY RESPONSE OF NORMAL CONTROL SYSTEMS

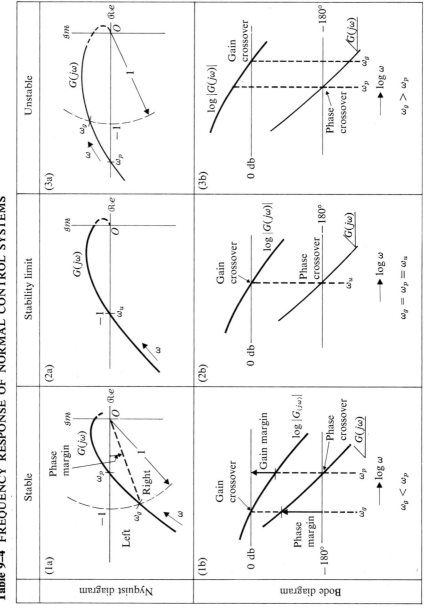

and (e) are stable when the gain is low. Two unstable closed-loop poles will appear (that is, $Z = 2$) when the gain of these systems is increased. By knowledge of the root locus we deduce that the unstable poles are a conjugate complex pair. System (d) in the table is unstable when open-loop. It becomes stable when the loop is closed, and the gain is sufficiently increased. We see in this case the system is stable when -1 is enclosed by the circle. In some systems the Nyquist diagram for the positive frequency range crosses the negative real axis more than once (see Fig. 9-13). If such multicrossing occurs, the system will be stable (or unstable) for some finite range of k. The system is then said to be conditionally stable (or unstable). A conditionally stable system is characterized by a pair of conjugate root-locus branches that will cross the imaginary axis in the root-locus plane more than once.

The majority of control problems in engineering deal with objects which possess no poles with positive real parts. In general, we do not introduce any poles with positive real parts into an open-loop transfer function by controller (and feedback) action. Hence, as noted earlier, $P = 0$ in Eq. (9-30) is the typical condition in engineering practice. For well-behaved (or "normal") stable open-loop engineering systems the portion of the Nyquist diagram crucial for stability investigation reduces to a small section in some vicinity of the $-1 + 0j$ point, as shown in (1a) through (3a) in Table 9-4. A closed-loop system is stable if the $-1 + 0j$ point is seen to the left of the open-loop system's Nyquist diagram when its frequency is increased [(1a) of the table]. The closed-loop system is at its stability limit when the open-loop Nyquist diagram threads the $-1 + 0j$ point [(2a) of the table]. For this case the characteristic equation of the closed-loop system has an imaginary root (closed-loop pole) $j\omega_u$. Thus,

$$1 + G(j\omega_u) = 0, \qquad (9\text{-}32)$$

and the closed-loop system will have an undamped oscillation of period

$$P_u = 2\pi/\omega_u.$$

As a concluding note to this section on the Nyquist stability criterion, we will cite the findings of L. S. Dzung [17], sometimes referred to as the Dzung stability criterion. In order for a system to be unstable, $F(s)$ must have at least one zero *inside* the contour in the *s*-plane shown in Table 9-3(a). Existence of such a zero will be exhibited by the origin of the $F(s)$-plane falling *inside* the $F(s)$ plot. Alternatively, the -1 point will fall *inside* the $G(j\omega)H(j\omega)$ plot for an unstable system. By identifying the "inside" of the plot as we did earlier, by looking to the right of the increasing-frequency direction, we may immediately determine the stability of a system. By this approach, however, the number of unstable roots is not determined as it is by application of the Nyquist stability theorem. As an example, consider Table 9-3(d). For low gain, the -1 point (point A) lies "inside" the plot and the system is unstable. At high gain, the -1 point (point B) lies "outside" the plot and the system is stable. These findings corroborate the results of the earlier Nyquist tabulation in Table 9–3(d).

9-3 OPEN-LOOP APPROACH TO CONTROL-SYSTEM DESIGN

It was observed in the preceding section that the Nyquist diagram of a typical engineering system, stable both for open- and closed-loop conditions, crosses the unit circle about the origin in the third quadrant (between $-90°$ and $-180°$) [Table 9–4 (1a)]. The Bode diagram for this case is shown in (1b) of the table. It can be deduced, then, that the gain and phase margins indicated in this figure represent a degree of stability of normal control systems. The *phase margin* is defined at a point of gain crossover. It is the phase angle measured upward from $-180°$ at the frequency ω_g where the magnitude ratio (called gain) $|G(j\omega)|$ is 0 db:

$$\text{phase margin} = 180° + \angle G(j\omega_g). \tag{9-33}$$

The *gain margin* is measured at the frequency ω_p of phase crossover, that is, at the point where $\angle G(j\omega) = -180°$. Thus,

$$\text{gain margin} = -|G(j\omega_p)| \text{ db}. \tag{9-34}$$

Note that these margins are defined to be positive when a control system is stable. Both margins vanish simultaneously when a system reaches its stability limit [(2b) of Table 9–4]. The word "margin" loses its original meaning when a system is unstable [(3b) of the table] where the values defined by Eqs. (9–25) and (9–26) both become negative.

The following rule-of-thumb design has been recommended by Oldenburger [1]: "The phase margin should be at least $30°$ and the gain margin at least 8 db." That is:

$$\text{phase margin} \geqslant 30° \quad \text{and} \quad \text{gain margin} \geqslant 8 \text{ db}. \tag{9-35}$$

This condition is usually met when one of the two expressions in (9–35) assumes an equality sign, as shown in Fig. 9–14(a) and (c). A situation like (b) in the figure does not happen unless by coincidence an open-loop transfer function satisfies the two equalities simultaneously. Since a compromise between stability (lower gain) and effectiveness of control (higher gain) is usually a design aim, allowing for excessive margins is generally not recommended, since the gain would become inordinately low and the resulting steady-state error unnecessarily high.

Example 9-3 The Bode diagram of the transfer function,

$$G(s) = k \frac{1 + 0.5s}{s(1 + 0.8s + s^2)},$$

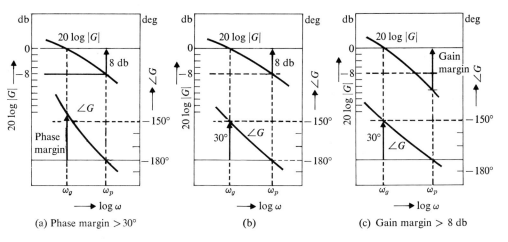

Fig. 9-14 Design by gain- and phase-margin criteria.

was determined in Fig. 9-4 for the value of the gain constant $k = \frac{1}{2}$. By visual inspection of the diagram we find that the phase margin is about 45°, while the gain margin is slightly below 8 db, but definitely over 6 db, where 6 db is the gain margin that corresponds to the second rule of Ziegler and Nichols (Eq. 8-66). The first equation of (8-66) states that $k_c = \frac{1}{2}k_u$ is about optimal, corresponding, in decibel units, to -6 db. Therefore, this rule is equivalent to recommending a 6-db gain margin. We therefore conclude that a closed-loop system with this $G(s)$ as its open-loop transfer function will have adequate stability when $k = \frac{1}{2}$. For this simple system the analytical approach presented in Section 8-3 yields a quick answer to the closed-loop system's stability limit. We have the closed-loop system's characteristic equation $1 + G(s) = 0$ in the following form:

$$s^3 + 0.8s^2 + (1 + 0.5k)s + k = 0 .$$

By Eq. (8-53), we find:

$$k_u = \frac{4}{3} \quad \text{and} \quad \omega_u = \frac{1}{\sqrt{6}} = 0.407 .$$

Therefore, the phase-crossover frequency is 0.407 rad/sec.

A $-180°$ phase angle will be produced by a minimal phase system with a -2 slope of its log magnitude (Section 9-1). Therefore, if the gain crossover of a minimal-phase system occurs at a point interior to a long range of slope -2, the phase margin will be either small or nonexistent. The situation will be worse if a nonminimum phase behavior, such as dead time, is involved in an open-loop transfer function. To ensure an adequate phase margin, it is usually desired to have -1 slope of the log magnitude curve at and near gain crossover. It is this thought that forms the basis for reshaping the Bode diagram of a system by the addition of suitable compensating elements.

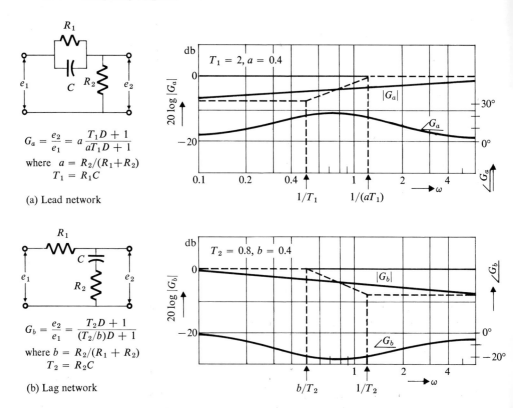

Fig. 9-15 Compensation networks.

The passive elements shown in Fig. 9-15 are often used in servomechanisms [13] to shape a desirable frequency-response pattern (see [14] for a transistorized lead/lag circuit). The series *lead network* [(a) in the figure] has a transfer function

$$G_a(s) = \frac{a(T_1 s + 1)}{a T_1 s + 1}, \tag{9-36}$$

where the value of the parameter a is usually in the range 0.05 to 0.1. The main function of this network, when inserted in a control loop, is to advance the phase and thus improve the phase margin. Another network [(b) in the figure], is called a *lag network*. When coupled in series to a main component, it increases gain at low frequency without an adverse effect on gain margin. Its transfer function is

$$G_b(s) = \frac{T_2 s + 1}{(T_2/b)s + 1}, \tag{9-37}$$

9-3 Open-loop approach to control system design

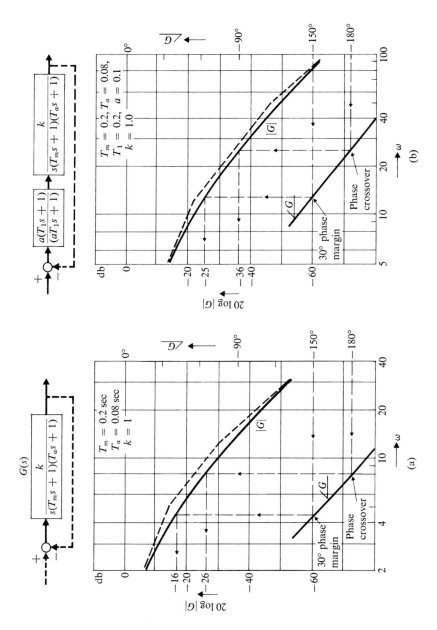

Fig. 9-16 Lead network compensation of a servomechanism.

where $b = 0.1$ to 0.05 and T_2 is usually greater than ten times the major time constant of the controlled system.

Example 9-4 Suppose that a servomechanism has an open-loop transfer function

$$G(s) = \frac{k}{s(T_m s + 1)(T_a s + 1)}, \qquad (9\text{-}38)$$

where $T_m = 0.2$ sec is the main time constant of the first example in Section 8-2, and $T_a = 0.08$ sec is the time constant of an amplifier, ignored in Section 8-2. If this system has no compensation network, the gain k is determined, as shown in Fig. 9-16(a), by the 30° phase margin condition. We find:

$$k = +16 \text{ db} = 6.3, \quad \text{phase crossover } \omega_p = 8. \qquad (9\text{-}39)$$

A lead network increases both gain and ω_p of this system, as shown in Fig. 9-16(b). The values of the network parameters in this plot are intuitively chosen as:

$$T_1 = T_m, \quad a = 0.1.$$

The intuition is based on experience formed after several trial and error attempts. The pole cancellation by the zero of the lead network introduces no hazard because the resulting uncontrollable mode is sufficiently stable (see Example 8-10). The 30° phase margin condition in the compensated system gives the following result:

$$\text{loop gain } ak = 26 \text{ db} = 20, \quad \text{phase crossover } \omega_p = 25. \qquad (9\text{-}40)$$

Compared to the original data (9-39), the loop gain and the phase crossover frequency of the compensated system are both three times higher. This means a marked improvement in static accuracy (due to high loop gain) and speed of response (due to high value of the phase crossover frequency).

Example 9-5 We try a lag network on the same servomechanism. Intuitively (as above) choosing the following values for the network parameters,

$$T_2 = 3, \quad b = 0.1,$$

we obtain the Bode diagram shown in Fig. 9-17. The 30° phase margin is again the determining factor for the gain k, which is found to be:

$$k = 30 \text{ db} = 56, \quad \text{and} \quad \omega_p = 7.6. \qquad (9\text{-}41)$$

The gain is nine times higher than the original one, but the speed of response, which is represented by ω_p, becomes slightly lower.

The counterparts of the lead and lag networks in conventional process controls are D- and I-actions added to a P-action, respectively. This will be shown in the next example.

Example 9-6 Consider the following process transfer function:

$$G_p(s) = \frac{e^{-0.5s}}{(s+1)(0.5s+1)(0.1s+1)}. \qquad (9\text{-}42)$$

9-3 Open-loop approach to control system design

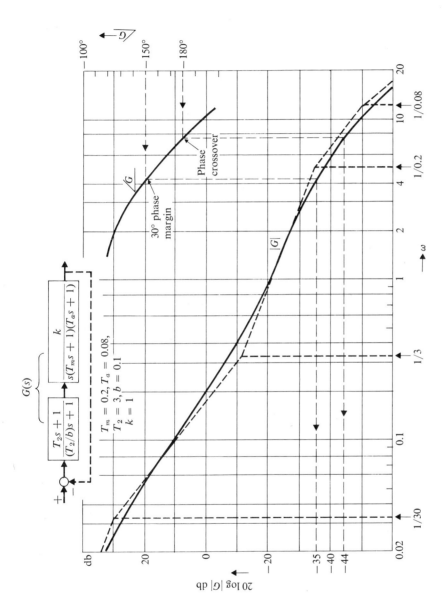

Fig. 9-17 Lag network compensation of a servomechanism.

The Bode diagram for this process (Fig. 9-18a) serves to determine the gain k_c of a P-control. As shown in the figure, the gain and phase margin conditions are almost simultaneously satisfied by the following value of k_c:

$$k_c = +2 \text{ db} = 1.25 , \tag{9-43}$$

In the following PID-control law written in the product form,

$$G_c(s) = k'_c \left(1 + \frac{1}{T'_i s}\right)(T'_d s + 1) , \tag{9-44}$$

we select the values of T'_i and T'_d in such a way that the two dominant lags of the process will be canceled:

$$T'_i = 1 , \qquad T'_d = 0.5 .$$

As a consequence of the pole cancellation, two uncontrollable modes will appear. There is no hazard, however, for the modes are sufficiently stable. (See Chapters 3 and 10 for further discussion of the effects of this pole cancellation.) A long stretch of -1 slope characterizes the log magnitude diagram of the compensated open-loop transfer function (Fig. 9-18b), although the phase angle maintains a steep descent due to the nonminimal phase behavior of the dead time. We fix the controller gain k'_c by the 8-db gain margin condition as

$$k'_c = +1 \text{ db} = 1.12 ,$$

and compute the controller adjustments in its canonical form (see Eq. 5-79)

$$G_c(s) = 1.12 \left(1 + \frac{1}{s}\right)(0.5s + 1) = 1.68 \left(1 + \frac{1}{1.5s} + 0.332s\right) ,$$

or

$$k_c = 1.68 , \qquad T_i = 1.5 \quad \text{and} \quad T_d = 0.332 . \tag{9-45}$$

The gain k_c is 30% higher than the P-control gain given by (9-43). The phase crossover frequency is higher than that of the P-control system. This is due to the D-action which contributes to a speedup in the recovery transient. The I-action, on the other hand, produces an indefinitely increasing gain in the low frequency range, and thus eliminates static error (or offset).

The frequency-response method can be extended from an undamped sinusoidal to a damped sinusoidal input. This provides us with the possibility of designing a control system for a prescribed pattern of dominant closed-loop oscillation, such as $\frac{1}{4}$ decay, which was discussed in Chapter 8. Consider, for the purpose of introduction, the second-order system of Fig. 9-19(a). The Nyquist diagram of the system's open-loop transfer function,

$$G(s) = \frac{k}{s(s+1)} ,$$

is shown in part (b) of the figure. We now substitute a complex frequency for s:

$$s = -\sigma + j\omega ,$$

9-3 Open-loop approach to control system design

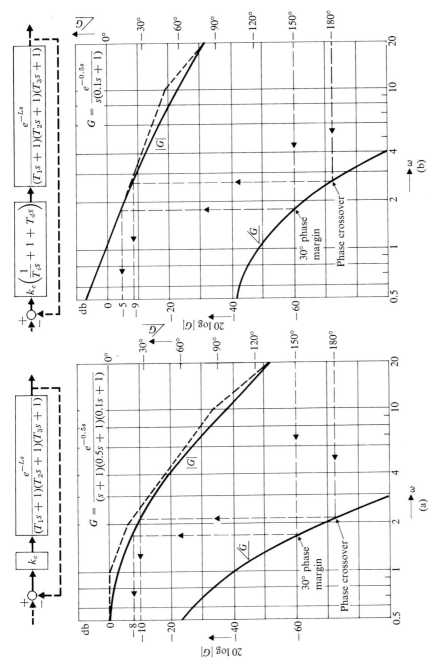

Fig. 9-18 P- and PID-control of a process.

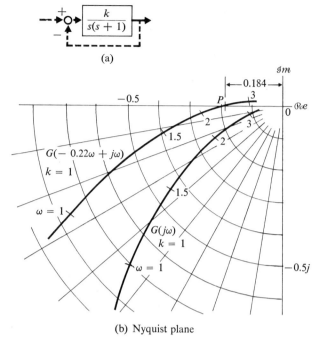

Fig. 9-19 Conformal mapping relation between the s-plane and the Nyquist plane.

where, for a $\frac{1}{4}$ decay (Eq. 8-62), the real part is related to the imaginary part by

$$\sigma = 0.22\omega .$$

The Nyquist diagram, modified for the complex frequency, $G(-0.22\omega + j\omega)$, is also shown in Fig. 9-19(b). It crosses the negative real axis at P, where

$$\omega = 2.28 \quad \text{and} \quad |G(-0.22\omega + j\omega)| = 0.184 .$$

The curve will therefore cross the real axis at -1 if the gain is set at $k = 1/0.184 = 5.44$. This is the value of gain that produces the $\frac{1}{4}$-decay mode of oscillation. For this simple system it is possible to confirm the result by directly solving the closed-loop system's characteristic equation,

$$1 + \frac{5.44}{s(s+1)} = 0 ,$$

for the roots

$$s_1, s_2 = -0.22\beta \pm j\beta , \quad \beta = 2.28 .$$

Since manual computation of the complex frequency response is usually tedious, use of templates (made for a specified decay such as $\frac{1}{4}$) helps to decrease the amount of computation necessary [15].

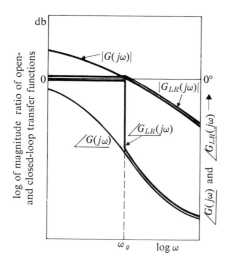

Fig. 9–20 First approximation of closed-loop frequency response $G_{LR}=G/(1+G)$.

9–4 FREQUENCY RESPONSE OF CLOSED-LOOP SYSTEMS

The closed-loop transfer functions defined by Eq. (8–12) yield the following closed-loop frequency responses:

$$G_{LR}(j\omega) = \frac{G(j\omega)}{1+G(j\omega)}, \quad G_{LV}(j\omega) = \frac{N(j\omega)}{1+G(j\omega)}, \tag{9-46}$$

where G_{LR} and G_{LV} are the closed-loop transfer functions for reference and disturbance inputs, respectively. Since an ideal limit of control is

$$E(j\omega) = 0 \quad \text{for all } \omega,$$

that is

$$G_{LR}(j\omega) = 1, \quad G_{LV}(j\omega) = 0, \tag{9-47}$$

it is desirable to have a high open-loop gain,

$$|G(j\omega)| \to \infty,$$

provided that the closed-loop system is stable. In general, a compromise between stability and high loop gain is sought in the design of control systems, as we saw in the preceding section. The ideal closed-loop frequency response (Eq. 9–47) is approximately realized only up to a certain frequency. A quick estimate of this frequency range (often termed the system *bandwidth*) is shown in Fig. 9–20. This estimate is based on a rough approximation where $|G(j\omega)|$ is compared with 1.

$$\begin{aligned} G_{CL}(j\omega) &\approx 1 & \text{for} \quad |G(j\omega)| > 1, \\ G_{CL}(j\omega) &\approx G(j\omega) & \text{for} \quad |G(j\omega)| < 1, \end{aligned} \tag{9-48}$$

where G_{CL} is the closed loop frequency response.

392 Frequency response

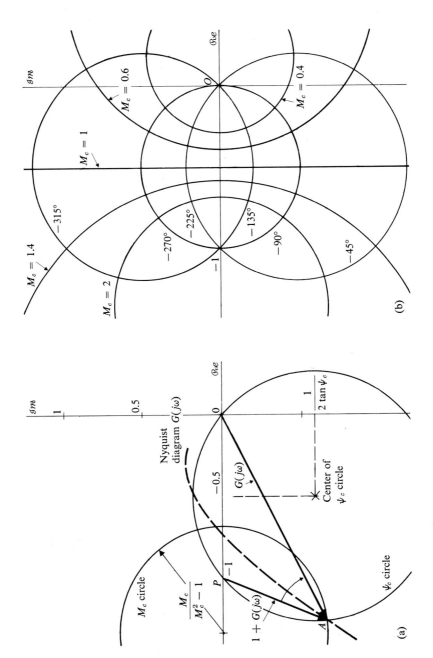

Fig. 9-21 M_c and ψ_c contours in the Nyquist diagram.

9-4 Frequency response of closed-loop systems

We saw examples in the preceding discussion where a pair of dominant closed-loop (oscillatory) poles appear when the open-loop gain is made high to attain effective control. Due to these poles, the closed-loop system will behave very much like a second-order, damped oscillatory system in the frequency range where gain and phase crossover take place. A peak in $G_{CL}(j\omega)$, like the resonance peak in a vibration system, shows up in a normal control system when its Nyquist diagram passes by on the stable side of the -1 point in the $G(j\omega)$ plot.

A geometric interpretation of this phenomenon is given by the complex-phasor relation in Fig. 9–21(a), where OP is $-1 + 0j$; thus PO is the $+1$ phasor and the phasor PA is $1 + G(j\omega)$. We therefore have

$$G_{CL}(j\omega) = \frac{\text{phasor } OA}{\text{phasor } PA}$$

or

$$|G_{CL}(j\omega)| = \frac{\overline{OA}}{\overline{PA}}, \qquad \angle G_{CL}(j\omega) = \angle PAO,$$

where $\angle PAO$ is negative when measured clockwise from the phasor PA. Since $\overline{PA}$ becomes short in the neighborhood of the -1 point, a peak in $|G_{CL}(j\omega)|$ appears in its frequency-response diagram.

The closed-loop frequency response $G_{CL}(j\omega)$ can be directly determined from its open-loop frequency response $G(j\omega)$ plotted on the Nyquist plane. For this purpose we superimpose curves $M_c = \text{const}$ and $\psi_c = \text{const}$ on the Nyquist plane, where M_c and ψ_c are the closed-loop magnitude ratio and phase angle, respectively. That is,

$$|G_{CL}(j\omega)| = M_c, \qquad \angle G_{CL}(j\omega) = \psi_c. \tag{9-49}$$

By virtue of the relations $M_c = \overline{OA}/\overline{PA}$ and $\psi_c = \angle PAO$ in Fig. 9–21(a), it can be proved geometrically that the locus of $M_c = \text{const}$ is a circle of radius $M_c/(M_c^2 - 1)$ with its center at $-M_c^2/(M_c^2 - 1)$ on the real axis. It can also be shown that $\psi_c = \text{const}$ is a circle which passes through the -1 point and the origin in the Nyquist plane, with its center at $-\frac{1}{2} + j/(2 \tan \psi_c)$. Figure 9–21(b) shows sets of these circles. The magnitude ratio (gain) and phase angle of the closed-loop frequency response $G_{CL}(j\omega)$ can be read directly at intersections of the open-loop frequency-response plot (the Nyquist diagram) and these circles. This graphical method serves to find the closed-loop frequency response quickly over a crucial frequency range. There are also available graphical techniques [16] for determining (on the Nyquist plane) the system gain required to yield a specified maximum value of M_c.

The Nichols chart is used to find the closed-loop frequency response of a system whose open-loop frequency response is given by the Bode diagram representation. The Nichols chart has as its two axes the open-loop

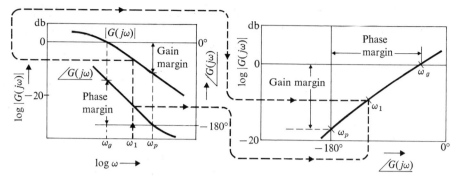

Fig. 9-22 Bode diagram vs. Nichols chart.

magnitude ratio $|G(j\omega)|$ in decibels and the phase angle $\angle G(j\omega)$ in degrees. The Bode diagram must be replotted onto the Nichols chart axes, as indicated by dashed lines in Fig. 9-22. Constant M_c and ψ_c contours are printed on the Nichols chart (Fig. 9-23). Reference [16] presents a method for transferring these constant M_c and ψ_c contours onto the Bode plot. The method is not generally recommended, since the transference will be different for each new system considered. These M_c and ψ_c curves on the Nichols chart are determined by means of the following relations:

$$M_c = \frac{1}{\sqrt{1 + (2\cos\psi)/M + 1/M^2}}, \qquad \psi_c = \frac{\sin\psi}{M + \cos\psi}, \qquad (9\text{-}50)$$

where

$$M = |G(j\omega)| \quad \text{and} \quad \psi = \angle G(j\omega).$$

Equation (9-50) is derived by the following computation where polar form is used for $G_{CL}(j\omega)$ and $G(j\omega)$:

$$G_{CL}(j\omega) = M_c e^{j\psi_c}$$
$$= \frac{G(j\omega)}{1 + G(j\omega)} = \frac{1}{1 + G^{-1}(j\omega)}$$
$$= \frac{1}{1 + (e^{-j\psi})/M} = \frac{1}{1 + (\cos\psi - j\sin\psi)/M}.$$

Since gain and phase margin appear as indicated on the right-hand side in Fig. 9-22, it is obvious that a control system is stable when its gain–phase plot runs below point P in Fig. 9-24(a). If it passes above P, as shown in Fig. 9-24(c), the control system is unstable.

The value M_c of a stable control system is maximum at point Q in Fig. 9-24(a) where the gain–phase plot becomes tangent to the smallest (in size) M_c locus. The following design rule, which is in common use, applies to

9–4 Frequency response of closed-loop systems 395

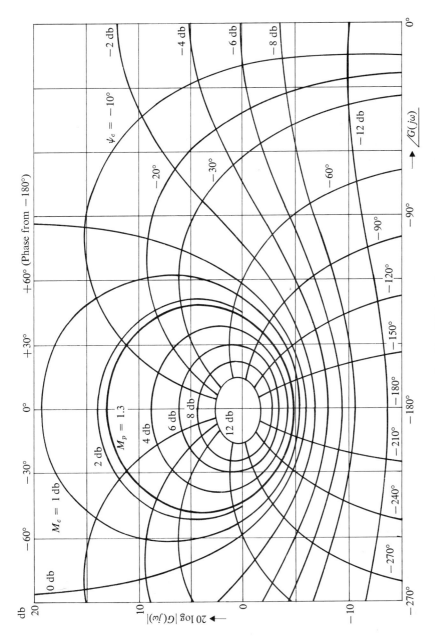

Fig. 9-23 The Nichols chart.

Frequency response

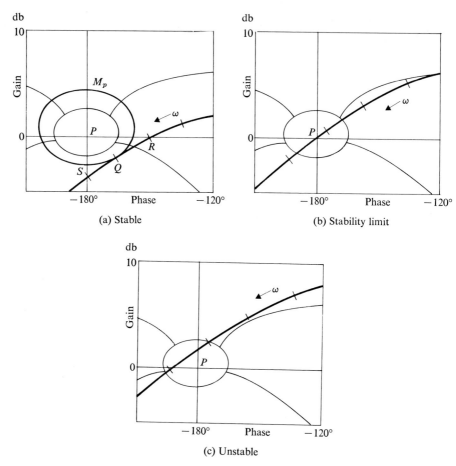

Fig. 9-24 Behavior of normal systems on a Nichols chart.

such maximum values of M_c:

The maximum value of M_c should be less than 2 (6 db), and should be about 1.3 in normal control systems. The M_c contour for 1.3 is shown in Fig. 9-23, labeled

$$M_p = 1.3. \tag{9-51}$$

Note that 1.3 is a numeric and is not in decibels. Since gain and phase margin of a normal control system, *PS* and *PR* in Fig. 9-24(a), form a triangular pattern *PRS*, there is a close relation between the $M_p = 1.3$ rule and the design criteria based on gain and phase margins (Eq. 9-35). The design of a control system based on the M_p criterion is performed on the Nichols chart by shaping a gain-phase plot so that it becomes tangent to the desired M_p locus.

9–4 Frequency response of closed-loop systems   397

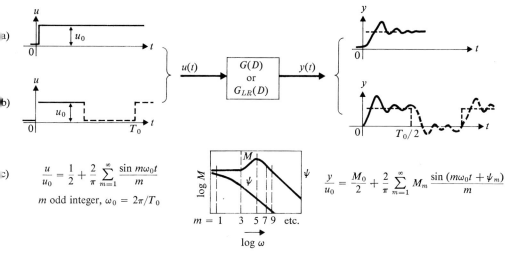

Fig. 9–25 Step-input response approximated by a Fourier series.

A signal in the frequency domain is related to the same signal in the time domain by a Fourier integral. The Fourier integral of a signal $u(t)$ is given by

$$u(t) = \left[\frac{1}{2\pi} \int_{s=-\infty}^{s=\infty} U(s) e^{st} \, ds \right]_{s=j\omega},$$

where

$$[U(s)]_{s=j\omega} = \left[\int_{t=-\infty}^{t=\infty} u(t) e^{-st} \, dt \right]_{s=j\omega}$$

is the Fourier transform of the signal. A Fourier integral can be approximated by a Fourier series for convenience of computation. This requires approximation of a continuous spectrum by a line spectrum. We can apply this approximation to estimate a time response from data in the frequency domain. Consider, for instance, an open- or closed-loop linear system subject to a step input of magnitude u_0 (Fig. 9–25a). We approximate the step signal by the periodic rectangular pattern shown in part (b) of the figure, for which we can write the following Fourier series (refer to any text on advanced mathematics for engineers for a discussion of Fourier series):

$$u(t) = u_0 \left[\frac{1}{2} + \frac{2}{\pi} \sum_{m=1}^{\infty} \frac{\sin(m\omega_0 t)}{m} \right]. \tag{9-52}$$

In Eq. (9–52)

$$m = \text{odd integers} \quad \text{and} \quad \omega_0 = \frac{2\pi}{T_0}, \quad \text{where } T_0 = \text{period}.$$

398 Frequency response

Since we can determine sinusoidal response components for each sinusoidal input separately, by use of the sinusoidal transfer function evaluated at the appropriate frequency, we may employ superposition. The total response $y(t)$ is given as a sum of the responses to the various input harmonics:

$$y(t) = u_0 \left[\frac{M_0}{2} + \frac{2}{\pi} \sum_{m=1}^{\infty} Mm \frac{\sin(m\omega_0 t + \phi_m)}{m} \right]. \quad (9\text{-}53)$$

The frequency-response transfer functions in Eq. (9-53) are given by

$$M_m = G(jm\omega_0), \quad \phi_m = \underline{/G(jm\omega_0)} \quad \text{and} \quad M_0 = \lim_{\omega \to 0} |G(j\omega)|,$$

as indicated in Fig. 9-25(c). If ω_0 is selected according to the rule

$$\omega_0 = (\omega \text{ for max } M)/(4 \text{ to } 10),$$

Eq. (9-53), when truncated, will give a fairly good approximation to a step-input response over the period

$$0 < t < \tfrac{1}{2} T_0.$$

The scheme of computation just shown is instructive rather than practical. It shows, for instance, how frequency components in the vicinity of a resonance peak dominate the system's time response.

9-5 FREQUENCY RESPONSE IN STATE-SPACE

As a final step, we consider the relation between the frequency and time domains, and shall briefly interpret the frequency response of a system in state-space. (For those readers who have not yet studied state-space vector techniques, we suggest that reading of this section be postponed until at least Chapters 1 through 3 have been covered.) Our reason for following this path is to take advantage of the clarification of the internal state of a system which the state-space approach affords. Due to the fact that the internal state of a system is represented by a vector, it is to be expected that the state-space approach will be more mathematically complex than the classical scalar input–output frequency-response technique. Although no design techniques have yet emerged from this new point of view that were not available in the classical frequency-response approach, the insight we gain into the internal state of linear systems in sinusoidal input–output steady state may eventually lead to new design and analysis approaches. The following treatment is offered with the aim of narrowing the gap between the classical control viewpoint based on the frequency-response technique and the contemporary (so-called "modern") approach using state-space vector methods.

9-5 Frequency response in state space

Let us consider a linear lumped system described by the following state and output equations:

$$\frac{d\mathbf{x}}{dt} = \mathbf{A}\mathbf{x} + \mathbf{B}\mathbf{u} \quad \text{or} \quad \frac{d\mathbf{x}}{dt} = \mathbf{A}\mathbf{x} + \mathbf{b}u, \tag{9-54}$$

and

$$\mathbf{y} = \mathbf{C}\mathbf{x} \quad \text{or} \quad y = \mathbf{c}\mathbf{x}, \tag{9-55}$$

where $\mathbf{x}(t)$ is an n-dimensional state, $\mathbf{u}$ is an r-dimensional input, and $\mathbf{y}$ is an m-dimensional output. The system is characterized by an $n \times n$ matrix $\mathbf{A}$, input matrix $\mathbf{B}$ which is $n \times r$, and output matrix $\mathbf{C}$ which is $m \times n$. For the special case of scalar input and scalar output, $\mathbf{B}$ and $\mathbf{C}$ reduce to an n-column vector $\mathbf{b}$ and an n-row vector $\mathbf{c}$, respectively.

As before, the derivative operator d/dt or D is replaced by $j\omega$ to deal with sinusoidal steady state (or frequency response), where ω (rad/sec) is the angular frequency of the sinusoidal input and $j = \sqrt{-1}$. This in turn means that we replace s in the Laplace domain by $j\omega$ to get frequency-domain relations. Since we have

$$\mathscr{I}m\,(e^{j\omega t}) = \sin \omega t,$$

the complex-frequency domain and the time domain are related to each other by the following conventions:

1. If $U(j\omega) = e^{j\omega t}$, then $u(t) = \sin \omega t = \mathscr{I}m\,(U(j\omega))$.
2. If $X(j\omega) = A e^{j(\omega t + \phi)}$, then

$$x(t) = \mathscr{I}m\,(X(j\omega)) = \mathscr{I}m\,(\cos \omega t + j \sin \omega t)(A \cos \phi + jA \sin \phi)$$
$$= A \sin \omega t \cos \phi + A \cos \omega t \sin \phi = A \sin (\omega t + \phi). \tag{9-56}$$

Although it is possible to compute matrix and vector relations in the complex domain, state space relations in the real domain permit us to interpret directly the observed behavior of physical systems. For this reason we develop frequency-response relations without introducing imaginary quantities into our computations. We write the following form for the ith state variable in sinusoidal steady state:

$$x_i(\omega) = \alpha_i \sin \omega t + \beta_i \cos \omega t. \tag{9-57}$$

Comparing (9-57) and (9-56), we see that

$$\alpha_i = A_i \cos \phi_i, \quad \beta_i = A_i \sin \phi_i, \tag{9-58}$$

where A_i is amplitude and ϕ_i is phase of the ith state variable. Thus, if α_i and β_i can be determined from the system of Eqs. (9-54) and (9-55), then the amplitude and phase shift of the ith state can be predicted. In vector form (9–57) becomes

$$\mathbf{x}(\omega) = [\boldsymbol{\alpha}, \boldsymbol{\beta}] \cdot \begin{bmatrix} \sin \omega t \\ \cos \omega t \end{bmatrix}, \tag{9-59}$$

where

$$\boldsymbol{\alpha} = \begin{bmatrix} \alpha_1 \\ \cdot \\ \cdot \\ \cdot \\ \alpha_n \end{bmatrix}, \quad \boldsymbol{\beta} = \begin{bmatrix} \beta_1 \\ \cdot \\ \cdot \\ \cdot \\ \beta_n \end{bmatrix}. \tag{9-60}$$

The time derivative of $\mathbf{x}(\omega)$ is then given by

$$\frac{d}{dt}\mathbf{x}(\omega) = [\boldsymbol{\alpha}, \boldsymbol{\beta}] \begin{bmatrix} \omega \cos \omega t \\ -\omega \sin \omega t \end{bmatrix} = [-\omega\boldsymbol{\beta}, \omega\boldsymbol{\alpha}] \begin{bmatrix} \sin \omega t \\ \cos \omega t \end{bmatrix}. \tag{9-61}$$

Let us suppose that the sinusoidal forcing input is scalar with unity amplitude;

$$u = \sin \omega t. \tag{9-62}$$

Substituting Eqs. (9–59), (9–61), and (9–62) into Eq. (9–54), we find that

$$[-\omega\boldsymbol{\beta}, \omega\boldsymbol{\alpha}] \begin{bmatrix} \sin \omega t \\ \cos \omega t \end{bmatrix} = \mathbf{A}[\boldsymbol{\alpha}, \boldsymbol{\beta}] \begin{bmatrix} \sin \omega t \\ \cos \omega t \end{bmatrix} + [\mathbf{b}, \mathbf{0}] \begin{bmatrix} \sin \omega t \\ \cos \omega t \end{bmatrix}$$

or

$$[-\omega\boldsymbol{\beta}, \omega\boldsymbol{\alpha}] = \mathbf{A}[\boldsymbol{\alpha}, \boldsymbol{\beta}] + [\mathbf{b}, \mathbf{0}]. \tag{9-63}$$

Therefore

$$-\omega\boldsymbol{\beta} = \mathbf{A}\boldsymbol{\alpha} + \mathbf{b}, \quad \omega\boldsymbol{\alpha} = \mathbf{A}\boldsymbol{\beta},$$

and, solving for $\boldsymbol{\alpha}$ and $\boldsymbol{\beta}$,

$$\boldsymbol{\alpha} = -(\omega^2\mathbf{I} + \mathbf{A}^2)^{-1}\mathbf{A}\mathbf{b}, \quad \boldsymbol{\beta} = -(\omega^2\mathbf{I} + \mathbf{A}^2)^{-1}\omega\mathbf{b}. \tag{9-64}$$

In the process of deriving Eq. (9–64) we note that the inverse matrix and $\mathbf{A}$ commute in their product.

$$\mathbf{A}(\omega^2\mathbf{I} + \mathbf{A}^2)^{-1} = (\omega^2\mathbf{I} + \mathbf{A}^2)^{-1}\mathbf{A}.$$

The relation can be immediately generalized to vector input. Let

$$\mathbf{u}(\omega) = [\mathbf{u}_1, \mathbf{u}_2] \begin{bmatrix} \sin \omega t \\ \cos \omega t \end{bmatrix}, \tag{9-65}$$

9-5   Frequency response in state space

where $\mathbf{u}_1$ and $\mathbf{u}_2$ are prescribed r-vectors. Converting these into n-vectors $\mathbf{b}_1$ and $\mathbf{b}_2$ by

$$\mathbf{Bu}_1 = \mathbf{b}_1, \qquad \mathbf{Bu}_2 = \mathbf{b}_2, \tag{9-66}$$

we obtain

$$[-\omega\boldsymbol{\beta}, \omega\boldsymbol{\alpha}] = \mathbf{A}[\boldsymbol{\alpha}, \boldsymbol{\beta}] + [\mathbf{b}_1, \mathbf{b}_2],$$

from which $\boldsymbol{\alpha}$ and $\boldsymbol{\beta}$ can be found,

$$\begin{aligned}\boldsymbol{\alpha} &= (\omega^2\mathbf{I} + \mathbf{A}^2)^{-1}(-\mathbf{A}\mathbf{b}_1 + \omega\mathbf{b}_2), \\ \boldsymbol{\beta} &= (\omega^2\mathbf{I} + \mathbf{A}^2)^{-1}(-\mathbf{A}\mathbf{b}_2 - \omega\mathbf{b}_1).\end{aligned} \tag{9-67}$$

Substituting Eq. (9-59) into Eq. (9-55), we see that the scalar response $y(\omega)$, as well as the vector response $\mathbf{y}(\omega)$, are given by

$$y(\omega) = \mathbf{cx}(\omega), \qquad \mathbf{y}(\omega) = \mathbf{Cx}(\omega), \tag{9-68}$$

where, by Eq. (9-64),

$$\mathbf{x}(\omega) = -(\omega^2\mathbf{I} + \mathbf{A}^2)^{-1}[\mathbf{Ab}, \omega\mathbf{b}]\begin{bmatrix}\sin\omega t\\ \cos\omega t\end{bmatrix} \tag{9-69}$$

for scalar input, and

$$\mathbf{x}(\omega) = -(\omega^2\mathbf{I} + \mathbf{A}^2)^{-1}[(-\mathbf{Ab}_1 + \omega\mathbf{b}_2), (-\mathbf{Ab}_2 - \omega\mathbf{b}_1)]\begin{bmatrix}\sin\omega t\\ \cos\omega t\end{bmatrix} \tag{9-70}$$

for vector input, and we have achieved our purpose of predicting $\mathbf{x}(\omega)$. Note that the inverse matrix $(\omega^2\mathbf{I} + \mathbf{A}^2)^{-1}$ exists when $\mathbf{A}$ has no eigenvalues which are equal to $\pm j\omega$.

It is possible to obtain Eq. (9-64) by substitution of $j\omega$ for d/dt in Eq. (9-54). We then have

$$j\omega\mathbf{X}(j\omega) = \mathbf{AX}(j\omega) + \mathbf{b}U(j\omega),$$

and hence

$$\mathbf{X}(j\omega) = (j\omega\mathbf{I} - \mathbf{A})^{-1}\mathbf{b}U(j\omega).$$

Recalling Eq. (9-56), we define

$$\boldsymbol{\alpha} = \mathscr{R}e\,(j\omega\mathbf{I} - \mathbf{A})^{-1}\mathbf{b}, \qquad \boldsymbol{\beta} = \mathscr{I}m\,(j\omega\mathbf{I} - \mathbf{A})^{-1}\mathbf{b}. \tag{9-71}$$

The identity

$$\boldsymbol{\alpha} + j\boldsymbol{\beta} = (j\omega\mathbf{I} - \mathbf{A})^{-1}\mathbf{b}$$

can be confirmed by substituting Eq. (9-64) into the left-hand side of this expression. We now consider several examples to illustrate the application of Eq. (9-64).

Example 9-7 Consider a system with distinct real eigenvalues having a diagonal A-matrix:

$$\mathbf{A} = \begin{bmatrix} \lambda_1 & & & 0 \\ & \lambda_2 & & \\ & & \ddots & \\ 0 & & & \lambda_n \end{bmatrix}, \quad \mathbf{b} = \begin{bmatrix} b_1 \\ b_2 \\ \vdots \\ b_n \end{bmatrix}.$$

By Eq. (9-64) we have

$$\boldsymbol{\alpha} = -\begin{bmatrix} \dfrac{\lambda_1 b_1}{\omega^2 + \lambda_1^2} \\ \dfrac{\lambda_2 b_2}{\omega^2 + \lambda_2^2} \\ \vdots \\ \dfrac{\lambda_n b_n}{\omega^2 + \lambda_n^2} \end{bmatrix}, \quad \boldsymbol{\beta} = -\begin{bmatrix} \dfrac{\omega b_1}{\omega^2 + \lambda_1^2} \\ \dfrac{\omega b_2}{\omega^2 + \lambda_2^2} \\ \vdots \\ \dfrac{\omega b_n}{\omega^2 + \lambda_n^2} \end{bmatrix}.$$

Example 9-8 Next we consider a system with multiplicity in real eigenvalues (A-matrix in Schwartz form).

a) Double eigenvalue:

$$\mathbf{A} = \begin{bmatrix} \lambda & 1 \\ 0 & \lambda \end{bmatrix}, \quad \mathbf{b} = \begin{bmatrix} b_1 \\ b_2 \end{bmatrix},$$

$$\boldsymbol{\alpha} = -\begin{bmatrix} \dfrac{\lambda b_1}{\omega^2 + \lambda^2} + \left(\dfrac{1}{\omega^2 + \lambda^2} - \dfrac{2\lambda^2}{(\omega^2 + \lambda^2)^2} \right) b_2 \\ \dfrac{\lambda b_2}{\omega^2 + \lambda^2} \end{bmatrix} = -\begin{bmatrix} \dfrac{\lambda b_1}{\omega^2 + \lambda^2} + \dfrac{\omega^2 - \lambda^2}{(\omega^2 + \lambda^2)^2} b_2 \\ \dfrac{\lambda b_2}{\omega^2 + \lambda^2} \end{bmatrix},$$

$$\boldsymbol{\beta} = -\begin{bmatrix} \dfrac{\omega b_1}{\omega^2 + \lambda^2} - \dfrac{2\lambda \omega b_2}{(\omega^2 + \lambda^2)^2} \\ \dfrac{\omega b_2}{\omega^2 + \lambda^2} \end{bmatrix}.$$

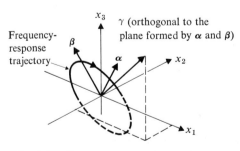

Fig. 9-26 Frequency-response trajectory in state space.

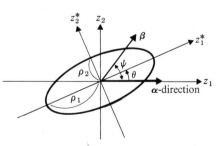

Fig. 9-27 Lissajous figure.

b) Triple eigenvalue:

$$A = \begin{bmatrix} \lambda & 1 & 0 \\ 0 & \lambda & 1 \\ 0 & 0 & \lambda \end{bmatrix}, \quad b = \begin{bmatrix} b_1 \\ b_2 \\ b_3 \end{bmatrix},$$

$$\alpha = -\begin{bmatrix} \dfrac{\lambda b_1}{\omega^2 + \lambda^2} + \dfrac{\omega^2 - \lambda^2}{(\omega^2 + \lambda^2)^2} b_2 - \dfrac{\omega^2 - \lambda^2}{(\omega^2 + \lambda^2)^3} b_3 \\ \dfrac{\lambda b_2}{\omega^2 + \lambda^2} + \dfrac{\omega^2 - \lambda^2}{(\omega^2 + \lambda^2)^2} b_3 \\ \dfrac{\lambda b_3}{\omega^2 + \lambda^2} \end{bmatrix},$$

$$\beta = -\begin{bmatrix} \dfrac{\omega b_1}{\omega^2 + \lambda^2} - \dfrac{2\lambda\omega b_2}{(\omega^2 + \lambda^2)^2} + \dfrac{3\lambda^2 - \omega^2}{(\omega^2 + \lambda^2)^3} \omega b_3 \\ \dfrac{\omega b_2}{\omega^2 + \lambda^2} - \dfrac{2\lambda\omega b_3}{(\omega^2 + \lambda^2)^2} \\ \dfrac{\omega b_3}{\omega^2 + \lambda^2} \end{bmatrix}.$$

Example 9-9 We now investigate a second-order system with conjugate complex eigenvalues (**A**-matrix in symmetric form):

$$A = \begin{bmatrix} \sigma & \Omega \\ -\Omega & \sigma \end{bmatrix}, \quad b = \begin{bmatrix} b_1 \\ b_2 \end{bmatrix},$$

$$\alpha = -\frac{1}{\varDelta} \begin{bmatrix} (\omega^2 + \sigma^2 + \Omega^2)\sigma b_1 + (\omega^2 - \sigma^2 - \Omega^2)\Omega b_2 \\ (-\omega^2 + \sigma^2 + \Omega^2) b_1 + (\omega^2 + \sigma^2 + \Omega^2)\sigma b_2 \end{bmatrix},$$

$$\beta = -\frac{1}{\varDelta} \begin{bmatrix} (\omega^2 + \sigma^2 - \Omega^2)\omega b_1 - 2\sigma\Omega\omega b_2 \\ 2\sigma\Omega\omega b_1 + (\omega^2 + \sigma^2 - \Omega^2)\omega b_2 \end{bmatrix},$$

where

$$\varDelta = (\sigma^2 + \Omega^2)^2 + 2\omega^2(\sigma^2 - \Omega^2) + \omega^4.$$

Since the frequency-response trajectory of Eq. (9-59) is composed of two vectors, one in the α-direction and the other in the β-direction, the trajectory lies in a plane (two-dimensional manifold), and it is an ellipse (Fig. 9-26). However, the analytical expression for the ellipse (a Lissajous figure) is not simple. In order to present a unified analytical result, let us define a dimensionless number r, which is smaller than 1, such that

$$\frac{||\alpha||^2}{||\beta||^2} \left(\text{or } \frac{||\beta||^2}{||\alpha||^2}\right) = r < 1.$$

Also let ψ = acute angle between α and β (Fig. 9-27), and put

$$\cos \psi = p.$$

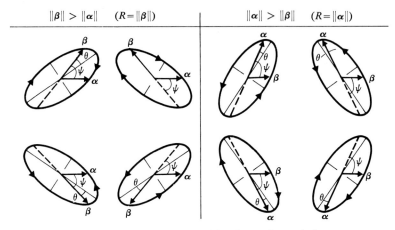

Fig. 9-28 Various combinations of α and β.

Then the axes of the ellipse will be given by

$$\left(\frac{\rho_1}{R}\right)^2 = \frac{1}{2}(1+r)(1+q) \quad \text{for the major axis,}$$
$$\left(\frac{\rho_2}{R}\right)^2 = \frac{1}{2}(1+r)(1-q) \quad \text{for the minor axis,} \quad (9\text{-}72)$$

where

$$q = \frac{\sqrt{(1+r)^2 + 4pr^2(1-p)}}{(1+r) + 2r(1-p)},$$

$R = \|\alpha\|$ if $\|\alpha\| > \|\beta\|$, $\qquad R = \|\beta\|$ if $\|\alpha\| < \|\beta\|$.

The angle θ for the major axis, measured from the longer of α and β (that is R), is

$$\tan \theta = \sqrt{\frac{1-p}{p}} - \sqrt{\frac{1-q}{1+q}} \frac{1}{\sqrt{rp}}. \quad (9\text{-}73)$$

Note that p ($=|\cos \psi|$) in these expressions is given by the following normalized absolute value of inner product:

$$p = \frac{|\alpha'\beta|}{\|\alpha\| \cdot \|\beta\|}. \quad (9\text{-}74)$$

As the frequency of the input changes, so do the vectors α and β. The planar trajectory in state space of the frequency response thus rotates and warps. Figure 9-28 depicts the planar trajectory for various combinations of α and β. Projections of this trajectory on a plane made up of any two state variables will, in general, yield an ellipse which changes shape and size as the forcing frequency changes.

9-5 Frequency response in state space

The trajectory plane in a third-order system rotates around a fixed axis when the forcing frequency ω is changed. Let $\mathbf{h}$ be the direction vector of this axis; then the elements of $\mathbf{h}$ (which are independent of ω) must be uniquely fixed by the following vector equation up to a proportionality constant:

$$f_1(\omega)\boldsymbol{\alpha} + f_2(\omega)\boldsymbol{\beta} = \mathbf{h}, \qquad (9\text{-}75)$$

where $f_1(\omega)$ and $f_2(\omega)$ are scalar functions of ω yet to be determined.

To prove this property we consider a third-order system with scalar input, described by the general form of transfer function

$$Y(s) = \left(\frac{c_1 + c_2 s + c_3 s^2}{a_1 + a_2 s + a_3 s^2 + s^3} + d\right) U(s)$$

or

$$Y(s) = (c_1 X_1 + c_2 s X_1 + c_3 s^2 X_1 + d) U(s).$$

Letting

$$sX_1 = X_2, \qquad s^2 X_1 = X_3,$$

we obtain the main canonical state-space formulation:

$$D_z \mathbf{x} = \mathbf{A}\mathbf{x} + \mathbf{b}u, \qquad y = \mathbf{c}\mathbf{x} + du,$$

where

$$\mathbf{A} = \begin{bmatrix} 0 & 1 & 0 \\ 0 & 0 & 1 \\ -a_1 & -a_2 & -a_3 \end{bmatrix}, \quad \mathbf{b} = \begin{bmatrix} 0 \\ 0 \\ 1 \end{bmatrix}, \quad \mathbf{c} = [c_1, c_2, c_3] \qquad (9\text{-}76)$$

in which the output equation, and hence also $\mathbf{c}$ and d, do not appear in Eqs. (9-64). We eliminate $\boldsymbol{\alpha}$ and $\boldsymbol{\beta}$ from Eq. (9-75) by the substitutions (9-64), and obtain

$$f_1(\omega)\mathbf{A}\mathbf{b} + f_2(\omega)\omega\mathbf{b} = -(\omega^2 \mathbf{I} + \mathbf{A}^2)\mathbf{h}; \qquad (9\text{-}77)$$

and, applying (9-76), we have

$$\begin{bmatrix} 0 \\ f_1(\omega) \\ -a_3 f_1(\omega) + \omega f_2(\omega) \end{bmatrix} = -\begin{bmatrix} \omega^2 & 0 & 1 \\ -a_1 & \omega^2 - a_2 & -a_3 \\ a_1 a_3 & -a_1 + a_2 a_3 & \omega^2 - a_2 + a_3^2 \end{bmatrix} \mathbf{h}.$$

Let

$$\mathbf{h} = \begin{bmatrix} 0 \\ 1 \\ 0 \end{bmatrix}.$$

Then

$$f_1(\omega) = a_2 - \omega^2, \qquad f_2(\omega) = \frac{a_1 - a_3\omega^2}{\omega}. \tag{9-78}$$

The invariant axis for the system of (9–76) is therefore the x_2-axis, and projections of the trajectory onto the x_1x_3-plane will be straight lines.

Example 9-10 We consider a numerical example to illustrate the fixed-axis result just derived. Let

$$\mathbf{A} = \begin{bmatrix} 0 & 1 & 0 \\ 0 & 0 & 1 \\ -6 & -11 & -6 \end{bmatrix}, \qquad \mathbf{b} = \begin{bmatrix} 0 \\ 0 \\ 1 \end{bmatrix}.$$

The eigenvalues of the system are -1, -2 and -3.

In order to apply the results of Example 9-7, we diagonalize $\mathbf{A}$ by a coordinate change $\mathbf{x} = \mathbf{T}\mathbf{x}^*$ such that

$$\mathbf{T} = \begin{bmatrix} 1 & 1 & 1 \\ -1 & -2 & -3 \\ 1 & 4 & 9 \end{bmatrix}, \qquad \mathbf{T}^{-1} = \begin{bmatrix} 3 & 2.5 & 0.5 \\ -3 & -4 & -1 \\ 1 & 1.5 & 0.5 \end{bmatrix}.$$

The diagonalized relation is

$$\frac{d\mathbf{x}^*}{dt} = \Lambda \mathbf{x}^* + \mathbf{b}^* u,$$

where

$$\Lambda = \begin{bmatrix} -1 & 0 & 0 \\ 0 & -2 & 0 \\ 0 & 0 & -3 \end{bmatrix}, \qquad \mathbf{b}^* = \mathbf{T}^{-1}\mathbf{b} = \begin{bmatrix} \tfrac{1}{2} \\ -1 \\ \tfrac{1}{2} \end{bmatrix}.$$

From Example 9-7 we have

$$\boldsymbol{\alpha}^* = \begin{bmatrix} \dfrac{0.5}{\omega^2+1} \\ \dfrac{-2}{\omega^2+4} \\ \dfrac{1.5}{\omega^2+9} \end{bmatrix}, \qquad \boldsymbol{\beta}^* = \begin{bmatrix} \dfrac{-0.5\omega}{\omega^2+1} \\ \dfrac{\omega}{\omega^2+4} \\ \dfrac{-0.5\omega}{\omega^2+9} \end{bmatrix}.$$

By inspection we find the direction vector $\boldsymbol{\gamma}^*$ which is orthogonal to both $\boldsymbol{\alpha}^*$ and $\boldsymbol{\beta}^*$:

$$\boldsymbol{\gamma}^* = \begin{bmatrix} \omega^2+1 \\ \omega^2+4 \\ \omega^2+9 \end{bmatrix}.$$

If the trajectory plane ($\boldsymbol{\alpha}^*\boldsymbol{\beta}^*$-plane) has an invariant axis of rotation $\mathbf{h}^*$, then it must satisfy for all ω the orthogonality condition

$$(\mathbf{h}^*)' \cdot \boldsymbol{\gamma}^* = 0.$$

For $(\mathbf{h}^*)' = [h_1^*, h_2^*, h_3^*]$, using the result just found for $\boldsymbol{\gamma}^*$, we obtain the condition

$$(\mathbf{h}^*)' \cdot \boldsymbol{\gamma}^* = (h_1^* + h_2^* + h_3^*)\omega^2 + (h_1^* + 4h_2^* + 9h_3^*) = 0 .$$

Since this equation must hold for all ω, we require

$$h_1^* + h_2^* + h_3^* = 0 \quad \text{and} \quad h_1^* + 4h_2^* + 9h_3^* = 0 .$$

Choosing $h_3^* = 1.5$, we obtain

$$(\mathbf{h}^*)' = [2.5, \;\; -4, \;\; 1.5] ,$$

and confirm that

$$\mathbf{h} = \mathbf{T}\mathbf{h}^* = \begin{bmatrix} 0 \\ 1 \\ 0 \end{bmatrix} .$$

To determine $f_1(\omega)$ and $f_2(\omega)$, we solve Eq. (9–77) in the modal domain

$$f_1(\omega)\mathbf{A}^*\mathbf{b}^* + f_2(\omega)\omega\mathbf{b}^* = -(\omega^2\mathbf{I} + \mathbf{A}^{*2})\mathbf{h}^*$$

and obtain

$$f_1(\omega) = 11 - \omega^2 , \quad f_2(\omega) = \frac{6 - 6\omega^2}{\omega} ,$$

These are also the functions we obtain by substituting $a_1 = 6$, $a_2 = 11$, $a_3 = 6$ in Eq. (9–78).

We now generalize the fixed-axis theorem to higher-order systems. For an nth-order system, consider two conditions for the vector $\mathbf{h}$:

$$\mathbf{p} = [\mathbf{b}, \mathbf{Ab}, \mathbf{A}^2\mathbf{b}, \ldots, \mathbf{A}^{n-1}\mathbf{b}]^{-1}\mathbf{h} ,$$

and $\hspace{10cm}$ (9–79)

$$\mathbf{q} = [\mathbf{b}, \mathbf{Ab}, \mathbf{A}^2\mathbf{b}, \ldots, \mathbf{A}^{n-1}\mathbf{b}]^{-1}\mathbf{A}^2\mathbf{h} .$$

The inverse matrix on the right-hand side exists when the system is controllable. We fix $\mathbf{h}$ in such a way that the following $n - 1$ conditions are satisfied.

$$p_3 = p_4 = \cdots = p_n = 0 \quad \text{and} \quad q_n = 0 . \tag{9–80}$$

This implies that $\mathbf{h}$ is orthogonal to the last $n - 2$ row vectors of $[\mathbf{b}, \mathbf{Ab}, \ldots, \mathbf{A}^{n-1}\mathbf{b}]^{-1}$, and also orthogonal to the last row vector of $[\mathbf{b}, \mathbf{Ab}, \ldots]^{-1}\mathbf{A}^2$. Since $\mathbf{h}$ is an n-dimensional direction vector (the spinning axis), $n - 1$ conditions are necessary and sufficient to fix $\mathbf{h}$.

From Eqs. (9–79) and (9–80) we obtain

$$[\mathbf{b}, \mathbf{Ab}, \ldots, \mathbf{A}^{n-1}\mathbf{b}](\omega^2\mathbf{p} + \mathbf{q}) = (\omega^2\mathbf{I} + \mathbf{A}^2)\mathbf{h}$$

or

$$(p_1\omega^2 + q_1)\mathbf{b} + (p_2\omega^2 + q_2)\mathbf{Ab} + q_3\mathbf{A}^2\mathbf{b} + \cdots + q_{n-1}\mathbf{A}^{n-2}\mathbf{b} = (\omega^2\mathbf{I} + \mathbf{A}^2)\mathbf{h} .$$

Letting

$$p_1^2 + q_1 = -f_2(\omega) \cdot \omega, \qquad p_2\omega^2 + q_2 = -f_1(\omega),$$

and substituting $\boldsymbol{\alpha}$ and $\boldsymbol{\beta}$ of (9-64), we obtain

$$f_1(\omega)\boldsymbol{\alpha} + f_2(\omega)\boldsymbol{\beta} + (\omega^2\mathbf{I} + \mathbf{A}^2)^{-1}(q_3\mathbf{A}^2\mathbf{b} + \cdots + q_{n-1}\mathbf{A}^{n-1}\mathbf{b}) = \mathbf{h}.$$

In a fourth-order system, for example, a 3-dimensional subspace that consists of the $(\boldsymbol{\alpha} - \boldsymbol{\beta})$-plane and a vector $(\omega^2\mathbf{I} + \mathbf{A}^2)^{-1}q_3\mathbf{A}^2\mathbf{b}$ spins around a fixed $\mathbf{h}$ line.

REFERENCES

1. Oldenburger, R. (ed.), *Frequency Response*. New York: Macmillan, 1956.
2. Truxal, J. G. (ed.), *Control Engineers Handbook*. New York: McGraw-Hill, 1958.
3. Ausman, J. S., "Amplitude frequency response analysis and synthesis of unfactored transfer functions," *Trans. ASME, Series D, J. of Basic Eng.*, **86**, No. 1, March 1964, pp. 32–36.
4. Bode, H. W., *Network Analysis and Feedback Amplifier Design*. Princeton, N. J.: Van Nostrand, 1945.
5. Meadows, N. G., "Transfer functions from frequency response"' *Control Eng.* **11**, No. 6, June 1964, pp. 95–96.
6. Chen, C. F., Philip, B. L., "Accurate determination of complex root transfer functions from frequency response data." *1965 JACC Preprint*, pp. 467–472.
7. Karam, J., Franke, M. E., "The frequency response of pneumatic lines." *Trans. ASME, J. of Basic Eng.*, **89**, No. 2, 1967, pp. 371–378.
8. Ball, S. J., "Approximate models for distributed-parameter heat-exchanger systems." *1963 JACC Preprint*, pp. 131–139.
9. Esterson, G. L., "Fluid-filled conduit frequency response." *1963 JACC Preprint*, pp. 328–339.
10. Esterson, G. L., et al., "Dynamic response of a continuous stirred tank." *1964 JACC Preprint*, p. 155.
11. Hale, J. C., "Frequency-domain models for extraction columns." *1966 JACC Preprint*, p. 321.
12. Nyquist, H., "Regeneration theory." *Bell System Tech. J.*, **11**, 1932, pp. 126–147.
13. Shucker, S., "Error coefficients ease servo response analysis." *Control Eng.*, **10**, No. 5, May 1963, pp. 119–123.
14. Scharf, W., "Don't overlook positive feedback." *Control Eng.*, **7**, No. 11, November 1960, pp. 115–118.
15. Caldwell, W. I., Coon, G. A., Zoss, L. M., *Frequency Response for Process Control*. New York: McGraw-Hill, 1959.

16. Chestnut, H., Mayer, R. W., *Servomechanisms and Regulating System Design*, Vol. I, 2nd ed. New York: Wiley, 1960, Chapter 9.
17. Dzung, L. S., *The Stability Criterion*. London: Butterworths Scientific Publications, 1952; "Automatic and Manual Control," *Proceedings of the 1951 Cranfield Conference*, pp. 13–23.

PROBLEMS

9-1 A temperature u (°C) is changing sinusoidally with a frequency of 0.2 cps. The response x (°C) of a thermometer is described by $TDx + x = u$, where $T = 3$ sec. If the amplitude of x is 2°C, what is the true amplitude?

9-2 Data of a frequency-response test are shown below. Estimate the transfer function.

ω (1/sec)	0.1	0.2	0.4	0.8	1	2	4	6
Magnitude ratio	5	4.8	3.8	2.6	2.3	1.25	0.63	0.40
Phase lag (deg)	12	25	45	73	83	122	173	218

9-3 Prove that the Nyquist plot of

$$G(s) = a + \frac{b}{1 + Ts}$$

is a semicircle, for positive or negative real values of a and b. Make a table for all possible Nyquist plots. Also sketch the unit-step-input response for each case.

9-4 Consider a second-order system,

$$G(s) = \frac{1}{a_0 s^2 + a_1 s + a_2},$$

where a_0, a_1, and a_2 are positive, real, and finite. (a) Prove that $1/(a_1 \omega)$ is tangent to $|G(j\omega)|$ in the Bode plot. (b) Find the condition under which $G(j\omega)$ has a resonance peak. Also determine the magnitude of the peak.

9-5 An experimentally determined Bode plot has the following characteristics:

(i) an initial slope of -1 log units/dec,

(ii) a break, up to 0 log units/dec slope, at $\omega = 1$ rad/sec, at which frequency the magnitude is $+2$ log units, and

(iii) a break, down to -1 log units/dec, at $\omega = 100$ rad/sec.

(a) Sketch the straight-line approximation to the Bode plot and the minimum-phase plot.

(b) From the above plot, sketch the Nyquist diagram. Label several points on both plots to show equivalence.

(c) Write the transfer function represented by this data.

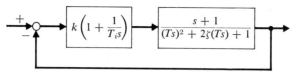

Fig. P9-7

9-6 An unknown parameter k of a dynamic system

$$\frac{d}{dt}\begin{bmatrix} x_1 \\ x_2 \\ x_3 \end{bmatrix} = \begin{bmatrix} -1 & 1 & 0 \\ -2 & -(k+1) & -1 \\ 0 & k & -2 \end{bmatrix} \begin{bmatrix} x_1 \\ x_2 \\ x_3 \end{bmatrix}$$

is to be investigated for stability by the frequency-response method. Find the canonical form of $G(s)$ for this purpose. Obtain the slope of gain asymptotes in the Bode diagram.

9-7 In the PI-control system shown in Fig. P9-7, T_i instead of k is the unknown parameter to be determined by the frequency-response method. Find the canonical form of $G(s)$ that will fit the purpose.

9-8 Determine the gain and phase of the irrational transfer function (see Example 7-5)

$$G(s) = e^{-\sqrt{s}}$$

and sketch the Nyquist plot.

9-9 Given that

$$G(s) = \frac{k}{(s+1)(s+2)(s+3)},$$

(a) what value of k yields a gain margin of 6 db? Solve this problem using an analytical, and not graphical, approach; and

(b) sketch the $G(j\omega)$ plot, carefully indicating the gain and phase margins on the Nyquist plot.

9-10 An open-loop transfer function of a feedback control system is

$$G(s) = \frac{k}{s(0.1s+1)(s^2+0.2s+1)}.$$

Determine k which satisfies the gain and phase margin criteria.

9-11 A mapping function (that is, the open-loop transfer function) is

$$G(s) = \frac{1}{s(s^2+25)}.$$

(a) Sketch the closed conformal map of the closed contour C consisting of the imaginary axis and the infinite semicircle enclosing the right half-plane. Indent as necessary into the right half of the s-plane along the C contour. Label several corresponding points on the contour C and its conformal map C'. Comment on the stability of a unity feedback system with such a feedforward $G(s)$.

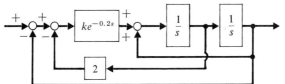

Fig. P9-15

(b) Repeat the mapping, but now indent into the left half-plane so that the open-loop poles are now included in the right half-plane. Show that the Nyquist theorem gives the same result in both cases.

9-12 Resketch the polar plots of Table 9–3(c) and (e) when the indent *efa* of Table 9–3(a) is to the left rather than to the right. Show that the Nyquist theorem gives the same results in both cases.

9-13 Apply Dzung's criterion to the Nyquist plots of Table 9–3. Repeat the procedure for the Nyquist plots of Problem 9–12.

9-14 Obtain the Bode plot for

$$G(s) = \frac{k}{s(0.1s + 1)^2}, \quad k = 1,$$

where a straight-line approximation is allowed for the gain plot. Superimpose on the Bode diagram the $M = 1.10$ contour from a Nichols chart and find the value of gain k to achieve $M_{max} = 1.10$.

9-15 An open-loop transfer function (see Fig. P9–15) of a conditionally stable feedback control system is:

$$G(s) = \frac{k(2s + 1)e^{-sL}}{s^2 - 1},$$

where $L = 0.2$. (a) Obtain the Bode plot. (b) Find the value of k for $M_p = 1.3$.

9-16 Prove that the M_c and ϕ_c contours in the G-plane are circles (shown in Fig. 9–21). Also derive the following M' and N' contours in the G^{-1}-plane:

$$\frac{1}{|1 + G(j\omega)|} = M_c, \quad \angle \frac{1}{1 + G(j\omega)} = \phi_c.$$

9-17 A system is described by the matrix equation $dx/dt = \mathbf{A}x + \mathbf{b}u$, where

$$\mathbf{A} = \begin{bmatrix} 0 & 1 & 0 \\ 0 & 0 & 1 \\ -4 & -6 & -4 \end{bmatrix}, \quad \mathbf{b} = \begin{bmatrix} 0 \\ 0 \\ 1 \end{bmatrix}.$$

(a) Obtain vectors $\boldsymbol{\alpha}$ and $\boldsymbol{\beta}$ for frequency response in the state space.

(b) Determine the output (or measurement) direction vector $\mathbf{c}$ in which forcing sinusoidal change is completely suppressed at some prescribed frequency $\omega = \omega_1$.

(c) For this $\mathbf{c}$ define output $y = \mathbf{c}x$ and compute the transfer function $G(s) = Y(s)/U(s)$.

10

MULTIVARIABLE CONTROL SYSTEMS

Linear multivariable feedback control systems may be represented in the same block-diagram form as single-variable, single-loop feedback control systems. When this is done, however, scalar variables and transfer functions must be replaced by vector variables and matrix transfer functions, respectively. Because of this mathematical complication, the design methods of the previous two chapters are not directly applicable to the multivariable problem. The classical control-theory approach to the problem was to decouple the system intentionally (in the physical sense) and treat each loop independently. As a result, final system design was limited in complexity by the then available mathematical techniques. In Section 10-1 this physical decoupling procedure, making use of transfer-function notation, is reviewed, and the feedforward design principle is presented.

The state-space approach to multivariable control problems is far more powerful than the scalar-variable method in that more design flexibility is maintained by the new mathematical formulation of the problem. The mathematical framework developed in Section 10-2 is amenable to digital solution while retaining all dependent relationships between the variables. To bridge the gap between the two approaches, we further discuss in this section a single-variable control system, using state-space techniques.

The dynamic state of a lumped-parameter multivariable system is uniquely represented by a state vector of finite dimensions. If all elements of this vector are measurable, we can expect improved feedback performance, as well as new approaches to process identification and control. When only a partial state-vector measurement is available, the problem becomes more complex. Both of these cases are investigated in Section 10-3 by means of an independent or decoupled modal control approach. The decoupling procedure of this method is a mathematical convenience and, unlike the method of Section 10-1, in no way constrains the structure of the final design. It is emphasized that state-variable feedback is the most effective way to maintain a desired process balance by achieving desired closed-loop-system eigenvalues.

The lumped-parameter-system modal control methods of Section 10-3 are extended in the next section to modal control of distributed-parameter systems. Here, final control elements impress distributed controlling inputs on the control object. The last section of the chapter deals with a class of optimal controls realizable by state-variable feedback, a topic to which we

10-1 Transfer-function approach to multivariable systems

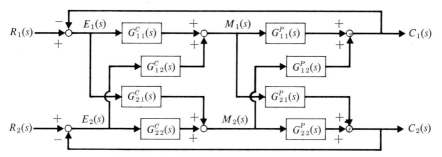

Fig. 10-1 Two-variable control system.

return in Chapter 14. In this section the performance index chosen to be minimized is a weighted sum of squared errors and controlling inputs.

10-1 TRANSFER-FUNCTION APPROACH TO MULTIVARIABLE SYSTEMS

Multivariable control systems are those which have a multiplicity of inputs and/or outputs. Feedback control loops in such systems are usually coupled to each other [1]. If this cross coupling or interaction is weak relative to the desired control performance, we may apply the design methods presented in the preceding two chapters as an initial step in the design of each control loop. If, on the other hand, the interaction is strong enough to be a dominant effect, we must consider a multivariable system as a single entity.

Let us consider a control object which has two inputs m_1, m_2 (manipulated variables) and two outputs c_1, c_2 (controlled variables). Referring to the block diagram shown in Fig. 10-1, we write the following s-domain relations for a linear plant:

$$C_1(s) = G_{11}^P(s)M_1(s) + G_{12}^P(s)M_2(s),$$
$$C_2(s) = G_{21}^P(s)M_1(s) + G_{22}^P(s)M_2(s).$$
(10-1)

The superscript P in the equations designates the control object, or plant. The relation can be expressed by the vector equation [2]

$$\mathbf{C}(s) = \mathbf{G}^P(s)\mathbf{M}(s),$$
(10-2)

where usually $\mathbf{C}(s)$ is an m-vector, $\mathbf{M}(s)$ is a p-vector, and the matrix transfer function $\mathbf{G}^P(s)$ has m rows and p columns (Fig. 10-2).

We write a similar equation for a set of controllers expressed by a matrix transfer function $\mathbf{G}^C(s)$ that relates the error vector $\mathbf{E}(s)$ of dimension m and manipulated variable vector $\mathbf{M}(s)$:

$$\mathbf{M}(s) = \mathbf{G}^C(s)\mathbf{E}(s),$$
(10-3)

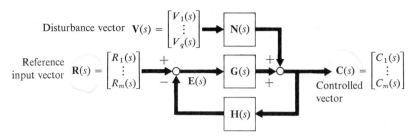

Fig. 10-2 Multivariable control system.

where $G^c(s)$ has p rows and m columns. To derive an open-loop matrix transfer function $G(s)$, we substitute $M(s)$ from Eq. (10-3) into Eq. (10-2):

$$C(s) = G^P(s)G^c(s)E(s) = G(s)E(s),$$

where

$$G(s) = G^P(s)G^c(s). \tag{10-4}$$

$G(s)$ is an $m \times m$ square matrix. Introducing the m-vector $R(s)$ as the reference input vector

$$E(s) = R(s) - C(s),$$

for a unity-feedback system we get

$$C(s) = G(s)R(s) - G(s)C(s)$$

or

$$C(s) = (I + G(s))^{-1}G(s)R(s). \tag{10-5}$$

The characteristic equation of the closed-loop system is

$$\det(I + G(s)) = 0, \tag{10-6}$$

and the closed-loop matrix transfer function for reference vector input is:

$$G_{LR}(s) = (I + G(s))^{-1}G(s). \tag{10-7}$$

These relations are exactly analogous to those of a scalar-variable, single-loop system, except that matrix inversion is now involved. The analogy holds for the general case where a disturbance $V(s)$ (q-vector) affects the controlled vector via a plant matrix transfer function $N(s)$ (m rows, q columns, Fig. 10-3) and an $m \times m$ matrix transfer function $H(s)$ is inserted as a feedback compensator (Fig. 10-2). The dimensionality of the R- and C-vectors must be the same since there can be only as many reference set points as there are measurements. However, the number of manipulated variables (P above) need not be the same as the number of measurements.

10-1 Transfer-function approach to multivariable systems

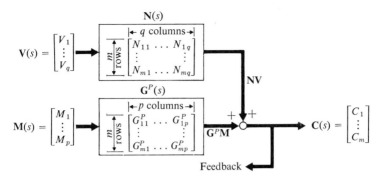

Fig. 10-3 Multivariable plant.

Although the vector equations in terms of matrix transfer functions are compact, the actual relations in terms of scalar transfer functions are quite complex. For instance, the closed-loop characteristic equation for the two variable systems of Fig. 10-1 is

$$[1 + G_{11}(s)][1 + G_{22}(s)] - G_{12}(s)G_{21}(s) = 0, \qquad (10\text{-}8)$$

where

$$\begin{aligned}
G_{11}(s) &= G_{11}^C(s)G_{11}^P(s) + G_{21}^C(s)G_{12}^P(s), \\
G_{12}(s) &= G_{12}^C(s)G_{11}^P(s) + G_{22}^C(s)G_{12}^P(s), \\
G_{21}(s) &= G_{11}^C(s)G_{21}^P(s) + G_{21}^C(s)G_{22}^P(s), \\
G_{22}(s) &= G_{12}^C(s)G_{21}^P(s) + G_{22}^C(s)G_{22}^P(s).
\end{aligned} \qquad (10\text{-}9)$$

In the classical viewpoint, there was no generally satisfactory approach known for designing all elements of the controller matrix $G^C(s)$ for a given plant $G^P(s)$. The decoupling or diagonalization approach was developed [3, 4, 5, and more recently, 18] to achieve simplicity in the design problem. Decoupling in this case means that each input $R_i(s)$, must affect one, and only one, output $C_i(s)$. For this condition to be satisfied the off-diagonal elements of matrix $G(s)$, and thus $G_{LR}(s)$, must be zero. In Eq. (10-9) this condition would be satisfied if, for instance,

$$G_{12}(s) = 0, \qquad G_{21}(s) = 0.$$

Two elements of the controller matrix are therefore fixed by the following conditions:

$$G_{12}^C(s) = -\frac{G_{12}^P(s)}{G_{11}^P(s)} G_{22}^C(s), \qquad G_{21}^C(s) = -\frac{G_{21}^P(s)}{G_{22}^P(s)} G_{11}^C(s). \qquad (10\text{-}10)$$

When (10-10) is satisfied in a two-variable system, the decoupled open-loop transfer functions are obtained by substituting (10-10) in (10-9). These functions, are:

$$G_{11}(s) = \frac{\det \mathbf{G}^P(s)}{G_{22}^P(s)} G_{11}^C(s), \qquad G_{22}(s) = \frac{\det \mathbf{G}^P(s)}{G_{11}^P(s)} G_{22}^C(s). \qquad (10\text{--}11)$$

Since the system is now reduced to two noninteracting single loops, $G_{11}^C(s)$ and $G_{22}^C(s)$ can be designed, for instance, by the gain and phase margin criteria applied to the frequency transfer function equivalents $G_{11}(j\omega)$ and $G_{22}(j\omega)$ of Eq. (10–11). In the general case of Fig. 10–3, we design for only p of the outputs if there are more outputs than manipulated variables ($m > p$), while we utilize only m of the manipulated variables (or define additional fictitious controlled variables) if $p > m$.

In the state-space formulation of a linear system in terms of $\mathbf{A}$, $\mathbf{B}$, and $\mathbf{C}$ matrices, the necessary and sufficient condition for input–output decouplability is known [18]. The condition is

$$\det \mathbf{B}^* \neq 0,$$

where $\mathbf{B}^*$ is an $m \times m$ matrix for m-vector input and output, and

$$\mathbf{B}^* = \begin{bmatrix} \mathbf{c}_1' \mathbf{A}^{N_1} \mathbf{B} \\ \mathbf{c}_2' \mathbf{A}^{N_2} \mathbf{B} \\ \vdots \\ \mathbf{c}_m' \mathbf{A}^{N_m} \mathbf{B} \end{bmatrix}.$$

$\mathbf{c}_1', \mathbf{c}_2', \ldots, \mathbf{c}_m'$ are row vectors of the $m \times n$ output matrix $\mathbf{C}$. N_1 through N_m are integer numbers chosen by the following rule shown for N_1:

$N_1 = 0 \quad$ if $\mathbf{c}_1'\mathbf{B} \neq 0$,

$N_1 = 1 \quad$ if $\mathbf{c}_1'\mathbf{B} = 0 \quad$ and $\quad \mathbf{c}_1'\mathbf{AB} \neq 0$,

$\vdots$

$N_1 = N \quad$ if $\mathbf{c}_1'\mathbf{B}, \mathbf{c}_1'\mathbf{AB}, \ldots, \mathbf{c}_1'\mathbf{A}^{N-1}\mathbf{B}$ are all $0 \quad$ and $\quad \mathbf{c}_1'\mathbf{A}^N\mathbf{B} \neq 0$,

$\vdots$

$N_1 = n \quad$ if $\mathbf{c}_1'\mathbf{A}^N\mathbf{B} = 0 \quad$ for all $N = 0, \ldots, (n-1)$, where n is the order of the system.

N_i, $i = 2, \ldots, m$ are chosen by similar arguments.

When $\mathbf{B}^*$ satisfies the condition $\det \mathbf{B}^* \neq 0$, the state-space analysis proceeds to obtaining conditions for the state-vector feedback matrix and the command-input processing matrix that will decouple the input–output relation while specifying some of the closed-loop poles. However, much remains to be done to develop a practical design technique via this approach.

The noninteracting control, where a one-to-one relation is established between a reference input and a controlled variable, has certain advantages in some applications [6, 7]. However an important drawback of this approach is the loss of controllability due to decoupling compensation (that

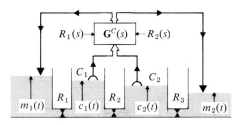

Fig. 10-4 Liquid-level control by supply-head manipulation.

is, pole cancellation; see example below). In many cases a better (or more effective) control is possible when the interaction is retained, so that system optimization is not subordinated to obtaining simplicity in design. Design simplification by diagonalization of $\mathbf{G}(s)$ may lead to worse problems than suboptimal performance. For instance, we may have trouble if $G_{11}^P(s)$, which appears in the denominator of $G_{12}^C(s)$ (Eq. 10–10), has a positive zero. In some cases a decoupled system may have an unrealizable transfer function.

Example 10-1 The two levels in Fig. 10-4 are controlled variables, while upstream and downstream heads are manipulated variables. The resistances and capacitances are

$$R_1 = 1, \quad R_2 = \tfrac{1}{2}, \quad R_3 = 2, \quad C_1 = 1, \quad C_2 = \tfrac{1}{2},$$

so that the plant is described by:

$$\frac{d}{dt}\begin{bmatrix} c_1 \\ c_2 \end{bmatrix} = \mathbf{A}\begin{bmatrix} c_1 \\ c_2 \end{bmatrix} + \begin{bmatrix} m_1 \\ m_2 \end{bmatrix}, \quad \text{where} \quad \mathbf{A} = \begin{bmatrix} -3 & 2 \\ 4 & -5 \end{bmatrix}.$$

Then we have

$$\mathbf{G}^P(s) = (s\mathbf{I} - \mathbf{A})^{-1} = \frac{1}{(s+1)(s+7)}\begin{bmatrix} s+5 & 2 \\ 4 & s+3 \end{bmatrix}.$$

By Eq. (10–10), we find the following decoupling controllers:

$$G_{12}^C(s) = -\left(\frac{2}{s+5}\right)G_{22}^C(s), \quad G_{21}^C(s) = -\left(\frac{4}{s+3}\right)G_{11}^C(s).$$

With these cross controllers the open-loop transfer functions of the decoupled system (Eq. 10–11) reduce to:

$$G_{11}(s) = \frac{G_{11}^C(s)}{s+3}, \quad G_{22}(s) = \frac{G_{22}^C(s)}{s+5}.$$

The original system's eigenvalues of -1 and -7 are completely missing in these open-loop transfer functions. This means that these eigenvalues (poles) were canceled by the cross controllers. To investigate the effect of this cancellation from the viewpoint of controllability and observability (Section 3–3), let us simplify the following computation by assuming a proportional control $G_{11}^C = k$. Further, we focus

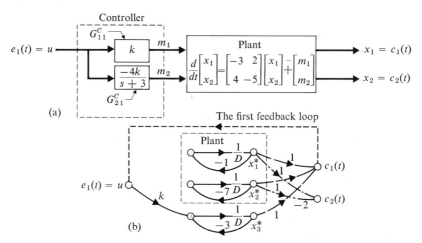

Fig. 10-5 Controllability and observability of a noninteracting control system.

our attention only on the signal transmission from the first error element $e_1(t)$ [that is, output of top summing junction in Fig. 10-1 with $e_2(t)$ set at zero] to plant outputs $c_1(t)$ and $c_2(t)$. Letting $e_1(t) = u(t) =$ input, $c_1(t) = x_1(t)$ and $c_2(t) = x_2(t)$ be state variables, we introduce a third state variable, $x_3(t)$, as the output of the cross controller $G_{21}^C(s) = -4k/(3+s)$, such that

$$m_2(t) = x_3(t), \quad Dx_3 = -3x_3 - 4ku.$$

We thus write the following vector equation for the data channel of our interest (Fig. 10-5a):

$$D\begin{bmatrix} x_1 \\ x_2 \\ x_3 \end{bmatrix} = \begin{bmatrix} -3 & 2 & 0 \\ 4 & -5 & 1 \\ 0 & 0 & -3 \end{bmatrix} \begin{bmatrix} x_1 \\ x_2 \\ x_3 \end{bmatrix} + \begin{bmatrix} k \\ 0 \\ -4k \end{bmatrix} u.$$

The m-vector of the plant in Fig. 10-5(a) appears as the $(k)(u)$ product (the m_1-term), and the $(1)(x_3)$ product (the m_2-term). With the coordinate change $\mathbf{x} = \mathbf{T}\mathbf{x}^*$, the relation becomes diagonal:

$$D\begin{bmatrix} x_1^* \\ x_2^* \\ x_3^* \end{bmatrix} = \begin{bmatrix} -1 & 0 & 0 \\ 0 & -7 & 0 \\ 0 & 0 & -3 \end{bmatrix} \begin{bmatrix} x_1^* \\ x_2^* \\ x_3^* \end{bmatrix} + \begin{bmatrix} 0 \\ 0 \\ k \end{bmatrix} u, \quad \text{where } \mathbf{T} = \begin{bmatrix} 1 & 1 & 1 \\ 1 & -2 & 0 \\ 0 & 0 & -4 \end{bmatrix}.$$

The outputs are now given in terms of the modal variables x_1^*, x_2^*, x_3^* in the following form:

$$\begin{bmatrix} c_1 \\ c_2 \end{bmatrix} = \begin{bmatrix} 1 & 0 & 0 \\ 0 & 1 & 0 \end{bmatrix} \begin{bmatrix} x_1 \\ x_2 \\ x_3 \end{bmatrix} = \begin{bmatrix} 1 & 0 & 0 \\ 0 & 1 & 1 \end{bmatrix} \mathbf{T} \begin{bmatrix} x_1^* \\ x_2^* \\ x_3^* \end{bmatrix}$$

$$= \begin{bmatrix} x_1^* + x_2^* + x_3^* \\ x_1^* - 2x_2^* \end{bmatrix}.$$

10-1 Transfer-function approach to multivariable systems

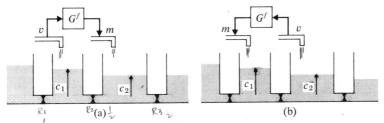

Fig. 10-6 Feed-forward compensation of load disturbance.

The signal-flow diagram of these relations, given in Fig. 10-5(b), shows that the system is decoupled in such a way that both x_1^* and x_2^* (the modes of the plant that affect the second controlled variable c_2) are not controllable by the first error signal e_1. Thus the input R_1 affects only c_1. This decoupling makes the main modes of the plant (x_1^* and x_2^*) unreachable by the first controller $G_{11}^C(s)$ in its main control loop. Similar reasoning shows that R_2 affects only c_2. Fortunately, the plant of this example is stable so that there is no danger of instability due to the uncontrolled modes. However, since this method of control does not affect the open-loop poles, it does not speed up the recovery process.

The block-diagram structure shown in Fig. 10-2 is not the only possible form of multivariable systems. On the contrary, a variety of structures exist [1]. For example, feedback control systems connected in series appear in cascade control systems and are often effective in reducing load upsets. Also, there can be a pattern of parallel feedback loops inside a major feedback loop [8]. In particular, in the process control field, feedforward control, which compensates for measured load upsets [9, 10], is considered effective.

The basic principle of feedforward compensation is shown in Fig. 10-6(a), where a disturbance $v(t)$ is measured, and a corrective action is produced by a feedforward controller $G^f(s)$. Let us suppose that $c_2(t)$, the level of the second tank, is the variable of major interest. For the same values of system parameters we assumed in the preceding example (Fig. 10-4), the equations for the two tank levels are:

$$\frac{dc_1}{dt} = -3c_1 + 2c_2 + v, \qquad \frac{dc_2}{dt} = 4c_1 - 5c_2 + 2m.$$

Eliminating c_1 from the pair of equations, we obtain

$$[(D+3)(D+5) - 8]c_2 = 4v + 2(D+3)m.$$

Therefore, if the load upset $v(t)$ is measured, and a compensating input $m(t)$ is generated by the feedforward controller $G^f(D)$ such that

$$4v + 2(D+3)m = 0$$

or

$$G^f(D) = \frac{m}{v} = -\frac{2}{D+3},$$

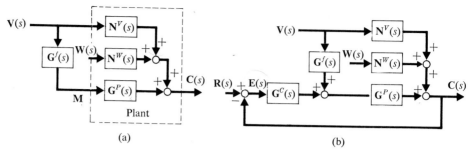

Fig. 10-7 Multivariable system with feed-forward compensation.

no deviation would appear in the main variable $c_2(t)$ due to the disturbance $v(t)$. With the configuration of the figure, a reversal of flow m [corresponding to the minus sign in $G^f(D)$] is impossible, but the relation is true for deviations about a nominal flow level. This is exact compensation. Such exact compensation, however, is not always realizable. For instance the feedforward controller for exact compensation of the system of Fig. 10-6(b) will be:

$$G^f(D) = \frac{m}{v} = \frac{D+3}{2},$$

which is physically unrealizable.

The feedforward principle just shown is generalized to a multivariable system in Fig. 10-7(a), where the vector input $\mathbf{V}(s)$ is a measurable disturbance input. For a linear plant we have

$$\mathbf{C}(s) = \mathbf{N}^V(s)\mathbf{V}(s) + \mathbf{G}^P(s)\mathbf{M}^f(s) + \mathbf{N}^W(s)\mathbf{W}(s),$$

where $\mathbf{N}^V(s)$, $\mathbf{N}^W(s)$, and $\mathbf{G}^P(s)$ are matrix transfer functions of the plant for measurable disturbance $\mathbf{V}(s)$, unmeasurable disturbance $\mathbf{W}(s)$, and controlling (or compensating) input vector $\mathbf{M}^f(s)$, respectively. For the matrix feedforward transfer function $\mathbf{G}^f(s)$,

$$\mathbf{M}^f(s) = \mathbf{G}^f(s)\mathbf{V}(s)$$

and thus

$$\mathbf{C}(s) = [\mathbf{N}^V(s) + \mathbf{G}^P(s)\mathbf{G}^f(s)]\mathbf{V}(s) + \mathbf{N}^W(s)\mathbf{W}(s).$$

Exact dynamic compensation for $\mathbf{V}(s)$ is possible if $[\mathbf{G}^P(s)]^{-1}$ exists and if

$$\mathbf{G}^f(s) = -[\mathbf{G}^P(s)]^{-1}\mathbf{N}^V(s) \tag{10-12}$$

is realizable.

Since feedforward compensation is not feasible for the unmeasurable disturbance $\mathbf{W}(s)$, and since, in general, exact compensation is impossible, it is advisable to design a combined feedback-feedforward system as shown in Fig. 10–7(b). The closed-loop response of the combined system is given by

$$\mathbf{C}(s) = [\mathbf{I} + \mathbf{G}^P(s)\mathbf{G}^c(s)]^{-1}[\mathbf{G}^P(s)\mathbf{G}^c(s)\mathbf{R}(s) + \mathbf{N}^W(s)\mathbf{W}(s) + \mathbf{\Delta}(s)\mathbf{V}(s)], \quad (10\text{--}13)$$

where

$$\mathbf{\Delta}(s) = \mathbf{G}^P(s)\mathbf{G}^f(s) + \mathbf{N}^V(s) \quad (10\text{--}14)$$

introduces a compensation error due to the imperfection of the feedforward controller $\mathbf{G}^f(s)$. In the ideal case, $\mathbf{G}^f(s)$ would be chosen such that $\mathbf{\Delta}(s) = \mathbf{0}$ for all $\mathbf{V}(s)$ and for all time. Note that the feedforward principle involves an open-loop operation and therefore does not influence the closed-loop poles [the closed-loop poles are determined from Eq. (10–6)].

10–2 STATE-SPACE FORMULATION

The state-variable viewpoint provides a direct approach to multivariable control system design and analysis. The design is straightforward if the inverses of both **B** and **C** exist in a linear, lumped control object described by

$$\frac{d\mathbf{x}(t)}{dt} = \mathbf{A}\mathbf{x}(t) + \mathbf{B}\mathbf{u}, \qquad \mathbf{y} = \mathbf{C}\mathbf{x}. \quad (10\text{--}15)$$

To simplify the problem somewhat, we assume no direct transmission from **u** to **y** (that is, $\mathbf{D} = \mathbf{0}$). If $\mathbf{B}^{-1}$ and $\mathbf{C}^{-1}$ exist, then **B** and **C** must be square $n \times n$ matrices. Consequently, the control vector **u** and the output vector **y** have the same dimension (n) as the state vector **x**. When a desired plant behavior is specified by an $n \times n$ matrix $\mathbf{A}_d$, a vector feedback control law **K** in the form

$$\mathbf{u} = -\mathbf{K}\mathbf{y} \quad (10\text{--}16)$$

can be uniquely determined such that Eqs. (10–15) and (10–16) will produce the desired closed-loop plant dynamics

$$\frac{d\mathbf{x}}{dt} = \mathbf{A}_d\mathbf{x}. \quad (10\text{--}17)$$

To fix **K** we first substitute the second of Eqs. (10–15) into Eq. (10–16) to find the control vector as a function of the state,

$$\mathbf{u} = -\mathbf{K}\mathbf{C}\mathbf{x}. \quad (10\text{--}18)$$

If Eq. (10–18) is now substituted into the first of Eqs. (10–15) and the result equated to Eq. (10–17), we get

$$\frac{d\mathbf{x}}{dt} = \mathbf{A}\mathbf{x} - \mathbf{B}\mathbf{K}\mathbf{C}\mathbf{x} = \mathbf{A}_d\mathbf{x},$$

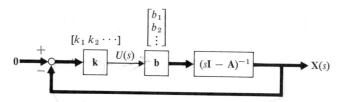

Fig. 10-8 Proportional control of the state vector.

from which follows

$$\mathbf{A} - \mathbf{BKC} = \mathbf{A}_d$$

or

$$\mathbf{K} = \mathbf{B}^{-1}(\mathbf{A} - \mathbf{A}_d)\mathbf{C}^{-1}. \qquad (10\text{-}19)$$

The ability to prescribe completely a desired $n \times n$ matrix $\mathbf{A}_d$ is lost if either $\mathbf{B}^{-1}$ or $\mathbf{C}^{-1}$ does not exist. We shall show next, however, that by using a properly manipulated scalar variable $u(t)$ we can design a multivariable feedback control system in which all closed-loop poles are specified. The necessary conditions involved are (1) that a direct and complete state variable measurement is possible (that is, $\mathbf{C}^{-1}$ exists) and (2) that the control object is controllable. The equation of the control object for this case is

$$\frac{d\mathbf{x}}{dt} = \mathbf{A}\mathbf{x}(t) + \mathbf{b}u(t), \qquad (10\text{-}20)$$

where $\mathbf{x}$ is an n-state vector, $\mathbf{A}$ is a constant $n \times n$ matrix, and $\mathbf{b}$ is an n-vector. Since $\mathbf{C}^{-1}$ exists by assumption, we simplify what follows by taking $\mathbf{C} = \mathbf{I}$, or $\mathbf{y} = \mathbf{x}$ so that a direct measurement of the state vector is assumed. The control law for this case can be represented by a *row* vector $\mathbf{k}'$, yielding a scalar feedback control

$$u(t) = -\mathbf{k}'\mathbf{x}(t). \qquad (10\text{-}21)$$

The closed-loop system obtained by Laplace transformation of Eqs. (10-20) and (10-21) is graphically depicted in Fig. 10-8 for zero reference input and and zero initial conditions.

In the following discussion we treat the relation between open- and closed-loop poles. The error vector $\mathbf{E}(s)$ is $-\mathbf{X}(s)$ when the reference input is zero; this vector is operated on by a constant row vector $\mathbf{k}'$ to produce a scalar control $U(s)$. The row vector is a set of proportional controls:

$$\mathbf{k}' = [k_1, k_2, \ldots, k_n]. \qquad (10\text{-}22)$$

The scalar controlling input is therefore given by:

$$U(s) = -\mathbf{k}'\mathbf{X}(s), \quad \text{or} \quad u(t) = -\mathbf{k}'\mathbf{x}(t). \qquad (10\text{-}23)$$

Substituting the linear control law (Eq. 10–23) into the equation of the control object,

$$\frac{d\mathbf{x}(t)}{dt} = \mathbf{A}\mathbf{x}(t) + \mathbf{b}u(t),$$

we obtain the following equation for the free, closed-loop system:

$$\frac{d\mathbf{x}(t)}{dt} = (\mathbf{A} - \mathbf{b}\mathbf{k}')\mathbf{x}(t). \tag{10-24}$$

Hence the characteristic polynomial of the closed-loop system is:

$$\begin{aligned}\Delta_c(s) &= \det(s\mathbf{I} - \mathbf{A} + \mathbf{b}\mathbf{k}') \\ &= \det\{(s\mathbf{I} - \mathbf{A})[\mathbf{I} + (s\mathbf{I} - \mathbf{A})^{-1}\mathbf{b}\mathbf{k}']\} \\ &= \det(s\mathbf{I} - \mathbf{A}) \cdot \det[\mathbf{I} + (s\mathbf{I} - \mathbf{A})^{-1}\mathbf{b}\mathbf{k}'] \\ &= \Delta(s) \cdot \det\{\mathbf{I} + \mathbf{h}(s)\mathbf{k}'\},\end{aligned} \tag{10-25}$$

where $\Delta(s)$ is the characteristic polynomial of the open-loop system given by

$$\Delta(s) = |s\mathbf{I} - \mathbf{A}| = a_0 + a_1 s + \cdots + a_{n-1} s^{n-1} + a_n s^n, \qquad a_n = 1, \tag{10-26}$$

and

$$\mathbf{h}(s) = (s\mathbf{I} - \mathbf{A})^{-1}\mathbf{b}$$

is an n-vector. Bass [12] has shown that the second determinant of (10–25) reduces to:

$$\det(\mathbf{I} + \mathbf{h}(s)\mathbf{k}') = 1 + \mathbf{k}'\mathbf{h}(s). \tag{10-27}$$

To derive Eq. (10–27), we denote the i-th column of $\mathbf{I}$ by $\mathbf{e}^i$ and compute:

$$\begin{aligned}\Delta &= \det(\mathbf{I} + \mathbf{h}\mathbf{k}') = \det[\mathbf{e}^1 + k_1\mathbf{h},\ \mathbf{e}^2 + k_2\mathbf{h},\ \ldots,\ \mathbf{e}^n + k_n\mathbf{h}] \\ &= \det[\mathbf{e}^1,\ \mathbf{e}^2 + k_2\mathbf{h},\ \ldots] + k_1 \cdot \det[\mathbf{h},\ \mathbf{e}^2 + k_2\mathbf{h},\ \ldots],\end{aligned}$$

where

$$\det[\mathbf{h},\ \mathbf{e}^2 + k_2\mathbf{h},\ \ldots,\ \mathbf{e}^n + k_n\mathbf{h}] = \det[\mathbf{h},\ \mathbf{e}^2,\ \ldots,\ \mathbf{e}^n].$$

The last statement is true because, by properties of determinants, we can subtract $k_2\mathbf{h}, \ldots, k_n\mathbf{h}$ (which are all linearly dependent on the first column) from other columns. Hence,

$$\begin{aligned}\Delta &= \det[\mathbf{e}^1, \ldots, \mathbf{e}^n] + k_1 \cdot \det[\mathbf{h}, \mathbf{e}^2, \ldots, \mathbf{e}^n] + \cdots + k_n \cdot \det[\mathbf{e}^1, \ldots, \mathbf{h}] \\ &= 1 + k_1 h_1 + \cdots + k_n h_n = 1 + \mathbf{k}'\mathbf{h}.\end{aligned}$$

With Eq. (10–27), the closed-loop characteristic polynomial (Eq. 10–25) becomes

$$\Delta_c(s) = \Delta(s)(1 + \mathbf{k}'\mathbf{h}(s)) = \Delta(s) + \Delta(s)\mathbf{k}'\mathbf{h}(s) . \tag{10–28}$$

In order to relate $\Delta_c(s)$, the characteristic polynomial of the closed-loop system, and $\Delta(s)$, its open-loop characteristic equation, we will make use of Eq. (10–26).

In Chapter 3 we saw the important role played by the solution matrix

$$e^{\mathbf{A}t} = \mathscr{L}^{-1}[(s\mathbf{I} - \mathbf{A})^{-1}] .$$

The inverse matrix $(s\mathbf{I} - \mathbf{A})^{-1}$ is called the resolvent of $(s\mathbf{I} - \mathbf{A})$. Since the inverse of a matrix is given by its transposed cofactor (or adjoint) matrix divided by its determinant (Section 2–2), the order in s of the numerator polynomials of the resolvent must be equal to or less than $n - 1$ for an nth-order system. Therefore, the resolvent can be expressed in the following form:

$$(s\mathbf{I} - \mathbf{A})^{-1} = \frac{1}{\Delta(s)}\{\mathbf{S}_1 + s\mathbf{S}_2 + \cdots + s^{n-1}\mathbf{S}_n\} = \frac{1}{\Delta(s)}\sum_{i=1}^{n} s^{i-1}\mathbf{S}_i . \tag{10–29}$$

It is possible to determine the square $n \times n$ matrices $\mathbf{S}_i$, $i = 1, \ldots, n$, in terms of $\mathbf{A}$ and the coefficients $a_0, \ldots, a_{n-1}$ of Eq. (10–26). We shall show the process of this computation for a fourth-order system. Letting $n = 4$ in Eq. (10–29), we have [using Eq. (10–26)]

$$(a_0 + a_1 s + a_2 s^2 + a_3 s^3 + a_4 s^4)\mathbf{I}$$
$$= (s\mathbf{I} - \mathbf{A})(\mathbf{S}_1 + s\mathbf{S}_2 + s^2\mathbf{S}_3 + s^3\mathbf{S}_4)$$
$$= -\mathbf{A}\mathbf{S}_1 + s(\mathbf{S}_1 - \mathbf{A}\mathbf{S}_2) + s^2(\mathbf{S}_2 - \mathbf{A}\mathbf{S}_3)$$
$$+ s^3(\mathbf{S}_3 - \mathbf{A}\mathbf{S}_4) + s^4\mathbf{S}_4 .$$

Coefficients of like powers of s on both sides of this equation must agree with each other. Therefore,

$$\mathbf{S}_4 = a_4 \mathbf{I} = \mathbf{I} \quad \text{since} \quad a_4 = 1,$$

and

$$a_0 \mathbf{I} = -\mathbf{A}\mathbf{S}_1 , \quad a_1 \mathbf{I} = \mathbf{S}_1 - \mathbf{A}\mathbf{S}_2 , \quad a_2 \mathbf{I} = \mathbf{S}_2 - \mathbf{A}\mathbf{S}_3 , \quad a_3 \mathbf{I} = \mathbf{S}_3 - \mathbf{A} .$$

By successive substitution we find:

$$\mathbf{S}_3 = a_3 \mathbf{I} + a_4 \mathbf{A} ,$$
$$\mathbf{S}_2 = a_2 \mathbf{I} + \mathbf{A}\mathbf{S}_3 = a_2 \mathbf{I} + a_3 \mathbf{A} + a_4 \mathbf{A}^2 ,$$
$$\mathbf{S}_1 = a_1 \mathbf{I} + \mathbf{A}\mathbf{S}_2 = a_1 \mathbf{I} + a_2 \mathbf{A} + a_3 \mathbf{A}^2 + a_4 \mathbf{A}^3 .$$

10-2 State-space formulation

Generalizing the computation to the nth-order case, we obtain

$$\mathbf{S}_i = \sum_{j=i}^{n} a_j \mathbf{A}^{j-i}, \qquad \mathbf{S}_n = \mathbf{I}, \qquad a_n = 1. \tag{10-30}$$

Using (10-30) in Eq. (10-29), we can now express the resolvent as a matrix polynomial. For simplicity let us first consider a fourth-order system. By Eq. (10-29) we have

$$\Delta(s)(s\mathbf{I} - \mathbf{A})^{-1} = \mathbf{S}_1 + s\mathbf{S}_2 + s^2\mathbf{S}_3 + s^3\mathbf{S}_4, \qquad \mathbf{S}_4 = \mathbf{I}.$$

Substituting (10-30) into this expression, we obtain

$$\Delta(s)(s\mathbf{I} - \mathbf{A})^{-1} = a_1\mathbf{I} + a_2\mathbf{A} + a_3\mathbf{A}^2 + a_4\mathbf{A}^3$$
$$+ a_2 s\mathbf{I} + a_3 s\mathbf{A} + a_4 s\mathbf{A}^2$$
$$+ a_3 s^2\mathbf{I} + a_4 s^2\mathbf{A}$$
$$+ a_4 s^3\mathbf{I}.$$

Collecting terms in each column, we have

$$(s\mathbf{I} - \mathbf{A})^{-1} = F_0(s)\mathbf{I} + F_1(s)\mathbf{A} + F_2(s)\mathbf{A}^2 + F_3(s)\mathbf{A}^3,$$

where

$$F_0(s) = \frac{a_1 + a_2 s + a_3 s^2 + a_4 s^3}{\Delta(s)},$$

$$F_1(s) = \frac{a_2 + a_3 s + a_4 s^2}{\Delta(s)},$$

$$F_2(s) = \frac{a_3 + a_4 s}{\Delta(s)},$$

$$F_3(s) = \frac{a_4}{\Delta(s)}.$$

The general form for an nth-order case is now deduced to be:

$$(s\mathbf{I} - \mathbf{A})^{-1} = \sum_{i=0}^{n-1} F_i(s)\mathbf{A}^i, \tag{10-31}$$

where

$$F_i(s) = \frac{a_{i+1} + a_{i+2}s + \cdots + a_n s^{n-(i+1)}}{\Delta(s)}. \tag{10-32}$$

Consequently the solution matrix in the time domain is expressed by the following matrix polynomial:

$$\mathscr{L}^{-1}[(s\mathbf{I} - \mathbf{A})^{-1}] = e^{\mathbf{A}t} = \sum_{i=0}^{n-1} f_i(t)\mathbf{A}^i, \tag{10-33}$$

where

$$f_i(t) = \mathscr{L}^{-1}[F_i(s)]. \quad (10\text{–}34)$$

The above development is similar to the derivation of Sylvester's theorem (see pp. 28–82 of [26]).

Returning to the original problem of relating $\Delta_c(s)$ to $\Delta(s)$, we see that the second term on the right-hand side of Eq. (10–28) may now be expressed as the following explicit polynomial in s [using Eq. (10–29) and the expression for $\mathbf{h}(s)$ following Eq. (10–26)]:

$$\Delta(s)\mathbf{k}'\mathbf{h}(s) = \Delta(s)\mathbf{k}'(s\mathbf{I} - \mathbf{A})^{-1}\mathbf{b} = \Delta(s)\mathbf{k}' \frac{1}{\Delta(s)} \sum_{i=1}^{n} s^{i-1} \mathbf{S}_i \mathbf{b}$$

$$= \mathbf{k}' \sum_{i=1}^{n} s^{i-1} \mathbf{S}_i \mathbf{b} \;;$$

hence

$$\Delta_c(s) = \Delta(s) + \mathbf{k}' \sum_{i=1}^{n} s^{i-1} \mathbf{S}_i \mathbf{b}. \quad (10\text{–}35)$$

In Eq. (10–26) the open-loop characteristic polynomial was expressed in terms of coefficients $a_0, a_1, \ldots, a_n$. Restated, it becomes

$$\Delta(s) = a_0 + a_1 s + \cdots + a_n s^n, \quad \text{where} \quad a_n = 1.$$

Let the closed-loop characteristic equation be

$$\Delta_c(s) = a_{0c} + a_{1c} s + \cdots + a_{nc} s^n, \quad \text{where} \quad a_{nc} = 1. \quad (10\text{–}36)$$

Then the coefficients of the two characteristic polynomials are related as follows:

$$a_{jc} = a_j + \mathbf{k}' \mathbf{S}_{j+1} \mathbf{b}, \quad j = 0, \ldots, n. \quad (10\text{–}37)$$

Example 10–2 Consider the system shown in Fig. 10-9, with parameter values as given in Section 10-1. The control object is described by

$$\frac{d}{dt} \begin{bmatrix} x_1 \\ x_2 \end{bmatrix} = \begin{bmatrix} -3 & 2 \\ 4 & -5 \end{bmatrix} \begin{bmatrix} x_1 \\ x_2 \end{bmatrix} + \begin{bmatrix} 1 \\ 0 \end{bmatrix} u.$$

The linear control law is

$$u = -[k_1, k_2] \begin{bmatrix} x_1 \\ x_2 \end{bmatrix}.$$

The open-loop characteristic polynomial is

$$\Delta(s) = |(s\mathbf{I} - \mathbf{A})| = 7 + 8s + s^2,$$

and thus

$$a_0 = 7, \quad a_1 = 8, \quad a_2 = 1.$$

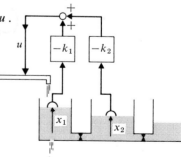

Fig. 10-9 Example of state variable feedback control.

Equation (10-30) gives

$$S_1 = a_1 I + a_2 A = \begin{bmatrix} 5 & 2 \\ 4 & 3 \end{bmatrix}, \qquad S_2 = a_2 I = I;$$

hence

$$a_{0c} = a_0 + k'S_1 b = 7 + 5k_1 + 4k_2,$$
$$a_{1c} = a_1 + k'S_2 b = 8 + k_1,$$
$$a_{2c} = a_2 = 1,$$

and the closed-loop characteristic polynomial becomes

$$\Delta_C(s) = (7 + 5k_1 + 4k_2) + (8 + k_1)s + s^2.$$

If, for example, we desire our system to exhibit two negative real closed-loop poles, both -10, we must then have

$$\Delta_C(s) = (s + 10)^2 = 100 + 20s + s^2.$$

We can fix k_1 and k_2 for this condition by solving

$$8 + k_1 = 20, \qquad 7 + 5k_1 + 4k_2 = 100;$$

thus

$$k_1 = 12, \qquad k_2 = \tfrac{33}{4}.$$

In the following discussion we shall show that a unique solution for the control law k' exists, and that it produces a prescribed set of closed-loop poles when a control object is controllable.

A formal statement for this problem is:

Given: (a) a controllable control object, hence A, b, and the coefficients $a_0, \ldots, a_n$ of its characteristic polynomial $\Delta(s)$; and (b) a complete set of desired closed-loop poles represented by coefficients $a_{0c}, a_{1c}, \ldots, a_{nc}$ of the closed-loop characteristic polynomial $\Delta_c(s)$,

Find: the control law k'.

We derive a matrix equation from Eq. (10-37) as the first step of the problem solution. Rewriting Eq. (10-37) in the following form:

$$a_{0c} - a_0 = k'S_1 b = (S_1 b)'k,$$
$$\vdots$$
$$a_{(n-1)c} - a_{n-1} = (S_n b)'k,$$

we collect the set into the vector expression

$$\begin{bmatrix} (S_1 b)' \\ \vdots \\ (S_n b)' \end{bmatrix} k = \begin{bmatrix} a_{0c} - a_0 \\ \vdots \\ a_{(n-1)c} - a_{n-1} \end{bmatrix}. \tag{10-38}$$

Let
$$\mathbf{D} = \begin{bmatrix} (\mathbf{S}_1\mathbf{b})' \\ \vdots \\ (\mathbf{S}_n\mathbf{b})' \end{bmatrix}.$$

Equation (10–38) can be solved for **k** if $\mathbf{D}^{-1}$ is known. To find $\mathbf{D}^{-1}$, we introduce a constant *n*-vector **p** such that

$$\mathbf{p}'\mathbf{b} = 0, \quad \mathbf{p}'\mathbf{A}\mathbf{b} = 0, \quad \ldots, \quad \mathbf{p}'\mathbf{A}^{n-2}\mathbf{b} = 0, \quad \mathbf{p}'\mathbf{A}^{n-1}\mathbf{b} = 1.$$

It then follows that

$$\mathbf{p}'[\mathbf{b}, \mathbf{A}\mathbf{b}, \ldots, \mathbf{A}^{n-1}\mathbf{b}] = \mathbf{p}'\mathbf{P} = [0, 0, \ldots, 0, 1]$$

or, transposing,

$$\mathbf{P}'\mathbf{p} = \mathbf{e}^n,$$

where $\mathbf{e}^n$ is the *n*th column of **I**, and **P** is defined by

$$\mathbf{P} = [\mathbf{b}, \mathbf{A}\mathbf{b}, \ldots, \mathbf{A}^{n-1}\mathbf{b}]. \tag{10-39}$$

This is the matrix that was crucial in the discussion of controllability in Chapter 3. If the object is controllable, $|\mathbf{P}| \neq 0$ and we can fix **p** by

$$\mathbf{p} = (\mathbf{P}^{-1})'\mathbf{e}^n. \tag{10-40}$$

Let us now compute the matrix product,

$$[\mathbf{p}, \mathbf{A}'\mathbf{p}, \ldots, (\mathbf{A}')^{n-1}\mathbf{p}] \begin{bmatrix} (\mathbf{S}_1\mathbf{b})' \\ \vdots \\ (\mathbf{S}_n\mathbf{b})' \end{bmatrix} \mathbf{b}$$

$$= [\mathbf{p}, \mathbf{A}'\mathbf{p}, \ldots, (\mathbf{A}')^{n-1}\mathbf{p}]\mathbf{D}\mathbf{b}$$
$$= \mathbf{p}(\mathbf{S}_1\mathbf{b})'\mathbf{b} + \mathbf{A}'\mathbf{p}(\mathbf{S}_2\mathbf{b})'\mathbf{b} + \cdots + (\mathbf{A}')^{n-1}\mathbf{p}(\mathbf{S}_n\mathbf{b})'\mathbf{b}$$
$$= [\mathbf{S}_1\mathbf{b}\mathbf{p}' + \mathbf{S}_2\mathbf{b}\mathbf{p}'\mathbf{A} + \cdots + \mathbf{S}_n\mathbf{b}\mathbf{p}'(\mathbf{A})^{n-1}]'\mathbf{b}$$
$$= \mathbf{I}\mathbf{b}. \tag{10-41}$$

The reduction is based on the equality of the products of column vector and row vector: $\mathbf{p}(\mathbf{S}_i\mathbf{b})' = [(\mathbf{S}_i\mathbf{b})\mathbf{p}']'$; on the property of the vector **p**,

$$\mathbf{p}'\mathbf{A}^i\mathbf{b} = 0 \quad \text{for all } i \text{ except } i = (n-1);$$

and on the fact that

$$\mathbf{S}_n\mathbf{b}(\mathbf{p}'(\mathbf{A})^{n-1}\mathbf{b}) = \mathbf{S}_n\mathbf{b} = \mathbf{I}\mathbf{b}$$

because $\mathbf{S}_n = \mathbf{I}$.

From Eq. (10-41) we get

$$[\mathbf{p}, \mathbf{A}'\mathbf{p}, \ldots, (\mathbf{A}')^{n-1}\mathbf{p}]\mathbf{D} = \mathbf{I};$$

therefore,

$$\mathbf{D}^{-1} = [\mathbf{p}, \mathbf{A}'\mathbf{p}, \ldots, (\mathbf{A}')^{n-1}\mathbf{p}],$$

and Eq. (10-38) is solved:

$$\mathbf{k} = [\mathbf{p}, \mathbf{A}'\mathbf{p}, \ldots, (\mathbf{A}')^{n-1}\mathbf{p}] \begin{bmatrix} a_{0c} & -a_0 \\ a_{1c} & -a_1 \\ \vdots & \vdots \\ a_{(n-1)c} & -a_{n-1} \end{bmatrix}$$

$$= (a_{0c} - a_0)\mathbf{p} + (a_{1c} - a_1)\mathbf{A}'\mathbf{p} + \cdots + (a_{(n-1)c} - a_{n-1})(\mathbf{A}')^{n-1}\mathbf{p}$$

$$= \sum_{i=0}^{n-1} (a_{ic} - a_i)(\mathbf{A}')^i \mathbf{p}. \tag{10-42}$$

The state-variable feedback control law (row vector) $\mathbf{k}$ can be computed by Eq. (10-42) (usually using a digital computer) for scalar control of a given object when all closed-loop poles are specified.

Example 10-3 Reconsider the system we solved in the last example.

Control object: $\mathbf{A} = \begin{bmatrix} -3 & 2 \\ 4 & -5 \end{bmatrix}, \quad \mathbf{b} = \begin{bmatrix} 1 \\ 0 \end{bmatrix}.$

Control law : $\mathbf{k} = \begin{bmatrix} k_1 \\ k_2 \end{bmatrix}.$

If the desired closed-loop poles are a double pole at -10, we have

$$\varDelta_C(s) = (s + 10)^2 = a_{0c} + a_{1c}s + s^2,$$

where $a_{0c} = 100$, $a_{1c} = 20$. We compute:

$$\mathbf{P} = [\mathbf{b}, \mathbf{Ab}] = \begin{bmatrix} 1 & -3 \\ 0 & 4 \end{bmatrix}, \quad \mathbf{P}^{-1} = \begin{bmatrix} 1 & \frac{3}{4} \\ 0 & \frac{1}{4} \end{bmatrix}, \quad (\mathbf{P}^{-1})' = \begin{bmatrix} 1 & 0 \\ \frac{3}{4} & \frac{1}{4} \end{bmatrix},$$

$$\mathbf{p} = (\mathbf{P}^{-1})' \mathbf{e}^n = \begin{bmatrix} 1 & 0 \\ \frac{3}{4} & \frac{1}{4} \end{bmatrix} \begin{bmatrix} 0 \\ 1 \end{bmatrix} = \begin{bmatrix} 0 \\ \frac{1}{4} \end{bmatrix},$$

$$\mathbf{k} = (a_{0c} - a_0)\mathbf{p} + (a_{1c} - a_1)\mathbf{A}'\mathbf{p} = (100 - 7)\mathbf{p} + (20 - 8)\begin{bmatrix} -3 & 4 \\ 2 & -5 \end{bmatrix}\mathbf{p} = \begin{bmatrix} 12 \\ \frac{33}{4} \end{bmatrix}.$$

The results are the same as in the preceding example, but the solution is now in closed form suitable for direct machine computation.

Example 10-4 Let us find the control law $\mathbf{k}$ in order to obtain specified closed-loop poles for a position control system (Fig. 10-10). The equation of the control object is

$$\frac{d}{dt}\begin{bmatrix} x_1 \\ x_2 \end{bmatrix} = \begin{bmatrix} 0 & 1 \\ -1/CI & -1/CR \end{bmatrix} \begin{bmatrix} x_1 \\ x_2 \end{bmatrix} + \begin{bmatrix} 0 \\ 1/CI \end{bmatrix} u,$$

430 Multivariable control systems

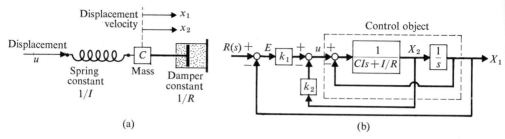

Fig. 10-10 Control of position with velocity feedback.

where x_1 = output displacement, x_2 = velocity, u = input displacement, C = mass, $1/I$ = spring constant, and $1/R$ is damper constant. Let $e = r - x_1$ = position error, where r = desired displacement (reference input), and

$$u = (r - x_1) k_1 - x_2 k_2 = rk_1 - \mathbf{k}'\mathbf{x}.$$

Let $I = 1$, $C = 1$, and $1/R = 0$ so that the control object is an undamped oscillator whose characteristic polynomial is

$$\Delta(s) = \frac{1}{CI} + \frac{1}{CR} s + s^2 = 1 + s^2, \qquad a_0 = 1, \qquad a_1 = 0.$$

If a double pole, -10, is desired for the closed-loop system, then $a_{0c} = 100$ and $a_{1c} = 20$. We compute

$$\mathbf{A} = \begin{bmatrix} 0 & 1 \\ -1 & 0 \end{bmatrix}, \quad \mathbf{b} = \begin{bmatrix} 0 \\ 1 \end{bmatrix}, \quad \mathbf{P} = \begin{bmatrix} 0 & 1 \\ 1 & 0 \end{bmatrix}, \quad \mathbf{P}^{-1} = \begin{bmatrix} 0 & 1 \\ 1 & 0 \end{bmatrix},$$

and

$$(\mathbf{P}^{-1})' = \begin{bmatrix} 0 & 1 \\ 1 & 0 \end{bmatrix}, \quad \mathbf{p} = \begin{bmatrix} 1 \\ 0 \end{bmatrix}.$$

Thus, by Eq. (10-42),

$$\mathbf{k} = (100 - 1)\begin{bmatrix} 1 \\ 0 \end{bmatrix} + 20 \begin{bmatrix} 0 & -1 \\ 1 & 0 \end{bmatrix} \begin{bmatrix} 1 \\ 0 \end{bmatrix} = \begin{bmatrix} 99 \\ 20 \end{bmatrix}.$$

The closed-loop transfer function of the system (Fig. 10-10b) for reference input is

$$\frac{x_1(s)}{R(s)} = \frac{k_1/IC}{s^2 + \left(\frac{I}{R} + k_2\right)\frac{s}{CI} + \frac{1 + k_1}{CI}} = \frac{99}{(s + 10)^2}.$$

The steady-state error (offset or droop) is 1 %.

10-3 MODAL CONTROL OF LUMPED-PARAMETER OBJECTS

Two cases of state-variable feedback controls were discussed in the preceding section. The first was the ideal case of vector control [see Eqs. (10-15) through (10-19)] where $\mathbf{B}^{-1}$ and $\mathbf{C}^{-1}$ both exist. The second, and the more thoroughly covered, was the case of scalar control based on complete state-vector measurement when C^{-1} exists. In many practical control problems, however, there often are situations in which a complete state-vector measurement is impossible *and* more than one but less than n manipulated variables ("control valves") can be applied to the control object. Thus the number of measurements m and/or the number of controls r are less than the order of the object n. The controller matrix **K** will then become rectangular, with $r \times m$ elements to be fixed. An obvious difficulty arises in the search for these controlling parameters when a designer stays in the original state space, because with $r \times m$ elements only a partial correction in the original n^2 elements of **A** can be achieved, and it is difficult to see the effect of such control in the original state-space formulation:

$$\frac{d\mathbf{x}}{dt} = (\mathbf{A} - \mathbf{BKC})\mathbf{x}.$$

A great reduction in the number of elements of **A** is obtained when **A** is reduced to the diagonal or the Jordan canonical form Λ by a suitable coordinate transformation $\mathbf{x} = \mathbf{Tx^*}$ such that

$$\frac{d\mathbf{x^*}}{dt} = \Lambda \mathbf{x^*} + \mathbf{T^{-1}Bu}, \qquad \mathbf{y} = \mathbf{CTx^*}. \tag{10-43}$$

When all eigenvalues of the control object are real and distinct, only these eigenvalues appear as diagonal elements of Λ, and Λ is diagonal. If duplications exist in the eigenvalues, it may be necessary to use the Schwartz form. To avoid complex numbers in Λ, we shall apply the modified transformation (see Chapter 3) and use the resulting modified canonic form when the system has conjugate complex eigenvalues (that is, oscillatory modes).

A multivariable control system design in the modal domain was first proposed by Rosenbrock [13]. The concept was extended by M. A. Murray-Lasso, L. A. Gould, and F. M. Schlaefer [14, 15, and 16] to include linear distributed-parameter systems (covered in the next section). Apparently independently, J. H. Wykes and A. S. Mori [17] made a feasibility study involving control of the elastic modes of a supersonic aircraft. The presentation of this section closely follows that of [19].

Our aim in the following discussion is to provide guidance in the engineering design of multivariable control systems which have dominant dynamic modes. The conditions to be satisfied for various degrees of modal control are more restrictive than observability and controllability conditions. The discussions in this section are limited to zero input conditions. The state corresponding to the fixed operating point will be regarded as the origin of state space. As shown in Fig. 10-11, the modally controlled object can be viewed as a subsystem of a cascade control system. The overall system may also include, in addition to the modal controller **K**, a feedforward compensator for measurable disturbances. Since the modal control in

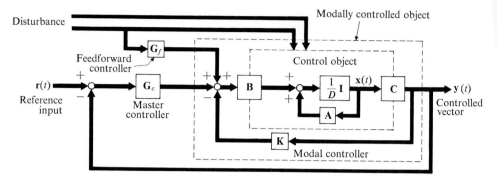

Fig. 10-11 Modal control as a subsystem of a cascade or computing control system.

this overall scheme will act to relocate (or speed up) the original modes of the control object, should we wish to move the operating point by changing the reference vector input **r**, the output **y** would exhibit improved dynamics in following such changes.

It should be noted that the modal control scheme proposed in this chapter involves proportional control only. As depicted in Fig. 10-11, the modal controller may be viewed as an inner-loop feedback in addition to the outer-loop controller G_c, which may be PI or PID in form. From this point of view, we will use the inner-loop modal controller to alter the dynamics (eigenvalues) of the system, while the outer-loop controller ensures the attainability of any desired equilibrium state.

To introduce the basic principle of modal control, let us first consider an ideal case where all elements of state vector $\mathbf{x}(s)$ are directly measurable ($\mathbf{C} = \mathbf{I}$) and the control vector $\mathbf{u}_c(t) = \mathbf{B}\mathbf{u}(t)$ is produced by a nonsingular $\mathbf{B}$ from an n-vector $\mathbf{u}(t)$. The control object is described by:

$$\frac{d\mathbf{x}}{dt} = \mathbf{A}\mathbf{x} + \mathbf{u}_c, \quad \mathbf{u}_c = \mathbf{B}\mathbf{u}. \tag{10-44}$$

To design an $n \times n$ matrix proportional controller $\mathbf{K}$ (Fig. 10-12a), we convert Eq. (10-44) into the modal domain by $\mathbf{x} = \mathbf{T}\mathbf{x}^*$, so that

$$\frac{d\mathbf{x}^*}{dt} = \mathbf{\Lambda}\mathbf{x}^* + \mathbf{u}^*, \tag{10-45}$$

where $\mathbf{u}^* = \mathbf{T}^{-1}\mathbf{u}_c$. Assuming for the moment that all eigenvalues of the control object are real and distinct, hence the $\mathbf{\Lambda}$-matrix of Eq. (10-45) is diagonal, let us suppose that

$$\mathbf{\Lambda}_d = \begin{bmatrix} p_{1d} & & 0 \\ & \ddots & \\ 0 & & p_{nd} \end{bmatrix}$$

10-3   Modal control of lumped-parameter objects

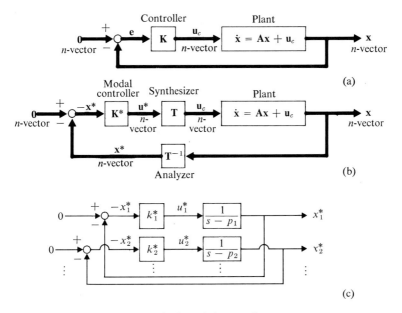

Fig. 10-12 Ideal modal control system.

is the desired or prescribed closed-loop modal-domain matrix of the control object with modal control, where $p_{1d}, \ldots, p_{nd}$ are the prescribed closed-loop eigenvalues of the controlled system. By a modal-domain vector feedback (for zero-reference-input vector)

$$\mathbf{u}^* = -\mathbf{K}^*\mathbf{x}^*,$$

we have the closed-loop system's equation

$$\frac{d\mathbf{x}^*}{dt} = \Lambda\mathbf{x}^* - \mathbf{K}^*\mathbf{x}^* = (\Lambda - \mathbf{K}^*)\mathbf{x}^*$$

$$= \Lambda_d\mathbf{x}^*.$$

Hence $\mathbf{K}^*$, the modal-domain feedback control matrix, can be fixed by

$$\mathbf{K}^* = \Lambda - \Lambda_d. \tag{10-46}$$

These steps are summarized in Fig. 10-12(b) from which we can see that

1. $\mathbf{x}^* = \mathbf{T}^{-1}\mathbf{x}$ is generation of the modal-domain state vector $\mathbf{x}^*$, via "mode analyzer" $\mathbf{T}^{-1}$, from measured state vector $\mathbf{x}$;
2. $\mathbf{u}^* = -\mathbf{K}^*\mathbf{x}^*$ is production of the modal-domain control vector $\mathbf{u}^*$; and
3. $\mathbf{u}_c = \mathbf{T}\mathbf{u}^*$ is conversion of $\mathbf{u}^*$ into the control vector $\mathbf{u}_c$ via the "mode synthesizer" $\mathbf{T}$.

434 Multivariable control systems

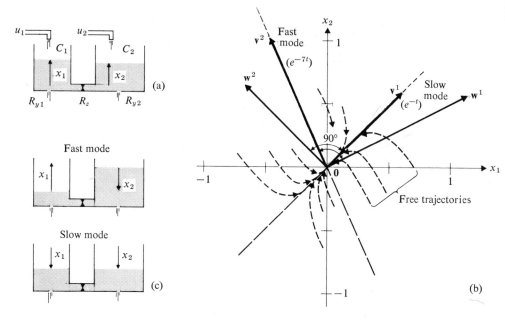

Fig. 10-13 Free motion of a second-order CR-system.

The control law **K** for $\mathbf{u}_c = -\mathbf{Kx}$ in the original state space (Fig. 10-12a) can be fixed by incorporating the three steps:

$$\mathbf{K} = \mathbf{T}\mathbf{K}^*\mathbf{T}^{-1}. \tag{10-47}$$

The results of the decoupled controlling action are shown in Fig. 10-12(c).

Example 10-5 We now apply the foregoing theory to modal cotrol of a second-order CR system (Fig. 10-13a). Let the system be

$$\frac{d\mathbf{x}}{dt} = \begin{bmatrix} -3 & 2 \\ 4 & -5 \end{bmatrix} \mathbf{x}(t) + \begin{bmatrix} u_1(t) \\ u_2(t) \end{bmatrix}.$$

For the eigenvalues $p_1 = -1$ and $p_2 = -7$, the eigenvectors are fixed by $(\mathbf{A} - p_i\mathbf{I})\mathbf{v}^i = \mathbf{0}$:

$$\mathbf{v}^1 = \begin{bmatrix} 0.5 \\ 0.5 \end{bmatrix}, \quad \mathbf{v}^2 = \begin{bmatrix} -0.5 \\ 1 \end{bmatrix},$$

and thus

$$\mathbf{T} = \begin{bmatrix} 0.5 & -0.5 \\ 0.5 & 1 \end{bmatrix} \quad \text{and} \quad \mathbf{T}^{-1} = \frac{4}{3} \begin{bmatrix} 1 & 0.5 \\ -0.5 & 0.5 \end{bmatrix}.$$

These vectors are shown in Fig. 10-13(b). The first mode, which corresponds to $p_i = -1$ (the slower motion), takes place in the direction of $\pm \mathbf{v}^1$. As shown in part (c) of the figure, x_1 and x_2 are in-phase for this mode (see Chapter 3). The second

10-3 Modal control of lumped-parameter objects

mode for $p_2 = -7$ decays much faster than the first mode, and x_1 and x_2 are out of phase because motion takes place in the direction of $\pm \mathbf{v}^2$. The row vectors $\mathbf{w}^1$ and $\mathbf{w}^2$ of $\mathbf{T}^{-1}$,

$$\mathbf{T}^{-1} = \begin{bmatrix} \mathbf{w}^1 \\ \mathbf{w}^2 \end{bmatrix}, \quad \mathbf{w}^1 = \tfrac{4}{3}[1 \quad 0.5], \quad \mathbf{w}^2 = \tfrac{4}{3}[-0.5 \quad 0.5],$$

are also indicated in Fig. 10-13(b). Since the vector $\mathbf{w}^1$ is orthogonal to $\mathbf{v}^2$, the second mode does not appear in $\langle \mathbf{w}^1, \mathbf{x} \rangle$ so that the inner product is $x_1^* = (\mathbf{w}^1)'\mathbf{x}$, which shows how $\mathbf{T}^{-1}$ acts as a mode analyzer. The first element of the control vector $\mathbf{u}^*$, $u_1^* = -k_1 x_1^*$, must act on the first mode. Therefore u_1^* is assigned to the direction $\mathbf{v}^1$ by the synthesizer; similarly, u_2^* is directed to $\mathbf{v}^2$. The overall control action $\mathbf{K}$ in $\mathbf{u}_c = -\mathbf{K}\mathbf{x}$ is

$$\mathbf{K} = \mathbf{T}\mathbf{K}^*\mathbf{T}^{-1} = \frac{4}{3}\begin{bmatrix} 0.5 & -0.5 \\ 0.5 & 1 \end{bmatrix}\begin{bmatrix} k_1^* & 0 \\ 0 & k_2^* \end{bmatrix}\begin{bmatrix} 1 & 0.5 \\ -0.5 & 0.5 \end{bmatrix}$$

$$= \frac{4}{3}\begin{bmatrix} 0.5k_1^* + 0.25k_2^* & 0.25k_1^* - 0.25k_2^* \\ 0.5k_1^* - 0.5k_2^* & 0.25k_1^* + 0.5k_2^* \end{bmatrix}.$$

This control system requires two controlling inputs, u_1 and u_2. It is interesting to note that the same result (from Eq. 10-46),

closed loop poles at $p_{1c} = -1 - k_1^*$, $p_{2c} = -7 - k_2^*$,

will be obtained by the controller vector $\mathbf{k}$ of Eq. (10-42) where only one controlling variable (either u_1 or u_2) is needed, provided that the object is controllable with the chosen controlling variable. From a practical point of view, however, the modal control scheme is likely to be the better of the two (that is, scalar or vector controlling signal), for the total control load would not be assigned to a single manipulator, as in the scalar case, but would be shared by several manipulators under a vector controlling signal.

Our next design problem is the independent control of q modes of an nth-order object, where $q < n$ and

$q =$ number of measurements (dimension m of $\mathbf{y}$) if $m \leqslant r$ or

$q =$ number of controls (dimension r of $\mathbf{u}$) if $r \leqslant m$.

A pair of modal variables x_i^* and x_{i+1}^* that corresponds to a conjugate complex pair of eigenvalues must be contained in the q controlled variables as a pair, or deleted from the set as a pair. If a system has repeated eigenvalues, there is no reason for a modal control to control only some of them. In other words, a high-order eigenvalue

$$p_i = p_{i+1} = \cdots = p_{i+k}$$

must be either all included in or all excluded from modal control. For the purpose of modal control (to "speed up" the dynamics of the control object, especially the slower modes) it is logical to consider control of the first q

modes such that

$$\mathscr{R}e\, p_1 \geqslant \mathscr{R}e\, p_2 \geqslant \mathscr{R}e\, p_3 \geqslant \cdots$$

We shall number the eigenvalues below in such descending order.

The modal-domain vector $\mathbf{x}_q^*$ in this case is q-dimensional. It must be "produced" from the measured m-vector $\mathbf{y}$. The measured vector $\mathbf{y}$ and the n-vector $\mathbf{x}^*$ in the modal space are related by

$$\mathbf{y} = \mathbf{F}\mathbf{x}^*, \quad \mathbf{F} = \mathbf{CT}. \tag{10-48}$$

The $m \times n$ rectangular matrix $\mathbf{F}$ consists of a square matrix $\mathbf{F}_m$ and an $m \times (n-m)$ rectangular matrix $\Delta\mathbf{F}_m$ such that

$$\mathbf{F} = [\mathbf{F}_m, \Delta\mathbf{F}_m].$$

A mode-oriented design is practical when $\mathbf{F}_m^{-1}$ exists, that is when

$$\det \mathbf{F}_m \neq 0. \tag{10-49}$$

We consider a measurement system to be ideal when, in addition to Eq. (10-49), the following condition is met:

$$\Delta\mathbf{F}_m = \mathbf{0}. \tag{10-50}$$

The second condition implies $\mathbf{x}_{m+1}^*, \ldots, \mathbf{x}_n^*$ are unobservable. In the mode-oriented design of a control system it is important to select measuring points and hence $\mathbf{C}$ in such a way that condition (10-49) will hold and also that elements of $\Delta\mathbf{F}_m$ will be as small as possible. This implies that the measurements must be sensitive to the first m modes while differentiating them from each other and, at the same time, insensitive to the remaining neglected modes. A certain mode is made unobservable to a sensing element when the sensor is located at a node of that mode. If the eigenvectors $\mathbf{T}$ of a system are well known, then the measurement matrix $\mathbf{C}$ may be chosen (that is, instrumentation located) such that Eq. (10-50) is very nearly satisfied.

It is also possible to state conditions on manipulation. The control vector $\mathbf{u}$ in the original state space, and the control vector $\mathbf{u}^*$ in the modal domain defined by

$$\frac{d\mathbf{x}^*}{dt} = \Lambda\mathbf{x}^* + \mathbf{u}^*$$

are related to each other by the equations

$$\mathbf{u}^* = \mathbf{H}\mathbf{u}, \quad \mathbf{H} = \mathbf{T}^{-1}\mathbf{B}. \tag{10-51}$$

The $n \times r$ rectangular matrix $\mathbf{H}$ has an $r \times r$ matrix $\mathbf{H}_r$ and an $(n-r) \times r$ matrix $\Delta\mathbf{H}_r$ such that

$$\mathbf{H} = \begin{bmatrix} \mathbf{H}_r \\ \Delta\mathbf{H}_r \end{bmatrix}. \tag{10-52}$$

10-3 Modal control of lumped-parameter objects 437

For effective modal control, it is essential that a control vector $\mathbf{u}^*$ in the modal domain be converted to $\mathbf{u}$ in the original state space, and for this to be possible, $\mathbf{H}_r^{-1}$ must exist; hence

$$\det \mathbf{H}_r \neq 0 . \tag{10-53}$$

The control will not cause any disturbance in $x_{r+1}^*, \ldots, x_n^*$ if

$$\Delta \mathbf{H}_r = \mathbf{0} , \tag{10-54}$$

that is if all modes other than the first r are made uncontrollable. We consider control (or manipulation) to be ideal when conditions (10-53) and (10-54) are both satisfied. For effective execution of modal control the first condition must be met and the second condition must be met at least approximately; the better the approximation, the better the control. This provides a guide for manipulator design. Controls must be chosen in such a way that, between them, they control all the dominant modes.

If measurement and control are both ideal, in the sense of Eqs. (10-49), (10-50), (10-53), and (10-54), it is possible to determine a control law $\mathbf{K}$ which will modify the motion of the first q modal variables $x_1^*, \ldots, x_q^*$, into a prescribed pattern in the modal domain, while not influencing the other modal variables. As stated above, the number q is taken to be the smaller of either m or r. However, the excess number of controls $(r - q) > 0$, or the excess number of measurements $(m - q) > 0$, turns out to be redundant for control of the first q modes. We therefore assume that $r = m = q$, where the smaller of r or m determines the indicial value that all three take on. In this case the q-dimensional modal-domain control matrix $\mathbf{K}_q^*$ can be fixed in exactly the same way as in the previous case of complete modal

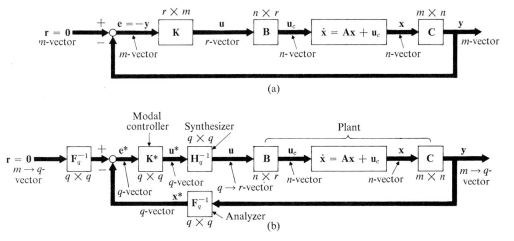

Fig. 10-14 Non-purely-ideal modal control system ($q = r \leqslant m$ or $q = m < r$).

control (Fig. 10–12b) with the $n \times n$ matrix $\mathbf{K}_q^*$ having the last $(n - q)$ rows and columns made up of zeros.

Figure 10–12(a) and (b) may now be extended to the non-purely-ideal case (when m and/or $r < n$). Figure 10–14(a) shows the system in the original state space, and Fig. 10–14(b) shows it in the modal domain (see analysis below). The subscripts r and m have been replaced by q, where q is the smaller of the two. Although the $q \times q$ matrix $\mathbf{F}_q^{-1}$ is not conformable with the m-vector $\mathbf{y}$ if $q = r < m$, we interpret the figure to mean that only the first q of the m measurements ($\mathbf{y}$) are utilized and the remaining $(m - q)$ are disregarded as being redundant. Similarly, when $q = m < r$, we discard the last $(r - q)$ controllers as being redundant.

We specify a desired form of the q-dimensional modal-domain matrix Λ_{qd}. For instance, it could be

$$\Lambda_{qd} = -\rho \mathbf{I}_q, \qquad \mathbf{I}_q = q \times q \text{ identity matrix,}$$

if a star-pattern trajectory (9 in Table 3–4) is desired for testing modal control on the first q variables $x_1^*, \ldots, x_q^*$. The state-space relation of the multi-variable control system,

$$\frac{d\mathbf{x}}{dt} = \mathbf{A}\mathbf{x} - \mathbf{B}\mathbf{K}\mathbf{C}\mathbf{x},$$

converted into the modal domain, yields (for ideal measurement and control)

$$\frac{d\mathbf{x}^*}{dt} = \Lambda \mathbf{x}^* - \begin{bmatrix} \mathbf{H}_q \\ 0 \end{bmatrix} \mathbf{K}[\mathbf{F}_q, 0]\mathbf{x}^*.$$

Let $\mathbf{x}_q^*$ be a q-vector for the first q modal variables $x_1^*, \ldots, x_q^*$, and Λ_q be a $q \times q$ square matrix that has the first q rows and columns of Λ; then

$$\frac{d\mathbf{x}_q^*}{dt} = \Lambda_q \mathbf{x}_q^* - \mathbf{H}_q \mathbf{K} \mathbf{F}_q \mathbf{x}_q^* \equiv \Lambda_{qd} \mathbf{x}_q^*,$$

so that

$$\mathbf{K}_q^* = \Lambda_q - \Lambda_{qd} = \mathbf{H}_q \mathbf{K} \mathbf{F}_q,$$

whence

$$\mathbf{K} = \mathbf{H}_q^{-1}(\Lambda_q - \Lambda_{qd})\mathbf{F}_q^{-1} = \mathbf{H}_q^{-1}\mathbf{K}_q^*\mathbf{F}_q^{-1}. \tag{10–55}$$

An ideal modal control up to the qth order can be realized by the $q \times q$ control law $\mathbf{K}$ of this expression.

By comparing Figs. 10–14(a) and (b) we can directly relate the modal controller $\mathbf{K}^*$ to the original state-space controller $\mathbf{K}$ (see Eq. 10–55). Such a relation is equivalent to Eq. (10–47) and the three steps described leading to Eq. (10–47) for the purely ideal case. Even when either the measurement is nonideal ($\Delta F_m \neq 0$) or the control is nonideal ($\Delta H_r \neq 0$), Fig. 10-14 still applies, though, as will be shown shortly, disturbances which couple the modal responses will then be introduced.

When control is ideal ($\Delta H_r = 0$) but the measurement is not ideal ($\Delta F_m \neq 0$), the modal-domain relation will be:

$$\frac{d\mathbf{x}^*}{dt} = \Lambda \mathbf{x}^* - \begin{bmatrix} \mathbf{H}_r \\ 0 \end{bmatrix} \mathbf{K}[\mathbf{F}_m, \Delta \mathbf{F}_m]\mathbf{x}^*$$

$$= \Lambda \mathbf{x}^* - \begin{bmatrix} \mathbf{I}_r \\ 0 \end{bmatrix} \mathbf{K}^*[\mathbf{I}_m, \Delta \mathbf{F}]\mathbf{x}^*,$$

where

$$\mathbf{H}_r \mathbf{K} \mathbf{F}_m = \mathbf{K}^*, \qquad \Delta \mathbf{F} = \mathbf{F}_m^{-1} \Delta \mathbf{F}_m$$

and $\mathbf{I}_r$, $\mathbf{I}_m$ are $r \times r$ and $m \times m$ identity matrices, respectively. Let $m > r = q$, or $r > m = q$. Modal control of up to q variables is realizable for both cases, but the controlled modes will be disturbed by the uncontrolled modes. To see how the disturbance occurs, let us investigate the first case: $m > r = q$.

$\mathbf{K}^*$ in this case is a $q \times m$ rectangular matrix. Let the first q columns of $\mathbf{K}^*$ be expressed by a $q \times q$ square matrix $\mathbf{K}_q^*$, while making the rest of the columns zero:

$$\mathbf{K}^* = [\mathbf{K}_q^*, \Delta \mathbf{K}^*], \qquad \Delta \mathbf{K}^* = 0.$$

Then the matrix product for the control action in the modal domain will reduce to

$$\begin{bmatrix} \mathbf{I}_r \\ 0 \end{bmatrix} [\mathbf{K}_q^*, 0][\mathbf{I}_m, \Delta \mathbf{F}] = \begin{bmatrix} \overbrace{\mathbf{K}_q^*}^{q \text{ cols}}, & \overbrace{0}^{(m-q) \text{ cols}}, & \overbrace{\mathbf{K}_q^* \Delta_1 \mathbf{F}}^{(n-m) \text{ cols}} \\ 0, & 0, & 0 \end{bmatrix} \begin{matrix} \} q \text{ rows} \\ \} (n-q) \text{ rows} \end{matrix} \qquad (10\text{-}56)$$

where $\Delta_1 \mathbf{F}$ is that part of $\Delta \mathbf{F}$ which consists of the first q rows of the latter. Letting

$$\mathbf{K}_q^* = \Lambda_q - \Lambda_{qd}, \qquad (10\text{-}57)$$

we can determine the control law $\mathbf{K}$ for independent modal control of the first q modal variables, $x_1^*, \ldots, x_q^*$. The matrix $\mathbf{K}_q^* \Delta_1 \mathbf{F}$ in the upper right corner of Eq. (10-56) means the last $(n - m)$ modal variables $x_{n-m+1}^*, \ldots, x_n^*$ will disturb the controlled modes. This disturbance is due to the nonideal measurement which senses not only the dominant modes but also the neglected modes. A similar discussion can be given for the second case, $r > m = q$; it is worked out in detail in [19].

Example 10-6 Given:

$$\mathbf{A} = \begin{bmatrix} -1 & 1 & -3 \\ 1 & -1 & -1 \\ 0 & 1 & -4 \end{bmatrix}, \quad \mathbf{T} = \begin{bmatrix} 1 & 2 & -1 \\ 3 & -2 & 0 \\ 1 & -1 & -1 \end{bmatrix},$$

so that

$$\mathbf{T}^{-1} = \frac{1}{9}\begin{bmatrix} 2 & 3 & -2 \\ 3 & 0 & -3 \\ -1 & 3 & -8 \end{bmatrix}, \quad \Lambda = \begin{bmatrix} -1 & 1 & 0 \\ 0 & -1 & 0 \\ 0 & 0 & -4 \end{bmatrix}.$$

For the measuring matrix $\mathbf{C}$,

$$\mathbf{C} = \begin{bmatrix} 0 & 1 & 0 \\ 0 & 0 & 1 \end{bmatrix},$$

we get

$$\mathbf{CT} = \mathbf{F} = \begin{bmatrix} 3 & -2 & 0 \\ 1 & -1 & -1 \end{bmatrix}, \quad \mathbf{F}_m = \begin{bmatrix} 3 & -2 \\ 1 & -1 \end{bmatrix}, \quad \mathbf{F}_m^{-1} = \begin{bmatrix} 1 & -2 \\ 1 & -3 \end{bmatrix},$$

and

$$\mathbf{F}_m^{-1}\mathbf{F} = \begin{bmatrix} 1 & 0 & 2 \\ 0 & 1 & 3 \end{bmatrix} = [\mathbf{I}_m, \Delta \mathbf{F}].$$

Also given is the controlling matrix $\mathbf{B}$,

$$\mathbf{B} = \begin{bmatrix} 3 & -2 \\ 1 & 2 \\ 0 & 1 \end{bmatrix},$$

so that

$$\mathbf{T}^{-1}\mathbf{B} = \mathbf{H} = \begin{bmatrix} 1 & 0 \\ 1 & -1 \\ 0 & 0 \end{bmatrix},$$

and hence

$$\mathbf{H}_r = \begin{bmatrix} 1 & 0 \\ 1 & -1 \end{bmatrix} \quad \text{and} \quad \mathbf{H}_r^{-1} = \begin{bmatrix} 1 & 0 \\ 1 & -1 \end{bmatrix}.$$

Let

$$\mathbf{K}_q^* = \begin{bmatrix} k_1^* & 1 \\ 0 & k_2^* \end{bmatrix}.$$

Then (by Eq. 10-57) the eigenvalues will be $\lambda_{1d} = -1 - k_1^*$, $\lambda_{2d} = -1 - k_2^*$, $\lambda_3 = -4$ (unchanged), and

$$\mathbf{K} = \mathbf{H}_r^{-1}\mathbf{K}_q^*\mathbf{F}_m^{-1} = \begin{bmatrix} k_1^* + 1 & -2k_1^* - 3 \\ k_1^* + 1 - k_2^* & -2k_1^* - 3 + 3k_2^* \end{bmatrix}.$$

To show the disturbance caused by the third mode in the first two controlled modes due to nonideal measurement, we reconsider the modal equation

$$\frac{d\mathbf{x}^*}{dt} = \Lambda\mathbf{x}^* - \begin{bmatrix} \mathbf{I}_r \\ 0 \end{bmatrix}\mathbf{K}^*[\mathbf{I}_m, \Delta\mathbf{F}]\mathbf{x}^*.$$

10-3　Modal control of lumped-parameter objects

Using the results obtained above, after multiplication and subtraction we get

$$\frac{d\mathbf{x}^*}{dt} = \begin{bmatrix} -1-k_1^* & 0 & -(2k_1^*+3) \\ 0 & -1-k_2^* & -3k_2^* \\ 0 & 0 & -4 \end{bmatrix} \mathbf{x}^*.$$

From this last equation it is clear that due to nonideal measurement the uncontrolled third mode x_3^* is contributing to the motion of the first two, even though the desired closed-loop eigenvalues have been achieved.

A situation complementary to the cases of nonideal measurement arises when Eq. (10–50) is satisfied while Eq. (10–54) does not hold. This situation corresponds to ideal measurement but nonideal control. We have the modal-domain relation:

$$\frac{d\mathbf{x}^*}{dt} = \Lambda\mathbf{x}^* - \begin{bmatrix} \mathbf{H}_r \\ \Delta\mathbf{H}_r \end{bmatrix} \mathbf{K}[\mathbf{F}_m, \mathbf{0}]\mathbf{x}^*$$

$$= \Lambda\mathbf{x}^* - \begin{bmatrix} \mathbf{I}_r \\ \Delta\mathbf{H} \end{bmatrix} \mathbf{K}^*[\mathbf{I}_m, \mathbf{0}]\mathbf{x}^*,$$

where

$$\mathbf{H}_r\mathbf{K}\mathbf{F}_m = \mathbf{K}^* \quad \text{or} \quad \mathbf{K} = \mathbf{H}_r^{-1}\mathbf{K}^*\mathbf{F}_m^{-1},$$

$$\Delta\mathbf{H} = \Delta\mathbf{H}_r\mathbf{H}_r^{-1}.$$

As before, modal control of up to q variables is realizable, where $q = r$ when $m > r$, and $q = m$ when $m < r$. No disturbance will appear in the controlled modes (the first q), while the controlled modes disturb a part of or all the remaining modes due to nonideal manipulator action. If the number of measurements m is greater than $r (= q)$, $(m - q)$ sensors may be removed without causing any loss in control quality. For $r > m = q$, with

$$\mathbf{K}^* = \begin{bmatrix} \mathbf{K}_q^* \\ \mathbf{0} \end{bmatrix}$$

we obtain

$$\begin{bmatrix} \mathbf{I}_r \\ \Delta\mathbf{H} \end{bmatrix} \begin{bmatrix} \mathbf{K}_q^* \\ \mathbf{0} \end{bmatrix} [\mathbf{I}_q, \mathbf{0}] = \begin{bmatrix} \mathbf{K}_q^* & \mathbf{0} \\ \mathbf{0} & \mathbf{0} \\ \Delta_1\mathbf{H}\mathbf{K}_q^* & \mathbf{0} \end{bmatrix},$$

where $\Delta_1\mathbf{H}$ is the part of $\Delta\mathbf{H}$ that consists of the first q columns of the latter. The disturbance, due to $\Delta_1\mathbf{H}\mathbf{K}_q^*$ in this case, will appear only in the last $(n - r)$ modal variables. In the limiting case when $r = m = q$,

$$\begin{bmatrix} \mathbf{I}_q \\ \Delta\mathbf{H} \end{bmatrix} \mathbf{K}_q^*[\mathbf{I}_q, \mathbf{0}] = \begin{bmatrix} \mathbf{K}_q^* & \mathbf{0} \\ \Delta\mathbf{H}\mathbf{K}_q^* & \mathbf{0} \end{bmatrix}.$$

Example 10-7 Consider a system represented by

$$A = \begin{bmatrix} -2 & 1 & 2 \\ -1 & -2 & 2 \\ -2 & 0 & 2 \end{bmatrix},$$

where the eigenvalues are $\pm j$ and -2. We have

$$T = \begin{bmatrix} 1 & 2 & 2 \\ 0 & 1 & -2 \\ 0 & 2 & 1 \end{bmatrix}, \quad T^{-1} = \frac{1}{5}\begin{bmatrix} 5 & 2 & -6 \\ 0 & 1 & 2 \\ 0 & -2 & 1 \end{bmatrix},$$

and the modified canonical form we derive is

$$T^{-1}AT = \begin{bmatrix} 0 & 1 & 0 \\ -1 & 0 & 0 \\ 0 & 0 & -2 \end{bmatrix}.$$

Let

$$C = \begin{bmatrix} 1 & 1 & 0 \\ 0 & 1 & 2 \end{bmatrix}$$

so that

$$F = CT = \begin{bmatrix} 1 & 3 & 0 \\ 0 & 5 & 0 \end{bmatrix}, \quad F_m = \begin{bmatrix} 1 & 3 \\ 0 & 5 \end{bmatrix}, \quad F_m^{-1} = \frac{1}{5}\begin{bmatrix} 5 & -3 \\ 0 & 1 \end{bmatrix}$$

with

$$B = \begin{bmatrix} 0 & 0 \\ 0 & 5 \\ 5 & 0 \end{bmatrix}.$$

We have

$$H = T^{-1}B = \begin{bmatrix} -6 & 2 \\ 2 & 1 \\ 1 & -2 \end{bmatrix}, \quad H_r = \begin{bmatrix} -6 & 2 \\ 2 & 1 \end{bmatrix}, \quad H_r^{-1} = \frac{1}{10}\begin{bmatrix} -1 & 2 \\ 2 & 6 \end{bmatrix},$$

and

$$HH_r^{-1} = \begin{bmatrix} 1 & 0 \\ 0 & 1 \\ -\frac{1}{2} & -2 \end{bmatrix} = \begin{bmatrix} I_r \\ \Delta H \end{bmatrix}.$$

Let

$$K_q^* = \begin{bmatrix} k_1^* & 1 \\ -1 & k_2^* \end{bmatrix}.$$

Then the control law is found to be

$$K = H_r^{-1} K_q^* F_m^{-1} = \frac{1}{50}\begin{bmatrix} -5k_1^* - 10 & 3k_1^* + 2k_2^* + 5 \\ 10k_1^* - 30 & -6k_1^* + 6k_2^* + 20 \end{bmatrix}.$$

If $k_1^* = 2$, $k_2^* = 2$, then

$$K = \begin{bmatrix} -0.4 & 0.3 \\ -0.2 & 0.4 \end{bmatrix},$$

and the three closed-loop eigenvalues will all be -2.

As in the previous example, we may evaluate the disturbance introduced by the nonideal conditions. From the modal equation for this case,

$$\frac{dx^*}{dt} = \Lambda x^* - \begin{bmatrix} I_r \\ \Delta H \end{bmatrix} K^* [I_m, 0] x^*,$$

we find after substitution and algebraic manipulation that

$$\frac{dx^*}{dt} = \begin{bmatrix} -k_1^* & 0 & 0 \\ 0 & -k_2^* & 0 \\ \frac{1}{2}k_1^* - 2 & 2k_2^* + \frac{1}{2} & -2 \end{bmatrix} x^*.$$

Again, the desired closed-loop eigenvalues have been achieved. However, due to the nonideal control ($\Delta H \neq 0$), the first and second modes contribute to the motion of the third mode.

It is worth comparing the discussion in the last paragraph of Example 10-6 with the foregoing paragraph, to ascertain how and why the disturbances are introduced in the two different cases. Note carefully in Example 10-7 that even though there are only two manipulators (see the **B**-matrix), the control is nonideal in that all three of the states in the modal domain experience manipulation. Thus, it is *not how many* controllers are introduced in this problem that causes the nonideal control, but *how and where* they are introduced. Similarly, in Example 10-6, although there were two measurements (see the **C**-matrix), they were nonideal. It was *how and where* the measurements were made that created the disturbance.

If measurement and control are both nonideal, controlled modes and uncontrolled modes will interact with each other. When $m = r = q$, for instance, the modal-domain matrix takes the form

$$\begin{bmatrix} I_q \\ \Delta H \end{bmatrix} [K_q^*][I_q \quad \Delta F] = \begin{bmatrix} K_q^* & K_q^* \cdot \Delta F \\ \Delta H \cdot K_q^* & \Delta H \cdot K_q^* \cdot \Delta F \end{bmatrix}. \quad (10\text{-}58)$$

An inspection of this result will show that due to the combined cross-coupling effect of ΔH and ΔF, the eigenvalues of the controlled system are not those we wanted. Therefore, as mentioned earlier, the proper design of measuring and controlling schemes (**B** and **C**) is of the utmost importance in minimizing the effects of coupling. Unless these conditions are met and the coupling effects are made small, modal control may not be effective. However, for given **B**- and **C**-matrices the interaction effect may be reduced by the addition of off-diagonal terms to the **K*** matrix.

It is possible to condition *all n* modes of a system (even when r and/or m are less than n) by a recursive application of the foregoing techniques. For each repetition, the eigenvalues in the Λ-matrix and the eigenvectors in the **T**-matrix must be recorded so that the previously uncontrolled modes are shifted up to the first q columns and rows of Λ and **T**, and the remaining modes are shifted down. The details of this repetitive approach are presented in [20].

An alternative design approach to modal control of systems when measurement and control are both nonideal involves use of the pseudo-inverse of a rectangular matrix (see Appendix). When the number of measurements m is smaller than the number of modal variables n, use of the right pseudoinverse will yield a minimum-norm measurement. Let

$$\mathbf{y} = \mathbf{Fx}^* \quad \text{and} \quad \mathbf{F} = \mathbf{CT},$$

where $\mathbf{y}$ is an m-vector and $\mathbf{x}^*$ is an n-vector, $m < n$. The "solution" $\mathbf{x}^{*0}$ of the previous equation,

$$\mathbf{x}^{*0} = \mathbf{F}^{RM}\mathbf{y}, \qquad \mathbf{F}^{RM} = \mathbf{F}'(\mathbf{FF}')^{-1}, \tag{10-59}$$

will exist when

$$\text{rank } \mathbf{F} = m,$$

and $\|\mathbf{x}^{*0}\|$ is minimal among all possible $\mathbf{x}^*$ which satisfy the original relation. The norm $\|\mathbf{x}^{*0}\|$ is, as in Chapter 4, the square root of the sum of the squares of the elements of $\mathbf{x}^{*0}$.

Suppose that we produce a control vector $\mathbf{u}^*$ in the modal domain based on the "measured" n-vector $\mathbf{x}^{*0}$,

$$\mathbf{u}^* = -\mathbf{K}^* \mathbf{x}^{*0}. \tag{10-60}$$

The n-vector $\mathbf{u}^*$ must be converted into the r-vector $\mathbf{u}$, where $r < n$. The relation to be satisfied is

$$\mathbf{Hu} = \mathbf{u}^*.$$

This equation can be "solved" for the unknown vector $\mathbf{u}$ by the left inverse of $\mathbf{H}$ (see Appendix):

$$\mathbf{u}^0 = \mathbf{H}^{LM}\mathbf{u}^*, \qquad \mathbf{H}^{LM} = (\mathbf{H}'\mathbf{H})^{-1}\mathbf{H}'. \tag{10-61}$$

The solution has the property that $\|\mathbf{Hu}^0 - \mathbf{u}^*\|$ will be minimal among all $\mathbf{u}$ that satisfy $\mathbf{Hu} = \mathbf{u}^*$.

We now specify

$$\mathbf{K}^* = \mathit{\Lambda} - \mathit{\Lambda}_d. \tag{10-62}$$

Then the control law $\mathbf{K}$ in the original state space is given by

$$\mathbf{K} = \mathbf{H}^{LM}\mathbf{K}^*\mathbf{F}^{RM} = \mathbf{H}^{LM}(\mathit{\Lambda} - \mathit{\Lambda}_d)\mathbf{F}^{RM}. \tag{10-63}$$

In the ideal cases (ideal measurement or ideal control) this approach agrees with what was discussed before. Also, as in the nonideal case treated in the preceding paragraphs, this approach will not eliminate the adverse effects caused by cross coupling. If the cross-coupling effect is not serious, the approach based on the pseudoinverse will give a control that is satisfactory to the extent the former approach will be. Since control quality

Modal control of lumped-parameter objects

depends on specific conditions (**A**, **B**, **C**, and n, m, r, q), it is usually not possible to draw a general conclusion as to which approach is better for the nonideal case.

Example 10-8 Consider a control object

$$\mathbf{A} = \begin{bmatrix} -2 & 1 & 0 \\ 2 & -3 & 0 \\ 0 & 4 & -5 \end{bmatrix}, \quad \mathbf{B} = \begin{bmatrix} 1 & 0 \\ 0 & 1 \\ 0 & 0 \end{bmatrix}, \quad \mathbf{C} = \begin{bmatrix} 1 & 0 & 0 \\ 0 & 0 & 1 \end{bmatrix}.$$

For this system

$$\mathbf{T} = \begin{bmatrix} 1 & -\tfrac{1}{2} & 0 \\ 1 & 1 & 0 \\ 1 & 4 & 1 \end{bmatrix}, \quad \mathbf{T}^{-1} = \frac{2}{3}\begin{bmatrix} 1 & \tfrac{1}{2} & 0 \\ -1 & 1 & 0 \\ 3 & -\tfrac{9}{2} & \tfrac{3}{2} \end{bmatrix}, \quad \Lambda = \begin{bmatrix} -1 & 0 & 0 \\ 0 & -4 & 0 \\ 0 & 0 & -5 \end{bmatrix}.$$

For control-law design by the first method,

$$\mathbf{H} = \mathbf{T}^{-1}\mathbf{B} = \frac{2}{3}\begin{bmatrix} 1 & \tfrac{1}{2} \\ -1 & 1 \\ 3 & -\tfrac{9}{2} \end{bmatrix}, \quad \mathbf{F} = \mathbf{CT} = \begin{bmatrix} 1 & -\tfrac{1}{2} & 0 \\ 1 & 4 & 1 \end{bmatrix}.$$

Using

$$\mathbf{K}_q^* = \begin{bmatrix} k_1^* & 0 \\ 0 & k_2^* \end{bmatrix},$$

we find that **K** becomes

$$\mathbf{K} = \mathbf{H}_q^{-1}\mathbf{K}_q^*\mathbf{F}_q^{-1} = \frac{2}{9}\begin{bmatrix} 4k_1^* + 0.5k_2^* & 0.5k_1^* - 0.5k_2^* \\ 4k_1^* - k_2^* & 0.5k_1^* + k_2^* \end{bmatrix}.$$

The modal-domain control matrix takes the form:

$$(\mathbf{HH}_q^{-1})\mathbf{K}_q^*(\mathbf{F}_q^{-1}\mathbf{F}) = \begin{bmatrix} k_1^* & 0 & \tfrac{1}{9}k_1^* \\ 0 & k_2^* & \tfrac{2}{9}k_2^* \\ -k_1^* & -4k_2^* & -\tfrac{1}{9}k_1^* - \tfrac{8}{9}k_2^* \end{bmatrix} = \mathbf{T}^{-1}\mathbf{BKCT},$$

from which we can clearly see the disturbance in the first two modes by the third due to nonideal measurement and the disturbance in the third by the first two due to nonideal control. Further, the desired closed-loop eigenvalues have not been achieved.

For a control-law design by the second, or pseudoinverse, method;

$$\mathbf{H}^{LM} = (\mathbf{H}'\mathbf{H})^{-1}\mathbf{H}' = \frac{1}{9}\begin{bmatrix} 9.5 & -2.5 & 0.5 \\ 6.5 & -1 & -2.5 \end{bmatrix},$$

$$\mathbf{F}^{RM} = \mathbf{F}'(\mathbf{FF}')^{-1} = \frac{1}{43}\begin{bmatrix} 38 & 4.5 \\ -10 & 9 \\ 2 & 2.5 \end{bmatrix},$$

and

$$\mathbf{K}^* = \begin{bmatrix} k_1^* & 0 & 0 \\ 0 & k_2^* & 0 \\ 0 & 0 & k_3^* \end{bmatrix}.$$

By Eq. (10-63), the control matrix in the original state space is given by:

$$\mathbf{K} = \frac{1}{774} \begin{bmatrix} 722k_1^* + 50k_2^* + 2k_3^* & 85.5k_1^* - 45k_2^* + 2.5k_3^* \\ 494k_1^* + 20k_2^* - 10k_3^* & 58.5k_1^* - 18k_2^* - 12.5k_3^* \end{bmatrix}.$$

To investigate the coupling disturbances due to nonideal condititions, we evaluate (after lengthy algebra)

$$\mathbf{M} = \mathbf{T}^{-1}\mathbf{BKCT}$$

$$= \begin{bmatrix} (0.933k_1^* + 0.005k_2^* - 0.006k_3^*) & (-0.022k_1^* - 0.212k_2^* - 0.012k_3^*) & (0.099k_1^* - 0.047k_2^* - 0.003k_3^*) \\ (-0.220k_1^* - 0.003k_2^* - 0.023k_3^*) & (0.005k_1^* + 0.106k_2^* - 0.047k_3^*) & (-0.023k_1^* + 0.023k_2^* - 0.013k_3^*) \\ (-0.055k_1^* + 0.055k_2^* + 0.099k_3^*) & (0.001k_1^* - 0.212k_2^* + 0.198k_3^*) & (-0.006k_1^* - 0.047k_2^* - 0.042k_3^*) \end{bmatrix}.$$

It would appear that the desired eigenvalues have not been achieved *and* that all modes are coupled. However, the off-diagonal terms do not predominate, and a reasonably close approximation to Λ_d can be obtained. To illustrate this, suppose that ideally we desire a star pattern with three repeated roots at $s = -6$. Then, by Eq. (10-62), we try

$$\mathbf{K}^* = \begin{bmatrix} 5 & 0 & 0 \\ 0 & 2 & 0 \\ 0 & 0 & 1 \end{bmatrix},$$

and evaluate $\mathbf{A}_d$ in $\dot{\mathbf{x}}^* = \mathbf{A}_d \mathbf{x}^*$ as

$$\mathbf{A}_d = \Lambda - \mathbf{M} = \begin{bmatrix} -5.669 & 1.536 & -0.398 \\ 1.129 & -4.190 & 0.082 \\ 0.166 & 0.221 & -4.834 \end{bmatrix}.$$

The closed-loop roots are then given by

$$\det(s\mathbf{I} - \mathbf{A}_d) = s^3 + 14.693s^2 + 69.725s + 105.766$$
$$= (s + 5.963)(s + 5.42)(s + 3.188),$$

and we see that a reasonable design has been achieved. An analog simulation of this system [19] does indeed indicate that a star pattern is obtainable by suitable choice of the elements in the $\mathbf{K}^*$-matrix. The $\mathbf{K}^*$-matrix of the pseudoinverse approach and the $\mathbf{K}_q^*$-matrix of the first method are, of course, different. If necessary, they may be related by somewhat lengthy algebraic manipulations which will be omitted here.

10-4 MODAL CONTROL OF DISTRIBUTED-PARAMETER SYSTEMS

Application of modal control to linear distributed-parameter objects has been reported by Gould, Lasso, and Schlaefer [14, 15, and 16]. In order to illustrate the principle, a heat conduction system (Fig. 10-15) will be considered next. Using normalized quantities for temperature y, distance z, and time t, we can describe the process by a normalized diffusion equation

$$\frac{\partial^2 y(t, z)}{\partial z^2} = \frac{\partial y(t, z)}{\partial t}. \qquad (10\text{-}64)$$

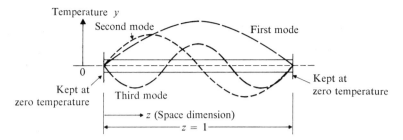

Fig. 10-15 A diffusion process and its space eigenmodes.

If both ends of the object are kept at zero temperature, $y(t, 0) = 0$, $y(t, 1) = 0$, the solution of (10-64) will take the form:

$$y(t, z) = \sum_{i=1}^{\infty} x_i^*(t) \sin(i\pi z), \tag{10-65}$$

or

$$y(t, z) = \sum_{i=1}^{\infty} x_i^*(t) f_i(z), \quad f_i(z) = \sin(i\pi z), \tag{10-66}$$

where $x_i^*(t)$ is a function of time only. In general, the function $f_i(z)$ appearing in the solution is called a *space eigenfunction*. Because of the prescribed boundary conditions (zero temperature at both ends), the space eigenfunctions of our problem are given by sine functions; the first three of them are shown in Fig. 10-15.

To control the amplitude (function of time) of the space modes, a distributed heating or cooling input $u(t, z)$ will be applied to the object, so that

$$\frac{\partial y(t, z)}{\partial t} = \frac{\partial^2 y(t, z)}{\partial z^2} + u(t, z). \tag{10-67}$$

In independent modal control, the amplitude of the first n modes, $x_i^*(t)$, $i = 1, \ldots, n$, will be controlled separately. As shown in Fig. 10-16, the control will require a mode analyzer, a diagonal controller matrix, and a synthesizer.

If measurement of a distributed response $y(t, z)$ is available (perhaps by a quick scanning of thermal radiation), the analyzer will compute Fourier coefficients of the measured distribution by the equation:

$$\int_0^1 f_i(z) y(t, z) dz = \sum_{j=1}^{\infty} x_j^*(t) \int_0^1 f_i(z) f_j(z) dz = c_i x_i^*(t). \tag{10-68}$$

The reduction in Eq. (10-68) is due to the following property which an independent set of space eigenmodes will satisfy:

$$\int_0^1 f_i(z) f_j(z) dz = \begin{cases} c_i \text{ (constant)} & \text{if } i = j, \\ 0 & \text{for } i \neq j. \end{cases} \tag{10-69}$$

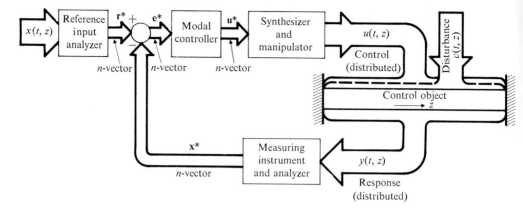

Fig. 10-16 *n*-mode control of a distributed-parameter object.

The first n modes, $x_i^*(t)$, $i = 1, \ldots, n$, detected by Eq. (10–68), will produce an error signal

$$e_i^*(t) = r_i^*(t) - x_i^*(t), \tag{10-70}$$

where $r_i^*(t)$ is a Fourier coefficient of the desired distribution $r(z, t)$. If, for example, a constant uniform temperature θ_c is desired,

$$r(t, z) = \theta_c = \frac{4\theta_c}{\pi} \left\{ \sin(\pi z) + \frac{\sin(3\pi z)}{3} + \frac{\sin(5\pi z)}{5} + \cdots \right\},$$

then

$$r_1^* = \frac{4\theta_c}{\pi}, \quad r_2^* = 0, \quad r_3^* = \frac{4\theta_c}{3\pi}, \quad r_4^* = 0, \quad \text{etc.}$$

The error signal $e_i(t)$ is fed into a proportional controller of gain k_i to produce $u_i^*(t) = k_i^*(r_i^* - x_i^*)$, by which a distributed controlling input $u(t, z)$ is constructed as

$$u(t, z) = \sum_{i=1}^{\infty} u_i^*(t) f_i(z). \tag{10-71}$$

Restated, the set of relations becomes:

plant: $\quad \dfrac{\partial y(t, z)}{\partial t} = \dfrac{\partial^2 y(t, z)}{\partial z^2} + u(t, z),$

distributed response: $\quad y(t, z) = \sum_{i=1}^{\infty} x_i^*(t) f_i(z),$

distributed control: $\quad u(t, z) = \sum_{i=1}^{\infty} u_i^*(t) f_i(z),$

reference input: $\quad r(t, z) = \sum_{i=1}^{\infty} r_i^*(t) f_i(z),$

proportional control: $\quad u_i^*(t) = k_i^* \bigl(r_i^*(t) - x_i^*(t)\bigr),$

where

$$k_i^* > 0 \quad \text{for} \quad i = 1, \ldots, n, \qquad k_i^* = 0 \quad \text{for} \quad i > n.$$

Substituting Fourier-series expressions for each variable into the plant equation, we obtain

$$\sum_{i=1}^{\infty} \frac{dx_i^*}{dt} f_i(z) = \sum_{i=1}^{\infty} x_i^*(-i^2\pi^2)f_i(z) + \sum_{i=1}^{\infty} k_i^*(r_i^* - x_i^*)f_i(z), \qquad (10\text{–}72)$$

where $f_i(z)$ is $\sin(i\pi z)$ so that $\partial^2 f_i(z)/\partial z^2 = -(i\pi)^2 f_i(z)$. To make use of the orthogonal property (Eq. 10–69) of the space eigenmodes, all terms of Eq. (10–72) will be multiplied by $f_j(z)$ and then integrated with respect to z from 0 to 1. The result is:

$$\frac{dx_i^*(t)}{dt} = -(i\pi)^2 x_i^*(t) + k_i^*\big(r_i^*(t) - x_i^*(t)\big), \qquad (10\text{–}73)$$

where

$$k_i^* > 0 \qquad \text{for} \qquad i = 1, \ldots, n.$$

Therefore, the closed-loop poles are

$$p_{ci} = p_i - k_i^*,$$

where

$$p_i = i\text{th open-loop pole} = -(i\pi)^2.$$

Equation (10–73) thus represents an independent control of the first n space modes.

The practical limitations of measuring and manipulating instruments will make it difficult to realize the control just proposed. To investigate the nature of these difficulties, and to formulate a practical design approach, we idealize the plant by assuming it to be "band limited." (A "band-limited" time signal can be reconstructed from its sampled data, as shown in Chapter 11.) This idealization takes the form of terminating the infinite Fourier series after n terms:

$$y(t, z) = \sum_{i=1}^{n} x_i^*(t) f_i(z), \qquad u(t, z) = \sum_{i=1}^{n} u_i^*(t) f_i(z). \qquad (10\text{–}74)$$

When $y(t, z)$ is band limited, a set of sensing elements sampled in space, at $z = z_1, \ldots, z_n$, are necessary and sufficient for the mode detection. Let

$$y_k = y(t, z_k), \qquad f_i^k = f_i(z_k).$$

Then

$$y_k = f_1^k x_1^* + f_2^k x_2^* + \cdots + f_n^k x_n^* = [f_1^k, f_2^k, \ldots, f_n^k]\mathbf{x}^*.$$

Introducing the matrix notation,

$$\mathbf{y} = \begin{bmatrix} y_1 \\ \vdots \\ y_n \end{bmatrix}, \qquad \mathbf{x}^* = \begin{bmatrix} x_1^* \\ \vdots \\ x_n^* \end{bmatrix}, \qquad \mathbf{F} = \begin{bmatrix} f_1^1 & \cdots & f_n^1 \\ \vdots & & \vdots \\ f_1^n & \cdots & f_n^n \end{bmatrix},$$

we have

$$\mathbf{y}(t) = \mathbf{F}\mathbf{x}^*(t), \tag{10-75}$$

so that $\mathbf{x}(t)$ can be detected by

$$\mathbf{x}^*(t) = \mathbf{F}^{-1}\mathbf{y}(t). \tag{10-76}$$

The location of the sensors, $z_1, \ldots, z_n$, must be properly chosen so that $\mathbf{F}^{-1}$ will exist.

In general, it is almost impossible to design a manipulator whose output distribution is $f_i(z)$, $i = 1, \ldots, n$. Let us suppose that each manipulator has an output distribution $w_i(z)$. In our example, $w_i(z)$ represents heat-flux distribution of a heating unit, such as a thermal radiator, an electric heater element, or a burner. Since our plant is assumed to be band limited, we require that each $w_i(z)$ be expressible in terms of n space eigenmodes:

$$w_i(z) = h_{1i} f_1(z) + \cdots + h_{ni} f_n(z) = [f_1, \ldots, f_n] \begin{bmatrix} h_{1i} \\ \vdots \\ h_{ni} \end{bmatrix}.$$

Collecting $w_1(z), \ldots, w_n(z)$ in a row vector $\mathbf{w}(z)$, we have

$$\mathbf{w}(z) = [w_1(z), \ldots, w_n(z)] = [f_1, \ldots, f_n]\mathbf{H} = \mathbf{f}\mathbf{H},$$

where

$$\mathbf{f}(z) = [f_1, f_2, \ldots, f_n], \qquad \mathbf{H} = \begin{bmatrix} h_{11} & \cdots & h_{1n} \\ \vdots & & \vdots \\ h_{n1} & \cdots & h_{nn} \end{bmatrix}.$$

The intensity of each distribution $w_i(z)$ is modulated by a time signal $m_i(t)$, so that the total controlling input will be given by

$$u(t, z) = \sum_{i=1}^{n} m_i(t) w_i(z) = [w_1, \ldots, w_n] \begin{bmatrix} m_1 \\ \vdots \\ m_n \end{bmatrix} = \mathbf{w}\mathbf{m},$$

where $\mathbf{m}$ is an n-vector with column elements $m_1(t), \ldots, m_n(t)$.

The control $u(t, z)$ must be equal to

$$\sum_{j=1}^{n} f_j(z) u_j^*(t) = \sum_{j=1}^{n} f_j(z) u_j^*(t) = [f_1, \ldots, f_n] \begin{bmatrix} u_1^* \\ \vdots \\ u_n^* \end{bmatrix} = \mathbf{f}\mathbf{u}^*, \tag{10-77}$$

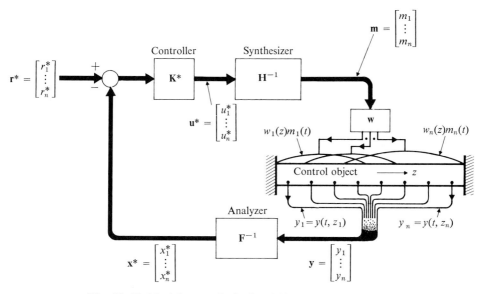

Fig. 10-17 Modal control of a band-limited control object.

and $\mathbf{w} = \mathbf{fH}$. We therefore get

$$\mathbf{fHm} = \mathbf{fu}^*$$

or

$$\mathbf{m}(t) = \mathbf{H}^{-1}\mathbf{u}^*(t). \qquad (10\text{--}78)$$

In this expression $\mathbf{u}^*(t)$ is the n-vector output of the matrix controller (Fig. 10-17) and $\mathbf{H}^{-1}$ is the synthesizer matrix. For $\mathbf{H}^{-1}$ to exist, n manipulators must be designed in such a way that the row vectors of $\mathbf{H}$ are linearly independent. Each manipulator output is indicated by an arc in Fig. 10-17, where the arc may symbolize a radiant heater. The action of the manipulators must be linear in the sense that the total control $u(t, z)$ is given by the sum of all manipulator outputs.

It would be desirable in many applications to simplify the instrumentation by reducing the number of control loops to some number m (such as 3 or 4, for instance). The dimension of the vectors $\mathbf{x}$, $\mathbf{r}$, and $\mathbf{u}$ (Fig. 10-17) then becomes m. The number of manipulators may be reduced to m, provided that the distribution $w_i(z)$ of each element is designed broadly enough to assure controllability. There will be no problem if n sensors are still installed.

If, however, the number of sensors is also reduced to m, a problem arises. A set of m sensors will have an m-vector $\mathbf{y}$ as its output, while $\mathbf{x}$ is an n-vector, $n > m$. Therefore $\mathbf{F}$ in the vector equation

$$\mathbf{y} = \mathbf{Fx}^*$$

becomes an $m \times n$ rectangular matrix, which has no inverse. A pseudoinverse (see Appendix) must be introduced to yield a unique solution for $\mathbf{x}^*$. Using the right pseudoinverse as in Eq. (10–59), we have

$$\mathbf{x}^{*0} = \mathbf{F}^{RM}\mathbf{y}, \qquad \mathbf{F}^{RM} = \mathbf{F}'(\mathbf{FF}')^{-1}. \tag{10–79}$$

This will introduce an error in the mode measurement. To reduce this error, careful engineering study of measuring points (relative to the space mode pattern) seems desirable.

10–5 LINEAR OPTIMAL FEEDBACK CONTROL FOR QUADRATIC CRITERIA

Design of an optimal control depends on the choice of a performance index to be minimized (or maximized). The following discussion will be limited to minimization of an index in quadratic form:

$$J = \int_0^\infty \left(\sum_{i=1}^m w_i y_i^2 + u^2 \right) dt, \tag{10–80}$$

where y_i is an element of the response vector $\mathbf{y} = \mathbf{Cx}$ of a linear, stationary, lumped control object

$$\frac{d\mathbf{x}}{dt} = \mathbf{Ax} + \mathbf{b}u,$$

and u is a scalar input. The weighting factors w_i in Eq. (10–80) are nonnegative scalars, and at least one is nonzero. It is known [21] that a linear control law

$$u_0 = -\mathbf{k}'\mathbf{x} \tag{10–81}$$

will generate the optimal control. The discussion which follows is based on Salukvanze's approach [24] in which a Lyapunov function plays an important role. Although some other approaches [12, 21, 22, 23] can provide a direct algorithm to compute $\mathbf{k}'$, while the Salukvanze method does not, the optimal-control theorem based on a Lyapunov function is instructive as well as flexible enough for application to time-varying linear systems and/or systems with constraints. The problem statement in the general case is:

Given: (a) a control object described by

$$\frac{dx_i(t)}{dt} = f_i(\mathbf{x}, \mathbf{u}, t), \qquad i = 1, \ldots, n, \tag{10–82}$$

(b) a constraint in admissible control $\mathbf{u}$, $\mathbf{u}(t) \in U$ for all t, and

(c) a performance index J in terms of a positive definite function $f_0(\mathbf{x}, \mathbf{u}, t)$,

$$J = \int_0^\infty f_0(\mathbf{x}, \mathbf{u}, t) dt. \tag{10–83}$$

Find: an optimal control $\mathbf{u} = \mathbf{u}^o$ for which J is minimal.

10-5   Linear optimal feedback control for quadratic criteria

The sufficient condition for a control $\mathbf{u} = \mathbf{u}^o$ to be optimal is expressed by the following equation:

$$\min_{\mathbf{u} \in U} H(\mathbf{u}) = H(\mathbf{u}^o) = 0, \tag{10-84}$$

where

$$H(\mathbf{u}) = \left(\frac{dV}{dt}\right)_{\mathbf{u}} + f_0(\mathbf{x}, \mathbf{u}, t). \tag{10-85}$$

$V(t)$ in Eq. (10–85) is a positive definite Lyapunov function related to the control object, and $(dV/dt)_{\mathbf{u}}$ is its time derivative along the control $\mathbf{u}$. A rigorous statement and proof of the theorem of Eq. (10–84) appears in [25]. Since $H(\mathbf{u}^o) = 0$,

$$\left(\frac{dV}{dt}\right)_{\mathbf{u}^o} = -f_0(\mathbf{u}^o, \mathbf{x}, t)$$

and

$$\min J = \int_0^\infty f_0(\mathbf{u}^o, \mathbf{x}, t)\, dt = V(\mathbf{x}_0) - V(\mathbf{x}(\infty)),$$

where $V(\mathbf{x}(\infty))$ will vanish when a system under optimal control moves to the origin in state space, since, then, $\mathbf{x}(\infty) = \mathbf{0}$. We must have $\mathbf{x}(\infty) = \mathbf{0}$ when Eq. (10–80) is chosen as a performance index. Therefore, for our problem, the Lyapunov function represents a minimal value of J for each initial state $\mathbf{x}_0$ in the state space.

To solve the problem for a linear lumped-parameter object, let us assume a quadratic form for the Lyapunov function:

$$V(\mathbf{x}) = \mathbf{x}'\mathbf{Y}\mathbf{x}, \tag{10-86}$$

where $\mathbf{Y}$ is a positive definite symmetric matrix (Section 4–3) yet to be determined. Its time derivative along a scalar control u is

$$\left(\frac{dV}{dt}\right)_u = \frac{d\mathbf{x}'}{dt}\mathbf{Y}\mathbf{x} + \mathbf{x}'\mathbf{Y}\frac{d\mathbf{x}}{dt},$$

where

$$\frac{d\mathbf{x}}{dt} = \mathbf{A}\mathbf{x} + \mathbf{b}u.$$

Thus

$$\frac{d\mathbf{x}'}{dt} = \mathbf{x}'\mathbf{A}' + \mathbf{b}'u,$$

and

$$\left(\frac{dV}{dt}\right)_u = (\mathbf{x}'\mathbf{A}' + \mathbf{b}'u)\mathbf{Y}\mathbf{x} + \mathbf{x}'\mathbf{Y}(\mathbf{A}\mathbf{x} + \mathbf{b}u)$$

$$= \mathbf{x}'(\mathbf{A}'\mathbf{Y} + \mathbf{Y}\mathbf{A})\mathbf{x} + (\mathbf{b}'\mathbf{Y}\mathbf{x} + \mathbf{x}'\mathbf{Y}\mathbf{b})u. \tag{10-87}$$

Since **Y** is symmetric, $\mathbf{Y}' = \mathbf{Y}$; hence $\mathbf{x}'\mathbf{Yb} = \mathbf{b}'\mathbf{Y}'\mathbf{x} = \mathbf{b}'\mathbf{Yx}$, and the second term on the right-hand side of Eq. (10–87) reduces to $2\mathbf{b}'\mathbf{Yx}u$.

For the performance index of Eq. (10–80) we have, by definition of J,

$$-u^2 + f_0(\mathbf{x}, u) = \mathbf{y}'\mathbf{Wy} = (\mathbf{Cx})'\mathbf{W}(\mathbf{Cx}) = \mathbf{x}'\mathbf{C}'\mathbf{WCx},$$

where **W** is an $m \times m$ diagonal matrix having weighting factors $w_1, \ldots, w_m$ as its diagonal elements. With $\mathbf{C}'\mathbf{WC} = \mathbf{F}_x$, $\mathbf{F}_x$ is an $n \times n$ symmetric matrix, and

$$-u^2 + f_0(\mathbf{x}, u) = \mathbf{x}'\mathbf{F}_x\mathbf{x}. \tag{10–88}$$

Equations (10–87) and (10–88) are now substituted into Eq. (10–85) to obtain

$$H(u) = \mathbf{x}'(\mathbf{A}'\mathbf{Y} + \mathbf{YA} + \mathbf{F}_x)\mathbf{x} + 2\mathbf{b}'\mathbf{Yx}u + u^2. \tag{10–89}$$

Letting $\partial H(u)/\partial u = 0$, we find that $u = u^\circ$ for the minimum of $H(u)$ is

$$u^\circ = -\mathbf{b}'\mathbf{Yx}. \tag{10–90}$$

This will yield a feedback control law,

$$u^\circ = -\mathbf{k}'\mathbf{x}$$

if **Y** is known so that

$$\mathbf{k}' = \mathbf{b}'\mathbf{Y}. \tag{10–91}$$

We apply the second condition of Eq. (10–84), that is

$$H(u^\circ) = 0, \tag{10–92}$$

to derive a condition for **Y**. Substituting Eq. (10–90) into Eq. (10–89), we obtain

$$H(u^\circ) = \mathbf{x}'(\mathbf{A}'\mathbf{Y} + \mathbf{YA} + \mathbf{F}_x)\mathbf{x} + (2\mathbf{b}'\mathbf{Yx} + u^\circ)u^\circ,$$

where $u^\circ = -\mathbf{b}'\mathbf{Yx}$ so that the second term on the right-hand side of the last equation reduces to

$$-(2\mathbf{b}'\mathbf{Yx} + u^\circ)u^\circ = \mathbf{b}'\mathbf{Yxb}'\mathbf{Yx} = \mathbf{x}'\mathbf{Ybb}'\mathbf{Yx}$$

and

$$H(u^\circ) = \mathbf{x}'(\mathbf{A}'\mathbf{Y} + \mathbf{YA} - \mathbf{Ybb}'\mathbf{Y} + \mathbf{F}_x)\mathbf{x} = 0.$$

Since **x** is an arbitrary state vector, for $H(u^\circ) = 0$ in the last expression to be true, we must have

$$\mathbf{A}'\mathbf{Y} + \mathbf{YA} - \mathbf{Ybb}'\mathbf{Y} + \mathbf{F}_x = 0. \tag{10–93}$$

Equation (10-93) must be "solved" for **Y**. From Eq. (10-93) can be obtained nonlinear relations for the elements of **Y**. From these relations we choose

10-5 Linear optimal feedback control for quadratic criteria

a positive definite **Y**, and this will give the required control law **k**' of Eq. (10-91). The procedure is illustrated in the following example.

Example 10-9 Let us consider a purely inertial object,

$$\frac{d}{dt}\begin{bmatrix} x_1 \\ x_2 \end{bmatrix} = \begin{bmatrix} 0 & 1 \\ 0 & 0 \end{bmatrix}\begin{bmatrix} x_1 \\ x_2 \end{bmatrix} + \begin{bmatrix} 0 \\ 1 \end{bmatrix} u,$$

where x_1, x_2, u are the displacement, velocity, and force, respectively, and the mass is unity. Suppose that the performance index to be minimized is given as

$$J = \int_0^\infty (x_1^2 + u^2)\,dt,$$

so that

$$f_0(\mathbf{x}, u) = x_1^2 + u^2;$$

then, for $\mathbf{C} = \mathbf{I}$

$$\mathbf{F}_x = \begin{bmatrix} 1 & 0 \\ 0 & 0 \end{bmatrix}.$$

Equation (10-93) for the three unknown elements a, b, and c of

$$\mathbf{Y} = \begin{bmatrix} a & b \\ b & c \end{bmatrix}$$

is

$$\begin{bmatrix} 0 & 0 \\ 1 & 0 \end{bmatrix}\begin{bmatrix} a & b \\ b & c \end{bmatrix} + \begin{bmatrix} a & b \\ b & c \end{bmatrix}\begin{bmatrix} 0 & 1 \\ 0 & 0 \end{bmatrix}$$

$$- \begin{bmatrix} a & b \\ b & c \end{bmatrix}\begin{bmatrix} 0 \\ 1 \end{bmatrix}[0 \ 1]\begin{bmatrix} a & b \\ b & c \end{bmatrix} + \begin{bmatrix} 1 & 0 \\ 0 & 0 \end{bmatrix} = \begin{bmatrix} 0 & 0 \\ 0 & 0 \end{bmatrix}.$$

This reduces to

$$\begin{bmatrix} -b^2 + 1 & a - bc \\ a - bc & 2b - c^2 \end{bmatrix} = \mathbf{0}.$$

We thus have

$$1 - b^2 = 0, \qquad 2b - c^2 = 0, \qquad a - bc = 0,$$

from which the following set,

$$a = \sqrt{2}, \qquad b = 1, \qquad c = \sqrt{2},$$

is chosen, since the other set does not yield a real, positive definite **Y**. The feedback control law is

$$\mathbf{k}' = \mathbf{b}'\mathbf{Y} = [0 \ 1]\begin{bmatrix} a & b \\ b & c \end{bmatrix} = [b \ c] = [1 \ \sqrt{2}].$$

The Lyapunov function for this example is, by Eq. (10-86),

$$V = x_1^2 + \sqrt{2}\,x_1 x_2 + x_2^2.$$

456 Multivariable control systems

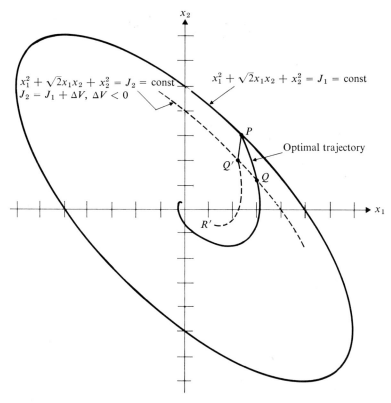

Fig. 10-18 Lyapunov contour for optimal trajectory.

The ellipse in Fig. 10-18 is the locus of initial states $\mathbf{x}_0$ starting from which, with optimal control, the performance index will take the value

$$J = J_1 = \text{const}.$$

The ellipse comes directly from the equation;

$$x_1^2 + \sqrt{2}\, x_1 x_2 + x_2^2 = J_1 = \text{const}.$$

To see why Eq. (10-84) yields an optimal control (that is, J_1 is minimized for u°), let us suppose that the trajectory PQ in the figure is an optimal trajectory for an initial state P. The value of an ellipse

$$x_1^2 + \sqrt{2}\, x_1 x_2 + x_2^2 = J_2$$

that passes through state Q must be such that for

$$J_1 = \int_0^\infty f_0(u^\circ)\, dt, \qquad J_2 = \int_{t_1}^\infty f_0(u^\circ)\, dt,$$

10-5 Linear optimal feedback control for quadratic criteria

so that

$$\Delta V = J_2 - J_1 = -\int_0^{\Delta t_1} f_0(u^\circ)\, dt \approx -f_0(u^\circ)\Delta t_1$$

or

$$\Delta V + f_0(u^\circ)\Delta t_1 = 0, \qquad (10\text{-}94)$$

where Δt_1 is the time required for the motion from P to Q. At the limit $\Delta t_1 \to dt$, $\Delta V \to dV$, Eq. (10-94) yields the condition:

$$H(u^\circ) = \left(\frac{dV}{dt}\right)_{u^\circ} + f_0(u^\circ) = 0. \qquad (10\text{-}95)$$

This is the right-hand side of Eq. (10-84).

To derive the first part of the conditions of Eq. (10-84), we consider a suboptimal trajectory (PQ' in Fig. 10-18), with a control $\hat{u}$ which is not optimal. If Δt_2 is the time required for the motion from P to Q', there must exist an inequality

$$f_0(\hat{u})\Delta t_2 > f_0(u^\circ)\Delta t_1, \qquad (10\text{-}96)$$

because otherwise a control equal to or better than the optimal could be realized by continuing the optimal trajectory $Q'R'$ from point Q'. Adding ΔV to both sides of Eq. (10-96), and applying Eq. (10-94) to the right-hand side, we get:

$$\Delta V + f_0(\hat{u})\Delta t_2 > \Delta V + f_0(u^\circ)\Delta t_1 = 0. \qquad (10\text{-}97)$$

Letting $\Delta V \to dV$ and $\Delta t_2 \to dt$ in Eq. (10-97), we find that:

$$\Delta H(\hat{u}) = \left(\frac{dV}{dt}\right)_{\hat{u}} + f_0(\hat{u}) > 0. \qquad (10\text{-}98)$$

This leads to the minimum condition of Eq. (10-84).

It is also possible to deal with the more general case in which the quadratic performance index is specified as:

$$J = \int_0^\infty \{\mathbf{y}'\mathbf{F}_y\mathbf{y} + \mathbf{u}'\mathbf{F}_u\mathbf{u}\}dt. \qquad (10\text{-}99)$$

The weighting matrix $\mathbf{F}_u$ for control vector $\mathbf{u}$ must be positive definite, but the weighting matrix $\mathbf{F}_y$ for the response vector $\mathbf{y}$ may be either positive definite or positive semidefinite. In general, the weight on the squared control u^2 is a restriction designed to avoid excessively large magnitudes of control (which may cause saturation). The control object for this general case is considered to be:

$$\frac{d}{dt}\mathbf{x} = \mathbf{A}\mathbf{x} + \mathbf{B}\mathbf{u}, \qquad \mathbf{y} = \mathbf{C}\mathbf{x}.$$

458 Multivariable control systems

The optimal state-vector feedback control law for this case is given by:

$$\mathbf{u}^o = -\mathbf{Kx},$$

where $\mathbf{K}$ is the $r \times n$ gain matrix,

$$\mathbf{K} = \mathbf{F}_u^{-1}\mathbf{B}'\mathbf{Y}, \tag{10-100}$$

and $\mathbf{Y}$ is defined by:

$$\mathbf{A}'\mathbf{Y} + \mathbf{YA} - \mathbf{YBF}_u^{-1}\mathbf{B}'\mathbf{Y} + \mathbf{F}_x = \mathbf{0}, \tag{10-101}$$

where $\mathbf{F}_x$ is an $n \times n$ symmetric matrix defined by:

$$\mathbf{F}_x = \mathbf{C}'\mathbf{F}_y\mathbf{C}.$$

Equation (10-101) is a generalization of Eq. (10-93). The solution $\mathbf{Y}$ of Eq. (10-101) must be either positive definite or positive semidefinite. We shall generalize the condition (Eq. 10-101) further in Section 14-4.

REFERENCES

1. Mesarović, M. D., *The Control of Multivariable Systems.* New York: Wiley, 1960.
2. Kavanaugh, R. J., "The application of matrix methods to multivariable control systems." *J. Franklin Inst.*, **262**, 1956, p. 349.
3. Boksenbom, A. S., Hood, R., "General algebraic method applied to control analysis of complex engine types." *NACA Report No. 980*, 1950.
4. Kavanaugh, R. J., "Noninteracting controls in linear multivariable systems," *AIEE Trans.*, **76**, Part II, May 1957, pp. 95-100; "Multivariable control system synthesis," *AIEE Trans.*, **77**, Part II, November 1958, pp. 425-429.
5. Freeman, H., "A Synthesis method for multivariable control systems," *AIEE Trans.*, **76**, Part II, March 1957, pp. 28-31; "Stability and physical realizability consideration in the synthesis of multiple control systems," *Trans. AIEE*, **77**, Part II, March 1958, pp. 1-5.
6. Ohien, K. L., et al., "The noninteracting controller for steam-generating systems." *Control Eng.*, **5**, Oct. 1958, pp. 95-101.
7. Valstar, J. E., "How to reduce interaction between control loops." *Control Eng.*, **6**, June 1959, pp. 112-113.
8. Cook, R. P., "Gain and phase boundary routine for two loop feedback systems." *1966 JACC Preprint*, pp. 762-775.
9. Bollinger, R. E., Lamb, D. E., "Multivariable systems," *IEC Fundamentals*, **1**, No. 4, November 1962, pp. 245-252.
10. Woolverton, P. F., Murrill, P. W., "An evaluation of four ideas in feedback control." *Instrumentation Technology*, **14**, No. 1, January 1967, pp. 35-40.

11. Kalman, R. E., "When is a linear control system optimal? *J. Basic Eng.*, *Trans. ASME*, **86**, Series D., No. 1, March 1964, pp. 51-60.
12. Bass, R. W., Gura, I., "High-order systems design via state-space considerations." *1965 JACC Preprint*, pp. 311-318.
13. Rosenbrock, H. H., "Distinctive problems of process control." *Chem. Eng. Progress*, **58**, No. 9, September 1962, pp. 43-50.
14. Murray-Lasso, M. A., "The modal analysis and synthesis of linear distributed control systems." Ph. D. Thesis, M. I. T., Cambridge, November 1966.
15. Gould, L. A., Murray-Lasso, M. A., "On the modal control of distributed systems with distributed feedback." *IEEE Trans. on Automatic Control*, AC-11, No. 4, October 1966, pp. 729-737.
16. Schlaefer, F. M., "Modal control of an ammonia reactor." *Report ESL-R-293*, M. I. T. Project DSR 74994, December 1966.
17. Wykes, J. H., Mori, A. S., "An analysis of flexible aircraft structural mode control," Part 1. *North American Adviation Technical Report AFFDL-TR-65-190*, June 1966.
18. Falb, P. L., Wolovich, W. A., "Decoupling in the design and synthesis of multivariable control systems," IEEE Trans. on Automatic Control, AC-12, December 1967, pp. 651-659.
19. Takahashi, Y., *et al.*, "Mode oriented design viewpoint for linear, lumped-parameter multivariable control systems," *J. Basic Eng.*, *Trans. ASME*, **90**, June 1968, pp. 222-230.
20. Loscutoff, W., "Modal analysis and synthesis of linear lumped-parameter systems," Ph. D. Thesis, Univ. of Calif., Berkeley, 1968.
21. Letov, A. M., "Analytical controller design I," *Avtomatika i Telemekhanika*, **21**, No. 4, April 1960, pp. 436-441.
22. Tyler, J. S., "Use of optimal control theory in multivariable control systems design." *1964 JACC Preprint*, pp. 40-51.
23. Willis, R. H., Brockett, R. W., "Frequency-domain solution of regulator problems," *1965 JACC Preprint*, pp. 228-234.
24. Salukvanze, M. E., "On the analytic design of an optimal controller," *Automatic and Remote Control*, **24**, No. 4, April 1963, pp. 409-417.
25. Krasovskii, N. N., "On the global stability of a system of nonlinear differential equations," (in Russian), *Prikl. mat. i. mekh.*, **18**, 1954, pp. 735-737.
26. Pipes, L. A., *Matrix Methods for Engineering*. Englewood Cliffs, N. J.: Prentice-Hall, 1963.

PROBLEMS

10-1 A two-variable feedback control system is described by the general matrix block diagram of Fig. 10-2, where the vector variables are

$$\mathbf{R}(s) = \begin{bmatrix} R_1(s) \\ R_2(s) \end{bmatrix}, \quad \mathbf{V}(s) = \begin{bmatrix} V_1(s) \\ V_2(s) \end{bmatrix}, \quad \mathbf{C}(s) = \begin{bmatrix} C_1(s) \\ C_2(s) \end{bmatrix},$$

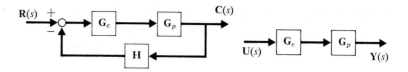

Fig. P10-2

and the matrix transfer functions are

$$G(s) = \begin{bmatrix} G_{11}(s) & G_{12}(s) \\ G_{21}(s) & G_{22}(s) \end{bmatrix}, \quad N(s) = \begin{bmatrix} N_1(s) & 0 \\ 0 & N_2(s) \end{bmatrix}.$$

Determine $C_1(s)$, $C_2(s)$, and the characteristic equation of the closed-loop system.

10-2 Open- and closed-loop configurations of a multivariable control system are shown in Fig. P10-2, where

G_p = matrix transfer function of the control object,
G_c = multivariable feedback controller matrix, and
G_e = equivalent open-loop control matrix.

G_e is defined by the equality condition

$$G_{CL} = G_{OP},$$

where G_{CL} and G_{OP} are the closed-loop and open-loop overall matrix transfer functions, respectively. These are given by

$$G_{CL}R = C, \quad G_{OP}U = Y.$$

As shown in the figure, **R** is the reference input vector to the closed-loop system with output **C** representing the controlled vector. **U** and **Y** represent the input-output vector pair of the open-loop system.

Suppose that the matrix transfer function of the control object changes from G_p to $G_p + \Delta G_p$, and as a consequence, the response vectors become C_e and Y_e. Define error vectors by

$$\Delta C = C - C_e, \quad \Delta Y = Y - Y_e.$$

Show that the sensitivity matrix **S** such that, for $R = U$,

$$S \cdot (\Delta Y) = (\Delta C),$$

is given by

$$S = [I + (G_p + \Delta G_p)G_c H]^{-1}.$$

If we let $G = G_p G_c H$ = the open-loop matrix transfer function, the final result to be proved will correspond to the sensitivity relation given by Eq. (8-28) for scalar-variable feedback control systems.

10-3 The thermal system, treated in Example 5-6 (Section 5-3), is shown in Fig. P10-3 with a multivariable controller G_c. Design G_c as a classical independent pro-

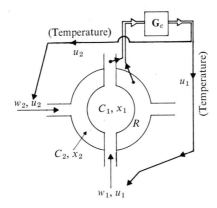

Fig. P10-3

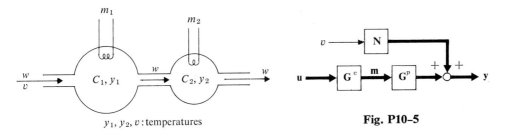

Fig. P10-5

portional controller. Assume that the system parameters are

$$\frac{w_1}{C_1} = \frac{w_2}{C_2} = 1, \qquad C_1 R = C_2 R = 2.$$

10-4 The overall resistance R in Problem 10-3 is doubled due to scale buildup. Does the G_c-matrix determined in Problem 10-3 still yield independent control? For this change in R, and $G_{11} = G_{22} = 8$, compute the sensitivity matrix **S** which was given in Problem 10-2.

10-5 The two mixing heaters shown in Fig. P10-5 have heat inputs m_1 and m_2 respectively. The heaters are completely insulated against heat loss to the environment. Heat capacitances are $C_1 = 1$ and $C_2 = \frac{1}{2}$, and the fluid flowrate specific heat product is $w = 1$. Inlet temperature (deviation) v is the disturbance. (a) Find the **N**- and G_p-matrices shown in the figure. (b) Determine the decoupling control matrix G_c where $G_{11} = k_1$ and $G_{22} = k_2$ (both P-action). (c) Obtain the scalar block diagram of the open-loop system that includes G_c, G_p, and **N**. (d) Is this (open- or closed-loop) system controllable or observable?

10-6 State-variable feedback control is applied to the object of the preceding problem (restated in Fig. P10-6). The desired closed-loop poles are λ_1 and λ_2 (both negative, of course). (a) Obtain the control law **k**. (b) Is this approach limited to control systems with real eigenvalues? (c) For this scheme of control, can we choose m_2 of Problem 10-5 as a controller output? (d) Suppose that the flowrate w varies. For a control system with control law **k** fixed by computation as in part (a) of this

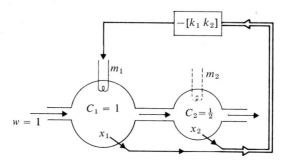

Fig. P10-6

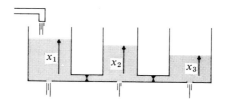

Fig. P10-7

problem (for $w = 1$), can we trace the effect of variations in w by the root-locus methods?

10-7 A three-tank system is described by

$$\frac{d}{dt}\mathbf{x} = \begin{bmatrix} -3 & 1 & 0 \\ 2 & -3 & 2 \\ 0 & 1 & -3 \end{bmatrix} \mathbf{x} + \begin{bmatrix} 1 \\ 0 \\ 0 \end{bmatrix} u$$

[see Fig. P10-7, and Eq. (3-32)]. (a) Obtain $\mathbf{S}_1$ and $\mathbf{S}_2$ in the expansion

$$(s\mathbf{I} - \mathbf{A})^{-1} = \frac{1}{\Delta(s)} \{\mathbf{S}_1 + s\mathbf{S}_2 + s^2\mathbf{S}_3\},$$

where $\mathbf{S}_3 = \mathbf{I}$, and $\Delta(s) = |s\mathbf{I} - \mathbf{A}|$. (b) Find $F_0(s)$, $F_1(s)$, and $F_2(s)$ in the expansion

$$(s\mathbf{I} - \mathbf{A})^{-1} = F_0(s)\mathbf{I} + F_1(s)\mathbf{A} + F_2(s)\mathbf{A}^2.$$

(c) Design the state-variable feedback controller $\mathbf{k}$ that will give closed-loop poles at $(-1 - c)$, $(-3 - c)$, and $(-5 - c)$, where c is a positive constant.

10-8 All vector variables in the matrix block diagram (see Fig. P10-8) are n-dimensional. The control object is described by $d\mathbf{x}/dt = \mathbf{A}\mathbf{x} + \mathbf{B}\mathbf{u}$, $\mathbf{y} = \mathbf{C}\mathbf{x} + \mathbf{D}\mathbf{u}$. We are to design the control matrix $\mathbf{G}$ in such a way that the closed-loop system will satisfy $d\mathbf{x}/dt = \mathbf{A}_d\mathbf{x}$ when the control system is free, where $\mathbf{A}_d$ is a prescribed $n \times n$ matrix. Obtain $\mathbf{G}$ in terms of $(\mathbf{A} - \mathbf{A}_d) = \Delta\mathbf{A}$, $\mathbf{B}$, $\mathbf{C}$, and $\mathbf{D}$. Assume that all matrix inverses necessary to derive the solution exist.

10-9 All C's and R's of the control object shown in Fig. P10-9 are 1. Obtain the control matrix $\mathbf{G}$ such that with feedback control, $\mathbf{u} = \mathbf{G}\mathbf{y}$, the controlled system

Problems 463

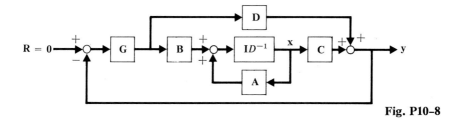

Fig. P10-8

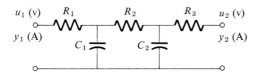

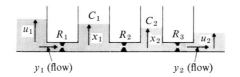

y_1 (flow) $\qquad\qquad\qquad$ y_2 (flow) $\qquad$ Fig. P10-9

will have an **A**-matrix

$$\mathbf{A}_d = \begin{bmatrix} \lambda_1 & 0 \\ 0 & \lambda_2 \end{bmatrix},$$

where λ_1 and λ_2 are negative constants.

10-10 In the three-tank system of Problem 10-7, two levels are measured:

$$y_1 = x_1, \qquad y_2 = x_2.$$

Obtain the minimum-norm solution for the modal-domain vector $\mathbf{x}^*$.

10-11 Suppose that the three-tank system of Problem 10-7 has two control inflows: u_1 for the left tank, and u_2 for the tank in the middle. We are to impress a three-dimensional modal-domain control vector $\mathbf{u}^*$ on the system via the two-dimensional control vector

$$\mathbf{u} = \begin{bmatrix} u_1 \\ u_2 \end{bmatrix}.$$

Express $\mathbf{u}$ in terms of $\mathbf{u}^*$. If this is not an exact relation, what property does it have?

10-12 The control object is similar to that of Problem 10-5, described by

$$\frac{d}{dt}\mathbf{x} = \begin{bmatrix} -1 & 0 \\ 2 & -2 \end{bmatrix} \mathbf{x} + \begin{bmatrix} 1 & 0 \\ 0 & 2 \end{bmatrix} \mathbf{u},$$

where x_1, x_2 are the temperatures of tanks C_1 and C_2, and u_1, u_2, respectively, are heat inputs to these tanks. The first measurement is x_2, $y_1 = x_2$; and the second output is the temperature difference between the two tanks, $y_2 = x_1 - x_2$. The desired eigenvalues of the modally controlled system are specified to be p_{c1} and p_{c2}

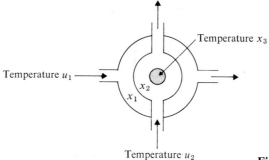

Fig. P10-13

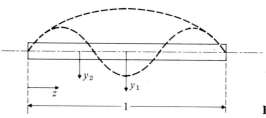

Fig. P10-15

(both negative real). Obtain the analyzer, synthesizer, modal-domain control matrix **K***, and control matrix **K** in the original variable space.

10-13 The thermal system shown in Fig. P10-13 consists of three heat capacitances. Two inputs are applied to the object as temperatures of incoming flows, so that (assuming numerical values for the C's and R's) the object is described by

$$\frac{d}{dt}\begin{bmatrix} x_1 \\ x_2 \\ x_3 \end{bmatrix} = \begin{bmatrix} -2 & 1 & 0 \\ 2 & -3 & 0 \\ 0 & 18 & -10 \end{bmatrix} \begin{bmatrix} x_1 \\ x_2 \\ x_3 \end{bmatrix} + \begin{bmatrix} 1 & 0 \\ 0 & \frac{1}{2} \\ 0 & 0 \end{bmatrix} \begin{bmatrix} u_1 \\ u_2 \end{bmatrix}.$$

Two temperatures are measured: $y_1 = x_1$ and $y_2 = x_2$. By modal control it is desired that we make all three eigenvalues -10. Determine the control law **K*** in the modal domain, **K** in the original space, the analyzer and synthesizer matrices, and discuss the effect caused by imperfection of either measurement or control.

10-14 The two control signals, u_1 and u_2 in the preceding problem, converted into heat units, are also applied directly to the third (inner) heat capacitance so that the equation for x_3 takes the form:

$$\frac{dx_3}{dt} = 18x_2 - 10x_3 + b_{31}u_1 + b_{32}u_2.$$

(a) Obtain the gain constants b_{31} and b_{32} which will make the control matrix **B** ideal in the sense of decoupled modal control. (b) Suppose that temperature x_2 in the preceding problem is impossible to measure. On the other hand, x_3 is measurable so that $y_1 = x_1$ and $y_2 = x_3$. Using **B** obtained above, design a **K**-matrix that will

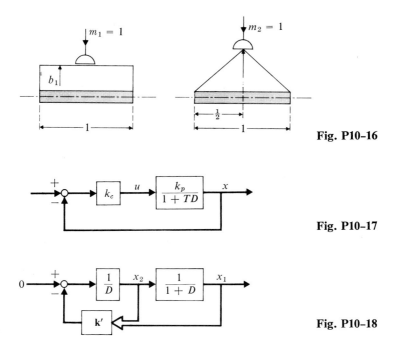

Fig. P10-16

Fig. P10-17

Fig. P10-18

make the first two modes e^{-10t}, as in the preceding problem, and discuss the effect of any imperfection.

10-15 The first two modes of a distributed-parameter object are the sinusoidals shown in Fig. P10-15. Measurements are made at $z = \frac{1}{2}$ and $\frac{1}{4}$. Obtain the analyzer matrix for the modes.

10-16 To control the two modes of the preceding problem, two manipulators are installed on the control object. The first manipulator emits a uniformly distributed controlling input of intensity b_1 for control signal $m_1(t) = 1$, and the second unit has a triangular distribution with peak value b_2 for $m_2(t) = 1$ (see Fig. P10-16). Obtain the synthesizer matrix.

10-17 For the control system given in Fig. P10-17, determine an optimal gain $k_c = k^o$ for which the performance index,

$$J = \int_0^\infty (x^2 + wu^2)\, dt, \qquad w > 0,$$

will be minimal.

10-18 A state-variable feedback control system is of second order, as shown in Fig. P10-18. Determine the optimal feedback control law $\mathbf{k}'$ for which the performance index

$$J = \int_0^\infty (u^2 + x_1^o)\, dt$$

will be minimal.

10-19 Consider an nth-order system in the modal domain,

$$\frac{d}{dt}\mathbf{x}^* = \Lambda \mathbf{x}^* + \mathbf{u}^*,$$

where the modal-domain control vector $\mathbf{u}^*$ is n dimensional, and

$$\Lambda = \begin{bmatrix} p_1 & & 0 \\ & p_2 & \\ & & \ddots \\ 0 & & p_n \end{bmatrix},$$

and where $p_1, \ldots, p_n$ are real and negative. Design a modal-domain control law $\mathbf{K}^*$ such that, with feedback control $\mathbf{u}^* = -\mathbf{K}^*\mathbf{x}^*$, the following performance index will be minimized:

$$J = \int_0^\infty \sum_{i=1}^n (w_{xi}^* x_i^{*2} + w_{ui}^* u_i^{*2}) \, dt, \quad w_{xi}^* \geq 0, \quad w_{ui}^* > 0 \quad \text{for all} \quad i.$$

10-20 The quadratic performance index minimized in the modal domain (see Problem 10-19) can be expressed in the general form

$$J^* = \int_0^\infty [(\mathbf{x}^*)' \mathbf{W}_x^* \mathbf{x}^* + (\mathbf{u}^*)' \mathbf{W}_u^* \mathbf{u}^*] \, dt,$$

where $\mathbf{W}_x^*$ is a positive semidefinite diagonal matrix with diagonal elements w_{xi}^*, $i = 1, \ldots, n$, and $\mathbf{W}_u^*$ is a positive definite diagonal matrix with elements w_{ui}^*, $i = 1, \ldots, n$. The control object in the original state space is expressed by

$$\frac{d\mathbf{x}}{dt} = \mathbf{A}\mathbf{x} + \mathbf{B}\mathbf{u}, \quad \mathbf{y} = \mathbf{C}\mathbf{x},$$

and the control law is $\mathbf{u} = -\mathbf{K}\mathbf{y}$. Suppose that $\mathbf{K}$ is designed in the modal domain to minimize J^*, and the latter is to be expressed in terms of the original state vector and control vector by

$$J = \int_0^\infty (\mathbf{x}'\mathbf{F}_x\mathbf{x} + \mathbf{u}'\mathbf{F}_u\mathbf{u}) \, dt.$$

Relate $\mathbf{F}_x$ to $\mathbf{W}_x^*$, and $\mathbf{F}_u$ to $\mathbf{W}_u^*$, respectively, to satisfy the equality $J = J^*$. Determine $\mathbf{F}_x$ and $\mathbf{F}_u$ for the system of Problem 10-12.

11

LINEAR DIGITAL CONTROL

A digital controller generates output by discrete computation. Digital control differs from analog control in two important respects: (1) the input to a digital controller must be quantized (analog-to-digital conversion is necessary if the original signal is analog), and (2) digital computation is performed only at discrete times instead of continuously, so that a sampler is required for the input side and a holder on the output side of the computer. Although quantization is a nonlinear process, linearity can be assumed without serious error if a sufficiently small quantization step is chosen. Sampling, on the other hand, produces some phenomena which have no counterparts in continuous systems. The theory of linear sampled-data control systems, based either on difference equations [1] or on the z-transformation [2 through 7], attracted the interest of control theorists more than a decade before the engineering realization of digital control. The importance and usefulness of some of this theory has been confirmed by recent progress in direct digital control (DDC). In typical DDC installations, a large number of analog control loops are replaced by a single digital computer. The rapid growth of DDC is also introducing many new engineering problems.

We begin this chapter with a brief introduction to DDC, emphasizing the differences between digital and analog (or conventional) controls. An introduction to DDC and the sampled-data domain is followed by a discussion of the design and analysis of linear, single-loop, sampled-data control systems. The theory is then extended to multivariable systems and some optimal feedback controls.

11-1 INTRODUCTION TO DIGITAL CONTROL

In previous chapters we were concerned with the design and operation of analog controllers. These controllers produce a continuous (in time) output in response to a continuous input. Digital computers are sequential devices: they produce a sequence of output states in response to a sequence of inputs. Typical analog controllers (pneumatic, electronic, or hydraulic) are relatively small devices which control single loops, and the overwhelming majority of them use the the P, PI, or PID control laws. Digital computers used in direct digital control (DDC) applications, on the other hand, usually control many loops "simultaneously" (50, 100, or more). The control computer does this by scanning (usually with a multiplexor) a number of plant variables and generating control signals on a time-sharing basis. The flexibility of stored-program digital computers allows the use of a wide variety of control algorithms, and digital versions of the PI and PID algorithms are widely used [14].

Much of the instrumentation used in control systems is analog in nature so that input signals from analog instrumentation must be sampled and quantized by an analog-to-digital converter for use in the control computer. Likewise, output signals from the computer are intermittent, and unless final control elements that are designed to operate from intermittent signals (such as stepping motors) are used, the output must be converted to analog form and held in a staircase fashion until the next signal is generated.

The advantages of DDC over conventional analog control lie in the intercommunication capabilities of computers, and the possibility of programming control computers to share time between direct single-loop control activities and other higher-level activities. Single-loop or *first-level* controls can be integrated into cascade, ratio, feedforward, and multivariable controls, which are considered to be the *second level* in the control hierarchy. In addition, sequential control for start-up, shutdown, or emergency operation can be incorporated, or adaptive action such as gain tuning can be included. The *third level* includes overall supervisory control and plant optimization; these functions can be carried out by a large-scale computer which has direct communication with the smaller first- and second-level control computers. In this chapter we are interested in first- and second-level controls.

Sampling period is one of the most important of the new parameters introduced into the control problem by the use of control computers. By choosing longer sampling periods we can decrease the cost of control since the reduced speed requirements for the multiplexing and computation allows us to increase the number of loops under control, increase the complexity of the control algorithms, buy less expensive equipment, or make possible some combination of these. On the other hand, the sampling period must be short enough to permit effective control. The dynamics of the control object, the types of reference input changes, expected disturbances, and the control algorithm to be used all affect the choice of sampling time. Typical sampling times for process control, as reported in [10], show that the fastest rates are required for liquid-flow loops (about 1 sec/sample); sample times of about 5 sec/sample for pressure and level controls and as much as 20 sec/sample or more for temperature and composition control are representative.

In control theory, data-sampling is considered to be an amplitude modulation of an impulse train. To formulate a sampling process for this interpretation, we must first derive the equation for an impulse train as the limiting case of a pulse train. The Fourier series for a pulse train (see

11-1 Introduction to digital control

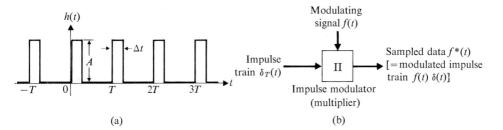

Fig. 11-1 Sampled data as a modulation of an impulse train.

Fig. 11-1a) is:

$$h(t) = \frac{A}{\pi}\left[\frac{\Omega \Delta t}{2} + \sum_{n=1}^{\infty}\frac{1}{n}(\sin n\Omega \Delta t)\cos n\Omega t\right.$$
$$\left.+ \sum_{n=1}^{\infty}\frac{1}{n}(1 - \cos n\Omega \Delta t)\sin n\Omega t\right]. \quad (11\text{-}1)$$

Ω in this expression is $2\pi/T$, where T is the period, A = pulse amplitude, and Δt = pulse width. Keeping the product $A\Delta t$ equal to unity, $A\Delta t = 1$, as we let $\Delta t \to 0$, we obtain an expression for the impulse train $\delta_T(t)$:

$$\delta_T(t) = \frac{\Omega}{2\pi} + \frac{\Omega}{\pi}\sum_{n=1}^{\infty}\cos n\Omega t = \frac{\Omega}{2\pi}\left[1 + \sum_{n=1}^{\infty}(e^{jn\Omega t} + e^{-jn\Omega t})\right]$$
$$= \frac{1}{T}\sum_{n=-\infty}^{+\infty}e^{j(2\pi nt/T)}. \quad (11\text{-}2)$$

The sampled signal $f^*(t)$ of a continuous signal $f(t)$ as an amplitude-modulated impulse train (Fig. 11-1b) is:

$$f^*(t) = \frac{1}{T}\sum_{n=-\infty}^{+\infty}f(t)e^{j(2\pi nt/T)}. \quad (11\text{-}3)$$

Making use of the following Laplace transform pair,

$$\mathscr{L}[e^{\lambda t}f(t)] = F(s - \lambda),$$

we find that the Laplace transform of Eq. (11-3) is:

$$\mathscr{L}[f^*(t)] \triangleq F^*(s) = \frac{1}{T}\sum_{n=-\infty}^{+\infty}F(s - jn\Omega) = \frac{1}{T}\sum_{n=-\infty}^{+\infty}F(s + jn\Omega). \quad (11\text{-}4)$$

$F(s)$ of the original continuous signal, $f(t)$, is related to $F^*(s)$ of its sampled data $f^*(t)$ by Eq. (11-4). Substituting $j\omega$ for s, the frequency-domain relation follows:

$$F^*(j\omega) = \frac{1}{T}\sum_{n=-\infty}^{+\infty}F(j\omega + jn\Omega). \quad (11\text{-}5)$$

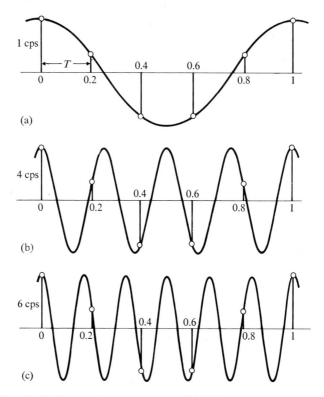

Fig. 11-2 Various harmonics that yield identical time series.

This expression shows a periodicity due to Ω of the sampling process. Let us consider, for instance, a set of sampled data of a 1-cps sinusoidal signal, taken by a sampling process with a period $T = \frac{1}{5}$ sec (Fig. 11-2a). Since $\Omega/2\pi = 1/T = 5$ cps, an identical set of sampled data can be obtained by sampling $5 - 1 = 4$ cps (Fig. 11-2b) or $5 + 1 = 6$ cps (Fig. 11-2c) sinusoidal signals. Similarly, for $n = 2, 3, \ldots$ in Eq. (11-5), identical sampled data will be obtained for sinusoidal signals at frequencies of $(n\Omega \pm \omega)$.

Example 11-1 Consider an exponential decay $f(t) = e^{-t}$. Since $F(s) = 1/(1 + s)$, $F(j\omega)$ for $\omega = 0$ to ∞, yields a semicircle which is shown in Fig. 11-3(a) For a sampling period $T = \pi/2$, $F^*(j\omega)$ of the sampled data can be determined from $F(j\omega)$ by graphical phasor addition, as shown for a frequency $\omega = 1$. The resulting $F^*(j\omega)$ has a periodicity of $\Omega = 4$. As shown, in parts (b) and (c) of the figure for two different sampling rates, a set of *spurious components*,

$$F(j\omega + jn\Omega), \quad n = 1, 2, \ldots \quad \text{and} \quad n = -1, -2, \ldots$$

will appear due to the periodicity. When a continuous signal is sampled, part of the information will be lost and the noise effect will be aggravated due to these spurious components.

11-1　Introduction to digital control

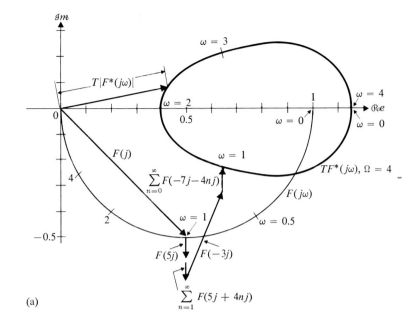

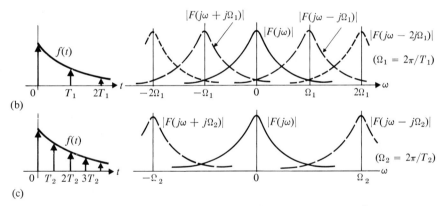

Fig. 11-3 Frequency-domain relation between $F(s)$ and $F^*(s)$.

The Someya-Shannon *sampling theorem* is instructive on the basic relation between a continuous signal and its sampled data. The theorem gives an answer to the question: When and how can a continuous signal $f(t)$ be reconstructed from its sampled data? An intuitive answer to this question is the following: the original signal $F(j\omega)$ may be regained (a) if there is no overlapping of $F(s)$ and its spurious components (Fig. 11-3c), and (b) if it is possible to filter out $F(s)$ without any phase and gain distortion. The first condition implies that the frequency of the original signal $F(j\omega)$ must be

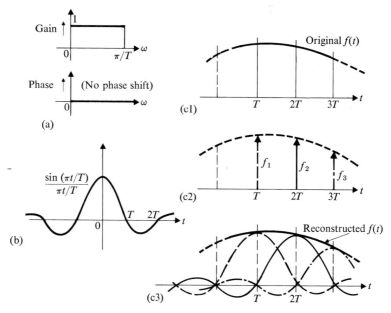

Fig. 11-4 Frequency response (a) and unit impulse response (b) of an ideal filter; reconstruction of a continuous signal $f(t)$ from its sampled data via an ideal filter (c1, c2, c3).

limited to the range

$$-\Omega/2 < \text{frequency range of } F(j\omega) < +\Omega/2 . \qquad (11\text{--}6)$$

In practice, therefore, the sampling period T must be shorter than a half-period of the highest-frequency signal component. The second condition implies that an ideal filter $G_i(s)$ (Fig. 11-4a) such that

$$G_i(j\omega) = \begin{cases} 1 & \text{for } 0 < \omega < \Omega/2 , \\ 0 & \text{for } \Omega/2 \leqslant \omega , \end{cases} \qquad (11\text{--}7)$$

is necessary to pick up $F(j\omega)$ without any distortion while rejecting all spurious components. Figure 11-4 shows how an original signal $f(t)$ is reconstructed from its sampled data when these two conditions are met. A band-limited (that is, a "smooth") signal $f(t)$ [(c1) in the figure] is sampled, with a sampling period T [(c2) in the figure]. Each sample $f_1, f_2, \ldots$ (modulated impulse) is then passed through an ideal filter, whose unit impulse response is shown in Fig. 11-4(b). The filtered outputs, superimposed as shown in (c3) of the figure, reproduce the original signal $f(t)$. Looking at Fig. 11-4(b), we see that an ideal filter responds in advance of the input application; therefore, it is physically unrealizable. Hence, complete reconstruction of a continuous signal from its sampled data is in general

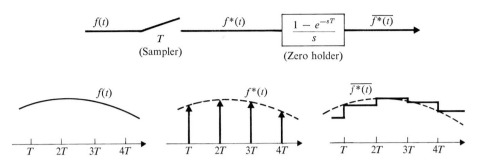

Fig. 11-5 Smoothing sampled data with a zero holder.

impossible. However, approximate reconstruction (or smoothing) can be accomplished, for instance, by a zero holder. The transfer function of a zero holder (see Section 1–5) is

$$G_h(s) = \frac{1 - e^{-sT}}{s}. \tag{11-8}$$

Figure 11–5 shows how a smooth signal $f(t)$ is approximately reconstructed from its sampled data by a zero holder. It is also possible to see the approximate reconstruction process in the frequency domain. A series coupling (phasor multiplication) of $F^*(j\omega)$ and the zero-holder frequency response $G_h(j\omega)$ (Table 9–2c) will produce an approximation of $F(j\omega)$ in the complex plane of Fig. 11–3(a).

A minor contamination of a measured signal by high-frequency noise is in general not very serious in many analog control systems, mainly because the dynamic components act as low-pass filters and thus attenuate the noise. However, if sampling is involved, high-frequency noise may produce low-frequency fluctuations due to spurious-component superposition [Fig. 11–3(b) and (c)]. The phenomenon is known as *aliasing*. The low-frequency noise component thus produced has the same amplitude as the original noise, and its frequency is the *difference* between the original noise and a multiple of the sampling frequency (see Fig. 11–2). The aliasing effect of sampling in DDC systems must be reduced by the insertion of an appropriate filter between the measuring element and the input to the digital control algorithm.

A simple first-order filter, in the form

$$G_f(s) = \frac{1}{1 + T_f s}, \tag{11-9}$$

seems satisfactory for many industrial applications. The following approximate computation [12] may serve as a guide to choosing an appropriate filter time constant T_f. According to Eq. (13–34), the variance (mean-

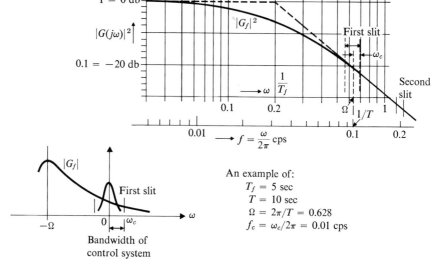

Fig. 11-6 Single lag filter reducing a white-noise aliasing.

squared fluctuation) σ^2 of a filtered noise is given by:

$$\sigma^2 = \frac{1}{2\pi} \int_{-\infty}^{\infty} |G_f(j\omega)|^2 \, \Phi_{nn}(j\omega) \, d\omega , \quad (11\text{–}10)$$

where $\Phi_{nn}(j\omega)$ is the power spectral density of the noise. Let us assume a system with a closed-loop cutoff frequency ω_c (that is, the bandwidth, or point at which the system gain falls below -3 db). Due to the aliasing effects of sampling at frequency Ω, spurious components within the system bandwidth will appear because of noise at the frequency slits $\Omega + \omega_c$, $2\Omega + \omega_c$, etc. A filter (Eq. 11–9) should be inserted (as part of the sampling process) which attenuates the noise without affecting the original system bandwidth (Fig. 11-6). $|G_f(j\omega)|^2$ in Eq. (11–10) may be approximated over the slit by its value at $\omega = \Omega$; thus,

$$|G_f(j\omega)|^2_{\text{average for slit}} \approx \frac{1}{1 + (2\pi T_f/T)^2} . \quad (11\text{–}11)$$

For $T_f = T/2$ the average value is $1/(1 + \pi^2) \approx 1/10$; hence the aliasing effect as a variance in the controlled variable will be reduced to 10% for white noise [$\Phi_{nn}(j\omega) = $ const]. If the noise bandwidth is limited, the filter time constant can be made even shorter [12]. The additional time constant T_f, in the control system produces a loss in control performance, and hence an excessively large time constant for filtering is not advisable.

Analog filters made of passive elements are adequate for filter time constants up to a few seconds. If T_f is longer than a few seconds, combined

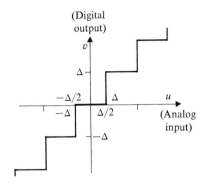
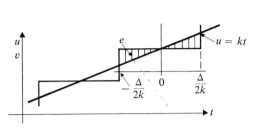

Fig. 11-7 Input-output of an encoder.

Fig. 11-8 Quantization error of a ramp signal.

digital and analog filtering is generally preferred. The numerical approximation to the first-order lag

$$T_f \frac{dx}{dt} + x = u$$

for an input u and output x is given by Eq. (1-25) as:

$$x_{n+1} = (e^{-T/T_f})x_n + (1 - e^{-T/T_f})u_n . \tag{11-12}$$

If a digital system has one output sample for m input samples, the inputs may be averaged over m input sample periods.

Both the input and the output of a digital controller must be digital; hence any analog input must be quantized. It is desirable to choose a quantization step Δ that will preserve the basic sensitivity of the measuring element. A quantization size of one part in a thousand ($2^{10} = 1024$ or 10-bit word length) is usually *sufficient* to ensure that no significant quantization error be introduced by the analog-to-digital converter ("encoder"). The output quantization level depends on the control algorithm, the manipulator (control valve) actuation, and the noise level in the loop. An output quanta of 1 : 250 ($2^8 = 256$ or 8-bit word length) to 1 : 500 ($2^9 = 512$) seems to be acceptable for many process systems.

The encoder is a nonlinear device (Fig. 11-7). The error it introduces,

$$e(t) = u(t) - v(t) ,$$

where $u(t)$ = analog quantity, $v(t)$ = its digital equivalent, gives the following mean-squared value if the analog signal is a ramp (Fig. 11-8):

$$\overline{e^2} = \frac{1}{\Delta/k} \int_{-\Delta/2k}^{\Delta/2k} (kt)^2 \, dt = \frac{\Delta^2}{12} , \tag{11-13}$$

where Δ is the size of the quantization step, and k is the slope of the ramp.

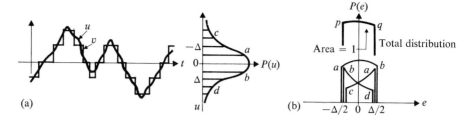

Fig. 11-9 Probability distribution of quantization error.

Although the analog signal may not be a ramp, it can be shown that Eq. (11-13) is a good representation of the mean-squared error produced by encoding. Let us suppose that an analog signal $u(t)$ is stationary random, with a probability distribution of amplitude as shown in Fig. 11-9(a). The amplitude distribution also represents the quantization error distribution in the following way: a–b for the error about $v = 0$, c–a for error about $v = -\Delta$, and so on. Therefore the probability distribution of quantization error is determined by summing these sections up (Fig. 11-9b) [5 and 13]. Since the total distribution p–q is almost flat, the mean-squared error is computed by

$$\overline{e^2} = \frac{1}{\Delta} \int_{-\Delta/2}^{\Delta/2} e^2 \, de = \Delta^2/12 \,.$$

The digital control algorithm is the last topic in this section. PI and PID algorithms are typical for first level DDC [11, 12, 14]. There are two ways to operate a final control element (for example, control valve): a position algorithm or a velocity algorithm. The position algorithm requires the control computer to command the full value of valve position m_N at each calculation, while only the required change in valve position is calculated in the *velocity algorithm*. For practical considerations such as safety, simpler computation, and easier manual-to-automatic transfer, the velocity algorithm is generally preferred. An analog PID control law, written in the form

$$m(t) = k_c \left[e(t) + \frac{1}{T_i} \int_{-\infty}^{t} e(t) \, dt + T_d \frac{de(t)}{dt} \right], \qquad (11\text{-}14)$$

can be written in the finite difference form

$$m_N = k_c \left[e_N + \frac{1}{T_i} \sum_{j=-\infty}^{N} e_j \cdot T + T_d \frac{e_N - e_{N-1}}{T} \right], \qquad (11\text{-}15)$$

where $T =$ sampling period. Applying Eq. (11-15) to the $(N - 1)$th sample, and subtracting the resulting equation from both sides of Eq. (11-15), we can determine an output increment

$$m_N - m_{N-1} = \Delta m_N$$
$$= k_c \left[(e_N - e_{N-1}) + \left(\frac{T}{T_i}\right) e_N + \left(\frac{T_d}{T}\right)(e_N - 2e_{N-1} + e_{N-2}) \right]. \quad (11\text{--}16)$$

Equation (11–16) shows that we can make adjustments on the integral- and derivative-action parameters not only with T_i and T_d, but also by changing the sampling period T. The range of adjustment can be made extremely wide.

The error signal e_N (direct digital controller input) in a feedback control system is the difference between the reference input (or set point) r_N and the value c_N of the controlled variable:

$$e_N = r_N - c_N. \quad (11\text{--}17)$$

If we tentatively assume r to be a constant, and substitute Eq. (11–17) in Eq. (11–16), we get

$$\Delta m_N = k_c[(c_{N-1} - c_N) + k_i(r - c_N) + k_d(2c_{N-1} - c_{N-2} - c_N)], \quad (11\text{--}18)$$

where

$$k_i = T/T_i, \quad k_d = T_d/T.$$

The set point signal r in Eq. (11–18) can be made into r_N in the computer program. This form of the control law has the advantage of avoiding a derivative "kick" for a sudden set-point change. Since r_N does not appear in the derivative term, a change in set point will not cause a large, sudden change in the output. We can rewrite the algorithm so that the P, I, and D adjustments are independent:

$$\Delta m_N = K_p(c_{N-1} - c_N) + K_i(r_N - c_N) + K_d(2c_{N-1} - c_{N-2} - c_N). \quad (11\text{--}19)$$

Since the set point r appears only in the integral action, this term must always be included. If the feedback signal is subject to sudden changes, a provision can be made in the controller program for limiting the output Δm_N per computation step to some maximum value Δm_{max}; the limiting, however, must be applied with caution so that it will not degrade the control quality. If the measured signal is noisy, a four-point difference algorithm may be advisable for the D-action [11]. This algorithm is defined by (see Fig. 11–10)

$$\frac{\Delta c}{\Delta t} = \frac{1}{4}\left(\frac{c_N - \bar{c}}{1.5T} + \frac{\bar{c} - c_{N-3}}{1.5T} + \frac{c_{N-1} - \bar{c}}{0.5T} + \frac{\bar{c} - c_{N-2}}{0.5T}\right),$$

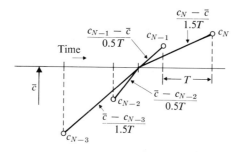

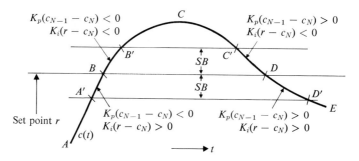

Fig. 11-10 Four-point difference.

Fig. 11-11 Sign combinations of PI algorithm.

where

$$\bar{c} = \tfrac{1}{4}(c_N + c_{N-1} + c_{N-2} + c_{N-3});$$

thus

$$\frac{\Delta c}{\Delta t} = \frac{c_N - c_{N-3} + 3c_{N-1} - 3c_{N-2}}{6T}. \tag{11-20}$$

Since DDC algorithms are flexible, various modifications are possible to improve control quality. (A DDC algorithm for the second canonical form (Table 5-3b) was discussed by DeBolt and Powell [14].) For instance, a program can be added to a position algorithm to prevent integral "windup" from occurring when the final control element is at saturation. Integral overshoot may be prevented by allowing the proportional action to depend on the magnitude of the error. As shown in Fig. 11-11, the P-action term and the I-action term in Eq. (11-19) have the same sign when the controlled variable is moving away from the set point (*BC* and *DE* in the figure), but have opposite signs when the controlled variable is moving toward the set point (*AB* and *CD*). Noting this property, Cox *et al.* made a suggestion [11] to put in a "set-point band" (*SB* in Fig. 11-11) such that K_p in Eq. (11-19) is made zero when the response is outside *SB*, and the P-term sign is opposite

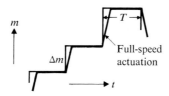

Fig. 11-12 Actuation of final control element by velocity algorithm.

to that of the I-term (AA' and CC' in Fig. 11-11). A further improvement in control quality may be expected by making K_p higher outside the band ($B'C$ and $D'E$) and much lower within the band ($A'B'$ and $C'D'$) [15].

The computer output of the velocity algorithm goes to a stepping motor or a similar device where an integration is performed:

$$m_N = m_{N-1} + \Delta m_N = m_0 + \sum_{i=1}^{N} \Delta m_i . \tag{11-21}$$

In general, the real position of the final control element (valve opening, for instance) cannot exactly follow this relation, for there are lags and thus instantaneous change of position is impossible. As shown in Fig. 11-12, which is exaggerated, the final control element moves with finite speed. So far as control quality is concerned, the highest possible speed gives the best quality.

11-2 SINGLE-LOOP DIGITAL CONTROL SYSTEMS

The optimum-tuning problem for DDC systems is more complicated than the same problem for analog PID control, because the DDC algorithm has more adjustable parameters than the equivalent analog control system. Suppose that the object of control is given; for instance, its dynamics might be represented by a dead time L and a major time constant T_p, or by R and L in the Ziegler-Nichols approximation (Section 8-4). A linear, invariant three-term DDC algorithm involves the three parameters, K_p, K_i, and K_d; and in addition to these, the size of quanta Δ and sampling period T are also system parameters. For a free response (zero reference input) from a unity initial state, an error criterion J, such as integrated absolute error (IAE; see Section 8-4), will be

$$J = \text{function } (\Delta, T, K_i, K_p, K_d, T_p, L) .$$

No rule of thumb is known to determine some of these in terms of others for the minimal J condition. The size Δ of quantization may be ignored for the moment if a sufficient word length is provided for data processing. Based on a criteria of 15% increase in settling time due to sampling, Goff [12] published a chart for T as a function of L and T_p, according to which the following rule of thumb seems to hold:

$$T \cong 0.3L.$$

These rules of thumb seem to suggest that there is an "optimum" sampling interval T. Our intuition, on the other hand, tells us that the more information we have, the better we can control, that is, T should be as small as possible. The reason for the seeming contradiction is that the rules of thumb presented are based on the assumption that a PID or similar algorithm is used. If no algorithm has been decided on, however, we can always achieve better control with a smaller T. Goff [12] also reports that the number of T per I-time constant ($K_i/T = K_1$) ranges from approximately 2 for predominantly dead-time processes to about 6 for processes having little dead time. This seems to approximately confirm the Ziegler-Nichols rule (Eq. 8–65) which gives $K_i k_c \approx 3L$ for PI-action. He also points out that

$$\text{(derivative time constant)}/T \approx 5 \text{ to } 10$$

is desirable if the PID algorithm is to obtain the full benefit of the D-action.

To introduce analytical formulations for a first-level linear DDC loop, we shall assume a first-order control object, ignore the quantization error, and consider a PI-control velocity algorithm:

$$\Delta m_k = K_p(y_{k-1} - y_k) + K_i(r_k - y_k) \quad \text{for PI control}, \tag{11-22}$$

and

$$y_{k+1} = a y_k + b m_k \quad \text{for control object}, \tag{11-23}$$

where k = sampling index, y = controlled variable, m = manipulated variable, and r = reference input. K_p and K_i are controller parameters, and a and b are plant parameters.

Since we have

$$\Delta m_{k+1} = m_{k+1} - m_k = K_p(y_k - y_{k+1}) - K_i y_{k+1}$$
$$= K_1 y_k - K_2 y_{k+1},$$

where r_k is assumed to be zero, and

$$K_1 = K_p, \qquad K_2 = K_p + K_i,$$

it follows that

$$m_{k+1} = m_k + K_1 y_k - K_2 y_{k+1}. \tag{11-24}$$

Substituting Eq. (11–23) for y_{k+1} in the right-hand side of Eq. (11–24), we obtain

$$m_{k+1} = m_k + K_1 y_k - K_2(a y_k + b m_k)$$
$$= (K_1 - a K_2) y_k + (1 - b K_2) m_k. \tag{11-25}$$

11-2 Single-loop digital control systems

Equations (11–23) and (11–25) combined yield the vector equation for the closed-loop system:

$$\begin{bmatrix} y_{k+1} \\ m_{k+1} \end{bmatrix} = \begin{bmatrix} a & b \\ (K_1 - aK_2) & (1 - bK_2) \end{bmatrix} \begin{bmatrix} y_k \\ m_k \end{bmatrix}. \qquad (11\text{–}26)$$

The characteristic equation of the closed-loop system is

$$\begin{vmatrix} z - a & -b \\ aK_2 - K_1 & z - (1 - bK_2) \end{vmatrix} = 0$$

or

$$z^2 - (a + 1 - bK_2)z + (a - bK_1) = 0. \qquad (11\text{–}27)$$

If the control object is a first-order plant described by the transfer function

$$G_p(s) = \frac{1}{1 + T_p s},$$

we will have (by Eq. 1–25)

$$a = e^{-T/T_p}, \qquad b = 1 - a. \qquad (11\text{–}28)$$

If a dead-beat response is desired [16], we specify the roots of Eq. (11–27) to be

$$z_1 = 0, \qquad z_2 = \beta, \qquad (11\text{–}29)$$

where $0 < \beta < 1$. Then, it follows that

$$z^2 - (a + 1 - bK_2)z + (a - bK_1) = z(z - \beta),$$

and thus

$$K_1 = \frac{a}{b}, \qquad K_2 = \frac{a + 1 - \beta}{b},$$

or, in terms of original controller and system parameters [where a and b are given by Eq. (11–28)],

$$K_p = \frac{1}{e^{T/T_p} - 1}, \qquad K_i = \frac{1 - \beta}{1 - e^{-T/T_p}}. \qquad (11\text{–}30)$$

If the conventional approach is preferred, we first determine the pulse transfer function of the plant and a zero hold by

$$G_p(z) = (1 - z^{-1}) \mathscr{Z} \frac{G_p(s)}{s}. \qquad (11\text{–}31)$$

For $G_p(s) = 1/(1 + T_p s)$ we find that

$$G_p(z) = (1 - a)/(z - a), \qquad (11\text{–}32)$$

where $a = e^{-T/T_p}$.

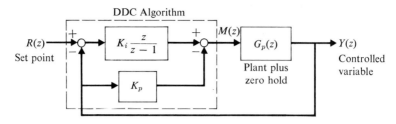

Fig. 11-13 Block diagram of a PI control system.

The PI algorithm in its original form in the z-domain is

$$M = -\left(K_p + \frac{K_i z}{z-1}\right) Y + K_i \left(\frac{z}{z-1}\right) R. \tag{11-33}$$

The block diagram of Fig. 11-13 is constructed from these relations.

The closed-loop response in the z-domain is

$$Y(z) = \frac{K_i[z/(z-1)]G_p(z)}{1 + \{K_p + K_i[z/(z-1)]\}G_p(z)} R(z). \tag{11-34}$$

The characteristic equation is

$$1 + \left(K_p + K_i \frac{z}{z-1}\right) G_p(z) = 0, \tag{11-35}$$

which agrees with Eq. (11-27) when Eq. (11-32) is substituted for $G_p(z)$.

In general, a control system is asymptotically stable if the roots of Eq. (11-35) are all inside the unit circle centered at the origin. The response will be dead-beat (nonoscillatory) if all roots are real and

$$0 \leqslant \text{roots} < 1.$$

Moreover, the sampled-data output of a control system for a step reference input will settle down to a steady-state level within a finite number of sampling periods if all closed-loop poles are at the origin of the z-plane. We shall discuss a control of this kind later.

The root-locus and frequency-response design methods apply to sampled-data systems, with appropriate modifications. We shall show in the following discussion the root-locus approach to a second-order control system. The control object with zero hold is expressed by Eq. (11-32), and restated:

$$G_p(z) = \frac{1-a}{z-a}.$$

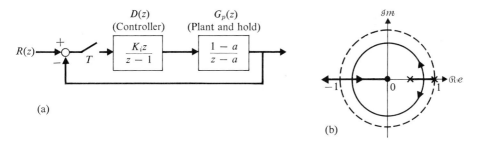

Fig. 11-14 I-control of first-order object (a), and its root locus (b).

The control law is an I-action:

$$D(z) = \frac{K_i z}{z - 1},$$

where $D(z)$ is the pulse transfer function of the control algorithm. The characteristic equation of the closed-loop system, $1 + D(z)G_p(z) = 0$, in general, becomes

$$1 + \left(\frac{kz}{z-1}\right)\left(\frac{1-a}{z-a}\right) = 0, \quad \text{where} \quad k = K_i. \quad (11\text{-}36)$$

The open-loop poles of the system are at $z = a$ and $z = 1$, and an open-loop zero is at the origin ($z = 0$). Applying the rule of root locus, we find the locus shown in Fig. 11-14(b). The two closed-loop poles are both positive, real, and less than 1 when

$$0 < k \leq k_c, \quad \text{where} \quad k_c = \frac{1 - \sqrt{a}}{1 + \sqrt{a}}.$$

The closed-loop response is nonoscillatory when k is less than k_c. A critical damping condition appears at $k = k_c$, followed by a damped oscillation when k becomes greater than k_c. The conjugate complex roots of the damped oscillation are on a circle of radius $\sqrt{a}$. Then the circle and the negative real axis intersect, and the latter becomes part of the root locus. The system reaches its stability limit when one of the closed-loop poles becomes -1.

An approximate I-control can be realized by pulse-duration control of a reversible on–off electric motor which drives a final control element, where the pulse duration is made proportional to actuating error. An analog simulation test of such a system by Zalkind [18] indicated that nonlinear action (a limit cycle) occurred when the roots were located in the left half of the z-plane, because of various imperfections, aggravated by high gain.

A digital control system is said to be finite-time settling when its closed-loop transfer function for a reference input is a finite polynomial in

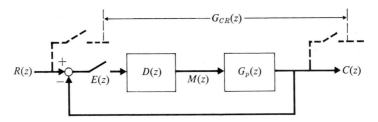

Fig. 11-15 Closed-loop pulse transfer function.

z^{-1}. The latter condition, in turn, requires, all closed-loop poles to be at the origin $z = 0$. The closed-loop pulse-transfer function of a typical linear sampled-data system, shown in Fig. 11-15, is given by

$$G_{CR}(z) = \frac{C(z)}{R(z)} = \frac{G(z)}{1 + G(z)}, \tag{11-37}$$

where $G(z) = D(z)G_p(z) =$ open-loop pulse-transfer function. If $G_{CR}(z)$ is specified for a given control object $G_p(z)$, the digital control algorithm $D(z)$ can be determined by solving Eq. (11-37) for $D(z)$:

$$D(z) = \frac{1}{G_p(z)} \left[\frac{G_{CR}(z)}{1 - G_{CR}(z)} \right]. \tag{11-38}$$

The digital control law $D(z)$ in its general form (see Problem 11-9) is

$$D(z) = \frac{b_0 + b_1 z^{-1} + \cdots + b_i z^{-i}}{1 + a_1 z^{-1} + \cdots + a_j z^{-j}} \tag{11-39}$$

where the "1" as the first term in the denominator polynomial assures the physical realizability of $D(z)$ because the output of a realizable pulse-transfer function must not depend on any future input or output.

Let us consider, as an example of this design approach, a control object whose unit-step-input response has a delay and lag pattern, as shown in Fig. 11-16(a). Approximating the process dynamics by a dead time L and a single time constant T_p, we write

$$G_p(s) = \frac{e^{-Ls}}{1 + T_p s}$$

for the process. The pulse-transfer function of a single-time-constant process and a zero hold without a dead time was given by Eq. (11-32). If a sampling period T is chosen in such a way that the condition,

$$L = NT, \quad N = \text{integer},$$

will hold, the pulse-transfer function of the process with zero hold and the

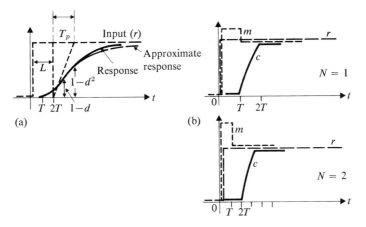

Fig. 11-16 Minimal settling time control of a plant with a dead time.

dead time will be given by

$$G_p(z) = z^{-N}\left(\frac{1-a}{z-a}\right), \quad \text{where} \quad a = e^{-T/T_p} < 1. \tag{11-40}$$

We now specify the following closed-loop pulse-transfer functions:

$$G_{CR}(z) = z^{-(N+1)}. \tag{11-41}$$

Since the object has a delay $L = NT$, a finite-time settling at the $(N+1)$th sampling seems to be the shortest possible choice. Applying Eqs. (11-40) and (11-41) in (11-38), we find

$$D(z) = \frac{1}{1-a}\left[\frac{1-az^{-1}}{1-z^{-(N+1)}}\right]. \tag{11-42}$$

The step-reference-input responses of the system are shown in Fig. 11-16(b) for $N = 1$ and $N = 2$. The response in the manipulated variable,

$$\frac{M(z)}{R(z)} = \frac{G_{CR}(z)}{G_p(z)} = \frac{1-az^{-1}}{1-a}$$

is also shown in the figure. The pattern of change of the manipulated variable looks quite similar to Smith's Posicast Control ([14] of Chapter 8). This control has appeal [19] for start-up (and shutdown) operation(s), provided that the final control element has the capacity to accommodate the first peak value $1/(1-a)$. However, since $D(z)$ cancels the pole of the plant, uncontrollable modes will appear. A signal-flow diagram of the original system ($N = 1$), with a load upset v hitting the control object at the control valve (Fig. 11-17a), when diagonalized (part (b) of the figure), reveals two modes of the object to be uncontrollable. In the next section we shall discuss a finite settling time controller design, keeping controllability in mind.

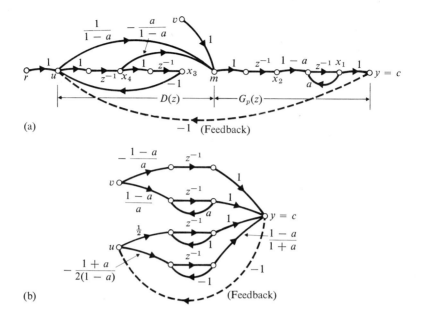

Fig. 11-17 Controllability of the system in Fig. 11-16.

11-3 CONTROL ALGORITHM FOR FINITE-TIME SETTLING

The general-purpose control algorithms, such as PI and PID, will give satisfactory results for a variety of control objects when the operating state is not very far from the desired equilibrium state. However, for a transient from an initial state which is far from the desired end state, a different control algorithm, which is tailored for a given process, will be more efficient. In the following discussion, we shall discuss a control algorithm for this purpose.

Let us consider a linear, time-invariant, sampled-data control object where a scalar controlling input u_k is held constant (zero hold) for each sampling period. The state vector equation for such a system can be written as

$$\mathbf{x}_{k+1} = \mathbf{P}\mathbf{x}_k + \mathbf{q}u_k, \qquad (11\text{-}43)$$

where $\mathbf{x}, \mathbf{q}$ are n-vectors and $\mathbf{P}$ is an $n \times n$ state transition matrix. The problem is to determine a control sequence $u_0, u_1, \ldots, u_k$, which will take the system from an arbitrary initial state $\mathbf{x}_0$ to the origin $\mathbf{0}$ of the state space in the minimum number of sampling periods (Fig. 11-18a). Mullin and Barbeyrac [20] solved this problem and concluded that it was possible to reach the origin in at most n sampling periods, provided that the system is controllable and state-variable feedback is possible.

11-3 Control algorithm for finite-time settling

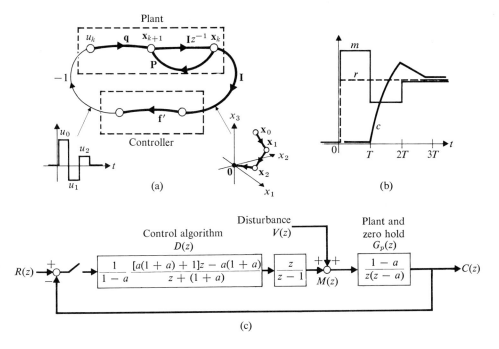

Fig. 11-18 Finite-time settling control.

The control algorithm for this purpose is given by

$$u_k = -\mathbf{f}'\mathbf{x}_k, \qquad \mathbf{f}' = [f_1, \ldots, f_n]. \tag{11-44}$$

To obtain the control law $\mathbf{f}'$ we compute the state at successive sampling periods. Starting from an arbitrary initial state $\mathbf{x}_0$, the next state $\mathbf{x}_1$ with this control will be

$$\mathbf{x}_1 = \mathbf{P}\mathbf{x}_0 + \mathbf{q}u_0 = \mathbf{P}\mathbf{x}_0 - \mathbf{q}\mathbf{f}'\mathbf{x}_0 = (\mathbf{P} - \mathbf{q}\mathbf{f}')\mathbf{x}_0.$$

Similarly,

$$\mathbf{x}_2 = \mathbf{P}\mathbf{x}_1 + \mathbf{q}u_1 = \mathbf{P}\mathbf{x}_1 - \mathbf{q}\mathbf{f}'\mathbf{x}_1 = (\mathbf{P} - \mathbf{q}\mathbf{f}')\mathbf{x}_1;$$

and, substituting the preceding equation into this one, we have

$$\begin{aligned}
\mathbf{x}_2 &= (\mathbf{P} - \mathbf{q}\mathbf{f}')^2\mathbf{x}_0 = \{[\mathbf{P}(\mathbf{P} - \mathbf{q}\mathbf{f}') - \mathbf{q}\mathbf{f}'(\mathbf{P} - \mathbf{q}\mathbf{f}')\}\mathbf{x}_0 \\
&= \{\mathbf{P}^2 - \mathbf{P}\mathbf{q}\mathbf{f}' - \mathbf{q}\mathbf{f}'(\mathbf{P} - \mathbf{q}\mathbf{f}')\}\mathbf{x}_0, \\
\mathbf{x}_3 &= (\mathbf{P} - \mathbf{q}\mathbf{f}')\mathbf{x}_2 = (\mathbf{P} - \mathbf{q}\mathbf{f}')^3\mathbf{x}_0 = \mathbf{P}\mathbf{x}_2 - \mathbf{q}\mathbf{f}'(\mathbf{P} - \mathbf{q}\mathbf{f}')^2\mathbf{x}_0 \\
&= \{\mathbf{P}^3 - \mathbf{P}^2\mathbf{q}\mathbf{f}' - \mathbf{P}\mathbf{q}\mathbf{f}'(\mathbf{P} - \mathbf{q}\mathbf{f}') - \mathbf{q}\mathbf{f}'(\mathbf{P} - \mathbf{q}\mathbf{f}')^2\}\mathbf{x}_0 \\
&= \left\{\mathbf{P}^3 - [\mathbf{P}^2\mathbf{q}, \mathbf{P}\mathbf{q}, \mathbf{q}]\begin{bmatrix}\mathbf{f}' \\ \mathbf{f}'(\mathbf{P} - \mathbf{q}\mathbf{f}') \\ \mathbf{f}'(\mathbf{P} - \mathbf{q}\mathbf{f}')^2\end{bmatrix}\right\}\mathbf{x}_0.
\end{aligned}$$

Proceeding in like manner, we obtain the nth state which must be at the origin of the state space:

$$\mathbf{x}_n = 0 = \left\{ \mathbf{P}^n - [\mathbf{P}^{n-1}\mathbf{q}, \ldots, \mathbf{Pq}, \mathbf{q}] \begin{bmatrix} \mathbf{f}' \\ \mathbf{f}'(\mathbf{P} - \mathbf{qf}') \\ \vdots \\ \mathbf{f}'(\mathbf{P} - \mathbf{qf}')^{n-1} \end{bmatrix} \right\} \mathbf{x}_0 .$$

Thus

$$\begin{bmatrix} \mathbf{f}' \\ \mathbf{f}'(\mathbf{P} - \mathbf{qf}') \\ \vdots \\ \mathbf{f}'(\mathbf{P} - \mathbf{qf}')^{n-1} \end{bmatrix} = [\mathbf{P}^{n-1}\mathbf{q}, \ldots, \mathbf{Pq}, \mathbf{q}]^{-1} \mathbf{P}^n . \qquad (11\text{-}45)$$

The inverse matrix on the right-hand side of Eq. (11 − 45) exists if the system is controllable [Eq. (3–58), restated]:

$$|\mathbf{P}^{n-1}\mathbf{q}, \ldots, \mathbf{Pq}, \mathbf{q}| \neq 0 .$$

Therefore, the required control law $\mathbf{f}'$ is given by the first row of the right-hand-side matrix of Eq. (11–45) when the system is controllable. In general, a set of n steps is needed for the control. However, depending on the initial state, fewer steps than n may complete the control.

Example 11-2 Consider a single-time-constant system with a delay (Fig. 11-16a), where $L = T$. The pulse-transfer function of the plant

$$G_p(z) = \frac{1 - a}{z(z - a)}$$

may be expressed by the following state-space equations:

$$x_{1,k+1} = (1 - a)u_k , \qquad x_{2,k+1} = x_{1,k} + ax_{2,k} ;$$

thus,

$$\mathbf{P} = \begin{bmatrix} 0 & 0 \\ 1 & a \end{bmatrix}, \qquad \mathbf{q} = \begin{bmatrix} 1 - a \\ 0 \end{bmatrix},$$

We have:

$$\mathbf{Pq} = \begin{bmatrix} 0 \\ 1 - a \end{bmatrix}, \qquad [\mathbf{Pq}, \mathbf{q}] = \begin{bmatrix} 0 & 1 - a \\ 1 - a & 0 \end{bmatrix},$$

and

$$[\mathbf{Pq}, \mathbf{q}]^{-1}\mathbf{P}^2 = \begin{bmatrix} a/(1 - a) & a^2/(1 - a) \\ 0 & 0 \end{bmatrix};$$

thus,

$$\mathbf{f}' = \begin{bmatrix} \dfrac{a}{1 - a}, & \dfrac{a^2}{1 - a} \end{bmatrix}.$$

The control law, however, has two defects: (1) it has a steady-state error (offset)—in other words, if the system is started from a zero state with a unit-step reference input, the final state will be less than 1; and (2) it requires state-variable feedback.

Example 11-3 To avoid the offset we found in Example 11-2, we apply the velocity algorithm instead of the position algorithm. Since an integrator, $z/(z-1)$ relates the controller output and manipulator position (valve opening) (Fig. 11-18c), we consider the original plant and an integrator combined,

$$G'_p(z) = \left(\frac{z}{z-1}\right)\left(\frac{1-a}{z(z-a)}\right) = \frac{1-a}{(z-1)(z-a)},$$

as a new plant transfer function. The state equations for this case are

$$x_{1,k+1} = x_{1,k} + (1-a)u_k, \qquad x_{2,k+1} = x_{1,k} + ax_{2,k},$$

or

$$\mathbf{P} = \begin{bmatrix} 1 & 0 \\ 1 & a \end{bmatrix}, \qquad \mathbf{q} = \begin{bmatrix} 1-a \\ 0 \end{bmatrix}.$$

Computing $(\mathbf{Pq}, \mathbf{q})^{-1}$, we obtain

$$\mathbf{f}' = \begin{bmatrix} \dfrac{1+a}{1-a}, & \dfrac{a^2}{1-a} \end{bmatrix}. \tag{11-46}$$

To avoid the second difficulty raised in Example 11-2, let us develop an algorithm which does not require state-variable feedback. Suppose that the available measurement is a scalar quantity y:

$$y_k = \mathbf{c}\mathbf{x}_k, \qquad \text{where} \qquad \mathbf{c} = [c_1, \ldots, c_n]. \tag{11-47}$$

For brevity let us suppose that the system is of second order. The initial measurement is

$$y_0 = \mathbf{c}\mathbf{x}_0,$$

and it is impossible to determine two state variables $x_{1,0}$ and $x_{2,0}$ from one measurement. So we take the next measurement,

$$y_1 = \mathbf{c}\mathbf{x}_1 = \mathbf{c}\mathbf{P}\mathbf{x}_0 + \mathbf{c}\mathbf{q}u_0.$$

Combining the two measurements in a vector form, we have

$$\begin{bmatrix} y_0 \\ y_1 \end{bmatrix} = \begin{bmatrix} \mathbf{c} \\ \mathbf{c}\mathbf{P} \end{bmatrix} \mathbf{x}_0 + \begin{bmatrix} 0 \\ \mathbf{c}\mathbf{q} \end{bmatrix} u_0,$$

or

$$\mathbf{x}_0 = \begin{bmatrix} \mathbf{c} \\ \mathbf{c}\mathbf{P} \end{bmatrix}^{-1} \left\{ \begin{bmatrix} y_0 \\ y_1 \end{bmatrix} - \begin{bmatrix} 0 \\ \mathbf{c}\mathbf{q} \end{bmatrix} u_0 \right\},$$

where the matrix inverse exists if the system is observable. Therefore, the initial state of a second-order system can be determined at $t = T$. In general, the initial state of an nth-order system can be identified at the $(n-1)$th step by a scalar measurement if the system is observable.

The control algorithm of the second-order system is

$$u_1 = -\mathbf{f}'\mathbf{x}_1 = -\mathbf{f}'\mathbf{P}\mathbf{x}_0 - \mathbf{f}'\mathbf{q}u_0$$

$$= -\mathbf{f}'\mathbf{P}\begin{bmatrix} \mathbf{c} \\ \mathbf{cP} \end{bmatrix}^{-1} \left\{ \begin{bmatrix} y_0 \\ y_1 \end{bmatrix} - \begin{bmatrix} 0 \\ \mathbf{cq} \end{bmatrix} u_0 \right\} - \mathbf{f}'\mathbf{q}u_0 , \qquad (11\text{-}48)$$

which is a difference equation in terms of present and past measurements and past control; hence the problem is solved. It is possible to derive a similar difference equation for an nth-order system, for which $n-1$ steps are required for identification and then n steps for control, or with a total of $(2n-1)$ periods.

Example 11-4 We reconsider the system of Example 11-3 with a measurement

$$\mathbf{c} = [0 \;\; 1] \quad \text{or} \quad y = x_2 .$$

We have

$$\begin{bmatrix} \mathbf{c} \\ \mathbf{cP} \end{bmatrix} = \begin{bmatrix} 0 & 1 \\ 1 & a \end{bmatrix};$$

hence

$$\begin{bmatrix} \mathbf{c} \\ \mathbf{cP} \end{bmatrix}^{-1} = \begin{bmatrix} -a & 1 \\ 1 & 0 \end{bmatrix}, \quad \text{and} \quad \mathbf{cq} = 0 ,$$

and thus

$$\mathbf{x}_0 = \begin{bmatrix} -a & 1 \\ 1 & 0 \end{bmatrix} \begin{bmatrix} y_0 \\ y_1 \end{bmatrix} = \begin{bmatrix} y_1 - ay_0 \\ y_0 \end{bmatrix} .$$

$\mathbf{f}'$ was determined in Example 11-3, and $\mathbf{f}'\mathbf{q} = 1 + a$. Therefore,

$$u_1 = -\begin{bmatrix} \dfrac{1+a}{1-a}, & \dfrac{a^2}{1-a} \end{bmatrix} \begin{bmatrix} 1 & 0 \\ 1 & a \end{bmatrix} \begin{bmatrix} y_1 - ay_0 \\ y_0 \end{bmatrix} - (1+a)u_0 .$$

Letting $u_1 \to zU$, $u_0 \to U$, $y_1 \to zY$, $y_0 \to Y$, we obtain

$$-D(z) = \frac{U(z)}{Y(z)} = \frac{1}{(1-a)} \frac{a(1+a) - (a^2 + a + 1)z}{z + (1+a)} . \qquad (11\text{-}49)$$

Since there is no pole cancellation between $D(z)$ and $G'_p(z)$, controllability is assured. By Eq. (11-37) the closed-loop pulse-transfer function is found to be

$$\frac{C(z)}{R(z)} = \frac{[a(1+a) + 1]z - a(1+a)}{z^3} . \qquad (11\text{-}50)$$

According to the final-value theorem in the z-domain [3, 4],

$$\lim_{n \to \infty} f(nT) = \lim_{z \to 1} (z-1)F(z) .$$

11-4 SOME MULTIVARIABLE AND OPTIMAL-CONTROL SYSTEMS

The right-hand side of Eq. (11-50) is 1 when $z = 1$; hence there is no offset. The response of the system for a unit-step reference input (with zero initial state) is shown in Fig. 11-18(b). Since controllability is assured, the controller also works to counteract a load upset.

The finite-time-settling control presented in the last section can be generalized for a vector (or multiple) input system described by

$$\mathbf{x}_{k+1} = \mathbf{P}\mathbf{x}_k + \mathbf{Q}\mathbf{u}_k, \tag{11-51}$$

where $\mathbf{u}_k$ is an r-vector and $\mathbf{Q}$ is an $n \times r$ matrix. If $m = n/r$ is an integer and if the controllability condition [Eq. (11-54) below] is satisfied, an arbitrary state $\mathbf{x}_0$ can be brought to the origin in m steps. The control law for this case is an $r \times n$ matrix $\mathbf{F}$ such that

$$\mathbf{u}_k = -\mathbf{F}\mathbf{x}_k. \tag{11-52}$$

We find $\mathbf{F}$ in much the same way as we did in the preceding section, by computing the sequence of states until the final state $\mathbf{x}_m = \mathbf{0}$ is reached. We have:

$$\mathbf{x}_1 = \mathbf{P}\mathbf{x}_0 + \mathbf{Q}\mathbf{u}_0 = (\mathbf{P} - \mathbf{Q}\mathbf{F})\mathbf{x}_0,$$

$$\vdots$$

$$\mathbf{x}_m = \mathbf{0} = \left\{ \mathbf{P}^m - [\mathbf{P}^{m-1}\mathbf{Q}, \ldots, \mathbf{P}\mathbf{Q}, \mathbf{Q}] \begin{bmatrix} \mathbf{F} \\ \mathbf{F}(\mathbf{P} - \mathbf{Q}\mathbf{F}) \\ \vdots \\ \mathbf{F}(\mathbf{P} - \mathbf{Q}\mathbf{F})^{m-1} \end{bmatrix} \right\} \mathbf{x}_0.$$

Solving the last equation for the control sequence, we obtain

$$\begin{bmatrix} \mathbf{F} \\ \mathbf{F}(\mathbf{P} - \mathbf{Q}\mathbf{F}) \\ \vdots \\ \mathbf{F}(\mathbf{P} - \mathbf{Q}\mathbf{F})^{m-1} \end{bmatrix} = [\mathbf{P}^{m-1}\mathbf{Q}, \ldots, \mathbf{P}\mathbf{Q}, \mathbf{Q}]^{-1}\mathbf{P}^m. \tag{11-53}$$

The inverse matrix in this expression exists when

$$|\mathbf{P}^{m-1}\mathbf{Q}, \ldots, \mathbf{P}\mathbf{Q}, \mathbf{Q}| \neq 0, \tag{11-54}$$

and, if this condition is satisfied, the required control law $\mathbf{F}$ will be given by the first r rows of Eq. (11-53). If n/r is not an integer, however, a modification in the control algorithm becomes necessary. We shall show the modification by a simple example.

Example 11-5 Let us consider a third-order system ($n = 3$) with two controls ($r = 2$):

$$\mathbf{x}_{k+1} = \mathbf{P}\mathbf{x}_k + [\mathbf{q}_1, \mathbf{q}_2] \begin{bmatrix} u_k \\ v_k \end{bmatrix}.$$

We let $u_0 = -\mathbf{f}'\mathbf{x}_0$, $v_0 = -\mathbf{g}'\mathbf{x}_0$, for the first step, so that

$$\mathbf{x}_1 = (\mathbf{P} - \mathbf{q}_1\mathbf{f}' - \mathbf{q}_2\mathbf{g}')\mathbf{x}_0 .$$

We then apply only one of the controls in the next (and final) step; for instance,

$$u_1 = -\mathbf{f}'\mathbf{x}_1 \quad \text{and} \quad v_1 = 0,$$
$$\mathbf{x}_2 = 0 = (\mathbf{P} - \mathbf{q}_1\mathbf{f}')\mathbf{x}_1$$
$$= \left\{\mathbf{P}^2 - [\mathbf{Pq}_1, \mathbf{Pq}_2, \mathbf{q}_1] \begin{bmatrix} \mathbf{f}' \\ \mathbf{g}' \\ \mathbf{f}'(\mathbf{P} - \mathbf{q}_1\mathbf{f}' - \mathbf{q}_2\mathbf{g}') \end{bmatrix}\right\} \mathbf{x}_0 ;$$

hence

$$\begin{bmatrix} \mathbf{f}' \\ \mathbf{g}' \\ \mathbf{f}'(\mathbf{P} - \mathbf{q}_1\mathbf{f}' - \mathbf{q}_2\mathbf{g}') \end{bmatrix} = [\mathbf{Pq}_1, \mathbf{Pq}_2, \mathbf{q}_1]^{-1}\mathbf{P}^2 , \tag{11-55}$$

where the following condition must be satisfied:

$$|\mathbf{Pq}_1, \mathbf{Pq}_2, \mathbf{q}_1| \neq 0 .$$

The necessary algorithms, $\mathbf{f}'$, $\mathbf{g}'$, are given by the first two rows of Eq. (11–55), but $\mathbf{g}'$ must be switched to $\mathbf{0}'$ in the last sampling period.

In some cases, the amplitude of the controlling pulse(s) in a minimal-time control may become excessively large for a given system. It is then possible to increase the number of finite-time-settling periods from a minimum number (which is equal to the order n of the system) to some specified number N, which is greater than n, and choose a control sequence with the constraint that minimizes the following performance index.

$$J = \sum_{i=0}^{N-1} u_i^2 . \tag{11-56}$$

The terminal state $\mathbf{x}_N$ is required to be at the origin, that is,

$$\mathbf{x}_N = 0 = \mathbf{P}^N\mathbf{x}_0 + [\mathbf{P}^{N-1}\mathbf{q}, \ldots, \mathbf{Pq}, \mathbf{q}]\mathbf{u} , \tag{11-57}$$

where the N-vector $\mathbf{u}$ represents the control sequence:

$$\mathbf{u}' = [u_0, u_1 \ldots, u_{N-1}] .$$

Rearranging Eq. (11–57) and left-multiplying by $\mathbf{P}^{-N}$, we obtain

$$\mathbf{x}_0 = \mathbf{Fu} , \tag{11-58}$$

where

$$\mathbf{F} = -\mathbf{P}^{-N}[\mathbf{P}^{N-1}\mathbf{q}, \ldots, \mathbf{Pq}, \mathbf{q}] .$$

Since $\mathbf{F}$ is an $n \times N$ rectangular matrix, $\mathbf{F}^{-1}$ does not exist; hence the pseudoinverse (Appendix A–4) is required. Equation (11–58), left-multiplied by an $n \times n$ identity

$$\mathbf{F}\mathbf{F}'(\mathbf{F}\mathbf{F}')^{-1} = \mathbf{I}_n,$$

becomes

$$\mathbf{F}\mathbf{F}'(\mathbf{F}\mathbf{F}')^{-1}\mathbf{x}_0 = \mathbf{I}_n\mathbf{F}\mathbf{u} = \mathbf{F}\mathbf{u};$$

and canceling out* the left-end $\mathbf{F}$, we obtain

$$\mathbf{u} = \mathbf{F}'(\mathbf{F}\mathbf{F}')^{-1}\mathbf{x}_0. \tag{11-59}$$

As shown in Appendix A–4, $\mathbf{u}$ given by this pseudoinverse is the solution for which the norm $\|\mathbf{u}\|$ is minimal and thus satisfies the performance requirement [21].

It must be noted that Eq. (11–59) does *not* yield a feedback control law. The optimal-control sequence $\mathbf{u}$ is fixed once and for all at the initial instant, when a complete knowledge of the initial state $\mathbf{x}_0$ is necessary for computation. If we adopt the first row vector of $\mathbf{F}'(\mathbf{F}\mathbf{F}')^{-1}$ as a feedback control law $\mathbf{f}$, as we did before, the control sequence will not take the Nth state to the origin. In the limiting case, $n = N$, the minimum right inverse reduces to $\mathbf{F}^{-1}$ (which exists when the system is controllable),

$$\mathbf{F}'(\mathbf{F}\mathbf{F}')^{-1} = \mathbf{F}^{-1} \quad \text{if} \quad N = n.$$

The solution of Eq. (11–59) in this special case agrees with what we discussed earlier in this section, and feedback control is possible.

Example 11-6 If the control object is an integrator, then

$$G_p(s) = \frac{1}{s}.$$

For an input u_k held constant for the period $kT < t < (k+1)T$, the difference equation is

$$x_{k+1} = x_k + Tu_k.$$

Therefore,

$$\mathbf{P} = 1 \quad \text{and} \quad \mathbf{q} = T.$$

If we specify $N = 3$, then

$$\mathbf{F} = -[T, T, T]$$

and

$$\mathbf{u} = -\begin{bmatrix} T \\ T \\ T \end{bmatrix} \left\{ [T, T, T] \begin{bmatrix} T \\ T \\ T \end{bmatrix} \right\}^{-1} x_0 = -\frac{1}{3T} \begin{bmatrix} 1 \\ 1 \\ 1 \end{bmatrix} x_0;$$

---

* Such cancellation is not always allowed, but in this case we get a pseudoinverse solution.

that is,
$$u_0 = u_1 = u_2 = -\frac{1}{3T}.$$

The transient response is
$$x_1 = \tfrac{2}{3}x_0, \qquad x_2 = \tfrac{1}{3}x_0, \qquad x_3 = 0.$$

It is obvious that a feedback control law $u_k = -x_k/3T$ will not make $x_N = 0$ at $N = 3$.

If the number of permissible control steps (N) is less than the order (n) of the control object, $n > N$, then it is, in general, not possible to expect $\mathbf{x}_N$ to be $\mathbf{0}$. The most we can do to optimize the control seems to be to minimize the distance of terminal state $\mathbf{x}_N$ from the origin. Thus

$$J = \|\mathbf{x}_N\|^2 = \mathbf{x}_N' \mathbf{x}_N \tag{11-60}$$

is the performance index to be minimized. For the control sequence $u_0, u_1, \ldots, u_{N-1}$ we have

$$\mathbf{x}_N = \mathbf{P}^N \mathbf{x}_0 + \mathbf{H}\mathbf{u}, \tag{11-61}$$

where
$$\mathbf{H} = [\mathbf{P}^{N-1}\mathbf{q}, \ldots, \mathbf{P}\mathbf{q}, \mathbf{q}]$$

is an $n \times N$ matrix. Cadzow [21] gives the following solution to this problem:

$$\mathbf{u} = -(\mathbf{H}'\mathbf{H})^{-1}\mathbf{H}'\mathbf{P}^N \mathbf{x}_0. \tag{11-62}$$

The control sequence $\mathbf{u}$ must be computed once and for all at the initial instant, when $\mathbf{x}_0$ must be identified. In other words, it is an open-loop (or program) control with a complete state measurement.

The algorithm of Eq. (11-62) is based on the minimum left inverse of $\mathbf{H}$ which is given in Appendix A-5, restated

$$\mathbf{H}^{LM} = (\mathbf{H}'\mathbf{H})^{-1}\mathbf{H}',$$

and the control is
$$\mathbf{u}^0 = -\mathbf{H}^{LM} \mathbf{P}^N \mathbf{x}_0.$$

For the proof that
$$\|\mathbf{x}_N\| = \|\mathbf{H}\mathbf{u} + \mathbf{P}^N \mathbf{x}_0\| \tag{11-63}$$

is minimal when $\mathbf{u} = \mathbf{u}^0$, let $\mathbf{P}^N \mathbf{x}_0 = \mathbf{y}$ and pose the problem of approximately solving the equation

$$\mathbf{H}\mathbf{u} = -\mathbf{y} \tag{11-64}$$

for **u**, where **u** is an N-dimensional vector, and **y** is an n vector, $N < n$. This problem is treated in Appendix A-5.

A complete state-vector measurement was assumed in all the preceding discussions of this section. In practice, however, this measurement is usually hard to accomplish. The modal-control approach, presented in Section 10-3 for continuous-time, lumped-parameter objects, also applies to discrete-time control systems, and provides us with a design guide to deal with systems that have imperfections in state measurement and/or control application [22]. For the ideal case presented in the first part of Section 10-3 [Eqs. (10-44) through (10-47)] a discrete-time modal-control law exists that will bring a state $\mathbf{x}_0$ into the origin **0** with one control, that is, within one sampling period. To obtain the control law, consider an ideal system

$$\mathbf{x}_{k+1} = \mathbf{P}\mathbf{x}_k + \mathbf{Q}\mathbf{u}_k, \qquad \mathbf{y}_k = \mathbf{C}\mathbf{x}_k, \qquad (11\text{-}65)$$

where both $\mathbf{Q}^{-1}$ and $\mathbf{C}^{-1}$ exist. The control law **K** is an $n \times n$ matrix such that

$$\mathbf{u}_k = -\mathbf{K}\mathbf{y}_k \qquad (11\text{-}66)$$

and, upon substitution into Eq. (11-65), yields

$$\mathbf{x}_1 = \mathbf{P}\mathbf{x}_0 - \mathbf{QKC}\mathbf{x}_0 = \mathbf{0};$$

hence $\mathbf{P} = \mathbf{QKC}$, or

$$\mathbf{K} = \mathbf{Q}^{-1}\mathbf{P}\mathbf{C}^{-1}. \qquad (11\text{-}67)$$

If the number of measurements (dimension m of $\mathbf{y}_k$) and/or the number of controls (dimension r of $\mathbf{u}_k$) are less than the order n of the control object, we convert Eq. (11-65) into the modal domain to design a mode-oriented control algorithm. With the modal matrix **T**, such that $\mathbf{x}_k = \mathbf{T}\mathbf{x}_k^*$, we obtain the modal-domain equations

$$\mathbf{x}_{k+1}^* = \mathbf{P}^*\mathbf{x}_k^* + \mathbf{Q}^*\mathbf{u}_k, \qquad \mathbf{y}_k = \mathbf{C}^*\mathbf{x}_k^*, \qquad (11\text{-}68)$$

where $\mathbf{Q}^* = \mathbf{T}^{-1}\mathbf{Q}$, $\mathbf{C}^* = \mathbf{CT}$ are modal-domain input and output matrices, respectively, and $\mathbf{P}^* = \mathbf{T}^{-1}\mathbf{PT}$ is in the diagonal or Jordan canonical form, that has eigenvalues $z_1, z_2, \ldots$ of the system as its elements in descending order of their norms:

$$|z_1| \geqslant |z_2| \geqslant |z_3| \geqslant \cdots$$

As in Eqs. (10-48) and (10-51), the rectangular matrices for observation and/or control are expressed in terms of the square matrix $[\mathbf{F}_m, \mathbf{H}_r]$ and the rectangular remainder $[\Delta\mathbf{F}_m, \Delta\mathbf{H}_r]$:

$$\mathbf{CT} = \mathbf{C}^* = [\mathbf{F}_m, \Delta\mathbf{F}_m], \qquad \mathbf{T}^{-1}\mathbf{Q} = \mathbf{Q}^* = \begin{bmatrix} \mathbf{H}_r \\ \Delta\mathbf{H}_r \end{bmatrix}. \qquad (11\text{-}69)$$

For the ideal case, where $m = r = q$ and

$$\Delta F_m = 0, \quad \Delta H_r = 0, \quad (11\text{-}70)$$

a control law K_q exists (if $|F_m| \neq 0$ and $|H_r| \neq 0$) that will bring the first q modes (that is, q-dimensional vector x_q^*) into the origin faster, while leaving the remaining $n - q$ modes for natural decay. K_q is a matrix that can be determined by

$$x_{q,1}^* = P_q^* x_{q,0}^* + H_r u_0, \quad u_0 = -K_q y_0 = -K_q F_m x_{q,0}^*,$$

and $x_{q,1}^* = 0$ for one-shot control. Hence

$$K_q = H_r^{-1} P_q^* F_m^{-1}, \quad (11\text{-}71)$$

where P_q^* is the $q \times q$ square matrix in the upper left-hand corner of P^*.

If the control is ideal ($\Delta H_r = 0$) but the measurement is not ($\Delta F_m \neq 0$), then we consider an $r \times m$ rectangular matrix control law K to produce the control vector q_0,

$$u_0 = -K y_0 = -K[F_m \; \Delta F_m] x_0^*. \quad (11\text{-}72)$$

Substituting Eq. (11-72) into the modal-domain state equation, we obtain

$$x_1^* = \left\{ P^* - \begin{bmatrix} H_r \\ 0 \end{bmatrix} K[F_m \; \Delta F_m] \right\} x_0^*. \quad (11\text{-}73)$$

Taking the first r rows of this expression, we have

$$x_{r,1}^* = (P_r^* - H_r K F) x_0^* = H_r(H_r^{-1} P_r^* - K F) x_0^*, \quad (11\text{-}74)$$

where P_r^* consists of the first r rows of P^*, and $F = [F_m, \Delta F_m]$. If H_r^{-1} exists, then K can be fixed as an approximate solution of

$$(H_r^{-1} P_r^* - KF)' = (P_r^*)'(H_r^{-1})' - F'K' \approx 0.$$

To solve this equation, we take the left inverse of F', which turns out to be the right inverse of F,

$$[(F')^{LM}]' = [(FF')^{-1}F]' = F'(FF')^{-1} = F^{RM},$$

and obtain

$$K' = (F')^{LM} P_r^{*\prime} (H_r^{-1})'$$

or

$$K = H_r^{-1} P_r^* F^{RM}. \quad (11\text{-}75)$$

With this control the norm of each row vector of $(H_r^{-1} P_r^* - KF)$ will be minimized.

A situation complementary to the preceding case exists if measurement is ideal ($\Delta F_m = 0$) but control is not ideal ($\Delta H_r \neq 0$). In place of Eq.

11-4

(11-73), we now have

$$\mathbf{x}_1^* = \left\{ \mathbf{P}^* - \begin{bmatrix} \mathbf{H}_r \\ \Delta \mathbf{H}_r \end{bmatrix} \mathbf{K}[\mathbf{F}_m, \mathbf{0}] \right\} \mathbf{x}_0^*, \qquad (11\text{-}76)$$

where $\mathbf{K}$, the control law, is $r \times m$. To approximately solve

$$\mathbf{P}^* - \mathbf{HKF} = \mathbf{0}$$

for $\mathbf{K}$, we apply the left inverse of

$$\mathbf{H} = \begin{bmatrix} \mathbf{H}_r \\ \Delta \mathbf{H}_r \end{bmatrix},$$

$$\mathbf{H}^{LM} = (\mathbf{H}'\mathbf{H})^{-1}\mathbf{H}'.$$

The left inverse exists if the rank of $\mathbf{H}$ is r, and we obtain

$$\mathbf{H}^{LM}\mathbf{P}^* = \mathbf{K}[\mathbf{F}_m, \mathbf{0}];$$

thus

$$\mathbf{K} = \mathbf{H}^{LM}\mathbf{P}_m^*\mathbf{F}_m^{-1}, \qquad (11\text{-}77)$$

where $\mathbf{P}_m^*$ is the $n \times m$ matrix that consists of the first m columns of $\mathbf{P}^*$. With this control the norm of each column vector of $(\mathbf{P}_m^* - \mathbf{HKF}_m)$ will be minimized. If both measurement and control are nonideal, interactions (as noted in Section 10-3) will appear and, depending on the magnitude of the interactions, performance will be affected. As in continuous-time systems, the control law may be determined by

$$\mathbf{K} = \mathbf{H}^{LM}\mathbf{P}^*\mathbf{F}^{RM}. \qquad (11\text{-}78)$$

As noted in Section 11-3, the velocity algorithm (I-action) may be applied to decoupled control loops in place of the P-algorithm, if we so desire.

Example 11-7 Let

$$\mathbf{A} = \begin{bmatrix} -3 & 1 & 0 \\ 2 & -3 & 2 \\ 0 & 1 & -3 \end{bmatrix}, \quad \mathbf{B}_1 = \begin{bmatrix} 1 & 0 \\ 0 & 0 \\ 0 & 1 \end{bmatrix}, \quad \mathbf{B}_2 = \begin{bmatrix} 1 & 0 \\ 1 & 1 \\ 0 & 1 \end{bmatrix},$$

$$\mathbf{C}_1 = \begin{bmatrix} 0 & 1 & 0 \\ 0 & 0 & 1 \end{bmatrix}, \quad \mathbf{C}_2 = \begin{bmatrix} 1 & 0 & -1 \\ 2 & 1 & 0 \end{bmatrix}.$$

The eigenvalues are

$$p_1 = -1, \quad p_2 = -3, \quad p_3 = -5$$

and

$$\mathbf{T} = \begin{bmatrix} 1 & 1 & -1 \\ 2 & 0 & 2 \\ 1 & -1 & -1 \end{bmatrix}, \quad \mathbf{T}^{-1} = \frac{1}{4}\begin{bmatrix} 1 & 1 & 1 \\ 2 & 0 & -2 \\ -1 & 1 & -1 \end{bmatrix}.$$

For a sampling period $T_s = 0.5$ sec, we obtain

$$\mathbf{P}^* = \begin{bmatrix} e^{-T_s} & 0 & 0 \\ 0 & e^{-3T_s} & 0 \\ 0 & 0 & e^{-5T_s} \end{bmatrix} = \begin{bmatrix} 0.606 & 0 & 0 \\ 0 & 0.223 & 0 \\ 0 & 0 & 0.082 \end{bmatrix}.$$

Case 1. For ideal measurement and ideal control, $\mathbf{B} = \mathbf{B}_2$ and $\mathbf{C} = \mathbf{C}_2$, we have

$$\mathbf{F} = \mathbf{CT} = \begin{bmatrix} 0 & 2 & 0 \\ 4 & 2 & 0 \end{bmatrix}, \quad \mathbf{F}_m = \begin{bmatrix} 0 & 2 \\ 4 & 2 \end{bmatrix},$$

$$\mathbf{H} = \mathbf{T}^{-1}\mathbf{Q} = (\mathbf{P}^* - \mathbf{I})\boldsymbol{\Lambda}^{-1}\mathbf{T}^{-1}\mathbf{B}_2 = \begin{bmatrix} 0.197 & 0.197 \\ 0.129 & -0.129 \\ 0 & 0 \end{bmatrix},$$

$$\mathbf{H}_r = \begin{bmatrix} 0.197 & 0.197 \\ 0.129 & -0.129 \end{bmatrix},$$

and, by Eq. (11-71),

$$\mathbf{K} = \begin{bmatrix} 0.045 & 0.394 \\ -0.83 & 0.394 \end{bmatrix}.$$

Case 2. For ideal control and nonideal measurement, $\mathbf{B} = \mathbf{B}_2$, $\mathbf{C} = \mathbf{C}_1$, we have

$$\mathbf{F} = \mathbf{CT} = \begin{bmatrix} 2 & 0 & -2 \\ 1 & -1 & -1 \end{bmatrix},$$

and hence

$$\mathbf{F}^{RM} = \mathbf{F}'(\mathbf{FF}')^{-1} = \begin{bmatrix} \frac{1}{4} & \frac{1}{2} \\ 0 & -\frac{1}{2} \\ -\frac{1}{4} & \frac{1}{2} \end{bmatrix};$$

and, by Eq. (11-75), we obtain

$$\mathbf{K} = \begin{bmatrix} 0.394 & 0.35 \\ 0.394 & 1.22 \end{bmatrix}.$$

Case 3. For nonideal control and ideal measurement, $\mathbf{B} = \mathbf{B}_1$, $\mathbf{C} = \mathbf{C}_2$, we have

$$\mathbf{H} = (\mathbf{P}^* - \mathbf{I})\boldsymbol{\Lambda}^{-1}\mathbf{T}^{-1}\mathbf{B}_1 = \begin{bmatrix} 0.098 & 0.098 \\ 0.13 & -0.13 \\ -0.046 & -0.046 \end{bmatrix},$$

and

$$\mathbf{H}^{LM} = (\mathbf{H}'\mathbf{H})^{-1}\mathbf{H}' = \begin{bmatrix} 4.2 & 3.85 & -1.97 \\ 4.2 & -3.85 & -1.97 \end{bmatrix}.$$

By Eq. (11-77), we find that

$$\mathbf{K} = \begin{bmatrix} -0.19 & 0.62 \\ -1.05 & 0.62 \end{bmatrix}.$$

We mostly focused our attention on sampled data in the preceding treatment where the following assumptions were made:

(1) sampling period is fixed ($T = $ const);
(2) all sampling operations are performed synchronously; and
(3) all sampling operations are "ideal" in the sense that they are described by an impulse modulation (Eq. 11–3).

A linear digital control sysytem consists of continuous-time elements (control object, etc.), discontinuous-time elements (digital controller, etc.), sampler(s), and (zero-order) holder(s). Kalman and Bertram [23] pioneered in the state-space solution of linear, digital control systems where the solution (or transition) matrices (Section 3–1) are first determined for the continuous-time elements, then for the discrete-time elements, and then for the sample holders, and finally combined for an overall system. In following this procedure, the listed assumptions are not necessary.

REFERENCES

1. Oldenbourg, R. C., Sartorius, H., *Dynamik selbsttätiger Regelungen*. Munich: Oldenbourg, 1944.
2. Barker, R. H., "The pulse transfer function and its applications to sampling servo systems," *Proc. IEE*, Part IV, No. 43, July 1952, pp. 302–317.
3. Ragazzini, J. R., Franklin, G. F., *Sampled Data Control Systems*. New York: McGraw-Hill, 1958.
4. Jury, E. I., *Sampled Data Control Systems*. New York: Wiley, 1958.
5. Tou, J. T., *Digital and Sampled Data Control Systems*. New York: McGraw-Hill, 1958.
6. Kuo, B. C., *Analysis and Synthesis of Sampled Data Control Systems*. Englewood Cliffs, N. J.: Prentice Hall, 1963.
7. Jury, E. I., *Theory and Application of the Z-Transform Method*. New York: Wiley, 1964.
8. Centner, R. M., "How to design low-cost digital controls." *Control Eng.*, **11**, No. 2, February 1964, pp. 75–80.
9. Mergler, H. W., Hubbard, K. H., "Digital control for the single loop," *Control Eng.*, **12**, No. 2, February 1965, pp. 61–64.
10. "Guidelines and General Information on User Requirements Concerning Direct Digital Control," *First Users' Workshop on Direct Digital Control*, Princeton, New Jersey, April 1963.
11. Cox, J. B., Hellums, L. J., Williams, T. J., Banks, R. S., Kirk, G. J., Jr., "A practical spectrum of DDC chemical process control algorithms." *ISA-J*, October 1966, pp. 65–72.

12. Goff, K. W., "Dynamics in direct digital control," I and II. *ISA-J*, November 1966, pp. 45–49; December 1966, pp. 44–54.
13. Larson, R. E., "Optimum quantization in dynamic systems." *1966 JACC Preprint*, pp. 586–593.
14. DeBolt, R. R., Powell, R. E., "A natural 3-mode controller algorithm for DDC." *ISA-J*. September 1966, pp. 43–53.
15. Takahashi, Y., Harris, R., "Performance improvement of process controller by gain switching." *ASME Preprint 61-WA-238*.
16. Koepcke, R. W., "A discrete design method for digital control," *Control Eng.*, **13,** No. 6, June 1966, pp. 83–89.
17. Cundall, C. M., Latham, V., "Designing digital computer control systems," *Control Eng.*, **9,** No. 10, October 1962, pp. 82–86; **10,** No. 1, January 1963, pp. 109–113.
18. Zalkind, C. S., "Pulse duration control." *Control Eng.*, **14,** No. 2, February 1967, pp. 74–77.
19. Mori, M., "Discrete compensator controls dead-time process." *Control Eng.*, **9,** No. 1, January 1962, pp. 57–60.

    Soliman, J. L., Al-Shaikh, A., "Extension of the state-space method to the analysis of discrete systems with finite delays." *Electronics Letters*, **2,** No. 4, April 1966.
20. Mullin, F. J., DeBarbeyrac, J., "Linear digital control." *J. Basic Eng.*, *ASME Trans.*, **86,** March 1964, pp. 61–66.
21. Cadzow, J. A., "A study of minimal norm control for sampled data systems." *1965 JACC* Preprint, pp. 545–550.
22. Takahashi, Y., *et al.*, "Mode oriented design viewpoint for linear, lumped-parameter multivariable control systems." *J. Basic Eng.*, *ASME Trans.*, **90,** June 1968.
23. Kalman, R. E., Bertram, J. E., "A unified approach to the theory of sampling systems." *J. Franklin Inst.*, **265,** No. 5, May 1959, pp. 405–436.

PROBLEMS

11-1 The sampling rate of the servomechanism shown in Fig. P11-1 is 10 cps. Since the sampling period, $T = 0.1$ sec, is short relative to the speeds involved in normal operation of the servo, the discrete-time system may be approximated by a continuous-time system over some bandwidth. A 12-cps noise is introduced in the position detector. (a) What is the highest frequency of the reference input the sampling servo can handle? (b) Does the noise cause any difficulty?

11-2 A continuous-time pressure-control system is shown in Fig. P11-2, where the gain k of I-control is set at half of its stability limit value. Introduce a sampler/holder ahead of the open-loop transfer function and convert it to a sampled-data system of a sampling period $T = 5$ sec. Show, on the Bode plot of $|G(j\omega)|$, the first "slit" of the noise band that must be filtered to avoid noise aliasing.

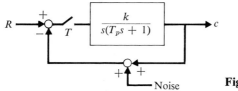

Fig. P11-1

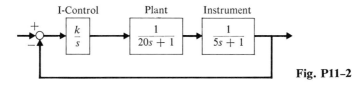

Fig. P11-2

11-3 There are two ways to determine the pulse-transfer function $G(z)$ of a continuous-time system $dx/dt = \mathbf{A}x + \mathbf{b}u$, $y = \mathbf{c}x$, with zero hold:

1. (*The classical approach*) $G(z) = (1 - z^{-1})\mathscr{Z}[G(s)/s]$, where

$$G(s) = \mathbf{c}(s\mathbf{I} - \mathbf{A})^{-1}\mathbf{b}$$

and

$$\mathscr{Z}[G(s)/s] = h_0 + h_1 z^{-1} + \cdots + h_k z^{-k} + \cdots,$$

and where

$$h_k = h(kT) = \mathscr{L}^{-1}\left[\frac{G(s)}{s}\right]_{t=kT}.$$

2. (*The modern approach*) $\mathbf{x}_{k+1} = \mathbf{P}\mathbf{x}_k + \mathbf{q}u_k$ for staircase-pattern input u_k, and $G(z) = \mathbf{c}(z\mathbf{I} - \mathbf{P})^{-1}\mathbf{q}$.

Assuming a diagonal form for the A-matrix,

$$\mathbf{A} = \begin{bmatrix} p_1 & & 0 \\ & p_2 & \\ 0 & & \ddots \end{bmatrix},$$

show that the two approaches give the same result.

11-4 Apply the two methods discussed in the preceding problem to determine $G(z)$ of the process shown in Fig. P11-4. Also obtain the state-space formulation. What kind of process is this?

11-5 For the open-loop discrete-time system shown in Fig. P11-5: (a) obtain the discrete-time state equation, and (b) determine the pulse-transfer functions $G_1(z)$ and $G_2(z)$ indicated in the Figure. Is $G_1(z)$ equal to the z-transform of $1/(s + 2)$?

11-6 A family tree is shown in Fig. P11-6, where each circle represents a couple. The time unit (sampling interval) is called a "year." According to this tree, a couple produces two new couples at age 1 (1-year old), and perishes shortly after

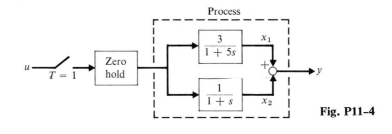

Fig. P11-4

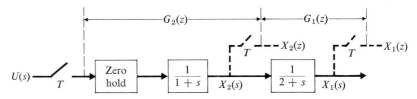

Fig. P11-5

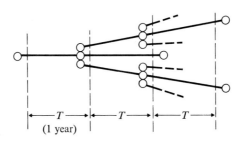

Fig. P11-6

two years. At $t = kT$, let x_{1k} = number of newly born couples x_{2k} = number of 1-year-old couples, x_{3k} = number of 2-year-old couples, and y_k = total population. Obtain the state-space formulation of the genealogical process. Is this reaction autocatalytic?

11-7 A second-order, oscillatory, discrete-time system is described by

$$\mathbf{x}_{k+1} = \mathbf{P}\mathbf{x}_k + \begin{bmatrix} q_1 \\ q_2 \end{bmatrix} u_k, \qquad y_k = [c_1 \quad c_2]\mathbf{x}_k,$$

where **P** is the matrix directly derived from the continuous-time system's **A**-matrix in its canonical form,

$$\mathbf{A} = \begin{bmatrix} -\sigma & \omega \\ -\omega & -\sigma \end{bmatrix}.$$

(See Problem 3–8.) State the conditions for the system to be controllable/observable.

11-8 In the PID-DDC algorithm,

$$\Delta m_k = K_p(c_{k-1} - c_k) + K_i(r_k - c_k) + K_d(2c_{k-1} - c_{k-2} - c_k) = m_k - m_{k-1},$$

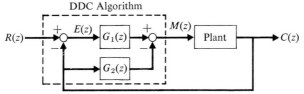

Fig. P11-8

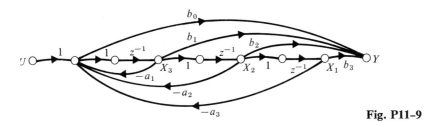

Fig. P11-9

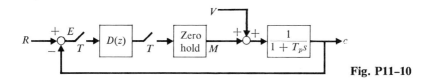

Fig. P11-10

c_k and r_k are taken as inputs, and m_k is considered to be the output. Obtain (a) the signal-flow diagram of the algorithm, (b) the state and output equations, and (c) $G_1(z)$ and $G_2(z)$ such that

$$M(z) = G_1(z)E(z) - G_2(z)C(z).$$

(See Fig. P11-8.) (d) Also compute the unit-step-input response of $G_1(z)$, $G_2(z)$, and a modified $G_2(z)$, where the original D-action is replaced by the four-point difference algorithm.

11-9 A signal-flow diagram in the canonical form of Fig. 2–10(b), with a slight modification, is shown in Fig. P11-9. Find the input-output relationship by the equation $D(z) = Y(z)/U(z)$. Also obtain the state-space formulation.

11-10 A control object is first order such that $e^{-T/T_p} = \frac{1}{2}$ (see Fig. P11-10). $D(z)$ is to be designed for minimal finite-time settling, and the closed-looped response is specified to be

$$G_{CR}(z) = z^{-1}.$$

(a) Obtain $D(z)$. (b) Determine e_k, m_k, and c_k for a unit-step input $R(z) = z/(z-1)$.

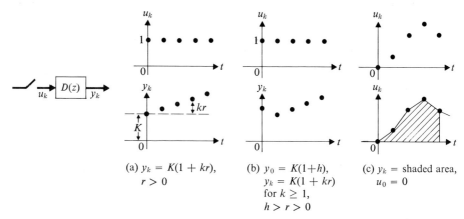

(a) $y_k = K(1 + kr)$, $r > 0$

(b) $y_0 = K(1+h)$, $y_k = K(1 + kr)$ for $k \geq 1$, $h > r > 0$

(c) y_k = shaded area, $u_0 = 0$

Fig. P11-11

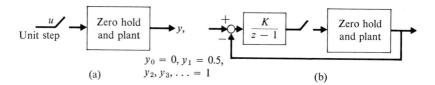

(a) $y_0 = 0, y_1 = 0.5,$ $y_2, y_3, \ldots = 1$

(b)

Fig. P11-12

(c) Investigate controllability and observability of the control system in the original state space and on the modal-domain signal-flow diagram.

11-11 Obtain $D(z)$, the digital control algorithm in the z-domain, for the following input–output pairs; (a) $y_k = K(1 + k_r)$, K and r constants, for a unit-step input, $u_k = 1$, $k = 0, 1, 2, \ldots$, (b) $y_0 = K(1 + h)$, $y_k = K(1 + kr)$, for $k = 1, 2, \ldots$ for a unit-step input, (c) y_k = area shown in Fig. P11-11(c).

11-12 A control object with a zero-holder is characterized by the unit-step-input response time series shown in Fig. P11-12. The control algorithm is a delayed I-action,

$$D(z) = \frac{k}{z - 1}.$$

Determine the stability condition of the control system in terms of k. Sketch the root locus in the z-plane.

11-13 We consider sampled-data control systems similar to the continuous-control system shown in part (a) of Fig. P11-13. The optimal gain of the continuous-control system is $k_{\text{opt}} = 1/RL = 1$ by Ziegler-Nichols' first rule, and $k_{\text{opt}} = \frac{1}{2}\pi/2 = 0.785$ by the ultimate sensitivity method [see Eqs. (8-65) and (8-66)], where $\pi/2$ is the stability limit gain. Three sampled-data systems are shown in parts (b) through (d) of the figure, where $T = 1$. Determine the stability limit gain and the optimal gain for the minimum of $J = \sum_{k=0}^{\infty} y_k^2$ for the free response of the three systems with

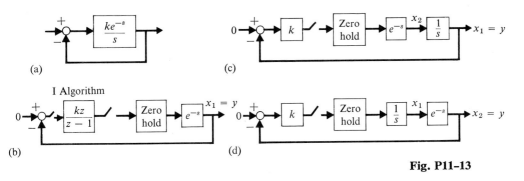

(a)

(b) I Algorithm

(c)

(d)

Fig. P11-13

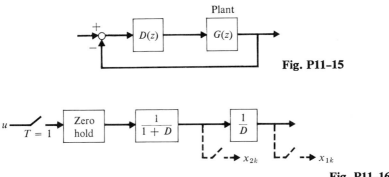

Fig. P11-15

Fig. P11-16

initial state

$$\mathbf{x}_0 = \begin{bmatrix} 1 \\ 1 \end{bmatrix}.$$

11-14 The control object is an integrator ($1/s$) with zero hold. The control algorithm is PI. (a) Draw a block diagram in the z-domain. (b) Take $x_1 =$ controlled variable, and $x_2 =$ controlling (or manipulated) variable, and obtain the P-matrix. (c) Determine the stability condition in terms of K_p and K_i in the control algorithm. (d) Discuss the response pattern at the stability limit.

11-15 A control object is described by

$$\mathbf{x}_{k+1} = \begin{bmatrix} 1 & 2 \\ 0 & 1 \end{bmatrix} \mathbf{x}_k + \begin{bmatrix} 1 \\ 2 \end{bmatrix} u_k, \quad y_k = \begin{bmatrix} 1 & 0 \end{bmatrix} \mathbf{x}_k,$$

(a) Determine $\mathbf{f}$ for state-variable feedback, minimal-time-settling control. (b) Obtain $D(z)$ for minimal-time settling via y_k feedback. (c) Check the result by computing $1 + D(z)G(z) = 0$ of the system shown in Fig. P11-15.

11-16 A control object is shown in Fig. P11-16, where the state variables x_1 and x_2 are both measured for the purpose of control. Determine the feedback control law for minimal-time settling. Also obtain the control law that will minimize the norm of the state vector in one sampling period.

11-17 Let

$$\mathbf{x}_{k+1} = \begin{bmatrix} \frac{1}{4} & -\frac{1}{2} & \frac{1}{2} \\ -\frac{1}{4} & \frac{1}{2} & -\frac{1}{4} \\ \frac{1}{4} & \frac{1}{4} & 0 \end{bmatrix} \mathbf{x}_k + \begin{bmatrix} 2 & -1 \\ -1 & 1 \\ -1 & 0 \end{bmatrix} \mathbf{u}_k, \quad \mathbf{y}_k = \begin{bmatrix} 3 & 2 & -2 \\ 1 & -1 & 1 \end{bmatrix} \mathbf{x}_k.$$

The system has eigenvalues $z_1 = \frac{3}{4}$, $z_2 = -\frac{1}{4}$, and $z_3 = \frac{1}{4}$. Determine the control law $\mathbf{K}$ such that with $\mathbf{u}_k = -\mathbf{K}\mathbf{y}_k$ the first two modes are brought to the origin in one sampling period.

PART 4

NONLINEAR, STOCHASTIC, OPTIMAL CONTROL AND LOGIC SYSTEMS

Chapter 12
Nonlinear Systems and Nonlinear Controls

Chapter 13
Linear Stochastic Systems

Chapter 14
Optimal Control

Chapter 15
Switching Control

12

NONLINEAR SYSTEMS

Nonlinearities in control systems may be either incidental or intentional. The incidental type of nonlinearities that are often forced on the designer includes saturation, Coulomb friction, backlash, parameter aging, and various nonlinear relations between system variables which occur in most physical systems. On the other hand, the designer can introduce intentional nonlinearities to either improve system performance or to effect economy in component selection.

In either case, the designer of nonlinear systems is usually confronted with relations for which no general mathematical solutions exist. The problem is compounded by the peculiar behavior of nonlinear systems: superposition no longer applies, the response of a nonlinear system often depends on its initial state, and the nature of the system transient usually changes at different nominal operating points in the state space. For all of these reasons, we are unable to present a unified and generalized method of nonlinear system analysis.

There are available, however, various broad avenues of approach that may be systematized, although, even within the scope of one approach, the solution methods vary greatly from problem to problem. We will thus have to resort to frequent examples to illustrate the techniques involved. The ultimate choice of analytical tools must be compatible with the design objectives of the particular problem.

We initially consider piecewise linear systems which, when second-order, may be conveniently analyzed on the state (or phase) plane. Here we will use both graphical and analytical techniques to investigate on–off control action. This is followed by an approximate analysis of nonlinear systems in the frequency domain based on the describing-function approach to investigate the existence of limit cycles (sustained oscillations of a specific period and amplitude).

To consider the stability of nonlinear systems, we shall briefly discuss the direct method of Lyapunov (Chapter 4). Although the method is general enough to cover all nonlinear systems, there is no assurance that an appropriate Lyapunov function can be constructed for a given system. To present a guaranteed, but unfortunately only sufficient, test for asymptotic stability, we will review some recent work by Popov and others in the frequency domain.

510 Nonlinear systems

In the Chapter 14 we will see that, according to Pontryagin's maximum principle, in order to achieve optimal control for a particular performance index a nonlinear mode of control becomes necessary. This method, along with the Posicast approach and the problem of adaptive control, comes under the general heading of positive use of nonlinear action (that is, the intentional nonlinearities mentioned above). The chapter ends with a discussion of hill-climbing and peak-holding methods of control.

12-1 ON-OFF CONTROL SYSTEMS IN THE STATE PLANE

An nth-order dynamic system is in general described by the vector differential equation

$$\frac{d\mathbf{x}}{dt} = \mathbf{f}(\mathbf{x}, \mathbf{u}), \tag{12-1}$$

where $\mathbf{x}$ is an n-dimensional state vector, and $\mathbf{u}$ is an r-dimensional input vector; $\mathbf{f}$ is a nonlinear vector function of $\mathbf{x}$ and $\mathbf{u}$ when the system is nonlinear. In this section we consider a special case of nonlinear systems, for which the control object is piecewise linear, expressed by

$$\frac{d\mathbf{x}}{dt} = \mathbf{A}\mathbf{x} + \mathbf{b}m, \tag{12-2}$$

and where m is a controlling input applied over a specified range of $\mathbf{x}$. Throughout this section we shall study the system via state-space trajectory analysis, or its analytical equivalent.

There is a class of control laws where the controlling signal can take on only two or three values. A typical example of this type of control is a relay control with dead zone, as shown in Fig. 12–1(a). Suppose that the desired state of the plant is constant for all time, and this state is taken as the origin of the state space. The input to the contactor, called a switching function, is then a function of measured variables, which in turn are a function of the state. If the dead (or inactive) zone of the relay is sufficiently small, the relay action may be expressed as

$$m(t) = M \operatorname{sgn} f(\mathbf{x}), \tag{12-3}$$

where sgn is the sign function, and M is a scalar magnitude (Fig. 12–1a). For a form of the switching function $f(\mathbf{x})$,

$$f(\mathbf{x}) = -k_1 x_1 - \cdots - k_n x_n = -\mathbf{k}'\mathbf{x}, \tag{12-4}$$

where

$$\mathbf{k}' = [k_1, \cdots, k_n],$$

12-1 On-off control systems in the state plane

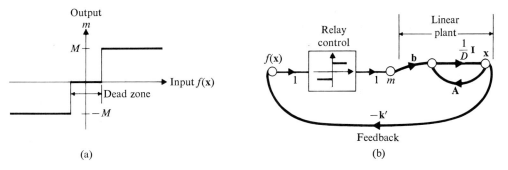

Fig. 12-1 Relay control system.

Eq. (12-2) becomes

$$\frac{dx}{dt} = \mathbf{A}\mathbf{x} - \mathbf{b}M \operatorname{sgn}(\mathbf{k'}\mathbf{x}). \tag{12-5}$$

The sign function in this equation is $+1$ on one side of the hyperplane

$$\mathbf{k'x} = 0, \tag{12-6}$$

and it is -1 on the other side. In other words, the state space can be divided by the hyperplane into two halves, and the system will be linear in each. Such a system is said to be piecewise linear. The trajectory of such a system can be determined by combining linear trajectories on each side of the hyperplane. If a piecewise linear system is second-order, or can be approximated as such, the isocline method (Section 3–4) applies to each linear zone of the state plane. This fact greatly simplifies the graphical solution of the system's motion.

Example 12-1 Let us consider the simple servomotor-type plant described by

$$\frac{d^2x}{dt^2} + \frac{dx}{dt} = m,$$

where $x =$ displacement and $m =$ force (or torque). Letting $x = x_1$ and $dx_1/dt = x_2$, we obtain the state-space expression

$$\frac{dx_1}{dt} = x_2, \qquad \frac{dx_2}{dt} = -x_2 + m, \tag{12-7}$$

where

$$m = -\operatorname{sgn}(k_1 x_1 + k_2 x_2) \cdot M.$$

Dividing both sides of the second equation of (12-7) by the first (see Section 3–4), we find the relation for the isocline method to be

$$\frac{dx_2}{dx_1} = -1 + \frac{(\pm M)}{x_2},$$

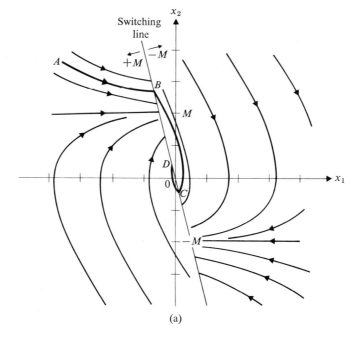

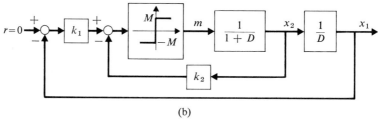

Fig. 12-2 Contactor servomechanism with rate feedback.

where we take $+M$ on the left-hand side of the switching line $k_1 x_1 + k_2 x_2 = 0$, and $-M$ on the right-hand side (Fig. 12-2). Trajectories can be easily determined on both sides of the switching line, as shown in Fig. 12-2(a), where $ABCD$ is a sample trajectory. It is obvious that all trajectories lead to the origin regardless of the initial state; hence the system is asymptotically stable in the large (see Chapter 4).

Example 12-2 Let us reconsider the CR control object introduced in Fig. 10-13 and restated in Fig. 12-3(a). The object is expressed by

$$\frac{dx_1}{dt} = -3x_1 + 2x_2, \quad \frac{dx_2}{dt} = 4x_1 - 5x_2 + m. \tag{12-8}$$

We now take as the control law an on–off action with a differential gap (or hysteresis), as shown in Fig. 12-3(b). Let us assume that the switching function in this example is simply an error signal,

$$f(\mathbf{x}) = r - x_1,$$

12-1 On-off control systems in the state plane

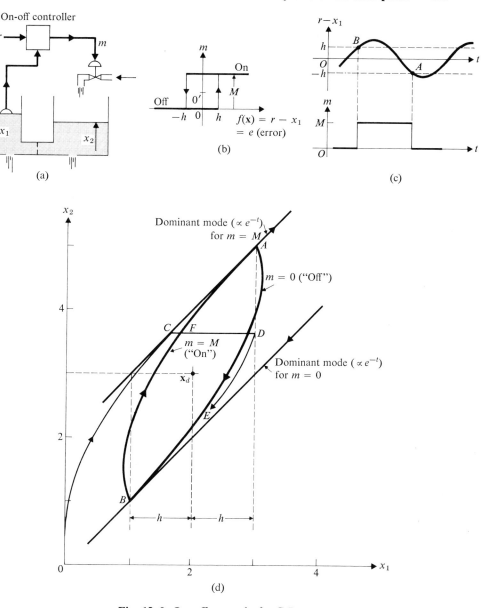

Fig. 12-3 On-off control of a C-R system.

where r = reference input and x_1 = level of the first tank. Because of the differential gap, the fluid supply (controlling input) m is cut off when the error $e = r - x_1$ decreases and reaches $-h$, where h = differential gap. Similarly, the control valve opens when error is increasing and it reaches $+h$ as indicated at point B in Fig. 12-3(c).

If $m = M = 14$ for "ON" and $m = 0$ for "OFF", by Eq. (12-8) the equilibrium state for the "ON" condition is $x_{1_{\text{on}}} = 4$, $x_{2_{\text{on}}} = 6$. This point becomes the origin of the state plane for the "ON" period, and all the trajectories given in Fig. 10-13 also apply for the filling (or "ON") process relative to this new origin. Let us suppose that the reference input (desired level x_{1d}) is $r = 2$ so that the desired state is

$$\mathbf{x}_d = \begin{bmatrix} 2 \\ 3 \end{bmatrix}.$$

For a differential gap $h = 1$ we first determine an "OFF" period trajectory starting at A of Fig. 12-3(d), where A is on the dominant eigenvector (or dominant mode) of the "ON" process and is located at $x_1 = r + h = 3$. (See also Example 12-4.) As we saw in Fig. 10-13, the "OFF" trajectory almost coincides with the dominant mode of the "OFF" period at B, where $x_1 = r - h$. According to Fig. 12-3(c), a switching from "OFF" to "ON" takes place at B, and the "ON" trajectory closely approximates the dominant mode of the "ON" period at A. Therefore the "OFF-ON" cycle we just followed from A to B and back to A is the limit cycle.

If initially both tanks are empty ($\mathbf{x} = \mathbf{0}$), the start-up trajectory from the origin via C closely approaches A during the first "ON" period, and the limit cycle then begins. The geometric pattern of the trajectories is such that all motions (both inside and outside the limit cycle), regardless of the initial state, quickly converge to the limit cycle. Therefore the limit cycle is "stable," and the system is stable in the sense of Lyapunov, but not asymptotically stable.

To note the flexibility of the phase-plane method, suppose that the second tank overflows at $x_2 = 3.6$. The start-up trajectory for this case first goes up to C, then an overflowing process (with x_2 constant at 3.6) takes place from C to D, where the supply flow is then turned off. The "OFF" trajectory DE converges on point B. The next "ON" process takes the state from (approximately) B via F to D. Thus the limit cycle of the overflowing-tank system will be $DEBF$.

Although the phase-plane method of analysis and design [1 through 7] is powerful for nonlinear systems of first and second orders, it can become quite tedious when a system is not piecewise linear. The method can be extended to systems of order higher than two [8 and 9], but complexity then quickly mounts. It is interesting to note, however, that first- and second-order systems with transportation lag (dead time) sometimes yield rather simple phase-plane trajectory patterns, as shown in the following example.

Example 12-3 The system is given in Fig. 12-4(a) where time and signal scales are normalized, so that without losing generality we may take the process time constant and magnitude of m to be unity. The response for the "ON" signal obeys the relation

$$\frac{dc}{dt} + c = +k, \tag{12-9}$$

where the transportation lag will be accommodated in the switching analysis given below. The response trajectory in the phase plane of Fig. 12-4(b), where c and dc/dt are taken as axes, is the straight line AC of -1 slope that crosses the dc/dt axis at $+k$. Time is implicit in the plot. To recover it from the diagram, we write the solution of Eq. (12-9) for zero initial value (corresponding to point P in the figure).

12-1 On-off control systems in the state plane 515

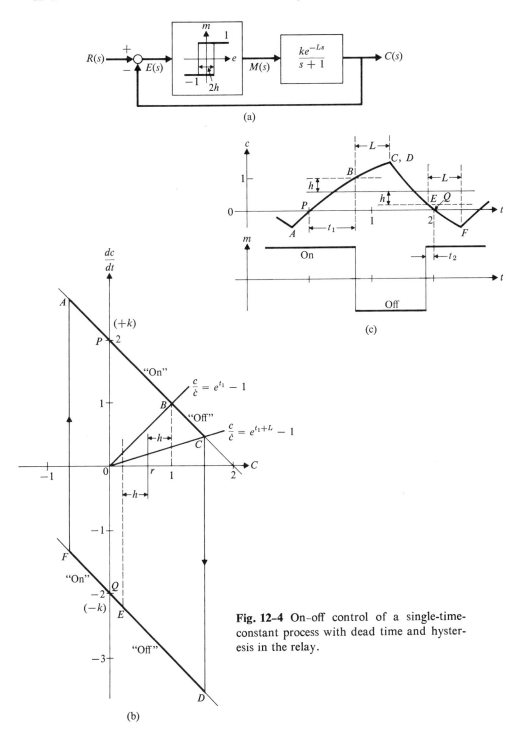

Fig. 12-4 On–off control of a single-time-constant process with dead time and hysteresis in the relay.

The solution is

$$c = k(1 - e^{-t}),$$

so that

$$\frac{dc}{dt} = ke^{-t}$$

and

$$c/\dot{c} = e^t - 1. \tag{12-10}$$

Making use of this relation, we can compute the time elapsed from P to B, t_1, from $(c/\dot{c})$ at B,

$$(c/\dot{c})_B = e^{t_1} - 1 = \tan(\angle POB).$$

The switching from "ON" to "OFF" takes place at time t_1, since c is crossing $(r + h)$ at B and the error signal, e is then at $-h$. Since the process has a delay L (taken to be $\frac{1}{2}$ in this example), the response for "OFF" input does not appear until time $t_1 + L$. The state C where the switching appears in the response can be fixed by the relation

$$(c/\dot{c})_{C \text{ at}} = e^{t_1+L} - 1 = \tan(\angle COP).$$

The response trajectory for the "OFF" input is the straight line DEF of -1 slope that crosses the $\dot{c}$-axis at $-k$. The system state jumps from C to D at time $t_1 + L$, which means a discontinuous change in dc/dt (Fig. 12-4c). The next switching, from "OFF" to "ON," takes place at point E where the output c equals $(r - h)$. By Eq. (12-10) we compute time t_2 from Q to E (negative) and fix F such that the time E to F is L. The switching from "OFF" to "ON" appears in the response at F, at which point the state jumps to A. Thus $ACDF$ in Fig. 12-4(b) is the limit cycle. The effect of the transportation lag is to maintain the response on a switched trajectory (AB or DE) for L units of time after the switching has occurred.

The following analytical development holds for a limit cycle which consists of piecewise linear trajectories. Let us suppose, for brevity, that a limit cycle consists of two linear trajectories which are symmetric to each other about the origin (Fig. 12-5). A linear control object,

$$\frac{d\mathbf{x}}{dt} = A\mathbf{x} + \mathbf{b}m,$$

is at the initial state $\mathbf{x}_0$ when switching takes place. Let

$$m(0^-) = -M, \quad \text{and} \quad m(t) = M \text{ for } 0 \leqslant t < t_1,$$

where t_1 is the half-period of the limit cycle. The linear trajectory will be given by

$$\mathbf{x}(t) = e^{At}\mathbf{x}_0 + e^{At} \int_0^t e^{-A\tau}\mathbf{b}m(\tau)\,d\tau$$

$$= e^{At}\mathbf{x}_0 + (e^{At} - I)A^{-1}\mathbf{u}_1, \tag{12-11}$$

12-1 On-off control systems in the state plane

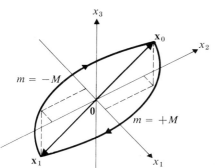

Fig. 12-5 A symmetric limit cycle.

where $\mathbf{A}^{-1}$ is assumed to exist, and

$\mathbf{u}_1 = \mathbf{b}M$ = a constant vector .

For a symmetric limit cycle, the state $\mathbf{x}(t_1) = \mathbf{x}_1$ at the next switching instant (from M to $-M$) must satisfy the equation

$$\mathbf{x}_1 = -\mathbf{x}_0 .$$

Therefore, by Eq. (12-11), we obtain

$$\mathbf{x}_0 = (\mathbf{I} + e^{\mathbf{A}t_1})^{-1}(\mathbf{I} - e^{\mathbf{A}t_1})\mathbf{A}^{-1}\mathbf{u}_1 . \qquad (12\text{-}12)$$

In this equation, t_1 is unknown, but we have one more condition available from the switching control law: the fact that $\mathbf{x}_0$ is on a switching hyperplane. It is therefore possible to analytically determine a limit cycle by Eq. (12-12) when $\mathbf{A}^{-1}$ exists. However, the computation usually gets quite involved. In an analytical approach developed by Tsypkin [10] a limit cycle is determined by a frequency plot which is based on a Fourier-series expression of the limit cycle. The main complication in the time-domain approach is the necessity of solving a set of transcendental equations [15 and 16]. In the example that follows the transcendental equations are easily solved, but this is not generally true.

Example 12-4 Let us reconsider the system of Example 12-2 (Fig. 12-3) from an analytical point of view. Since

$$\mathbf{A} = \begin{bmatrix} -3 & 2 \\ 4 & -5 \end{bmatrix},$$

the solution matrix $e^{\mathbf{A}t}$ is found to be

$$e^{\mathbf{A}t} = \mathscr{L}^{-1}[(s\mathbf{I} - \mathbf{A})^{-1}] = \begin{bmatrix} \tfrac{2}{3}e^{-t} + \tfrac{1}{3}e^{-7t} & \tfrac{1}{3}e^{-t} - \tfrac{1}{3}e^{-7t} \\ \tfrac{2}{3}e^{-t} - \tfrac{2}{3}e^{-7t} & \tfrac{1}{3}e^{-t} + \tfrac{2}{3}e^{-7t} \end{bmatrix}.$$

We take $\mathbf{x}_d$ in Fig. 12-3(d) as a new origin, and let $M = \pm 7$ to get a symmetric relation. In Fig. 12-3(b), this means that the origin for the relay action is shifted up from point 0 to point 0'. The switching from $m = +M$ to $-M$ takes place when

518 Nonlinear systems

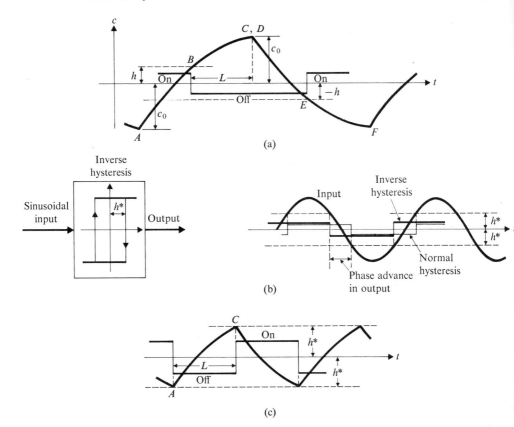

Fig. 12-6 Inverse hysteresis (b) and controlled response with regular hysteresis (a) vs. with inverse hysteresis (c).

$x_1 = +h$ in the new coordinate [A in Fig. 12-3(d)]. Recalling the fact that B (the next switching point) was practically in the direction of the dominant eigenmode (e^{-t}), we ignore all terms for the second mode (e^{-7t_1}) in the solution matrix at t_1. Thus, Eq. (12-12) reduces to

$$\begin{bmatrix} 1 + \tfrac{2}{3}e^{-t_1} & \tfrac{1}{3}e^{-t_1} \\ \tfrac{2}{3}e^{-t_1} & 1 + \tfrac{1}{3}e^{-t_1} \end{bmatrix} \begin{bmatrix} x_{10} \\ x_{20} \end{bmatrix} = \begin{bmatrix} 1 - \tfrac{2}{3}e^{-t_1} & -\tfrac{1}{3}e^{-t_1} \\ -\tfrac{2}{3}e^{-t_1} & 1 - \tfrac{1}{3}e^{-t_1} \end{bmatrix} \begin{bmatrix} 2 \\ 3 \end{bmatrix},$$

from which, letting $x_{10} = h = 1$, we find

$$x_{20} = 2,$$

and this is the ordinate of state A (Fig. 12-3d) in the new coordinate system where $\mathbf{x}_d$ is the origin.

Example 12-5 We return to the on-off control system of a single-time-constant process with a delay (Example 12-3, Fig. 12-4) to formulate the analytical solution of the problem. The limit-cycle pattern shown in Fig. 12-4(c) is not symmetric. Therefore a computation over a full cycle is necessary to determine the periodic pattern.

12-1 On-off control systems in the state plane

A symmetric pattern is expected when $r = 0$, and we compute period and amplitude of the limit cycle for this condition.

The system's equation for the "ON" period was given by Eq. (12-9), restated

$$\frac{dc}{dt} + c = +k.$$

Solution of this equation for an initial value $-c_0$ [A in Fig. 12-6(a)] is

$$c = -c_0 e^{-t} + k(1 - e^{-t}).$$

Let $t = t_1$ at C where $c(t_1) = +c_0$ (the condition for symmetry); then

$$c_0 = -c_0 e^{-t_1} + k(1 - e^{-t_1}).$$

The switching takes place at time $t_1 - L$ when $c(t_1 - L) = h$ [point B in Fig. 12-6(a)], so that

$$h = -c_0 e^{-(t_1-L)} + k(1 - e^{-(t_1-L)}).$$

Eliminating t_1 from the last two equations, we obtain an expression for the half-amplitude

$$c_0 = k - (k - h)e^{-L}. \tag{12-13}$$

Substituting Eq. (12-13) back into one of the last two expressions, we find the relation for the half-period of the limit cycle, t_1, to be

$$e^{t_1} = \frac{2ke^L - (k - h)}{k - h}. \tag{12-14}$$

Equations (12-13) and (12-14) may represent the control performance of this system [11]. However, a single time constant and a delay are often only an approximate model of a multilag system. For this reason, the sharp corner in the response of Fig. 12-6(a) is usually fictitious.

Compensation to improve the performance of discontinuous control systems may take the same form as the series or feedback compensators discussed in Chapters 8 and 9. In the presence of noise, however, multivariable feedback of a state vector measurement (Fig. 12-1b) to determine the switching hyperplane will, in general, yield improved performance. The nonlinearity itself may contribute to system compensation if it is suitably chosen. Yamashita [12] suggests an interesting application of inverse hysteresis to introduce phase advance. Inverse hysteresis can be realized, for example, by a logic network (see Example 15-8). The phase advance of inverse (or negative) hysteresis subjected to an open-loop sinusoidal input is illustrated in Fig. 12-6(b). The effect of phase advance in improving system performance will be shown in the following example.

Example 12-6 Example 12-5 is now reworked for the case of inverse hysteresis. The minimum half-amplitude c_0 of Eq. (12-13) is obtained when the differential gap (or "positive" hysteresis) h is zero:

$$c_{0 \text{ min}} = k(1 - e^{-L}). \tag{12-15}$$

The most effective "tuning" of the inverse hysteresis is shown in Fig. 12-6(c), where h^* of the hysteresis equals the half-amplitude. Note that $L = t_1$ (of Example 12-3) in this case. Computing the transient response from A to C in the figure, we obtain

$$h^* = -h^* e^{-L} + k(1 - e^{-L});$$

hence,

$$h^* = \frac{k(e^L - 1)}{e^L + 1}. \tag{12-16}$$

The performance index of the inverse hysteresis follows from Eqs. (12-15) and (12-16):

$$\eta = \frac{h^*}{c_{0\,\min}} = \frac{1}{1 + e^{-L}}. \tag{12-17}$$

At large values of L, η approaches unity and it has a minimum value of $\frac{1}{2}$ as L approaches zero.

An interaction of on–off action and sampling takes place in discrete-time relay control systems, because the controlling signals are then restricted not only in amplitude but also in time. As a consequence, an infinite number of stable limit cycles may exist in such a system. Nelson [13] points out that any nth order relay control system of the type described by Eqs. (12-2) through (12-5), which exhibits asymptotic stability with continuous control, will not necessarily be asymptotically stable when subjected to sampling. For a sampled-data system with a relay, as shown in Fig. 12-1(a), the only way the system can be in equilibrium at the origin is for the state to be at the origin at a time coinciding with a sampling instant.

Let us consider a linear control object described by Eq. (12-2), restated:

$$\frac{d\mathbf{x}}{dt} = \mathbf{A}\mathbf{x} + \mathbf{b}m.$$

If the plant itself is asymptotically stable, there must exist a pair of equilibrium states for constant values of the controlling input m:

$$m = +M \quad \text{and} \quad -M.$$

These states are

$$\mathbf{x}_e = \pm \mathbf{A}^{-1} \mathbf{b} M, \tag{12-18}$$

and they are either stable nodes [Table 3-3 (1a)] or foci [Table 3-5 (12a)]. Since limit cycles are formed by matching segments of trajectories toward the two symmetric equilibrium states $\mathbf{x}_e$, it seems logical to consider only symmetric limit cycles if the control object is stable. If, on the other hand, a control object has a pole(s) at zero [Table 3-3 (4a), Table 3-4 (7a)], $\mathbf{A}$ is singular and the $\pm M$ equilibrium states are not finite. The limit cycles which can exist in this case are not, in general, symmetric. The following discussion is limited to systems with symmetric limit cycles, and therefore the conclusions may not apply to control objects which are not asymptotically stable.

Let hT coincide with the half-period of a limit cycle, where h is the integer number of samples per half-period and T is the sampling period. We assume that the controlling input to the process, m_k, changes polarity only at the beginning of each half-period of the limit cycle. If a sign change of the switching function occurs at the zeroth sampling instant, from positive to negative, then

$$m_{-1} = +M \quad \text{and} \quad m_0 = -M.$$

For a linear control object described by the difference equation

$$\mathbf{x}_{k+1} = \mathbf{P}\mathbf{x}_k + \mathbf{q}m_k, \tag{12-19}$$

the half-period sequence is expressed by a set of equations,

$$\mathbf{x}_1 = \mathbf{P}\mathbf{x}_0 + \mathbf{q}(-M),$$
$$\vdots$$
$$\mathbf{x}_h = \mathbf{P}^h\mathbf{x}_0 + (\mathbf{P}^{h-1} + \cdots + \mathbf{P} + \mathbf{I})\mathbf{q}(-M). \tag{12-20}$$

When the states in each half-period are symmetric with respect to the origin, we must have

$$\mathbf{x}_{k+h} = -\mathbf{x}_k,$$

and thus

$$\mathbf{x}_h = -\mathbf{x}_0.$$

Therefore Eq. (12-20) can be solved for $\mathbf{x}_0$:

$$\mathbf{x}_0 = (\mathbf{P}^h + \mathbf{I})^{-1}(\mathbf{P}^{h-1} + \cdots + \mathbf{P} + \mathbf{I})\mathbf{q}M. \tag{12-21}$$

The starting state of a symmetric limit cycle (of half-period hT) can be determined by Eq. (12-21) provided that the sign-change condition,

$$m_{-1} = +M \quad \text{and} \quad m_0 = -M$$

or $\tag{12-22}$

$$m_{h-1} = -M \quad \text{and} \quad m_h = M,$$

is satisfied. The following example, given by Nelson [13], will show an application of these conditions and a design principle based on our knowledge of limit cycles.

Example 12-7 The system of Example 12-1 [Eq. (12-7) and Fig. 12-2], with a sampler following the relay (Fig. 12-7a), can be represented by the following difference equation:

$$\mathbf{x}_{k+1} = \begin{bmatrix} 1 & 1-d \\ 0 & d \end{bmatrix} \mathbf{x}_k + \begin{bmatrix} T-1+d \\ 1-d \end{bmatrix} m_k, \quad (12\text{-}23)$$

where $d = e^{-T}$. We have

$$\mathbf{P}^2 = \begin{bmatrix} 1 & 1-d^2 \\ 0 & d^2 \end{bmatrix}, \quad \ldots, \quad \mathbf{P}^h = \begin{bmatrix} 1 & 1-d^h \\ 0 & d^h \end{bmatrix};$$

hence

$$\mathbf{I} + \mathbf{P} + \cdots + \mathbf{P}^{h-1} = \begin{bmatrix} h & (h-1) - d(d^{h-1}-1)/(d-1) \\ 0 & (d^h-1)/(d-1) \end{bmatrix}.$$

Therefore Eq. (12-21) reduces to

$$\mathbf{x}_0 = M \begin{bmatrix} \tfrac{1}{2}hT - \tanh \tfrac{1}{2}hT \\ \tanh \tfrac{1}{2}hT \end{bmatrix}. \quad (12\text{-}24)$$

For $T = 1$ and $M = 1$, we obtain for various limit-cycle half-periods (number of samples h):

$$h = 1, \quad \mathbf{x}_0 = \begin{bmatrix} 0.03788 \\ 0.46212 \end{bmatrix}; \quad h = 2; \quad \mathbf{x}_0 = \begin{bmatrix} 0.23841 \\ 0.76159 \end{bmatrix};$$

$$h = 3; \quad \mathbf{x}_0 = \begin{bmatrix} 0.59485 \\ 0.90515 \end{bmatrix}; \quad \text{etc.}$$

These states are labeled P_1, P_2, and P_3 in Fig. 12-7(b). If a switching control law (Eq. 12-4) is chosen such that $\mathbf{x}_{(-1)}$ (labeled P'_1, P'_2, and P'_3 in the figure) is just to the other side of a switching line, then Eq. (12-22) is satisfied and the limit cycle does exist. Therefore, limit cycles for $h = 1$, 2, and 3 all exist if S_1–S_1 in the figure is the switching line. If, on the other hand, $\mathbf{k}'$ of Eq. (12-4) is modified in such a way that S_2–S_2 becomes the switching line, then these limit cycles can all be avoided, with the exception of the smallest one at $h = 1$. There is no way of

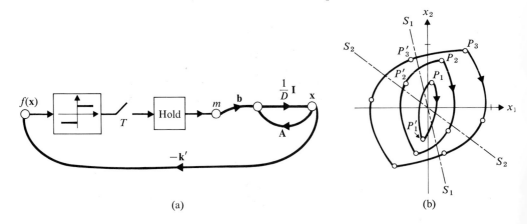

Fig. 12-7 Discrete-time discontinuous control (a) and some examples of limit cycles (b).

12-2 OSCILLATION OF NONLINEAR SYSTEMS

Among the characteristics of nonlinear system behavior which are not explained by linear theory, perhaps the most important is the self-excited oscillation, called the *limit cycle*. By assuming that the only significant output of a nonlinear element for a sinusoidal input is a sinusoidal component at the input frequency, we simplify the investigation of such oscillations [17, 18, 19]. The describing-function method [20] is based on this assumption and utilizes the frequency-response technique. Let us consider an amplitude-dependent, but frequency-insensitive, nonlinear input–output relation

$$v = f(u) .\tag{12-25}$$

For a sinusoidal input of an amplitude a and frequency ω,

$$u = a \sin \omega t ,\tag{12-26}$$

the response

$$\tilde{v} = f(a \sin \omega t)$$

has nonlinear distortion, but its periodicity is the same as the input for a variety of important practical nonlinear elements, such as those shown in

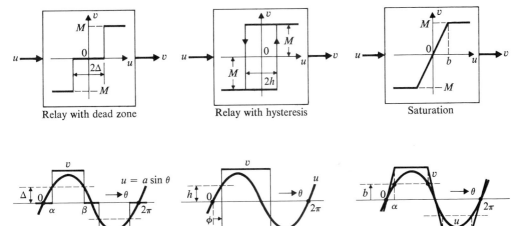

Fig. 12-8 Input-output of some nonlinear elements.

Fig. 12–8. The periodic output can be expressed as a Fourier series:

$$\tilde{v}(\theta) = c + \sum_{n=1}^{\infty} (f_n \sin n\theta + h_n \cos n\theta), \qquad \theta = \omega t, \qquad (12\text{–}27)$$

where the Fourier coefficients for the fundamental components are given by

$$f_1 = \frac{1}{\pi} \int_0^{2\pi} \tilde{v}(\theta) \sin\theta \, d\theta, \qquad h_1 = \frac{1}{\pi} \int_0^{2\pi} \tilde{v}(\theta) \cos\theta \, d\theta, \qquad (12\text{–}28)$$

and the dc-component is

$$c = \frac{1}{2\pi} \int_0^{2\pi} \tilde{v}(\theta) \, d\theta. \qquad (12\text{–}29)$$

The describing function of a nonlinear element $N(a)$ is defined by

$$N(a) = \frac{(f_1 + jh_1)}{a}, \qquad (12\text{–}30)$$

where a is defined by Eq. (12–26), and $N(a)$ is the output-over-input ratio of the nonlinear element in the frequency domain. The output of the non-linear element is taken as the fundamental (first) component of the Fourier-series representation of the output wave. $N(a)$ can involve both amplitude and phase information and may be a function of both input amplitude and frequency. It can be determined by analytical or numerical computation [21]. Reference [6] has an extensive table of describing functions, including a single generalized describing function that is applicable, by suitable choice of parameters, to most commonly encountered piecewise nonlinearities. Some examples follow:

Example 12–8 *Describing function for relay action with a dead band* (Fig. 12–8a). Since there is no phase shift between input and output, $h_1 = 0$, and (using symmetry) we have

$$f_1 = \frac{2}{\pi} \int_0^{\pi} \tilde{v}(\theta) \sin\theta \, d\theta = \frac{2M}{\pi} \int_\alpha^\beta \sin\theta \, d\theta = \frac{4M}{\pi} \cos\alpha,$$

where

$$a \sin\alpha = \Delta,$$

as shown in Fig. 12–8(a). Thus,

$$\cos\alpha = \sqrt{1 - (\Delta/a)^2}$$

and

$$N(a) = \frac{4M}{\pi a^2} \sqrt{a^2 - \Delta^2}. \qquad (12\text{–}31)$$

Equation (12–31) applies for $a \geqslant \Delta$, and $N(a)$ is zero for $a < \Delta$. In this example the output v is an odd function and hence involves only sine terms in the Fourier expansion ($h_n = 0$).

Example 12-9 *On–off action with hysteresis h* (Fig. 12-8b). The relation is not single valued. As in the case of all "memory-type" nonlinearities, there is a phase lag as a consequence. The rectangular output has a fundamental component of amplitude $\sqrt{f_1^2 + h_1^2}$, where

$$f_1 = \frac{2}{\pi} \int_0^\pi M \sin \theta \, d\theta = \frac{4M}{\pi} \cos \phi$$

and

$$h_1 = \frac{2}{\pi} \int_0^\pi M \cos \theta \, d\theta = -\frac{4M}{\pi} \sin \phi .$$

The amplitude ratio (or "gain") of $N(a)$ is therefore

$$|N(a)| = \frac{\sqrt{f_1^2 + h_1^2}}{a} = \frac{4M}{\pi a} .$$

The phase lag $[-\phi$ in Fig. 12-8(b) and in the previous equations for f_1 and $h_1]$ is determined by the equation $a \sin \phi = h$:

$$-\phi = -\sin^{-1} \frac{h}{a} .$$

The describing function is thus given in terms of its gain and phase by

$$N(a) = \frac{4M}{\pi a} \underline{/-\sin^{-1}(h/a)} \quad \text{for} \quad a \geqslant h . \tag{12-32}$$

Example 12-10 *Saturation* (Fig. 12-8c). The element is linear so long as the input amplitude is less than b:

$$N(a) = \frac{M}{b} = k \quad \text{for} \quad b > a .$$

For $a \geqslant b$ we compute (again $h_1 = 0$ since v is an odd function):

$$f_1 = \frac{4}{\pi} \int_0^\alpha \frac{aM}{b} (\sin \theta)^2 \, d\theta + \frac{4}{\pi} \int_\alpha^{\pi/2} M \sin \theta \, d\theta ,$$

where α is fixed by $a \sin \alpha = b$:

$$\sin \alpha = b/a, \quad \cos \alpha = \sqrt{1 - (b/a)^2}, \quad \text{and} \quad \alpha = \sin^{-1}(b/a) .$$

Therefore the describing function for $a \geqslant b$ is expressed by

$$N(a) = \frac{f_1}{a} = \frac{2M}{\pi b} \sin^{-1} \frac{b}{a} + \frac{2M}{\pi a} \sqrt{1 - \left(\frac{b}{a}\right)^2} . \tag{12-33}$$

The output of the nonlinearity is symmetric and single-valued; hence there is no phase shift in $N(a)$.

We can explore the possible existence of limit cycles in a control system which consists of a nonlinear element $N(a)$ and a linear element $G(s)$ (Fig. 12-9) by using a describing-function approximation of the non-

Fig. 12-9 Feedback control system with a nonlinear element

linearity. The input $m(t)$ to the linear element in the figure is not a pure sinusoidal signal. However, since the $G(s)$ of many engineering systems attenuates high-frequency signal components, the output of $G(s)$ [which is fed back into $N(a)$] may be assumed to be a sinusoid of the fundamental frequency when $m(t)$ is periodic. Therefore, a periodic (sinusoidal) motion will exist at the input to the nonlinearity in a free feedback control system when the following conditions are satisfied:

$$\frac{C(j\omega)}{M(j\omega)} = G(j\omega), \qquad \frac{-M(j\omega)}{C(j\omega)} = N(a),$$

where $N(a)$ is the describing function of the nonlinear element, and a is the amplitude of the controlled variable (which in turn is equal to the negative of the error signal when the reference input is zero). The two conditions combined will yield

$$G(j\omega) = \frac{-1}{N(a)}. \qquad (12\text{-}34)$$

If Eq. (12-34) is satisfied by a pair $\omega = \omega_1$ and $a = a_1$ (to the extent of the accuracy of the describing-function approximation), then there *will* exist a limit cycle which, in the error channel $e(t)$ of Fig. 12-9, is approximately described by $a_1 \sin \omega_1 t$. There are various graphical methods to solve Eq. (12-34) for a_1 and ω_1, among which the method in the Nyquist plane is most basic. We shall show the principle of the basic method by means of the next two examples.

Example 12-11 *On-off with hysteresis control of a single-time-constant process with a delay* (Fig. 12-4a). The linear plant is

$$G(s) = \frac{ke^{-sL}}{1+s} \qquad \text{where} \qquad k = 2, \quad L = \tfrac{1}{2}.$$

The Nyquist plot of $G(j\omega)$ is GG in Fig. 12-10. Since Eq. (12-34) will be satisfied at an intersection of $G(j\omega)$ (called a *frequency locus*) and $-1/N(a)$ (called an *amplitude locus*) in the Nyquist plane, we construct $-N(a)^{-1}$ for the describing function given by Eq. (12-32). Let us suppose OQ in Fig. 12-10 to be the phasor $-1/N(a_2)$ for some amplitude $a = a_2$; then

$$|OQ| = \frac{1}{|N(a_2)|} = \frac{\pi a_2}{4M}.$$

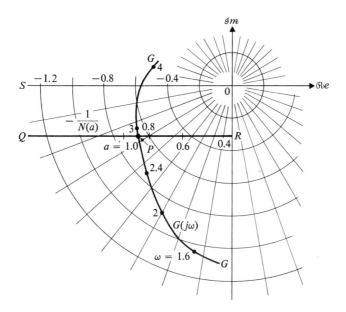

Fig. 12-10 Describing-function method applied to an on–off control system.

Angle QOS in the figure must be equal to $\sin^{-1}(h/a_2)$ due to inversion and negation; hence

$$OR = \frac{\pi h}{4M} = \text{const}.$$

The amplitude locus is therefore a straight line RQ which is parallel to the real axis; it starts at R where $a = h$. Let us assume that $h = 0.4$ and $M = 1$ so that $OR = \pi/10$.

The two loci intersect at P where a is about 0.9 and ω is about 2.9. We conclude that the limit-cycle condition (Eq. 12-34) is satisfied at P and the limit-cycle oscillation at the input side of the nonlinear element [$E(s)$ in Fig. 12-4(a)] will be approximately $0.9 \sin 2.9t$. Since the attenuation in the linear element depends on a mere single-time-constant "low-pass" filter $1/(1 + s)$, its output for on–off type input still has a considerable amount of nonlinear distortion, as indicated by the sharp corner C in Fig. 12-4(c). Therefore the accuracy of the describing-function method, especially regarding amplitude information, is not very high.

All the preceding examples dealt with intentional nonlinearities. Local instability is often caused by incidental nonlinearities in a system which was designed to be linear and stable. In a servomechanism, for instance, Coulomb friction and/or backlash may generate a limit cycle [17, 19, 22, 23]. We shall show an example below where a mechanical limit in the speed with which a valve can follow its control command is a cause of instability in a process control system. The process control system shown in Fig. 12-11(a) has two time constants (T_1, T_2), and a control valve in which the

528 Nonlinear systems

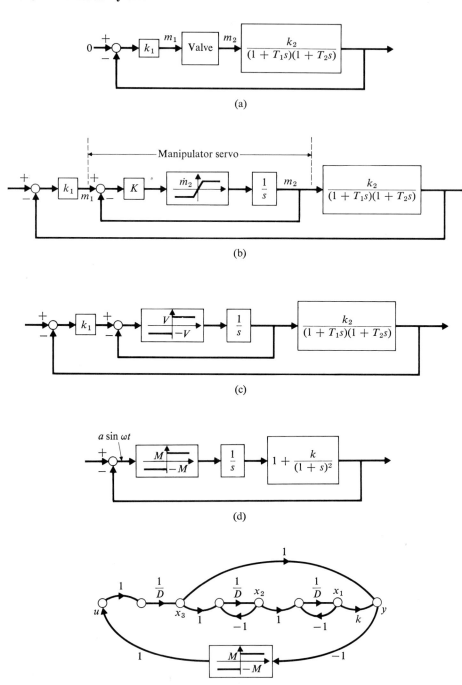

Fig. 12-11 Process control system with valve speed saturation.

following velocity saturation takes place:

$$\left|\frac{dm_2}{dt}\right| = V \quad \text{if} \quad \left|\frac{dm_1}{dt}\right| > V. \tag{12-35}$$

In Eq. (12-35), m_1 and m_2 are the input command and output response, respectively, of the valve stroke. A valve actuator of this kind may be represented by the minor feedback system shown in Fig. 12-11(b). Since $m_1 = m_2$ when saturation ends, the gain K of the minor loop system is infinity, and the system reduces to part (c) of the figure. The latter, after some block diagram algebra, is equivalent to Fig. 12-11(d) when $T_1 = T_2 = 1$. The relation is normalized in Fig. 12-11(d), where

$$M = VT_1 \quad \text{and} \quad k = k_1 k_2.$$

Example 12-12 *Limit cycles in the system of Fig. 12-11(d).* Since $\varDelta = 0$ in Eq. (12-31), the describing function is

$$N(a) = \frac{4M}{\pi a},$$

and the negative real axis becomes the amplitude locus $-1/N(a)$.

Let us assume one numerical value $k = 12$, so that $G(j\omega)$ crosses the negative real axis twice, as shown by P_1 and P_2 in Fig. 12-12. Analytical computation of these points is possible for this simple system. According to Eq. (12-34),

$$1 + \frac{K}{s}\left[1 + \frac{12}{(1+s)^2}\right] = 0,$$

where $K = N(a)$, must have imaginary roots $s = j\omega$ at the intersections; thus,

$$(j\omega)(1 + j\omega)^2 + K[(j\omega)^2 + 2(j\omega) + 13] = 0.$$

The imaginary part of this equation is

$$2K + 1 = \omega^2,$$

while the real part yields

$$(K + 2)\omega^2 = 13K.$$

Eliminating ω^2 from the two relations, we obtain

$$K^2 - 4K + 1 = 0,$$

which has the following roots:

$$K_1 = 2 - \sqrt{3} = 0.268, \quad K_2 = 2 + \sqrt{3} = 3.732,$$

where K_1, K_2 corresponds to P_1 and P_2, respectively, in the figure. These values of K_1 and K_2 (equal to $4M/(\pi a_1)$ and $4M/(\pi a_2)$, respectively) correspond to the two possible limit-cycle amplitudes a_1 and a_2. The frequencies at the points P_1 and P_2 on the Nyquist locus are:

$$\omega_1 = 1.24, \quad \omega_2 = 2.91.$$

530   Nonlinear systems

The low-frequency limit cycle ($\omega = \omega_1$) is said to be "stable" in this system because it tends to persist. Suppose, for instance, the amplitude a of the periodic input to the nonlinear element decreases from a_1 at point P_1 to a_1' which corresponds to P_1' on the amplitude locus (negative real axis) in the figure. Since the condition (12-34) can be interpreted as a stability limit in the Nyquist theorem, the enclosure of the point P_1' by the Nyquist locus $G(j\omega)$ means an instability. Therefore the amplitude a_1' tends to increase, and it will regain the original amplitude a_1. Similarly, an increase in amplitude a results in a stable transient which returns operation to point P_1. By the same logic we can deduce that the higher-frequency limit cycle (at ω_2) is "unstable," in the sense that the periodic oscillation with frequency ω_2 disappears with the slightest disturbance. If the disturbance is an increase in amplitude a, or a motion to the left of point P_2 on the negative real axis, operation will shift to the stable limit cycle at point P_1. A decrease in amplitude a, on the other hand, will result in a zero-amplitude infinite-frequency limit cycle corresponding to operation at the origin. This may be viewed as local asymptotic stability. We shall next show a state-space description of the limit cycles which will clarify the heuristic interpretations of "stable" and "unstable" limit cycles just given.

Example 12-13 We now derive an exact solution of the last example problem. Since the system is not too complicated, the method outlined in Section 12-1 can be used to determine the limit-cycle conditions exactly. Referring to Fig. 12-11(e), we write a vector equation for the linear portion of the system:

$$\frac{d\mathbf{x}}{dt} = \mathbf{A}\mathbf{x} + \mathbf{b}u, \qquad y = \mathbf{c}\mathbf{x},$$

$$\mathbf{A} = \begin{bmatrix} -1 & 1 & 0 \\ 0 & -1 & 1 \\ 0 & 0 & 0 \end{bmatrix}, \quad \mathbf{b} = \begin{bmatrix} 0 \\ 0 \\ 1 \end{bmatrix}, \quad \mathbf{c} = \begin{bmatrix} k & 0 & 1 \end{bmatrix},$$

where u is the output of the nonlinear element, and y is the input to it. The Laplace-domain solution, shown in Section 3-2 to be

$$\mathbf{X}(s) = (s\mathbf{I} - \mathbf{A})^{-1}\mathbf{x}_0 + (s\mathbf{I} - \mathbf{A})^{-1}\mathbf{b}U(s),$$

takes the following form:

$$\mathbf{X}(s) = \begin{bmatrix} \frac{1}{s+1} & \frac{1}{(s+1)^2} & \frac{1}{s(s+1)^2} \\ 0 & \frac{1}{s+1} & \frac{1}{s(s+1)} \\ 0 & 0 & \frac{1}{s} \end{bmatrix} \mathbf{x}_0 + \begin{bmatrix} \frac{1}{s(s+1)^2} \\ \frac{1}{s(s+1)} \\ \frac{1}{s} \end{bmatrix} U(s). \qquad (12\text{-}36)$$

Now $u(t) = M$ for a half-cycle of the limit-cycle oscillation from $t = 0$ to t_1. Thus, letting $U(s) = M/s$ in Eq. (12-36), we find the following inverse transform:

$$\mathbf{x}(t) = \begin{bmatrix} e^{-t} & te^{-t} & 1-(1+t)e^{-t} \\ 0 & e^{-t} & 1-e^{-t} \\ 0 & 0 & 1 \end{bmatrix} \mathbf{x}_0 + M \begin{bmatrix} (t-2)+(t+2)e^{-t} \\ t+e^{-t}-1 \\ t \end{bmatrix}. \qquad (12\text{-}37)$$

12-2

Applying a symmetry condition $x(t_1) = -x_0$, we find x_0 to be

$$\frac{x_{10}}{M} = -\frac{t_1 e^{-t_1}}{1+e^{-t_1}}\left(\frac{3}{2} + \frac{1-e^{-t_1}}{1+e^{-t_1}}\right) + \frac{2 - t_1/2 - 2e^{-t_1}}{1+e^{-t_1}},$$

$$\frac{x_{20}}{M} = -\frac{t_1}{2} + \frac{1-e^{-t_1}}{1+e^{-t_1}},$$

$$\frac{x_{30}}{M} = -\frac{t_1}{2}.$$

In addition to these three conditions, we have the switching condition

$$y(0) = kx_{10} + x_{30} = 0.$$

Eliminating x_0 from the four equations, we find [24],

$$k = \frac{\cosh t_1 + 1}{(4\sinh t_1)/t_1 - \cosh t_1 - 3}. \tag{12-38}$$

For $k = 12$ (the gain we assumed in Example 12-12), the half-periods t_1 of the limit cycles are found to be

$$t_1 = 2.42 \text{ and } 1.26.$$

The longer period, $2t_1 = 4.84$ (or $\omega_1 = \pi/2.42 = 1.30$), corresponds to the stable limit

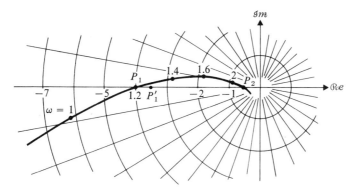

Fig. 12-12 Describing-function method analysis of the valve speed saturation system.

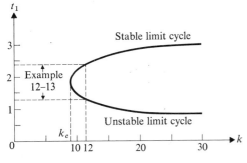

Fig. 12-13 Exact limit-cycle conditions of the velocity saturation system.

cycle (P_1 in Fig. 12-12), and the other period ($\omega_2 = \pi/1.26 = 2.50$) to the unstable limit cycle (see Example 12-14 below). The overall pattern of Eq. (12-38) is shown in Fig. 12-13. According to this result, absolute stability (asymptotic stability in the large, Section 4-2) is assured when

$$k < k_c, \qquad (12\text{-}39)$$

where k_c is about 9. In terms of the describing function method, the absolute stability condition of this system corresponds to

$$\angle G(j\omega) > -180°,$$

that is, the condition that the Nyquist locus will not cross the negative real axis in Fig. 12-12.

The describing-function method and the state-space formulation combined yield an approximate description of the stability boundary in the state space [25]. Such combined information clarifies the concept of stability and instability of a limit cycle discussed in the preceding examples. In some cases it may also serve as a guide for designing an adaptive control algorithm (in state space) to avoid a local instability. In the following discussion, we shall show the general principle of this approach by considering the valve-speed saturation system as an example.

Example 12-14 Since the describing-function method is a linearization technique in the vicinity of a limit cycle, a quadratic Lyapunov function [Eq. (4-13) restated]

$$V = \mathbf{x}'\mathbf{Y}\mathbf{x} \qquad (12\text{-}40)$$

may be used to determine the stability of the limit cycle. In this example we shall investigate the trajectory pattern in the neighborhood of the two limit cycles determined via the describing-function method in Example 12-12. For a free, stable, linear system, Eq. (12-40) describes a hyperellipsoid when V is kept at a constant value. The hyperellipsoid "explodes" and becomes a hypercylinder at an oscillatory stability limit of the linear system [see Table 3-6 (3a)]. The symmetric matrix $\mathbf{Y}$ in Eq. (12-40) is fixed for a linear system $d\mathbf{x}/dt = \mathbf{A}\mathbf{x}$ by the following matrix equation when the system is at the stability limit:

$$\mathbf{A}'\mathbf{Y} + \mathbf{Y}\mathbf{A} = \mathbf{0}. \qquad (12\text{-}41)$$

This is a homogeneous equation; hence a solution exists when

$$\det \mathbf{Y} = 0, \qquad (12\text{-}42)$$

and this is the condition for the quadratic function of Eq. (12-40) to become a cylinder. As shown in Table 3-6 (3a), the trajectory of a free system at the oscillatory stability limit lies on the cylinder surface. Since the system of Example 12-12 is nonlinear, but approximately linearized in the vicinity of the two limit cycles, two cylinders for these limit cycles can be fixed up to their size. The size depends on the value V we specify in Eq. (12-40); hence it is arbitrary if the system is linear. The cylinder is shown in Fig. 12-14 for L_1, where L_1 and L_2 are the unstable and stable limit cycles, respectively.

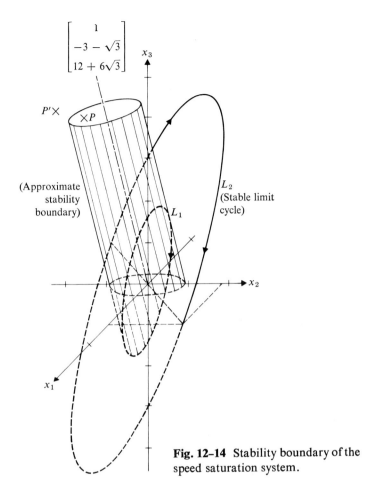

Fig. 12-14 Stability boundary of the speed saturation system.

The directions of the center lines of the cylinders are given by the eigenvectors for the real eigenvalue. The characteristic equation of our system, linearized by the describing function $N(a) = K$ in Fig. 12-11(e), is

$$\begin{vmatrix} s+1 & -1 & 0 \\ 0 & s+1 & -1 \\ kK & 0 & s+K \end{vmatrix} = 0,$$

or

$$s^3 + (2+K)s^2 + (1+2K)s + (K+kK) = 0,$$

where $k = 12$ and two values of K are fixed in Example 12-12 for the oscillatory stability limit. Since $\pm j\omega$ are the roots at the oscillatory stability limit, the third (real) root can be found by dividing the characteristic polynomial by $(s^2 + \omega^2)$ or by simply taking the negative of the second coefficient in the characteristic poly-

nomial. It is
$$s_3 = -(2+K),$$
and by $\mathbf{A}\mathbf{v}^3 = s_3\mathbf{v}^3$ the eigenvector for the real root is fixed as
$$\mathbf{v}^3 = \begin{bmatrix} 1 \\ -(1+K) \\ (1+K)^2 \end{bmatrix},$$
or
$$\begin{bmatrix} 1 \\ -3 - \sqrt{3} \\ 12 + 6\sqrt{3} \end{bmatrix} \quad \text{for the unstable limit cycle}$$
and
$$\begin{bmatrix} 1 \\ -3 + \sqrt{3} \\ 12 - 6\sqrt{3} \end{bmatrix} \quad \text{for the stable limit cycle}.$$

Once an unstable limit cycle (the ring L_1 in Fig. 12-14) is fixed for a sinusoidal steady state found by the describing-function method, the approximate stability boundary can be determined by drawing lines parallel to $\mathbf{v}^3$. Thus Eq. (12-41) need not be solved for $\mathbf{Y}$.

According to Fig. 12-14, the unstable limit cycle is realized only if an initial state is on the surface of the cylinder. All free trajectories which start from inside the cylinder (such as P in the figure) go to the origin. On the other hand, all trajectories with initial state outside the cylinder (for instance P' in the figure) tend to the stable limit cycle L_2. Therefore, the cylinder is the stability boundary; the system is asymptotically stable inside the cylinder, and the free response is bounded but not asymptotically stable for all initial states outside the cylinder.

The cylindrical stability boundary is an approximation of an exact manifold, with highest accuracy in the vicinity of the limit cycle. An exact determination of the boundary is extremely cumbersome even for this simple system [25].

An interesting but undesirable phenomenon called *jump resonance* may occur when a nonlinear system is subject to sinusoidal forcing. A typical example where it takes place is a servomechanism with a saturation amplifier (Fig. 12-15a). As we saw in Example 12-10, the system is linear if amplitude R of a sinusoidal reference input

$$r(t) = R \sin \omega t \quad \text{or} \quad R \exp(j\omega t)$$

is sufficiently small relative to the saturation limit. The magnitude ratio of the closed-loop frequency response, $C(j\omega)/R(j\omega)$, will then be given, for instance, by curve 1 in Fig. 12-15(b). If, on the other hand, the input amplitude is very large (relative to the saturation limit), we expect a drastic decrease in the resonance peak due to the low effective gain of the amplifier, curve 3 of the figure. The jump phenomenon may occur between these two extreme conditions, as shown by curve 2 in the figure, where the direction

12-2 Oscillation of nonlinear systems

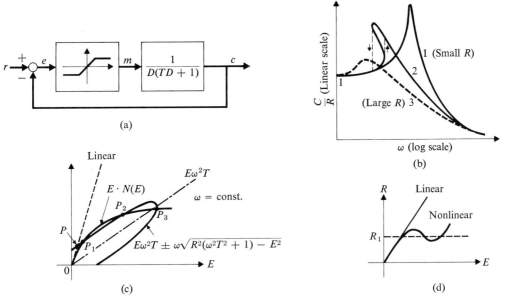

Fig. 12-15 Jump resonance.

of the jump is indicated by arrows as the forcing frequency is gradually increased or decreased.

In order to see why the jump occurs, let us consider the following sinusoidal steady state where the nonlinearity is approximated by a describing function $N(E)$:

$$r(t) = R \exp(j\omega t), \quad e(t) = E \exp(j\omega t + j\alpha), \quad m(t) = N(E)e(t).$$

The closed-loop frequency response is then given by

$$\frac{E}{R} \exp(j\alpha) = \frac{1}{1 + N(E)G(j\omega)},$$

where $G(s)$ is the transfer function of the linear element, which, in our example, is

$$G(j\omega) = \frac{1}{-\omega^2 T + j\omega}.$$

It then follows that

$$\left|\frac{E}{R}\right|^2 = \frac{\omega^2(1 + \omega^2 T^2)}{(N(E) - \omega^2 T)^2 + \omega^2},$$

whence

$$EN(E) = E\omega^2 T \pm \omega\sqrt{R^2(1 + \omega^2 T^2) - E^2}. \tag{12-43}$$

536 Nonlinear systems

The right-hand side of Eq. (12–43), plotted against E, is an ellipse, as shown in 12–15(c). If the servomechanism is linear, the left-hand side of Eq. (12–43) is a straight line in the same diagram; hence, only one intersection (P) will appear. For a saturation amplifier, however, the amplitude $EN(E)$ of the forced-frequency component of $m(t)$ will level off as E increases. Consequently, a multiplicity of intersections (P_1, P_2, and P_3 in the figure) will occur. (For lower values of ω, the ellipse in Fig. 12–15(c) slants down to the right more—leading to the single valued low frequency branch of curve 2 in Fig. 12–15(b)). Similarly, high values of ω lead to the single valued high frequency end of the curve). The jump-resonance phenomenon depends on such multivalued dependence of R on E. Hatanaka [26] makes an extensive analysis of the phenomenon in the frequency domain where a nonlinearity is replaced by its describing function. In Hatanaka's analysis, R is considered to be a function of both E and ω. For a fixed value of frequency ω, a multivalued relation exists between R and E if and only if there is a range of E where

$$\left(\frac{\partial R}{\partial E}\right)_{\omega=\text{const}} < 0 \qquad [\text{see Fig. 12–15(d) for } R = R_1].$$

In order for Hatanaka's approach to apply, we must assume that $N(E)$ is differentiable with respect to E and that its first derivative is continuous.

12–3 STABILITY OF NONLINEAR SYSTEMS

The stability of dynamic systems, linear or nonlinear, was *defined* in Chapter 4 by the "direct method" of Lyapunov. It is therefore possible to determine the stability of a nonlinear system, either locally or globally, without actually solving the system equation, *if* a suitable Lyapunov function is known. This big *IF* constitutes the major problem in applying the direct method of Lyapunov to nonlinear systems. The difficulty in finding a Lyapunov function that will give complete and exact information regarding a given nonlinear system's stability is comparable to the difficulty involved in determining an exact analytical solution of the system's nonlinear differential equation. In the following discussion we shall describe several approaches to stability analysis of nonlinear systems; some are based on the Lyapunov approach, and others are related to the Nyquist theorem.

Perhaps the first method of Lyapunov would be the simplest. Consider a nonlinear system expressed in terms of an n-vector $\mathbf{x}$,

$$\frac{d\mathbf{x}}{dt} = \mathbf{f}(\mathbf{x}), \qquad (12\text{–}44)$$

where the origin is chosen to be an equilibrium state $\mathbf{x}_e = \mathbf{0}$. If the system has more than one equilibrium state, the first method applies to each of them investigated separately.

We assume that the vector function $\mathbf{f}(\mathbf{x})$ is continuously differentiable in $x_1, \ldots, x_n$, and expand it in a Taylor series about the equilibrium state $\mathbf{0}$:

$$f_i(\mathbf{x}) = \sum_{j=1}^{n} \left(\frac{\partial f_i}{\partial x_j}\right)_{\mathbf{x}_j = 0} + \text{higher-order derivatives and residue}.$$

Then Eq. (12–44) can be expressed as

$$\frac{d\mathbf{x}}{dt} = \mathbf{A}_J \mathbf{x} + \mathbf{g}(\mathbf{x}), \qquad (12\text{–}45)$$

where $\mathbf{A}_J$ is an $n \times n$ Jacobian matrix

$$\mathbf{A}_J = \begin{bmatrix} \frac{\partial f_1}{\partial x_1} & \cdots & \frac{\partial f_1}{\partial x_n} \\ \vdots & & \vdots \\ \frac{\partial f_n}{\partial x_1} & \cdots & \frac{\partial f_n}{\partial x_n} \end{bmatrix}$$

with partial derivatives evaluated at $\mathbf{x} = \mathbf{0}$. The n-vector $\mathbf{g}(\mathbf{x})$, which may contain terms arising from high-order partial derivatives of the Taylor-series expansion, is considered "strictly nonlinear" in the sense that it is impossible to reduce it into a form

$$\mathbf{g}(\mathbf{x}) = \mathbf{A}_1 \mathbf{x} + \mathbf{g}_1 \mathbf{x}, \qquad \mathbf{A}_1 \neq \mathbf{0}.$$

The first part of Eq. (12–45),

$$\frac{d\mathbf{x}}{dt} = \mathbf{A}_J \mathbf{x}, \qquad (12\text{–}46)$$

is called the equation of the first approximation of the original equation (Eq. 12-44). It was shown by Lyapunov [27] and others [28] that the original equation (12-44) and its equation of the first approximation (12-46) are equivalent with respect to stability properties if (a) all the eigenvalues of $\mathbf{A}_J$ have nonzero real parts,* and if (b) for some $\lambda > 0$, which is sufficiently small, the following inequality holds:

$$\|\mathbf{g}(\mathbf{x})\| < \lambda \|\mathbf{x}\|. \qquad (12\text{–}47)$$

It therefore follows that the system of Eq. (12–44) is asymptotically stable in the neighborhood of the origin if (1) Eq. (12–46) describes a system which is asymptotically stable and (2) Eq. (12–47) holds. (The first method of Lyapunov gives information only on local stability. No indication is given by the method of how small the region is.)

---

* If at least one of the eigenvalues of $\mathbf{A}_J$ has a zero real part, the local stability behavior of the equilibrium state $\mathbf{0}$ cannot be determined from $\mathbf{A}_J$ (see Table 3-3, (4) and (5)).

Proof. If $d\mathbf{x}/dt = \mathbf{A}_J\mathbf{x}$ is asymptotically stable at $\mathbf{x} = \mathbf{0}$, then there exists, for every given real, positive definite symmetric matrix $\mathbf{Q}$, a real, positive definite quadratic form

$$V = \mathbf{x}'\mathbf{Y}\mathbf{x}, \qquad \mathbf{Y}' = \mathbf{Y}$$

such that

$$\frac{dV}{dt} = \mathbf{x}'(\mathbf{A}_J'\mathbf{Y} + \mathbf{Y}\mathbf{A}_J)\mathbf{x} = -\mathbf{x}'\mathbf{Q}\mathbf{x}.$$

The total derivative of V for the system of Eq. (12–45) is

$$\frac{dV}{dt} = (\mathbf{x}'\mathbf{A}_J' + \mathbf{g}')\mathbf{Y}\mathbf{x} + \mathbf{x}'\mathbf{Y}(\mathbf{A}_J\mathbf{x} + \mathbf{g})$$
$$= \mathbf{x}'(\mathbf{A}_J'\mathbf{Y} + \mathbf{Y}\mathbf{A}_J)\mathbf{x} + \mathbf{g}'\mathbf{Y}\mathbf{x} + \mathbf{x}'\mathbf{Y}\mathbf{g}$$
$$= -\mathbf{x}'\mathbf{Q}\mathbf{x} + 2\mathbf{g}'\mathbf{Y}\mathbf{x}. \tag{12–48}$$

Equation (12–48) is negative definite if

$$\mathbf{x}'\mathbf{Q}\mathbf{x} > 2\mathbf{g}'\mathbf{Y}\mathbf{x}. \tag{12–49}$$

Let b be the maximum of the absolute values of all elements of the $n \times n$ square matrix $\mathbf{Y}$; then

$$|\mathbf{g}'\mathbf{Y}\mathbf{x}| \leq nb\,||\mathbf{g}||\,||\mathbf{x}||,$$

where the bars on the right-hand side denote absolute value of the vectors $\mathbf{g}$ and $\mathbf{x}$ in the L_2-norm sense (that is, the square root of the sum of the squares of the vector elements). Now $||\mathbf{g}|| < \lambda\,||\mathbf{x}||$ by the second condition in the theorem statement, so that

$$|\mathbf{g}'\mathbf{Y}\mathbf{x}| < nb\lambda\,||\mathbf{x}||^2. \tag{12–50}$$

Since $\mathbf{x}'\mathbf{Q}\mathbf{x}$ is positive definite,

$$\left(\frac{\mathbf{x}'}{||\mathbf{x}||}\right)\mathbf{Q}\left(\frac{\mathbf{x}}{||\mathbf{x}||}\right) \geq c > 0, \tag{12–51}$$

where c is the smallest eigenvalue of $\mathbf{Q}$. If the positive number λ is so chosen that

$$\lambda < \frac{c}{2nb},$$

then by Eqs. (12–50) and (12–51) we have

$$\mathbf{x}'\mathbf{Q}\mathbf{x} > c\,||\mathbf{x}||^2 > 2nb\lambda\,||\mathbf{x}||^2 > 2\,|\mathbf{g}'\mathbf{Y}\mathbf{x}|;$$

hence Eq. (12–49) is true.

Example 12-15 [28] Consider a first-order system

$$\frac{dx}{dt} = -ax + |x|, \quad a > 0.$$

Let

$$V = x^2.$$

Then $b = 1$, and

$$\frac{dV}{dt} = 2x\frac{dx}{dt} = -2ax^2,$$

where $dx/dt = A_j x$, and $A_j = -a$ in this case. Therefore $Q = 2a > 0$ and $c = 2a$. Hence

$$\lambda < a;$$

but

$$|g(x)| = |x| < \lambda |x|$$

so that the system will be asymptotically stable if $a > 1$. This conclusion can be verified by direct solution of the differential equation for $x_0 < 0$ and $x_0 > 0$.

The Lyapunov function constructed by Krasovskii [29] for nonlinear systems reduces to the quadratic form presented in Section 4-3 when the system is linear. Therefore his approach gives a necessary and sufficient condition for stability if the system is linear, while for nonlinear systems the condition is sufficient for global asymptotic stability. Let a nonlinear system be described by

$$\frac{d\mathbf{x}}{dt} = \mathbf{f}(\mathbf{x})$$

with its equilibrium state at the origin, $\mathbf{f}(0) = 0$. The Jacobian matrix of the system is defined as before by

$$\mathbf{A}_J = \begin{bmatrix} \frac{\partial f_1}{\partial x_1} & \cdots & \frac{\partial f_1}{\partial x_n} \\ \vdots & & \vdots \\ \frac{\partial f_n}{\partial x_1} & \cdots & \frac{\partial f_n}{\partial x_n} \end{bmatrix},$$

where f_i must be differentiable. It must be noted that $\mathbf{A}_J$ depends on $\mathbf{x}$ in a nonlinear system.

Krasovskii's Lyapunov function has the general quadratic form

$$V = \mathbf{f}'\mathbf{Y}\mathbf{f}, \tag{12-52}$$

where $\mathbf{Y}$ is a symmetric matrix which must be positive definite in order for

540 Nonlinear systems

V to be positive definite. We have

$$\frac{\partial \mathbf{f}(\mathbf{x})}{\partial t} = \left(\frac{\partial \mathbf{f}(\mathbf{x})}{\partial \mathbf{x}}\right)\left(\frac{d\mathbf{x}}{dt}\right) = \mathbf{A}_J \mathbf{f}$$

and, similarly,

$$\frac{\partial \mathbf{f}'(\mathbf{x})}{\partial t} = \mathbf{f}' \mathbf{A}'_J .$$

Therefore the time derivative of V along a free trajectory is

$$\frac{dV}{dt} = \frac{\partial \mathbf{f}'}{\partial t} \mathbf{Y} \mathbf{f} + \mathbf{f}' \mathbf{Y} \frac{\partial \mathbf{f}}{\partial t}$$

$$= \mathbf{f}'(\mathbf{A}'_J \mathbf{Y} + \mathbf{Y} \mathbf{A}_J)\mathbf{f} . \tag{12-53}$$

Since V and dV/dt must have opposite signs to assure asymptotic stability when $\mathbf{Y}$ is positive definite, we require

$$\mathbf{W} = -(\mathbf{A}'_J \mathbf{Y} + \mathbf{Y} \mathbf{A}_J)$$

to be positive definite. The positive definiteness of $\mathbf{Y}$ and $\mathbf{W}$, expressed by Sylvester's inequalities (Eq. 4-16) yields a stability condition. The algebraic relation for stability simplifies when $\mathbf{Y}$ is replaced by a unit matrix, but this will limit the sharpness of the method.

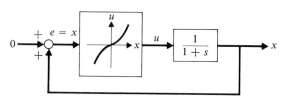

Fig. 12-16 Nonlinear positive-feedback system.

Example 12-16 We apply Krasovskii's approach to nonlinear positive feedback control of a first-order plant (Fig. 12-16). Let us suppose that the controller gain k is a function of x:

$$k = 0.5 + x^2 .$$

Then the control system's equation is

$$\frac{dx}{dt} = -0.5x + x^3 .$$

For this simple system we can draw a conclusion regarding stability without using a Lyapunov function. (An exact solution is unexpectly complex even for this simple case.) If $x_0 > 0$, the motion will converge to 0 from the positive side when $-0.5x + x^3 < 0$ or when $x^2 < \frac{1}{2}$. For $x_0 < 0$ we require $dx/dt > 0$, or $-0.5x + x^3 > 0$.

Applying Krasovskii's method with $\mathbf{Y} = \mathbf{I}$, we define a Lyapunov function

$$V = (x^3 - 0.5x)^2.$$

Then it follows that

$$\frac{dV}{dt} = 2(x^3 - 0.5x)^2(3x^2 - 0.5),$$

where the first factor on the right-hand side is positive definite. Therefore, for dV/dt to be negative definite, we require

$$x^2 < \tfrac{1}{6}.$$

This is far more restrictive than

$$x^2 < \tfrac{1}{2}.$$

Luré and Letov [30] suggest an approach for determining a system's Lyapunov function based on a "canonic form" of the diagonalized system equation. The method applies to a class of nonlinear control systems which consist of a nonlinear element and a linear, lumped-parameter control object (Fig. 12–21a). The Lyapunov function arrived at is in the form:

$$V = H + F + \int_0^z f(z)\, dz, \tag{12-54}$$

where H and F are functions of the state of the linear portion of the system. The last term on the right of Eq. (12–54) is an ingenious representation of a single-valued nonlinearity. It is positive definite for an arbitrary nonlinear relation $f(z)$ confined to the sector indicated in Fig. 12–21(a). For precise definitions of the H and F functions, and for a statement of the algebraic condition under which dV/dt is negative definite (all of which involves cumbersome algebra), the interested reader is referred to [30].

If a sufficient stability condition is needed for a specific type of nonlinear system, it is sometimes practical to construct a Lyapunov function in its simplest form by using inequality conditions. Johnson applies such conditions on the upper bound of error caused by quantization in a linear digital control system [31]. To determine the bound, Johnson formulates a discrete-time digital control system (Fig. 12–17) as a linear system with

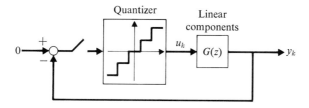

Fig. 12–17 Discrete-time system with a quantizer.

quantization error input:

$$\mathbf{x}_{k+1} = \mathbf{P}\mathbf{x}_k + \mathbf{q}u_k, \qquad y_k = \mathbf{C}\mathbf{x}_k, \qquad (12\text{-}55)$$

where u_k is quantization error, and y_k is scalar output, both at the kth sampling instant. It is assumed that the closed-loop pulse-transfer function $Y(z)/U(z)$ has an nth-order denominator polynomial in z, and an mth-order numerator polynomial where $m < n$. Introducing a quadratic Lyapunov function,

$$V_k = \mathbf{x}_k' \mathbf{Y} \mathbf{x}_k,$$

where $\mathbf{Y}$ is a symmetric and positive definite matrix that satisfies the condition

$$\mathbf{P}'\mathbf{Y}\mathbf{P} - \mathbf{Y} = -\mathbf{I} \qquad (12\text{-}56)$$

(see Eq. 4-36), the change ΔV_k along the system's motion,

$$\Delta V_k = V_{k+1} - V_k,$$

is computed for the forced system of Eq. (12-55). We have

$$\begin{aligned} V_{k+1} &= \mathbf{x}_{k+1}' \mathbf{Y} \mathbf{x}_{k+1} = (\mathbf{x}_k'\mathbf{P}' + \mathbf{q}'u_k)\mathbf{Y}(\mathbf{P}\mathbf{x}_k + \mathbf{q}u_k) \\ &= \mathbf{x}_k'\mathbf{P}'\mathbf{Y}\mathbf{P}\mathbf{x}_k + \mathbf{x}_k'\mathbf{P}'\mathbf{Y}\mathbf{q}u_k + \mathbf{q}'u_k\mathbf{Y}\mathbf{P}\mathbf{x}_k + \mathbf{q}'\mathbf{Y}\mathbf{q}u_k^2 \\ &= \mathbf{x}_k'\mathbf{P}'\mathbf{Y}\mathbf{P}\mathbf{x}_k + 2\mathbf{q}'\mathbf{Y}\mathbf{P}\mathbf{x}_k u_k + \mathbf{q}'\mathbf{Y}\mathbf{q} u_k^2. \end{aligned}$$

Thus

$$\begin{aligned} \Delta V_k &= \mathbf{q}'\mathbf{Y}\mathbf{q} u_k^2 + 2\mathbf{q}'\mathbf{Y}\mathbf{P}\mathbf{x}_k u_k + \mathbf{x}_k'(\mathbf{P}'\mathbf{Y}\mathbf{P} - \mathbf{Y})\mathbf{x}_k \\ &= \mathbf{q}'\mathbf{Y}\mathbf{q} u_k^2 + 2\mathbf{q}'\mathbf{Y}\mathbf{P}\mathbf{x}_k u_k - \mathbf{x}_k'\mathbf{I}\mathbf{x}_k. \end{aligned} \qquad (12\text{-}57)$$

According to Eq. (12-57), ΔV is negative when $||\mathbf{x}||$ is large enough so that $\mathbf{x}_k'\mathbf{I}\mathbf{x}_k$ dominates the other two terms. The question remains. How large should $||\mathbf{x}||$ be to keep ΔV negative? To estimate an upper bound for $||\mathbf{x}||$, above which a negative ΔV is assured, we introduce the worst possible error

$$|u_k| = \frac{\Delta}{2},$$

where Δ is the quantization step. Then ΔV_k is negative when

$$\mathbf{q}'\mathbf{Y}\mathbf{q}(\Delta^2/4) + |\mathbf{q}'\mathbf{Y}\mathbf{P}\mathbf{x}_k|\,\Delta \leqslant \mathbf{x}_k'\mathbf{I}\mathbf{x}_k = ||\mathbf{x}_k||^2. \qquad (12\text{-}58)$$

Noting that

$$|\mathbf{q}'\mathbf{Y}\mathbf{P}\mathbf{x}_k| \leqslant ||\mathbf{q}'\mathbf{Y}\mathbf{P}||\,||\mathbf{x}_k||,$$

we can convert Eq. (12-58) into the form

$$||\mathbf{q}'\mathbf{Y}\mathbf{P}||\,||\mathbf{x}_k||\,\Delta \leqslant ||\mathbf{x}_k||^2 - \mathbf{q}'\mathbf{Y}\mathbf{q}\left(\frac{\Delta^2}{4}\right). \qquad (12\text{-}59)$$

Let ρ = upper bound of $||\mathbf{x}_k||$ beyond which a negative ΔV is assured (for a stable control system). Then, replacing $||\mathbf{x}_k||$ in Eq. (12-59) by ρ and taking the equality sign as a limiting condition, we find that

$$\rho^2 - ||\mathbf{q}'\mathbf{YP}||\,\rho\Delta - \mathbf{q}'\mathbf{Yq}\left(\frac{\Delta^2}{4}\right) = 0. \qquad (12\text{-}60)$$

The positive root of Eq. (12-60) is

$$\rho = \tfrac{1}{2}\Delta(||\mathbf{q}'\mathbf{YP}|| + \sqrt{||\mathbf{q}'\mathbf{YP}||^2 + \mathbf{q}'\mathbf{Yq}'}). \qquad (12\text{-}61)$$

Since the system's response for quantization-error input is given by

$$y_k = \mathbf{c}\mathbf{x}_k,$$

the maximum possible magnitude of error, ε, that will appear in y due to quantization is

$$\varepsilon \leq ||\mathbf{c}||\,\rho. \qquad (12\text{-}62)$$

Applying Eq. (12-61) in (12-62), we obtain

$$\varepsilon \leq \tfrac{1}{2}\Delta(||\mathbf{c}||)(||\mathbf{q}'\mathbf{YP}|| + \sqrt{||\mathbf{q}'\mathbf{YP}||^2 + \mathbf{q}'\mathbf{Yq}}), \qquad (12\text{-}63)$$

where

$$\mathbf{c} = [c_1, \ldots, c_n] \quad \text{and} \quad ||\mathbf{c}|| = \sqrt{c_1^2 + \cdots + c_n^2}.$$

Equation (12-63) can be used either to estimate ε for a prescribed quantization step Δ or to determine Δ for a given ε.

Recent progress has returned the stability study of nonlinear systems to the frequency domain, where the resulting stability tests closely resemble the Nyquist criterion in both form and application. Unlike the describing-function approach in the frequency domain, there are no approximations involved, and asymptotic stability is investigated rather than the existence (or nonexistence) of limit cycles. Unfortunately, only sufficient conditions have been developed to date, but even these can be useful to the designer. In the remainder of this section we will study two of these stability tests: the Popov criterion and the circle criterion.

Perhaps the first success in this direction was achieved by V. M. Popov [32]. In his approach the nonlinear feedback control system whose stability is to be investigated is considered to consist of a nonlinear static relation N and a linear plant $G(s)$ as shown in Fig. 12-18(a). Certain assumptions about the system must be satisfied before applying the Popov criterion. They are:

1. $G(s)$ must be linear, time-invariant, and causal (that is, more poles than zeros). Also, $G(s)$ must be strictly stable and finite dimensional.

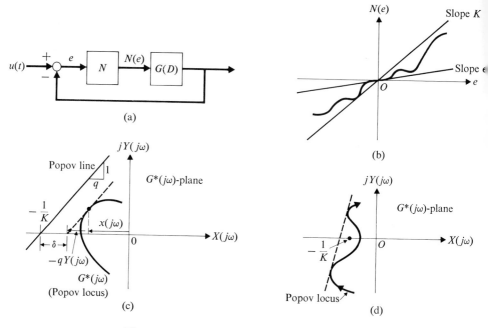

Fig. 12-18 Conditions in the Popov system.

2. The input $u(t)$ and any disturbances entering the system must be bounded, uniformly continuous (the size of the time interval across which continuity is investigated is not a function of time), and square integrable over the time interval $[0, \infty]$, for all initial states.

3. The output $N(e)$ of the nonlinear block must be time-invariant, memoryless, and a piecewise continuous function of e. Further, the slope $dN(e)/de$ must be bounded, and $N(e)$ must lie in the closed sector $[\varepsilon, K]$, where $\varepsilon > 0$ and $\varepsilon = 0$ if $G(s)$ has no poles on the imaginary axis. This last condition may be stated, as shown in Fig. 12–18(b), as

$$0 < \varepsilon < N(e)/e < K < \infty, \qquad (12\text{-}64)$$

where $N(e) = 0$ when $e = 0$.

Given these assumptions, the *sufficient* condition of the Popov theorem for asymptotic stability in the large may be stated as: *If, for the above system, there is any real number q and an arbitrarily small $\delta > 0$ such that*

$$\mathscr{R}e\,[(1 + j\omega q)G(j\omega)] + \frac{1}{K} \geqslant \delta > 0 \qquad (12\text{-}65)$$

for all $\omega \geqslant 0$, then for any initial state the output of the system is bounded and tends to zero as time goes to infinity.

The proof of this theorem is rather lengthy, and will be omitted here. The interested reader will find the derivation in either [33] or [35]. We note in passing, however, that satisfaction of the Popov criterion implies the existence of a Lyapunov function $V(\mathbf{x})$ for the system, and vice versa. We shall direct our attention below to the application of the Popov criterion.

The frequency-transfer function may be written:

$$G(j\omega) = \mathcal{Re}[G(j\omega)] + j\mathcal{Im}[G(j\omega)]. \qquad (12\text{-}66)$$

We now define a modified frequency-transfer function $G^*(j\omega)$ as

$$G^*(j\omega) = \mathcal{Re}[G(j\omega)] + j\omega\mathcal{Im}[G(j\omega)] \equiv X(j\omega) + jY(j\omega). \qquad (12\text{-}67)$$

The Popov locus is defined as the plot of $G^*(j\omega)$ in the complex plane as ω is varied from zero to infinity. Note that $G^*(j\omega)$ is entirely an even function of ω so that only positive values of ω need be plotted. By Eqs. (12-66) and (12-67), Eq. (12-65) becomes

$$X(j\omega) - qY(j\omega) + \frac{1}{K} \geq \delta > 0. \qquad (12\text{-}68)$$

The Popov line, defined by

$$Y(j\omega) = \frac{[X(j\omega) + 1/K]}{q}, \qquad (12\text{-}69)$$

is shown plotted at an arbitrary slope $1/q$ and through the point $-1/K$ on the real axis in Fig. 12-18(c). As indicated by the phasors in Fig. 12-18(c), the inequality of Eq. (12-68) is satisfied. The Popov condition is not satisfied in the case of Fig. 12-18(d), since there is no line through the $-1/K$ point that will not intersect the Popov locus. In summary, the Popov criterion is satisfied if the $G^*(j\omega)$ plot lies strictly to the right of any ($\pm q$) line which passes through the negative real axis at $-1/K$ and has a slope of $(1/q)$.

Although the Popov criterion is only a sufficient condition, resulting in conservative design, there is a separate necessary condition that may be checked. If we define the negative real axis to the left of the $-1/K$ point as the "forbidden zone," then a necessary condition for asymptotic stability of the system of Fig. 12-18(a) is that the Nyquist plot of the linear portion of the system does not cross the forbidden zone (see [50] for proof). This is obviously a weaker condition than the Popov criterion. The system of Fig. 12-18(d) passes this necessary test while it fails the Popov sufficiency test. The system *may*, however, still be asymptotically stable. What is required at this point, and is as yet unavailable, is a necessary *and* sufficient test. In a recent paper [52] Bergen and Sapiro have moved in this general direction by replacing the Popov line with a less stringent parabola. Another recent

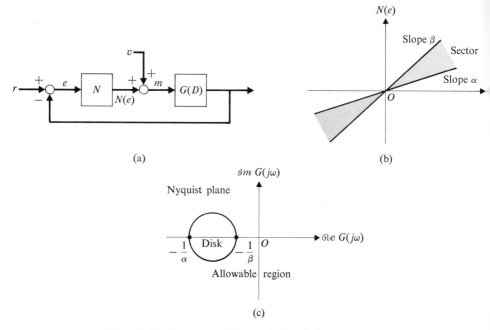

Fig. 12-19 Sector condition and the circle theorem.

development [53] involves transforming the allowable design region in the $G^*(j\omega)$ plane into a safe region on the Nichols chart.

Since Popov did his original work, considerable effort has been expended in attempts to extend and generalize his results [33, 34, and 35]. Of particular interest is Zames' circle criterion [34], which may be arrived at by setting $q = 0$ in Eq. (12-65) (see below). In this case, the proof is valid even for time-varying nonlinearities and, in fact, is valid for memory-type nonlinearities so long as the nonlinearity satisfies the condition

$$0 \leqslant \frac{N(e)}{e} \leqslant K.$$

The circle criterion is a sufficient test for asymptotic stability, and may be stated as follows:

If the nonlinearity N (Fig. 12-19a) lies inside a sector characterized by two slopes α and β (Fig. 12-19b), and if the frequency response $G(j\omega)$ of the linear element avoids a critical region [the disk or circle of Fig. 12-19(c)] in the Nyquist plane, then the closed-loop response is bounded, in the sense that for forcing input r or v in Fig. 12-19(a) whose squared integral over semi-infinite range of time $(\int_0^\infty r^2\, dt)$ or $\int_0^\infty v^2\, dt)$ is bounded, the same expression for loop variable e or m is also bounded.

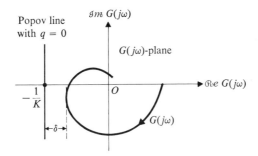

Fig. 12-20 Popov criterion with $q = 0$ in the Nyquist plane.

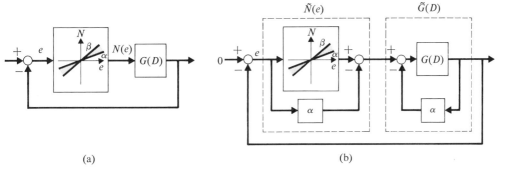

Fig. 12-21 Steps in arriving at the circle criterion.

Letting $q = 0$ in Eq. (12-65), we get

$$\mathcal{R}e\,[G(j\omega)] + \frac{1}{K} \geq \delta > 0. \tag{12-70}$$

This condition is depicted in Fig. 12-20, which indicates satisfaction of the Popov criterion. Note also that the plot is equivalent to Fig. 12-19(c) with $\alpha = 0$, $\beta = K$, and an infinite radius circle.

To arrive at the circle criterion, we convert the system of Fig. 12-21(a) with nonlinearity $N(e)$ in the sector $[\alpha, \beta]$ to the system of Fig. 12-21(b) with nonlinearity $\tilde{N}(e)$ in the sector $[0, K]$, where $K = (\beta - \alpha)$. This conversion is accomplished by adding and subtracting the same signal between the two feedforward blocks. The Popov criterion may now be applied to the system of Fig. 12-21(b) so long as $\tilde{G}(j\omega)$ is strictly stable [that is, so long as $G(s)$ with feedback gain α has closed-loop poles in the left half-plane]. Thus, Eq. (12-70) becomes

$$\mathcal{R}e\left[\frac{G(j\omega)}{1 + \alpha G(j\omega)}\right] + \frac{1}{\beta - \alpha} \geq \delta > 0,$$

or, after rearrangement,

$$\frac{1}{\beta - \alpha} \{\mathcal{R}e\,[H(j\omega)]\} \geq \delta > 0, \quad (12\text{-}71)$$

where

$$H(j\omega) = \frac{1 + \beta G(j\omega)}{1 + \alpha G(j\omega)}. \quad (12\text{-}72)$$

Equations (12–71) and (12–72) constitute the circle criterion, where the latter equation represents a conformal mapping of straight lines in the $H(j\omega)$-plane into circles in the $G(j\omega)$-plane. In particular, we consider the imaginary axis in the $H(j\omega)$-plane, since everything to the right of this line satisfies the inequality of Eq. (12–71). By setting $H(j\omega) = j\Omega$ in Eq. (12–72), and after rearranging, eliminating Ω, and completing the square, we arrive at the equation of a circle:

$$\left\{\mathcal{R}e\,[G(j\omega)] + \frac{\alpha + \beta}{2\alpha\beta}\right\}^2 + \{\mathcal{I}m\,[G(j\omega)]\}^2 = \frac{(\alpha - \beta)^2}{4\alpha^2\beta^2}. \quad (12\text{-}73)$$

The center of this circle is at $-(\alpha + \beta)/2\alpha\beta$ on the negative real axis, and the circle has radius $(\beta - \alpha)/(2\alpha\beta)$. This, of course, corresponds to a circle intersecting the negative real axis at $-1/\alpha$ and $-1/\beta$, as shown in Fig. 12–19(c).

All points to the right of the $j\Omega$-axis in the $H(j\omega)$-plane transform into points *outside* the circle. Thus, the sufficient test for asymptotic stability of the system of Fig. 12–19(a) is that the Nyquist plot of the linear portion of the system must not touch or encircle the disk of Fig. 12–19(c). If $\beta \to \alpha$, then the circle shrinks to the $-1/\alpha$ point, and the Nyquist criterion assures stability of the resulting linear system. If α is negative, the inequality of Eq. (12–71) reverses, so that asymptotic stability in this case requires that $G(j\omega)$ plot entirely within the resulting circle. This immediately rules out the possibility of any free integrators in $G(s)$ (that is, poles at the origin) when the α line in Fig. 12–19(b) has negative slope. As a final special case, consider a nonminimal phase $G(s)$, with α chosen such that $\hat{G}(s)$ of Fig. 12–21(b) is strictly stable. In this situation, asymptotic stability is assured when $G(j\omega)$ entirely encloses the circle (See Problem 12–16).

12-4 POSITIVE USE OF NONLINEAR ACTION

It is known that nonlinear action gives better control than linear action with respect to particular performance indices. According to the maximum principle of Pontryagin (see Chapter 14), for example, bang-bang control is the indicated operating mode for time-optimal performance. Pioneering research for control algorithms superior to linear action was initiated at a

12-4 Positive use of nonlinear action 549

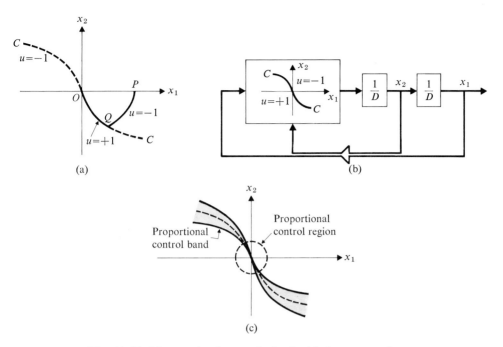

Fig. 12-22 Time-optimal control of a double-integrator plant.

time when the maximum principle was not yet known [36, 37, 38]. As a simple example, let us consider a second-order time-optimal position servo. The problem is to move a unit mass from one position to another in minimal time. The purely inertial system is described by

$$\frac{dx_1}{dt} = x_2 = \text{velocity},$$

$$\frac{dx_2}{dt} = u = \text{controlling force},$$

where x_1 = position and the available force magnitude is limited to (say)

$$|u| \leqslant 1.$$

The time-optimal control to move a mass from P to zero (Fig. 12–22a) is known to consist of two parts (see Section 14–1 for proof of optimality): $u = -1$ (full acceleration in the minus direction) from P to Q, and $u = +1$ (full deceleration) from Q to the origin. The locus of the force-switching point Q [CC in Fig. 12–22(a)], which is generally called the critical switching boundary, is identical with the trajectories going through the origin with either $u = +1$ or -1. Time-optimal control of an ideal inertial system can be realized by a feedback system which has a built-in critical switching boundary (Fig. 12–22b). However the switch of such an ideal

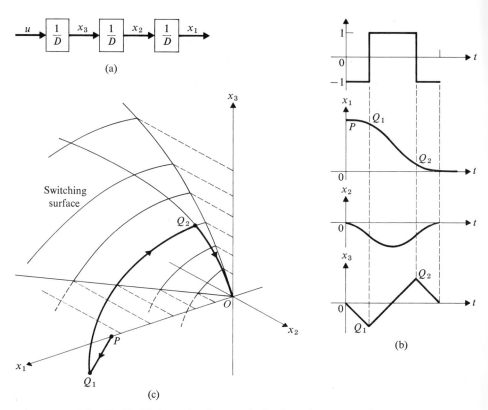

Fig. 12-23 Time-optimal control of a three-integrator plant.

servo would chatter (a high-frequency limit cycle, usually of low amplitude) when the error becomes zero. To avoid this chatter it is desirable to have a small continuous-control region (Fig. 12-22c) in the vicinity of the origin, although, strictly speaking, the system will become suboptimal with such an additional control mode.

The need for time-optimal control of inertial objects exists in various kinds of production and information equipment. The mechanical parts of a digital computer, an automatic machine tool, or an automatic assemblying machine, for example, should complete their motion in minimum time. However, the control objects in these systems are rarely ideal masses; Coulomb friction may exist, the order of the object may be higher than two, or the system parameters may be changing with time. It is also impossible to expect ideal action in the real control hardware.

Suppose, for instance, that the order of the control object is three instead of two, but otherwise the system is ideal. A time-optimal control for a normalized three-integrator system is shown in Fig. 12-23(a). In general, two switchings are required to bring an arbitrary initial state to zero. Since the state space is three dimensional, a trajectory cannot always hit the final "critical" trajectory without an intermediate switching. Starting from P (Fig. 12-23c), the trajectory first hits a switching surface at Q_1 in the ($+x_1$, $-x_2$, $-x_3$) octant. After the first switching at Q_1, the state will take the path Q_1Q_2 on the surface and meet the critical trajectory at Q_2, where the second switching will take place. To implement such a control is exceedingly cumbersome, if not impossible, for many practical applications. The complexity of an optimal-control algorithm rapidly increases when other deviations from ideal second-order systems are taken into account. Therefore, the simple control mode that has only one force reversal (Fig. 12-22) is generally preferred in many engineering applications, although it is suboptimal.

When a switching control is suboptimal, and hence the switching line does not exactly match the system's trajectory, there arises the possibility of chattering at the switching point [40]. Also, there is no assurance that the state will actually end at the target point (the origin in our examples). To avoid such unwanted phenomena, dual mode control is always desirable in the engineering realization of bang-bang type suboptimal controls [39]. One way to achieve this is to introduce a narrow proportional band in the vicinity of the switching line and/or target point (Fig. 12-22c). The result is a smoother transient.

The nonlinear suboptimal control discussed in the preceding paragraph is different from conventional feedback control in the sense that the pattern of the controlling signal (the manipulated variable) is predetermined by a knowledge of the control object and the specified task of control. For a specific job assignment, such as start-up of a batch process, a sequence of predetermined commands can be executed under feedback control of some selected signal. As an example, the Posicast control [41] originally proposed by Calvert* and named Posicast by Smith ([14], Chapter 8), when applied to a start-up (or shutdown) operation, follows a double-step command pattern to yield a smooth and fast transient for an oscillatory object.

The principle of Posicast can be readily explained. Let us suppose that the control object is second order with conjugate complex eigenvalues, described by a canonical state vector equation (see Section 2-4),

$$\frac{dx_1}{dt} = -\sigma x_1 + \omega x_2, \qquad \frac{dx_2}{dt} = -\omega x_1 - \sigma x_2 + u, \qquad (12\text{-}74)$$

or by transfer function

$$\frac{X_1(s)}{U(s)} = \frac{\omega}{(s+\sigma)^2 + \omega^2},$$

---

* J. F. Calvert, et al., U.S. Patent 2801351, 1957 and 3010035, 1961.

552 Nonlinear systems

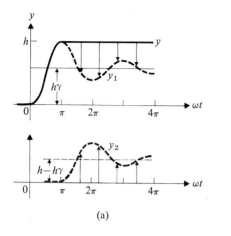

(a)

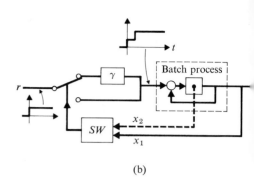

(b)

Fig. 12-24 Posicast control.

where ω is the imaginary part of the eigenvalues representing the damped natural frequency, and σ is the real part that produces an exponential decay. The unit-step-input response for zero initial state is:

$$x_1 = k\left[1 - e^{-\sigma t}\left(\frac{\sigma}{\omega}\sin \omega t + \cos \omega t\right)\right], \qquad k = \frac{\omega}{\omega^2 + \sigma^2}. \qquad (12\text{-}75)$$

The maximum overshoot (x_1 max) occurs when $dx_1/dt = 0$, or when $\sin \omega t = 0$. Thus $\omega t = \pi$ (Fig. 12-24a), and the peak value is

$$x_{1\ max} = \frac{\omega}{\omega^2 + \sigma^2}[1 + e^{-(\sigma/\omega)\pi}]. \qquad (12\text{-}76)$$

Suppose that a step change of x_1 from zero to h is required. We apply a step input of magnitude $(h/k)\gamma$ at $t = 0$, where

$$\gamma = \frac{1}{1 + e^{-\sigma t_1}}, \qquad t_1 = \frac{\pi}{\omega}. \qquad (12\text{-}77)$$

According to Eq. (12-75) and Eq. (12-76), the peak value of the response y_1 for this step occurs at $t = t_1$, and the peak value is equal to h. A second step input of height $(h/k)(1 - \gamma)$ must be superimposed on the first step at this instant. Since the second response lags the first response by exactly one half-period, the oscillation components ($\sin \omega t$ and $\cos \omega t$) of the response cancel each other; the cancellation is perfect when γ is chosen by Eq. (12-77). Therefore the total response, $x_1 = y_1 + y_2$ for $t > t_1$, stays constant at the desired level, $x_1 = h$ (Fig. 12-24).

The Posicast approach is applicable to any system which has a dominant pair of conjugate complex eigenvalues. A batch process with a feedback controller so tuned that it behaves almost like an oscillatory spring–mass–damper object can be connected to a Posicast controller as shown in

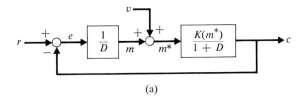

(a)

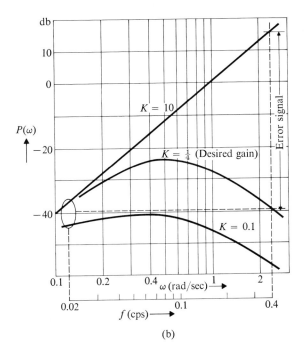

(b)

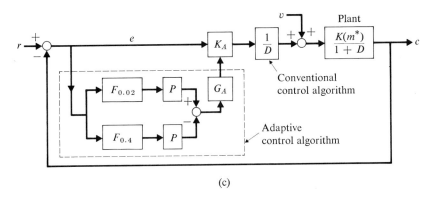

(c)

Fig. 12-25 Adaptive gain tuning.

Fig. 12–24(b), where SW is a switching logic that detects the switching instant when the gain must be changed from γ/k to $1/k$. A derivative network to detect $dx_1/dt = 0$ can be avoided by means of state-variable feedback. In any event, when the system is not purely second order (as in the case of the batch process of the figure), to produce approximate dead-beat response, the switching instant and the magnitude γ must be modified from what was given above.

There seems to be no precise definition of the term *adaptive control* in general acceptance. However, it is at least possible to state that the term applies to that class of automatic control systems which has the capability of automatic measurement of process dynamics *and* of automatic readjustment (or redesign) of the control algorithm [3]. Due mainly to the complexity involved in the realization of this type of control, initial investigations of the subject were purely theoretical. Recently, a new dimension of flexibility was added to control algorithm design by the on-line capability of digital computing controllers. In the following discussion we shall show two examples of adaptive process control which were developed and tested on real plants by Bakke [42 and 43].

According to conventional control theory (Chapters 8 and 9), favorable control performance (short settling time to set-point changes and rapid recovery from load disturbance) is obtained when controlling parameters (controller gain, etc.) are properly tuned to process dynamics. If, for instance, the gain of the control object changes as a function of load but the load upset is not measurable, then somehow the gain change must be detected and controlling parameter(s) must be readjusted to maintain satisfactory performance. One of the novelties of the adaptive scheme in Bakke's approach [42] is that the identification of the process gain is made without using any test signal. For purposes of illustration we shall consider I-control of a first-order plant (Fig. 12–25a), where the process gain $K(m^*)$ is a function of the load m^*. The power ratio $P(\omega)$ is generally defined as

$$P(\omega) = \left|\frac{E(j\omega)}{V(j\omega)}\right|^2, \qquad (12\text{–}78)$$

where $V(j\omega)$ is the load disturbance and $E(j\omega)$ is the error signal resulting from $V(j\omega)$, both in the frequency domain. Bakke found $P(\omega)$ to be a convenient measure for system gain identification. For our specific example $P(\omega)$ is given by

$$P(\omega) = \frac{\omega^2 K^2(m^*)}{[K(m^*) - \omega^2]^2 + \omega^2}. \qquad (12\text{–}79)$$

Equation (12–79) is plotted in Fig. 12–25(b) for three values of $K(m^*)$. The diagram shows high dependence of $P(\omega)$ on process gain $K(m^*)$ at a relatively high frequency, say, $f = 0.4$ cps, whereas it is practically gain independent at a low frequency such as $f = 0.02$ cps.

Suppose that $K(m^*) = \frac{1}{4}$ is the desired value of open-loop gain. (The closed-loop characteristic equation is $s^2 + s + K = 0$, so that $K = \frac{1}{4}$ is the "critical damping" condition.) The two frequencies we just noted in the figure have the following significance:

1. For $K(m^*) = \frac{1}{4}$, $P(\omega)$ at 0.02 cps is almost equal to the value at 0.4 cps.
2. This value of $P(\omega)$ (about -40 db) represents the forcing-energy intensity for a wide range of $K(m^*)$ at a frequency of 0.02 cps.

If these two frequency components are extracted from the error signal $e(t)$ via narrow-band-pass filters $F_{0.02}$ and $F_{0.4}$ tuned to 0.02 cps and 0.4 cps, respectively (Fig. 12–25c), and then squared or rectified by P in the figure, the difference will be a measure of the gain deviation from its desired value of $\frac{1}{4}$.

This difference (error signal in the adaptive loop) is fed into a conventional PI-control algorithm indicated by G_A in Fig. 12–25(c), and the gain K_A of the original loop is changed accordingly. Since a parametric change is involved, the system as a whole is nonlinear. However, when the adaptive loop is sufficiently slower than the original loop, the dynamic coupling between the two loops can be ignored in a preliminary design of the control algorithm. The load disturbance must have wide bandwidth (such as repeated steps) to ensure enough power at the two frequencies. Since all control algorithms in the block diagram (Fig. 12–25c) are carried out by a digital computer, no extra hardware is needed for the adaptive control. The overall range of gain change obtainable was reported to be about 20 : 1. This is also the approximate range of gain change in aircraft attitude-control systems.

The second example of adaptive control, also reported by Bakke [43], is concerned with a process having variable dead time. Consider a single-loop feedback control configuration of a process with dead time L (Fig. 12–26a). It is known from conventional control theory [44] that the effect of the dead time can be separated from the characteristic equation of the control system by finding a compensator transfer function $F(s)$ such that

$$\frac{C(s)}{R(s)} = \frac{G_c(s)G_p(s)F(s)e^{-Ls}}{1 + G_c(s)G_p(s)F(s)H(s)e^{-Ls}}$$

$$= \left[\frac{G_c(s)G_p(s)}{1 + G_c(s)G_p(s)H(s)}\right]e^{-Ls}. \qquad (12\text{–}80)$$

Solving Eq. (12–80) for $F(s)$, we find

$$F(s) = \frac{1}{1 + G_c(s)G_p(s)H(s)(1 - e^{-Ls})}. \qquad (12\text{–}81)$$

Equation (12–81), when placed in Fig. 12–26(a), can be realized by the minor loop configuration of Fig. 12–26(b).

556   Nonlinear systems

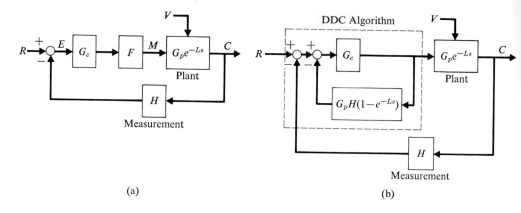

Fig. 12-26 Control of a process with dead time.

The control law $G_c(s)$, as well as $G_p(s)H(s)(1 - e^{-Ls})$, can be formulated into a DDC algorithm when the plant transfer functions $G_p(s)$, $H(s)$, and the dead time L are all known. The dead time produces no problem in computer instructions already written in a finite difference form.

Dead time in real process systems is quite often not constant, usually varying with operating conditions of the plant. Although we cannot always measure dead time directly, if we know the process structure, we can generally compute it from available measurements. If this is feasible, we can feed the computed dead time into the main control algorithm, which we can then make adaptive to a variable dead time.

The preceding examples of adaptive control indicate the necessity of on-line identification without using a test signal. The identification is possible by on-line digital data processing if an appropriate set of measurements is available [45]. The basic algorithm for an ideal case where the system is linear, autonomous, and all state variables are measurable is given below. Let

$$\frac{d\mathbf{x}}{dt} = \mathbf{A}\mathbf{x}$$

be the system for which the matrix $\mathbf{A}$ of known order is to be identified. The response of the system is

$$\mathbf{x}(t) = e^{\mathbf{A}t}\mathbf{x}_0 \quad \text{where} \quad \mathbf{x}_0 = \mathbf{x}(0),$$

and its sampled data for a sampling interval T will be denoted by $\mathbf{x}_i$, $i = 0, 1, \ldots$ A computer model of the system is constructed (Fig. 12-27), using an initial guess $\mathbf{A}_c(1)$ of $\mathbf{A}$, where the guess could be the zero matrix $\mathbf{0}$. The response of the model with the same initial state is

$$\mathbf{x}_c(t) = e^{\mathbf{A}_c(1)t}\mathbf{x}_0,$$

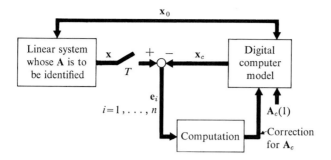

Fig. 12-27 On-line identification of **A**.

and its sampled data are denoted by $x_{c0}(1), x_{c1}(1), \ldots, x_{ci}(1), \ldots$ By defining an error vector $e_i(1)$ for the ith sample,

$$e_i(1) = x_i - x_{ci}(1) = (e^{A iT} - e^{A_c(1)iT})x_0, \quad i = 1, \ldots, n,$$

we derive a matrix relationship for a set of n samples;

$$E(1) = (e^{AT} - e^{A_c(1)T})X_0, \tag{12-82}$$

where

$$E(1) = [e_1(1), e_2(1), \ldots, e_n(1)] \quad \text{and} \quad X_0 = [x_0, x_1, \ldots, x_{n-1}]$$

are both $n \times n$ square matrices. If the system is in motion, all x_i, $i = 0, \ldots, n-1$, are different. Hence the inverse of X_0 exists, and

$$e^{AT} - e^{A_c(1)T} = E(1)X_0^{-1}. \tag{12-83}$$

In Eq. (12-83), X_0 is a collection of measurements and $E(1)$ is calculated. Theoretically it is possible to solve this equation for **A**. However, because of digital computer characteristics, it is faster and more accurate to use an iterative solution of Eq. (12-83). An iterative solution entails choosing an initial guess at $A_c(1)$ and then repeatedly making an approximation of Eq. (12-83) to improve $A_c(2)$, $A_c(3)$, ... If the matrix exponentials in Eq. (12-83) are expanded in series in terms of $A_c(1)$ and

$$\Delta A(1) = A - A_c(1), \tag{12-84}$$

we find that

$$\frac{1}{T}(e^{AT} - e^{A_c(1)T}) = \frac{1}{T} E(1)X_0^{-1} \approx \Delta A(1), \tag{12-85}$$

provided that $\|A(1)\|$ is small and/or T is small. The iteration equation to improve the model $A_c(k)$, $k = 1, 2, \ldots$, is the following:

$$A_c(k+1) = A_c(k) + \frac{1}{T} E(k)X_0^{-1}. \tag{12-86}$$

The iteration is terminated when

$$\|\mathbf{E}(k)\| < \text{Acc},$$

where Acc is a prescribed accuracy (such as 0.0005) and $\|\mathbf{E}\|$ is defined to be the sum of absolute errors in all elements of the matrix. The algorithm was tested on an electrohydraulic servo system with a smooth nonlinearity which is approximately third-order. Free response of the system was damped oscillatory, with a natural period of about 8 sec. The three state variables were sampled at an interval of $T = 1.25$ sec. Each identification required $nT = 3.75$ sec, and 6 to 8 iterations for an accuracy of Acc $=$ 0.0004. The matrix $\mathbf{A}$ to be identified in this experiment was 3×3, although, strictly speaking, the order of the servo was higher than three. Of all 34 tests, the identification was successfully made 24 times, while it diverged 10 times due mainly to the nonlinearity. The success or failure of this identification scheme depends on its convergence. Intuitively one would expect that the iteration will not converge if the sampling theorem (Section 11-1) is violated. The sufficient condition for convergence of the ideal case was found by Yore [46] to be

$$|a_{ij}|_{\max} T < \frac{(\log_e 2)}{n}, \qquad (12\text{-}87)$$

where n is the order of the linear system and a_{ij} is the largest element of $\mathbf{A}$. Yore [46] also deals with nonideal cases such as nonautonomous systems, noise effects, etc.

12-5 AUTOMATIC HILL-CLIMBING CONTROL

Static or steady-state performance of a plant—efficiency, product quantity, product quality, control quality, losses, etc.—in general varies as a function of operating condition or equilibrium state. Quite often there exists a static optimal point where a performance index assumes its extremum value (either maximum or minimum). Draper and Li's "optimalizing control" [47] was one of the first attempts to automatically maintain optimal performance by feedback control. Let us suppose that the performance index J to be maximized (Fig. 12-28a) is a function of a scalar controlling input u. Although the index J and the input u are considered to be the controlled variable and manipulated variable, respectively, an optimalizing control is basically different from conventional feedback control because the desired operating point is $J_{\max}$ instead of at some given reference input. Since the static relation between J and u is nonlinear with a peak of $J_{\max}$ at $u = u^0$, an algorithm is necessary to differentiate $u < u^0$ from $u > u^0$. Direct use of the slope $\partial J/\partial u$ for control is in general impractical due to inherent noise and dynamic lags. As shown in Fig. 12-28(b), a cosinusoidal test signal is

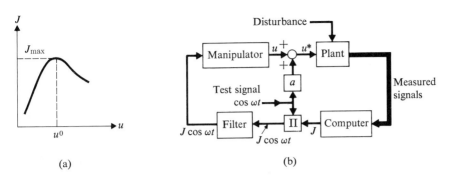

Fig. 12-28 Peak-holding control by sinusoidal forcing.

applied to the system in the Draper–Li optimalizing control approach. When the J–u nonlinear characteristic is considered to be quadratic in the vicinity of its apex point,

$$J = J_{\max} - k(u^* - u^0)^2 , \qquad (12\text{–}88)$$

where $k = $ const, the total input u^* is the sum of the controlling input u and a test signal

$$u^* = u + a \cos \omega t . \qquad (12\text{–}89)$$

Hence

$$J \cos \omega t = [J_{\max} - k(a \cos \omega t + u - u^0)^2] \cos \omega t ,$$

where

$$(a \cos \omega t + u - u^0)^2 \cos \omega t$$
$$= (u - u^0)^2 \cos \omega t + 2a(u - u^0) \cos^2 \omega t + a^2 \cos^3 \omega t .$$

Since $\cos^2 \omega t = \tfrac{1}{2} + \tfrac{1}{2} \cos 2\omega t$, and $\cos^3 \omega t = \tfrac{3}{4} \cos \omega t + \tfrac{1}{4} \cos 3\omega t$, the time average of $J \cos \omega t$ involves only the dc-component of $\cos^2 \omega t$. Thus

$$\overline{J \cos \omega t} = ka(u^0 - u) . \qquad (12\text{–}90)$$

Therefore, the filter output (Fig 12–28b) will serve as an actuating error of the feedback control system, and when $u = u^0$, the test signal will produce a limit-cycle oscillation of period π/ω in J. The magnitude a and frequency ω must be carefully chosen relative to the system's noise level and dynamics.

The Draper–Li optimalizing control is basically a static peak-holding method. Various hill-climbing algorithms have been developed since then, mostly for computational purposes. Some of these algorithms may be applicable to on-line static peak-holding controls.

If the static as well as the dynamic mathematical model of a control object is known, optimal control theory (Chapter 14) may apply to deter-

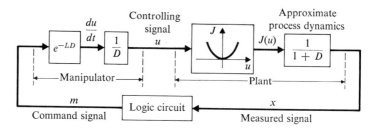

Fig. 12-29 Perret-Rouxel's extremal control system.

mine the optimal control policy $u(t)$, which will dynamically optimize a control transient for a given disturbance. In practice, however, a simpler mode of control which does not require a complete knowledge of the control object but still takes into account the approximate process dynamics to yield a compromise solution between static peak-holding and dynamic optimization is sometimes desirable. We introduce below a method in this category, the Perret–Rouxel method [48 and 49].

Consider a system (Fig. 12-29) that consists of a plant, logic circuit (the extremal computer), and a manipulator. For the normalized relation shown in the figure, where the plant dynamics are approximated by only a single lag, we have

$$\frac{dx}{dt} = -x + \tfrac{1}{2}u^2, \qquad \frac{du(t)}{dt} = m(t - L), \qquad (12\text{-}91)$$

where a normalized parabolic relation is assumed between the control u and the performance index J such that

$$J = \tfrac{1}{2}u^2.$$

The minimum point, $J = 0$ and $u = 0$, is the desired operating condition. A bang-bang control policy, $m = \pm 1$, is to be developed in such a way that any state in the xu-plane will be quickly brought to the vicinity of the origin.

We first study trajectory patterns forced by $m = 1$ and -1 without switching. The trajectories may be determined either analytically or by the isocline method. To derive an analytical expression for the trajectories, we eliminate t from Eq. (12-91):

$$\frac{dx}{du} = -\frac{x}{m} + \frac{1}{2}\frac{u^2}{m}, \qquad (12\text{-}92)$$

where $m = +1$ or -1. Solving Eq. (12-92), we have

$$x = \tfrac{1}{2}u^2 - mu + m^2 + Ce^{-u/m}.$$

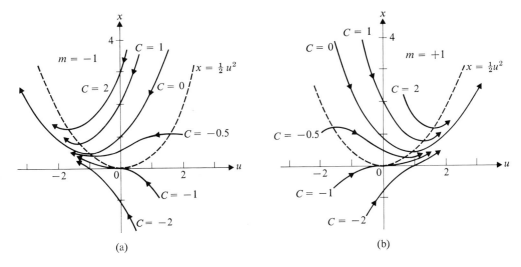

Fig. 12-30 Trajectory patterns.

Thus,

$$x = \tfrac{1}{2}(u+1)^2 + \tfrac{1}{2} + Ce^u \quad \text{for} \quad m = -1,$$
$$x = \tfrac{1}{2}(u-1)^2 + \tfrac{1}{2} + Ce^{-u} \quad \text{for} \quad m = +1, \qquad (12\text{-}93)$$

where C is a constant that represents initial state. The trajectories are shown in Fig. 12-30.

The Perret-Rouxel control law is based on the switching function:

$$S = k_1 \frac{dx}{dt} + k_2 \frac{d^2x}{dt^2} - \Delta, \qquad (12\text{-}94)$$

where $k_1 \geq 0$ and $k_2 > 0$ are both constant, and Δ is a constant threshold, positive or negative. The sign of the control command m is determined by the following logic-oriented statements:

1. $m(t)$ changes polarity (either from $+1$ to -1 or -1 to $+1$) when S crosses zero from negative to positive. This switching will be referred to below as "natural switching."

2. While S is positive, $m(t)$ is forced to change sign (from $+1$ to -1, or vice versa) with a preset period $2W$. We shall call this mode of switching "forced switching."

3. $m(t)$ does not change sign when S is negative.

To see how the above stated switching logic will drive the system toward the origin in the xu-space, let us first construct the switching line ($S = 0$) for the natural switching. For purposes of illustration we assume

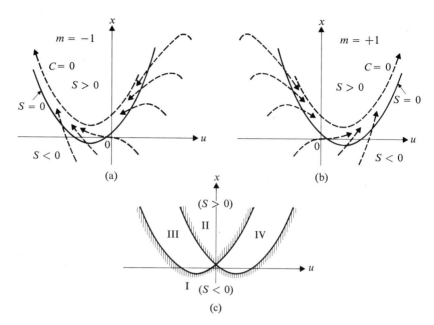

Fig. 12-31 Derivation of switching law.

a simpler form for the switching function:

$$S = \frac{d^2x}{dt^2} - \Delta, \tag{12-95}$$

which means that $k_1 = 0$ and $k_2 = 1$ in Eq. (12-94). By Eq. (12-91), we have

$$\frac{d^2x}{dt^2} = -\frac{dx}{dt} + u\frac{du}{dt},$$

where $dx/dt = -x + \frac{1}{2}u^2$ and $du/dt = m$; thus,

$$\frac{d^2x}{dt^2} = x - \frac{1}{2}u^2 + mu$$

and

$$S = x - \frac{1}{2}u^2 + mu - \Delta. \tag{12-96}$$

The switching condition $S = 0$ is satisfied by a pair x_s, u_s such that

$$x_s = \frac{1}{2}u_s^2 - mu_s + \Delta. \tag{12-97}$$

The parabolas for $m = \pm 1$ are shown in Fig. 12-31(a) and (b). The switching lines divide the state plane into four regions, I through IV (Fig. 12-31c).

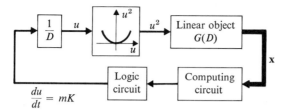

Fig. 12-32 Control of stable object.

No switching can occur in the region I, where all trajectories, for both $m = +1$ and -1, are directed upward (Fig. 12-30). If $m = -1$, the trajectories in this region will be bent to the left (Fig. 12-30a) and will cross the left-hand-side switching line (Fig. 12-31a) where natural switching will change the polarity of m from -1 to $+1$. Likewise, all trajectories for $m = +1$ in the region I are bent to the right (Fig. 12-30b), and they will cross the right-hand-side switching line (Fig. 12-31b) where the polarity will change from $+1$ to -1.

S is positive regardless of the sign of m in the region II (Fig. 12-31c), where forced switching will take place so long as the state remains in this zone. Since all trajectories are directed downward (Fig. 12-30), x will decrease, and eventually move into either zone III or zone IV. The natural and/or forced switching in the last two regions will take the state into a final mode which is a limit cycle in some vicinity of the origin. We shall shortly discuss the limit-cycle patterns.

The heuristic arguments in the preceding paragraphs imply that the system's motion under switching control is bounded (stable in the sense of Lyapunov). The boundedness of the response has been proved [48] for the general case of Fig. 12-32, where $G(D)$ is an asymptotically stable linear object,

$$G(D) = \frac{1}{1 + a_1 D + \cdots + a_n D^n}. \tag{12-98}$$

The normalized general relation that corresponds to Eq. (12-91) becomes

$$(a_n D^n + \cdots + a_1 D + 1)x = \tfrac{1}{2}u^2, \qquad Du = m. \tag{12-99}$$

Let λ_i, $i = 1, \cdots, n$, be the roots of the characteristic equation

$$a_n \lambda^n + \cdots + a_1 \lambda + 1 = 0,$$

and assume for brevity of discussion that all roots are distinct. The solution of Eq. (12-99) is

$$x = \tfrac{1}{2}u^2 - a_1 mu + (a_1^2 - a_2) + \sum_{i=1}^{n} C_i e^{m\lambda_i u}, \tag{12-100}$$

where C_i, $i = 1, \ldots, n$, are constants that depend on the initial state. We also have

$$\frac{dx}{dt} = mu - a_1 + \sum_{i=1}^{n} C_i \lambda_i e^{m\lambda_i u},$$

$$\frac{d^2x}{dt^2} = 1 + \sum_{i=1}^{n} C_i \lambda_i^2 e^{m\lambda_i u},$$

so that

$$S = k_1 mu - k_1 a_1 + k_2 - \Delta + \sum_{i=1}^{n} (k_1 + k_2 \lambda_i) C_i \lambda_i e^{m\lambda_i u}. \quad (12\text{--}101)$$

Suppose that no switching occurs. Then, as time goes to infinity, the product mu will tend to plus infinity, since $u \to +\infty$ when $m = +1$ and $u \to -\infty$ when $m = -1$. Since the control object itself is asymptotically stable by assumption, all exponential terms $e^{m\lambda_i u}$ in Eq. (12–101) will vanish as time tends to infinity. For a positive value of k_1, the asymptotic behavior of S will therefore be

$$\lim_{t \to \infty} S = k_1 mu - k_1 a_1 + k_2 - \Delta \to +\infty.$$

We can thus conclude that S will be positive. According to control-switching logic 2, at least the forced switching takes place so long as S stays positive. Since the control object is asymptotically stable, the limit cycle with forced switching must be bounded. Therefore, the controlled trajectory of the system cannot diverge.

The last item of investigation in the Perret-Rouxel control system is the pattern of limit cycles that can take place. There are three possible types: NF, NN, and FF, meaning alternating forced and natural switchings, natural switchings only, and forced switchings only, respectively. To determine the nature of the limit cycles, let us again consider the first-order object (Eq. 12-91) and the simplified switching function (Eq. 12-95). The effect of the dead time L (Fig. 12-29), which was implicit in the preceding discussion, must be explicitly taken into account in computing a limit-cycle pattern. The effect of switching at a state $x = x_s$ and $u = u_s$, Eq. (12-97), changes the direction of motion of u after L seconds, at a state $x = x_s'$ and $u = u_s'$, such that

$$\begin{aligned} u_s' - u_s &= mL, \\ x_s' &= \tfrac{1}{2} u_s'^2 - mu_s' + 1 - (1-\Delta)e^{-L}, \end{aligned} \quad (12\text{--}102)$$

where the expression for x_s' is derived by applying the trajectory equation (12-93) at the switching point (Eq. 12-97). Thus,*

$$x_s = \tfrac{1}{2} u_s^2 - mu_s + m^2 + Ce^{-mu_s} = \tfrac{1}{2} u_s^2 - mu_s + \Delta,$$

---

* Note that $e^{-u/m} = e^{-mu}$ for $m = \pm 1$.

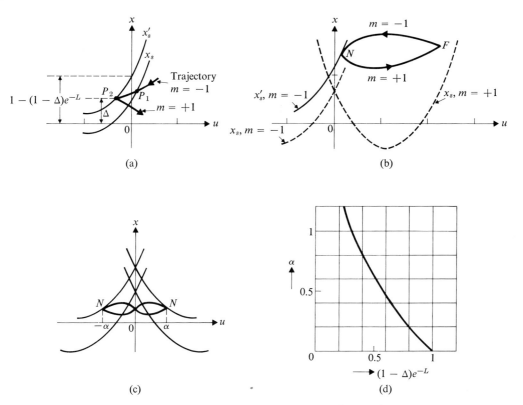

Fig. 12-33 Limit-cycle conditions of the extremal control system.

so that

$$\Delta = 1 + Ce^{-mu_s},$$
$$Ce^{-mu'_s} = Ce^{-m^2L}e^{-u_sm} = (\Delta - 1)e^{-L}$$

and

$$x'_s = \tfrac{1}{2}u'^2_s - mu'_s + 1 + Ce^{-mu'_s},$$

which reduces to Eq. (12-102). The dead time L caused by the effects represented in Eq. (12-102) produces a displaced switching curve, labeled x_s in Fig. 12-33(a). When the command signal $m = -1$ is changed to $+1$ at P_1 of the figure, the actual trajectory switch will appear at P_2.

An NF cycle is shown in Fig. 12-33(b). The limit cycle is not symmetric relative to the origin in the xu-plane. Hence, it is not desirable for our purpose of optimization. To avoid this unwanted limit cycle, an appropriate selection must be made in system parameters Δ (positive or negative threshold) and W (half-period of the forced switching) so that the point N in Fig. 12-33(b) will reach the left-most boundary of the right-hand-side switching parabola (labeled $m = +1$ in the figure).

The NN cycle shown in Fig. 12–33(c) is the limit the control system is designed for. The limit cycle amplitude of $u = \alpha$ in the figure can be found by a simple computation to satisfy the following condition:

$$\alpha = \tfrac{1}{2}(1 - \Delta)(e^{2\alpha} - 1)e^{-L} \ . \tag{12-103}$$

α as an explicit function of $(1 - \Delta)e^{-L}$ is shown in Fig. 12–33(d). The third kind of limit cycle, the FF cycle, is unstable for a stable object; it disappears quickly.

The implementation of the control system is straightforward. Natural switching can be realized by a simple sequential circuit (Chapter 15), and the same circuit can trigger a clock-pulse generation for forced switching. The time derivatives involved in the switching function S (Eq. (12–94) can be replaced by an appropriate linear combination of state variables when the control object has an order higher than one.

REFERENCES

1. Cosgriff, R. L., *Nonlinear Control Systems.* New York: McGraw-Hill, 1958.
2. West, J. G., *Analytical Techniques for Nonlinear Control Systems.* Princeton, N. J.: Van Nostrand, 1960.
3. Mishkin, E., et al., *Adaptive Control Systems.* New York: McGraw-Hill. 1961.
4. Graham, D., McRuer, D., *Analysis of Nonlinear Control Systems.* New York: Wiley, 1961.
5. Thaler, G. J., Pastel, M. P., *Analysis and Design of Nonlinear Feedback Control Systems.* New York: McGraw-Hill, 1962.
6. Gibson, J. E., *Nonlinear Automatic Control.* New York: McGraw-Hill, 1963.
7. Struble, R. A., *Nonlinear Differential Equations.* New York: McGraw-Hill, 1962.
8. Reis, G. C., "Phase trajectory construction of high-order, nonlinear, time-varying, nonautonomous systems." *J. Basic Eng., Trans. ASME,* **88,** June 1966, pp. 311–315. See also Whitbeck, R. F., "Special graphical construction methods for the study of nonlinear systems." *1965 JACC Preprint,* pp. 1–7.
9. Ku, Y. H., *Analysis and Control of Nonlinear Systems,* New York: Ronald Press, 1958.
10. Tsypkin, Y. Z., "A note on Tsypkin locus." *IEEE Trans.,* **AC-8,** No. 1, 1963, pp. 70–71; and his book published in USSR.
11. Woods, J. T., Roots, W. K., "On–off control performance." *Control Eng.,* December 1966, pp. 77–78.
12. Yamashita, S., "On–off control action improved by reverse hysteresis." *Instruments and Control* (Japan). **2,** No. 3, March 1963, pp. 175–181.
13. Nelson, W. L., "Pulse-width relay control in sampling systems." *J. of Basic Eng. Trans. ASME,* **83,** 1961, pp. 65–76.

14. Loos, C. H., "Time proportional control." *Control Eng.*, May 1965, pp. 65–70.
15. Kovatch, G., "On the computation of self-sustained oscillations in piecewise linear systems with two nonlinearities." *1964 JACC Preprint*, pp. 189–193.
16. Mufti, I. H., "A method for the exact determination of periodic motions in relay control systems." *1965 JACC Preprint*, pp. 8–14.
17. Tustin, A., "The effect of backlash and of speed-dependent friction on the stability of closed-cycle control systems." *J. Inst. Elec. Engrs.*, **94,** Part 2A, No. 1, 1947, pp. 143–151; "A method of analyzing the effect of certain kinds of nonlinearity in closed-cycle control systems," *Ibid.*, pp. 152–160.
18. Goldfarb, L. C., "On some nonlinear phenomena in regulatory systems." *Avtomatika i Telemekhanika*, **8,** No. 5, Sept.–Oct. 1947, pp. 349–383. (English translation in *Frequency Response*, R. Oldenburger (ed.). New York: Macmillan, 1956, pp. 239–259.)
19. Oppelt, W., "Über die Stabilität unstätiger Regelvorgänge," *Electrotechnik*, **2,** 1948, pp. 71–78; "über Ortskurvenverfahren bei Regelvorgängen mit Reibung," Z-VDI, **90,** No. 6, 1948, pp. 179–183.
20. Kochenburger, R., "Frequency-response method for analysis of a relay servomechanism." *Trans. AIEE*, **69,** Part I, 1950, pp. 270–284.
21. Deekshatulu, B. L., "How to evaluate harmonic response of nonlinear components." *Control Eng.*, October 1963, pp. 53–56.
22. Merritt, H. E., Stocking, G. L,, "How to find when friction causes instability." *Control Eng.*, December 1966, pp, 65–70.
23. Neal, C. B., Bunn, D. B. "The describing function for hysteresis." *1964 JACC Preprint*, pp. 185–188.
24. Takahashi, Y., Ziegler, J. G., Nichols, N. B., "Process control with a velocity limit," *Proc. First IFAC Congress* (Moscow, 1960), London: Butterworths, pp. 198–204.
25. Sarti, E., Takahashi, Y., "An extension of the describing-function method." *J. Basic Eng.*, *Trans. ASME*, **88,** No. 2, Series D, June 1966, pp. 469–474.
26. Hatanaka, H., "The frequency response and jump-resonance phenomena of nonlinear feedback control systems." *J. Basic Eng.*, *Trans. ASME*, **85,** No. 2, Series D, June 1963, pp. 236–242.
27. Lyapunov, M. A., Dissertation, Kharkov, USSR, 1892. Stability of motion, translated by F. Abramovici, M. Shimshoni, Academic Press, 1966.
28. Lehnigk, S. H., *Stability Theorems for Linear Motions.* Englewood Cliffs, N. J.: Prentice Hall, 1966.
29. Krasovskii, N. N., "On the global stability of a system of nonlinear differential equations," *Prikl. Mat. i. Mek.*, **18,** 1954, pp. 735–737.
30. Letov, A. M., *Stability in Nonlinear Control Systems* (translated from Russian by J. G. Adashko), Princeton. N. J.: Princeton University Press, 1961.
31. Johnson, G. W., "Upper bound of dynamic quantization error in digital control systems via the direct method of Lyapunov." *1965 JACC Preprint*, pp. 682–688.

32. Popov, V. M., "Absolute stability of nonlinear systems of automatic control." *Automatic and Remote Control*, **22**, No. 8, March 1962, pp. 857-875. (Original, August 1961).

33. Brocket, R. W., Willems, J. W., "Frequency-domain stability criteria." *1965 JACC Preprint*, pp. 735-747; *IEEE Trans.* **AC-10**, July 1965, pp. 262-267.

34. Zames, G., "On the input-output stability of time-varying nonlinear feedback systems," *IEEE Trans. of Automatic Control*, **AC-11**, No. 2, April 1966, pp. 228-238; *Ibid.*, No. 3, July 1966, pp. 465-476.

35. Desoer, C., "A generalization of the Popov criterion," *IEEE Trans. on A.C.*, **AC-10**, No. 2, April 1965, pp. 182-185.

36. Bushaw, D. W., "Differential equations with a discontinuous forcing term." Dissertation, Princeton University, June 1952; *Experimental Towing Tank Report No. 469*, Stevens Institute of Technology, New Jersey, January 1953.

37. Hopkin, A. M., "A phase-plane approach to the computation of saturating servomechanisms." *AIEE Trans.*, **70**, Part 1, 1951, pp. 631-639.

38. Oldenburger, R., *Optimal Control*. New York: Rinehart and Winston, 1966.

39. Fredriksen. T. R., "Practical bang-bang design." *Control Eng.*, October 1966, pp. 78-84.

40. O'Donnell, J. J., "Bounds on limit cycle in two-dimensional bang-bang control systems with an almost time-optimal switching curve." *1964 JACC Preprint*, pp. 199-206.

41. Gorbatenko, G. G., "Posicast control by delayed gain." *Control Eng.*, February 1965, pp. 74-77.

42. Bakke, R. M., "Adaptive gain tuning applied to process control." *ISA Paper, No. 32-1-64*, October 1964.

43. Bakke, R. M., "Direct digital control algorithms for processes with dead time." *A. I. Ch. E. Preprint*, March 1966.

44. Reswick, J. B., "The design and application of the delay-line synthesizer." Dissertation, M.I.T., 1954.

45. Yore, E. E., Takahashi, Y., "Identification of dynamic systems by digital computer modelling in state space," *Trans. ASME*, **89**, No. 2, Series D, June 1967, pp. 295-299.

46. Yore, E. E., "Identification of dynamic systems by digital computer modelling in state space and component parameter identification of static and dynamic systems," Dissertation, University of California, Berkeley, June 1966.

47. Draper, C. S., Li, Y. T., "Principles of optimalizing control systems and an application to the internal combustion engines." *ASME Publication*, September 1951.

48. Perret, R., Rouxel, R. R., "Principle and application of an extremal computer." Proc. 2nd IFAC, 1963, pp. 527-540, Butterworths-Oldenboung.

49. Carmassi, M., Helein, J., Rouxel, R., "Optimization of a sulphur recovery plant." *1965 IFAC Tokyo Symposium Preprint Volume*, V-29-40.

50. Aizerman, M. A., Gantmacher, F. R., *Absolute stability of regulator Systems.* San Francisco: Holden, 1964.
51. Nichols, N. B., "Backlash in a velocity lag servomechanism." *Appl. and Ind. AIEE.,* No. 10, January 1954, pp. 462–467.
52. Bergen, A. R., Sapiro, M. A., "The parabola test for absolute stability." *IEEE Trans. on AC,* **12**, June 1967, pp. 312–314.
53. Murphy, G. J., "A method for the design of stable nonlinear feedback Systems." *Proc. National Electronics Conference,* **XXIII**, 1967, pp. 108–113.

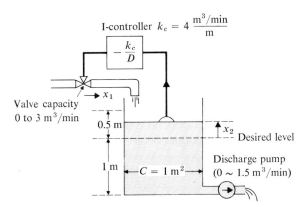

Fig. P12–1

PROBLEMS

12-1 The second-order control system shown in Fig. P12–1 is linear except for limiting effects; the control valve hits stops, and overflow takes place in the tank. For the controlling input (that is, controller output), take the first state variable x_1 as supply flowrate, with $x_1 = 0$ when total supply flowrate is 1.5 m³/min (which is equal to the discharge pump delivery rate). Liquid level is measured by x_2 (in meters) with $x_2 = 0$ at the desired level shown in the figure. Assume that the control valve is completely closed $[x_1(0) = -1.5]$ and the tank is empty $[x_2(0) = -1]$ at $t = 0$, but the discharge pump is in operation so that all liquid supply will be sucked out until x_1 reaches 1.5. Obtain the state-plane trajectory for start-up transient, label crucial states on the trajectory, and show these points on $x_1(t)$ and $x_2(t)$ response curves plotted against time. For this problem it is convenient to choose a scale ratio $x_1/x_2 = 2$ for state-space coordinate scaling. Why?

12-2 Chattering (or dither) is a limit cycle (or "self excited oscillation") seen, for instance, in machine tool operation or in a servomechanism moving at a creeping speed. Consider a mass pulled by a spring at a constant speed $u = vt$ (v = constant velocity) as a simple model of this phenomenon. The mass moves when the friction (or sticking) force reaches a magnitude F_0. Assume that there is no friction while the mass is in motion. Let l_0 = length l, shown in Fig. P12–2, when the spring is in its free state (no compression or tension acting on the spring), and introduce state variables $x_1 = l - l_0$, $dx_1/dt = x_2$. Sketch the limit-cycle pattern in the state plane.

570 Nonlinear systems

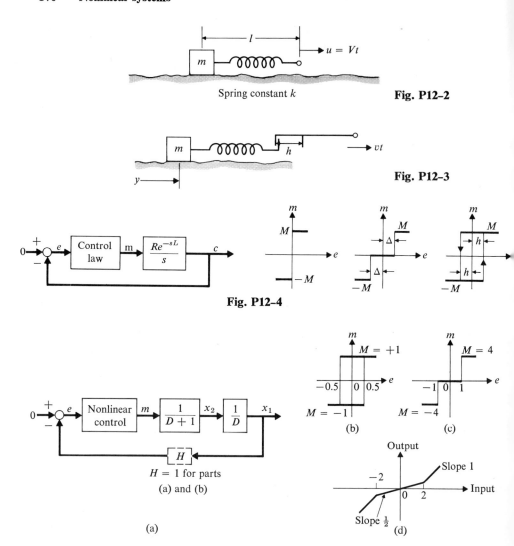

12-3 The system of Problem 12-2 has a backlash h added as shown in Fig. P12-3. What effect does the backlash have on the limit cycle pattern? Sketch both the state-plane trajectory and $x_1(t)$ and $y(t)$ curves versus time, where $y(t)$ = position of the mass.

12-4 The control object to be considered is an integrator with delay, where R is gain and L is dead time (see Fig. P12-4). The controller is of the bang-bang type. Determine the limit-cycle amplitude and period for the three types of control laws shown in the figure, and sketch the phase-plane (c vs. dc/dt) pattern for each control law.

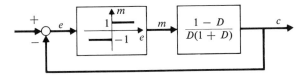

Fig. P12-6

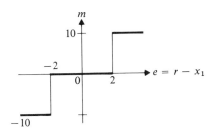

Fig. P12-7

12-5 A control system (servomechanism) is shown in Fig. P12-5(a) where the control law is either bang-bang action with hysteresis (Fig. P12-5b) or bang-bang action with dead zone (Fig. P12-5c). (a) Obtain the trajectory pattern (and limit cycle if there is any) for the bang-bang control with hysteresis. (b) Find the trajectory for the bang-bang control with dead zone, for the initial state $x_{10} = -2$, $x_{20} = +5$. Is this system asymptotically stable in the large, locally asymptotically stable, stable in the sense of Lyapunov, or unstable? (c) It is desired, in the system of question (b), to expand the region of x_1 represented by $x_{1\,max}$ such that, starting from an initial state $0 < x_{10} \leqslant x_{1\,max}$ and $x_{20} = 0$, the motion will reach the equilibrium state with only one switching. Is the feedback element shown in Fig. P12-5(d) an improvement over the original system for this purpose? (Design problems of this general nature arise, for instance, in machine tool control systems).

12-6 Bang-bang control of a nonminimum phase plant is shown in Fig. P12-6. (a) Sketch the unit-step-input response pattern of the plant for zero initial state. (b) Obtain the stability limit gain and the period of oscillation for P-control of the plant instead of bang-bang control. (c) Determine the limit cycle of the nonlinear control system by an analytical approach. (d) Confirm the results of part (c) by sketching state-plane trajectories. (e) Apply the describing-function method to the problem of part (c).

12-7 The control object is linear, described by

$$\frac{dx_1}{dt} = -x_1 + 3x_2, \qquad \frac{dx_2}{dt} = -3x_1 - x_2 + m(t),$$

where $m(t)$ is the controller output and x_1 is the plant output. The control law is bang-bang with a dead zone, as shown in Fig. P12-7. (a) Determine the trajectory of the control system for $r = 0$. The initial state is

$$\mathbf{x}_0 = \begin{bmatrix} -6 \\ 0 \end{bmatrix}.$$

(b) Discuss the system's stability using the describing-function concept.

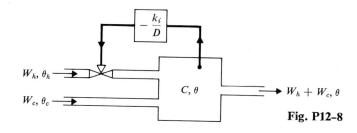

Fig. P12-8

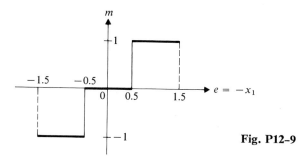

Fig. P12-9

12-8 Figure P12-8 shows a mixing type thermal system, where θ is the mixed temperature to be controlled. Thermal capacitance C of the mixing tank is 20. Inlet temperatures θ_h and θ_c of hot and cold flow are kept constant, $\theta_h = 20$ and $\theta_c = 0$. The product of the flowrate and the specific heat of the hot flow, w_h, is considered to be the controller output. The product for the cold flow w_c is kept constant at $w_c = 10$. Desired temperature of the mixture (set point) is 10, and x_1 is temperature deviation,

$$x_1 = \theta - 10.$$

The controller action is I, with $k_i = \frac{1}{2}$. Obtain the state-plane trajectory (θw_h-plane) for the start-up transient from an initial state $\theta = 0$, $w_h = 0$. In the state plane, compare the results with the trajectory of a linear model (linearized in the vicinity of the final equilibrium state).

12-9 The control object is $G_p(s) = 1/(1 + T_p s)$, where T_p is the plant time constant. The state variable x_1 is the plant output (that is, the controlled variable). Control is DDC, with an I-algorithm; $G_z(z) = K_i z/(z - 1)$, where K_i is the controller gain. $G_c(z)$ (with x_{2k} as the digital integrator state) is connected to the plant via a zero order holder. The DDC algorithm is preceded by an A-to-D converter, part of which is shown in Fig. P12-9. (a) Determine the condition under which a limit cycle of period $2T$ and h equal only to ± 1 exists (where T is sampling period). (b) Is this limit cycle stable?

12-10 There is saturation in the manipulator of a discrete-time control system shown in Fig. P12-10, so that

$$m(t) = M \quad \text{if} \quad u > h,$$
$$= ku \quad \text{if} \quad -h \leqslant u \leqslant h,$$
$$= -M \quad \text{if} \quad u < -h,$$

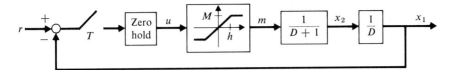

Fig. P12-10

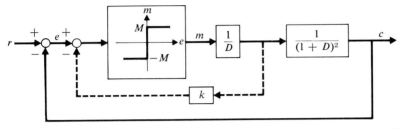

Fig. P12-11

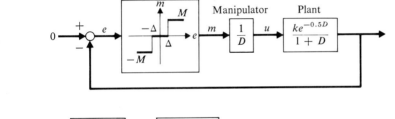

Fig. P12-12

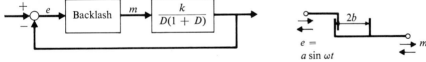

Fig. P12-13

where $k = M/h$ and h is a positive constant. Obtain $\mathbf{P}$ and $\mathbf{q}$ in the difference equation of the control system,

$$\mathbf{x}_{k+1} = \mathbf{P}\mathbf{x}_k + \mathbf{q}r_k,$$

that will serve for successive numerical computation of the closed-loop response.

12-11 A bang-bang control system is shown in Fig. P12-11. Using the describing-function method, (a) determine the limit-cycle period and amplitude when $k = 0$. (b) Local feedback (around the nonlinear element) serves to squelch the limit cycle. Find the condition on k by which the control system will be made asymptotically stable.

12-12 Using the describing-function method for the system shown in Fig. P12-12, (a) obtain the amplitude and period of the stable limit cycle for $kM/\varDelta = 5$. (b) Find the condition necessary in order for the control system to be asymptotically stable.

12-13 Backlash often exists in the drive mechanism of a servo (see Fig. P12-13). Obtain the amplitude locus $-1/N(h)$ of the backlash (using $b/a = h$ as the plotting

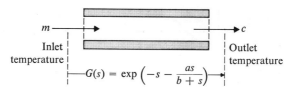

Fig. P12-14

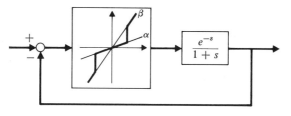

Fig. P12-15

parameter, where $2b =$ backlash and a is the input amplitude to the backlash). Determine the nature of limit cycle(s) resulting from a servo gain of $k = 5$ [51].

12-14 Consider control of a percolation process, where inlet temparature is the controlling input (see Fig. P12-14). Assume that $a = 1$ and $b = 1$. (a) Estimate, using the describing-function method, the amplitude and period of the limit cycle for a single-speed floating control (that is, process input is a bang-bang action of magnitude ± 1 followed by an integrator). Does the describing-function approach effectively apply to bang-bang control of the process without the integrator? (b) Using the Popov method, obtain the maximum slope (see Fig. 12-21) of a proportional type nonlinear control below which asymptotic stability of the percolation process control system is assured.

12-15 For the control system given in Fig. P12-15, where $\alpha = 1$, obtain the stability limit value of β by (a) the Popov theorem and (b) the circle theorem.

12-16 The system of Fig. 12-22 has $\alpha = 2$, $\beta = 10$ and $G(s) = 1/(s - 1)$. Analyze the stability of this nonminimum phase plant by the circle criterion.

12-17 In Fig. 12-21, let $G(s) = (s + 1)/(s^2 + 1)$ and determine by the Popov criterion the range of slope K of $N(e)$ for asymptotic stability. Repeat for the case $G(s) = 1/((s^2 + 1)(s + 1))$.

13

LINEAR STOCHASTIC SYSTEMS

There are dynamic systems in which input, state, and output variables randomly change. If the randomness is characterized by some rule, such as a Gaussian distribution, it is often possible to theoretically relate these variables. In linear, stationary systems, the input–output relation in terms of a transfer function can be expressed in a form that closely parallels those we saw in the preceding chapters on deterministic signals. Basic time-domain relations for such systems are introduced in the first section. This is followed by frequency-domain considerations in Section 13-2, where the frequency spectrum concept is introduced. The time-domain relations for stochastic variables are generalized to vector variables in Section 13-3. The vector form fits the use of the state-space approach to linear systems, and develops into the optimum filter design problem which is introduced in Section 13-4. Various forms of the optimum filtering algorithm are presented in the last section, including a case where the inversion in the original algorithm does not exist. Application of the filter to linear, optimal-feedback control will be treated in the last section of the next chapter.

13-1 STATISTICAL CONSIDERATIONS

Time variables of engineering significance fall into two categories, *deterministic* and *random*. Up to now we have considered deterministic inputs and outputs. In practice, however, there are cases where time variables vary randomly so that, at most, they can be characterized by a probability function. Such variables are called stochastic or random. Although most authors use these two words synonymously, a "stochastic" variable is one that is not entirely random but contains a degree of randomness [1]. The necessity for taking a statistical viewpoint to deal with time variables arises in various problems related to control, communications, acoustics, vibrations, strength of materials, materials processing, traffic, biology, business administration, etc. Probably the earliest work to apply the statistical approach to control problems, that of James, Nichols, and Phillips [2], was published in 1947. This was followed by a book devoted to the subject [3], and the subject now appears in a number of books on control. We shall briefly review below basic relations in probability theory, and then apply these to random time variables.

In probability theory, a random variable x (weight of men, for instance) is considered as the outcome of an experiment in a sample space which represents a collection of possible outcomes. To describe certain characteristics of such a sample space, we define the *probability distribution function* $F(x)$ of a random variable x by the relation:

$$F(x_1) = P(x \leqslant x_1) . \tag{13-1}$$

In this expression $P(x \leqslant x_1)$ is the probability that the random variable is less than or equal to a prescribed value x_1. If $F(x)$ is differentiable and $dF(x)/dx$ exists, the *first probability-density function* $p(x)$ is defined by $p(x) = dF(x)/dx$. It then follows that

$$p(x_1) \, dx = P(x_1 < x \leqslant x_1 + dx) .$$

This implies that the probability P of a random variable taking on a value between x and $x + dx$ is $p(x) \, dx$. The probability that the random variable lies between particular values x_1 and x_2 is then given by

$$P(x_1 < x \leqslant x_2) = \int_{x_1}^{x_2} p(x) \, dx . \tag{13-2}$$

If the range of possible values of x is bounded by a and b,

$$a \leqslant x \leqslant b ,$$

then $p(x) \geqslant 0$ over the range and

$$\int_a^b p(x) \, dx = 1 ,$$

where it is possible that $a = -\infty$ and $b = +\infty$. (Note in Eq. (13-1) that $F(-\infty) = 0$ and $F(+\infty) = 1$.) The probability-distribution function (13-1) is now expressed in terms of the first probability-density function as:

$$P(x \leqslant x_1) = \int_a^{x_1} p(x) \, dx . \tag{13-3}$$

We define the *mean* or *expected value* μ of the random variable x (written as $E\{x\}$), by the following equation:

$$\int_a^b (x - \mu) p(x) \, dx = 0 , \quad \text{or} \quad E\{x\} = \mu = \int_a^b x p(x) \, dx . \tag{13-4}$$

The *n*th *central moment* μ_n (the expected value of $(x - \mu)^n$) is defined by an expression which is similar to (13-4):

$$E\{(x - \mu)^n\} = \int_a^b (x - \mu)^n p(x) \, dx = \mu_n . \tag{13-5}$$

The preceding discussion extends to the problem where two samples of x are taken from a sample space, or two variables x and y are jointly taken from

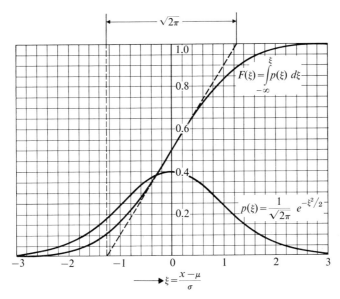

Fig. 13-1 The Gaussian distribution (normalized).

two sample spaces, and so on. For instance, the expected value of a product term xy of two random variables x and y is expressed by

$$E\{xy\} = \phi_{xy} = \int_{-\infty}^{\infty} \int_{-\infty}^{\infty} (xy)p(x, y)\, dx\, dy , \qquad (13\text{-}6)$$

where $p(x, y)$ is called the joint probability-density function.

The first probability-density function of the *Gaussian* or *normal* distribution is expressed by the following analytical form [4]:

$$p(x) = \frac{1}{\sigma\sqrt{2\pi}} e^{-(x-\mu)^2/2\sigma^2} . \qquad (13\text{-}7)$$

Since x is involved in this expression only in the quadratic form $(x - \mu)^2$, the distribution is symmetric about $x = \mu$ (Fig. 13-1), and μ is the mean value. The other parameter involved in this expression, σ, is called the standard deviation, and σ^2, the variance. It is a measure of scatter. As σ decreases, the curve becomes narrower and more highly peaked. The variance is the second central moment (defined by (13-5) with $n = 2$). It can be shown [3] that

$$E\{(x - \mu)^2\} = \int_{-\infty}^{\infty} (x - \mu)^2 p(x)\, dx = \sigma^2 , \qquad (13\text{-}8)$$

where $p(x)$ is the normal distribution given by (13-7).

Because the Gaussian distribution function has several unique attributes (see below and [1] through [6]), it is often convenient to assume a Gaussian distribution for analytical work. Standard tests can sometimes be applied to check the validity of the assumption.

The above discussion of the probability-density function was based on an *ensemble*, that is, a collection of possible outcomes. We now wish to apply the technique to a single, random, *time* variable $x(t)$ in place of an ensemble $x^1(t), x^2(t), \ldots$. This, in turn, means replacing averaging processes over an ensemble (given by Eqs. (13-3) through (13-6), for instance) by *time* averages. In Eq. (13-4), for example, we must assume the equality:

Ensemble average = Time average,

or

$$\int_{-\infty}^{\infty} xp(x)\, dx = E\{x(t)\} = \lim_{T \to \infty} \frac{1}{2T} \int_{-T}^{+T} x(t)\, dt. \tag{13-9}$$

The set of assumptions of this general nature that is necessary to deal with a random time variable $x(t)$ in the time domain is called the *ergodic hypothesis*, and the time variable is said to come from an ergodic process. As a necessary condition for this hypothesis, it is required that the probability-density functions of a random time variable, $x(t)$, must not change by a shift of the time origin. The variable is then said to be *stationary random*. To deal with random signals by assuming them to be stationary and ergodic simplifies the engineering approach. Fortunately, this assumption normally does not cause trouble. Again, the assumptions may be checked when the problem is completed [3].

By Eq. (13-9) the mean or expected value of a stationary random ergodic signal $x(t)$ is given by an averaging process in the time domain:

$$E\{x(t)\} = \overline{x(t)} = \lim_{T \to \infty} \frac{1}{2T} \int_{-T}^{T} x(t)\, dt, \tag{13-10}$$

where the bar designates a time average. The expected value of the product $x(t)x(t + \tau)$, with a time shift τ, involves the joint probability-density (Eq. (13-6)), defined in the time domain as:

$$E\{x(t)x(t + \tau)\} = \phi_{xx}(\tau) = \lim_{T \to \infty} \frac{1}{2T} \int_{-T}^{T} x(t)x(t + \tau)\, dt. \tag{13-11}$$

This is called the *autocorrelation function*. It represents a correlation of two samples of the same signal taken τ time units apart. Similarly, a correlation between two variables $x(t)$ and $y(t)$, an input and an output, for instance, is represented by the *crosscorrelation function*:

$$E\{x(t)y(t + \tau)\} = \phi_{xy}(\tau) = \overline{x(t)y(t + \tau)}$$

$$= \lim_{T \to \infty} \frac{1}{2T} \int_{-T}^{T} x(t)y(t + \tau)\, dt. \tag{13-12}$$

13-1　Statistical considerations

Table 13-1 SOME AUTOCORRELATION FUNCTIONS

	Time function $x(t)$	Autocorrelation function $\phi_{xx}(\tau)$
(a)	$x = A = $ const	$\phi_{xx} = A^2$
(b)	$x = A \sin(\omega t + \varphi)$ $\varphi = $ fixed phase	$\phi_{xx} = \dfrac{A^2}{2} \cos \omega \tau$
(c)	Binary random noise	
(d)	White noise	$\phi_{xx} = \sigma^2 \delta(\tau)$
(e)	A pseudo-random binary signal (the M-series) 1 cycle = 15 time units	

The variance or the mean square value $x(t)^2$ of a random signal $x(t)$ is given, as the special case of $\tau = 0$ in its autocorrelation function (13–11), by

$$E\{x(t)^2\} = \phi_{xx}(0) = \overline{x(t)^2}.$$

Since the self-correlation is strongest at $\tau = 0$ where agreement is complete,

$$|\phi_{xx}(\tau)| \leqslant \phi_{xx}(0) \qquad \text{for all } \tau. \tag{13–13}$$

From the equality $\overline{x(t)x(t+\tau)} = \overline{x(t-\tau)x(t)}$, it follows that the autocorrelation function is even in τ:

$$\phi_{xx}(\tau) = \phi_{xx}(-\tau). \tag{13–14}$$

The crosscorrelation function does not have these properties. However, $\phi_{xy}(\tau) = \phi_{yx}(-\tau)$. Some examples of the autocorrelation function are shown in Table 13–1. In general, the correlation functions are found by numerical computation according to the defining equations (13–11) or (13–12) from measured signals (allowing for the approximation of integrating over a finite length of time), or by relations equivalent to these equations in the discrete time domain. The following example is one of the few cases in which the autocorrelation function can be determined analytically in a straightforward manner for a probabilistic process. In this example, the autocorrelation function of binary white noise is derived by a limiting process.

Example 13-1 *The autocorrelation function of a random binary time series.*

A random signal $x(t)$ is generated by flipping a coin once each second, according to the rule $x = +1$ for a head, $x = -1$ for a tail. The pattern of the resulting signal may appear as shown in Table 13-1(c). The mean value is zero. It is also obvious that

$$\phi_{xx}(0) = 1.$$

Since there exists no correlation between two plays, we have

$$\phi_{xx}(\tau) = 0 \qquad \text{for } |\tau| \geqslant 1.$$

If $|\tau| < 1$, there exists a 100 percent correlation for the time interval $1 - |\tau|$ where $x(t)$ and $x(t + \tau)$ belong to an identical play, whereas absolutely no correlation exists for the rest of the time interval. Therefore,

$$\phi_{xx}(\tau) = 1 - |\tau|, \qquad \text{for } |\tau| < 1.$$

We thus get the triangular function shown in (c) of the table. To generalize the result we obtained, let Δ seconds be the interval of each play, and c be the score, $+c$ and $-c$ for heads and tails, respectively. The result will be:

$$\phi_{xx}(\tau) = c^2 \left(1 - \frac{|\tau|}{\Delta}\right) \qquad \text{for } |\tau| \leqslant \Delta,$$

$$\phi_{xx}(\tau) = 0 \qquad\qquad\qquad \text{otherwise}.$$

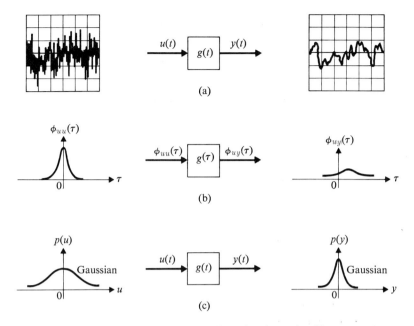

Fig. 13-2 Stationary, random, ergodic noise through a linear system.

If c and Δ in the above example are so chosen that $c^2\Delta = N =$ a constant (note that this is not a variance), at the limit of $\Delta \to 0$ we obtain a delta-function expression for $\phi_{xx}(\tau)$:

$$E\{x(t)x(t+\tau)\} = \phi_{xx}(\tau) = N\delta(\tau). \tag{13-15}$$

A random signal $x(t)$, whose autocorrelation function is given by (13–15), is called *white noise*. Since an infinite magnitude of power is required to generate a white-noise signal (see next section), such a signal does not actually exist. However, an approximate (band-limited) white-noise signal exists, and it plays an important role in the problem that follows.

Let us apply the correlation technique to the input–output relation of a linear system shown in Fig. 13–2, where the system is represented by its unit-impulse input-response $g(t)$. Note that this is an open-loop system (Thal-Larsen [7] analyzed a linear feedback-control system relative to correlations of various loop variables). The input-to-output crosscorrelation function is

$$\phi_{uy}(\tau) = \lim_{T\to\infty} \frac{1}{2T} \int_{-T}^{T} u(t-\tau)y(t)\,dt.$$

The output $y(t)$ in this expression for a deterministic input $u(t)$ is given by the convolution integral (8–41), restated as:

$$y(t) = \int_{-\infty}^{+\infty} u(t-\eta)g(\eta)\,d\eta. \tag{13-16}$$

Substituting the latter relation into the former, and rearranging the terms, we get

$$\phi_{uy}(\tau) = \int_{-\infty}^{+\infty} g(\eta) \left[\lim_{T \to \infty} \frac{1}{2T} \int_{-T}^{T} u(t-\eta) u(t-\tau) \, dt \right] d\eta \, .$$

In this expression the integral inside the parenthesis with the limiting process $T \to \infty$ yields the autocorrelation function $\phi_{uu}(\tau - \eta)$; and thus:

$$\phi_{uy}(\tau) = \int_{-\infty}^{+\infty} \phi_{uu}(\tau - \eta) g(\eta) \, d\eta \, . \tag{13-17}$$

Equation (13-17) is identical with (13-16) in its form. We can therefore say that if the input autocorrelation function $\phi_{uu}(\tau)$ is fed into a linear system as an input time function, the input-output crosscorrelation function $\phi_{uy}(\tau)$ would come out as a response (Fig. 13-2(b)). In particular, if $\phi_{uu}(\tau)$ were given by a delta function (the autocorrelation function for a white-noise input), $\phi_{uy}(\tau)$ would take on the pattern of the system's impulse response!

It is now clear that the principle of process identification presented in Sec. 8-2 for a deterministic input-output also applies when the input is ergodic. We simply replace deterministic input and output by $\phi_{uu}(\tau)$ and $\phi_{uy}(\tau)$, respectively, and apply the same algorithm of identification. Since a white-noise test signal produces a system's impulse response through the input-output crosscorrelation, noise generators for on-line identification purposes are often built to give an approximation to white noise. For example, a periodic binary signal which repeats the following fixed sequence,

$$----+++-++--+-+ \quad \text{(15 time units per period)},$$

that can be easily generated by a sequential logic system, has been found useful for process identification. As shown in Table 13-1(e), the autocorrelation function of this series resembles that of a white noise. A deterministic signal of this kind is said to be pseudorandom. The particular time series shown here belongs to the general class of "maximum period sequences," abbreviated as the M-Series [8].

In general, the amplitude distribution pattern of a stationary random input is different from the pattern of its output. For instance, a binary input does not produce a binary output. An exception arises, however, when the amplitude distribution is normal (Gaussian). The output, $y(t)$ in Fig. 13-2, has Gaussian amplitude distribution when a system is linear and its input $u(t)$ has Gaussian distribution. This statement can be justified by the following sequence of logic: If $u(t)$ has a normal amplitude distribution, the quantity $u(t - \eta)g(\eta)$ in Eq. (13-16) is also normally distributed, since $g(\eta)$ depends only on η and not on t. Statistical theory states that the sum of a set of Gaussian variables is also Gaussian [4]. Since a convolution integral

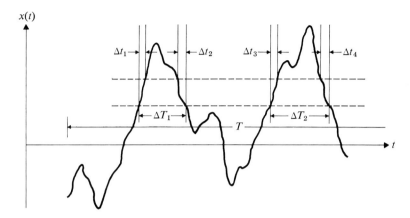

Fig. 13-3 Amplitude distribution of ergodic, stationary random signal.

is a summation process, $y(t)$ given by Eq. (13-16) must also have a Gaussian distribution. Moreover, although it cannot be shown rigorously, the central-limit theorem [3] seems to strongly indicate a tendency for a non-Gaussian stationary random signal to approach a Gaussian amplitude distribution when it is passed through those linear systems we commonly encounter in practice (band-pass type).

Referring to Fig. 13-3, we see that the probability density function $p(x)$ of an amplitude $x(t)$ is given by

$$p(x_1)\,dx = P(x_1 < x \leqslant x_1 + dx) = \lim_{\substack{T \to \infty \\ \Delta x \to 0}} \frac{\text{sum of } \Delta t}{T}$$

and

$$P(x_1 < x) = \lim_{T \to \infty} \frac{\text{sum of } \Delta T}{T} = \% \text{ of the time that } x \text{ becomes greater than } x_1, \qquad (13\text{--}18)$$

where the Δt's refer to the times for which $x(t)$ is between x_1 and $x_1 + \Delta x$ and the ΔT's refer to the times for which the signal $x(t)$ is above x_1.

If a given random signal is assumed to be stationary, ergodic, and Gaussian, and if its variance and mean value are known, it is possible to give an estimate of the percent of time given by Eq. (13-18) from a plot of the Gaussian function (Fig. 13-1). Suppose that $u(t)$ in Fig. 13-2 represents a product quality and it has a normal distribution. If the quality of a delivery flow, $y(t)$, is specified either by its variance or by the percent of the time $y(t)$ is allowed to exceed a certain limit, c, it is possible to design a smoothing device (a mixing tank, for instance) with an impulse response $g(t)$ that will give the required amount of reduction in the variance (see Example 13-2 below).

13-2 STATISTICAL RELATIONSHIPS IN THE FREQUENCY DOMAIN

The stationary random variable was discussed from the time-domain viewpoint in the preceding section. We describe it below in terms of its frequency spectrum. The transformations between the time and frequency domain are given by the well-known Fourier transform pair:

$$F(j\omega) = \int_{-\infty}^{\infty} f(t)e^{-j\omega t} \, dt, \qquad f(t) = \frac{1}{2\pi} \int_{-\infty}^{\infty} F(j\omega)e^{j\omega t} \, d\omega. \qquad (13\text{-}19)$$

In the following we assume that the correlation functions and the linear system's impulse-response function $g(t)$ are all Fourier-transformable. The autocorrelation function $\phi_{xx}(\tau)$ of a random variable $x(t)$ is transformed into the frequency domain by

$$\Phi_{xx}(j\omega) = \int_{-\infty}^{\infty} \phi_{xx}(\tau)e^{-j\omega \tau} \, d\tau, \qquad (13\text{-}20)$$

and the inverse transformation is

$$\phi_{xx}(\tau) = \frac{1}{2\pi} \int_{-\infty}^{\infty} \Phi_{xx}(j\omega)e^{j\omega \tau} \, d\omega. \qquad (13\text{-}21)$$

Recalling the equation following Eq. (13-12), we let $\tau = 0$ in (13-21) to obtain the expression for the mean-square value $\overline{x(t)^2}$:

$$\overline{x(t)^2} = \phi_{xx}(0) = \frac{1}{2\pi} \int_{-\infty}^{\infty} \Phi_{xx}(j\omega) \, d\omega. \qquad (13\text{-}22)$$

This expression shows that the frequency function $\Phi_{xx}(j\omega)$ is a representation of the $\overline{x(t)^2}$ distribution along the frequency axis; that is, the area under the $\Phi_{xx}(j\omega)$ curve is 2π times the mean square value $\overline{x(t)^2}$. To avoid the $1/2\pi$ factor on the right of Eq. (13-22), $\Phi(j\omega)$ in some literature is defined to be $1/2\pi$ times the Fourier transform of the correlation function. Since the squared value of a signal is sometimes associated with energy (see bottom of Table 6-1), $\Phi_{xx}(j\omega)$ is termed the *power-density spectrum* (or power-density function). The power-density spectrum is a real function of ω; it has no phase shift. To see this, we substitute $e^{-j\omega \tau} = \cos \omega\tau - j \sin \omega\tau$ in (13-20) and obtain

$$\Phi_{xx}(j\omega) = \int_{-\infty}^{\infty} \phi_{xx}(\tau) \cos \omega\tau \, d\tau - j \int_{-\infty}^{\infty} \phi_{xx}(\tau) \sin \omega\tau \, d\tau.$$

Since $\phi_{xx}(\tau)$ is an even function of τ, the product term $\phi_{xx}(\tau) \sin \omega\tau$ is odd in τ, and its integral over the infinite range in τ is zero. Equation (13-20) thus reduces to

$$\Phi_{xx}(j\omega) = \int_{-\infty}^{\infty} \phi_{xx}(\tau) \cos \omega\tau \, d\tau. \qquad (13\text{-}23)$$

13-2 Statistical relationships in the frequency domain

Table 13-2 EXAMPLES OF POWER DENSITY SPECTRA

	Autocorrelation function	Power density spectrum
(a)	$\phi_{xx}(\tau) = A^2$	$\Phi_{xx}(j\omega) = 2\pi A^2 \delta(\omega)$
(b)	$\phi_{xx}(\tau) = \dfrac{A^2}{2} \cos \omega_1 \tau$	$\Phi_{xx}(j\omega) = \dfrac{\pi A^2}{2} \delta(|\omega| - \omega_1)$
(c)	$\phi_{xx}(\tau) = 1 - |\tau|,\ |\tau| \leq 1$	$\Phi_{xx}(j\omega) = \left(\sin \dfrac{\omega}{2}\right)^2 \Big/ \left(\dfrac{\omega}{2}\right)^2$
(d)	$\phi_{xx}(\tau) = \delta(\tau)$ (white noise)	$\Phi_{xx}(j\omega) = 1$
(e)	$\phi_{xx}(\tau) = e^{-a|\tau|},\ a > 0$	$\Phi_{xx}(j\omega) = \dfrac{2a}{a^2 + \omega^2}$
(f)	$\phi_{xx}(\tau) = e^{-a|\tau|} \cos(b|\tau| + c),$ $a \cos c = -b \sin c$	$\Phi_{xx}(j\omega) = \dfrac{2a\{2(a^2 + b^2) + \omega^2\} \cos c}{\{a^2 + (b - \omega)^2\}\{a^2 + (b + \omega)^2\}}$

Since the right side of (13–23) is an even function of ω, $\Phi_{zz}(j\omega) = \Phi_{zz}(-j\omega)$ and

$$\Phi_{zz}(j\omega) = 2\int_0^\infty \phi_{zz}(\tau) \cos \omega\tau \, d\tau . \qquad (13\text{–}24)$$

Assuming an electrical model where a filtered noise voltage is impressed across a resistor, it can be generally shown that the power-density spectrum is nonnegative over all frequencies.

Some pairs of autocorrelation functions and their power-density spectra are shown in Table 13–2, where (a) through (d) are related to time variables listed in Table 13–1. It is interesting to note that *white noise* (d) has a constant power distribution over the entire frequency range. So-called Gaussian white-noise generators are built to produce an approximately constant power distribution over a finite frequency range. *Exact* white-noise signals cannot be generated in the laboratory.

When experimental data of a random signal are available, it is often convenient first to compute the autocorrelation function, approximate the plot by a mathematical function, and then take the Fourier transform to obtain the power spectrum. The last two pairs (e) and (f) of Table 13–2 are useful for this purpose, where (f) may be replaced by the following simpler pair [9]:

$$\phi_{zz}(\tau) = e^{-a|\tau|} \cos b\tau ,$$

$$\Phi_{zz}(j\omega) = \frac{2a(a^2 + b^2 + \omega^2)}{(a^2 + (b+\omega)^2)(a^2 + (b-\omega)^2)} .$$

It was noted in the preceding section that the following relation,

$$\phi_{uy}(\tau) = \int_{-\infty}^\infty \phi_{uu}(\tau - \eta) g(\eta) \, d\eta , \qquad (13\text{–}25)$$

is exactly like the deterministic input–output relationship of a linear system. Recalling the Laplace transformation pair of an input–output relationship,

$$Y(s) = G(s)U(s) , \qquad y(t) = \int_{-\infty}^\infty u(t-\eta) g(\eta) \, d\eta ,$$

where $u(t)$ is the input, $y(t)$ is the output, and $g(t) = \mathscr{L}^{-1}[G(s)]$, we write the following frequency-domain relation of (13–25):

$$\Phi_{uy}(s) = G(s)\Phi_{uu}(s) . \qquad (13\text{–}26)$$

In Eq. (13–26),

$$\Phi_{uy}(s) = \int_{-\infty}^\infty \phi_{uy}(\tau) e^{-s\tau} \, d\tau ,$$

$$\Phi_{uu}(s) = \int_{-\infty}^\infty \phi_{uu}(\tau) e^{-s\tau} \, d\tau . \qquad (13\text{–}27)$$

13-2 Statistical relationships in the frequency domain

The bilateral Laplace transform is required for our problem because time functions extend on both sides of the time origin indefinitely. (See [10] for more information on the bilateral Laplace transform.)

For the real frequency ω we let $s = j\omega$ and obtain

$$\Phi_{uy}(j\omega) = G(j\omega)\Phi_{uu}(j\omega), \qquad (13\text{-}28)$$

where $\Phi_{uu}(j\omega)$ is the power-density spectrum of the input autocorrelation function, and

$$\Phi_{uy}(j\omega) = \int_{-\infty}^{\infty} \phi_{uy}(\tau)e^{-j\omega\tau}\, d\tau \qquad (13\text{-}29)$$

is referred to as the cross-power-density spectrum of the input and output signals. It follows from Eq. (13-28) that, if $\Phi_{uu}(j\omega)$ and $\Phi_{uy}(j\omega)$ are both determined by experiment (input–output record), the system transfer function can be identified by

$$G(j\omega) = \frac{\Phi_{uy}(j\omega)}{\Phi_{uu}(j\omega)}. \qquad (13\text{-}30)$$

The time-domain relation (13-25) involves a convolution, whereas the frequency-domain relation is a division process and therefore may be easier for numerical evaluation. However, a computation is needed to convert correlation functions into power-density spectra [11].

Another important relationship is known between the power-density spectra of an input and output pair. To derive the second relationship we begin with the convolution integral for an output $y(t)$,

$$y(t) = \int_{-\infty}^{\infty} u(t-\eta)g(\eta)\, d\eta,$$

$$y(t+\tau) = \int_{-\infty}^{\infty} u(t+\tau-\xi)g(\xi)\, d\xi.$$

By the definition of the autocorrelation function, we have

$$\phi_{yy}(\tau) = \lim_{T\to\infty} \frac{1}{2T} \int_{-T}^{T} y(t)y(t+\tau)\, dt$$

$$= \lim_{T\to\infty} \frac{1}{2T} \int_{-T}^{T}\int_{-\infty}^{\infty}\int_{-\infty}^{\infty} u(t-\eta)g(\eta)u(t+\tau-\xi)g(\xi)\, d\eta\, d\xi\, dt.$$

$$= \int_{-\infty}^{\infty}\int_{-\infty}^{\infty} \left\{\lim_{T\to\infty} \frac{1}{2T}\int_{-T}^{T} u(t-\eta)u(t+\tau-\xi)\, dt\right\} g(\eta)g(\xi)\, d\eta\, d\xi,$$

where the integral in brackets is by definition the input autocorrelation function $\phi_{uu}(\tau+\eta-\xi)$. We therefore have

$$\phi_{yy}(\tau) = \int_{-\infty}^{\infty}\int_{-\infty}^{\infty} \phi_{uu}(\tau+\eta-\xi)g(\eta)g(\xi)\, d\eta\, d\xi.$$

Multiplying both sides of this expression by $e^{-s\tau}$ and integrating over the infinite range in τ yields

$$\left\{\int_{-\infty}^{\infty} \phi_{yy}(\tau)e^{-s\tau}\,d\tau\right\}$$
$$= \left\{\int_{-\infty}^{\infty} e^{s\eta}g(\eta)\,d\eta\right\}\left\{\int_{-\infty}^{\infty} e^{-s\xi}g(\xi)\,d\xi\right\}\left\{\int_{-\infty}^{\infty} \phi_{uu}(\tau+\eta-\xi)e^{-s(\tau+\eta-\xi)}\,d\tau\right\}.$$

The arrangements of the exponentials are adjusted in this expression so that factors in the parentheses are replaced by frequency functions of input and output autocorrelations and the system transfer function, to obtain the final result:

$$\Phi_{yy}(s) = G(-s)G(s)\Phi_{uu}(s). \qquad (13\text{-}31)$$

In the frequency domain $s = j\omega$,

$G(-j\omega)$ is the conjugate of $G(j\omega)$,

so that

$$G(-j\omega)G(j\omega) = |G(j\omega)|^2.$$

We therefore have

$$\Phi_{yy}(j\omega) = |G(j\omega)|^2\,\Phi_{uu}(j\omega). \qquad (13\text{-}32)$$

This is the second basic relation between the power-density spectra of the input and output. Although Eq. (13-32) is applicable to system transfer-function identification, it gives gain only,

$$|G(j\omega)| = \sqrt{\frac{\Phi_{yy}(j\omega)}{\Phi_{uu}(j\omega)}}. \qquad (13\text{-}33)$$

The application of (13-32) to the mean-square output computation is more interesting. For this purpose we derive the following relation from (13-32):

$$\overline{y(t)^2} = \frac{1}{2\pi}\int_{-\infty}^{\infty} \Phi_{yy}(j\omega)\,d\omega = \frac{1}{2\pi}\int_{-\infty}^{\infty} |G(j\omega)|^2\,\Phi_{uu}(j\omega)\,d\omega. \qquad (13\text{-}34)$$

Equation (13-34) is the basis of various analytical design methods in which mean-square error, $\overline{e(t)^2}$, is to be minimized. To deal with practical problems, analog-computer simulation often works much faster than the analytical approach [12]. The mean-absolute-error criterion, $|e(t)|$, is useful and easily simulated when a problem is treated on an analog computer. For an application of Eq. (13-34) to a distributed-parameter system, see [13]. To illustrate the application of Eq. (13-34) to design problems, two examples will be treated.

13-2 Statistical relationships in the frequency domain 589

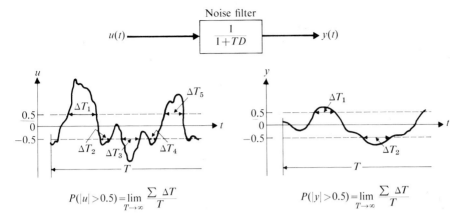

Fig. 13-4 Noise filter for smoothing a random signal.

Example 13-2 In the process industry, a product flow of quality $u(t)$ is passed through a mixing tank (a low-pass filter) of time constant T for the purpose of smoothing fluctuations in the output. Suppose that the zero-average input signal $u(t)$ has a Gaussian distribution of variance $\sigma_u^2 = 0.90$. Referring to a handbook table (or chart) of Gaussian functions (or Fig. 13-1), we find the probability that the signal exceeds a chosen limit of ± 0.5 to be

$$P(|u| > 0.5) = 0.6 \;.$$

By the ergodic hypothesis, the probability is considered equal to the fraction of the total time the signal exceeds the limit (Fig. 13-4). Let us suppose that the relation was favorably confirmed by test data.

It is desired to find the time constant such that the output fluctuation will satisfy the following condition:

$$P(|y| > 0.5) \leqslant 0.1 \;.$$

To solve this problem, we must know the autocorrelation function $\phi_{uu}(\tau)$ which, let us assume, was found to be

$$\phi_{uu}(\tau) = 0.90 e^{-|\tau|} \cos \tau \;.$$

Then (f) of Table 13-2 gives the power-density spectrum:

$$\Phi_{uu}(j\omega) = 0.90 \, \frac{2(2 + \omega^2)}{(2 + 2\omega + \omega^2)(2 - 2\omega + \omega^2)} \;.$$

Also we have

$$|G(j\omega)|^2 = \frac{1}{1 + (\omega T)^2} \;.$$

Substituting these expressions into the right side of Eq. (13-34), we use the tabular integration formula given in the appendix of [1] or [14] or the state-space approach

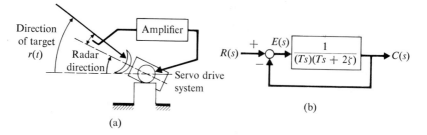

Fig. 13-5 A radar tracking station.

(see next section, Example 13-6) and obtain

$$\overline{y(t)^2} = 0.9 \frac{1+T}{1+2T+2T^2},$$

Since the input is Gaussian and the system is linear, the output is Gaussian. The variance $\bar{y}^2$ for the prescribed probability (equal to or less than 0.1 for the probability that $|y|$ exceeds 0.5) of a Gaussian process is found to be 0.09. Hence

$$0.09 \geqslant 0.9 \frac{1+T}{1+2T+2T^2},$$

and we solve for

$$T \geqslant 4.82 .$$

Example 13-3 Let us consider a radar system [9] where the tracking error is amplified and fed back to a drive mechanism (Fig. 13-5(a)). For the purposes of illustration the servo follow-up system is assumed to be second-order [(b) of the figure], such that

$$\frac{C(s)}{R(s)} = \frac{1}{1 + 2\zeta Ts + (Ts)^2} .$$

Let us assume that the reference input has both message component and noise; the message is a ramp of slope B (a typical situation when a target appears on the horizon) and the noise is Gaussian white noise, $\Phi_{nn}(j\omega) = N$, where N is the constant that appears in Eq. (13-15).

The value of the mean-square output due to the noise input is

$$\overline{c_n^2} = \frac{1}{2\pi} \int_{-\infty}^{\infty} \left\{ \frac{\Phi_{nn}(j\omega)}{|1 + 2\zeta(j\omega T) + (j\omega T)^2|^2} \right\} d\omega .$$

Referring to the table of integrals in [1] or [14], or by using the state-space approach (see next section, Example 13-4), we evaluate the integral and obtain

$$\overline{c_n^2} = \frac{N}{4\zeta T} .$$

On the other hand, the steady-state error for the ramp input is found by Eq. (8-21) as

$$e_{ss} = 2\zeta TB .$$

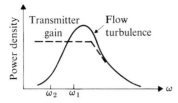

Fig. 13-6 Noise problem in a flow system.

Now we are faced with a compromise decision. To determine the best compromise we define a performance index J involving weighting factors w_1 and w_2 in the following form:

$$J = w_1 e_{ss}^2 + w_2 \overline{c_n^2}.$$

It is now possible to determine the optimal parameter values (which are combined in a product form, ζT, in this problem) by the condition $\partial J/\partial(\zeta T) = 0$. The answer is

$$\zeta T = \frac{1}{4}\left(\frac{2w_2 N}{w_1 B^2}\right)^{1/3}.$$

A compromise solution to a noise-filtering problem was shown in the above example. There are many problems of a similar nature in process instrumentation. In flow measurement and control, for instance, noise appears due to turbulence, Fig. 13-6. If a message (significant change of flow rate) has a high-frequency component such as ω_1, where the noise intensity is high, it is practically impossible to separate this message component from noise; a wide-band flow transmitter would pick up the message as well as the noise. If, on the other hand, a crucial message is in the lower-frequency range, ω_2 in the figure, then we must design a filter which will effectively cut down noise at the high-frequency end while retaining maximum information carried by the message at low frequency. In mathematical theory such observation takes us to the problems of optimum filter design which will appear in Section 13-4. There exist real problems of an almost endless variety for which engineering development in noise-filtering and/or predictor design is needed. This is especially true in some man-machine problems (for instance, see [15], [16] for an application of fast time-scale plant models for prediction purposes).

13-3 STATISTICAL RELATIONSHIPS IN THE STATE SPACE

We dealt with scalar random variables in the preceding sections. In this section the time-domain relations of Section 13-1 will be generalized to account for random vector variables. In view of the useful attributes of a Gaussian distribution (see Section 13-1), all random vectors in the following will be assumed Gaussian, and, in addition, to have zero mean. A random n-vector variable **x** is said to be Gaussian when its probability distribution has the following form:

$$p(\mathbf{x}) = \frac{1}{\sqrt{\det \mathbf{X}}(\sqrt{2\pi})^n} \exp\left\{-\frac{1}{2}(\mathbf{x} - E\{\mathbf{x}\})'(\mathbf{X})^{-1}(\mathbf{x} - E\{\mathbf{x}\})\right\},$$
(13-35)

where $E\{\mathbf{x}\} = \mathbf{0}$ for zero mean $\mathbf{x}$. The expression closely parallels the Gaussian distribution function given by Eq. (13-7) for a scalar variable, except for the variance term which is now replaced by a square matrix $\mathbf{X}$. The matrix $\mathbf{X}$, called the covariance matrix of $\mathbf{x}$, is defined by

$$\mathbf{X} = E\{(\mathbf{x} - E\{\mathbf{x}\})(\mathbf{x} - E\{\mathbf{x}\})'\}.$$
(13-36)

The covariance matrix for n-vector $\mathbf{x}$ is $n \times n$ square, symmetric, and nonnegative. It is symmetric because of the equality, $E\{x_i x_j\} = E\{x_j x_i\}$. It is nonnegative because, in its diagonalized form, all diagonal elements are variances and therefore nonnegative.

Let us consider a discrete-time system

$$\mathbf{x}_{k+1} = \mathbf{P}\mathbf{x}_k + \mathbf{Q}\mathbf{u}_k,$$
(13-37)

where the input r-vector $\mathbf{u}_k$ is zero-mean, ergodic, Gaussian random noise that has no correlation between different samples. Hence,

$$E\{\mathbf{u}_i \mathbf{u}_j'\} = \begin{cases} \mathbf{0} & \text{if } i \neq j, \\ \mathbf{U} & \text{if } i = j. \end{cases}$$
(13-38)

$\mathbf{U}$ in Eq. (13-38) is the covariance matrix of $\mathbf{u}_k$. Although, strictly speaking, a white-noise vector is defined only for the continuous-time case, a discrete-time noise vector in which each sample is independent will be informally called "white" in the following.

To compute the covariance matrix $\mathbf{X}$ of the state vector $\mathbf{x}_k$ in Eq. (13-37), we right multiply by the transposed relation, $\mathbf{x}_{k+1}' = \mathbf{x}_k'\mathbf{P}' + \mathbf{u}_k'\mathbf{Q}'$ on both sides of Eq. (13-37), and take the expected value of each vector product. Since the state $\mathbf{x}_k$ is not correlated with the input $\mathbf{u}_k$, $E\{\mathbf{u}_k \mathbf{x}_k'\} = \mathbf{0}$ and we get

$$\mathbf{X} = \mathbf{P}\mathbf{X}\mathbf{P}' + \mathbf{Q}\mathbf{U}\mathbf{Q}'.$$
(13-39)

The covariance matrix $\mathbf{X} = E\{\mathbf{x}_k \mathbf{x}_k'\}$ can be determined by Eq. (13-39). If a scalar response y_k of the system is given in terms of the row vector $\mathbf{c}$ by the output equation

$$y_k = \mathbf{c}\mathbf{x}_k,$$

the variance of the response is

$$E\{y_k^2\} = \mathbf{c}E\{\mathbf{x}_k \mathbf{x}_k'\}\mathbf{c}' = \mathbf{c}\mathbf{X}\mathbf{c}'.$$
(13-40)

It is also possible to express the matrix $E\{\mathbf{x}_{k+m}\mathbf{x}'_k\}$ that corresponds to the autocorrelation function in terms of scalar variables. For this purpose we right multiply by $\mathbf{x}'_k$ on both sides of Eq. (13-37), and take the expected value of each vector product. Noting $E\{\mathbf{u}_k\mathbf{x}'_k\} = 0$ as before, we obtain

$$E\{\mathbf{x}_{k+1}\mathbf{x}'_k\} = \mathbf{PX},$$

and, likewise,

$$E\{\mathbf{x}_{k+m}\mathbf{x}'_k\} = \mathbf{P}^m\mathbf{X}, \tag{13-41}$$

where m is a positive integer. If m is negative, Eq. (13-41) must be transposed;

$$E\{\mathbf{x}_k\mathbf{x}'_{k+m}\} = E\{\mathbf{x}_{k-m}\mathbf{x}'_k\} = \mathbf{X}(\mathbf{P}')^m. \tag{13-42}$$

The continuous-time equivalences of Eqs. (13-39) through (13-41) are the following:

$$\mathbf{AX} + \mathbf{XA}' + \mathbf{BUB}' = \mathbf{0}, \tag{13-43}$$

$$E\{y(t)^2\} = \mathbf{cXc}', \tag{13-44}$$

and

$$E\{\mathbf{x}(t+\tau)\mathbf{x}'(t)\} = \mathbf{S}(\tau)\mathbf{X}, \qquad 0 \leqslant \tau. \tag{13-45}$$

Equations (13-43) through (13-45) apply to a system described by

$$\frac{d\mathbf{x}}{dt} = \mathbf{Ax} + \mathbf{Bu}, \qquad y = \mathbf{cx}, \tag{13-47}$$

where $\mathbf{u}(t)$ is a zero-mean Gaussian white-noise vector defined by

$$E\{\mathbf{u}(t)\mathbf{u}'(t+\tau)\} = \mathbf{U}\delta(\tau). \tag{13-48}$$

The Dirac delta function appears in Eq. (13-48) for the reason given in Section 13-1. We use Eq. (13-43) to compute the covariance matrix of the state vector $\mathbf{x}(t)$,

$$E\{\mathbf{x}(t)\mathbf{x}'(t)\} = \mathbf{X}. \tag{13-49}$$

The diagonal elements of the matrix (13-45) are autocorrelation functions of the state variables, and the off-diagonal elements are their crosscorrelation functions. $\mathbf{S}(\tau)$ in this expression is the solution matrix of the system, $e^{\mathbf{A}\tau}$. Equations (13-43) through (13-45) are derived from Eqs. (13-39) through (13-41), respectively, by taking the following limiting process that applies when the sampling interval, T, is made infinitesimally small, dt,

$$\mathbf{P} \to \mathbf{I} + \mathbf{A}\,dt, \qquad \mathbf{Q} \to \mathbf{B}\,dt. \tag{13-50}$$

594   Linear stochastic systems

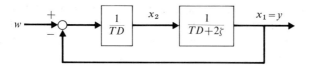

Fig. 13-7 A servomechanism with white-noise input w.

The defining equation of the delta function,

$$\int_{-\varepsilon}^{+\varepsilon} \delta(t)\,dt = 1, \qquad \varepsilon \to 0^+,$$

must also be taken into account in the limiting process (see Eq. (13-82) in the next section).

The relations presented in this section are useful for computing mean-squared values of zero-mean Gaussian random variables in the time domain, as shown by the next example.

Example 13-4 The mean-square output of a radar system subject to Gaussian white-noise input was evaluated in the frequency domain in Example 13-3. Let us apply Eq. (13-43) and confirm the result.

The radar system of Fig. 13-5(b) is restated in Fig. 13-7 with minor changes in variable symbols. The state-space expression of the system yields

$$\mathbf{A} = \frac{1}{T}\begin{bmatrix} -2\zeta & 1 \\ -1 & 0 \end{bmatrix}, \quad \mathbf{B} = \begin{bmatrix} 0 \\ \frac{1}{T} \end{bmatrix}, \quad \mathbf{c} = \begin{bmatrix} 1 & 0 \end{bmatrix},$$

and the zero-mean, Gaussian white-noise input is characterized by

$$E\{u(t)u(t+\tau)\} = N\delta(\tau),$$

where N is a nonnegative constant. Substituting these into Eq. (13-43), we get

$$\frac{1}{T}\begin{bmatrix} -2\zeta & 1 \\ -1 & 0 \end{bmatrix}\begin{bmatrix} X_{11} & X_{12} \\ X_{21} & X_{22} \end{bmatrix} + \frac{1}{T}\begin{bmatrix} X_{11} & X_{12} \\ X_{21} & X_{22} \end{bmatrix}\begin{bmatrix} -2\zeta & -1 \\ 1 & 0 \end{bmatrix}$$
$$+ \frac{1}{T^2}\begin{bmatrix} 0 & 0 \\ 0 & N \end{bmatrix} = \begin{bmatrix} 0 & 0 \\ 0 & 0 \end{bmatrix}.$$

The matrix equation reduces to three scalar relations for three unknowns, X_{11}, X_{22}, and $X_{12} = X_{21}$. By Eq. (13-44), it follows that

$$E\{y(t)^2\} = X_{11} = \frac{N}{4\zeta T},$$

which agrees with the value of c_n^2 which we found in Example 13-3.

The input noise vector must be "white" if all the relations from Eq. (13-39) through Eq. (13-45) are to hold. Therefore, if the input noise has a color, as in Example 13-2, a shaping filter that will produce a prescribed colored noise from a white noise must be included in the system. A shaping filter is determined in the next example for the colored noise input of Example 13-2, followed by Example 13-6 where the filter is applied in the time-domain solution of Example 13-2.

Example 13-5 We shall construct a shaping filter which, subject to a zero-mean Gaussian white-noise input w, will produce an output u whose autocorrelation function is

$$\phi_{uu}(\tau) = 0.90 e^{-|\tau|} \cos \tau .$$

Let the filter be described by

$$\frac{d\mathbf{x}}{dt} = \mathbf{A}\mathbf{x} + \mathbf{b}w, \qquad u = \mathbf{c}\mathbf{x},$$

where

$$E\{w(t)w(t+\tau)\} = N\delta(\tau), \qquad 0 \leqslant N .$$

By Eqs. (13-44) and (13-45) we have

$$\phi_{uu}(\tau) = E\{u(t+\tau)u(t)\} = \mathbf{c}E\{\mathbf{x}(t)\mathbf{x}'(t+\tau)\}\mathbf{c}' = \mathbf{c}\mathbf{S}(\tau)\mathbf{X}\mathbf{c}' . \tag{13-51}$$

Recalling the solution matrix for the symmetric form of the $\mathbf{A}$-matrix,

$$\mathbf{A}_m = \begin{bmatrix} -\sigma & \omega \\ -\omega & -\sigma \end{bmatrix}; \qquad e^{\mathbf{A}_m \tau} = \mathbf{S}(\tau) = e^{-\sigma \tau} \begin{bmatrix} \cos \omega \tau & -\sin \omega \tau \\ \sin \omega \tau & \cos \omega \tau \end{bmatrix},$$

we let

$$\mathbf{A} = \begin{bmatrix} -1 & 1 \\ -1 & -1 \end{bmatrix}, \qquad \mathbf{c} = \begin{bmatrix} 1 & 0 \end{bmatrix} \quad \text{and} \quad \mathbf{X} = \begin{bmatrix} 0.90 & 0 \\ 0 & X_{22} \end{bmatrix}, \tag{13-52}$$

so that (13-51) will satisfy the prescribed condition,

$$\phi_{uu}(\tau) = \mathbf{c}\mathbf{S}(\tau)\mathbf{X}\mathbf{c}' = 0.90 e^{-\tau} \cos \tau, \qquad 0 \leqslant \tau .$$

Equation (13-43), for the known set of $\mathbf{A}$ and $\mathbf{X}$, yields a condition for input vector

$$\mathbf{b} = \begin{bmatrix} b_1 \\ b_2 \end{bmatrix}.$$

We have

$$\mathbf{A}\mathbf{X} = \begin{bmatrix} -0.90 & X_{22} \\ -0.90 & -X_{22} \end{bmatrix},$$

so that Eq. (13-43) becomes

$$\begin{bmatrix} -0.90 & X_{22} \\ -0.90 & -X_{22} \end{bmatrix} + \begin{bmatrix} -0.90 & -0.90 \\ X_{22} & -X_{22} \end{bmatrix} + N \begin{bmatrix} b_1^2 & b_1 b_2 \\ b_1 b_2 & b_2^2 \end{bmatrix} = \mathbf{0} .$$

Taking $b_1 = 1$ and $N = 1.8$, it can be found that

$$b_2 = \sqrt{2} - 1, \qquad \text{and} \qquad X_{22} = 0.9 b_2^2 = 0.9(3 - 2\sqrt{2}) .$$

The shaping filter is thus fixed, as shown in the left side of Fig. 13-8.

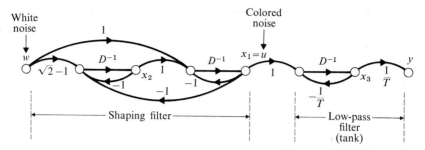

Fig. 13-8 State-space formulation of Example 13-2.

Example 13-6 Using the shaping filter of the preceding example as part of the system, the problem of Example 13-2 will be treated in the time domain. The system of Fig. 13-8 is described by

$$\frac{d\mathbf{x}}{dt} = \mathbf{A}\mathbf{x} + \mathbf{b}w, \quad y = \mathbf{c}\mathbf{x},$$

where

$$\mathbf{A} = \begin{bmatrix} -1 & 1 & 0 \\ -1 & -1 & 0 \\ 1 & 0 & -\frac{1}{T} \end{bmatrix}, \quad \mathbf{b} = \begin{bmatrix} 1 \\ \sqrt{2}-1 \\ 0 \end{bmatrix}, \quad (13\text{-}53)$$

$$\mathbf{c} = \begin{bmatrix} 0 & 0 & \frac{1}{T} \end{bmatrix}.$$

Since four elements X_{11} through X_{22} of $\mathbf{X}$ were already found in Example 13-5, we write the following form for $\mathbf{X}$:

$$\mathbf{X} = \begin{bmatrix} 0.90 & 0 & X_{13} \\ 0 & 0.9(3-2\sqrt{2}) & X_{23} \\ X_{31} & X_{32} & X_{33} \end{bmatrix}.$$

The unknown elements of $\mathbf{X}$ are determined by Eq. (13-43) for $\mathbf{A}$ and $\mathbf{b}$ of (13-53) as:

$$X_{13} = X_{31} = \frac{0.9T(1+T)}{1+2T+2T^2},$$

$$X_{23} = X_{32} = -\frac{0.9T^2}{1+2T+2T^2},$$

$$X_{33} = \frac{0.9T^2(1+T)}{1+2T+2T^2}.$$

By Eq. (13-44) the variance of the output is

$$E\{y(t)^2\} = \frac{1}{T^2} X_{33} = \frac{0.9(1+T)}{1+2T+2T^2},$$

which confirms the condition that was used in Example 13-2.

13-4 THE PRINCIPLE OF OPTIMUM FILTERING

In this section we will discuss the following filtering problem: Let a system be described by the difference equation of state,

$$\mathbf{x}_{k+1} = \mathbf{P}\mathbf{x}_k + \mathbf{Q}\mathbf{u}_k,$$

and an output (or measurement) equation

$$\mathbf{y}_k = \mathbf{C}\mathbf{x}_k + \mathbf{v}_k,$$

where $\mathbf{u}_k$ and $\mathbf{v}_k$ are zero-mean Gaussian white noise vectors. Suppose we know $\mathbf{P}, \mathbf{Q}, \mathbf{C}$, and the noise intensity of both $\mathbf{u}$ and $\mathbf{v}$. Also suppose that the measurement has been made so that the vector time series $\mathbf{y}_0, \ldots, \mathbf{y}_k$ is known. Then, what is the best estimate of $\mathbf{x}_k$ we can make at time $t = kT$? The answer to this question is useful for various problems such as unknown state estimation, prediction, stochastic optimal-control system design, noise-filtering, interception of a hostile object, etc. In classical control theory the Wiener filter (based mainly on continuous-time, linear systems) affords an answer to this problem. In "modern" control theory, on the other hand, the Kalman filter serves this purpose even better because the algorithm can be programmed on a digital computer without going through mathematical analysis (which was unavoidable in Wiener theory). Essentially, the Kalman filter is an adaptive, gain-tuning technique in which the prediction of the state at any instant is weighted between the extrapolated past value and the observed value.

To introduce the so-called Kalman filter theory in its simplest form, let us consider, for the time being, a first-order, stationary, discrete-time system subject to stationary random inputs:

$$x_{k+1} = px_k + qu_k, \qquad y_k = cx_k + v_k, \qquad (13\text{-}54)$$

where u_k and v_k are zero-mean Gaussian noise, independent of each other (that is, uncorrelated), and moreover, the expected values of the products, $E\{u_k u_j\}$ and $E\{v_k v_j\}$, are zero for all integer subscripts except for $k = j$,

$$E\{u_k u_j\} = 0, \qquad E\{v_k v_j\} = 0, \qquad \text{for all} \quad k \neq j,$$

and

$$E\{u_k^2\} = \sigma_u^2, \qquad E\{v_k^2\} = \sigma_v^2.$$

In other words, the noise signals u and v are "white" sampled data with rms values σ_u and σ_v, respectively.

To obtain the best estimate of x_k at time $t = kT$, denoted by $\hat{x}_k$, the set of information available for the computation is the output time series $y_0, y_1, y_2, \ldots, y_k$ (to be denoted by the set Y_k of the time series for brevity in the following), the variances σ_u^2 and σ_v^2, and system constants p, q, and c.

The estimation error e_k is then expressed by

$$e_k = x_k - \hat{x}_k. \tag{13-55}$$

We define the best estimate to be an estimate for which the mean squared error will be minimal. Let μ_k and σ_k^2 be the mean and variance of the random variable x_k:

$$E\{x_k\} = \mu_k, \qquad E\{(x_k - \mu_k)^2\} = \sigma_k^2. \tag{13-56}$$

In Eq. (13-56) x_k is a random *variable*; for formality a distinction in symbols should be made between a variable and its value, but we shall abandon formality in this tutorial presentation. Squaring Eq. (13-55), we obtain

$$e_k^2 = \hat{x}_k^2 - 2\hat{x}_k x_k + x_k^2,$$

where $\hat{x}_k$ in this expression is the best estimate, yet to be determined, and hence a value. The expected value of e_k^2 (the mean-squared error) is computed by putting $E\{\ |\ Y_k\}$ on the stochastic variable x_k, where $E\{\ |\ Y_k\}$ means the expectation (average) conditional of a set of observations $Y_k = \{y_k, \ldots, y_1, y_0\}$:

$$E\{e_k^2 | Y_k\} = \hat{x}_k^2 - 2\hat{x}_k E\{x_k | Y_k\} + E\{x_k^2 | Y_k\}.$$

Since

$$\sigma_k^2 \equiv E\{(x_k - \mu_k)^2 | Y_k\} = E\{x_k^2 | Y_k\} - 2\mu_k E\{x_k | Y_k\} + \mu_k^2$$

where

$$\mu_k = E\{x_k | Y_k\},$$

thus

$$\sigma_k^2 = E\{x_k^2 | Y_k\} - 2\mu_k^2 + \mu_k^2$$

or

$$E\{x_k^2 | Y_k\} = \mu_k^2 + \sigma_k^2.$$

It follows that

$$E\{e_k^2 | Y_k\} = \hat{x}_k^2 - 2\hat{x}_k \mu_k + \mu_k^2 + \sigma_k^2 = (\hat{x}_k - \mu_k)^2 + \sigma_k^2. \tag{13-57}$$

Therefore, the mean-squared error of the estimate will be minimized when $\hat{x}_k = \mu_k\ (=E\{x_k | Y_k\})$, for which the following relation will hold:

$$E\{x_k^2 | Y_k\} = E\{\hat{x}_k^2 | Y_k\} + E\{e_k^2 | Y_k\}.$$

In the general case of vector state estimation (next section),

$$\hat{\mathbf{x}}_k = E\{\mathbf{x}_k | Y_k\},$$
$$\mathbf{e}_k = \mathbf{x}_k - \hat{\mathbf{x}}_k,$$

and the minimization of mean-squared error applies to all elements of $\mathbf{e}_k = [e_{1k}, \ldots, e_{nk}]$, hence

$$J = E\{\mathbf{e}'_k \mathbf{e}_k | Y_k\}$$

will be minimized.

For the $(k+1)$st sample of the first order system the optimal estimate is:

$$\hat{x}_{k+1} = E\{x_{k+1} | Y_{k+1}\}. \tag{13-58}$$

This is the estimate which we shall determine below. For this optimal value, $\hat{x}$, the mean-squared error (variance) of estimation is given by σ_k^2; $E\{e_k^2 | Y_k\} = \sigma_k^2$ in Eq. (13-57). Since σ_k^2 is an important performance index of the optimum filter, we shall designate a symbol z_k for it: that is,

$$E\{e_k^2 | Y_k\} = E\{(x_k - \hat{x}_k)^2 | Y_k\} = z_k. \tag{13-59}$$

It can be shown that the final form of the optimum estimation algorithm reduces to the following set of conditions:

$$\hat{x}_{k+1} = p\hat{x}_k + K_{k+1}(y_{k+1} - cp\hat{x}_k), \tag{13-60}$$

$$K_{k+1} = \frac{m_{k+1} c}{c^2 m_{k+1} + \sigma_v^2}, \tag{13-61}$$

$$m_{k+1} = p^2 z_k + q^2 \sigma_u^2, \tag{13-62}$$

$$z_k = (1 - K_k c) m_k. \tag{13-63}$$

Equation (13-60) is the main condition of the optimum filtering, where K_{k+1} is the weight put on the latest observation y_{k+1}. The extrapolation of the past best estimate $\hat{x}_k$ is $p\hat{x}_k$ because, by Eq. (13-54),

$$E\{x_{k+1}\} = pE\{x_k\} + qE\{u_k\},$$

where u_k is assumed to be zero-mean noise, $E\{u_k\} = 0$. The output that will correspond to $p\hat{x}_k$ is $cp\hat{x}_k$, and the difference $(y_{k+1} - cp\hat{x}_k)$ in Eq. (13-60) is a correction for updating the data. If y_{k+1} (which is the observation of the state x_{k+1} with noise v_{k+1}) is considered highly reliable, a weight close to $K_{k+1} = 1/c$ will be justified. If, on the other hand, y_{k+1} is not reliable due to measurement noise v_{k+1},

$$K_{k+1} \to 0 \quad \text{in the limiting case},$$

and we must rely mostly on $p\hat{x}_k$, which is a prediction of $\hat{x}_{k+1}$ based only on $\hat{x}_k$. We initialize Eq. (13-60) with

$$\hat{x}_{-1} = 0, \tag{13-64}$$

and compute $\hat{x}_0 = K_0 y_0$, $\hat{x}_1$, ...

The filter weights $K_0, K_1, \ldots$ are given by Eq. (13–61), where m_{k+1} represents the variance of the one-step prediction error,

$$m_{k+1} = E\{(x_{k+1} - E\{x_{k+1}\})^2 \mid Y_k\}. \tag{13-65}$$

$E\{x_{k+1} \mid Y_k\}$ in this expression is the optimal estimate of x_{k+1} given measurements $y_0, y_1, \ldots, y_k$. Since Y_{-1} is an empty set (no available measurement), the initial condition for m_k is

$$m_0 = E\{(x_0 - E\{x_0 \mid Y_{-1}\})^2\} = E\{x_0^2\}, \tag{13-66}$$

where the initial state, x_0, is considered to be a zero-mean Gaussian random variable; m_0 is the variance of x_0.

Eliminating z_k from Eqs. (13–62) and (13–63), we obtain

$$m_{k+1} = p^2(1 - K_k c)m_k + q^2 \sigma_u^2, \tag{13-67}$$

where, by Eq. (13–61),

$$1 - K_k c \quad \text{may be replaced by} \quad 1 - K_k c = \frac{\sigma_v^2}{c^2 m_k + \sigma_v^2}.$$

Therefore, starting with the initial value m_0 given by Eq. (13–66), $K_0, K_1, \ldots$ and $m_1, m_2, \ldots$ can be computed recursively. However, the set of recursive relations given by Eqs. (13–60) through (13–63) is generally preferred, because z_k (Eq. (13–59)) gives the statistical quality of the optimum filter (which is of interest to us). Note that z_k is different from m_k.

Example 13-7 Let $c = 1$ in Eq. (13–54), so that

$$x_{k+1} = px_k + qu_k, \qquad y_k = x_k + v_k,$$

where

$$E\{u_k^2\} = \sigma_u^2 \quad \text{and} \quad E\{v_k^2\} = \sigma_v^2.$$

If the rms value of x_k is also σ_u,

$$E\{x_{k+1}^2\} = p^2 E\{x_k^2\} + 2pq E\{x_k u_k\} + q^2 E\{u_k^2\}$$

where

$$E\{x_{k+1}^2\} = E\{x_k^2\} = \sigma_u^2, \qquad E\{u_k^2\} = \sigma_u^2, \qquad E\{x_k u_k\} = 0;$$

hence we must have

$$1 - p^2 = q^2.$$

Let us choose the following set of constants, which will satisfy $q = \sqrt{1 - p^2}$:

$$p = 0.6, \qquad q = 0.8, \qquad \text{and} \qquad \sigma_u^2 = 5, \qquad \sigma_v^2 = 1.$$

The initial value, m_0, for Eq. (13–66) is then

$$m_0 = 5,$$

and we find (by Eq. (13-61)),

$$K_0 = \frac{m_0}{m_0 + 1} = 0.833 .$$

Recalling Eq. (13-64), we obtain, by Eq. (13-60)

$$\hat{x}_0 = 0.833 y_0 .$$

Next we find m_1 by Eq. (13-67),

$$m_1 = 0.36(1 - K_0)m_0 + 3.2 = 3.50 ,$$

and

$$K_1 = \frac{m_1}{m_1 + 1} = 0.777 ,$$

$$\hat{x}_1 = 0.6\hat{x}_0 + 0.777(y_1 - 0.6\hat{x}_0) = 0.222(0.6\hat{x}_0) + 0.777 y_1 .$$

The sequence quickly converges to a set of steady-state values (in practice, the steady-state condition of the filter is achieved at about

$$t = (5 \text{ to } 10)T_p$$

where T_p is the dominant-process time constant). Letting both m_{k+1} and m_k in Eq. (13-67) equal a steady-state value, m_s, we obtain

$$m_s = p^2(1 - K_s c)m_s + q^2 \sigma_u^2 ,$$

where the steady-state value of the weighting gain is

$$K_s = \frac{m_s c}{c^2 m_s + \sigma_v^2} .$$

Eliminating K_s from the two conditions, for our example, we obtain

$$m_s^2 - 2.56 m_s - 3.2 = 0 .$$

Since m is a variance, and hence nonnegative, we take the positive root of the quadratic equation $m_s = 3.48$; thus $K_s = 0.775$.

For our example, we see (from Eqs. (13-61) and (13-63)) that $K_s = z_s$. Therefore the variance of the estimation error at steady state is

$$z_s = 0.775 .$$

If we relied completely on observation (without using the filter) for state estimation, the estimation error would have given a variance

$$E\left\{\left(x_k - \frac{y_k}{c}\right)^2\right\} = \frac{1}{c} E\{v_k^2\} = \frac{\sigma_v^2}{c} = 1 .$$

The measurement noise assumed in our example is much lower ($\sigma_v^2 = 1$) than the input noise ($\sigma_u^2 = 5$) which perturbs the state. This is why the weight $K_s = 0.775$ is high. The effect of the noise ratio on K_s for our example is shown in Fig. 13-9. Here K_s changes from $K_s = 0$ to $K_s = 1/c = 1$ as the rms value of measurement noise σ_v is decreased relative to σ_u.

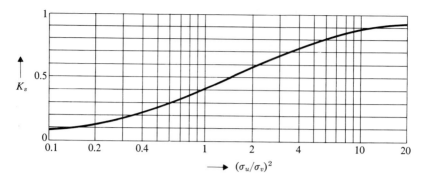

Fig. 13-9 Steady-state gain K_s as a function of noise-magnitude ratio in Example 13-8.

The structure of the optimal filter given by Eqs. (13-60) is shown by the block diagram of Fig. 13-10, where the recursive relation (Eqs. (13-61) through (13-63)), used to compute the time-varying "gain" K_k during the transient of the filter, is indicated by a flow chart in informal, condensed form. Let us now derive this set of conditions for optimal filtering.

In general, a conditional probability-density function has the following property:

$$p(a|b, c, \ldots, z) = \frac{p(a, b, c, \ldots, z)}{p(b, c, \ldots, z)}. \tag{13-68}$$

We can verify the following equality of conditional probability-density functions by substituting from Eq. (13-68) for each of them:

$$p(x_{k+1}|Y_{k+1}) = \frac{p(y_{k+1}|Y_k, x_{k+1}) \cdot p(x_{k+1}|Y_k)}{p(y_{k+1}|Y_k)}. \tag{13-69}$$

In Eq. (13-54) we assumed u_k and v_k to be Gaussian. In general, Gaussian random variables have the important property we saw in Fig. 13-2(c). In our system of Eq. (13-54), therefore, both x_k and y_k are also Gaussian for all k, and the form of Eq. (13-7) applies to all p's of Eq. (13-69); thus

$$\frac{1}{\sqrt{2\pi}\sigma_{k+1}} \exp\left\{-\frac{1}{2}\frac{(x_{k+1} - \mu_{k+1})^2}{\sigma_{k+1}^2}\right\} = \frac{1}{\sqrt{2\pi}(\sigma_1\sigma_3/\sigma_2)}$$
$$\times \exp\left\{-\frac{1}{2}\left(\frac{(x_{k+1} - \mu_1)^2}{\sigma_1^2} - \frac{(y_{k+1} - \mu_2)^2}{\sigma_2^2} + \frac{(y_{k+1} - \mu_3)^2}{\sigma_3^2}\right)\right\}, \tag{13-70}$$

or, equating terms on both sides,

$$\frac{(x_{k+1} - \mu_{k+1})^2}{\sigma_{k+1}^2} = \frac{(x_{k+1} - \mu_1)^2}{\sigma_1^2} - \frac{(y_{k+1} - \mu_2)^2}{\sigma_2^2} + \frac{(y_{k+1} - \mu_3)^2}{\sigma_3^2},$$

$$\sigma_{k+1} = \frac{\sigma_1\sigma_3}{\sigma_2}. \tag{13-71}$$

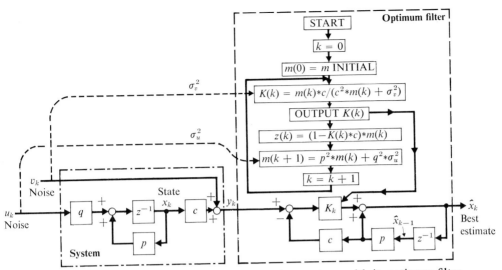

Fig. 13-10 Time-invariant, first-order, discrete-time system, with its optimum filter and symbolic flowchart for gain computation.

The symbols used in Eqs. (13-70) and (13-71) are:

Left side of the equation: μ_{k+1} and σ^2_{k+1} for $p(x_{k+1} | Y_{k+1})$,

and

Right side of the equation: μ_1, σ^2_1 for $p(x_{k+1} | Y_k)$,
μ_2, σ^2_2 for $p(y_{k+1} | Y_k)$,
μ_3, σ^2_3 for $p(y_{k+1} | Y_k, x_{k+1})$.

Note that $\mu_{k+1} = \hat{x}_{k+1}$ by Eq. (13-58) and $\sigma^2_{k+1} = z_{k+1}$ by Eq. (13-59). We now compute these in terms of the statistical quantities on the right side of Eq. (13-71):

i) We want to find μ_1 and σ_1 for $p(x_{k+1} | Y_k)$. Since $x_{k+1} = px_k + qu_k$,

$$\mu_1 = E\{x_{k+1} | Y_k\} = pE\{x_k | Y_k\} + qE\{u_k | Y_k\},$$

where $E\{x_k | Y_k\} = \hat{x}_k$ by Eq. (13-58) and $E\{u_k | Y_k\} = 0$ for the independent zero-mean noise u_k; hence

$$\mu_1 = p\hat{x}_k. \tag{13-72}$$

The variance σ^2_1 has already been denoted by m_{k+1} in Eq. (13-65):

$$\sigma^2_1 = E\{(x_{k+1} - E\{x_{k+1}\})^2 | Y_k\} = m_{k+1}.$$

Substituting $x_{k+1} = px_k + qu_k$ and $E\{x_{k+1} | Y_k\} = p\hat{x}_k$,

$$m_{k+1} = E\{(px_k + qu_k - p\hat{x}_k)^2 | Y_k\}$$
$$= p^2 E\{(x_k - \hat{x}_k)^2 | Y_k\} + q^2 E\{u_k^2\} + 2pq E\{(x_k - \hat{x}_k)u_k | Y_k\},$$

where

$$E\{(x_k - \hat{x}_k)^2 \mid Y_k\} = z_k \quad \text{by Eq. (13-59)},$$

$$E\{u_k^2\} = \sigma_u^2$$

and

$$E\{(x_k - \hat{x}_k)u_k \mid Y_k\} = 0$$

for the independent noise u_k; hence

$$\sigma_1^2 = m_{k+1} = p^2 z_k + q^2 \sigma_u^2. \tag{13-73}$$

This is identical to Eq. (13-62).

ii) We can now find μ_2 and σ_2 for $p(y_{k+1} \mid Y_k)$. Recalling that $E\{v_{k+1}\} = 0$ in the output equation $y_{k+1} = cx_{k+1} + v_{k+1}$, we have

$$\mu_2 = E\{y_{k+1} \mid Y_k\} = cE\{x_{k+1} \mid Y_k\},$$

where, by Eq. (13-72),

$$E\{x_{k+1} \mid Y_k\} = \mu_1 = p\hat{x}_k.$$

Thus

$$\mu_2 = c\mu_1 = cp\hat{x}_k. \tag{13-74}$$

The variance is

$$\sigma_2^2 = E\{(y_{k+1} - \mu_2)^2 \mid Y_k\}$$
$$= E\{(cx_{k+1} + v_{k+1} - c\mu_1)^2 \mid Y_k\}$$
$$= c^2 E\{(x_{k+1} - \mu_1)^2 \mid Y_k\} + E\{v_{k+1}^2\} + 2cE\{v_{k+1}(x_{k+1} - \mu_1) \mid Y_k\},$$

where

$$E\{(x_{k+1} - \mu_1)^2 \mid Y_k\} = m_{k+1},$$
$$E\{v_{k+1}^2\} = \sigma_v^2,$$

and the last term is zero for independent noise v_{k+1}. Therefore:

$$\sigma_2^2 = c^2 m_{k+1} + \sigma_v^2. \tag{13-75}$$

iii) μ_3 and σ_3 for $p(y_{k+1} \mid Y_k, x_{k+1})$ will now be determined. Since $y_{k+1} = cx_{k+1} + v_{k+1}$, the set of past observations, Y_k, is not needed to determine y_{k+1} when x_{k+1} is given as a condition. Thus $p(y_{k+1} \mid Y_k, x_{k+1}) = p(y_{k+1} \mid x_{k+1})$ and

$$\mu_3 = E\{y_{k+1} \mid x_{k+1}\} = cx_{k+1} + E\{v_{k+1}\} = cx_{k+1}, \tag{13-76}$$

$$\sigma_3^2 = E\{(y_{k+1} - \mu_3)^2 \mid x_{k+1}\}$$
$$= E\{[(cx_{k+1} + v_{k+1}) - cx_{k+1}]^2 \mid x_{k+1}\}$$
$$= E\{v_{k+1}^2\} = \sigma_v^2. \tag{13-77}$$

13-4 The principle of optimum filtering

Substituting Eqs. (13–72) through (13–77) in the right side of Eq. (13–71), and collecting terms on x_{k+1}, we obtain

$$\frac{(x_{k+1} - p\hat{x}_k)^2}{m_{k+1}} - \frac{(y_{k+1} - cp\hat{x}_k)^2}{c^2 m_{k+1} + \sigma_v^2} + \frac{(y_{k+1} - cx_{k+1})^2}{\sigma_v^2}$$

$$= \left(\frac{1}{m_{k+1}} + \frac{c^2}{\sigma_v^2}\right) x_{k+1}^2 - 2\left(\frac{p\hat{x}_k}{m_{k+1}} + \frac{cy_{k+1}}{\sigma_v^2}\right) x_{k+1}$$

$$+ \left\{\frac{p^2 \hat{x}_k^2}{m_{k+1}} + \frac{y_{k+1}^2}{\sigma_v^2} - \frac{(y_{k+1} - cp\hat{x}_k)^2}{c^2 m_{k+1} + \sigma_v^2}\right\}$$

$$= \left(\frac{\sigma_v^2 + c^2 m_{k+1}}{\sigma_v^2 m_{k+1}}\right) \left\{x_{k+1} - \left(\frac{p\hat{x}_k \sigma_v^2 + c m_{k+1} y_{k+1}}{\sigma_v^2 + c^2 m_{k+1}}\right)\right\}^2. \qquad (13-78)$$

Equating the right side of Eq. (13–78) with the left side of Eq. (13–71), we find:

$$\mu_{k+1} = \frac{p\hat{x}_k \sigma_v^2 + c m_{k+1} y_{k+1}}{\sigma_v^2 + c^2 m_{k+1}} = p\hat{x}_k + K_{k+1}(y_{k+1} - cp\hat{x}_k)$$

where

$$K_{k+1} = \frac{m_{k+1} c}{c^2 m_{k+1} + \sigma_v^2}.$$

These are the first two conditions of the filter, Eqs. (13–60) and (13–61). Equation (13–62) has already been derived, as Eq. (13–73). To obtain the last condition, Eq. (13–63), we recall the variance condition, $\sigma_{k+1} = (\sigma_1 \sigma_3)/\sigma_2$, which was stated in Eq. (13–71). Substituting Eqs. (13–73), (13–75), and (13–77) into the right side, it follows that:

$$z_{k+1} = \sigma_{k+1}^2 = \frac{m_{k+1} \sigma_v^2}{c^2 m_{k+1} + \sigma_v^2} = (1 - K_{k+1} c) m_{k+1},$$

which, shifting $(k + 1)$ back to k, is Eq. (13–63). This completes the derivation.

It was hinted before that $\mu_1 = p\hat{x}_k$ is a one-step prediction of the state (and m_{k+1} is the variance of the prediction error). From this we deduce that the *optimal prediction* of a state which is N sampling periods ahead is given by

$$E\{x_{k+N} \mid Y_k\} = p^N \hat{x}_k, \qquad (13-79)$$

where $\hat{x}_k$ is the optimum filter output at the present instant, $t = kT$.

It is possible to adapt Eqs. (13–60) through (13–63) to use on a linear, time-invariant continuous-time system whose equations are,

$$\frac{dx}{dt} = ax + bu, \qquad y = cx + v, \qquad (13-80)$$

where u and v are independent white noise. For this conversion we apply the following limiting process to Eq. (13–54):

$$\frac{dx}{dt} = \lim_{\Delta t \to dt} \frac{x(kT + \Delta t) - x(kT)}{\Delta t}$$

$$= \lim_{T \to dt} \left\{ \frac{p-1}{T} x_k + \frac{q}{T} u_k \right\} = ax_k + bu_k \,;$$

that is,

$$p \to 1 + a\,dt, \qquad q \to b\,dt. \tag{13–81}$$

The variances σ_{uc}^2, σ_{vc}^2 for $u(t)$ and $v(t)$ in the continuous domain are given in terms of σ_u^2 and σ_v^2 for the sampled-data system by:

$$\sigma_{uc}^2 = \lim_{T \to dt} \sigma_u^2 T = \sigma_u^2\,dt, \qquad \sigma_{vc}^2 = \lim_{T \to dt} \sigma_v^2 T = \sigma_v^2\,dt,$$

that is,

$$\sigma_u^2 \to \sigma_{uc}^2/dt, \qquad \sigma_v^2 \to \sigma_{vc}^2/dt. \tag{13–82}$$

Substituting Eqs. (13–81) and (13–82) into the filter equations, and neglecting terms involving $(dt)^2$, we obtain:

$$\frac{d\hat{x}}{dt} = a\hat{x} + \frac{mc}{\sigma_{vc}^2}(y - c\hat{x}),$$

$$\frac{dm}{dt} = 2am - \frac{c^2 m^2}{\sigma_{vc}^2} + b^2 \sigma_{uc}^2. \tag{13–83}$$

The system of Eqs. (13–80) and (13–83) is shown in Fig. 13–11, where the upper half of the optimal filter becomes static when $m(t)$ settles down to the stationary value m_s, where m_s is the positive root of the quadratic equation

$$2am_s - \frac{c^2}{\sigma_{vc}^2} m_s^2 + b^2 \sigma_{uc}^2 = 0\,;$$

that is,

$$m_s = \left(\frac{\sigma_{vc}}{c}\right)^2 \left\{ a + \sqrt{a^2 + \left(\frac{bc\sigma_{uc}}{\sigma_{vc}}\right)^2} \right\}. \tag{13–84}$$

The gain of the filter, $K = mc/(\sigma_{vc})^2$, becomes constant at $m(t) = m_s$ and the filter transfer function at this state is given by

$$\frac{\hat{x}}{y} = \frac{1}{c} \frac{a + \sqrt{a^2 + (bc\sigma_{uc}/\sigma_{vc})^2}}{D + \sqrt{a^2 + (bc\sigma_{uc}/\sigma_{vc})^2}}. \tag{13–85}$$

This is a low-pass filter, which is stable regardless of the eigenvalue a of the original system. The Wiener theory of optimum filtering yields the same transfer function only when the original system is assumed stable ($a < 0$).

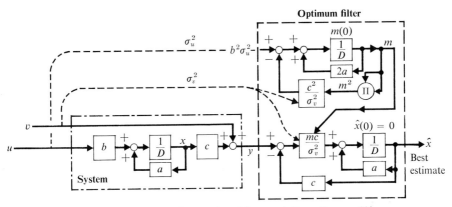

Fig. 13-11 A linear, first-order object and its optimum filter.

13-5 OPTIMUM FILTER ALGORITHMS

The filtering algorithm just derived for stationary first-order systems can now be generalized. We shall consider continuous- or discrete-time, linear, nonstationary systems of an arbitrary (but finite) order, with multiple (or vector) random-variable inputs. For this purpose we replace the variance by the covariance matrix, Eq. (13-36), restated as:

$$E\{(\mathbf{x} - E\{\mathbf{x}\})(\mathbf{x} - E\{\mathbf{x}\})'\} = \mathbf{X}.$$

If, for instance, a zero-mean random vector is $\mathbf{u} = \begin{bmatrix} u_1 \\ u_2 \end{bmatrix}$ and $u_2 = 0$, its covariance matrix $\mathbf{U}$ will be

$$\mathbf{U} = E\left\{\begin{bmatrix} u_1 \\ 0 \end{bmatrix} [u_1 \quad 0]\right\} = E\left\{\begin{bmatrix} u_1^2 & 0 \\ 0 & 0 \end{bmatrix}\right\}.$$

The probability-density function, $p(x)$, for a scalar random Gaussian variable, Eq. (13-7), must be replaced by Eq. (13-35) for a random-vector variable.

If a dynamic process is given by a set of vector difference equations with time-dependent coefficient matrices

$$\mathbf{x}_{k+1} = \mathbf{P}_k \mathbf{x}_k + \mathbf{Q}_k \mathbf{u}_k, \qquad \mathbf{y}_k = \mathbf{C}_k \mathbf{x}_k + \mathbf{v}_k, \tag{13-86}$$

it can be shown [18, 19, 20, 21] that the optimum filter for the best estimate $\hat{\mathbf{x}}_k$ of $\mathbf{x}_k$ in the least mean-squared error sense will be given by the following set of conditions:

$$\begin{aligned}
\hat{\mathbf{x}}_{k+1} &= \mathbf{P}_k \hat{\mathbf{x}}_k + \mathbf{K}_{k+1}(\mathbf{y}_{k+1} - \mathbf{C}_{k+1}\mathbf{P}_k \hat{\mathbf{x}}_k), \qquad \hat{\mathbf{x}}_{-1} = \mathbf{0}, \\
\mathbf{K}_{k+1} &= \mathbf{M}_{k+1}\mathbf{C}'_{k+1}(\mathbf{C}_{k+1}\mathbf{M}_{k+1}\mathbf{C}'_{k+1} + \mathbf{V}_{k+1})^{-1}, \\
\mathbf{M}_{k+1} &= \mathbf{P}_k \mathbf{Z}_k \mathbf{P}'_k + \mathbf{Q}_k \mathbf{U}_k \mathbf{Q}'_k, \qquad \mathbf{M}_0 = E\{\mathbf{x}_0 \mathbf{x}'_0\}, \\
\mathbf{Z}_k &= (\mathbf{I} - \mathbf{K}_k \mathbf{C}_k)\mathbf{M}_k.
\end{aligned} \tag{13-87}$$

Here:

1. $\mathbf{x}_k$ is an *n*-vector of state at time $t = kT$, $E\{\mathbf{x}_0\} = 0$ and $E\{\mathbf{x}_0\mathbf{x}_0'\} = \mathbf{M}_0$ is a known (or assumed) covariance matrix.

2. $\{\mathbf{u}_k\}$ is an uncorrelated, zero-mean, random-sequence, *r*-vector of Gaussian distribution; that is,

$$E\{\mathbf{u}_k\} = 0, \qquad E\{\mathbf{u}_k\mathbf{u}_j'\} = 0 \qquad \text{for all} \quad k \neq j,$$

and

$$E\{\mathbf{u}_k\mathbf{u}_k'\} = \mathbf{U}_k,$$

with the property that $E\{\mathbf{u}_k\mathbf{x}_0\} = 0$. Its covariance matrix $\mathbf{U}_k$ may depend on k.

3. $\mathbf{y}_k$ is an *m*-vector of measurements at time $t = kT$.

4. $\{\mathbf{v}_k\}$ is a zero-mean uncorrelated Gaussian random-sequence *m*-vector,

$$E\{\mathbf{v}_k\} = 0, \qquad E\{\mathbf{v}_k\mathbf{v}_j'\} = 0 \qquad \text{for all} \quad k \neq j, \qquad E\{\mathbf{v}_k\mathbf{v}_k'\} = \mathbf{V}_k,$$

where the covariance matrix $\mathbf{V}_k$ may vary with k.

5. There is no correlation between $\{\mathbf{u}_k\}$ and $\{\mathbf{v}_k\}$.

Since the vector difference equation, Eq. (13–86), is first-order, a new state, $\mathbf{x}_{k+1}$, is influenced only by the preceding state, $\mathbf{x}_k$, and the Gaussian random input $\mathbf{u}_k$ at $t = kT$. Such a process is called a discrete, Gaussian, Markov process. Derivation of the optimal filter conditions (Eq. (13–87)) closely parallels the steps shown previously.

A further generalization of Eq. (13–87) is possible for the case when $\mathbf{u}_k$ and $\mathbf{v}_k$ are correlated. For this case we define the matrices $\mathbf{R}_k$ and $\mathbf{R}_k'$ (rectangular if $r \neq m$) by

$$E\left\{\begin{bmatrix}\mathbf{u}_k \\ \mathbf{v}_k\end{bmatrix}[\mathbf{u}_k' \quad \mathbf{v}_k']\right\} = \begin{bmatrix}\mathbf{U}_k & \mathbf{R}_k \\ \mathbf{R}_k' & \mathbf{V}_k\end{bmatrix}.$$

The final result for this case will be:

$$\left.\begin{aligned}\hat{\mathbf{x}}_{k+1} &= \mathbf{P}_k\hat{\mathbf{x}}_k + \mathbf{K}_{k+1}(\mathbf{y}_{k+1} - \mathbf{C}_{k+1}\mathbf{P}_k\hat{\mathbf{x}}_k), \qquad \hat{\mathbf{x}}_{-1} = 0, \\ \mathbf{K}_{k+1} &= (\mathbf{M}_{k+1}\mathbf{C}_{k+1}' + \mathbf{Q}_k\mathbf{R}_k)(\mathbf{C}_{k+1}\mathbf{M}_{k+1}\mathbf{C}_{k+1}' + \mathbf{V}_k)^{-1}, \\ \mathbf{M}_{k+1} &= \mathbf{P}_k\mathbf{Z}_k\mathbf{P}_k' + \mathbf{Q}_k\mathbf{U}_k\mathbf{Q}_k', \\ \mathbf{Z}_k &= (\mathbf{I} - \mathbf{K}_k\mathbf{C}_k)\mathbf{M}_k - \mathbf{K}_k\mathbf{R}_{k-1}'\mathbf{Q}_{k-1}'.\end{aligned}\right\} \qquad (13\text{–}88)$$

The structure of the second of Eqs. (13–88) is similar to the minimal norm solution of the output equation

$$\mathbf{y} = \mathbf{C}\mathbf{x}$$

13-5   Optimum filter algorithms

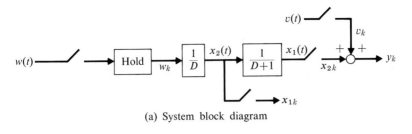

(a) System block diagram

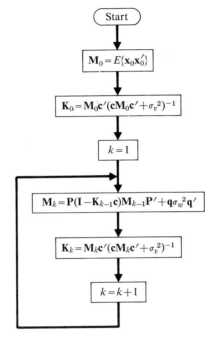

(b) Flow diagram for computation of time-varying gain

Fig. 13-12 Computation of optimum-filter gains.

for a deterministic system. According to Eq. (A-33) of the Appendix, norm $\|x^\circ\|$ of the "solution" x° is minimized when

$$x^\circ = C'(CC')^{-1}y.$$

Example 13-8 We would like to develop an optimal (Kalman) filter to estimate the state of the system shown in Fig. 13-12a. The state equations for the continuous-time part of the system are:

$$\frac{d}{dt}x = \begin{bmatrix} -1 & 1 \\ 0 & 0 \end{bmatrix} x + \begin{bmatrix} 0 \\ 1 \end{bmatrix} w,$$

$$y = [1 \quad 0]x + v,$$

where w and v are uncorrelated Gaussian white noise and

$$\mathbf{x} = \begin{bmatrix} x_1 \\ x_2 \end{bmatrix}.$$

These equations can be transformed to the following discrete-time, difference equations:

$$\mathbf{x}_{k+1} = \underbrace{\begin{bmatrix} e^{-T} & (1-e^{-T}) \\ 0 & 1 \end{bmatrix}}_{\mathbf{P}} \mathbf{x}_k + \underbrace{\begin{bmatrix} e^{-T} + T - 1 \\ T \end{bmatrix}}_{\mathbf{q}} w_k,$$

$$y_k = \underbrace{[1 \quad 0]}_{\mathbf{c}} \mathbf{x}_k + v_k,$$

where T is the time between samples.

The optimum filter is used to get the "best" estimate for the state of the system. The basic filter equation for the estimated state at the $(k + 1)$st time interval is (from Eq. (13-88) with $\mathbf{P}$ and $\mathbf{C}$ constant):

$$\hat{\mathbf{x}}_{k+1} = \mathbf{P}\hat{\mathbf{x}}_k + \mathbf{K}_{k+1}(\mathbf{y}_{k+1} - \mathbf{cP}\hat{\mathbf{x}}_k).$$

Our major problem is to find the time-varying gains, $\mathbf{K}_k$, such that we do get the best possible estimate. We use the relations in Eq. (13-88) recursively to generate these gains. To start the computation we note that

$$\mathbf{M}_0 = E\{\mathbf{x}_0 \mathbf{x}_0'\} = E\left\{\begin{bmatrix} x_{1_0}^2 & x_{1_0} x_{2_0} \\ x_{1_0} x_{2_0} & x_{2_0}^2 \end{bmatrix}\right\}.$$

If we take $x_{1_0} = x_{2_0} = 0$, then

$$\mathbf{M}_0 = \begin{bmatrix} 0 & 0 \\ 0 & 0 \end{bmatrix}.$$

The starting gain can be found from the equation for $\mathbf{K}(k)$,

$$\mathbf{K}_0 = \mathbf{M}_0 \mathbf{c}'(\mathbf{cM}_0 \mathbf{c}' + \sigma_v^2)^{-1},$$

where we have reduced $\mathbf{V}$ to σ_v^2, and noted that $\mathbf{cMc}'$ is also a scalar for this problem. The remainder of the gains are generated by direct application of the equations, as shown in the flow chart in Fig. 13-12(b) (note the similarity of the flow diagram to the one shown in Fig. 13-10). This algorithm can be implemented directly on a digital computer; the results are shown on Fig. 13-13. The input noise intensity, σ_w^2, was kept at unity for all the cases shown since it is only the ratio, (σ_v^2/σ_w^2), that affects the solution. A sampling interval of $T = 0.5$ was used, giving values for $\mathbf{P}$ and $\mathbf{q}$ of:

$$\mathbf{P} = \begin{bmatrix} 0.6 & 0.4 \\ 0 & 1 \end{bmatrix}, \quad \mathbf{q} = \begin{bmatrix} 0.1 \\ 0.5 \end{bmatrix}.$$

As we expect, as the noise contamination of the measurement gets stronger relative to the input noise, the filter gains go down, showing more reliance on the predicted value than on the measured one. If the input noise were to disappear entirely

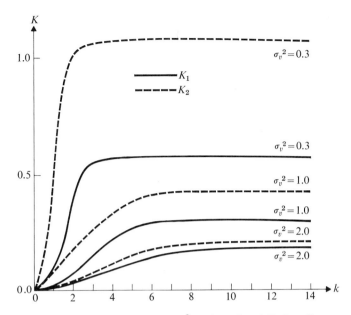

Fig. 13-13 Kalman filter gains, $\sigma_w^2 = 1.0$, $T = 0.5$, for all cases.

($\sigma_w^2 = 0$), then the equations show that **K** goes to zero—a logical result since the system is not perturbed at all, just the measurement.

An important final point is that the system analyzed here has a Gaussian, white-noise input, w, that is sampled and held before it reaches the system. It is *not* correct to apply these results to a similar system without the input sampler. Hybrid computer simulation of this system shows clearly that though the filter derived here is optimal for the system of Fig. 13-12(a), it is not optimal for the equivalent system *without* the sampler. See the limiting process below, which takes us from a discrete-time (T) to a continuous-time (dt) system.

If a dynamic system is stationary and the input-noise intensity is time-invariant, in Eq. (13-87) we let

$$\mathbf{P}_k \to \mathbf{P}, \quad \mathbf{Q}_k \to \mathbf{Q}, \quad \mathbf{C}_k \to \mathbf{C} \quad \text{and} \quad \mathbf{V}_k \to \mathbf{V}, \quad \mathbf{U}_k \to \mathbf{U}.$$

As noted in Example 13-7, the filter will then assume a steady state after a start-up transient. Equation (13-87) for the steady state will reduce to:

$$\left.\begin{aligned}
\hat{\mathbf{x}}_{k+1} &= \mathbf{P}\hat{\mathbf{x}}_k + \mathbf{K}(\mathbf{y}_{k+1} - \mathbf{CP}\hat{\mathbf{x}}_k), \\
\mathbf{K} &= \mathbf{MC}'(\mathbf{CMC}' + \mathbf{V})^{-1}, \\
\mathbf{PMP}' &- \mathbf{M} - \mathbf{PMC}'(\mathbf{CMC}' + \mathbf{V})^{-1}\mathbf{CMP}' + \mathbf{QUQ}' = \mathbf{0}.
\end{aligned}\right\} \quad (13\text{-}89)$$

A nonnegative **M** must be chosen from the third condition of (13-89). In digital-computer solutions of this type of matrix Riccati equation, the authors recommend that **M** be determined as the steady-state solution of the difference equation,

$$\mathbf{M}_{k+1} = \mathbf{P}\mathbf{M}_k\mathbf{P}' - \mathbf{P}\mathbf{M}_k\mathbf{C}'(\mathbf{C}\mathbf{M}_k\mathbf{C}' + \mathbf{V})^{-1}\mathbf{C}\mathbf{M}_k\mathbf{P}' + \mathbf{Q}\mathbf{U}\mathbf{Q}', \qquad (13\text{-}90)$$

initializing with, say, $\mathbf{M}_0 = 0$.

When the state $\mathbf{x}_k$ at time kT is estimated by Eq. (13-89), the predictor algorithm for a state at a future time point N steps ahead, $(k + N)T$, is given by

$$(\mathbf{x}_{k+N})_{\text{optimum prediction}} = \mathbf{P}^N \hat{\mathbf{x}}_k .$$

The relations which apply to discrete-time linear systems can be transformed so as to apply to continuous-time linear systems. For this case, we replace Eq. (13-86) by

$$\frac{d\mathbf{x}(t)}{dt} = \mathbf{A}(t)\mathbf{x}(t) + \mathbf{B}(t)\mathbf{u}(t), \qquad \mathbf{y}(t) = \mathbf{C}(t)\mathbf{x}(t) + \mathbf{v}(t). \qquad (13\text{-}91)$$

We obtain, from Eq. (13-88), the following algorithm for the continuous-time system:

$$\left.\begin{aligned}
\frac{d\hat{\mathbf{x}}(t)}{dt} &= \mathbf{A}(t)\hat{\mathbf{x}}(t) + \mathbf{K}(t)\big(\mathbf{y}(t) - \mathbf{C}(t)\hat{\mathbf{x}}(t)\big), \\
\mathbf{K}(t) &= \big(\mathbf{M}(t)\mathbf{C}'(t) + \mathbf{B}(t)\mathbf{R}(t)\big)\big(\mathbf{V}(t)\big)^{-1}, \\
\frac{d\mathbf{M}(t)}{dt} &= \mathbf{A}(t)\mathbf{M}(t) + \mathbf{M}(t)\mathbf{A}'(t) \\
&\quad - \mathbf{K}(t)\big(\mathbf{C}(t)\mathbf{M}(t) + \mathbf{R}'(t)\mathbf{B}'(t)\big) + \mathbf{B}(t)\mathbf{U}(t)\mathbf{B}'(t),
\end{aligned}\right\} \qquad (13\text{-}92)$$

where $\mathbf{M}(t)$ is the covariance matrix of the estimation error, $\mathbf{x}(t) - \hat{\mathbf{x}}(t)$, and

$$E\left\{\begin{bmatrix}\mathbf{u}(t)\\ \mathbf{v}(t)\end{bmatrix}[\mathbf{u}'(t+\tau) \quad \mathbf{v}'(t+\tau)]\right\} = \begin{bmatrix}\mathbf{U}(t) & \mathbf{R}(t)\\ \mathbf{R}'(t) & \mathbf{V}(t)\end{bmatrix}\delta(\tau). \qquad (13\text{-}93)$$

In the derivation of (13-92) from (13-88), we make use of Eqs. (13-50) and (13-82), and note that

$$\mathbf{K}_k \to \mathbf{K}(t)\, dt .$$

If the system and the noise are both stationary, and there is no crosscorrelation between $\mathbf{u}(t)$ and $\mathbf{v}(t)$, $\mathbf{R} = 0$ in Eq. (13-93), then the last condition of (13-92) will reduce to the following Riccati equation.

$$\frac{d\mathbf{M}(t)}{dt} = \mathbf{A}\mathbf{M}(t) + \mathbf{M}(t)\mathbf{A}' - \mathbf{M}(t)\mathbf{C}'(\mathbf{V})^{-1}\mathbf{C}\mathbf{M}(t) + \mathbf{B}\mathbf{U}\mathbf{B}' . \qquad (13\text{-}94)$$

When the start-up transition is over, $d\mathbf{M}(t)/dt = 0$ in Eq. (13-94), and thus

$$\mathbf{M}(t) \to \mathbf{M}, \qquad \mathbf{K}(t) \to \mathbf{K},$$

and we obtain a steady-state set of equations similar to Eq. (13-89).

The optimal filtering algorithm, Eqs. (13–87) through (13–92), must be modified if the system has a deterministic forcing input. In addition to the present discussion, we will discuss such a case, that of linear stochastic optimal control, in the next chapter.

In place of Eq. (13–86), we now consider the following system:

$$\mathbf{x}_{k+1} = \mathbf{P}_k\mathbf{x}_k + \mathbf{Q}_{u,k}\mathbf{u}_k + \mathbf{Q}_{w,k}\mathbf{w}_k, \qquad \mathbf{y}_k = \mathbf{C}_k\mathbf{x}_k + \mathbf{v}_k, \qquad (13\text{–}95)$$

where the r_1-vector $\mathbf{u}_k$ is a *deterministic* input, and the r_2-vector $\mathbf{w}_k$ is zero-mean Gaussian noise, as in Eq. (13–86). For $\mathbf{w}_k$ and $\mathbf{v}_k$ we assume the following covariance matrix:

$$E\left\{\begin{bmatrix}\mathbf{w}_k \\ \mathbf{v}_k\end{bmatrix}[\mathbf{w}'_k \quad \mathbf{v}'_k]\right\} = \begin{bmatrix}\mathbf{W}_k & 0 \\ 0 & \mathbf{V}_k\end{bmatrix}. \qquad (13\text{–}96)$$

The filter algorithm, Eq. (13–87), modified for this case, is:

$$\begin{aligned}
\hat{\mathbf{x}}_{k+1} &= (\mathbf{P}_k\hat{\mathbf{x}}_k + \mathbf{Q}_{u,k}\mathbf{u}_k) \\
&\quad + \mathbf{K}_{k+1}\{\mathbf{y}_{k+1} - \mathbf{C}_{k+1}(\mathbf{P}_k\hat{\mathbf{x}}_k + \mathbf{Q}_{u,k}\mathbf{u}_k)\}, \quad \hat{\mathbf{x}}_{-1} = 0, \\
\mathbf{K}_{k+1} &= \mathbf{M}_{k+1}\mathbf{C}'_{k+1}(\mathbf{C}_{k+1}\mathbf{M}_{k+1}\mathbf{C}'_{k+1} + \mathbf{V}_{k+1})^{-1}, \\
\mathbf{M}_{k+1} &= \mathbf{P}_k\mathbf{Z}_k\mathbf{P}'_k + \mathbf{Q}_{w(k)}\mathbf{W}_k\mathbf{Q}'_{w,k}, \quad \mathbf{M}_0 = E\{\mathbf{x}_0\mathbf{x}'_0\}, \\
\mathbf{Z}_k &= (\mathbf{I} - \mathbf{K}_k\mathbf{C}_k)\mathbf{M}_k.
\end{aligned} \qquad (13\text{–}97)$$

A major change from the original form is seen in the first (or main) equation of (13–97). We shall derive it, for brevity, for the scalar, stationary case. Since the state x_{k+1} is forced by two inputs, one deterministic (u_k) and one stochastic (w_k), it is convenient to consider two components of x_k,

$$x_k = x_k^d + x_k^n,$$

where the superscripts d and n mean deterministic, d, and noise, n, respectively. It then follows that:

$$x_{k+1}^d = px_k^d + q_u u_k, \qquad x_{k+1}^n = px_k^n + q_w w_k.$$

Similarly, for the output, y_k,

$$y_k = y_k^d + y_k^n,$$

where

$$y_k^d = cx_k^d, \qquad y_k^n = cx_k^n + v_k.$$

The optimal estimate of the stochastic component, x_k^n, is given by the main equation of optimal filtering, Eq. (13–60) for our scalar system, restated as:

$$\hat{x}_{k+1}^n = p\hat{x}_k^n + K_{k+1}(y_{k+1}^n - c p\hat{x}_k^n).$$

Here we make the following substitutions:

Left side: $\hat{x}^n_{k+1} = \hat{x}_{k+1} - x^d_{k+1} = \hat{x}_{k+1} - px^d_k - q_u u_k$,
Right side: $p\hat{x}^n_k = p\hat{x}_k - px^d_k$,

$$y^n_{k+1} = y_{k+1} - y^d_{k+1} = y_{k+1} - cx^d_{k+1} = y_{k+1} - c(px^d_k + q_u u_k),$$
$$cp\hat{x}^n_k = cp(\hat{x}_k - x^d_k),$$

and, rearranging terms, we obtain

$$\hat{x}_{k+1} = (p\hat{x}_k + q_u u_k) + K_{k+1}\{y_{k+1} - c(p\hat{x}_k + q_u u_k)\}.$$

This is the scalar form of the main equation in (13–97).

As was done before, Eq. (13–97) can be transformed for use with a continuous-time system. Let the system be described by:

$$\frac{d\mathbf{x}(t)}{dt} = \mathbf{A}(t)\mathbf{x}(t) + \mathbf{B}_u(t)\mathbf{u}(t) + \mathbf{B}_w(t)\mathbf{w}(t),$$
$$\mathbf{y}(t) = \mathbf{C}(t)\mathbf{x}(t) + \mathbf{v}(t),$$
(13–98)

where $\mathbf{u}(t)$ is deterministic, and, for independent noise vectors $\mathbf{w}(t)$ and $\mathbf{v}(t)$,

$$E\{\mathbf{w}(t)\mathbf{w}'(t+\tau)\} = \mathbf{W}(t)\delta(\tau), \qquad E\{\mathbf{v}(t)\mathbf{v}'(t+\tau)\} = \mathbf{V}(t)\delta(\tau).$$

The optimum filter algorithm for $\hat{\mathbf{x}}(t) = E\{\mathbf{x}(t)\,|\,\mathbf{y}(\tau), 0 \leq \tau \leq t\}$ is

$$\left.\begin{aligned}
\frac{d\hat{\mathbf{x}}(t)}{dt} &= \mathbf{A}(t)\hat{\mathbf{x}}(t) + \mathbf{B}_u(t)\mathbf{u}(t) + \mathbf{K}(t)\bigl(\mathbf{y}(t) - \mathbf{C}(t)\hat{\mathbf{x}}(t)\bigr), \\
\mathbf{K}(t) &= \mathbf{M}(t)\mathbf{C}'(t)\bigl(\mathbf{V}(t)\bigr)^{-1}, \\
\frac{d\mathbf{M}(t)}{dt} &= \mathbf{A}(t)\mathbf{M}(t) + \mathbf{M}(t)\mathbf{A}'(t) \\
&\quad - \mathbf{M}(t)\mathbf{C}'(t)\bigl(\mathbf{V}(t)\bigr)^{-1}\mathbf{C}(t)\mathbf{M}(t) + \mathbf{B}_w(t)\mathbf{W}(t)\mathbf{B}'_w(t), \\
\mathbf{M}(0) &= E\{\mathbf{x}(0)\mathbf{x}'(0)\}, \quad \hat{\mathbf{x}}(0) = \mathbf{0}.
\end{aligned}\right\} \quad (13\text{–}99)$$

Input noise was assumed white in all of the preceding discussions. Since any physical system is band-limited, "white noise" (see Table 13–2(d)) must also be band-limited, and hence not *purely white*. If, however, the noise is not white in the frequency band of importance, we construct a shaping filter that will convert the white noise into the prescribed colored noise. This was done in Example 13–5. By including the filter dynamics in the *P*- or *A*-matrix, as was shown in Example 13–6, the problem can be handled by the preceding algorithms. A serious difficulty arises, however, especially in continuous-time systems where the covariance matrix $\mathbf{V}$ must be inverted, when there is *no noise* in some elements of the measurement $\mathbf{y}$, or the measurement noise is colored. There is a way to avoid this difficulty [22, 23]. To

13-5 Optimum filter algorithms

introduce the basic principle of this procedure, let us first treat a simple example.

Example 13-9 A stationary, continuous-time first-order system is described by

$$\frac{dx}{dt} = Ax + u_1,$$

where u_1 is white noise. The measurement

$$y = Cx + v,$$

however, has a colored noise v. Let us assume that the noise is characterized by the following shaping-filter dynamics:

$$\frac{dv}{dt} = Fv + Fu_2,$$

where F is a constant, and u_2 is white noise. The algorithm (13-92) does not apply for this system because $(V)^{-1}$ becomes singular.

To avoid this difficulty, we *tentatively* differentiate the output with respect to time, so that white noise will appear in the right side,

$$\frac{dy}{dt} = C\frac{dx}{dt} + \frac{dv}{dt} = C(Ax + u_1) + (Fv + Fu_2).$$

Substituting $v = y - Cx$, and rearranging terms, we obtain

$$\frac{dy}{dt} - Fy = (CA - FC)x + (Cu_1 + Fu_2),$$

or, introducing the equivalent measurement $y^* = (dy/dt) - Fy$ and the equivalent noise $v^* = Cu_1 + Fu_2$,

$$y^* = (CA - FC)x + v^*.$$

Since v^* is white, y^* can be used in the optimum filtering algorithm. The main filter equation (top of Eq. (13-92)) for this output is:

$$\frac{d\hat{x}(t)}{dt} = A\hat{x}(t) + K\{y^*(t) - (CA - FC)\hat{x}(t)\}$$

$$= A\hat{x}(t) + K\left\{\frac{dy}{dt} - Fy(t) - (CA - FC)\hat{x}(t)\right\}.$$

We now move the "tentative" term, dy/dt, in the last expression to the left side to obtain the final form for a steady-state condition ($K = $ const.),

$$\frac{d}{dt}\{\hat{x}(t) - Ky\} = \{A - K(CA - FC)\}\hat{x}(t) - KFy(t).$$

As shown in Fig. 13-14, the final form does not involve the time derivative dy/dt of the output y.

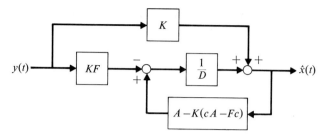

Fig. 13-14 Optimum filter of Example 13-9.

A general procedure outlined in the following (introduced for a stationary case) has been developed by Bryson [22] to deal with stationary and time-varying linear systems for which the matrix inversion in the filter algorithm does not exist.

1. Including a shaping filter as part of the system, write the state equation

$$\frac{d}{dx}\mathbf{x}(t) = \mathbf{A}\mathbf{x}(t) + \mathbf{u}(t), \quad (13\text{-}100)$$

where $\mathbf{u}(t)$ is zero-mean Gaussian white noise with the covariance matrix $\mathbf{U}$,

$$E\{\mathbf{u}(t)\mathbf{u}'(t+\tau)\} = \mathbf{U}\delta(\tau).$$

2. Divide the output $\mathbf{y}$ into two parts, $\mathbf{y}_0$ and $\mathbf{y}_1$, where $\mathbf{y}_0$ is an ideal measurement that has no noise,

$$\mathbf{y}_0 = \mathbf{C}_0\mathbf{x},$$

and $\mathbf{y}_1$ is perturbed by a white-noise vector $\mathbf{v}_1(t)$,

$$\mathbf{y}_1 = \mathbf{C}_1\mathbf{x} + \mathbf{v}_1.$$

3. Differentiate each element y_{0i} of $\mathbf{y}_0$ with respect to time, and replace the resulting dx_j/dt in the right side by the right side of Eq. (13-100). White-noise components (elements of $\mathbf{u}$ in Eq. (13-100)) may or may not eventually appear in the right side of the output derivatives.

4. Let $\mathbf{x}_2$ be the collection of linearly independent elements of $\mathbf{y}_0$ and its time derivatives that do not have white-noise components, and $\mathbf{y}_2$ be the collection of time derivatives of $\mathbf{y}_0$—elements that involve linearly independent white-noise components in the right side. The time derivatives of the output elements that belong to $\mathbf{x}_2$ must be generated by approximate differentiation of the output; for these, the absorption of dy/dt, as in Example 13-9, is not possible.

5. Following the definitions of $\mathbf{x}_2$ and $\mathbf{y}_2$, obtain the matrices $\mathbf{M}_2$, $\mathbf{C}_2$, and $\mathbf{D}$ for

$$\mathbf{x}_2 = \mathbf{M}_2\mathbf{x}, \qquad \mathbf{y}_2 = \mathbf{C}_2\mathbf{x} + \mathbf{D}\mathbf{u}.$$

If $\mathbf{x}_2$ is an m-vector and $\mathbf{x}$ is an n-vector, $m < n$, obtain the $\mathbf{M}_1$ matrix for the $(n - m)$-vector $\mathbf{x}_1$ such that

$$\mathbf{x}_1 = \mathbf{M}_1 \mathbf{x} .$$

The matrices $\mathbf{M}_1$ and $\mathbf{M}_2$ must be so chosen that, in the combined form

$$\begin{bmatrix} \mathbf{x}_1 \\ \mathbf{x}_2 \end{bmatrix} = \mathbf{M}\mathbf{x}, \qquad \mathbf{M} = \begin{bmatrix} \mathbf{M}_1 \\ \mathbf{M}_2 \end{bmatrix}, \tag{13-101}$$

the matrix $\mathbf{M}$ will be nonsingular.

6. Substitute Eq. (13–101) into (13–100) and obtain

$$\frac{d}{dt}\begin{bmatrix} \mathbf{x}_1 \\ \mathbf{x}_2 \end{bmatrix} = \begin{bmatrix} \mathbf{A}_{11} & \mathbf{A}_{12} \\ \mathbf{A}_{21} & \mathbf{A}_{22} \end{bmatrix} \begin{bmatrix} \mathbf{x}_1 \\ \mathbf{x}_2 \end{bmatrix} + \begin{bmatrix} \mathbf{B}_1 \\ \mathbf{B}_2 \end{bmatrix} \mathbf{u} . \tag{13-102}$$

Also combine $\mathbf{y}_1$ and $\mathbf{y}_2$ of the preceding expressions into one to obtain

$$\begin{bmatrix} \mathbf{y}_1 \\ \mathbf{y}_2 \end{bmatrix} = \begin{bmatrix} \mathbf{C}_{11} & \mathbf{C}_{12} \\ \mathbf{C}_{21} & \mathbf{C}_{22} \end{bmatrix} \begin{bmatrix} \mathbf{x}_1 \\ \mathbf{x}_2 \end{bmatrix} + \begin{bmatrix} \mathbf{v}_1 \\ \mathbf{D}\mathbf{u} \end{bmatrix},$$

and convert it into the following form which can be used in the optimal-filter algorithm:

$$\mathbf{y} = \mathbf{C}\mathbf{x}_1 + \mathbf{v}, \tag{13-103}$$

where

$$\mathbf{y} = \begin{bmatrix} \mathbf{y}_1 - \mathbf{C}_{12}\mathbf{x}_2 \\ \mathbf{y}_2 - \mathbf{C}_{22}\mathbf{x}_2 \end{bmatrix}, \qquad \mathbf{C} = \begin{bmatrix} \mathbf{C}_{11} \\ \mathbf{C}_{21} \end{bmatrix} \quad \text{and} \quad \mathbf{v} = \begin{bmatrix} \mathbf{v}_1 \\ \mathbf{D}\mathbf{u} \end{bmatrix}.$$

7. Since it is not necessary to estimate $\mathbf{x}_2$ (for which a noise-free measurement is available), write a (tentative) filter equation for $\mathbf{x}_1$. The state equation for $\mathbf{x}_1$ is given by Eq. (13–102) as

$$\frac{d}{dt}\mathbf{x}_1 = \mathbf{A}_{11}\mathbf{x}_1 + \mathbf{A}_{12}\mathbf{x}_2 + \mathbf{B}_1\mathbf{u} .$$

The filter equation for $\mathbf{x}_1$ is a hybrid of Eqs. (13–92) and (13–99):

$$\frac{d}{dt}\hat{\mathbf{x}}_1 = \mathbf{A}_{11}\hat{\mathbf{x}}_1 + \mathbf{A}_{12}\mathbf{x}_2 + \mathbf{K}(t)(\mathbf{y} - \mathbf{C}\hat{\mathbf{x}}_1) ,$$

$$\mathbf{K}(t) = \left(\mathbf{M}(t)\mathbf{C}' + \mathbf{B}_1\mathbf{R}\right)(\mathbf{V})^{-1}, \tag{13-104}$$

$$\frac{d}{dt}\mathbf{M}(t) = \mathbf{A}_{11}\mathbf{M}(t) + \mathbf{M}(t)\mathbf{A}'_{11} - \mathbf{K}(t)\mathbf{V}\mathbf{K}'(t) + \mathbf{B}_1\mathbf{U}\mathbf{B}'_1 ,$$

where $\mathbf{V}$ in $E\{\mathbf{v}(t)\mathbf{v}'(t + \tau)\} = \mathbf{V}\delta(\tau)$ is positive definite, and

$$E\{\mathbf{u}(t)\mathbf{v}'(t + \tau)\} = \mathbf{R}\delta(\tau) .$$

8. The last step is the absorption of some of the output time derivatives into the left side of the main filter equation (the first of Eqs. (13–104)). Since y_0 is a member of x_2, and y_2 is derived by collecting time derivatives of y_0, it is possible to obtain the matrix H in the relation

$$y_2 = H \frac{dx_2}{dt}. \tag{13-105}$$

Substituting y_2 of Eq. (13–105) into y of the main filter equation, and moving the dx_2/dt component to the left side,

$$\frac{d}{dt}\left\{\hat{x}_1 - K(t)\begin{bmatrix} 0 \\ Hx_2 \end{bmatrix}\right\}$$
$$= A_{11}\hat{x}_1 + A_{12}x_2 + K(t)\begin{bmatrix} y_1 - C_{12}x_2 \\ -C_{22}x_2 \end{bmatrix} - K(t)C\hat{x}_1 - \frac{d}{dt}K(t)\begin{bmatrix} 0 \\ Hx_2 \end{bmatrix} \tag{13-106}$$

is obtained as the final form of the main equation.

REFERENCES

1. Newton, G. C., Gould, L. A., Kaiser, J. F., *Analytical Design of Linear Feedback Control*. New York: J. Wiley, 1957.
2. James, H. M., Nichols, N. B., Phillips, R. S., *Theory of Servomechanisms*. New York: McGraw-Hill, 1947.
3. Lanning, J. H., Jr., Battin, R. H., *Random Processes in Automatic Control*. New York: McGraw-Hill, 1956.
4. Davenport, W. B., Jr., Root, W. L., *Introduction to Random Signals and Noise*. New York: McGraw-Hill, 1956.
5. Mishkin, E., et al., *Adaptive Control Systems*. New York: McGraw-Hill, 1961.
6. Smith, G. L., "The scientific inferential relationships between statistical estimation, decision theory, and modern filter theory," *JACC Preprint 1965*, pp. 350–359.
7. Thal-Larsen, H., "Correlation functions and noise patterns in control analysis," *Trans. ASME*, **80**, No. 2, February 1958, pp. 479–489.
8. Huffmann, D. A., "The synthesis of linear sequential coding networks," *Information Theory*, New York: Academic Press, 1956.
9. Pastel, M. P., "Analyzing random signals in linear systems," *Control Eng.*, **8**, 12, December 1961, pp. 85–89.
10. Murphy, G. J., *Basic Automatic Control Theory*, second edition, Princeton, N. J.: D. v. Nostrand, 1966.
11. Henning, T., "Testing for plant transfer functions in presence of noise and nonlinearity," II, *Control Eng.*, **10**, 9, September 1963, pp. 119–124.

12. Vander Velde, W. E., "Making statistical studies on analog simulations," *Control Eng.*, **7**, 6, June 1960, pp. 127–130.
13. Pierre, D. A., "Minimum mean-square-error design of distributed-parameter control system," *JACC Preprint 1965*, pp. 381–389.
14. Seifert, W. W., Steeg, C. W., Jr., (eds.) *Control Systems Engineering*. New York: McGraw-Hill, 1960.
15. Kelley, C. R., "Predictor instruments look into the future," *Control Eng.*, **9**, 3, March 1962, pp. 86–90.
16. Fargel, L. C., Ulbrich, E. A., "Predictor displays extend manual operation," *Control Eng.*, **10**, 8, August 1963, pp. 57–60.
17. Wiener, N., *The Extrapolation, Interpolation and Smoothing of Stationary Time Series*. New York: J. Wiley, 1949.
18. Kalman, R. E., "A new approach to linear filtering and prediction problems." *J. Basic Eng., Trans. ASME*, Series D, **82**, March 1960, pp. 33–45.
19. Kalman, R. E., Bucy, R. S., "New results in linear filtering and prediction theory," *J. Basic Eng., ASME Trans.*, Series D, **83**, March 1961, pp. 95–101.
20. Richman, J., Thau, F., "Elements of the Kalman filtering technique," *Tech. News. Bulletin*, General Precision, Inc., Little Falls, New Jersey, **9**, No. 3, 3rd Quarter, 1966, pp. 13–18.
21. Sage, A. P., Optimum System Control, New Jersey: Prentice-Hall, 1968.
22. Bryson, A. E., Jr., Johansen, D. E., "Linear filtering for time-varying systems using measurements containing colored noise," *IEEE Trans. on Aut. Control*, January 1965, pp. 4–10.
23. Cox, H., "Estimation of state variables via dynamic programming." *JACC Preprint Volume*, 1964, pp. 376–381.

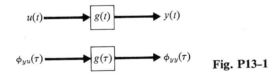

Fig. P13-1

PROBLEMS

13-1 Show that the following relation holds for a linear system's stationary random input-output correlation function (see Fig. P13-1):

$$\phi_{yy}(\tau) = \int_{-\infty}^{\infty} \phi_{yu}(\tau - \eta) g(\eta) \, d\eta$$

where

$$\phi_{yu}(\tau) = \phi_{uy}(-\tau) .$$

13-2 A first-order system is given by

$$G(s) = \frac{1}{s + a} \quad \text{where} \quad a > 0 .$$

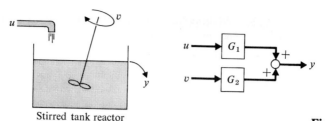

Stirred tank reactor

Fig. P13-3

Input to the system is white noise described by
$$\phi_{uu}(\tau) = \delta(\tau) .$$
Determine the correlation functions $\phi_{uy}(\tau)$ and $\phi_{yy}(\tau)$.

13-3 Two random inputs $u(t)$ and $v(t)$ are combined to produce a random output $y(t)$. For instance, $u(t)$ = fluctuation in feed concentration, $v(t)$ = fluctuation in mixing, and $y(t)$ = fluctuation in product concentration. A linear relation shown in the block diagram of Fig. P13-3 is assumed, and $u(t)$ and $v(t)$ have zero mean,
$$E\{u\} = 0 , \qquad E\{v\} = 0 .$$
Show that
$$\phi_{uy}(\tau) = \int_{-\infty}^{+\infty} g_1(\xi)\phi_{uu}(\tau - \xi) \, d\xi + \int_{-\infty}^{\infty} g_2(\xi)\phi_{uv}(\tau - \xi) \, d\xi .$$

13-4 Input u to a linear system $G(s)$ is a white noise such that
$$\phi_{uu}(j\omega) = 1 .$$
The input-output cross-correlation is
$$\phi_{ux}(\tau) = \frac{1}{T} e^{-\tau/T} \quad \text{for} \quad \tau > 0 ;$$
otherwise
$$\phi_{ux}(\tau) = 0 .$$
Identify $G(j\omega)$ in the frequency domain.

13-5 The autocorrelation function of a random noise has an oscillatory pattern,
$$\phi_{xx}(\tau) = e^{-a|\tau|} \cos (b |\tau| + c) .$$
Compute its power-density spectrum.

13-6 Input to a linear filter $G(s)$ is a white noise such that
$$\phi_{uu}(\tau) = \delta(\tau) .$$
Output x of the filter is an average of input u over the last one second:
$$x(t_1) = \int_{t_1-1}^{t_1} u(t) \, dt \quad \text{for all} \quad t_1 .$$

a) Find $G(s)$.
b) Determine the input-output cross-correlation function $\phi_{ux}(\tau)$.

c) Compute the output power-density spectrum $\phi_{xx}(j\omega)$, and mean squared output $\overline{x^2}$.

13-7 A first-order system is described by

$$x_{k+1} = 0.6x_k + 0.8u_k ,$$

where the rms value of u_k is σ_u and

$$E\{u_k u_j\} = 0 \quad \text{for all} \quad k \neq j .$$

Determine $E\{x_k x_{k+n}\}$ as a function of n (integer, plus or minus, including zero) and sketch the result.

13-8 Let $p(\mathbf{x})$ be the probability-density function for Gaussian random-vector variable $\mathbf{x}$ of dimension two, where x_1 and x_2 (the elements of $\mathbf{x}$) are independent, with mean values μ_1, μ_2 and rms values σ_1 and σ_2, respectively. Show that $p(\mathbf{x})$ decouples into the product of two functions,

$$p(\mathbf{x}) = p(x_1)p(x_2) .$$

13-9 Determine the output autocorrelation function $\phi_{yy}(\tau)$ of Problem 13-2 by using the state-space approach.

13-10 The input autocorrelation function of a first-order system,

$$G(s) = \frac{1}{s + a_2} ,$$

is

$$e^{-a_1|\tau|}/(2a_1) ,$$

where a_1, a_2 are positive constants. Determine the variance of the output.

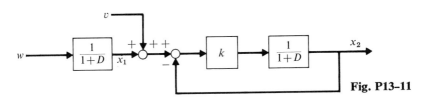

Fig. P13-11

13-11 In the system of Fig. P13-11, w and v are Gaussian white noise with zero mean, such that

$$E\{w(t)w(t + \tau)\} = 0.64\delta(\tau) , \quad E\{v(t)v(t + \tau)\} = 0.36\delta(\tau) ,$$
$$E\{v(t)w(t)\} = 0 .$$

The error e is defined to be

$$e = x_1 - x_2 .$$

Determine k for which $E\{e^2\}$ will be minimal.

13-12 A first-order system is described by $dx/dt = -x + w$, and its measurement is $y = x + v$, where w and v are zero-mean, Gaussian white noise for which the ex-

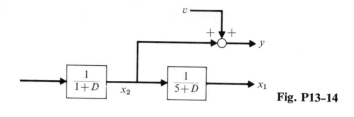

Fig. P13-14

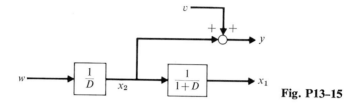

Fig. P13-15

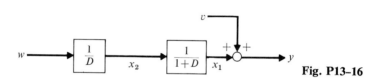

Fig. P13-16

pected values given in Problem 13-11 apply. Construct an optimal filter for the steady state, draw its block diagram, and discuss the relation of this problem to Problem 13-11.

13-13 A deterministic signal $u(t)$ also acts on the system of Problem 13-12, so that the state equation of the preceding problem is modified into the following form:

$$\frac{dx}{dt} = -x + u + w.$$

Determine the optimal filter for the steady state, and draw a block diagram of the system with the filter.

13-14 An unobservable system is given in Fig. P13-14, where w and v are white noise as specified in Problem 13-11. Obtain the K-matrix of the optimum filter for the steady state.

13-15 A system is given in Fig. P13-15, where

$$E\{w(t)w(t+\tau)\} = \sigma_w^2 \delta(\tau), \qquad E\{v(t)v(t+\tau)\} = \sigma_v^2 \delta(\tau), \qquad E\{w(t)v(t)\} = 0,$$

and $\sigma_w/\sigma_v = r > 0$. Obtain a block diagram for the optimal filter in the steady state, where r will appear as a parameter.

13-16 The system of Problem 13-15 is modified to the form shown in Fig. P13-16. Obtain the block diagram of the optimum filter in the steady state.

13-17 An optimal filter applied to a first-order object

$$x_{k+1} = 0.6k_k + 0.8u_k, \qquad y_k = x_k + v_k,$$

where

$$E\{u_k^2\} = 5, \qquad E\{v_k^2\} = 1,$$

is in a stationary state, that is, as shown Example 13-7,

$$\hat{x}_k = 0.6\hat{x}_{k-1} + 0.775(y_k - 0.6\hat{x}_{k-1}).$$

Let $T_p = NT =$ prediction time and $x^p =$ predicted state; obtain x^p for $N = 1$ and the variance of the prediction error. If y_k at $t = kT$ is used as x^p at $t = (k+1)T$, what would be the variance of the prediction error? Also find the variance of the prediction error for $N = 2$.

13-18 The continuous-time system of Fig. P13-16 is made into a discrete-time system by putting a sample-holder unit at the left end of the block diagram. For a fixed sampling period T, the discrete-time formulation of the system is

$$\mathbf{x}_{k+1} = \begin{bmatrix} e^{-T} & (1-e^{-T}) \\ 0 & 1 \end{bmatrix} \mathbf{x}_k + \begin{bmatrix} e^{-T} + T - 1 \\ T \end{bmatrix} w_k,$$

$$y_k = \begin{bmatrix} 1 & 0 \end{bmatrix} \mathbf{x}_k + v_k,$$

where

$$E\{w_k^2\} = \sigma_w^2, \qquad E\{w_i w_j\} = 0 \qquad \text{for all } i \neq j,$$

and there is no measurement noise, $v_k = 0$. Without using the algorithm (13-87), obtain $\mathbf{K}$ such that it will give a deterministic identification of the state.

13-19 If $T = 0.5$ in Problem 13-18,

$$e^{-T} = 0.60653 \approx 0.60$$

so that $\mathbf{P}$ and $\mathbf{q}$ in the preceding problem will reduce to

$$\mathbf{P} = \begin{bmatrix} 0.6 & 0.4 \\ 0 & 1 \end{bmatrix}, \qquad \mathbf{q} = \begin{bmatrix} 0.1 \\ 0.5 \end{bmatrix}.$$

Determine $\mathbf{K}$ and $\mathbf{Z}$ of Eq. (13-89) (the steady-state case) for this problem.

13-20 Apply steps (1) through (8) of Section 13-5 and obtain Eqs. (13-100) through (13-106) for the system of Example 13-9.

14

OPTIMAL CONTROL

The word "optimal" in its general sense means "best" or "most desirable." In controls, as in other endeavors, we are always searching for the best solution to the problem at hand. In earlier chapters, we spoke of the Ziegler-Nichols method (and other methods) for finding "optimum controller settings," which is a process of *parameter optimization*. In this chapter, we will deal with a more general problem: finding the best control *strategy* for a given system and cost function (an expression to be minimized or maximized for "best" system operation). In determining an optimal control strategy, we do not make any prior assumptions or commitments that would fix the controller structure; on the other hand, in the parameter-optimization problem, we first fix the structure of the controller (for example, a PI-controller); *then* we attempt to determine the optimum parameters. We assume in this chapter that the control object—its inputs, its outputs, and its dynamic structure and parameters—is known. Our aim is to find a *function*, which, when applied as the system input, will optimize the system's performance with respect to some cost criterion. Sometimes that function will depend on the system output; in that case we have a feedback solution. In other cases the function does not depend on a measurement of the system output.

Although the variational viewpoint is the basis from which we seek an optimal strategy, its mathematical difficulty and computational complexity are so great that real progress in this field started only after major breakthroughs by Bellman (dynamic programming) and Pontryagin (the maximum principle). Since then, the problem has attracted the attention of many control mathematicians, who have offered formal, rigorous, and general, but often very abstract, contributions. Even though we can often obtain solutions only by using simplified system models and simple performance indices (that is, cost-function integrals), the optimal control approach can nevertheless be an important tool for engineers, for example, by pointing the way to better controller structures, by indicating possible switching functions, and by providing a measure of performance so that a designer can determine whether there is any sizable room for improvement in his design. Engineers often prefer simplified system models rather than complex ones for realistic instrumentation design.

Since the main purpose of this chapter is to introduce optimal control theory, we will avoid mathematical formality and rigorous proofs in favor of a presentation of the basic principles: the maximum principle, its meaning, some applications, feedback realization of linear optimal control, and an introduction to stochastic optimal control.

14-1 PROBLEM STATEMENT AND PROPERTIES OF OPTIMAL TRAJECTORIES

A typical optimal control problem begins with the following statements (see [1] through [10]):

1. The mathematical model of a deterministic system is known to be

$$\frac{d\mathbf{x}}{dt} = \mathbf{f}(\mathbf{x}, \mathbf{u}), \qquad (14\text{-}1)$$

where the state $\mathbf{x}$ is an n-vector for an nth order lumped-parameter object and $\mathbf{f}$ is a vector function of $\mathbf{x}$ and the control vector $\mathbf{u}$.

2. The initial state $\mathbf{x}(0)$ is given, and the final state $\mathbf{x}(t_1)$ at some time t_1 is specified; the time t_1 is not known beforehand.

3. A performance index J is defined:

$$J = \int_0^{t_1} f_0(\mathbf{x}, \mathbf{u}) \, dt, \qquad (14\text{-}2)$$

where $f_0(\mathbf{x}, \mathbf{u})$ is a prescribed cost function.

4. If there are constraints in $\mathbf{u}$ and/or $\mathbf{x}$, the solution must not violate them. In other words, $\mathbf{u}$ and $\mathbf{x}$ must stay in the admissible regions U and X, respectively,

$$\mathbf{u} \in U, \qquad (14\text{-}3)$$

$$\mathbf{x} \in X. \qquad (14\text{-}4)$$

If, for instance, u_1 and u_2 are openings of two control valves, and the valve strokes are limited, then U in Eq. (14-3) is a rectangular region in the $u_1 u_2$-plane (Fig. 1-6).

5. It is required to determine an optimal control $\mathbf{u}^0$, and the time t_1, that will *minimize* the performance index Eq. (14-2), while moving the state from $\mathbf{x}(0)$ to the specified "target" $\mathbf{x}(t_1)$, subject to the constraints of Eqs. (14-3) and (14-4).

Since maximization problems can be treated with only slight modifications, we shall consider only minimization problems. We will also discuss some variations to the optimal control problem.

Example 14-1 The problem statement for time-optimal control of a second-order purely inertial object is:

1. A linear inertial object is described by

$$\frac{dx_1}{dt} = x_2, \qquad \frac{dx_2}{dt} = u, \qquad (14\text{-}5)$$

where x_1 = displacement, x_2 = velocity, u = controlling force, and the scales of variables are normalized so that the proportionality constant for mass is unity.

2. The system is initially in equilibrium $[x_2(0) = 0]$ with a deviation from zero position $[x_1(0) = h]$. The final state must be in equilibrium with zero deviation:

$$\mathbf{x}(t_1) = \mathbf{0} \ .$$

3. The performance index is time t_1. Since we have

$$J = \int_0^{t_1} f_0(\mathbf{x}, u) \, dt = \int_0^{t_1} dt = t_1 \ ,$$

$f_0(\mathbf{x}, \mathbf{u})$ for time-optimal control is

$$f_0(\mathbf{x}, \mathbf{u}) = 1 \ . \tag{14-6}$$

4. There is no constraint in $\mathbf{x}$, but the controlling force is limited to:

$$|u| \leqslant 1 \ . \tag{14-7}$$

5. The unknowns to be determined are $u^0(t)$ for $0 \leqslant t \leqslant t_1$ for which t_1 is minimal, and the value of t_1 itself.

Bellman's dynamic programming [9] is one of the best known methods for dealing with optimal control problems. The method is based on the "*principle of optimality*," which states:

"An optimal policy, or optimal control strategy, has the property that, whatever the initial state and the initial decision, the remaining decisions must form an optimal control strategy with respect to the state resulting from the first decision."

This principle can best be explained in the discretized state space. A set of discretized states is shown by $a, b_1, b_2, b_3, c_1, c_2$, etc., in Fig. 14-1. The problem is to bring the system from the initial state e to the final state a. We consider only those trajectories that pass through these states. Let us suppose that the trajectory segment $c_3 - b_2 - a$ is an optimal one. Then this segment remains optimal regardless of the motion that takes the initial state to c_3. If optimal trajectories from states d_1 and d_2 both go through the state c_3, then $d_1c_3b_2a$ and $d_2c_3b_2a$ are both optimal. It never happens that $d_2c_3b_2a$ is optimal for initial state d_2 while $d_1c_3b_1a$ is optimal for a different initial state d_1.

Because of this principle, the number of trials in a numerical search for an optimal trajectory can be drastically reduced [10]. In other words, dynamic programming in a discrete state space can be regarded as an ingenious way of programming a digital computer. The computations proceed in a backwards fashion, from the final state, as shown in the following example. In this particular example, however, a similar number of computations would result from a forward-moving scheme, due to the simplicity of the control problem.

Figure 14-1 shows a dynamic-programming approach to the time-optimal problem stated in Example 14-1, where the initial state is at e and

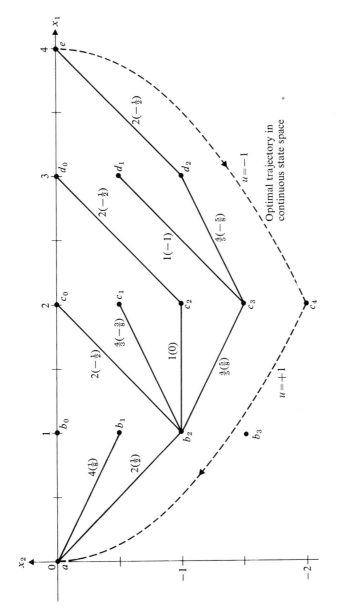

Fig. 14-1 Search for an optimal trajectory by dynamic programming in the discretized state space.

the final state is a (the origin). Time (incremental cost) and control for each incremental move are shown along each branch, outside and inside the parentheses, respectively. For instance, the motion from b_1 to a requires incremental time $\Delta t = 4$ and a constant control $u = \frac{1}{8}$ (see Example 14–2 below for the computation of these values). We begin with the last section of trajectories that hit the specified final state a. Out of four conceivable intermediate states, b_0, b_1, b_2, b_3, we reject b_0 because the time required to move from b_0 to a is infinity. Also, we reject b_3 because the control u for a move from b_3 to a is $\frac{9}{8}$ and the prescribed limit on u is $|u| \leqslant 1$. Therefore, the first computation comes at the last stage, and only two possibilities have to be computed: $b_1 a$ and $b_2 a$.

We next compute the two-stage decision process, from $x_1 = 2$ via $x_1 = 1$ to the origin. From state c_0 there are two possible routes, either via b_1 or via b_2. The time Δt from c_0 to b_1 is 4 (not shown in the figure) while it is 2 for $c_0 b_2$. The *total* time from c_0 to a is 8 for the trajectory $c_0 b_1 a$, while it is 4 for $c_0 b_2 a$. Therefore, $c_0 b_2$ is optimal, and we shall never consider the route $c_0 b_1$ again. In this particular example the incremental cost Δt is smaller on $c_0 b_2$ than the alternate, but this is not the reason we take $c_0 b_2$. Imagine, for example, a fictitious case where $\Delta J = 1$ for $c_0 b_1$ and it is 2 for $c_0 b_2$. The total cost of $c_0 b_1 a$ would be $1 + 4 = 5$ and the cost for $c_0 b_2 a$ would be $2 + 2 = 4$; thus the total cost would be smaller over the latter route and we would take it as an optimal, although $c_0 b_1$ is better than $c_0 b_2$ on the first step in this artificial case. We repeat the computations for other states, c_1, c_2, and c_3, and find that all the optimal trajectories go through b_2.

In the third stage of computation, the starting value of x_1 is 3. From d_0 we find that the total cost is minimized when the trajectory passes through c_2 rather than c_1, while c_3 is unreachable due to the constraint in u. For both d_1 and d_2 the trajectories via c_3 are preferred rather than those via c_2 or c_1.

Finally, starting from the prescribed initial state e, the total cost (time) is compared for the two possibilities, via d_1 and via d_2, and we find that the latter is better. Thus the optimal trajectory for the discretized states is e–d_2–c_3–b_2–a, and the optimal control consists of four stages; $u = -\frac{1}{2}$, $-\frac{5}{8}$, $+\frac{5}{8}$, and $+\frac{1}{2}$ (a control process like this is generally called a multistage decision process).

Example 14–2 Let us solve the time-optimal control of Example 14–1, in the discretized state space (Fig. 14–1) as a multistage decision process. Note that this is not a sampled-data system and also that the straight-line segments in Fig. 14–1 are not trajectories in the continuous state space.

Let $\mathbf{x}_k$ and $\mathbf{x}_{k+1}$ be the kth and $(k+1)$th discretized states, respectively, and let Δt_k be the time required for the transition from $\mathbf{x}_k$ to $\mathbf{x}_{k+1}$ with constant control u_k. Then we have, by integration of Eq. (14–5),

$$x_{2,k+1} = x_{2,k} + u_k \Delta t_k ,$$
$$x_{1,k+1} = x_{1,k} + x_{2,k}\Delta t_k + u_k(\Delta t_k)^2/2 .$$

14-1 Problem statement and properties of optimal trajectories

Solving for Δt_k and u_k, we have

$$\Delta t_k = \frac{2(x_{1,k+1} - x_{1,k})}{x_{2,k+1} + x_{2,k}}, \quad u_k = \frac{x_{2,k+1}^2 - x_{2,k}^2}{2(x_{1,k+1} - x_{1,k})}. \tag{14-8}$$

The values in Fig. 14-1 are computed from Eq. (14-8). The optimal trajectory for the continuous-state-space problem is shown by dashed lines in the same figure, for which $u = -1$ for the first 2 seconds, and $u = +1$ for the last 2 seconds. The optimal trajectory in the discretized space will approach the continuous-space optimal trajectory as the discretization steps are made finer.

Pontryagin's maximum principle gives a *necessary condition* that an optimal control must satisfy. Since it is a condition rather than an algorithm for computing an optimal control, the computation to determine an optimal control solution is generally not straightforward. Moreover, in some cases optimality of a chosen control must be confirmed by the J-value it yields because the maximum principle is really a local search for an optimal trajectory and, thus, a necessary but not sufficient condition. As an introduction to the maximum principle, we shall show its formulation for a second-order system (the general approach is given in the next section).

The first step is to write a vector equation in which the system's dynamics and the cost functional are combined. We assign the zeroth-state variable x_0 to the performance index given by Eq. (14-2) and write three scalar first-order differential equations for a second-order system:

$$\frac{dx_0}{dt} = f_0(\mathbf{x}, \mathbf{u}), \quad \frac{dx_1}{dt} = f_1(\mathbf{x}, \mathbf{u}), \quad \frac{dx_2}{dt} = f_2(\mathbf{x}, \mathbf{u}). \tag{14-9}$$

In the next step, we introduce a three-dimensional *covariant vector* $\mathbf{\Psi}$. This vector is defined by the following relation:

$$\begin{aligned}
\frac{d\psi_0}{dt} &= -\psi_0 \frac{\partial f_0}{\partial x_0} - \psi_1 \frac{\partial f_1}{\partial x_0} - \psi_2 \frac{\partial f_2}{\partial x_0}, \\
\frac{d\psi_1}{dt} &= -\psi_0 \frac{\partial f_0}{\partial x_1} - \psi_1 \frac{\partial f_1}{\partial x_1} - \psi_2 \frac{\partial f_2}{\partial x_1}, \\
\frac{d\psi_2}{dt} &= -\psi_0 \frac{\partial f_0}{\partial x_2} - \psi_1 \frac{\partial f_1}{\partial x_2} - \psi_2 \frac{\partial f_2}{\partial x_2}.
\end{aligned} \tag{14-10}$$

We then define the scalar product (or Hamiltonian) H by

$$H = f_0 \psi_0 + f_1 \psi_1 + f_2 \psi_2. \tag{14-11}$$

The control $\mathbf{u}$ is involved in the right-hand side of H; hence H depends on $\mathbf{u}$. We stress this dependency by writing $H(\mathbf{u})$ for H. When $\mathbf{u}$ is optimal, $\mathbf{u}^0$, then $H(\mathbf{u})$ takes a maximum (or supremum) value, and moreover, that maximum value is zero:

$$H(\mathbf{u}^0) = \max_{\mathbf{u} \in U} H(\mathbf{u}) = 0. \tag{14-12}$$

This is the maximum principle, and it gives a necessary condition for $\mathbf{u}^0$ to be optimal. In the next section we shall discuss the maximum principle in some detail. Here we concern ourselves with interpreting its importance by showing a simple application.

There is no boundary condition stated for the covariant vector $\boldsymbol{\Psi}$; however, since Eq. (14–10) is homogeneous in ψ_i, $i = 0, 1$, and 2, we can choose an arbitrary proportionality constant for them. Noting that x_0 is the performance index and that, therefore, it will not appear in the right-hand functions f_1, f_2, and f_0, it follows that

$$\frac{d\psi_0}{dt} = 0. \tag{14-13}$$

Therefore we can freely choose a constant value for ψ_0. As we shall see later, the maximum principle requires that ψ_0 be negative. Therefore we let

$$\psi_0 = -1. \tag{14-14}$$

If ψ_0 were chosen to be positive, the principle would have to be called the minimum principle (see Section 14–2).

Example 14-3 We shall again look at the system formulated in Example 14-2. We have $f_0 = 1$, $f_1 = x_2$, and $f_2 = u$. Hence $\psi_0 = -1$ and

$$\frac{d\psi_1}{dt} = 0, \quad \text{or} \quad \psi_1 = \text{const} = C_1,$$

$$\frac{d\psi_2}{dt} = -\psi_1, \quad \text{or} \quad \psi_2 = C_2 - C_1 t,$$

where C_2 is a constant. Applying these conditions in Eq. (14–11), we find H to be

$$H = -1 + C_1 x_2 + (C_2 - C_1 t) u.$$

According to the maximum principle, H must be maximized with respect to u; hence for $|u| \leqslant 1$,

$$u = u^0 = \text{sgn}\,(C_2 - C_1 t). \tag{14-15}$$

In other words, the optimal control must be bang-bang, and $\psi_2 = (C_2 - C_1 t)$ will act as the switching function.

The theorem also states that the maximum value of H is zero. To make use of this condition we consider the relation at $t = 0$. Since $x_2(0) = 0$, H at $t = 0$ reduces to

$$H\,|_{t=0} = -1 + C_2 u.$$

Intuitively, we can see that for this inertial system, in which u represents an acceleration, to drive a state

$$x_1(0) = 4, \quad x_2(0) = 0$$

to the origin in minimal time, we must first apply $u = -1$. We have no reason to

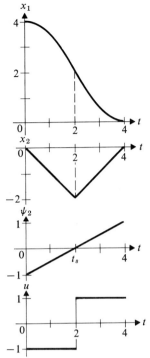

Fig. 14-2 Time optimal control of an inertial object.

use $u = +1$ at the beginning. With $u(0^+) = -1$ and $H = 0$, C_2 is fixed as

$$C_2 = -1.$$

Unfortunately there is no direct way to fix the other constant, C_1. We find it by the following reasoning: since

$$\psi_2 = -1 - C_1 t,$$

it can change sign only once if C_1 is negative. Let

$$t = t_s$$

be the switching instant; then

$$\psi_2(t_s) = -1 - C_1 t_s = 0,$$

or

$$C_1 = -\frac{1}{t_s}.$$

Thus C_1 can be fixed if t_s is known. For the trajectory shown by dashed lines in Fig. 14-1, both the initial and final conditions are satisfied if and only if

$$t_s = 2.$$

Thus the problem is completely solved (Fig. 14-2). The motion for $0 \leqslant t < t_s$ is

$$x_1 = 4 - \tfrac{1}{2}t^2, \qquad x_2 = -t, \tag{14-16}$$

and the remaining motion, when $u = +1$, is

$$x_1 = \tfrac{1}{2}(t-4)^2, \qquad x_2 = t - 4. \tag{14-17}$$

The final time t_1 is found to be:

$$t_1 = 4.$$

To confirm that max $H = 0$ over the entire period, $0 \leqslant t \leqslant 4$, we substitute the set $u = -1$ and $x_2 = -t$ and find

$$H = -1 - \tfrac{1}{2}x_2 + (\tfrac{1}{2}t - 1)u = 0.$$

Also $H = 0$ for the set $u = +1$ and $x_2 = t - 4$. Thus the switching control just determined satisfies the maximum principle, and, since there is no other control that brings the state to the origin with one switching, it is the optimal control.

The problem of the preceding example involved solving two first-order vector equations,

$$\frac{d\mathbf{X}}{dt} = \mathbf{F}(\mathbf{x}, \mathbf{u}), \qquad \frac{d\mathbf{\Psi}}{dt} = -\mathbf{M}'_J \mathbf{\Psi},$$

where

$$\mathbf{X} = \begin{bmatrix} x_0 \\ x_1 \\ x_2 \end{bmatrix}, \qquad \mathbf{\Psi} = \begin{bmatrix} \phi_0 \\ \phi_1 \\ \phi_2 \end{bmatrix}$$

and $\mathbf{M}_J$ is the Jacobian, defined by Eq. (14-22) below. Since $\mathbf{u}$ is fixed by the maximum condition, the two equations can be solved if six constants are known by boundary values. We have

$$\mathbf{x}(0) = \text{given}, \qquad \mathbf{x}(t_1) = \text{specified}, \qquad x_0(0) = 0$$

and

$$\phi_0 = -1;$$

thus six conditions are available (when $\mathbf{x}$ is two-dimensional) but t_1 is not known, for which one more condition,

$$\max H = 0 \quad \text{at} \quad t = 0 \quad \text{or} \quad t = t_1$$

becomes available. Therefore the number of unknowns matches the number of equations, and it is mathematically true that distinct solutions will exist if any solutions exist. Finding the solution(s) is a different matter, and is generally difficult. The problem of the existence of solutions is related to the existence of an optimal control. There are engineering (and theoretical) problems in which no optimal control exists. Such a system could be called degenerate. Although at present there is no clear-cut general answer to the problem of existence (of optimal control), engineers are usually motivated

14-1 Problem statement and properties of optimal trajectories   633

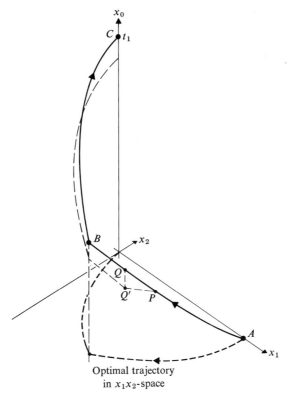

Fig. 14-3 Optimal trajectory in $(n + 1)$-dimensional space.

to find an optimal control because they feel that there should be an optimal way to control. This feeling is often related to some clue that a system will not be degenerate.

A three-dimensional vector space for the vector **X** was introduced for the second-order object in the preceding example. This implies that an optimal trajectory will be investigated in a state space in which the cost variable x_0 is added to the object's state variables. For instance, the time-optimal trajectory of the inertial object treated in Example 14-3 will appear as shown in Fig. 14-3, in a three-dimensional space, where the third axis x_0 represents time.

An important property of an optimal trajectory can be stated in this $(n + 1)$-dimensional space of an nth-order object. The property is: let ABC in Fig. 14-3 be an optimal trajectory; then any portion of it, such as PQ, is also optimal. The proof of this statement is straightforward. If PQ in the figure is not optimal, then there must exist a better control which will take the state from P to Q', where Q' has a smaller value of x_0 than Q. If this is the case, the rest of the trajectory, QBC, can be replaced by the same one

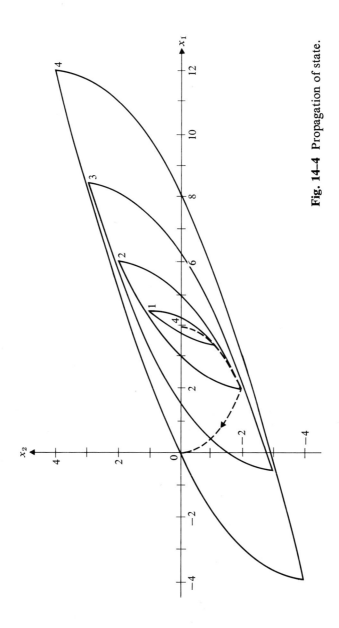

Fig. 14-4 Propagation of state.

starting from Q', by which we would obtain the result,

$$J = \min J - \overline{Q'Q} \ .$$

This violates the statement that ABC is an optimal trajectory. Therefore PQ must be optimal by itself.

We conclude this section with a remark on an analogy between control processes and wave-propagation processes. In the control of an inertial object by a control force subject to the constraint $|u| \leqslant 1$, the state of the object, which starts from a prescribed initial state, for instance

$$x_1(0) = 4, \qquad x_2(0) = 0 \ ,$$

will have a propagation "wave" in the state space. For example, all possible states at $t = 1$ will fill the leaf-pattern space labeled 1 in Fig. 14–4 (see Section 12–5 for a similar mapping concept). The reachable region will expand as time proceeds to 2, 3, and 4. The "wave fronts" for these instants are labeled 2, 3, and 4, respectively.

The boundary of the region 4 touches the origin (the target state). This provides a direct proof that $t_1 = 4$ is the minimal time in which the motion can reach the origin. The optimal trajectory (in the system's state space) can be determined by tracing the state from the origin (endpoint) backward in time, as indicated by dashed lines in the figure.

14–2 THE MAXIMUM PRINCIPLE

We shall now present the theorem statement of the maximum principle for the type of control problems discussed in the preceding section, and then give a heuristic explanation of the theorem instead of a rigorous theoretical proof.

Let us consider that an nth-order control object and a specified cost function are given by the following $(n + 1)$-dimensional vector equation

$$\frac{d\mathbf{X}}{dt} = \mathbf{F}(\mathbf{x}, \mathbf{u}) \tag{14–18}$$

or

$$\frac{dx_i}{dt} = f_i(x_1, \ldots, x_n, u_1, \ldots, u_r) \ , \qquad i = 0, 1, \ldots, n, \tag{14–19}$$

where the first state variable x_0 is for the performance index J,

$$J = \int_0^{t_1} f_0(\mathbf{x}, \mathbf{u}) \, dt \ . \tag{14–20}$$

The rest of the state variables $x_1, \ldots, x_n$ are elements of the state vector $\mathbf{x}$ of the control object. The control vector $\mathbf{u}$ is r-dimensional. It may have a constraint

$$\mathbf{u} \in U. \tag{14-21}$$

The problem is to find an optimal control $\mathbf{u}^0$ subject to the constraint of Eq. (14–21), which will transfer a given initial state $\mathbf{x}(0)$ to a specified final state $\mathbf{x}(t_1)$ at some time t_1, while minimizing the performance index J.

To formulate the theorem for this problem, we introduce an $(n+1)$-dimensional covariant (auxiliary, supplementary, costate, or adjoint are synonymously used) vector $\mathbf{\Psi}$ which is defined by the following linear and homogeneous relation:

$$\frac{d\mathbf{\Psi}}{dt} = -\begin{bmatrix} \frac{\partial f_0}{\partial x_0} & \frac{\partial f_1}{\partial x_0} & \cdots & \frac{\partial f_n}{\partial x_0} \\ \vdots & & & \vdots \\ \frac{\partial f_0}{\partial x_n} & \frac{\partial f_1}{\partial x_n} & \cdots & \frac{\partial f_n}{\partial x_n} \end{bmatrix} \mathbf{\Psi} . \tag{14-22}$$

Equation (14–18) and the costate vector $\mathbf{\Psi}$ of Eq. (14–22) are now combined into a scalar function by the Hamiltonian H such that

$$H = \mathbf{\Psi}'\mathbf{F} = \mathbf{\Psi}' \frac{d\mathbf{X}}{dt} = \sum_{i=0}^{n} \psi_i f_i(\mathbf{x}, \mathbf{u}) . \tag{14-23}$$

The Hamiltonian is regarded as a function of $\mathbf{u} \in U$. Let the supremum (maximum) value of this function with respect to $\mathbf{u}$ be denoted by $M(\mathbf{\Psi}, \mathbf{x})$:

$$M(\mathbf{\Psi}, \mathbf{x}) = \sup_{\mathbf{u} \in U} H(\mathbf{\Psi}, \mathbf{x}, \mathbf{u}) . \tag{14-24}$$

The statement of the maximum principle is: Let $\mathbf{u}(t)$, $0 \leq t \leq t_1$, be an admissible control that moves the state from a given initial state $\mathbf{x}(0)$ to a prescribed final state $\mathbf{x}(t_1)$ at some time t_1. In order that $\mathbf{u}^0(t)$ and the resulting trajectory $\mathbf{X}^0(t)$ in the $(n+1)$-space be optimal, it is *necessary* that there exist a nonzero continuous vector $\mathbf{\Psi}^0(t)$ that corresponds to $\mathbf{u}^0$ and $\mathbf{X}^0$ such that, for every t, $0 \leq t \leq t_1$,

1. the function H of the variable $\mathbf{u} \in U$ attains its maximum at the point $\mathbf{u} = \mathbf{u}^0$,

$$H(\mathbf{\Psi}^0(t), \mathbf{x}^0(t), \mathbf{u}^0(t)) = M(\mathbf{\Psi}^0(t), \mathbf{x}^0(t)) ; \tag{14-25}$$

2. the final value of M is zero, while the final value of ψ_0 is not positive. Furthermore, it turns out that both M and ψ_0 are constant over the entire period $0 \leq t \leq t_1$. Therefore,

$$M(\mathbf{\Psi}^0(t), \mathbf{x}^0(t)) = 0 , \qquad 0 \leq t \leq t_1 , \tag{14-26}$$

and since Eq. (14–22) is homogeneous, we can assign an arbitrary negative constant for ψ_0; thus

$$\psi_0 = -1, \quad 0 \leqslant t \leqslant t_1. \tag{14-27}$$

It is possible to rewrite the equations for **X** and **Ψ** in terms of the Hamiltonian H. We can immediately verify that the following expressions reduce to the original equations, (14–19) and (14–22), respectively:

$$\frac{dx_i}{dt} = \frac{\partial H}{\partial \psi_i}, \quad \frac{d\psi_i}{dt} = -\frac{\partial H}{\partial x_i}, \tag{14-28}$$

where $i = 0, 1, \ldots, n$. The formulation simplifies for a linear control object,

$$\frac{d\mathbf{x}}{dt} = \mathbf{Ax} + \mathbf{Bu}. \tag{14-29}$$

We denote the gradient (or derivative) of the scalar function f_0 of vector **x** by

$$\mathbf{grad}\, f_0(\mathbf{x}, \mathbf{u}) = \frac{\partial f_0(\mathbf{x}, \mathbf{u})}{\partial \mathbf{x}} = \begin{bmatrix} \frac{\partial f_0}{\partial x_1} \\ \frac{\partial f_0}{\partial x_2} \\ \vdots \\ \frac{\partial f_0}{\partial x_n} \end{bmatrix}.$$

With this notation the last n scalar equations in Eq. (14–22) can be replaced by

$$\frac{d\boldsymbol{\phi}}{dt} = \frac{\partial f_0}{\partial \mathbf{x}} - \mathbf{A}'\boldsymbol{\phi} \tag{14-30}$$

where the lower-case letter ϕ represents an n-vector having $\psi_1, \ldots, \psi_n$ as its elements. The first element of the original $(n+1)$-dimensional covariant vector, **Ψ**, is considered to be $\psi_0 = -1$. The Hamiltonian for this system is

$$H = -f_0(\mathbf{x}, \mathbf{u}) + \boldsymbol{\phi}'\{\mathbf{Ax} + \mathbf{Bu}\}. \tag{14-31}$$

Example 14–4 Consider the problem of achieving a smooth landing of a vehicle in a zero-g world, minimizing time and control-fuel consumption (Fig. 14–5). The control object is considered to be a pure inertial body,

$$\frac{d}{dt}\begin{bmatrix} x_1 \\ x_2 \end{bmatrix} = \begin{bmatrix} 0 & 1 \\ 0 & 0 \end{bmatrix} \mathbf{x} + \begin{bmatrix} 0 \\ 1 \end{bmatrix} u, \quad |u| \leqslant 1.$$

The initial state is

$$\mathbf{x}(0) = \begin{bmatrix} h \\ -v \end{bmatrix},$$

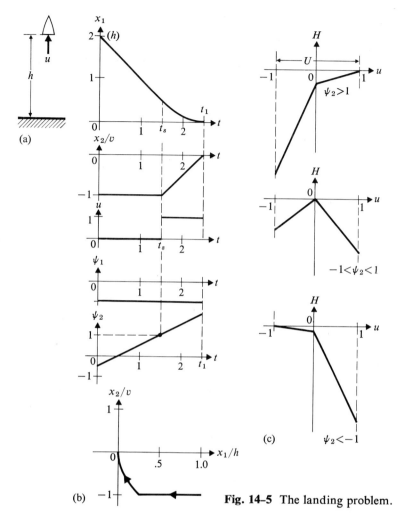

Fig. 14-5 The landing problem.

and the final, required, state is

$$\mathbf{x}(t_1) = \mathbf{0} \ .$$

The performance index J to be minimized is

$$J = \int_0^{t_1} (|u| + w) \, dt \ ,$$

or

$$f_0(\mathbf{x}, u) = |u| + w \ ,$$

where w is a weighting factor for time. By Eq. (14-30) we have

$$\frac{d\phi_1}{dt} = 0 \ , \qquad \frac{d\phi_2}{dt} = -\phi_1 \ ,$$

and the Hamiltonian (Eq. 14-31) is

$$H = -|u| - w + \phi_1 x_2 + \phi_2 u.$$

By application of the maximum condition, we find that the optimal control must be (Fig. 14-5c):

$$u^0 = +1 \quad \text{if } \phi_2 > 1,$$
$$u^0 = 0 \quad \text{if } -1 < \phi_2 < +1,$$
$$u^0 = -1 \quad \text{if } \phi_2 < -1.$$

By a simple computation of the vehicle's dynamics we find that a one-switching control,

$$u = 0 \quad \text{(coasting) from } t = 0 \text{ to some time } t_s,$$
$$u = +1 \quad \text{(retrorocket thrust) from } t_s \text{ to } t_1,$$

is possible when the following condition is satisfied for initial height h relative to initial velocity v:

$$\frac{v^2}{2} \leqslant h.$$

In order for this control to be optimal, the switching condition related to ϕ_2 requires an additional condition on w:

$$\frac{2}{(h/v^2) - \frac{1}{2}} \geqslant w.$$

By a computation similar to that of Example 14-3, we find the following solution when these two conditions are satisfied:

$$\phi_1 = -\frac{w}{v}, \quad \phi_2 = 1 - w\left(\frac{h}{v^2} - \frac{1}{2}\right) + \frac{w}{v}t,$$

and

$$x_1 = h - vt, \quad x_2 = -v \quad \text{with } u = 0, \quad \text{for } 0 \leqslant t \leqslant t_s,$$
$$x_1 = (h - vt_s) + \tfrac{1}{2}(t - t_s)^2 - v(t - t_s),$$
$$x_2 = (t - t_s) - v, \quad u = +1, \quad \text{for } t_s \leqslant t \leqslant t_1,$$

where

$$t_s = \frac{h}{v} - \frac{v}{2}, \quad t_1 = \frac{h}{v} + \frac{v}{2}.$$

The pattern of the optimal response is shown in Fig. 14-5(b). We shall call this the landing problem below.

The basis for the maximum principle is a local search for an optimal trajectory. Since the search is not global, the maximum principle is a necessary condition, not necessarily sufficient. In the local search we consider perturbations from the optimal control trajectory. Suppose PQ in Fig. 14-6(a) is an optimal trajectory. If the control is perturbed at time τ_1 from

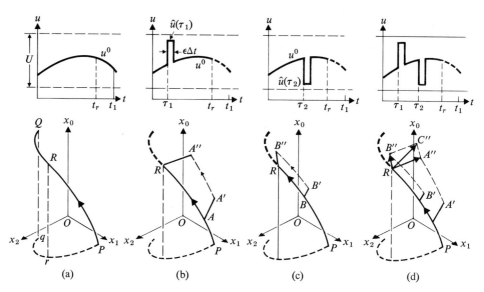

Fig. 14-6 Perturbation in optimal control and state deviation.

its optimal value $u^0(\tau_1)$ to some other value $\hat{u}(\tau_1)$ for a short time interval, it will produce the deviation AA' in the optimal trajectory (Fig. 14-6b). When control resumes the original optimal value shortly after τ_1, the perturbed trajectory $A'A''$ will reach the state A'' at some time t_r when the optimal trajectory is at point R. A similar argument holds for perturbation at a different time, τ_2, as shown in Fig. 14-6(c), where the deviation vector BB' at time τ_2 will become RB'' at time t_r. If these two perturbations are both involved in a trajectory, the state deviation at time t_r will be given by RC'' (Fig. 14-6d), which is the vector sum of RA'' and RB''. We assume the time duration of each perturbation, $\varepsilon\Delta t$, to be very short ($\varepsilon \to 0^+$) for linear superposition to hold.

Let us consider all the possible magnitudes of perturbed control ($\hat{u} \in U$) at every time from zero to t_r (except at instants where switching occurs in u^0; Δt must not include such discontinuities in u^0). Such control perturbations will produce state deviations RC'', RD'', etc., as shown in Fig. 14-7(a). Since the vector sum of any of these deviation vectors is also a deviation vector, the set of all possible vectors will fill a convex cone. The cone is convex because, if it has a concave edge, that portion will be filled by vector addition. The convex cone is called the *cone of attainability*. At any state R there is at least one half-line that does not belong to the cone, such as Rr in Fig. 14-7(a), because a state like R' in the figure represents a better-than-optimal performance and thus violates the optimality of trajectory PR. Therefore, the cone of attainability at any point R cannot span the entire $(n + 1)$-dimensional space and there must exist a hyperplane that supports the cone. Since we must also consider perturbations in time, $\pm \varepsilon \delta t$, at time

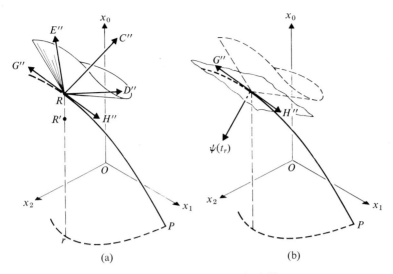

Fig. 14-7 The cone of attainability.

t_r, for which the deviation vectors RG'' and RH'' will be tangent to the trajectory (Fig. 14-7a), the cone will have the shape of a trough and the dividing hyperplane will include RG'' and RH'' as its members. Hence the hyperplane itself will be tangent to the optimal trajectory (Fig. 14-7b). We shall see below that $\Psi(t_r)$, the costate vector at time t_r, is orthogonal to the dividing hyperplane and pointed away from the cone of attainability.

We shall consider, for brevity, a second-order control object ($n = 2$, hence $n + 1 = 3$) in the following discussion. The basic structure of each equation remains the same for the nth-order case. Introducing a variational vector $\delta \mathbf{X}(t)$ [that will represent, for instance, AA' in Fig. 14-6(b)], we derive an equation of motion for $\delta \mathbf{X}(t)$. Substituting $\mathbf{X}(t) + \varepsilon \delta \mathbf{X}(t)$ for $\mathbf{X}(t)$ in Eqs. (14-18) and (14-19), we obtain

$$\frac{dx_i}{dt} + \varepsilon \frac{d\delta x_i}{dt} = f_i(\mathbf{x}, \mathbf{u}^0) + \varepsilon \sum_{j=0}^{n} \frac{\partial f_i}{\partial x_j} \delta x_j,$$

$$i = 0, 1, \ldots, n. \quad (14\text{-}32)$$

Here ε is assumed to be a small number of the first order; that is, ε^2 is negligible in this Taylor-series expansion. Since

$$\frac{dx_i}{dt} = f_i(\mathbf{x}, \mathbf{u}^0)$$

in Eq. (14-32), we obtain the following equation of motion for the variational vector:

$$\frac{d\delta x_i}{dt} = \sum_{j=0}^{n} \frac{\partial f_i}{\partial x_j} \delta x_j, \quad i = 0, 1, \ldots, n, \quad (14\text{-}33)$$

or, for $n = 2$,

$$\frac{d}{dt} \delta \mathbf{X}(t) = \begin{bmatrix} \frac{\partial f_0}{\partial x_0} & \frac{\partial f_0}{\partial x_1} & \frac{\partial f_0}{\partial x_2} \\ \frac{\partial f_1}{\partial x_0} & \frac{\partial f_1}{\partial x_1} & \frac{\partial f_1}{\partial x_2} \\ \frac{\partial f_2}{\partial x_0} & \frac{\partial f_2}{\partial x_1} & \frac{\partial f_2}{\partial x_2} \end{bmatrix} \delta \mathbf{X}(t) . \tag{14-34}$$

Note that ε has dropped out, so that the perturbed trajectory [such as $A'A''$ in Fig. 14-6(b)] need not be infinitesimally close to the original trajectory.

We next define a costate vector $\boldsymbol{\Psi}(t)$ by a vector differential equation with a matrix that is adjoint (meaning negative transpose) to the Jacobian matrix of Eq. (14-34):

$$\frac{d}{dt} \boldsymbol{\Psi}(t) = - \begin{bmatrix} \frac{\partial f_0}{\partial x_0} & \frac{\partial f_1}{\partial x_0} & \frac{\partial f_2}{\partial x_0} \\ \frac{\partial f_0}{\partial x_1} & \frac{\partial f_1}{\partial x_1} & \frac{\partial f_2}{\partial x_1} \\ \frac{\partial f_0}{\partial x_2} & \frac{\partial f_1}{\partial x_2} & \frac{\partial f_2}{\partial x_2} \end{bmatrix} \boldsymbol{\Psi}(t) .$$

For an adjoint system the following relation holds:

$$\frac{d}{dt} \{ \boldsymbol{\Psi}' \cdot \delta \mathbf{X} \} = \left(\frac{d}{dt} \boldsymbol{\Psi}' \right) \delta \mathbf{X} + \boldsymbol{\Psi}' \left(\frac{d}{dt} \delta \mathbf{X} \right) ,$$

where

$$\frac{d}{dt} \delta \mathbf{X} = \mathbf{M}_J \delta \mathbf{X} , \qquad \frac{d}{dt} \boldsymbol{\Psi} = -\mathbf{M}_J' \boldsymbol{\Psi} , \qquad \mathbf{M}_J = \text{the Jacobian matrix} ;$$

hence

$$\left(\frac{d}{dt} \boldsymbol{\Psi} \right)' = (-\mathbf{M}_J' \boldsymbol{\Psi})' = -\boldsymbol{\Psi}' \mathbf{M}_J$$

and

$$\frac{d}{dt} \{ \boldsymbol{\Psi}' \cdot \delta \mathbf{X} \} = -\boldsymbol{\Psi}' \mathbf{M}_J \delta \mathbf{X} + \boldsymbol{\Psi}' \mathbf{M}_J \delta \mathbf{X} = 0 .$$

We thus conclude that

$$\boldsymbol{\Psi}'(t) \cdot \delta \mathbf{X}(t) = \text{const} . \tag{14-35}$$

Eq. (14-35) applied to the deviation vectors AA' and RA'' in Fig. 14-6(b) will give:

$$\boldsymbol{\Psi}'(\tau_1) \delta \mathbf{X}(\tau_1) = \boldsymbol{\Psi}'(t_r) \delta \mathbf{X}(t_r) = C_1 , \tag{14-36}$$

14-2 The maximum principle 643

where C_1 is a constant, $\delta X(\tau_1)$ is AA' and $\delta X(t_r)$ is RA''. The initial deviation AA' is caused by the nonoptimal control $\hat{u}(\tau)$ at time τ_1 for a very short time interval $\varepsilon \Delta t$; thus by Eq. (14–18) we have:

$$\varepsilon \delta X(\tau_1) = \{F(x(\tau_1), \hat{u}(\tau_1)) - F(x(\tau_1), u^0(\tau_1))\} \varepsilon \Delta t,$$

hence

$$\delta X(\tau_1) = \{F(x(\tau_1), \hat{u}(\tau_1)) - F(x(\tau_1), u^0(\tau_1))\} \Delta t. \tag{14-37}$$

Substituting Eq. (14–37) into Eq. (14–36), and recalling the Hamiltonian of Eq. (14–23), we obtain:

$$H(\hat{u}) - H(u^0) = C_1. \tag{14-38}$$

Let us choose a negative value for C_1 for this deviation vector, zero for the dividing hyperplane, and positive values for all points below the hyperplane. Then Eq. (14–38) gives the maximum principle:

H for control other than optimal $\leqslant H$ for optimal control.

Since the tangent $G''H''$ to the optimal trajectory is a member of the dividing plane for which C_1 is defined to be zero (see Fig. 14–7), we must have

H for optimal control $= 0$.

Consider, lastly, the vertical half-line RR' in Fig. 14–7(a), for which δx_0 is obviously negative and $\delta x_1 = 0$, $\delta x_2 = 0$. When RR' is considered as the vector δX, we have

$$\Psi'(t_r) \cdot \delta X(t_r) = C_2 = \psi_0 \delta x_0,$$

where C_2 must be positive by the sign convention. Therefore,

$$\psi_0 < 0, \tag{14-39}$$

which determines the sign of all the components of Ψ [because Eq. (14–22) is homogeneous]. The sign convention is arbitrary; if the opposite sign were chosen, the theorem would have been called the minimum principle.

Example 14-5 Knowing the optimal trajectory of the smooth-landing system (see the preceding example), compute the deviation vector $\delta X(t)$, the cone of attainability, and the supporting plane for a time point t_r less than the switching instant $t_s = 1.5$. We assume $w = 1$, $v = -1$, $h = 2$, so that optimal control is $u = 0$ for $0 < t < t_s = 1.5$. The original equation of motion is

$$\frac{d}{dt} X(t) = F(x, u) = \begin{bmatrix} |u| + w \\ x_2 \\ u \end{bmatrix}.$$

For an optimal trajectory in the coasting period, the equation reduces to

$$F(x^0, u^0) = \begin{bmatrix} 1 \\ x_2 \\ 0 \end{bmatrix} = \begin{bmatrix} 1 \\ -1 \\ 0 \end{bmatrix} \quad \text{and} \quad X(t) = \begin{bmatrix} t \\ 2 - t \\ -1 \end{bmatrix}.$$

Equation (14–34) for the variational vector is

$$\frac{d}{dt}\delta\mathbf{X}(t) = \begin{bmatrix} 0 & 0 & 0 \\ 0 & 0 & 1 \\ 0 & 0 & 0 \end{bmatrix}\delta\mathbf{X}(t). \tag{14-40}$$

Let $u = \hat{u}$ at some time point τ less than 1.5; then

$$\mathbf{F}(\mathbf{x}(\tau), \hat{u}) = \begin{bmatrix} 1 + |\hat{u}| \\ -1 \\ \hat{u} \end{bmatrix},$$

and by Eq. (14–37),

$$\delta\mathbf{X}(\tau) = \begin{bmatrix} |\hat{u}| \\ 0 \\ \hat{u} \end{bmatrix}\Delta t. \tag{14-41}$$

Taking Eq. (14–41) as an initial condition, we solve Eq. (14–40) and obtain, for $\tau < t_r < 1.5$,

$$\delta\mathbf{X}(t_r) = \begin{bmatrix} |\hat{u}| \\ \hat{u}(t_r - \tau) \\ \hat{u} \end{bmatrix}\Delta t. \tag{14-42}$$

Taking all possible linear combinations of $\delta\mathbf{X}(t_r)$ at $t_r = 1$ for all admissible $\hat{u}$ (that is, $0 < |\hat{u}| \leq 1$) and $0 \leq \tau \leq 1$, we have the cone (a sector of a plane for this case) shown in Fig. 14-8. Since the cone of attainability must also have a component tangent to the optimal trajectory at $t = t_r$, it will take the form of a trough with the optimal trajectory as its edge. The trough is not shown in Fig. 14-8. The supporting plane of the cone is orthogonal to $\mathbf{\Psi}(t_r)$, where, from the preceding example, the costate vector $\mathbf{\Psi}(t)$, as a solution of

$$\frac{d}{dt}\mathbf{\Psi}(t) = -\begin{bmatrix} 0 & 0 & 0 \\ 0 & 0 & 0 \\ 0 & 1 & 0 \end{bmatrix}\mathbf{\Psi}(t),$$

was found to be

$$\mathbf{\Psi}(t) = \begin{bmatrix} -1 \\ -1 \\ t - \frac{1}{2} \end{bmatrix}.$$

To confirm that the cone is supported by the plane, we compute the scalar (inner) product:

$$\mathbf{\Psi}'(t_r)\cdot\delta\mathbf{X}(t_r) = \begin{bmatrix} -1 & -1 & t_r - \frac{1}{2} \end{bmatrix}\begin{bmatrix} |\hat{u}| \\ \hat{u}(t_r - \tau) \\ \hat{u} \end{bmatrix}\Delta t$$

$$= \{-|\hat{u}| + (\tau - \tfrac{1}{2})\hat{u}\}\Delta t. \tag{14-43}$$

Since τ is less than 1.5, $(\tau - \frac{1}{2}) < 1$, the right-hand side of Eq. (14–43) is therefore negative for all possible $\hat{u}$. Hence the cone is on the upper side of the plane.

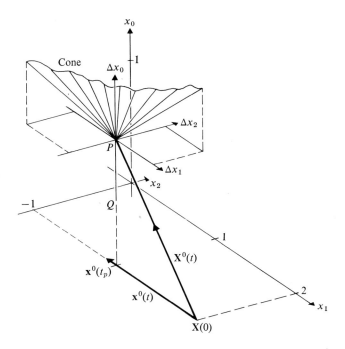

Fig. 14-8 The cone for deviations in control from its optimal.

The maximum principle and dynamic programming are mathematically related to each other. To derive the maximum principle from the dynamic-programming viewpoint, let us denote the optimal (that is, minimal) cost associated with the motion from a state $\mathbf{x}(t)$ to a fixed final state $\mathbf{x}(t_1)$ by $J^*(\mathbf{x})$,

$$J^*(\mathbf{x}) = \int_t^{t_1} f_0(\mathbf{x}, u^0) \, dt \,. \tag{14-44}$$

Differentiating $J^*(\mathbf{x})$ with respect to time, and letting

$$\psi_i = -\frac{\partial J^*}{\partial x_i}, \qquad i = 1, \ldots, n\,, \tag{14-45}$$

we obtain the relation

$$f_0(\mathbf{x}, u^0) = -\frac{dJ^*(\mathbf{x})}{dt} = \sum_{i=1}^n \left(-\frac{\partial J^*}{\partial x_i}\right) \frac{dx_i}{dt} = \sum_{i=1}^n \psi_i f_i \,. \tag{14-46}$$

Combining both sides of Eq. (14-46) into one expression, and introducing $\psi_0 = -1$, we write the last equation in the alternate form:

$$\sum_{i=0}^n \psi_i f_i = 0 \qquad \text{at} \quad u = u^0 \,. \tag{14-47}$$

Let us suppose that the control were nonoptimal, $\hat{u} \neq u^0(t)$, at time t for a short time interval dt; then by the principle of optimality, the following relation will hold:

$$J^*(\mathbf{x} + d\mathbf{x}) + f_0(\mathbf{x}, \hat{u})\, dt \geqslant J^*(\mathbf{x}). \tag{14-48}$$

By Taylor-series expansion, $J^*(\mathbf{x} + d\mathbf{x})$ in this expression can be converted into:

$$J^*(\mathbf{x} + d\mathbf{x}) = J^*(\mathbf{x}) + \sum_{i=1}^{n} \frac{\partial J^*}{\partial x_i}\, dx_i$$

where

$$dx_i = f_i(\mathbf{x}, \hat{u})\, dt\,;$$

hence

$$J^*(\mathbf{x} + d\mathbf{x}) - J^*(\mathbf{x}) = -\sum_{i=1}^{n} \psi_i f_i(\mathbf{x}, \hat{u})\, dt. \tag{14-49}$$

From Eqs. (14-48) and (14-49) follows the relation

$$\sum_{i=1}^{n} \psi_i f_i(\mathbf{x}, \hat{u}) \leqslant f_0(\mathbf{x}, \hat{u}),$$

which, with the identity $\psi_0 = -1$, becomes

$$\sum_{i=0}^{n} \psi_i f_i(\mathbf{x}, \hat{u}) \leqslant 0. \tag{14-50}$$

The statement of the maximum principle follows from Eqs. (14-47) and (14-50).

14-3 SOME APPLICATIONS OF THE MAXIMUM PRINCIPLE

A variety of simple examples will be treated in this section to demonstrate applications of the fundamental theorem of the maximum principle. Also, we shall introduce modifications of the fundamental theorem for the cases in which the final time is prescribed, or in which time appears explicitly on the right-hand side of a system's equations, and problems that require the transversality conditions.

Our first topic is concerned with modes of control. It is often possible to state the *mode of optimal control* by inspection of the maximum condition (Eq. 14-25) [11]. For the sake of brevity, let us assume that the control object is linear, so that Eq. (14-31) will hold. Equation (14-31) restated is:

$$H = -f_0(\mathbf{x}, \mathbf{u}) + \psi'(A\mathbf{x} + B\mathbf{u}).$$

It can be written as the sum of two functions, $H_1(\mathbf{u})$ which is a function of $\mathbf{u}$, and H_2 which is not a function of $\mathbf{u}$,

$$H(\mathbf{u}) = H_1(\mathbf{u}) + H_2.$$

14-3 Some applications of the maximum principle

For linear control objects $H_1(\mathbf{u})$ includes $\boldsymbol{\phi}'\mathbf{Bu}$ and some term(s) of $f_0(\mathbf{x}, \mathbf{u})$, which explicitly involve $\mathbf{u}$. The control mode depends on the terms contributed by f_0. We present three typical cases of practical interest below.

Case 1 (Bang-bang control with neutral or deadband region.) This is the case in which $|u|$ is involved in $f_0(\mathbf{x}, \mathbf{u})$, as in Example 14-4. For scalar control, H_1 takes the form

$$H_1 = -g_1|u| + g_2 u,$$

where g_1 is a constant and g_2 is $\boldsymbol{\phi}'\mathbf{b}$; $\mathbf{b}$ is the column vector in the linear object's equation

$$\frac{d\mathbf{x}}{dt} = \mathbf{Ax} + \mathbf{b}u.$$

With the control constraint

$$u_{\min} \leq u \leq u_{\max},$$

the mode of control is:

$$\begin{aligned} u &= u_{\max} & \text{if} \quad & g_1 < g_2, \\ u &= 0 & \text{if} \quad & -g_1 < g_2 < g_1, \\ u &= u_{\min} & \text{if} \quad & g_2 < -g_1, \end{aligned} \tag{14-51}$$

where g_1 is assumed to be positive, and $u_{\min}$ is considered negative.

Case 2 (Bang-bang control.) If $f_0(\mathbf{x}, u)$ does not involve u, or if it has a term which is proportional to u, then H_1 takes the form

$$H_1 = gu,$$

where g is a function of time. The mode of optimal control in this case is bang-bang, governed by the following condition:

$$\begin{aligned} u &= u_{\max} & \text{if} \quad g > 0, \\ u &= u_{\min} & \text{if} \quad g < 0. \end{aligned} \tag{14-52}$$

Example 14-3 belongs to this class of optimal control.

Case 3 (Continuous action with or without saturation.) This case occurs when $f_0(\mathbf{x}, u)$ involves u^2. Usually, the u^2 is intended to penalize large magnitudes of u (see Problem 14-15). For a quadratic form of H_1,

$$H_1 = -g_1 u^2 + g_2 u, \qquad g_1 > 0,$$

the maximum condition $\partial H_1/\partial u = 0$ yields the value

$$u^0 = \frac{g_2}{2g_1}, \tag{14-53}$$

if $u_{min} < u^0 < u_{max}$. We shall show an example of the third case, with saturation in control.

Example 14-6 The control object is first-order with unit time constant,

$$\frac{dx_1}{dt} = -x_1 + u .$$

The initial value is

$$x_1 = 1 \quad \text{at} \quad t = 0 .$$

We wish to bring the output x_1 to zero at some time t_1:

$$x_1(t_1) = 0 .$$

The performance index to be minimized in this transient is

$$J = \int_0^{t_1} (x_1^2 + u^2) \, dt . \tag{14-54}$$

We shall consider two cases: first, the case of no constraint in the magnitude of u; then the case in which u will saturate.

The first case is simple. We have

$$\frac{dx_0}{dt} = x_1^2 + u^2 ;$$

hence

$$\phi_0 = -1 , \quad \frac{d\phi_1}{dt} = \phi_1 + 2x_1 ,$$

and

$$H = -x_1^2 - u^2 + \phi_1(-x_1 + u) .$$

Applying Eq. (14-53), we find that $u^0 = \phi_1/2$. Substituting the last expression into the differential equations

$$\frac{dx_1}{dt} = -x_1 + \tfrac{1}{2}\phi_1 , \quad \frac{d\phi_1}{dt} = \phi_1 + 2x_1 ,$$

we find

$$x_1 = C_1 e^{\sqrt{2}\,t} + C_2 e^{-\sqrt{2}\,t} ,$$
$$u^0 = \tfrac{1}{2}\phi_1 = (\sqrt{2} + 1)C_1 e^{\sqrt{2}\,t} + (-\sqrt{2} + 1)C_2 e^{-\sqrt{2}\,t} . \tag{14-55}$$

The values of C_1 and C_2 in Eq. (14-55) must be fixed by the initial conditions:

$$x_1(0) = 1 \quad \text{and} \quad \sup H = 0 \quad \text{at} \quad t = 0 .$$

There are two possible pairs of values: $C_1 = 1$, $C_2 = 0$, or $C_1 = 0$, $C_2 = 1$. The first pair must be rejected because it will not take the system to the target point. Thus we obtain the answer

$$x_1 = e^{-\sqrt{2}\,t} , \quad u^0 = (1 - \sqrt{2})e^{-\sqrt{2}\,t} \quad \text{and} \quad t_1 = +\infty . \tag{14-56}$$

The solution of Eq. (14-56) serves as a guide in the computation of the second case. Let us assume that saturation in control will occur due to the constraint

$$|u| \leq \tfrac{1}{4}.\tag{14-57}$$

It seems likely that the control will be $u = -\tfrac{1}{4}$ for some time $0 \leq t \leq t_s$ and will then follow the exponential decay $u = C(1 - \sqrt{2})e^{-\sqrt{2}t}$. An approximation of the time t'_s at which the saturation will end can be determined from the condition

$$u = -\tfrac{1}{4} = (1 - \sqrt{2})e^{-\sqrt{2}t'_s},$$

and we find $t'_s = 0.357$. This is the time when saturation would end in the control system of Eq. (14-56) if its control had the constraint of Eq. (14-57) and if its initial state was $x_1(0) = 1$. However, this response is not truly optimal.

An exact solution for $\phi_1(t)$ in the saturation period is needed to determine the exact optimal control. To obtain $\phi_1(t)$, we first solve the system's equation with $u = -\tfrac{1}{4}$ and $x_1(0) = 1$. The solution is:

$$x_1(t) = 1.25e^{-t} - \tfrac{1}{4}.$$

With this solution, and the condition $H = 0$ at $t = 0$ in mind, we solve

$$\frac{d\phi_1}{dt} = \phi_1 + 2x_1$$

and obtain

$$\phi_1 = -1.25e^{-t} + \tfrac{1}{2} - 0.1e^t.\tag{14-58}$$

$\phi_1(t)$ of Eq. (14-58) must satisfy the maximum condition $u = \tfrac{1}{2}\phi_1$, with $u = -\tfrac{1}{4}$ at time t_s. Hence

$$\phi_1(t_s) = 2(-\tfrac{1}{4}) = -1.25e^{-t_s} + \tfrac{1}{2} - 0.1e^{t_s},$$

and we find that (Fig. 14-9),

$$t_s = 0.382.$$

The rest of the computation is similar to the first case. The results of the two cases are shown in Fig. 14-9. Note that optimal control (I) becomes suboptimal when saturation occurs.

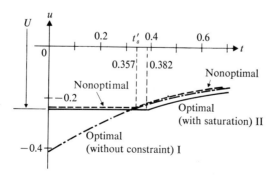

Fig. 14-9 Optimal control with saturation.

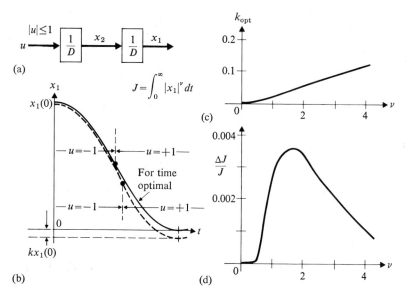

Fig. 14-10 A. T. Fuller's problem.

Although the mode of optimal control is determined by the maximum condition, the theorem must be applied with caution, mainly because it stipulates only a necessary condition. We shall show the difficulty of application in the next example, for which Eq. (14-52) applies.

Example 14-7 The control object is a pure inertial system (Fig. 14-10a), described by

$$\frac{dx_1}{dt} = x_2, \qquad \frac{dx_2}{dt} = u, \qquad |u| \leqslant 1,$$

where x_1 is displacement, x_2 is velocity, and u is the controlling force. The performance index to be minimized is:

$$J = \int_0^{t_1} |x_1|^\nu \, dt, \qquad \mathbf{x}(t_1) = \mathbf{0},$$

where ν is an integer, $\nu = 1, 2, \ldots$ Since H depends only on u, the optimal control must be bang-bang. In fact, it can be shown that the time-optimal switching [Fig. 12-22(a), (b), and Example 14-3] satisfies the necessary condition of optimality (the maximum principle) for any prescribed initial state, although this performance index is different from time-optimal except for the special case $\nu = 0$. According to A. T. Fuller [12], however, the optimal control is different from the time-optimal switching when $\nu \geqslant \frac{1}{2}$. The optimal response under bang-bang control has a constant half-cycle amplitude attenuation ratio k (Fig. 14-10b). As a consequence, the response oscillates (an infinite number of times; this phenomenon is called a "dither" or "chattering") with quickly decaying amplitude. The optimal value of the half-cycle amplitude ratio k is shown in Fig. 14-10(c) for various values of ν. The performance improvement due to such a dither pattern is plotted in Fig. 14-10(d). For instance, when $\nu = 2$ (the squared-error criterion) the performance of the optimal dither is about 0.35 percent better than the performance of the time-optimal switching. The difference therefore is only of theoretical interest. The dither produces some difficulty in the direct application of the maximum principle to this problem. To avoid the difficulty, D. M. Eggleston [13] proposed use of a reverse-time algorithm for this problem.

There are optimal control problems in which the time t_1 when the state will hit a specified target is also prescribed,

$$t_1 = T \quad \text{(given)}.$$

For this case, Eq. (14-26) must be modified as:

$$M(\boldsymbol{\Psi}^0(t), \mathbf{x}^0(t)) = \text{const}; \quad 0 \leqslant t \leqslant T, \tag{14-59}$$

this is the only change in the fundamental theorem.

To prove that the modified condition of Eq. (14-59) is appropriate for this type of problem, we add time t as the $(n+1)$th coordinate axis:

$$x_{n+1} = t, \quad \text{or} \quad \frac{dx_{n+1}}{dt} = 1, \tag{14-60}$$

The fundamental theorem (Section 14-2) holds for the revised system, for which the target coordinate is specified as

$$\begin{bmatrix} \mathbf{x} \\ x_{n+1} \end{bmatrix} = \begin{bmatrix} \mathbf{x}_t \\ T \end{bmatrix},$$

where $\mathbf{x}_t$ is the specified final state of the object. Since running time t is not explicitly involved in the right-hand side of the system's equation (Eq. 14-18), the additional costate variable will stay constant:

$$\frac{d\psi_{n+1}}{dt} = 0, \quad \psi_{n+1} = \text{const}.$$

The Hamiltonian H' of Eq. (14-23), in the new $(n+2)$-space, will have the form

$$H' = \sum_{i=0}^{n} \psi_i f_i + \psi_{n+1} = H + \psi_{n+1},$$

where H is the original Hamiltonian, $H = \sum_{i=0}^{n} \psi_i f_i$. Since

$$\sup_{\mathbf{u} \in U} H' = 0$$

by the fundamental theorem, it follows that

$$H = -\psi_{n+1} = \text{const}.$$

Example 14-8 Let us consider the first-order system of Example 14-6, with the same initial and final state as before. We now define the performance index to be

$$J = \int_0^T |u|\, dt, \quad |u| \leq 1,$$

where $T = 1$ is the specified final time. The differential equations are:

$$\frac{dx_0}{dt} = |u|, \quad \frac{dx_1}{dt} = -x_1 + u,$$

$$\phi_0 = -1, \quad \frac{d\phi_1}{dt} = \phi_1.$$

We have

$$H = -|u| + \phi_1(-x_1 + u);$$

hence the optimal control must be

$$u = +1 \quad \text{if } \phi_1 > 1,$$
$$u = 0 \quad \text{if } 1 > \phi_1 > -1,$$
$$u = -1 \quad \text{if } -1 > \phi_1.$$

Integration of the equation for ϕ_1 yields the solution

$$\phi_1 = Ce^t.$$

Because H is not equal to zero, we cannot use the initial condition $H = 0$ to find C (as we did in earlier cases). Since the initial state $x_1(0)$ is positive and the final state $x_1(T)$ must be zero, and also since ϕ_1 is monotonic, it is logical to consider C to be negative and $|C| < 1$, so that $u = 0$ for some time, and then $u = -1$. The solution for the first period ($u = 0$) is

$$x_1 = e^{-t}x_1(0), \quad 0 \leq t \leq t_s.$$

We use $x_1(t_s)$ as an initial condition to obtain the solution for the second period ($u = -1$):

$$x_1 = e^{-t}x_1(0) + e^{t_s - t} - 1 \quad \text{for } t_s \leq t \leq T.$$

Letting $x_1(T) = 0$ in the second solution, we find that

$$x_1(0) + e^{t_s} = e^T,$$

where $x_1(0) = 1$ and $T = 1$ (from the problem statement), so that

$$e^{t_s} = e - 1, \quad t_s = 0.541.$$

Since the switching condition

$$\phi_1(t) = Ce^t = -1$$

must be satisfied at $t = t_s$, we have $Ce^{t_s} = -1$ or

$$C = -e^{-t_s} = -1/(e - 1).$$

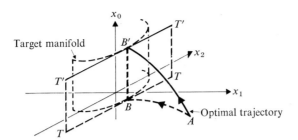

Fig. 14-11 The transversality condition.

Substituting the solutions into the Hamiltonian, we can confirm that

$$\sup H = -C$$

all through the period $0 \leqslant t \leqslant T$.

In some problems, time t will appear explicitly on the right-hand side of the system's vector equation,

$$\frac{d\mathbf{X}(t)}{dt} = \mathbf{F}(\mathbf{x}(t), \mathbf{u}(t), t). \qquad (14\text{-}61)$$

This can happen, for instance, if the control object is time-varying, if a disturbance (as a known function of time) is applied to a control object, or if a performance index has t in its integrand (for example, $J = \int tx_i^2 dt$). Such a system is then called *nonautonomous*. An $(n + 1)$-dimensional nonautonomous system (Eq. 14-61) can be represented as an $(n + 2)$-dimensional autonomous system whose $(n + 1)$th state variable is time,

$$x_{n+1} = t, \qquad \frac{dx_{n+1}}{dt} = 1. \qquad (14\text{-}62)$$

All the theorems and formulations given above can be applied to the $(n + 2)$-dimensional autonomous system (see Problem 14–10).

Sometimes the end state $\mathbf{x}(t_1)$, the target state, is only partially specified, or not specified at all. In a second-order object, for example, we might specify that the final state can lie anywhere on a circle in the vicinity of the origin. With the addition of an artificial coordinate x_0, the circle in the original (two-dimensional) state space will become a cylinder in the new three-dimensional space, as shown in Fig. 14–11. The graphical interpretation of an imperfectly (or partially) specified end state is that the target is in a subspace called the *manifold*.

Let AB' in Fig. 14–11 be an optimal trajectory. To derive a necessary condition for optimality, consider a perturbation in the vicinity of the endpoint B'. The cylindrical surface can be replaced by a tangent plane in the vicinity of B'. Let $T'T'$ be a horizontal tangent at B'. Then the lower half of the tangent plane (which has $T'T'$ as its upper edge) must not belong to the cone of attainability because, if it does, there must exist a trajectory which is better than optimal. Therefore $T'T'$ must belong to the "supporting plane." Since the plane is orthogonal to the costate vector $\Psi(t_1)$, $T'T'$ must also be orthogonal to it. If we represent the direction of $T'T'$ by the column vector

$$\begin{bmatrix} 0 \\ \eta_1 \\ \eta_2 \end{bmatrix},$$

we must have (for orthogonality)

$$\eta_1 \psi_1(t_1) + \eta_2 \psi_2(t_1) = 0. \tag{14-63}$$

In general, the n-dimensional column vector

$$\begin{bmatrix} \psi_1(t_1) \\ \vdots \\ \psi_n(t_1) \end{bmatrix}$$

must be orthogonal to the plane (or line) which is tangent to the (smooth) manifold in the n-dimensional state space. This is a statement of the *transversality* condition at the final point (the "right end") of a trajectory. Note that the transversality condition supplies the "missing" boundary condition that is necessary if we are to fix an optimal control for systems with partially specified end states. For instance, in a second-order control object, two end conditions are available if the target state is fixed: $x_1(t_1)$ and $x_2(t_1)$. Two conditions are also available in the case of Fig. 14–11: the equation of the circle in the $x_1 x_2$-plane on which the end state must lie, and the transversality condition (Eq. 14–63).

An important special case of the transversality condition at the final state arises when some coordinates of the target state are not specified. If only $x_1(t_1)$ is specified in a second-order system, for example, the target manifold is a plane parallel to the $x_0 x_2$-plane. Thus the transversality condition reduces to:

$$\psi_2(t_1) = 0.$$

In general, if a target's ith coordinate $x_i(t_1)$ is not specified, then

$$\psi_i(t_1) = 0 \tag{14-64}$$

is the transversality condition. If the end state at a given time T is not specified at all, the transversality condition is

$$\psi_i(T) = 0, \quad i = 1, \ldots, n, \quad \text{or} \quad \boldsymbol{\psi}(T) = \mathbf{0}. \tag{14-65}$$

14-3 Some applications of the maximum principle

We shall make use of this condition in the next section to derive a linear-feedback optimal control law.

If the *initial* state $\mathbf{x}(t_0)$ is incompletely specified, or not specified at all, we can apply the transversality condition to the initial point. The n-dimensional costate vector at initial time t_0,

column vector of $\psi_1(t_0)$, $\psi_2(t_0)$, ..., $\psi_n(t_0)$,

must be orthogonal to the tangent plane of the manifold which passes through the initial point $\mathbf{x}(t_0)$. We can use reasoning similar to that given above to apply the transversality condition to initial states.

Example 14-9 Consider a second-order object described by the differential equation

$$\frac{d^2x}{dt^2} + u(t)x = 0,$$

where $u(t)$ is the control, with the constraint $0 \leqslant u \leqslant 1$. The control u in this case is a system parameter: if $u = 0$, the system is a free inertial object, while a positive-valued u makes the system oscillatory. Such control is called *parametric* control (often seen in biological systems [14]). Letting $x = x_1$, we obtain the state-space formulation

$$\frac{dx_1}{dt} = x_2, \qquad \frac{dx_2}{dt} = -ux_1.$$

Its initial state is $x_1(0) = 1$, $x_2(0) = 0$, and the final state at some time t_1 is specified only for x_1:

$$x_1(t_1) = 0.$$

The cost function to be minimized is

$$J = \int_0^{t_1} (u + w)\, dt.$$

To solve this problem, we first write the equation for the costate,

$$\frac{d\psi_1}{dt} = u\psi_2, \qquad \frac{d\psi_2}{dt} = -\psi_1.$$

The Hamiltonian is

$$H = -w - u + \psi_1 x_2 - u x_1 \psi_2;$$

hence the optimal control is:

$$u^0 = 0 \quad \text{if} \quad (1 + x_1\psi_2) > 0, \qquad u^0 = 1 \quad \text{if} \quad (1 + x_1\psi_2) < 0.$$

Suppose that $u = 1$ for $0 \leqslant t \leqslant t_s$, followed by $u = 0$ for $t_s \leqslant t \leqslant t_1$. The solution for the first period is

$$x_1 = \cos t, \qquad x_2 = -\sin t$$

and
$$\psi_1 = C \sin(t + \phi), \qquad \psi_2 = C \cos(t + \phi),$$
where C and ϕ are unknown constants. The switching condition is
$$1 + x_1(t_s)\psi_2(t_s) = 0 = 1 + \cos t_s[C \cos(t_s + \phi)]. \tag{14-66}$$
The initial condition $H = 0$ at $t = 0$ yields
$$C \cos \phi = -(1 + w). \tag{14-67}$$

We have $u = 0$ in the second period ($t_s \leq t \leq t_1$); hence there is no change in x_2 or ψ_1,
$$x_2 = -\sin t_s, \qquad \psi_1 = C \sin(t_s + \phi).$$
Solving
$$\frac{dx_1}{dt} = x_2 = -\sin t_s,$$
we obtain
$$x_1(t) = \cos t_s + (t_s - t) \sin t_s,$$
where the integration constant is determined from continuity of x_1 at t_s. At time t_1 we require that $x_1(t_1) = 0$, that is,
$$\cos t_s + (t_s - t_1) \sin t_s = 0. \tag{14-68}$$
Solving
$$\frac{d\psi_2}{dt} = -\psi_1 = -C \sin(t_s + \phi)$$
for the boundary condition, $\psi_2(t_s) = C \cos(t_s + \phi)$, we obtain
$$\psi_2(t) = C \cos(t_s + \phi) + (t_s - t) \sin(t_s + \phi).$$
By the transversality condition, ψ_2 must satisfy
$$\psi_2(t_1) = 0 = C \cos(t_s + \phi) + (t_s - t_1) \sin(t_s + \phi). \tag{14-69}$$

We now have four conditions [Eqs. (14-66) through (14-69)] for four unknowns: t_s, t_1, C, and ϕ. To determine these unknowns, let us first eliminate $(t_1 - t_s)$ from the last two conditions:
$$\frac{\cos(t_s + \phi)}{\sin(t_s + \phi)} = \frac{\cos t_s}{\sin t_s};$$
hence
$$\phi = 0,$$
and
$$C = -(1 + w), \qquad \sin t_s = \sqrt{w/(1 + w)}, \qquad \cos t_s = 1/\sqrt{1 + w}.$$

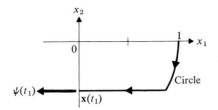

Fig. 14-12 A parametric control problem.

The final time t_1 is:

$$t_1 = 1/\sqrt{w} + \tan^{-1}\sqrt{w}\ .$$

It can be confirmed that $\max H = 0$ for $0 \leqslant t \leqslant t_1$. The optimal trajectory is shown in Fig. 14-12; it consists of a circle (simple harmonic motion) and a straight line (coasting).

A complication arises in the maximum principle if there is a constraint in the range of the state **x**. Such constraint in the state space may be caused, for instance, by overflow in a water tank system. If this happens, two conditions must be added to the fundamental theorem of the maximum principle. The first condition states that the value of $\psi_i(t)$ must jump when $x_i(t)$ hits or leaves such a boundary. If more than one coordinate variable, such as x_1 and x_2, form such a boundary, then the jump appears in the corresponding costate variables such as $\psi_1(t)$ and $\psi_2(t)$. The second condition applies when the state is moving along a constraint border. The maximum condition for this period must be modified to include a Lagrange multiplier (see [1] for details).

Compared with optimal control problems of lumped-parameter objects, mathematical sophistication of a higher degree is required to deal with problems of distributed-parameter objects. For instance, optimal control theory for a linear, lumped-parameter object is no longer simple once a dead time is introduced [15, 16]. The required mathematical sophistication increases for distributed-parameter objects governed by linear partial-differential equations [17, 18, 19]. For engineers, however, who are often interested in the numerical and practical aspects of optimal controls, some problems of less theoretical sophistication may already present difficulties in machine computation. One of the earliest contributors toward a repetitive algorithm was Y. C. Ho [20]. Recent publications [21 through 24] indicate a rapidly growing interest in the numerical approach.

14-4 LINEAR OPTIMAL FEEDBACK CONTROL

An automatic feedback-control-oriented approach to optimal control is the main theme of this section. We shall discuss a feedback realization of optimal control for a quadratic performance index. For the purpose of demonstrating basic principles rather than rigorous details, we shall first treat a first-order, linear, time-invariant control object, and then generalize the results for linear objects of finite order.

Let us derive a linear feedback-control law for linear control objects when the right-hand end of the trajectory is not specified and, in addition, when u^2 is involved in the performance index J without any constraint in u. Since the target state is not specified, we will use the transversality conditions (Eq. 14–65). We will assume a quadratic form for the performance index,

$$J = \int_0^T (wx^2 + u^2)\, dt,$$

and a first-order, linear, stationary control object

$$\frac{dx}{dt} = ax + bu,$$

where the weighting factor w is greater than (or equal to) zero and T is the prescribed final time. As we noted earlier, the squared magnitude of control u^2 involved in J penalizes excessive magnitudes of the controlling input.

We have

$$\psi_0 = -1, \qquad f_0 = wx^2 + u^2$$

and

$$H = -wx^2 - u^2 + (ax + bu)\psi,$$

where ψ must satisfy the differential equation

$$\frac{d\psi}{dt} = \frac{\partial f_0}{\partial x} - a\psi = 2wx - a\psi.$$

Since our control object is first-order, both x and ψ are scalar, so we shall omit the subscript 1 for brevity. By the maximum condition $\partial H/\partial u = 0$, and we find that the optimal control (that maximizes H) is:

$$u = \tfrac{1}{2}b\psi. \tag{14–70}$$

Substituting the control into the differential equation for x, and combining it with the differential equation for ψ, we obtain

$$\frac{d}{dt}\begin{bmatrix} x \\ \psi \end{bmatrix} = \begin{bmatrix} a & \tfrac{1}{2}b^2 \\ 2w & -a \end{bmatrix} \begin{bmatrix} x \\ \psi \end{bmatrix}. \tag{14–71}$$

Using the Laplace-transform method (Section 3–2), we find the solution matrix $\mathbf{S}(t)$ of this equation:

$$\mathbf{S}(t) = \mathscr{L}^{-1}\begin{bmatrix} s - a & -\tfrac{1}{2}b^2 \\ -2w & s + a \end{bmatrix}^{-1} = \begin{bmatrix} S_{11}(t) & S_{12}(t) \\ S_{21}(t) & S_{22}(t) \end{bmatrix}, \tag{14–72}$$

where

$$S_{11}(t) = \frac{(p+a)}{2p} e^{pt} + \frac{(p-a)}{2p} e^{-pt},$$

$$S_{12}(t) = \frac{b^2}{4p} (e^{pt} - e^{-pt}),$$

$$S_{21}(t) = \frac{w}{p} (e^{pt} - e^{-pt}), \qquad (14\text{--}73)$$

$$S_{22}(t) = \frac{(p-a)}{2p} e^{pt} + \frac{(p+a)}{2p} e^{-pt},$$

$$p = \sqrt{a^2 + b^2 w}.$$

It then follows that

$$\begin{bmatrix} x(t) \\ \psi(t) \end{bmatrix} = \mathbf{S}(t) \begin{bmatrix} x(0) \\ \psi(0) \end{bmatrix}, \quad \begin{bmatrix} x(T) \\ \psi(T) \end{bmatrix} = \mathbf{S}(T) \begin{bmatrix} x(0) \\ \psi(0) \end{bmatrix}; \qquad (14\text{--}74)$$

hence

$$\begin{bmatrix} x(T) \\ \psi(T) \end{bmatrix} = \mathbf{S}(T)\mathbf{S}^{-1}(t) \begin{bmatrix} x(t) \\ \psi(t) \end{bmatrix} = \mathbf{S}(T-t) \begin{bmatrix} x(t) \\ \psi(t) \end{bmatrix}, \qquad (14\text{--}75)$$

where

$$\mathbf{S}(T)\mathbf{S}^{-1}(t) = \mathbf{S}(T-t)$$

is a property of the solution matrix (as we saw in Section 3–1).

Since the end state $x(T)$ is not specified in the problem, we have the transversality condition

$$\psi(T) = 0. \qquad (14\text{--}76)$$

Therefore, by Eqs. (14–72) and (14–75), we have:

$$\psi(T) = S_{21}(T-t) \cdot x(t) + S_{22}(T-t) \cdot \psi(t) = 0,$$

or

$$\tfrac{1}{2}\psi(t) = -h(t) \cdot x(t), \qquad (14\text{--}77)$$

where

$$h(t) = \frac{1}{2} \frac{S_{21}(T-t)}{S_{22}(T-t)}. \qquad (14\text{--}78)$$

Since the optimal control is $u = \tfrac{1}{2}b\psi$ (as in Eq. 14–70), it follows that:

$$u = -bh(t)x(t) = -g(t)x(t),$$

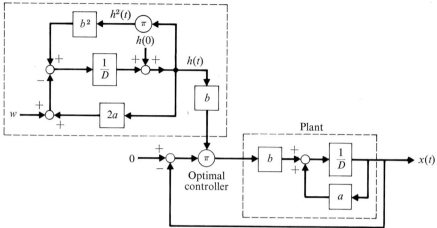

Fig. 14-13 Feedback optimal control of a first-order object.

where the time-varying feedback-control gain $g(t)$ is given by

$$g(t) = bh(t) = \frac{1}{2} b \left(\frac{S_{21}(T-t)}{S_{22}(T-t)} \right)$$

$$= wb \frac{e^{p(T-t)} - e^{-p(T-t)}}{(p-a)e^{p(T-t)} + (p+a)e^{-p(T-t)}} . \qquad (14\text{-}79)$$

A scheme of feedback realization of this optimal control is shown in Fig. 14-13. The control applies only for the prescribed time interval, from time 0 to T. To obtain an algorithm for the generation of the time-varying gain $g(t)$, which is shown in the figure, we derive a differential equation for either $g(t)$ or $h(t)$. Substituting $\frac{1}{2}\psi = -hx$ in the differential equation for ψ,

$$\frac{d\psi}{dt} = 2wx - a\psi \quad \rightarrow \quad -\frac{dh}{dt} x - h \frac{dx}{dt} = wx + ahx ,$$

where

$$\frac{dx}{dt} = ax + bu = ax - b^2 hx .$$

Therefore,

$$\frac{dh}{dt} x = (-2ah + b^2 h^2 - w)x .$$

Since this condition must hold for optimal control that starts from any initial state, that is, for all x, we conclude that

$$\frac{dh}{dt} = -2ah + b^2 h^2 - w , \qquad (14\text{-}80)$$

or, replacing h by $bh = g$, that

$$\frac{dg}{dt} = -2ag + bg^2 - wb .$$

Equation (14–79) is a solution of this Riccati-type equation. A computing network (or program) for Eq. (14–80) and Fig. 14–13 must be initialized by $h(0)$ or $g(0)$, which is determined by letting $t = 0$ in Eq. (14–79), or the computation must proceed backward in time, starting with $h(T) = 0$ (Eq. 14–76). An instability in the computation can be avoided by the reverse-time algorithm.

The optimal gain becomes constant if $T = \infty$. Constant gain g_s for this case can be obtained either by letting $T = \infty$ in Eq. (14–79) or by taking the positive root of the steady-state solution of the quadratic equation

$$bg_s^2 - 2ag_s - wb = 0 .$$

The gain is

$$g_s = \frac{p + a}{b} = \frac{wb}{p - a} .$$

The two expressions for g_s are equivalent because $p = \sqrt{a^2 + wb^2}$. Note that the optimal control is $u = 0$ ("do nothing") if $w = 0$ and $T = \infty$.

This feedback solution to optimal control of a first-order object can be generalized for an nth-order control object,

$$\frac{d\mathbf{x}}{dt} = \mathbf{Ax} + \mathbf{Bu} .$$

We redefine the performance index as the following quadratic form:

$$J = \int_0^T (\mathbf{x}'\mathbf{F}_x\mathbf{x} + \mathbf{u}'\mathbf{F}_u\mathbf{u}) \, dt ,$$

where $\mathbf{F}_x$ is a nonnegative $n \times n$ symmetric matrix, and $\mathbf{F}_u$ is a positive definite (zero penalty for excessive control is not allowed) $r \times r$ symmetric matrix when $\mathbf{u}$ is an r-vector. To derive a feedback-control law, we first write $\boldsymbol{\phi}(t)$ as:

$$\tfrac{1}{2}\boldsymbol{\phi}(t) = -\mathbf{H}(t)\mathbf{x}(t) , \qquad (14\text{–}81)$$

without really evaluating the solution matrix $\mathbf{S}(t)$. This is justified by the relation

$$\begin{bmatrix} \mathbf{x}(T) \\ \boldsymbol{\phi}(T) \end{bmatrix} = \mathbf{S}(T - t) \begin{bmatrix} \mathbf{x}(t) \\ \boldsymbol{\phi}(t) \end{bmatrix} = \begin{bmatrix} \mathbf{S}_{11}(T - t) & \mathbf{S}_{12}(T - t) \\ \mathbf{S}_{21}(T - t) & \mathbf{S}_{22}(T - t) \end{bmatrix} \begin{bmatrix} \mathbf{x}(t) \\ \boldsymbol{\phi}(t) \end{bmatrix}$$

and

$$\boldsymbol{\phi}(T) = \mathbf{0} ;$$

hence

$$\tfrac{1}{2}\varphi(t) = -\tfrac{1}{2}\mathbf{S}_{22}^{-1}(T-t)\mathbf{S}_{21}(T-t)\mathbf{x}(t) \equiv -\mathbf{H}(t)\mathbf{x}(t) \ .$$

The feedback optimal control law for this system is

$$\begin{aligned}\mathbf{u}^0(t) &= -\mathbf{G}(t)\mathbf{x}(t) \ , \\ \mathbf{G}(t) &= \mathbf{F}_u^{-1}\mathbf{B}'\mathbf{H}(t) \ , \end{aligned} \qquad (14\text{-}82)$$

and $\mathbf{H}(t)$ must satisfy a matrix differential equation of the Riccati type:

$$\frac{d\mathbf{H}(t)}{dt} = -\mathbf{A}'\mathbf{H} - \mathbf{H}\mathbf{A} + \mathbf{H}\mathbf{B}\mathbf{F}_u^{-1}\mathbf{B}'\mathbf{H} - \mathbf{F}_x \ . \qquad (14\text{-}83)$$

As mentioned before, Eq. (14–83) is "initialized" by $\mathbf{H}(T) = \mathbf{0}$, and the computation proceeds backwards in time. The equation reduces to the steady-state condition $d\mathbf{H}/dt = \mathbf{0}$ if $T \to \infty$, and we obtain Eq. (10–101). (In the latter equation, symbol $\mathbf{Y}$ was used for our present symbol $\mathbf{H}$.)

Example 14-10 The control object is a double-integrator system described by

$$\frac{d}{dt}\begin{bmatrix} x_1 \\ x_2 \end{bmatrix} = \begin{bmatrix} 0 & 1 \\ 0 & 0 \end{bmatrix}\begin{bmatrix} x_1 \\ x_2 \end{bmatrix} + \begin{bmatrix} 0 \\ 1 \end{bmatrix} u \ ,$$

and the performance index to be minimized is

$$J = \int_0^\infty (x_1^2 + wu^2)\, dt \ .$$

To determine the steady-state control law for this system, we solve the Riccati equation for elements h_{11}, $h_{12} = h_{21}$, and h_{22} of matrix $\mathbf{H}$:

$$\frac{d\mathbf{H}}{dt} = -\begin{bmatrix} 0 & h_{11} \\ h_{11} & 2h_{12} \end{bmatrix} + \frac{1}{w}\begin{bmatrix} h_{12}^2 & h_{12}h_{22} \\ h_{12}h_{22} & h_{22}^2 \end{bmatrix} - \begin{bmatrix} 1 & 0 \\ 0 & 0 \end{bmatrix} = \mathbf{0} \ .$$

The matrix equation yields three scalar conditions,

$$h_{12}^2 = w \ , \qquad h_{12}h_{22} = wh_{11} \ , \qquad h_{22}^2 = 2wh_{12} \ ,$$

from which we choose the following solution, which will give a stable feedback control:

$$h_{11} = \sqrt{2\sqrt{w}} \ , \qquad h_{12} = \sqrt{w} \ , \qquad h_{22} = \sqrt{2w\sqrt{w}} \ ;$$

hence

$$\mathbf{H} = \begin{bmatrix} \sqrt{2\sqrt{w}} & \sqrt{w} \\ \sqrt{w} & \sqrt{2w\sqrt{w}} \end{bmatrix} \ .$$

By Eq. (14–82), the control law is

$$\mathbf{G} = \frac{1}{w}\begin{bmatrix} 0 & 1 \end{bmatrix}\begin{bmatrix} h_{11} & h_{12} \\ h_{21} & h_{22} \end{bmatrix} = \begin{bmatrix} \dfrac{1}{\sqrt{w}} & \sqrt{\dfrac{2}{\sqrt{w}}} \end{bmatrix} \ .$$

The optimal control system is shown in Fig. 14-14. If $w = 1$, the characteristic

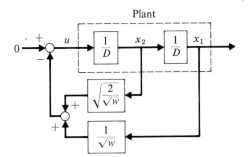

Fig. 14-14 A regulator system (Example 14–10).

equation of the control system will be:

$$s^2 + \sqrt{2}\,s + 1 = 0$$

which yields a damping factor $\zeta = 1/\sqrt{2} = 0.707$, which agrees with the usual rule of thumb that $\zeta = 0.7$ for adequate damping.

The preceding optimal-control conditions, Eqs. (14-81) through (14–83), must be modified if the m-vector output **y** of a control object includes a direct input-to-output coupling **Du**,

$$\mathbf{y} = \mathbf{Cx} + \mathbf{Du},$$

and the performance index is specified in terms of the output as

$$J = \int_0^T (\mathbf{y}'\mathbf{F}_y\mathbf{y} + \mathbf{u}'\mathbf{F}_u\mathbf{u})\, dt$$

where the $m \times m$ weighting matrix $\mathbf{F}_y$ is positive semidefinite.

We have

$$\begin{aligned}f_0(\mathbf{x}, \mathbf{u}) &= \mathbf{y}'\mathbf{F}_y\mathbf{y} + \mathbf{u}'\mathbf{F}_u\mathbf{u} = (\mathbf{x}'\mathbf{C}' + \mathbf{u}'\mathbf{D}')\mathbf{F}_y(\mathbf{Cx} + \mathbf{Du}) + \mathbf{u}'\mathbf{F}_u\mathbf{u}\\ &= \mathbf{x}'(\mathbf{C}'\mathbf{F}_y\mathbf{C})\mathbf{x} + (\mathbf{x}'\mathbf{C}'\mathbf{F}_y\mathbf{Du} + \mathbf{u}'\mathbf{D}'\mathbf{F}_y\mathbf{Cx}) + \mathbf{u}'(\mathbf{D}'\mathbf{F}_y\mathbf{D} + \mathbf{F}_u)\mathbf{u}\end{aligned}$$

where

$$\mathbf{x}'\mathbf{C}'\mathbf{F}_y\mathbf{Du} = \mathbf{u}'\mathbf{D}'\mathbf{F}_y\mathbf{Cx},$$

so that

$$f_0(\mathbf{x}, \mathbf{u}) = \mathbf{x}'\mathbf{F}_x\mathbf{x} + 2\mathbf{u}'\mathbf{F}_c\mathbf{x} + \mathbf{u}'\mathbf{F}_0\mathbf{u}, \tag{14-84}$$

in which

$$\mathbf{F}_x = \mathbf{C}'\mathbf{F}_y\mathbf{C}, \qquad \mathbf{F}_c = \mathbf{D}'\mathbf{F}_y\mathbf{C}, \qquad \mathbf{F}_0 = \mathbf{D}'\mathbf{F}_y\mathbf{D} + \mathbf{F}_u.$$

Equation (14–84), with Eq. (14–31), gives the Hamiltonian

$$H = -\mathbf{x}'\mathbf{F}_x\mathbf{x} - 2\mathbf{u}'\mathbf{F}_c\mathbf{x} - \mathbf{u}'\mathbf{F}_0\mathbf{u} + \boldsymbol{\phi}'(\mathbf{Ax} + \mathbf{Bu}).$$

Now $\partial H/\partial \mathbf{u} = \mathbf{0}$ for max H, so that

$$-2\mathbf{F}_c\mathbf{x} - 2\mathbf{F}_0\mathbf{u}^0 + \mathbf{B}'\boldsymbol{\phi} = \mathbf{0},$$

or

$$\mathbf{u}^0 = \mathbf{F}_0^{-1}[\mathbf{B}\tfrac{1}{2}\boldsymbol{\phi} - \mathbf{F}_c\mathbf{x}]. \tag{14-85}$$

Following the same logic as in the first-order case, but without really computing the solution matrix, we deduce the following vector form of Eq. (14-77):

$$\tfrac{1}{2}\boldsymbol{\phi}(t) = -\mathbf{H}(t)\mathbf{x}(t). \tag{14-86}$$

Substituting Eq. (14-86) into (14-85), the feedback control $\mathbf{G}(t)$ in

$$\mathbf{u}^0(t) = -\mathbf{G}(t)\mathbf{x}(t)$$

is given in terms of $\mathbf{H}(t)$ as

$$\mathbf{G}(t) = \mathbf{F}_0^{-1}(\mathbf{B}'\mathbf{H}(t) + \mathbf{F}_c). \tag{14-87}$$

In order to avoid the use of the solution matrix in the algorithm, we substitute Eq. (14-86) into Eq. (14-30) and obtain the following condition for $\mathbf{H}(t)$:

$$\frac{d}{dt}\mathbf{H} = -\mathbf{A}'\mathbf{H} - \mathbf{H}\mathbf{A} + \mathbf{H}\mathbf{B}\mathbf{F}_0^{-1}\mathbf{B}'\mathbf{H}$$
$$- \mathbf{F}_x + (\mathbf{H}\mathbf{B}\mathbf{F}_0^{-1}\mathbf{F}_c + \mathbf{F}_c'\mathbf{F}_0^{-1}\mathbf{B}'\mathbf{H}) + \mathbf{F}_c'\mathbf{F}_0^{-1}\mathbf{F}_c. \tag{14-88}$$

As before, machine computation of Eq. (14-88) is "initialized" by $\mathbf{H}(T) = \mathbf{0}$ and proceeds backward in time. The transient-solution approach is also recommended for the computation of a symmetric constant matrix $\mathbf{H}$ for the steady state that applies when $T \to \infty$.

Example 14-11 In Fig. 14-15(a), x_1 is a command signal to be tracked by the plant's output y_2. The tracking error y_1 is

$$y_1 = x_1 - y_2,$$

and an integrated squared error is involved in the performance index,

$$J = \int_0^\infty (y_1^2 + wu^2)\,dt.$$

Let us suppose that a value $w = \tfrac{1}{2}$ was found to be adequate in avoiding excessive magnitude of control, and that for this value of w we then seek to design an optimal control system.

The matrices of the system's equations

$$\frac{d\mathbf{x}}{dt} = \mathbf{A}\mathbf{x} + \mathbf{B}u, \qquad \mathbf{y} = \mathbf{C}\mathbf{x} + \mathbf{D}u$$

14-4 Linear optimal feedback control   665

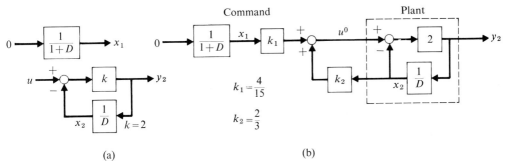

Fig. 14-15 A tracking system design (Example 14-11).

are found to be

$$\mathbf{A} = \begin{bmatrix} -1 & 0 \\ 0 & -2 \end{bmatrix}, \quad \mathbf{B} = \begin{bmatrix} 0 \\ 2 \end{bmatrix}, \quad \mathbf{C} = \begin{bmatrix} 1 & 2 \\ 0 & -2 \end{bmatrix}, \quad \mathbf{D} = \begin{bmatrix} -2 \\ 2 \end{bmatrix}.$$

The prescribed performance index is represented by

$$\mathbf{F}_y = \begin{bmatrix} 1 & 0 \\ 0 & 0 \end{bmatrix}, \quad \mathbf{F}_u = w = \tfrac{1}{2}.$$

Therefore it follows that

$$\mathbf{F}_z = \mathbf{C}'\mathbf{F}_y\mathbf{C} = \begin{bmatrix} 1 & 2 \\ 2 & 4 \end{bmatrix},$$
$$\mathbf{F}_c = \mathbf{D}'\mathbf{F}_y\mathbf{C} = [-2 \quad -4],$$
$$\mathbf{F}_0 = \mathbf{D}'\mathbf{F}_y\mathbf{D} + \mathbf{F}_u = 4.5.$$

Substituting these **F**-values into Eq. (14–88), and letting $d\mathbf{H}/dt = \mathbf{0}$, we obtain the following three conditions for the three unknown elements h_{11}, $h_{12} = h_{21}$, and h_{22}, of **H**:

$$18h_{11} + 8h_{12}^2 - 16h_{12} = 1,$$
$$27h_{12} + 8h_{12}h_{22} - 16h_{12} - 8h_{22} = 2,$$
$$36h_{22} + 8h_{22}^2 - 32h_{22} = 4.$$

Solving the third equation for h_{22}, we find $h_{22} = +\tfrac{1}{2}$ or -1, and, by the second equation,

$$h_{12} = \tfrac{2}{5} \text{ if } h_{22} = \tfrac{1}{2}, \qquad h_{12} = -2 \text{ if } h_{22} = -1.$$

By Eq. (14–87) the control law is

$$\mathbf{G} = \frac{2}{4.5}[(h_{12} - 1) \quad (h_{22} - 2)] = [-\tfrac{4}{15} \quad -\tfrac{2}{3}] \quad \text{or} \quad [-\tfrac{4}{3} \quad -\tfrac{4}{3}].$$

We reject the second solution because it makes the controlled plant (Fig. 14–15b) unstable, and conclude that the optimal control is

$$u^0(t) = -\mathbf{G}\mathbf{x} = k_1 x_1 + k_2 x_2, \qquad k_1 = \tfrac{4}{15}, \qquad k_2 = \tfrac{2}{3}.$$

666   Optimal control

If Eq. (14–88) were solved by computer, initialized by $\mathbf{H} = \mathbf{0}$, and reversed in time, the "wrong answer" will be ruled out automatically, and $\mathbf{H}(t)$ will converge to the correct steady-state values,

$$h_{12} = \tfrac{2}{5} \quad \text{and} \quad h_{22} = \tfrac{1}{2}.$$

The optimal control structure is shown in Fig. 14–15(b). Note that the structure was not assumed in the problem statement (Fig. 14–15a). If the structure is assumed to have k_1 and k_2 as unknown coefficients, it is possible to compute the quadratic performance index J by means of Eq. (4–26), where J will be a function of k_1, k_2, and initial state $\mathbf{x}(0)$. Then the values $k_1 = \tfrac{4}{15}$ and $k_2 = \tfrac{2}{3}$ will be determined by the condition that J must be minimal for all possible initial states. Such an approach is tedious, and in addition, there is no assurance that the assumed structure will be optimal. Equation (4–26), however, is useful to evaluate the value of J for an optimal control system.

In this example the command signal (an exponential decay of unit time constant) was replaced by a signal generator $1/(1 + D)$, and the optimal control solution was obtained for an equivalent free system. The approach closely parallels the "shaping filter" construction shown in Example 13–5 for an optimal filtering problem. The signal generator is decoupled from the system in the problem statement; hence the system of Fig. 14–15(a) is uncontrollable. This example, therefore, serves to demonstrate optimal control of an uncontrollable system.

14–5 DISCRETE-TIME OPTIMAL CONTROL

The maximum principle applies to discrete time systems but with some modifications [25]. Consider a control object described by the difference equation

$$\mathbf{x}_{k+1} = \mathbf{f}(\mathbf{x}_k, \mathbf{u}_k, k), \tag{14–89}$$

and the performance index

$$J = \sum_{k=0}^{N-1} f_0^*(\mathbf{x}_{k+1}, \mathbf{u}_k, k), \tag{14–90}$$

where N is the number of periods of the optimal control. The required unknown is a vector time series $\mathbf{u}_k$, where $k = 0, 1, \ldots, (N-1)$. Since $\mathbf{u}_{N-1}$ is the last control applied, and $\mathbf{x}_N$ is the final state, the cost increment f_0^* in Eq. (14–90) is expressed as a function of $\mathbf{x}_{k+1}$ and $\mathbf{u}_k$ for the period from kT to $(k+1)T$ (where T is the sampling period). However, $\mathbf{x}_{k+1}$ is the state the system will reach when the period ends; hence the condition for an optimal choice of $\mathbf{u}_k$ must be based upon $\mathbf{x}_k$ instead of $\mathbf{x}_{k+1}$. For this reason $\mathbf{x}_{k+1}$ in f_0^* must be replaced by the right-hand side of Eq. (14–89):

$$f_0^*(\mathbf{x}_{k+1}, \mathbf{u}_k, k) = f_0^*(\mathbf{f}(\mathbf{x}_k, \mathbf{u}_k, k), \mathbf{u}_k, k) = f_0(\mathbf{x}_k, \mathbf{u}_k, k). \tag{14–91}$$

14-5 Discrete-time optimal control

As before (in Eq. 14–90), the zeroth state variable is assigned to the performance index, so that

$$x_{0,k+1} = x_{0,k} + f_0(\mathbf{x}_k, \mathbf{u}_k, k) .$$

Combining $x_{0,k}$ and state vector $\mathbf{x}_k$ into an $(n+1)$-vector $\mathbf{X}_k$, we obtain the following equation of motion of the deviation vector $\delta \mathbf{X}_k$ (for a second-order object):

$$\delta \mathbf{X}_{k+1} = \begin{bmatrix} \delta x_{0,k+1} \\ \delta x_{1,k+1} \\ \delta x_{2,k+1} \end{bmatrix} = \begin{bmatrix} 1 & \partial f_0^k/\partial x_{1,k} & \partial f_0^k/\partial x_{2,k} \\ 0 & \partial f_1^k/\partial x_{1,k} & \partial f_1^k/\partial x_{2,k} \\ 0 & \partial f_2^k/\partial x_{1,k} & \partial f_2^k/\partial x_{2,k} \end{bmatrix} \delta \mathbf{X}_k , \qquad (14\text{–}92)$$

where f^k is an abbreviation of $f(\mathbf{x}_k, \mathbf{u}_k, k)$. The first element of the Jacobian matrix is 1 because $x_{0,k+1}$ has $x_{0,k}$ as its first term on the right-hand side. Since $x_{0,k}$ is not involved in $\mathbf{f}(\mathbf{x}_k, \mathbf{u}_k)$, other elements in the first column of the matrix are all zero. We next define the costate vector by the transposed Jacobian matrix,

$$\begin{bmatrix} \psi_{0,k} \\ \psi_{1,k} \\ \psi_{2,k} \end{bmatrix} = \begin{bmatrix} 1 & 0 & 0 \\ \dfrac{\partial f_0^k}{\partial x_{1,k}} & \dfrac{\partial f_1^k}{\partial x_{1,k}} & \dfrac{\partial f_2^k}{\partial x_{1,k}} \\ \dfrac{\partial f_0^k}{\partial x_{2,k}} & \dfrac{\partial f_1^k}{\partial x_{2,k}} & \dfrac{\partial f_2^k}{\partial x_{2,k}} \end{bmatrix} \begin{bmatrix} \psi_{0,k+1} \\ \psi_{1,k+1} \\ \psi_{2,k+1} \end{bmatrix} . \qquad (14\text{–}93)$$

Note that the sequence from k to $(k+1)$ is reversed, instead of using the negative transpose of the Jacobian matrix. Writing $\mathbf{M}_J$ for the Jacobian matrix of Eq. (14–92), we have

$$\delta \mathbf{X}_{k+1} = \mathbf{M}_J \delta \mathbf{X}_k , \qquad \mathbf{\Psi}_k = \mathbf{M}_J' \mathbf{\Psi}_{k+1} ,$$

so that

$$\mathbf{\Psi}_k' \delta \mathbf{X}_k = (\mathbf{M}_J' \mathbf{\Psi}_{k+1})' \delta \mathbf{X}_k = \mathbf{\Psi}_{k+1}'(\mathbf{M}_J \delta \mathbf{X}_k) = \mathbf{\Psi}_{k+1}' \delta \mathbf{X}_{k+1}$$

or

$$\mathbf{\Psi}_k' \delta \mathbf{X}_k = \text{const} . \qquad (14\text{–}94)$$

Equation (14–94) parallels Eq. (14–35) of continuous-time systems. Also, we see from Eq. (14–93) that $\psi_{0,k}$ does not change, so, as in Eq. (14–27), we choose -1 for this constant:

$$\psi_{0,k} = -1 \qquad \text{for all} \quad k .$$

The form for the Hamiltonian also parallels Eq. (14–23),

$$H = \mathbf{\Psi}_{k+1}' \mathbf{F}(\mathbf{x}_k, \mathbf{u}_k, k) = -f_0(\mathbf{x}_k, \mathbf{u}_k, k) + \boldsymbol{\phi}_{k+1}' \mathbf{f}(\mathbf{x}_k, \mathbf{u}_k, k) . \qquad (14\text{–}95)$$

Following almost the same procedure as for continuous-time systems, we arrive at the discrete maximum principle, which states that the Hamiltonian (Eq. 14–95) is maximum when $\mathbf{u}_k$ is optimal.

Despite the similarity between the two systems, however, there are subtle but important differences. Since changes that occur in adjacent sampling periods are finite, as opposed to the infinitesimally small state change $d\mathbf{x}$ per dt time interval of the continuous-time system, the discrete maximum principle does not apply to systems that have constraints on either control or state variables [26]. Also, the maximum value of H does not assume a constant (or zero) value.

If the control object is linear (but time-varying),

$$\mathbf{x}_{k+1} = \mathbf{P}_k \mathbf{x}_k + \mathbf{Q}_k \mathbf{u}_k, \qquad (14\text{–}96)$$

and if the performance index is quadratic in terms of symmetric, time-varying weighting matrices $\mathbf{F}_{x,k}$ and $\mathbf{F}_{u,k}$ for state $\mathbf{x}$ and control $\mathbf{u}$, respectively,

$$J = \sum_{k=0}^{N-1} [\mathbf{x}'_{k+1}\mathbf{F}_{x,k+1}\mathbf{x}_{k+1} + \mathbf{u}'_k \mathbf{F}_{u,k}\mathbf{u}_k], \qquad (14\text{–}97)$$

then Eqs. (14–93) and (14–95) will take the following form:

$$\boldsymbol{\phi}_k = -\frac{\partial f_0(\mathbf{x}_k, \mathbf{u}_k, k)}{\partial \mathbf{x}_k} + \mathbf{P}'_k \boldsymbol{\phi}_{k+1}, \qquad (14\text{–}98)$$

$$H = -f_0(\mathbf{x}_k, \mathbf{u}_k, k) + \boldsymbol{\phi}'_{k+1}(\mathbf{P}_k\mathbf{x}_k + \mathbf{Q}_k\mathbf{u}_k), \qquad (14\text{–}99)$$

where

$$\begin{aligned} f_0(\mathbf{x}_k, \mathbf{u}_k, k) &= \mathbf{x}'_{k+1}\mathbf{F}_{x,k+1}\mathbf{x}_{k+1} + \mathbf{u}'_k \mathbf{F}_{u,k}\mathbf{u}_k \\ &= (\mathbf{x}'_k\mathbf{P}'_k + \mathbf{u}'_k\mathbf{Q}'_k)\mathbf{F}_{x,k+1}(\mathbf{P}_k\mathbf{x}_k + \mathbf{Q}_k\mathbf{u}_k) + \mathbf{u}'_k \mathbf{F}_{u,k}\mathbf{u}_k \\ &= \mathbf{x}'_k\mathbf{P}'_k\mathbf{F}_{x,k+1}\mathbf{P}_k\mathbf{x}_k + \mathbf{u}'_k(\mathbf{Q}'_k\mathbf{F}_{x,k+1}\mathbf{Q}_k + \mathbf{F}_{u,k})\mathbf{u}_k \\ &\quad + 2\mathbf{x}'_k\mathbf{P}'_k\mathbf{F}_{x,k+1}\mathbf{Q}_k\mathbf{u}_k. \end{aligned} \qquad (14\text{–}100)$$

The equality relation for the scalar quantity,

$$\mathbf{u}'_k\mathbf{Q}'_k\mathbf{F}_{x,k+1}\mathbf{P}_k\mathbf{x}_k = \mathbf{x}'_k\mathbf{P}'_k\mathbf{F}_{x,k+1}\mathbf{Q}_k\mathbf{u}_k$$

is utilized in the reduction process. Recalling Eq. (4–8), we determine the gradient term in (14–98) from Eq. (14–100) as

$$\begin{aligned} \frac{\partial f_0(\mathbf{x}_k, \mathbf{u}_k, k)}{\partial \mathbf{x}_k} &= 2\mathbf{P}'_k\mathbf{F}_{x,k+1}\mathbf{P}_k\mathbf{x}_k + 2\mathbf{P}'_k\mathbf{F}_{x,k+1}\mathbf{Q}_k\mathbf{u}_k \\ &= 2\mathbf{P}'_k\mathbf{F}_{x,k+1}(\mathbf{P}_k\mathbf{x}_k + \mathbf{Q}_k\mathbf{u}_k) = 2\mathbf{P}'_k\mathbf{F}_{x,k+1}\mathbf{x}_{k+1}, \end{aligned}$$

so that Eq. (14–98) becomes

$$\tfrac{1}{2}\boldsymbol{\phi}_k = -\mathbf{P}'_k[\mathbf{F}_{x,k+1}\mathbf{x}_{k+1} - \tfrac{1}{2}\boldsymbol{\phi}_{k+1}]. \qquad (14\text{–}101)$$

Table 14-1 OPTIMAL CONTROL CONDITIONS FOR A LINEAR OBJECT WITH A QUADRATIC PERFORMANCE INDEX

	Discrete-time system	$T \to dt$	Continuous-time system
		Linear control object	
1	$\mathbf{x}_{k+1} = \mathbf{P}_k \mathbf{x}_k + \mathbf{Q}_k \mathbf{u}_k, \quad \mathbf{x}(0) = \mathbf{x}_0$	$\mathbf{P}_k \to \mathbf{I} + \mathbf{A}(t)\,dt,$ $\mathbf{Q}_k \to \mathbf{B}(t)\,dt$	$\dfrac{d}{dt}\mathbf{x}(t) = \mathbf{A}(t)\mathbf{x}(t) + \mathbf{B}(t)\mathbf{u}(t), \quad \mathbf{x}(0) = \mathbf{x}_0$
		Performance index	
2†	$J = \displaystyle\sum_{k=0}^{N-1} f_0^*(\mathbf{x}_{k+1}, \mathbf{u}_k, k)$ $= \displaystyle\sum_{k=0}^{N-1}\{\mathbf{x}'_{k+1}\mathbf{F}_{x,k+1}\mathbf{x}_{k+1} + \mathbf{u}'_k \mathbf{F}_{u,k}\mathbf{u}_k\}$	$\mathbf{F}_{x,k} \to \mathbf{F}_x(t)\,dt,$ $\mathbf{F}_{u,k} \to \mathbf{F}_u(t)\,dt$	$J = \displaystyle\int_0^T f_0(\mathbf{x}(t), \mathbf{u}(t), t)\,dt$ $= \displaystyle\int_0^T \{\mathbf{x}'(t)\mathbf{F}_x(t)\mathbf{x}(t) + \mathbf{u}'(t)\mathbf{F}_u(t)\mathbf{u}(t)\}\,dt$
		Equation for covariant n-vector	
3	$\tfrac{1}{2}\boldsymbol{\phi}_k = -\dfrac{1}{2}\dfrac{\partial f_0}{\partial \mathbf{x}_k} + \mathbf{P}'_k(\tfrac{1}{2}\boldsymbol{\phi}_{k+1})$ $= -\mathbf{P}'_k\{\mathbf{F}_{x,k+1}\mathbf{x}_{k+1} - \tfrac{1}{2}\boldsymbol{\phi}_{k+1}\}$ $\boldsymbol{\phi}_N = 0 \quad \text{if } \mathbf{x}_N \text{ is not specified}$	$\boldsymbol{\phi}_k \to \boldsymbol{\phi}(t),$ $\boldsymbol{\phi}_{k+1} \to \boldsymbol{\phi}(t) + d\boldsymbol{\phi}$	$\dfrac{d}{dt}\left(\tfrac{1}{2}\boldsymbol{\phi}(t)\right) = \dfrac{1}{2}\dfrac{\partial f_0}{\partial \mathbf{x}(t)} - \mathbf{A}'(t)(\tfrac{1}{2}\boldsymbol{\phi}(t))$ $= \mathbf{F}_x(t)\mathbf{x}(t) - \mathbf{A}'(t)(\tfrac{1}{2}\boldsymbol{\phi}(t))$ $\boldsymbol{\phi}(T) = 0 \quad \text{if } \mathbf{x}(T) \text{ is not specified}$
		Hamiltonian for the maximum principle	
4	$H = -f_0(\mathbf{x}_k, \mathbf{u}_k, k) + \boldsymbol{\phi}'_{k+1}(\mathbf{P}_k \mathbf{x}_k + \mathbf{Q}_k \mathbf{u}_k)$		$H = -f_0(\mathbf{x}(t), \mathbf{u}(t), t) + \boldsymbol{\phi}'(t)(\mathbf{A}(t)\mathbf{x}(t) + \mathbf{B}(t)\mathbf{u}(t))$
		Optimal control	
5	$\dfrac{\partial H}{\partial \mathbf{u}_k} = 0; \quad \mathbf{u}_k^0 = \mathbf{F}_{u,k}^{-1} \mathbf{Q}'_k (\tfrac{1}{2}\boldsymbol{\phi}_{k+1} - \mathbf{F}_{x,k+1}\mathbf{x}_{k+1})$		$\dfrac{\partial H}{\partial \mathbf{u}(t)} = 0; \quad \mathbf{u}^0(t) = \mathbf{F}_u^{-1}(t)\mathbf{B}'(t)(\tfrac{1}{2}\boldsymbol{\phi}(t))$

† $f_0^*(\mathbf{x}_{k+1}, \mathbf{u}_k, k) = f_0(\mathbf{P}_k \mathbf{x}_k + \mathbf{Q}_k \mathbf{u}_k, \mathbf{u}_k, k)$.

670   Optimal control

These relations are compared with corresponding relations of the continuous-time system in Table 14–1, #1 through #3. Note that the finite-difference relations on the left-hand side of the table are converted into the differential and integral relations on the right-hand side by the limiting process shown in the middle. The conversion, however, does not apply for the Hamiltonian, #4 of the Table.

According to the discrete maximum principle, an optimal control must satisfy $\partial H/\partial \mathbf{u}_k = 0$; hence

$$-\frac{\partial f_0(\mathbf{x}_k, \mathbf{u}_k, k)}{\partial \mathbf{u}_k} + \mathbf{Q}'_k \boldsymbol{\phi}_{k+1} = 0, \qquad (14\text{–}102)$$

where the gradient term is computed from Eq. (14–100) as

$$\frac{\partial f_0(\mathbf{x}_k, \mathbf{u}_k, k)}{\partial \mathbf{u}_k} = 2(\mathbf{Q}'_k \mathbf{F}_{x,k+1} \mathbf{Q}_k + \mathbf{F}_{u,k})\mathbf{u}_k + 2\mathbf{Q}'_k \mathbf{F}_{x,k+1} \mathbf{P}_k \mathbf{x}_k$$

$$= 2\mathbf{F}_{u,k}\mathbf{u}_k + 2\mathbf{Q}'_k \mathbf{F}_{x,k+1}(\mathbf{Q}_k \mathbf{u}_k + \mathbf{P}_k \mathbf{x}_k)$$

$$= 2\mathbf{F}_{u,k}\mathbf{u}_k + 2\mathbf{Q}'_k \mathbf{F}_{x,k+1}\mathbf{x}_{k+1}. \qquad (14\text{–}103)$$

Substituting Eq. (14–103) into Eq. (14–102), and solving for $\mathbf{u}_k$, we find the optimal control to be

$$\mathbf{u}^0_k = \mathbf{F}^{-1}_{u,k}\mathbf{Q}'_k(\tfrac{1}{2}\boldsymbol{\phi}_{k+1} - \mathbf{F}_{x,k+1}\mathbf{x}_{k+1}). \qquad (14\text{–}104)$$

Note that the weighting matrix $\mathbf{F}_u$ for the control vector must be positive definite in order for this optimal control to exist. Equation (14–104) also gives the continuous-time optimal control at the limit $T \to dt$, as shown in #5 of Table 14–1.

Example 14–12 The control object is first-order and stationary:

$$x_{k+1} = 0.5x_k + u_k \; .$$

The performance index to be minimized is specified as

$$J = F_{x,N}x_N^2 + \sum_{k=0}^{N-1}(F_x x_k^2 + F_u u_k^2),$$

where

$$F_x = 1, \qquad F_{x,N} = 6, \qquad F_u = 4 \quad \text{and} \quad N = 3 \; .$$

The form of this result, which is often seen in the literature, differs from Eq. (14–97). Here, a different (heavier) weight $F_{x,N}$ is placed only on the terminal state. It includes $F_x x_0^2$ as the first term of the summation, which contributes a constant bias to J since the initial state cannot be affected by control.

Since F_x is constant except for the terminal state, the problem can be solved as stationary. For this purpose we modify Eq. (14–100), and define the following stationary cost increment:

$$f_0(x_k, u_k) = F_{x,N}(x_{k+1}^2 - x_k^2) + (F_x x_k^2 + F_u u_k^2)$$

$$= F_{x,N}(Px_k + qu_k)^2 - (F_{x,N} - F_x)x_k^2 + F_u u_k^2 \; .$$

14-5 Discrete-time optimal control

For our problem

$$f_0(x_k, u_k) = 6(0.5x_k + u_k)^2 - 5x_k^2 + 4u_k^2$$
$$= -3.5x_k^2 + 6x_k u_k + 10u_k^2,$$

and the Hamiltonian is

$$H = 3.5x_k^2 - 6x_k u_k - 10u_k^2 + \psi_{k+1}(0.5x_k + u_k).$$

If we let $\partial H/\partial u_k = 0$, the optimal control is

$$u_k^0 = \tfrac{1}{20}(\psi_{k+1} - 6x_k).$$

By Eq. (14-98), the costate variable is governed by

$$\psi_k = 7x_k - 6u_k + 0.5\psi_{k+1}.$$

Substituting u_k^0 into the equations for x_{k+1} and ψ_k, and combining the two, we obtain

$$\begin{bmatrix} x_{k+1} \\ \psi_k \end{bmatrix} = \begin{bmatrix} \tfrac{1}{5} & \tfrac{1}{20} \\ \tfrac{44}{5} & \tfrac{1}{5} \end{bmatrix} \begin{bmatrix} x_k \\ \psi_{k+1} \end{bmatrix},$$

where $k = 0, 1, 2$, and

$$x_0 = \text{specified const } c.$$

Since terminal state x_3 is not specified in this problem, by the transversality condition (Eq. 14-65), which also applies to discrete time systems,

$$\psi_3 = 0.$$

Solving the two-point boundary-value problem, it can be found that

$$x_0 = c, \quad x_1 = \frac{14}{37}c, \quad x_2 = \frac{5}{37}c, \quad x_3 = \frac{1}{37}c,$$

$$\psi_0 = \frac{352}{37}c, \quad \psi_1 = \frac{132}{37}c, \quad \psi_2 = \frac{44}{37}c, \quad \psi_3 = 0,$$

and

$$u_0^0 = -\frac{4.5}{37}c, \quad u_1^0 = -\frac{2}{37}c, \quad u_2^0 = -\frac{1.5}{37}c.$$

The same result can be obtained more quickly by means of the feedback-control approach which we shall present next.

The optimal control of the linear object Eq. (14-96) for the quadratic performance index (Eq. 14-97) can be implemented by state-vector feedback. To derive the control algorithm, we introduce a time-varying symmetric matrix $\mathbf{H}_k$ that corresponds to $\mathbf{H}(t)$ of Eq. (14-81) in the continuous-time system. Letting

$$\mathbf{F}_{x,k+1}\mathbf{x}_{k+1} - \tfrac{1}{2}\boldsymbol{\psi}_{k+1} = \mathbf{H}_{k+1}\mathbf{x}_{k+1}, \tag{14-105}$$

in Eq. (14-104), we obtain the optimal control:

$$\mathbf{u}_k^0 = -\mathbf{F}_{u,k}^{-1}\mathbf{Q}_k'\mathbf{H}_{k+1}\mathbf{x}_{k+1} = -\mathbf{F}_{u,k}^{-1}\mathbf{Q}_k'\mathbf{H}_{k+1}(\mathbf{P}_k\mathbf{x}_k + \mathbf{Q}_k\mathbf{u}_k^0),$$

so that

$$(I + F_{u,k}^{-1} Q_k' H_{k+1} Q_k) u_k^0 = -F_{u,k}^{-1} Q_k' H_{k+1} P_k x_k$$

or

$$u_k^0 = -G_k x_k ,$$
$$G_k = (F_{u,k} + Q_k' H_{k+1} Q_k)^{-1} Q_k' H_{k+1} P_k . \quad (14\text{--}106)$$

The condition for H_k is derived by substituting ϕ of Eq. (14–105) into Eq. (14–101). The left-hand side of Eq. (14–101) is

$$\tfrac{1}{2}\phi_k = F_{x,k} x_k - H_k x_k .$$

The right-hand side of Eq. (14–101) is $-P_k' H_{k+1} x_{k+1}$, and

$$-P_k' H_{k+1} x_{k+1} = -P_k' H_{k+1} (P_k x_k + Q_k u_k^0) = -P_k' H_{k+1} (P_k - Q_k G_k) x_k ,$$

so that

$$(F_{x,k} - H_k) x_k = -P_k' H_{k+1} (P_k - Q_k G_k) x_k .$$

Since the equality must hold for all x_k, we have

$$F_{x,k} - H_k = -P_k' H_{k+1} (P_k - Q_k G_k)$$

or

$$H_k = P_k' H_{k+1} (P_k - Q_k G_k) + F_{x,k} .$$

Substituting G_k of Eq. (14–106), we find that the following Riccati equation holds for H:

$$H_k = P_k' H_{k+1} P_k$$
$$- P_k' H_{k+1} Q_k (F_{u,k} + Q_k' H_{k+1} Q_k)^{-1} Q_k' H_{k+1} P_k + F_{x,k} . \quad (14\text{--}107)$$

The boundary condition to "initialize" this recursive algorithm is given by the transversality condition

$$\phi_N = 0 \quad (14\text{--}108)$$

if the terminal state x_N is not specified. Equation (14–105) at $k + 1 = N$ then yields

$$H_N = F_{x,N} . \quad (14\text{--}109)$$

Starting with Eq. (14–109), the computation of Eq. (14–107) proceeds backwards:

$$k = N - 1, \ N - 2, \ \ldots, \ 2, \ 1, \ 0 .$$

As shown in Table 14–2, the feedback optimal control law of the discrete-time system reduces to the continuous-time case of the preceding section at the limit $T \to dt$.

Table 14-2 OPTIMAL LINEAR FEEDBACK CONTROL LAW FOR A DETERMINISTIC SYSTEM

	Discrete-time system	$T \to dt$	Continuous-time system
		Definition of $n \times n$ symmetric matrix $\mathbf{H}$	
6	$-\mathbf{F}_{x,k}\mathbf{x}_k + \tfrac{1}{2}\boldsymbol{\phi}_k = -\mathbf{H}_k\mathbf{x}_k$	$\mathbf{H}_k \to \mathbf{H}(t)$	$\tfrac{1}{2}\boldsymbol{\phi}(t) = -\mathbf{H}(t)\mathbf{x}(t)$
		Optimal feedback control law	
7	$\mathbf{u}_k^0 = -\mathbf{G}_k\mathbf{x}_k$ $\mathbf{G}_k = (\mathbf{F}_{u,k} + \mathbf{Q}_k'\mathbf{H}_{k+1}\mathbf{Q}_k)^{-1}\mathbf{Q}_k'\mathbf{H}_{k+1}\mathbf{P}_k$	$\mathbf{G}_k \to \mathbf{G}(t)$	$\mathbf{u}^0(t) = -\mathbf{G}(t)\mathbf{x}(t),\quad \mathbf{G}(t) = \mathbf{F}_u^{-1}(t)\mathbf{B}'(t)\mathbf{H}(t)$
		Matrix Riccati equation	
8	$\mathbf{H}_k = \mathbf{P}_k'\mathbf{H}_{k+1}\mathbf{P}_k$ $\quad - \mathbf{P}_k'\mathbf{H}_{k+1}\mathbf{Q}_k(\mathbf{F}_{u,k} + \mathbf{Q}_k'\mathbf{H}_{k+1}\mathbf{Q}_k)^{-1}\mathbf{Q}_k'\mathbf{H}_{k+1}\mathbf{P}_k + \mathbf{F}_{x,k}$ $\mathbf{H}_N = \mathbf{F}_{x,N}$		$\dfrac{d}{dt}\mathbf{H}(t) = -\mathbf{A}'(t)\mathbf{H}(t) - \mathbf{H}(t)\mathbf{A}(t)$ $\qquad + \mathbf{H}(t)\mathbf{B}(t)\mathbf{F}_u^{-1}(t)\mathbf{B}'(t)\mathbf{H}(t) - \mathbf{F}_x(t),$ $\mathbf{H}(T) = \mathbf{0}$

Example 14-13 The same problem as the preceding example, that is;

$P = 0.5$, $\quad Q = 1$,
$F_{x,k} = 1$ $\quad$ for $k = 1, 2$ $\quad$ and $\quad F_{x,N} = 6$ at $N = 3$.
$F_{u,k} = 4$ $\quad$ for all k.

Equation (14-107) is

$$H_k = \frac{1}{4} H_{k+1} - \frac{\frac{1}{4} H_{k+1}^2}{4 + H_{k+1}} + 1 \quad \text{for } k = 1, 2;$$

and by Eq. (14-109), $H_3 = 6$, so we find that

$$H_2 = \tfrac{8}{5}, \quad H_1 = 9/7.$$

The controller gain G_k is given by Eq. (14-106) as

$$G_k = \frac{\frac{1}{2} H_{k+1}}{4 + H_{k+1}};$$

hence

$$G_2 = \tfrac{3}{10}, \quad G_1 = \tfrac{1}{7}, \quad G_0 = \tfrac{9}{74},$$

and we obtain the same control sequence as in the preceding example.

Example 14-14 Here we discuss control of the same object,

$$x_{k+1} = 0.5 x_k + u_k,$$

for the performance index with weighting factors

$F_{x,k} = 1$, $\quad F_{u,k} = 4$ $\quad$ for all k,

and $N = 3$, with the terminal state specified:

$$x_N = x_3 = 0.$$

The specified target implies the limiting case in which the weight on the terminal state is infinite in the problem where the terminal state is not given:

$$F_{x,N} \to \infty;$$

this produces a difficulty in the initialization of Eq. (14-107). One way to avoid the difficulty is to choose a sufficiently large value for $F_{x,N}$, and proceed as in the last example. Letting

$$F_{x,N} = H_3 = 400,$$

for instance, we obtain

$$H_2 = \tfrac{201}{101}, \quad H_1 = \tfrac{806}{605};$$

hence

$$G_2 = \tfrac{50}{101} \approx \tfrac{1}{2}, \quad G_1 = \tfrac{201}{1206} \approx \tfrac{1}{6}, \quad G_0 = \tfrac{403}{3226} \approx \tfrac{1}{8}.$$

In fact, the target condition in the last stage does give $G_2 = \frac{1}{2}$;

$$x_3 = 0 = 0.5x_2 - G_2x_2 \ ; \quad \text{hence} \quad G_2 = \frac{1}{2} \ .$$

Since H_3 is indeterminate, however, we cannot proceed backward to determine G_1 and G_0 by Eq. (14–106).

An alternate approach is to first determine optimal control as in Example 14–12, and then obtain the feedback-control gain at each step. Applying $F_{x,N} = F_x = 1$ in the stationary cost increment $f_0(x_k, u_k)$ of example 14–12, we find that

$$f_0(x_k, u_k) = 0.25x_k^2 + x_k u_k + 5u_k^2 \ ,$$

and thus

$$H = -0.25x_k^2 - x_k u_k - 5u_k^2 + \psi_{k+1}(0.5x_k + u_k) \ ,$$
$$u_k^0 = (\psi_{k+1} - x_k)/10 \ ,$$
$$\psi_k = -0.5x_k - u_k + 0.5\psi_{k+1} = -0.4x_k + 0.4\psi_{k+1} \ ,$$

so that

$$\begin{bmatrix} x_{k+1} \\ \psi_k \end{bmatrix} = \begin{bmatrix} 0.4 & 0.1 \\ -0.4 & 0.4 \end{bmatrix} \begin{bmatrix} x_k \\ \psi_{k+1} \end{bmatrix}, \quad (k = 0, 1, 2) \ ,$$

where boundary values are

$$x_0 = c = \text{given} \ , \quad x_3 = 0 \ .$$

The solution of the equation is

$$x_0 = c \ , \quad x_1 = \tfrac{3}{8}c \ , \quad x_2 = \tfrac{1}{8}c \ , \quad x_3 = 0 \ ,$$
$$\psi_0 = -\tfrac{1}{2}c \ , \quad \psi_1 = -\tfrac{1}{4}c \ , \quad \psi_2 = -\tfrac{1}{4}c \ , \quad \psi_3 = -\tfrac{1}{2}c \ ,$$

and

$$u_k^0 = \frac{\psi_{k+1} - x_k}{10} \quad \text{for} \quad k = 0, 1, 2$$

are found to be

$$u_0^0 = -\tfrac{1}{8}c \ , \quad u_1^0 = -\tfrac{1}{16}c \ , \quad u_2^0 = -\tfrac{1}{16}c \ ,$$

or

$$G_0 = -\frac{u_0^0}{x_0} = \tfrac{1}{8} \ , \quad G_1 = -\frac{u_1^0}{x_1} = \tfrac{1}{6} \ , \quad G_2 = -\frac{u_2^0}{x_2} = \tfrac{1}{2} \ .$$

14–6 OPTIMAL FEEDBACK CONTROL OF LINEAR STOCHASTIC SYSTEMS

The optimal feedback-control algorithm presented in the preceding section for deterministic linear systems also applies to the optimal control of stochastic linear systems provided that we insert an optimal filter in the feedback loop (Fig. 14–16) [27 through 30]. As a stochastic-control object we consider a linear system subject to a zero-mean Gaussian "white-noise" vector input $\mathbf{w}_k$, with a measurement which is also perturbed by a zero-mean Gaus-

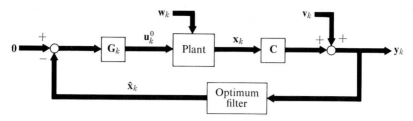

Fig. 14-16 Principle of stochastic optimal control.

sian white-noise vector $\mathbf{v}_k$;

$$\mathbf{x}_{k+1} = \mathbf{P}_k\mathbf{x}_k + \mathbf{Q}_k\mathbf{u}_k + \mathbf{Q}_{w,k}\mathbf{w}_k,$$
$$\mathbf{y}_k = \mathbf{C}_k\mathbf{x}_k + \mathbf{v}_k. \qquad (14\text{-}110)$$

Let us assume that $\mathbf{w}_k$ and $\mathbf{v}_k$ are not correlated, so that

$$E\left\{\begin{bmatrix}\mathbf{w}_k\\\mathbf{v}_k\end{bmatrix}[\mathbf{w}'_k\ \mathbf{v}'_k]\right\} = \begin{bmatrix}\mathbf{W}_k & 0\\0 & \mathbf{V}_k\end{bmatrix} \qquad (14\text{-}111)$$

where, as in Section 13–5, $\mathbf{W}_k$ and $\mathbf{V}_k$ are the covariance matrices of $\mathbf{w}_k$ and $\mathbf{v}_k$, respectively.

The performance index to be minimized is defined by

$$J = E\left\{\sum_{k=i}^{N-1}(\mathbf{x}'_{k+1}\mathbf{F}_{x,k+1}\mathbf{x}_{k+1} + \mathbf{u}'_k\mathbf{F}_{u,k}\mathbf{u}_k)\,\bigg|\,Y_i\right\}, \qquad (14\text{-}112)$$

where N is the prescribed number of steps for optimal control, $\mathbf{F}_x$ and $\mathbf{F}_u$ are weighting matrices (both symmetric), $\mathbf{F}_x$ is nonnegative, $\mathbf{F}_u$ is positive definite as in Eq. (14-97), and Y_i represents the set of observations made up to the ith sampling instant,

$$Y_i = \{\mathbf{y}_0, \mathbf{y}_1, \ldots, \mathbf{y}_i\}.$$

By minimizing J for all Y_i, $(i = 0, 1, \ldots, N-1)$, we also minimize the unconditional expectation.

The optimal control law for this system is given by

$$\mathbf{u}^0_k = -\mathbf{G}_k\hat{\mathbf{x}}_k, \qquad (14\text{-}113)$$

where $\hat{\mathbf{x}}_k$ is the best estimate of state $\mathbf{x}_k$ at the kth instant. According to Eq. (13-58), it is

$$\hat{\mathbf{x}}_k = E\{\mathbf{x}_k\,|\,Y_k\}, \qquad (14\text{-}114)$$

and $\mathbf{G}_k$ in Eq. (14-113) is the optimal-control matrix obtained for deterministic systems in the last section (Table 14-2). Since $\hat{\mathbf{x}}_k$ is the best estimate *determined* by the optimum filter, control vector $\mathbf{u}^0_k$ of Eq. (14-113) is also deterministic. Therefore the optimum filtering algorithm for the system of

Table 14-3 OPTIMAL LINEAR FEEDBACK CONTROL LAW FOR A STOCHASTIC SYSTEM

	Discrete-time system	$T \to dt$	Continuous-time system
		Control object and optimal control	
9	$\mathbf{x}_{k+1} = \mathbf{P}_k\mathbf{x}_k + \mathbf{Q}_k\mathbf{u}_k + \mathbf{Q}_{w,k}\mathbf{w}_k$ $\mathbf{y}_k = \mathbf{C}_k\mathbf{x}_k + \mathbf{v}_k, \quad \mathbf{u}_k = -\mathbf{G}_k\hat{\mathbf{x}}_k$	$\mathbf{Q}_{w,k} \to \mathbf{B}_w(t)\,dt$ $\mathbf{C}_k \to \mathbf{C}(t)$	$\dfrac{d}{dt}\mathbf{x}(t) = \mathbf{A}(t)\mathbf{x}(t) + \mathbf{B}(t)\mathbf{u}(t) + \mathbf{B}_w(t)\mathbf{w}(t)$ $\mathbf{y}(t) = \mathbf{C}(t)\mathbf{x}(t) + \mathbf{v}(t), \quad \mathbf{u}(t) = -\mathbf{G}(t)\hat{\mathbf{x}}(t)$
		Zero-mean Gaussian noise covariance matrix	
10	$E\left\{\begin{bmatrix}\mathbf{w}_k\\\mathbf{v}_k\end{bmatrix}[\mathbf{w}_k'\ \mathbf{v}_k']\right\} = \begin{bmatrix}\mathbf{W}_k & \mathbf{0}\\\mathbf{0} & \mathbf{V}_k\end{bmatrix}$	$\mathbf{W}_k \to \mathbf{W}(t)\delta(\tau)$ $\mathbf{V}_k \to \mathbf{V}(t)\delta(\tau)$	$E\left\{\begin{bmatrix}\mathbf{w}(t)\\\mathbf{v}(t)\end{bmatrix}[\mathbf{w}'(t+\tau)\ \mathbf{v}'(t+\tau)]\right\} = \begin{bmatrix}\mathbf{W}(t) & \mathbf{0}\\\mathbf{0} & \mathbf{V}(t)\end{bmatrix}\delta(\tau), \ 0 \leq \tau \leq t_i$
		Performance index	
11	$J = E\left\{\sum_{k=i}^{N-1}(\mathbf{x}_{k+1}'\mathbf{F}_{z,k+1}\mathbf{x}_{k+1} + \mathbf{u}_k'\mathbf{F}_{u,k}\mathbf{u}_k)\Big|Y_i\right\}$		$J = E\left\{\int_{t_i}^{T}(\mathbf{x}'(t)\mathbf{F}_z(t)\mathbf{x}(t) + \mathbf{u}'(t)\mathbf{F}_u(t)\mathbf{u}(t))\,dt\Big|\mathbf{y}(\tau),\ 0 \leq \tau \leq t_i\right\}$
		Optimum filter algorithm	
12	$\hat{\mathbf{x}}_{k+1} = (\mathbf{P}_k\hat{\mathbf{x}}_k + \mathbf{Q}_k\mathbf{u}_k)$ $+ \mathbf{K}_{k+1}\{\mathbf{y}_{k+1} - \mathbf{C}_{k+1}(\mathbf{P}_k\hat{\mathbf{x}}_k + \mathbf{Q}_k\mathbf{u}_k)\}$ $\hat{\mathbf{x}}_{-1} = \mathbf{0}$	$\mathbf{K}_k \to \mathbf{K}(t)\,dt$	$\dfrac{d}{dt}\hat{\mathbf{x}}(t) = (\mathbf{A}(t)\hat{\mathbf{x}}(t) + \mathbf{B}(t)\mathbf{u}(t)) + \mathbf{K}(t)(\mathbf{y}(t) - \mathbf{C}(t)\hat{\mathbf{x}}(t))$ $\hat{\mathbf{x}}(0) = \mathbf{0}$
		Filter gain	
13	$\mathbf{K}_k = \mathbf{M}_k\mathbf{C}_k'(\mathbf{V}_k + \mathbf{C}_k\mathbf{M}_k\mathbf{C}_k')^{-1}$	$\mathbf{M}_k \to \mathbf{M}(t)$	$\mathbf{K}(t) = \mathbf{M}(t)\mathbf{C}'(t)(\mathbf{V}(t))^{-1}$
		Matrix Riccati equation	
14	$\mathbf{M}_{k+1} = \mathbf{P}_k\mathbf{M}_k\mathbf{P}_k' - \mathbf{P}_k\mathbf{M}_k\mathbf{C}_k'(\mathbf{V}_{k+1} + \mathbf{C}_k\mathbf{M}_k\mathbf{C}_k')^{-1}\mathbf{C}_k\mathbf{M}_k\mathbf{P}_k'$ $+ \mathbf{Q}_{w,k}\mathbf{W}\mathbf{Q}_{w,k}',$ $\mathbf{M}_0 = E\{\mathbf{x}_0\mathbf{x}_0'\}$		$\dfrac{d}{dt}\mathbf{M}(t) = \mathbf{A}(t)\mathbf{M}(t) + \mathbf{M}(t)\mathbf{A}'(t) - \mathbf{M}(t)\mathbf{C}'(t)(\mathbf{V}(t))^{-1}\mathbf{C}(t)\mathbf{M}(t)$ $+ \mathbf{B}_w(t)\mathbf{W}(t)\mathbf{B}_w'(t),$ $\mathbf{M}(0) = E\{\mathbf{x}(0)\mathbf{x}'(0)\}$

Eq. (14–110), with deterministic input $\mathbf{u}_k = \mathbf{u}_k^0$, is given by Eq. (13–97), as restated in #12 through #14 of Table 14–3. The main relations for the continuous-time system are also included in the right-hand side of the table. Note that the Riccati equation for the optimal control (#8, Table 14–2) and the optimum filter (#14, Table 14–3) are dual.

Both **G** and **K** will reach a stationary state if the control object, input noise, and weighting factors are all stationary and the optimization period is indefinite. The performance index for the stationary case is expressed in terms of mean-squared values of variables as

$$J = E\{\mathbf{x}_k' \mathbf{F}_x \mathbf{x}_k + \mathbf{u}_k' \mathbf{F}_u \mathbf{u}_k\}$$

or

$$J = E\{\mathbf{x}(t) \mathbf{F}_x \mathbf{x}(t) + \mathbf{u}'(t) \mathbf{F}_u \mathbf{u}(t)\} \ .$$

Example 14–15 We shall next discuss stationary optimal control of a first-order system for minimization of

$$J = E\{F_x x(t)^2 + F_u u(t)^2\} \ .$$

The system's equation is

$$\frac{dx}{dt} = ax + bu + w \ , \qquad y = cx + v \ ,$$

for which the control law is $u = -g\hat{x}$, and the filter equation is

$$\frac{d\hat{x}}{dt} = a\hat{x} + bu + k(y - c\hat{x}) \ .$$

The controller gain is

$$g = \frac{bh}{F_u} \ ,$$

where h is the positive root of

$$-2ah + \frac{h^2 b^2}{F_u} - F_x = 0 \ ,$$

so that

$$g = \frac{1}{b} \{a + \sqrt{a^2 + b^2 (F_x / F_u)}\} \ .$$

The filter gain is

$$k = \frac{mc}{\sigma_v^2} \ ,$$

and m is the positive root of

$$2am - \frac{c^2 m^2}{\sigma_v^2} + \sigma_w^2 = 0 \ ;$$

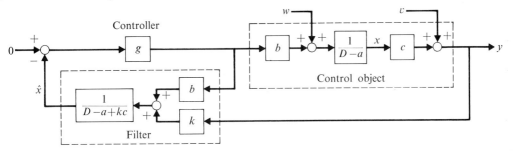

Fig. 14-17 Stationary optimal control of a first-order object.

hence

$$k = \frac{1}{c}\{a + \sqrt{a^2 + c^2(\sigma_w/\sigma_v)^2}\}.$$

The system's block diagram is shown in Fig. 14-17.

In order to show the basic steps involved in deriving the optimal control law, a first-order, time-invariant discrete-time system will be assumed in the following. Thus the control object is

$$x_{k+1} = px_k + qu_k + w_k, \quad k = 0, 1, \ldots, \quad (14\text{-}115)$$

where w_k is zero-mean Gaussian random noise, independent from one time to the next, and stationary, so that its variance $\sigma_w^2 = E\{w_k^2\}$ is constant. Output y_k of the system is

$$y_k = cx_k + v_k, \quad (14\text{-}116)$$

where v_k is zero-mean Gaussian noise, independent from one time to the next, also independent of x_0 and w_k, and its variance, σ_v^2, is constant:

$$\sigma_v^2 = E\{v_k^2\} = \text{const}.$$

The performance index to be minimized is quadratic:

$$J = E\left\{\sum_{k=i}^{N-1}(F_x x_{k+1}^2 + F_u u_k^2)\,\bigg|\,Y_i\right\}, \quad (14\text{-}117)$$

where the weighting factors are $F_x \geq 0$ and $F_u > 0$. Here N represents the specified number of samples of data to be optimized.

To apply the principle of optimality in the derivation, let us denote the minimal value of the performance index at the ith stage by

$$S(x_i) = \min_{u_i, u_{i+1}, \ldots, u_{N-1}} (J). \quad (14\text{-}118)$$

The relationship in time between the S's and the Y's is shown in Fig. 14-18. We shall show, by induction, that

$$S(x_i) = E\{(h_i x_i^2 + d_i)\,|\,Y_i\}, \quad i < N, \quad (14\text{-}119)$$

680 Optimal control

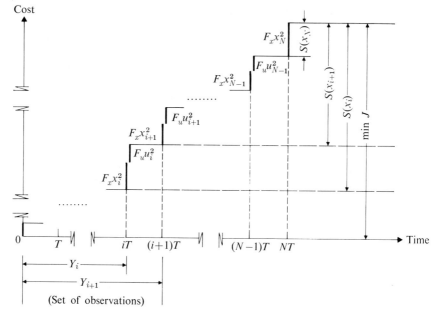

Fig. 14-18 Increments of cost function and set of observations.

where d_i, which will be defined later, satisfies:

$$E\{d_i \mid Y_i\} = E\{d_i\} \geqslant 0 ,$$

and h_i ($h_i \geqslant 0$) does not depend on Y_i.

Let us now assume that

$$S(x_{i+1}) = E\{(h_{i+1}x_{i+1}^2 + d_{i+1}) \mid Y_{i+1}\} . \tag{14-120}$$

Then, by the principle of optimality (a condition similar to Eq. 14-48), it follows that

$$S(x_i) = \min_{u_i} E\{F_x x_i^2 + F_u u_i^2 + S(x_{i+1}) \mid Y_i\}$$
$$= \min_{u_i} E\{F_x x_i^2 + F_u u_i^2 + E(h_{i+1}x_{i+1}^2 + d_{i+1} \mid Y_{i+1}) \mid Y_i\} .$$

Noting that Y_i is a subset of Y_{i+1} (see Fig. 14-18), we have

$$E[\{E( ) \mid Y_{i+1}\} \mid Y_i] = E[( ) \mid Y_i] ,$$

so that

$$S(x_i) = \min_{u_i} E\{(F_x x_i^2 + F_u u_i^2 + h_{i+1}x_{i+1}^2 + d_{i+1}) \mid Y_i\} . \tag{14-121}$$

Substituting the system's equation (Eq. 14–115) into (14–121) we obtain

$$S(x_i) = \min_{u_i} E\{F_z x_i^2 + F_u u_i^2 + h_{i+1}(px_i + qu_i + w_i)^2 + d_{i+1} | Y_i\}$$

$$= \min_{u_i} E\{(F_z + p^2 h_{i+1})x_i^2 + (F_u + h_{i+1}q^2)u_i^2 + 2h_{i+1}pqu_i x_i$$

$$+ 2h_{i+1}pw_i x_i + 2h_{i+1}qu_i w_i + h_{i+1}w_i^2 + d_{i+1} | Y_i\}. \quad (14\text{–}122)$$

Since x_i and w_i are independent, w_i has zero mean, and u_i is a deterministic function of Y_i (still unknown), we may rewrite Eq. (14–122) as:

$$S(x_i) = \min_{u_i} [E\{(p^2 h_{i+1} + F_z)x_i^2 | Y_i\} + (F_u + q^2 h_{i+1})u_i^2$$

$$+ 2h_{i+1}qpu_i \hat{x}_i + d_{i+1} + E\{(h_{i+1}w_i^2) | Y_i\}], \quad (14\text{–}123)$$

where

$$\hat{x}_i = E\{x_i | Y_i\} \quad (14\text{–}124)$$

is the optimum estimate of x_i at time $t = iT$, which is the output of the optimum filter.

The optimal control $u_i = u_i^0$ that will minimize the right side of Eq. (14–123) can be determined by the condition

$$\frac{\partial S}{\partial u_i} = 0;$$

that is,

$$2u_i(F_u + q^2 h_{i+1}) + 2h_{i+1}qp\hat{x}_i = 0.$$

Hence

$$u_i = u_i^0 = -\frac{qph_{i+1}}{F_u + q^2 h_{i+1}} \hat{x}_i$$

or

$$u_i^0 = -g_i \hat{x}_i, \quad (14\text{–}125)$$

where

$$g_i = \frac{qph_{i+1}}{F_u + q^2 h_{i+1}}. \quad (14\text{–}126)$$

We substitute Eqs. (14–125) and (14–126) back into Eq. (14–123), together with Eq. (14–124), to obtain:

$$S(x_i) = E\{(p^2 h_{i+1} + F_z)x_i^2 | Y_i\} + (F_u + q^2 h_{i+1})(g_i^2 \hat{x}_i^2)$$

$$+ 2h_{i+1}qp(-g_i \hat{x}_i)\hat{x}_i + d_{i+1} + E\{(h_{i+1}w_i^2) | Y_i\}. \quad (14\text{–}127)$$

For the second term on the right-hand side of Eq. (14–127), we have:

$$E\{(F_u + q^2h_{i+1})(g_i^2\hat{x}_i^2) \mid Y_i\}$$
$$= E\left\{(F_u + q^2h_{i+1})\left(\frac{q^2p^2h_{i+1}^2}{(F_u + q^2h_{i+1})^2}\right)(x_i^2 - e_i^2)\,\Big|\, Y_i\right\}$$
$$= \frac{q^2p^2h_{i+1}^2}{F_u + q^2h_{i+1}}[E\{x_i^2 \mid Y_i\} - E\{e_i^2 \mid Y_i\}], \qquad (14\text{–}128)$$

where

$$e_i = x_i - \hat{x}_i \qquad (14\text{–}129)$$

is the error of the optimal prediction, for which the following relation holds (see Section 13–4):

$$E\{e_i^2 \mid Y_i\} + E\{\hat{x}_i^2 \mid Y_i\} = E\{x_i^2 \mid Y_i\}. \qquad (14\text{–}130)$$

Both of these conditions (Eqs. 14–129 and 14–130) are used in the reduction process of Eq. (14–128). Similarly, for the third term on the right-hand side of Eq. (14–127), we compute

$$E\{2h_{i+1}qp(-g_i\hat{x}_i^2) \mid Y_i\} = -\left(\frac{2h_{i+1}^2q^2p^2}{F_u + q^2h_{i+1}}\right)E\{\hat{x}_i^2 \mid Y_i\}$$
$$= -2\left(\frac{q^2p^2h_{i+1}^2}{F_u + q^2h_{i+1}}\right)[E\{x_i^2 \mid Y_i\} - E\{e_i^2 \mid Y_i\}].$$
$$(14\text{–}131)$$

Equation (14–127), with relations Eqs. (14–128) and (14–131) substituted, reduces to

$$S(x_i) = E\left\{\left[p^2\left(h_{i+1} - \frac{q^2h_{i+1}^2}{F_u + q^2h_{i+1}}\right) + F_x\right]x_i^2 \right.$$
$$\left. + \left[\frac{p^2q^2h_{i+1}^2}{F_u + q^2h_{i+1}}\right]e_i^2 + d_{i+1} + h_{i+1}w_i^2 \,\Big|\, Y_i\right\}. \qquad (14\text{–}132)$$

Let us put, in Eq. (14–132),

$$\left[p^2\left\{h_{i+1} - \frac{q^2h_{i+1}^2}{F_u + q^2h_{i+1}}\right\} + F_x\right] = h_i, \qquad (14\text{–}133)$$

$$\left(\frac{p^2q^2h_{i+1}^2}{F_u + q^2h_{i+1}}\right)e_i^2 + d_{i+1} + h_{i+1}w_i^2 = d_i;$$

then Eq. (14–132) will reduce to:

$$S(x_i) = E\{(h_i x_i^2 + d_i) \mid Y_i\}. \qquad (14\text{–}134)$$

This completes the inductive proof of Eq. (14–119). Because of Eq. (14–124), an optimum filter enters into the optimal control system as a loop component.

14-6 Optimal feedback control of linear stochastic systems

According to Fig. 14–18, the cost increment at the last instant $t = NT$ is

$$S(x_N) = E\{F_x x_N^2 \mid Y_N\}. \tag{14-135}$$

Comparing Eq. (14–135) with Eq. (14–134), we obtain

$$h_N = F_x. \tag{14-136}$$

The recursion (Eq. 14–133) is "initialized" by Eq. (14–136), and proceeds backwards. Computation for an nth-order time-varying system closely parallels Eqs. (14–115) through Eq. (14–136). The optimal control law, #7 of Table 14–2, stems from Eqs. (14–125) and (14–126). Equation (14–133), with Eq. (14–136) in the general case, yields the Riccati equation and the boundary condition, #8 of Table 14–2. Since h_i in Eq. (14–134), and hence H_i in the general case, is the increment of cost, these must be nonnegative. This is why we took the positive root of h in Example 14–15.

It is possible to compute the statistical performance of a control system by the recursive algorithm. The analytical approach to deriving the algorithm is similar to what we saw in Section 13–3. For this purpose we combine the system's equation (Eq. 14–110) and the filter equation (#12 of Table 14–3):

$$\boldsymbol{\xi}_{k+1} = \boldsymbol{\Pi}_k \boldsymbol{\xi}_k + \boldsymbol{\Gamma}_k \boldsymbol{\eta}_k, \qquad E\{\boldsymbol{\xi}_0\} = 0, \tag{14-137}$$

where

$$\boldsymbol{\xi}_k = \begin{bmatrix} \mathbf{x}_k \\ \hat{\mathbf{x}}_k \end{bmatrix}, \qquad \boldsymbol{\eta}_k = \begin{bmatrix} \mathbf{w}_k \\ \mathbf{v}_{k+1} \end{bmatrix},$$

$$\boldsymbol{\Pi}_k = \begin{bmatrix} \mathbf{P}_k & -\mathbf{Q}_k \mathbf{G}_k \\ \mathbf{K}_{k+1} \mathbf{C}_{k+1} \mathbf{P}_k & \mathbf{L}_k \end{bmatrix}, \qquad \boldsymbol{\Gamma}_k = \begin{bmatrix} \mathbf{Q}_{w,k} & 0 \\ \mathbf{K}_{k+1} \mathbf{C}_{k+1} \mathbf{Q}_{w,k} & \mathbf{K}_{k+1} \end{bmatrix},$$

$$\mathbf{L}_k = (\mathbf{I} - \mathbf{K}_{k+1} \mathbf{C}_{k+1})(\mathbf{P}_k - \mathbf{Q}_k \mathbf{G}_k) - \mathbf{K}_{k+1} \mathbf{C}_{k+1} \mathbf{Q}_k \mathbf{G}_k.$$

Because of zero initial state $E\{\boldsymbol{\xi}_0\} = 0$, and zero-mean forcing $E\{\boldsymbol{\eta}_k\} = 0$, the following unconditional expectations are all zero;

$$E\{\boldsymbol{\xi}_k\} = 0; \quad \text{thus} \quad E\{\mathbf{x}_k\} = 0, \qquad E\{\hat{\mathbf{x}}_k\} = 0;$$

and

$$E\{\mathbf{u}_k\} = \mathbf{G}_k E\{\hat{\mathbf{x}}_k\} = 0.$$

To obtain second-order statistics we right-multiply

$$\boldsymbol{\xi}'_{k+1} = \boldsymbol{\xi}'_k \boldsymbol{\Pi}'_k + \boldsymbol{\eta}'_k \boldsymbol{\Gamma}'_k$$

on each side of Eq. (14–137), take the expected value of each term, and obtain

$$E\{\boldsymbol{\xi}_{k+1} \boldsymbol{\xi}'_{k+1}\} = \boldsymbol{\Pi}_k E\{\boldsymbol{\xi}_k \boldsymbol{\xi}'_k\} \boldsymbol{\Pi}'_k + \boldsymbol{\Gamma}_k E\{\boldsymbol{\eta}_k \boldsymbol{\eta}'_k\} \boldsymbol{\Gamma}'_k,$$

where $E\{\boldsymbol{\xi}_k\boldsymbol{\eta}_k'\} = \mathbf{0}$ is used. Letting

$$E\{\boldsymbol{\xi}_k\boldsymbol{\xi}_k'\} = \mathbf{J}_k, \qquad E\{\boldsymbol{\eta}_k\boldsymbol{\eta}_k'\} = \mathbf{N}_k$$

in the last expression, we obtain

$$\mathbf{J}_{k+1} = \boldsymbol{\Pi}_k \mathbf{J}_k \boldsymbol{\Pi}_k' + \boldsymbol{\Gamma}_k \mathbf{N}_k \boldsymbol{\Gamma}_k'. \tag{14-138}$$

This is the recursive relation used in computing the variance as a time series. The recursion is initialized by

$$\mathbf{J}_0 = \begin{bmatrix} E\{\mathbf{x}_0\mathbf{x}_0'\} & E\{\mathbf{x}_0\hat{\mathbf{x}}_0'\} \\ E\{\hat{\mathbf{x}}_0\mathbf{x}_0'\} & E\{\hat{\mathbf{x}}_0\hat{\mathbf{x}}_0'\} \end{bmatrix},$$

where, recalling the filter equation (Section 13-5),

$$E\{\mathbf{x}_0\mathbf{x}_0'\} = \mathbf{M}_0,$$

we have the filter equation

$$\hat{\mathbf{x}}_0 = \mathbf{L}_{-1}\hat{\mathbf{x}}_{-1} + \mathbf{K}_0\mathbf{C}_0\mathbf{x}_0 + \mathbf{K}_0\mathbf{v}_0,$$

where $\hat{\mathbf{x}}_{-1} = \mathbf{0}$, so that

$$\hat{\mathbf{x}}_0 = \mathbf{K}_0\mathbf{C}_0\mathbf{x}_0 + \mathbf{K}_0\mathbf{v}_0.$$

Since $\mathbf{v}_0$ and $\mathbf{x}_0$ are uncorrelated, it follows that

$$E\{\hat{\mathbf{x}}_0\mathbf{x}_0'\} = \mathbf{K}_0\mathbf{C}_0\mathbf{M}_0, \qquad E\{\mathbf{x}_0\hat{\mathbf{x}}_0'\} = \mathbf{M}_0\mathbf{C}_0'\mathbf{K}_0'$$

and

$$E\{\hat{\mathbf{x}}_0\hat{\mathbf{x}}_0'\} = \mathbf{K}_0\mathbf{C}_0\mathbf{M}_0\mathbf{C}_0'\mathbf{K}_0' + \mathbf{K}_0\mathbf{V}_0\mathbf{K}_0'.$$

If $\mathbf{w}_k$ and $\mathbf{v}_{k+1}$ are uncorrelated, the "forcing" covariance matrix of Eq. (14-138) is

$$\mathbf{N}_k = E\left\{\begin{bmatrix} \mathbf{w}_k \\ \mathbf{v}_{k+1} \end{bmatrix} [\mathbf{w}_k' \quad \mathbf{v}_{k+1}']\right\} = \begin{bmatrix} \mathbf{W}_k & \mathbf{0} \\ \mathbf{0} & \mathbf{V}_{k+1} \end{bmatrix}.$$

The recursive relation Eq. (14-138) in terms of $2n \times 2n$ matrices is general, in the sense that it is not limited to a control system with an optimum filter. If filter gain $\mathbf{K}$ in the above relations is optimum, the recursive relation can be reduced to an $n \times n$-matrix equation. Designating the first element of $\mathbf{J}_k$ in Eq. (14-138) by

$$\mathbf{J}_{11,k} = E\{\mathbf{x}_k\mathbf{x}_k'\},$$

we obtain (see Problem 14-22) the following recursion for $\mathbf{J}_{11}$:

$$\mathbf{J}_{11,k+1} = (\mathbf{P}_k - \mathbf{Q}_k\mathbf{G}_k)(\mathbf{J}_{11,k} - \mathbf{Z}_k)(\mathbf{P}_k' - \mathbf{G}_k'\mathbf{Q}_k')$$
$$+ \mathbf{P}_k\mathbf{Z}_k\mathbf{P}_k' + \mathbf{Q}_{w,k}\mathbf{W}_k\mathbf{Q}_{w,k}'. \tag{14-139}$$

In this expression $\mathbf{Z}_k$ is the covariance matrix of the estimation error,

$$\mathbf{Z}_k = E\{(\mathbf{x}_k - \hat{\mathbf{x}}_k)(\mathbf{x}_k - \hat{\mathbf{x}}_k)'\} ;$$

this is available from the optimum-filter computation (see Eq. 13–97). The covariance matrix of the control vector is given (see Problem 14–22) in terms of $\mathbf{J}_{11,k}$ by

$$E\{\mathbf{u}_k \mathbf{u}_k'\} = \mathbf{G}_k(\mathbf{J}_{11,k} - \mathbf{Z}_k)\mathbf{G}_k' . \qquad (14\text{–}140)$$

The theoretical possibility of realizing an optimal control by time-varying (or time-invariant) gain matrix feedback has a strong appeal to practicing engineers. Listed in the following are some remarks concerning the application of the optimal control approach as a useful guide to the design of control systems.

a) *System Model:* Linearization of nonlinearities and use of simple models are important. Because of constraints on available instrumentation or control-computer capacity, unreduced system models may lead to practically unacceptable control system designs [31].

b) *Performance Index:* Nonquadratic forms must be approximated by quadratic forms for which the optimal control solution exists. Selection of weighting matrices pertinent to a given problem requires insight, experience, and (often) trial-and-error search.

c) *Input Signal:* Actual noise must be approximated by Gaussian signals, and a shaping filter for the noise color must be built into the system. If the system transient is of importance, the initial state of the shaping filter must be carefully chosen to match a prescribed operating condition. If the command signal is deterministic in a tracking control problem, a "generator" of the signal is also considered as part of the system. The following simple but practical example is intended to illustrate some of these points.

Example 14-16 A single-actuator, one-dimensional earthquake simulator is shown in Fig. 14–19 where an electronic actuating signal, u, produced by a servoamplifier and a servovalve, operates a hydraulic actuator which, in turn, will "shake" a platform on which a test structure is erected. The control system must assure high fidelity in reproducing the "desired" earthquake acceleration which should be insensitive to test-structure dynamics. The command signal to the control system is generated from an earthquake record. Due to the nature of the test, there is no feedback available from the test structure. Measurements made on the actuator-platform complex (for example, platform displacement, velocity, and acceleration) are to be fed back to produce the actuating signal u by a suitable control law.

For a vibration-testing system of this nature a common practice of control has been the use of simple displacement feedback. For a typical set of plant parameters (with scale factors), including a poorly damped oscillatory test structure, the open-

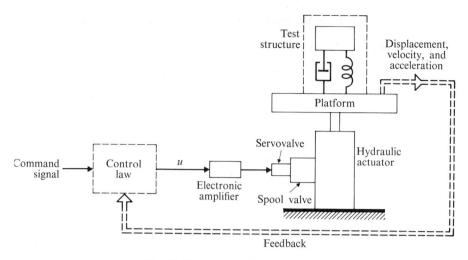

Fig. 14-19 Earthquake simulator.

loop transfer function for displacement output was found to be

$$G_d(s) = K_d \frac{1}{(s^2 + 2s + 5)} \frac{1}{(s^2 + 0.91s + 2.12)} \left(\frac{s^2 + 0.00625s + 0.1}{s^2 + 0.107s + 0.094}\right) \frac{1}{s},$$
(14-141)

where K_d is an overall open-loop gain, and 100 rad/sec is taken as a unit. The first factor on the right-hand side of $G_d(s)$ is the servovalve dynamics. The second quadratic factor involves interaction between the fluid system of the actuator and the mechanical load. The third factor is mainly due to the interaction of the platform and the test structure, producing a pair of open-loop zeros (roots of the numerator polynomial) close to a pair of open-loop poles. Because of the last-mentioned open-loop poles and zeros, persistent closed-loop oscillation occurs in the displacement-control system when the test structure is oscillatory. This is the objectionable feature of the displacement-feedback system, which is otherwise acceptable as long as the test structure is well damped.

Linear optimal-control theory provides us with a design principle whereby the difficulty just stated can be overcome and, in addition, a drastic improvement in output quality can be expected. A direct application of optimal-control theory, however, is not practical, because for the purpose of the earthquake simulator it is imperative to design a control system which makes full use of the readily available measurements of platform outputs but exhibits lowest sensitivity to test-structure dynamics. For this reason we shall derive an approximate system model for the platform-acceleration output.

The transfer function for displacement output (Eq. 14-141), converted into acceleration output by multiplying by s^2, produces the frequency-response pattern of Fig. 14-20, where the crucial corner point B comes from the second quadratic factor of Eq. (14-141). Since the control system's bandwidth is thus dominated by this factor, it is logical to consider the following form as a first approximation of the

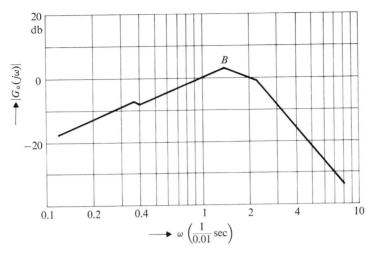

Fig. 14-20 Bode-diagram asymptotes of $|G_a(j\omega)|$.

open-loop transfer function for accleration output;

$$G_a(s) = K_a \left(\frac{s}{s^2 + a_1 s + a_0} \right), \qquad a_1 = 0.91, \qquad a_0 = 2.12. \tag{14-142}$$

In state-space, this is expressed as

$$\frac{d}{dt}\begin{bmatrix} x_1 \\ x_2 \end{bmatrix} = \begin{bmatrix} 0 & 1 \\ -a_0 & -a_1 \end{bmatrix}\begin{bmatrix} x_1 \\ x_2 \end{bmatrix} + \begin{bmatrix} 0 \\ K_a \end{bmatrix} u, \tag{14-143}$$

and

$$y_1 = \begin{bmatrix} 0 & 1 \end{bmatrix} \begin{bmatrix} x_1 \\ x_2 \end{bmatrix} \tag{14-144}$$

where y_1 is acceleration output.

For the purposes of design, we consider the acceleration command (x_3 in Fig. 14-21) as being generated by Gaussian white noise through a shaping filter [32]. Since our main concern is high-quality performance of the control system within the system's bandwidth, the acceleration-command signal is passed through a filter to produce a frequency-weighted signal x_4, and the error signal to be minimized is defined by

$$e = x_4 - x_2,$$

where x_2 is the system's acceleration output which was given by Eq. (14-143). Thus the performance index for this system is

$$J = E\{e^2(t)\} + F_u E\{u^2(t)\}.$$

The entire system is fourth-order, but, as in Example 14-11, the control object [A_1 of Eq. (14-143)] and the input processor [A_2 for Fig. 14-21] are decoupled so that

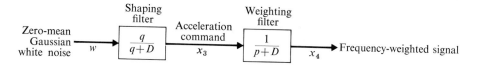

Fig. 14-21 Input processer for acceleration command.

the overall system equation takes the following form where the **A**-matrix is diagonal:

$$\frac{d}{dt}\mathbf{x} = \begin{bmatrix} \mathbf{A}_1 & 0 \\ 0 & \mathbf{A}_2 \end{bmatrix}\mathbf{x} + \begin{bmatrix} \mathbf{b}_1 \\ 0 \end{bmatrix} u + \begin{bmatrix} 0 \\ \mathbf{f}_2 \end{bmatrix} w . \tag{14-145}$$

For the performance index

$$J = E\{\mathbf{x}'\mathbf{F}_x\mathbf{x}\} + F_u E\{u^2\} ,$$

the matrix Riccati equation (#8, Table 14-2) for the steady state is

$$\begin{bmatrix} \mathbf{A}_1 & 0 \\ 0 & \mathbf{A}_2 \end{bmatrix}\begin{bmatrix} \mathbf{H}_{11} & \mathbf{H}_{12} \\ \mathbf{H}_{21} & \mathbf{H}_{22} \end{bmatrix} + \begin{bmatrix} \mathbf{H}_{11} & \mathbf{H}_{12} \\ \mathbf{H}_{21} & \mathbf{H}_{22} \end{bmatrix}\begin{bmatrix} \mathbf{A}_1 & 0 \\ 0 & \mathbf{A}_2 \end{bmatrix}$$

$$- \frac{1}{F_u}\begin{bmatrix} \mathbf{H}_{11} & \mathbf{H}_{12} \\ \mathbf{H}_{21} & \mathbf{H}_{22} \end{bmatrix}\begin{bmatrix} \mathbf{B} & 0 \\ 0 & 0 \end{bmatrix}\begin{bmatrix} \mathbf{H}_{11} & \mathbf{H}_{12} \\ \mathbf{H}_{21} & \mathbf{H}_{22} \end{bmatrix} + \begin{bmatrix} \mathbf{F}_{11} & \mathbf{F}_{12} \\ \mathbf{F}_{21} & \mathbf{F}_{22} \end{bmatrix} = 0 ,$$

where $\tag{14-146}$

$$\mathbf{F}_x = \begin{bmatrix} \mathbf{F}_{11} & \mathbf{F}_{12} \\ \mathbf{F}_{21} & \mathbf{F}_{22} \end{bmatrix}, \quad \mathbf{B} = \mathbf{b}_1\mathbf{b}_1' .$$

Since **A** is diagonal and **H** is symmetric ($\mathbf{H}'_{12} = \mathbf{H}_{21}$), Eq. (14-146) reduces to the following three conditions:

$$\mathbf{A}'_1\mathbf{H}_{11} + \mathbf{H}_{11}\mathbf{A}_1 - \frac{1}{F_u}\mathbf{H}_{11}\mathbf{B}\mathbf{H}_{11} + \mathbf{F}_{11} = 0 , \tag{14-147}$$

$$\mathbf{A}'_1\mathbf{H}_{12} + \mathbf{H}_{12}\mathbf{A}_2 - \frac{1}{F_u}\mathbf{H}_{11}\mathbf{B}\mathbf{H}_{12} + \mathbf{F}_{12} = 0 , \tag{14-148}$$

and

$$\mathbf{A}'_2\mathbf{H}_{22} + \mathbf{H}_{22}\mathbf{A}_2 - \frac{1}{F_u}\mathbf{H}_{21}\mathbf{B}\mathbf{H}_{12} + \mathbf{F}_{22} = 0 . \tag{14-149}$$

First $\mathbf{H}_{11}$ is fixed by Eq. (14-147). With this result applied in Eq. (14-148), we can determine $\mathbf{H}_{12}$, and $\mathbf{H}_{22}$ in Eq. (14-149) if it is also necessary. The decoupling of the Riccati equation thus simplifies the design procedure.

For our specific system we have

$$\mathbf{F}_{11} = \mathbf{F}_{22} = -\mathbf{F}_{12} = \begin{bmatrix} 0 & 0 \\ 0 & 1 \end{bmatrix},$$

and

$$\mathbf{A}_1 = \begin{bmatrix} 0 & 1 \\ -a_0 & -a_1 \end{bmatrix}, \quad \mathbf{A}_2 = \begin{bmatrix} -q & 0 \\ 1 & -p \end{bmatrix}, \quad \mathbf{b}_1 = \begin{bmatrix} 0 \\ K_a \end{bmatrix}, \quad \mathbf{f}_2 = \begin{bmatrix} q \\ 0 \end{bmatrix},$$

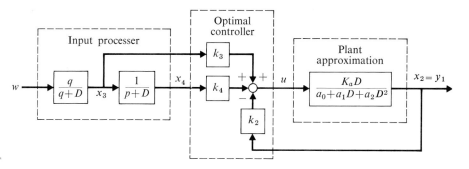

Fig. 14-22 Optimal acceleration control system.

so that, by Eq. (14-147), we obtain the following conditions for elements of $\mathbf{H}_{11}$:

$$-2a_0 h_{12} - \frac{1}{F_u} h_{12}^2 K_a^2 = 0, \quad \text{or} \quad h_{12} = 0,$$

$$-a_0 h_{22} - h_{11} - a_1 h_{12} - \frac{1}{F_u} h_{12} h_{22} K_a^2 = 0, \quad \text{or} \quad h_{11} = -a_0 h_{22},$$

$$2h_{12} - 2a_1 h_{22} - \frac{1}{F_u} K_a^2 h_{22}^2 + 1 = 0.$$

Solving the last expression for h_{22}, and rejecting the negative root, we find

$$h_{22} = \frac{\sqrt{a_1^2 s^2 + K_a^2 F_u} - a_1 F_u}{K_a^2}. \tag{14-150}$$

Two other elements of $\mathbf{H}$ that are necessary for the control law are found by Eq. (14-148) as

$$h_{23} = \frac{-(a_0/q)(1/p) + 1}{(a_0/q) + a_1 + q + (1/F_u) h_{22} K_a^2} h_{24}, \tag{14-151}$$

$$h_{24} = \frac{-1}{(a_0/p) + a_1 + p + (1/F_u) h_{22} K_a^2}. \tag{14-152}$$

The control law for the acceleration control is thus found to be

$$u^0_{(t)} = -\frac{1}{F_u} \mathbf{b}' \mathbf{H} \mathbf{x}(t) = -k_2 x_2 + k_3 x_3 + k_4 x_4,$$

where

$$k_2 = \frac{K_a}{F_u} h_{22}, \quad k_3 = -\frac{K_a}{F_u} h_{23}, \quad k_4 = -\frac{K_a}{F_u} h_{24}. \tag{14-153}$$

The control system is shown in Fig. 14-22. The numerical value of K_a is normally greater than other parameter values. If, in addition, F_u is chosen small, and p and q are so chosen that $a_0 = pq$, Eqs. (14-150) through (14-153) will reduce to the following approximate control gains:

$$k_2 = \frac{1}{2\sqrt{F_u}}, \qquad k_3 = 0, \qquad k_4 = \frac{2}{\sqrt{F_u}}.$$

The result indicates an insensitivity of the optimal-control gains to parameter values of the test structure. The value of the weighting factor F_u must be selected by experiment so that there will be no serious saturation in the servomechanism and at the same time the closed-loop system's high performance will be assured. In practice a weak displacement-feedback loop must be added to avoid drift of the platform position.

REFERENCES

1. Pontryagin, L. S., Boltyanskii, V. G., Gamkrelidze, R. V., Mishchenko, E. F., *Mathematical Theory of Optimal Processes.* New York: J. Wiley, 1962. (Authorized translation from Russian by K. N. Trirogoff.)
2. Fan, L. T., Wang, C. S., *The Discrete Maximum Principle.* New York: J. Wiley, 1964.
3. Fan, L. T., *The Continuous Maximum Principle.* New York: J. Wiley 1966.
4. Fel'dbaum, A. A., *Optimal Control Systems.* New York: Academic Press, 1966. (Translated from the Russian by A. Kraiman.)
5. Athans, M., Falb, P. L., *Optimal Controls.* New York: McGraw-Hill, 1966.
6. Sworder, D., *Optimal Adaptive Control Systems.* New York: Academic Press, 1966.
7. Leitmann, G., *An Introduction to Optimal Control.* New York: McGraw-Hill, 1966.
8. Bellman, R. E., *Dynamic Programming.* Princeton, N. J.: Princeton University Press, 1957.
9. Bellman, R. E., *Adaptive Control Processes.* Princeton, N. J.: Princeton University Press, 1961.
10. Elgerd, O. I., *Control Systems Theory.* New York: McGraw-Hill, 1967.
11. Boyadjieff, G., Eggleston, D., Jacques, M., Sutabutra, H., Takahashi, Y., "Some applications of the maximum principle to second-order systems, subject to input saturation, minimizing error, and effort," *J. of Basic Eng., Trans. ASME*, Series D, **86,** March 1964, pp. 11–22.
12. Fuller, A. T., "Relay control systems optimized for various performance criteria," *Proc. First IFAC* (Moscow 1960). London: Butterworths, pp. 510–519.
13. Eggleston, D. M., "On the application of the Pontryagin maximum principle using reverse time trajectories," *J. Basic Eng., Trans. ASME,* **85,** September 1963, pp. 478–480.
14. Milhorn, H. T., *The Application of Control Theory to Physiological Systems.* Philadelphia: Saunders, 1966.

15. Chang, J.-W., "The problem of synthesizing an optimal controller in systems with time delay," *Avtomatika i Telemekhanika*, **23** (2), February 1962, pp. 133–137.
16. Khatri, H. C., "Optimal control of linear systems with transportation lag," *J. Basic Eng., Trans. ASME*, Series D, **89** (2), 1967, pp. 385–392.
17. Khatri, H. C., Goodson, R. E., "Optimal feedback systems for a class of distributed systems," *J. Basic Eng., Trans. ASME*, Series D, June 1966, pp. 337–342.
18. Wiberg, D. M., "Feedback control of linear distributed systems," *J. Basic Eng., Trans. ASME*, Series D, **89** (2), 1967, pp. 379–384.
19. Uzgiris, S. O., "Optimal control of distributed-parameter systems," *J. Basic Eng., Trans. ASME*, **89**, Series D, 1967. (Also: Brogan, W. L., "Dynamic programming and distributed-parameter maximum principle," *1967 JACC Preprint*.)
20. Ho, Y.-C., "A successive-approximation technique for optimal control systems subject to input saturation," *J. Basic Eng., Trans. ASME*, Series D, **84** (1), March 1962, pp. 33–40.
21. Sutherland, J. W., et al., "A numerical trajectory optimization method suitable for a computation of limited memory," *1966 JACC Preprint*, pp. 177–185.
22. Schley, C. H., et al., "Optimal control computation by the Newton-Raphson method and the Riccati transformation," *1966 JACC Preprint*, pp. 186–192.
23. Noton, A. R. M., et al., "Numerical computation of optimal control," *1966 JACC Preprint*, pp. 193–204.
24. Sinnott, J. F., et al., "Solution of optimal control problems by the method of conjugate gradients," *1967 JACC Preprint*.
25. Katz, S., "A discrete version of Pontryagin's maximum principle," *J. of Electronics and Controls*, **13**, 179, 1962.
26. Sage, A. P., *Optimum Systems Control*. Englewood Cliffs, N. J.: Prentice-Hall, 1968.
27. Gunckel, T. L., II, Franklin, G. F., "A general solution for linear, sampled-data control." *J. Basic Eng., Trans. ASME*, Series D, **85**, June 1963, pp. 197–203.
28. Kalman, R. E., "When is a linear system optimal?" *Trans. ASME*, Series D, **86**, March 1964, pp. 51–60.
29. Schwartz, R. J., Friedland, B., *Linear Systems*. New York: McGraw-Hill, 1965.
30. Schultz, D. G., Melsa, J. L., *State Functions and Linear Control Systems*. New York: McGraw-Hill, 1967.
31. Widnall, W. S., *Application of Optimal Control Theory to Computer Controller Design*. Cambridge, Mass.: MIT Press, 1968.
32. Kanai, K., "Semi-empirical formula for the seismic characteristics of the ground," *Bull. Earthquake Res. Inst.*, Univ. of Tokyo, Japan, **35**, Pt. 2, June 1957, pp. 309–325; and Housner, T. W., and Jennings, P. C., "Generation of artificial earthquakes," *J. Eng. Mech. Div.*, A.S.C.E., **90**, No. E.M. 1, February 1964, pp. 113–150.

PROBLEMS

14-1 A first-order control object is described by

$$\frac{dx_1}{dt} = -(u+1)x_1 + u,$$

where the control u has the constraint $|u| \leq 1$. It is desired to move the system state from $x_1(0) = 2$ to $x_1(t_1) = 0$ in minimal time. Determine the optimal control, and confirm the result by computing the maximum value of H.

14-2 The control object is an integrator, $dx_1/dt = u$. The initial state is $x_1(0) = 1$, and the final state is specified to be $x_1(t_1) = 0$ at some time t_1. The performance index to be minimized is

$$J = \int_0^{t_1} (x_1^2 + u^2)\, dt .$$

The admissible range in u is $|u| \leq 1$. (a) Obtain the optimal trajectory in the $x_1 x_0$-plane. (b) Determine the cone of attainability at time $t = 1$ for all perturbations in control.

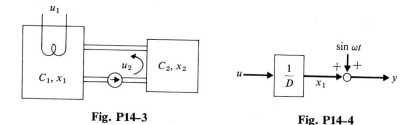

Fig. P14-3 **Fig. P14-4**

14-3 The control object is a thermal system of two tanks with a controlled rate of mixing (see Fig. P14-3). Lumped heat capacitances are $C_1 = 1$, $C_2 = \frac{1}{2}$. The flowrate for circulation is $0 \leq u_2 \leq 1$. Heat input has the constraint $0 \leq u_1 \leq 1$; there is no heat loss. It is desired to bring the temperature of the tanks from $x_1(0) = 0$, $x_2(0) = 0$ to $x_1(t_1) = 4$, $x_2(t_1) = 1$ at some time t_1, while minimizing time and mixing effort:

$$J = \int_0^t (w + u_2)\, dt ,$$

where the weight for time is $w \geq 0.05$. Find the optimal control and the time t_1.

14-4 A forced system is shown in Fig. P14-4, where $x_1(0) = 0$, $x_1(T) = 0$, and $T = 2\pi/\omega$. Find an optimal control for the minimization of

$$J = \int_0^T (y^2 + u^2)\, dt .$$

14-5 The population x_1 of a bacteria colony is described by

$$\frac{dx_1}{dt} = x_1 - u .$$

The reaction is autocatalytic, and u (the control) is a "poisoning." The initial state is $x_1(0) = 1$. Find an optimal control that will minimize

$$J = x_1(T)^2 + \int_0^T wu^2 \, dt \, ,$$

where T is a prescribed time, and the weight factor w is positive. Assume that $w = 1$. [*Hint:* For a terminal control problem of this nature we define an artificial coordinate x_0 by

$$\frac{dx_0}{dt} = 2x_1(t) \frac{dx_1}{dt} + wu^2$$

and substitute for dx_1/dt the right-hand side of the system's equation.]

14-6 Consider a first-order nonlinear process described by

$$\frac{dx_1}{dt} = 2u - u^2 - ux_1 \, ,$$

where u is positive but there is no upper constraint in control u. This equation may represent a reaction process where x_1 is a measure of product concentration, with $(2u - u^2)$ the rate of production, and ux_1 the carryover loss from a stirred tank reactor. The initial state $x_1(0)$ is zero. Find an optimal control that will bring the state x_1 up to the highest possible value in a prescribed time interval T.

14-7 Find an optimal control for a purely inertial object

$$\frac{dx_1}{dt} = x_2 \, , \qquad \frac{dx_2}{dt} = u$$

for minimization of the performance index

$$J = \int_0^\infty (wx_1^2 + u^2) \, dt \, .$$

The initial state is $x_1(0) = 1$, $x_2(0) = 0$. There is no constraint in u.

14-8 Consider a problem similar to that of cooling a cup of coffee in a minimal time using a limited supply of cream as a cooling agent. Suppose that temperature x_1 of the main tank in Fig. P14-8 is initially 100, and that the tank has heat loss to the zero-temperature surroundings, so that its dynamic equation is

$$\frac{dx_1}{dt} = -x_1 - 25u - 0.25ux_1 \, ,$$

where $-25u$ is due to the cold flow from the first tank, and $-0.25ux_1$ is for the heat

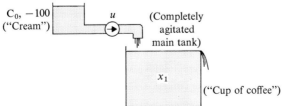

Fig. P14-8

carryover of the overflow liquid. The first tank is completely heat insulated. The initial storage of the cold fluid is limited to $C_0 = 1$. The capacity of the transfer pump produces a constraint in the control flow u so that

$$0 \leqslant u \leqslant 1.$$

Obtain the optimal control that will bring the temperature x_1 down to zero in minimal time. [*Hint:* For this control we make full use of the cooling agent.]

14-9 Only discrete values $(0, -1$ or $-2)$ are allowed for a control u. The control object is first-order, described by

$$\frac{dx_1}{dt} = -x_1 + u.$$

The initial state is $x_1(0) = 1$ and the specified final state is $x_1(T) = 0$ at $T = \frac{1}{2}$. The performance index to be minimized is quadratic,

$$J = \int_0^T u^2 \, dt.$$

Determine the optimal control.

14-10 The sale price v of a piece of merchandise is determined as the difference of demand q and rate of delivery u of the merchandise into a market: $v = q - u$. The demand is known to be growing according to the relation

$$q = 1 - e^{-t}.$$

There is no production going on, so that the equation for inventory is

$$\frac{dx_1}{dt} = -u.$$

The initial value of the inventory is $x_1(0) = \frac{1}{10}$, and it is desired to sell out the stock at some time t_1,

$$x_1(t_1) = 0,$$

while maximizing income J, where J is

$$J = \int_0^{t_1} uv \, dt - wt_1.$$

A fixed charge for inventory w is $w = \frac{1}{100}$. Determine an optimal sales strategy.

14-11 The control object is an inertial system,

$$\frac{dx_1}{dt} = x_2, \qquad \frac{dx_2}{dt} = u,$$

where control u has the constraint $|u| \leqslant 1$. The initial state is $x_1(0) = 2$, and $x_2(0) = -1$; and the final state must satisfy $x_1(t_1) = 0$ at some time t_1. Final velocity $x_2(t_1)$ is not specified. Find an optimal control that will minimize

$$J = \int_0^{t_1} (|u| + w) \, dt.$$

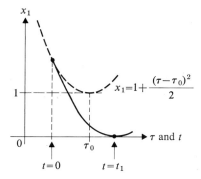

Fig. P14-12

where the weight on time w is less than $\frac{1}{2}$. Also determine the intersection of the supporting plane of the cone of attainability and the coordinate planes at t_1.

14-12 The trajectory of an inertial system,

$$\frac{dx_1}{dt} = x_2, \qquad \frac{dx_2}{dt} = u$$

must leave the parabola shown in Fig. P14-12, and reach a final state specified to be $\mathbf{x}(t_1) = \mathbf{0}$ at some t_1. Find an optimal control for which the following performance index is minimal:

$$J = \int_0^{t_1} (|u| + w) \, dt, \qquad 0 < w < 1, \qquad |u| \leqslant 1.$$

14-13 A target's motion is known to be

$$y_1 = \tfrac{1}{2}(\tau - 4)^2, \qquad y_2 = \tau - 4,$$

where τ is time, and y_1, y_2 are distance and velocity, respectively. A control object

$$\frac{dx_1}{d\tau} = x_2, \qquad \frac{dx_2}{d\tau} = u$$

is at $x_1 = 1$, $x_2 = 0$ until some time $\tau = \tau_0$, when it takes off. The object must intercept (or dock on) the target at some time τ_1, satisfying the condition

$$x_1(\tau_1) = y_1(\tau_1), \qquad x_2(\tau_1) = y_2(\tau_1).$$

For this maneuver, it is required to minimize the functional

$$J = \int_{\tau_0}^{\tau_1} (w + u^2) \, d\tau,$$

where the weighting factor w is assumed to be 1. There is no constraint on u. Obtain the optimal control and the times τ_0 and τ_1. [*Hint:* Replace time by $t = \tau - \tau_0$.]

14-14 The control object is an inertial system described by

$$\frac{dx_1}{dt} = x_2, \qquad \frac{dx_2}{dt} = u,$$

where control u has the constraint

$$|u| \leq 1 .$$

The initial state of the object is $x_1(0) = 2$, $x_2(0) = 0$. At the completion of the control effort the object must exhibit simple harmonic motion which, expressed by the target state x_1^T and x_2^T, is represented by a circle in the phase plane:

$$(x_1^T)^2 + (x_2^T)^2 = 1 .$$

Time t_1 when the object trajectory hits the circle is not known beforehand. The cost functional to be minimized is

$$J = \int_0^{t_1} (|u| + w) \, dt ,$$

where weight for time is $w = \frac{1}{2}$. Determine the optimal control, optimal trajectory, and time t_1.

14-15 The control object is

$$\frac{dx_1}{dt} = x_2, \qquad \frac{dx_2}{dt} = u ,$$

where u has no constraint. The initial and final states are specified to be $x_1(0) = 1$, $x_2(0) = 0$, and $x_1(t_1) = 0$, $x_2(t_1) = 0$, where t_1 is not known. The performance index is

$$J = \int_0^{t_1} (1 + u^4) \, dt .$$

Obtain the optimal control.

14-16 The control object is first-order, described by

$$\frac{dx}{dt} = ax + bu .$$

The performance index to be minimized is

$$J = hx(T)^2 + \int_0^T (wx^2 + u^2) \, dt ,$$

where $h \geq 0$, $w \geq 0$. The final time T is prescribed, and it is finite so that h is a weight placed on the terminal distance of the state from the origin. Determine the time-varying gain $k(t)$ of the optimal feedback control law $u^0 = -kx$.

14-17 The control object is first-order, described by

$$\frac{dx}{dt} = -x + u \qquad x(0) = 1 ;$$

and the performance index to be minimized is

$$J = x(T)^2 + \int_0^T u^2 \, dt , \qquad \text{where} \quad T = 1 .$$

Obtain the optimal control $u^0(t)$, optimal response $x(t)$, feedback gain $k(t)$ to generate optimal control by state-variable feedback, and plot the behavior of $k(t)$ over the period. Also confirm that $\sup(H) = \text{const}$.

14-18 The control object is first-order, described by

$$x_{k+1} = x_k + qu_k, \qquad q = \tfrac{1}{3},$$

with initial state $x(0) = 1$. End state x_N is not specified. The performance index is quadratic,

$$J = w_1 x_N^2 + \sum_{k=0}^{N-1} (w_1 x_k^2 + w_2 u_k^2), \qquad N = 3, \quad w_1 = 2, \quad w_2 = 1.$$

Determine the optimal control, its response, and the feedback control gain to produce this control. Compare the result with the following continuous time system:

$$\frac{dx}{dt} = u, \qquad J = \int_0^T (w_1 x^2 + w_2 u^2)\, dt, \qquad T = 1.$$

14-19 The performance index of a first-order sampled-data system

$$x_{k+1} = px_k + qu_k$$

is quadratic, as given in Prob. 14-18. The final state is specified to be $x_N = 0$. Initialize the feedback control gain. For the same system as in Prob. 14-18 ($p = 1$, $q = \tfrac{1}{3}$, $N = 3$, $w_1 = 2$, and $w_2 = 1$), obtain the response and feedback control gain that will take the state into the origin in three steps. Compare the result with the continuous-time system given in Prob. 14-18, with boundary (or terminal) conditions

$$x(0) = 1 \quad \text{and} \quad x(T) = 0 \quad \text{at } T = 1.$$

(Feedback optimal control for a specified terminal state is not always feasible by the algorithm of this chapter. See Chapter 11 for a different approach.)

14-20 The control object is shown in Fig. P14-20. Obtain k_1 and k_2 in the optimal feedback control law $u^0 = -[k_1\, k_2]\mathbf{x}$ for the minimization of the quadratic performance index

$$J = \int_0^\infty (wx_1^2 + u^2)\, dt, \qquad w > 0.$$

Discuss why $k_2 \neq 0$ for the uncontrollable mode.

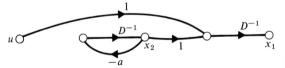

Fig. P14-20

14-21 The control object is an integrator subject to Gaussian, zero-mean, white-noise input w,

$$\frac{dx}{dt} = u + w,$$

and the output equation has a measurement noise v which is also Gaussian, zero-mean, and white,

$$y = x + v.$$

There is no correlation between w and v. Obtain the stationary optimal feedback-control system for the minimization of the following performance index:

$$J = E\{x(t)^2\} + F_u E\{u(t)^2\}.$$

Also determine the value of J.

14-22 The optimum filter has the following property:

$$E\{(\mathbf{x}_k - \hat{\mathbf{x}}_k)\hat{\mathbf{x}}_k'\} = \mathbf{0},$$

or, letting $\mathbf{x}_k - \hat{\mathbf{x}}_k = \mathbf{e}_k$ (error in the estimate),

$$E\{\hat{\mathbf{x}}_k \hat{\mathbf{x}}_k'\} = E\{\mathbf{x}_k \mathbf{x}_k'\} - E\{\mathbf{e}_k \mathbf{e}_k'\}.$$

Making use of this property, derive Eqs. (14–139) and (14–140) from Eq. (14–138).

15

SWITCHING SYSTEMS

The systems we consider in this chapter consist of components which we shall call logical elements, and which have only two possible states. These states are generally characterized as ONE and ZERO, but may just as well be called plus and minus, or on and off. The choice of 1 and 0 and the resulting theory to be developed, should not be confused with the binary counting procedure. The logical components that deliver these states may be fluidic, hydraulic, electrical or electronic. We shall start off by briefly discussing the physical operation of such widely diverse elements as transistors, relays, piston devices and fluidic gates. In the process we shall define some of the basic logic operations, such as OR and AND, that these devices are capable of producing.

The algebra of logic, developed by G. Boole in 1847 [13], was first applied to switching problems by C. E. Shannon in 1938 [14]. The fundamentals of Boolean or switching algebra are presented in Section 15-2. This algebra, in its basic form, yields a static (or time-independent) relation between input(s) and output(s). Physical switching systems that give such a static input–output relation are called *combinational* (or merely *logical*) systems. Some basic problems of logical-system design are presented in Section 15-3.

Although logical systems themselves have various engineering applications, such systems also constitute a basic building block in the construction of sequential switching control systems. In a sequential switching system, the output depends not only on the present input(s), but also on the past history of inputs (a combination lock is a good example). Sequential control systems have wide application to process-control problems, various types of automated machinery, and, in particular, to computer design.

A brief introduction to the concepts of sequential control analysis and synthesis is given, and the chapter concludes with a short discussion of finite-state systems, which include switching systems.

15-1 LOGIC ELEMENTS

Switching systems have become indispensable for automatic operations in various applications: aerospace navigation, production machinery, process-control, and information-handling, to name a few examples. Widely diverse types of elements are available as the building blocks of a switching system, for example, electromagnetic relays, transistor gates, integrated microcircuits,

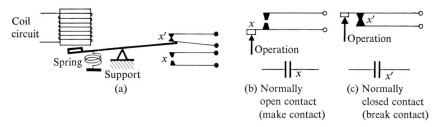

Fig. 15-1 Electromagnetic relay contacts.

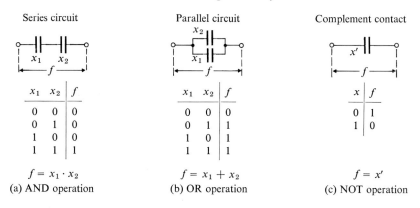

Fig. 15-2 Conducting paths associated with contact circuits.

fluidic devices, and piston logic circuits [6]. To help choose the most suitable building block for a particular application, we consider the operating characteristics of some of these devices in this section.

Figure 15-1(a) shows an electromagnetic relay that has two contacts, designated by x and x'. The contact x closes (operates) when the electromagnet is excited. The armature operation is symbolically indicated by an arrow in Fig. 15-1(b). The contact is called a normally open (or a make) contact. The other contact, x', is the complement of x; it is normally closed, as shown in Fig. 15-1(c). There is a dead time between a change in the coil current (excitation) and the resulting contact state (operation). As a consequence, both x and x' can be closed or open for a short time interval when the state of relay excitation is changed. Such imperfection may cause hazards, as we shall discuss later. The relay in Fig. 15-1(a) will break before making. There are also commercially available make-before-break relays. (See Problems 15-17 and 15-18.)

Let us introduce the following convention:

"0" for open path,   "1" for closed path,   (15-1)

and investigate the series and parallel connections of two contacts x_1 and x_2, Fig. 15-2. We consider x_1 and x_2 to be inputs, and the overall transmission

15-1 Logic elements 701

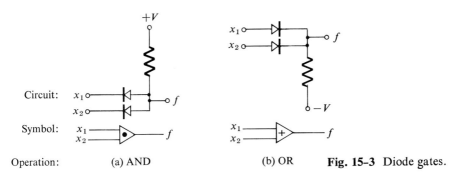

Fig. 15-3 Diode gates.

f of the network as the output. The input–output relation is shown for all possible input combinations in the tables of Fig. 15–2(a) and (b). Tables of this kind are called "truth tables." The truth table for the series connection of x_1 and x_2 shows that the overall transmission is ONE if and only if both contacts are closed. This is an AND circuit, expressed by the logic function

$$f = x_1 \cdot x_2. \tag{15-2}$$

The parallel structure (Fig. 15–2b) generates an output ONE if one or more inputs are ONE. This is an OR circuit, expressed by

$$f = x_1 + x_2. \tag{15-3}$$

If the transmission f via the normally closed contact x' is defined to be an output, f turns out to be the negation (or complement) of x (Fig. 15–2c). This is a NOT operation.

An element whose logical function is realized by opening or closing paths is classified as the *branch type*. An electromagnetic relay is a typical branch type (sometimes also called "contact") element. There is another class of elements where the logical information is represented by the level of a physical variable (a voltage, for example). Such elements are of the *gate type*.

Figure 15–3 shows the AND and OR functions implemented by diode gates. If we use the convention,

"0" for lower voltage,   "1" for higher voltage   (15-4)

for two distinct voltage levels (such as, arbitrarily, $+10$ V vs. 0 V or 0 V vs. -10 V), then the circuit (a) of the figure is an AND gate. The output voltage f is positive ("1") only if both inputs x_1 and x_2 are positive. A negative or zero voltage applied to either x_1 or x_2 (or both) causes that diode to conduct and the output potential to become the same ("0") as the input. The other circuit (Fig. 15–3(b)) has the diodes and bias potential reversed. A positive ("1") input on either x_1 or x_2 will cause conduction. Hence the circuit generates the OR function.

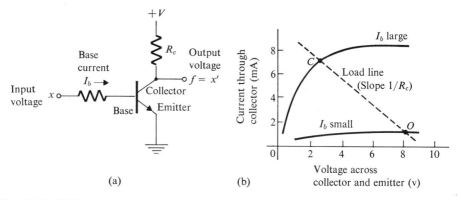

Fig. 15-4 NOT gate by an npn transistor; negative bias on the base side required to eliminate interaction with the collector current is not shown.

Diodes are passive components. Therefore the (small but finite) resistance in diodes and the interaction between stages causes signal deterioration unless amplifiers are added to the circuit. Because of their passiveness, diodes will not generate signal inversion; hence the NOT function cannot be generated by diodes alone. Transistor gates are active components. Therefore they do not have the above shortcomings of diode gates. There are two complementary types of transistors available, npn and pnp.

In an npn transistor, a positive voltage (for instance, $V = +10$ V) is applied at the collector side via a resistor R_c (for example, $R_c = 1.2$ kΩ). (See Fig. 15-4a.) The input voltage x is either at ground ("0") or some positive value ("1"). When the input is positive, the base current I_b flows and shorts the transistor, giving a low impedance between the output f and the grounded emitter, so that the output voltage is effectively zero. When the input voltage x is at ground ($x = 0$) the transistor "opens" (or blocks the output circuit), bringing the collector current to zero, so that the output voltage f is positive ("1"). We therefore conclude that the transistor acts as a switch, with its output always being the opposite of its input. It therefore produces the NOT function.

The characteristic curves of a typical transistor are shown in Fig. 15-4(b). Since R_c is the dominant resistance, the collector current is inversely proportional to R_c or the slope of the load line is $1/R_c$. The intersection C, where the base current is large, is the operating point at which the transistor "switch" is closed, producing an output voltage $f = $ "0." When the input is zero, and hence the base current is small, the transistor switch is open. The operating point is then at O in the figure, and the output voltage is high; $f = $ "1."

Let us suppose that the transistor gate in the preceding paragraph has more than one input, as shown by x_1 and x_2 in Fig. 15-5(a). It is clear that the transistor switch will open (and thus produce $f = $ "1") only when both

15-1
Logic elements 703

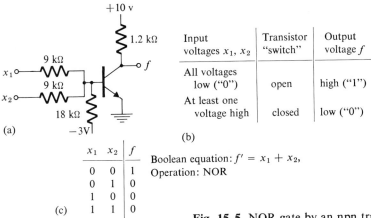

Fig. 15-5 NOR gate by an npn transistor.

Input voltages x_1, x_2	Transistor "switch"	Output voltage f
At least one voltage low ("0")	closed	high (ground, "1")
All high (ground)	open	low ("0")

(a)

$f' = x_1 \cdot x_2$, NAND

x_1	x_2	f
0	0	1
0	1	1
1	0	1
1	1	0

(b)

Fig. 15-6 NAND gate by a pnp transistor.

x_1 and x_2 are zero; otherwise the transistor switch will be closed, generating $f = $ "0". The statement is given in tabular form in Fig. 15-5(b), followed by a truth table (c). According to the truth table, the transistor gate of this figure is the complement (NOT) of an OR function. The resulting function is called a NOR function. Its Boolean expression is

$$f' = x_1 + x_2. \tag{15-5}$$

For a pnp transistor, the supply voltage (on the collector side) is negative, and input voltages are at either ground or some negative value. The relation becomes complementary to what is shown in Fig. 15-5(b). The input–output relation of a pnp transistor is shown in Fig. 15-6(a). For the convention of Eq. (15-4) (that is, "0" for low minus voltage, "1" for high ground voltage), the relation given by the truth table of Fig. 15-6(b) represents the Boolean expression

$$f' = x_1 \cdot x_2. \tag{15-6}$$

This is the complement (NOT) of the AND, called the NAND. If, however,

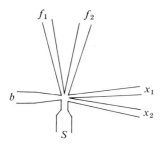

Fig. 15-7 Dual-input monostable fluid amplifier.

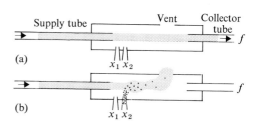

Fig. 15-8 Turbulence amplifier.

the association of voltage levels to the logical variables is reversed, so that

"1" represents a negative voltage,

and "0", the ground voltage, (15-7)

then the pnp gate (Fig. 15-6) will be a NOR element. Since npn and pnp transistors are not used simultaneously, no confusion can occur.

The wall-attachment amplifier and the turbulence amplifier are two typical fluidic devices suitable for logical operations [7, 8, 9]. They usually use air as the working fluid, but almost any other gas or liquid can be used. They have operating speeds (switching times) in the order of milliseconds, so they are much slower than solid-state devices (which have microsecond or shorter switching times). However, their principal advantages are simplicity and extreme reliability in severe environments. The disadvantage of a small steady-state power drain in such devices is offset by the distinct advantage of having no moving parts to wear out.

A wall-attachment device is shown in Fig. 15-7. It consists of a pneumatic or hydraulic power input duct S, a bias port b, control ports x_1 and x_2, and output ducts f_1 and f_2. With the convention

"1" for flow, "0" for no flow,

the output f_1 is "0" (and f_2 is "1") only if x_1 and x_2 are both "0"; thus

$$f_1 = f_2' = x_1 + x_2.$$

Therefore, the device is an OR gate for f_1 and a NOR gate for f_2. For one input x_1, the output f_2 performs the NOT operation. The device is basically a bistable flip-flop made monostable by a bias. The jet position is stable due to the tendency of a jet stream to attach to an adjacent wall (the Coanda effect described in Section 6-5).

The basic logic of the turbulence amplifier (Fig. 15-8) is the NOR operation. A supply tube vents a laminar stream of air into a collector tube

15-1 Logic elements

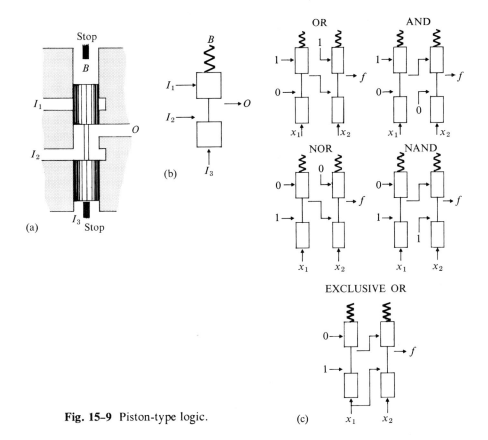

Fig. 15-9 Piston-type logic.

of the same diameter a few centimeters away. With no input ($x_1 = 0$, $x_2 = 0$), [part (a) of the figure] the laminar flow produces a "high" output pressure, $f = $ "1." An input flow [for instance $x_2 = $ "1" in Fig. 15-8(b)] causes the air stream to become turbulent and the output pressure to drop; $f = $ "0." This is a NOR operation. It performs the NOT function for a single input.

There are various other components available for producing binary logic, such as piston logic elements [10, 11] and pneumatic logic systems [12]. The building block of piston-type logic is a piston–cylinder pair (Fig. 15-9a). It has three input ports, I_1, I_2, and I_3, one of which is for the input variable x; the others are for high ("1") or low ("0") pressure supply. B in the figure is a medium (bias) pressure which is constantly applied on one side of the piston. This pressure is indicated by a spring in the symbolic representation of the unit (Fig. 15-9b). Some basic logic elements realized by two cascaded cylinders are shown in Fig. 15-9(c); the last one, labeled EXCLUSIVE OR (also called a "half-adder"), has the Boolean expression

$$f = x_1 x_2' + x_1' x_2 . \tag{15-8}$$

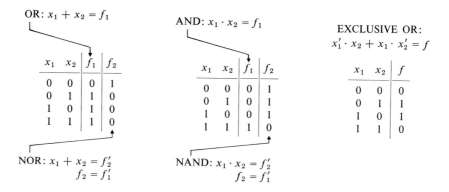

Fig. 15-10 A summary of basic logic functions.

The truth table for the EXCLUSIVE OR is shown in Fig. 15-10, where other logical functions of two variables considered in this section are also summarized.

15-2 ELEMENTS OF SWITCHING ALGEBRA*

The algebra of logic (also called Boolean or switching algebra) was developed by G. Boole [13] and applied to switching circuits by C. E. Shannon [14]. This algebra provides a theoretical tool [1 through 5] for dealing with systems in which only two values (0 or 1) of a variable are involved. These are *not*, however, the binary variables associated with binary counting. The effectiveness of Boolean algebra in switching-system design is not unlimited. For the design of extremely simple systems, intuition may work even faster than the algebraic approach, while in less simple systems the use of auxiliary tables (see below) becomes advisable. In complicated systems, trained intuition with the support of machine computation (the latter is yet to be fully developed) seems essential. In all these situations, however, switching algebra constitutes a basis for a systematic and consistent mode of expression. These expressions do not include time as a factor. Some attempts have been made [15, 16] to introduce the time element in switching algebra. In Section 15-4, when we discuss sequential switching circuits, we shall see how time is accounted for when it becomes an important variable in switching-circuit operation. Since the variables in switching algebra can have only two values, 0 or 1, some rules and expressions will be entirely different from the operations of ordinary algebra and arithmetic, although for the most part the differences between Boolean and ordinary algebra are not excessive.

---

* In this and subsequent sections the authors are indebted to Professor R. Perret of the University of Grenoble, France, for his lectures at the University of California, 1966.

Table 15-1 GENERALLY ACCEPTED CODING

State	Relay		Transistor gate	Fluidics	Magnetic core	Logic
	contact open or closed	current through coil	output voltage	output flow or pressure	polarity	true or false
0	open	not operated	small or zero / zero	small, zero or negative	normal	false
1	closed	operated	large or positive / negative	large or positive	reverse	true

Some examples of assigning two values for the state or condition of a switching variable were given in the preceding section. Table 15-1 shows the generally accepted coding for these and some other systems. For a variable x that can take only two values, one at a time, we can postulate:

$x = 0$ if $x \neq 1$,
$x = 1$ if $x \neq 0$.

This postulate is stated in a dual form which we shall repeatedly use in most of the relations we shall develop. We shall adopt a format in which the dual of any relation will be stated immediately after the relation in question, but indented and denoted as necessary with a primed equation number. Thus, complementation (negation or NOT operation denoted by a primed variable) of a variable x is expressed as:

$$\text{If } x = 0, \text{ then } x' = 1. \tag{15-9}$$

$$\text{If } x = 1, \text{ then } x' = 0. \tag{15-9'}$$

However, there are some neutral (self-dual) relations, for example,

$$(x')' = x. \tag{15-10}$$

In those cases where a relation is derived, the derivation of its dual will generally be left as an exercise for the reader.

The framework of switching algebra is built on the three fundamental operations:
   complementation or negation (NOT),
   logical sum (OR),
   logical product (AND).
As shown by the truth table for OR and AND (Fig. 15-10), the following

Table 15-2 ALL POSSIBILITIES OF A FUNCTION OF A SINGLE VARIABLE

x	$f_0 = 0$	$f_1 = x$	$f_2 = x'$	$f_3 = 1$
0	0	0	1	1
1	0	1	0	1

relations hold for these operations:

$$\text{OR:} \quad 0 + 0 = 0,$$
$$0 + 1 = 1 + 0 = 1, \tag{15-11}$$
$$1 + 1 = 1.$$

$$\text{AND:} \quad 1 \cdot 1 = 1,$$
$$1 \cdot 0 = 0 \cdot 1 = 0, \tag{15-11'}$$
$$0 \cdot 0 = 0.$$

Note that to arrive at (15-11'), we may replace 1 by 0, 0 by 1, and "+" by "." in its dual (15-11).

Since a logic variable can assume only two values, there exists only a *limited* number of functions ($= 2^{2^n}$ where n is the number of variables) for a finite number of variables. As shown in Table 15-2, there are only four possibilities for a function $f(x)$ of a single variable x. The basic relations that hold for expressions related to one variable x are the following:

$$x + 0 = x, \quad x + 1 = 1,$$
$$x + x + \cdots + x = x, \tag{15-12}$$
$$x + x' = 1.$$

$$x \cdot 1 = x, \quad x \cdot 0 = 0,$$
$$x \cdot x \cdot \cdots \cdot x = x, \tag{15-12'}$$
$$x' \cdot x = 0.$$

The validity of these relations and others that will follow can be verified by substituting, in turn, the two possible values of the variable, 0 and 1. This procedure is equivalent to constructing a truth table for all possible combinations. Such a process of perfect induction often constitutes the simplest proof of a statement. Using this approach, it is possible to prove the following general relation for a function of one variable:

$$f(x) = x \cdot f(1) + x' \cdot f(0), \tag{15-13}$$
$$f(x) = \{x + f(0)\} \cdot \{x' + f(1)\}. \tag{15-13'}$$

Table 15-3 VENN DIAGRAMS FOR SOME BASIC LOGIC RELATIONS

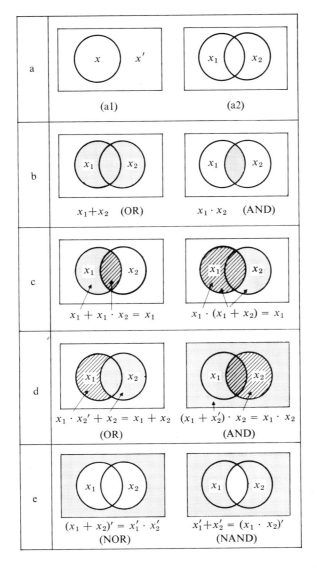

In arriving at Eq. (15–13′) from its dual (15–13), we note that in addition to the steps listed after Eq. (15–11′), x and x' exchange roles. Although Eqs. (15–13) and (15–13′) appear to be quite different, in fact they give the same result for any one of the four functions of Table 15–2. (Again, corroborate by substituting first $x = 0$ and then $x = 1$ for each case.) Equation (15–13) is written in the first canonical form, the canonical *sum* of product terms; and its dual Eq. (15–13′) is in the second canonical form, the canonical *product*

of sum terms. Although there is no basis for preferring one canonical form over the other, we shall direct our attention primarily to results expressed in the first canonical form.

Simple logical relations can be visualized in a diagram where a set is shown as a region. In Table 15–3(a1) the region inside the circle represents x, and everything outside the circle x'. This is known as a Venn diagram [17]. For a system of two variables, two circles will appear [Table 15–3(a2)]. When two circles overlap, the logical sum (OR) and product (AND) can be represented by the regions shown in Table 15–3(b). The sum and product are sometimes called union and intersection, and are expressed by cup ($\cup$) and cap ($\cap$), respectively.

Logical sum (OR), $x_1 + x_2$, is a union: $x_1 \cup x_2$.
Logical product (AND), $x_1 \cdot x_2$, is an intersection: $x_1 \cap x_2$.
Simple and useful two-variable relations are given in Table 15–3(c) and (d):

$$x_1 + x_1 \cdot x_2 = x_1, \tag{15-14}$$

$$x_1 \cdot (x_1 + x_2) = x_1, \tag{15-14'}$$

$$x_1 \cdot x_2' + x_2 = x_1 + x_2, \tag{15-15}$$

$$(x_1 + x_2') \cdot x_2 = x_1 \cdot x_2. \tag{15-15'}$$

The two basic sum and product forms are related to each other as shown in Table 15–3(e). They are called the first and second DeMorgan relations respectively. The relations signify that a logic function can be written either by sum or by product (and its complement). In other words, a system can be built entirely from OR and NOT elements, or from AND and NOT elements. Symbolically, the DeMorgan relations are:

$$(x_1 + x_2)' = \text{NOR} = x_1' \cdot x_2' \quad \text{(the first relation)}, \tag{15-16}$$

$$(x_1 \cdot x_2)' = \text{NAND} = x_1' + x_2' \quad \text{(the second)}. \tag{15-16'}$$

Negating both sides of each relation, we also have:

$$(x_1 + x_2) = \text{OR} = (x_1' \cdot x_2')', \tag{15-17}$$

$$x_1 \cdot x_2 = \text{AND} = (x_1' + x_2')'. \tag{15-17'}$$

There are 16 ($=2^{2^2}$) possible functions of two variables, as listed in Table 15–4. The table is symmetric with respect to its centerline in the sense that the functions on the right of the table are complementary to those on the left:

$$f_i(x_1, x_2) = f'_{15-i}(x_1, x_2). \tag{15-18}$$

The first and the last functions are trivial,

$$f_0 = 0, \quad f_{15} = 1.$$

15-2 Elements of switching algebra

Table 15-4 SIXTEEN POSSIBLE FUNCTIONS OF TWO VARIABLES

x_1	x_2	f_0	f_1	f_2	f_3	f_4	f_5	f_6	f_7	f_8	f_9	f_{10}	f_{11}	f_{12}	f_{13}	f_{14}	f_{15}
0	0	0	1	0	1	0	1	0	1	0	1	0	1	0	1	0	1
0	1	0	0	1	1	0	0	1	1	0	0	1	1	0	0	1	1
1	0	0	0	0	0	1	1	1	1	0	0	0	0	1	1	1	1
1	1	0	0	0	0	0	0	0	0	1	1	1	1	1	1	1	1

By inspection we can also find other trivial functions,

$$f_3 = x_1', \qquad f_{12} = x_1,$$
$$f_5 = x_2', \qquad f_{10} = x_2.$$

The OR, AND, and related functions are:

$$f_{14} = x_1 + x_2 \quad \text{(OR)}; \qquad f_1 = f_{14}' = x_1' \cdot x_2' = (x_1 + x_2)' \quad \text{(NOR)};$$
$$f_8 = x_1 \cdot x_2 \quad \text{(AND)}; \qquad f_7 = f_8' = x_1' + x_2' = (x_1 \cdot x_2)' \quad \text{(NAND)};$$
$$f_{11} = x_1' + x_2; \qquad f_{13} = x_1 + x_2';$$
$$f_4 = f_{11}' = x_1 \cdot x_2'; \qquad f_2 = f_{13}' = x_1' \cdot x_2.$$

The pair $f_6 = f_9'$ takes the most complicated form:

$$f_6 = f_9' = x_1 x_2' + x_1' x_2. \tag{15-19}$$

This is called the disjunctive function (EXCLUSIVE OR), for which a special symbol (circled plus) is sometimes used.

$$f_6 = x_1 \oplus x_2. \tag{15-20}$$

Its realization by piston-type logic was shown in Fig. 15-9(c). According to a Venn-diagram interpretation (Fig. 15-11), a complementary relation applies to the f_6, f_9 pair:

$$f_6 = x_1 x_2' + x_1' x_2 \quad \text{(EXCLUSIVE OR)}, \tag{15-21}$$

$$f_9 = (x_1 + x_2') \cdot (x_1' + x_2). \tag{15-21'}$$

Just as a function of a single variable (Table 15-2) was summarized in Eq. (15-13), so we may also categorize Table 15-4 by a single expression.

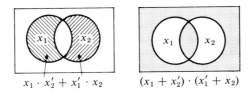

Fig. 15-11 EXCLUSIVE OR and its complement.

To arrive at the first canonical (standard-sum) form of a function of two variables, we merely apply Eq. (15-13) twice. Thus,

$$f(x_1, x_2) = x_1 \cdot f(1, x_2) + x_1' \cdot f(0, x_2),$$

where

$$f(1, x_2) = x_2 \cdot f(1, 1) + x_2' \cdot f(1, 0)$$

and

$$f(0, x_2) = x_2 \cdot f(0, 1) + x_2' \cdot f(0, 0).$$

Hence,

$$f(x_1, x_2) = x_1 \cdot x_2 \cdot f(1, 1) + x_1 \cdot x_2' \cdot f(1, 0)$$
$$+ x_1' \cdot x_2 \cdot f(0, 1) + x_1' \cdot x_2' \cdot f(0, 0). \qquad (15\text{-}22)$$

In general, it is possible to express a logical function by a sum of products, as in Eq. (15-22), or by a product of sums (the second canonical form). To derive the dual of Eq. (15-22), we apply Eq. (15-13′) twice and obtain (see Problem 15-12)

$$f(x_1, x_2) = \{x_1 + x_2 + f(0, 0)\} \cdot \{x_1 + x_2' + f(0, 1)\}$$
$$\cdot \{x_1' + x_2 + f(1, 0)\} \cdot \{x_1' + x_2' + f(1, 1)\}. \qquad (15\text{-}22')$$

The relay circuit equivalents of Eqs. (15-22) and (15-22′) are shown in Fig. 15-12. Later in this section we shall discuss NOR and NAND gate implementation.

Let us assume that the function $f_i(x_1, x_2)$ in Eq. (15-22) is known for all possible combinations of x_1 and x_2. The following rule may then be applied to simplify the results:

a) All product terms in the sum for which $f_i = 0$ disappear,
b) All product terms in the sum for which $f_i = 1$ appear without f_i.

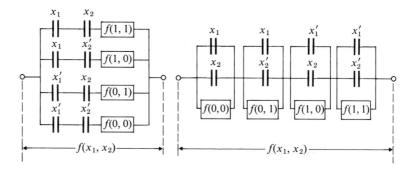

Fig. 15-12 Relay circuit equivalent of the canonical forms.

Example 15-1 Determine the expression for f_6 in Table 15-4 by using Eq. (15-22). From the table, we obtain

$$f_6(0, 0) = 0, \quad f_6(0, 1) = 1, \quad f_6(1, 0) = 1, \quad f_6(1, 1) = 0.$$

Applying rules (a) and (b) above, we get

$$f_6(x_1, x_2) = x_1 x_2' + x_1' x_2 \quad \text{(EXCLUSIVE OR)}.$$

When working in the second canonical form (see Problem 15–12), the rules become:

 a′) All sum terms in the product for which $f_i = 0$ appear without f_i,
 b′) All sum terms in the product for which $f_i = 1$ disappear.

It must be noted that, in general, these canonical forms are not (necessarily) the simplest. We shall discuss circuit simplification later.

It is possible to generalize the preceding relations into functions of n variables. In general the sum ($+$) and the product ($\cdot$) relations are commutative, associative, and distributive. The associative and distributive laws shown for three variable relations are as follows.

Associative law

$$x_1 + x_2 + x_3 = (x_1 + x_2) + x_3 = x_1 + (x_2 + x_3),$$

$$x_1 \cdot x_2 \cdot x_3 = x_1 \cdot (x_2 \cdot x_3) = (x_1 \cdot x_2) \cdot x_3.$$

Distributive law

$$x_1 \cdot x_2 + x_1 \cdot x_3 = x_1 \cdot (x_2 + x_3), \tag{15–23}$$

$$(x_1 + x_2) \cdot (x_1 + x_3) = x_1 + x_2 \cdot x_3. \tag{15–23′}$$

Proofs of these relations using Venn diagrams are shown in Fig. 15–13.

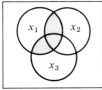
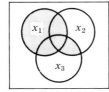

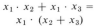

$x_1 \cdot x_2 + x_1 \cdot x_3 =$ $\quad$ $(x_1 + x_2) \cdot (x_1 + x_3) =$
$x_1 \cdot (x_2 + x_3)$ $\quad\quad\quad$ $x_1 + x_2 \cdot x_3$

Fig. 15–13 Venn diagrams for functions of three variables.

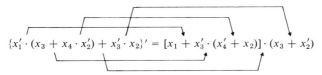

Fig. 15–14 Example of the Shannon rule on duality.

The DeMorgan relations [Eqs. (15–16) and (15–16′)] generalized for n-variable functions are:

$$(x_1 + x_2 + \cdots + x_n)' = x_1' \cdot x_2' \cdot \cdots \cdot x_n' = \text{No}(x_1, \ldots, x_n), \quad (15\text{-}24)$$

$$(x_1 \cdot x_2 \cdot \cdots \cdot x_n)' = x_1' + x_2' + \cdots + x_n'$$
$$= \text{Na}(x_1, \ldots, x_n), \quad (15\text{-}24')$$

where No signifies the NOR operation, and Na the NAND operation.

A further generalization of the DeMorgan theorem was suggested by Shannon, in the general symbolic form

$$[f(x_1, \ldots, x_n, +, \cdot, 0, 1)]' = f(x_1', \ldots, x_n', \cdot, +, 1, 0). \quad (15\text{-}25)$$

The duality expressed by Eq. (15–25) signifies that a complement of any function f is obtained by replacing each variable x_i by x_i', and by interchanging the symbols for summation $(+)$ and multiplication $(\cdot)$. An example is shown in Fig. 15–14.

A function of n variables can be expanded about any of its variables. For instance, expansion about x_1 takes the following form:

$$f(x_1, x_2, \ldots, x_n) = x_1 \cdot f(1, x_2, \ldots, x_n) + x_1' \cdot f(0, x_2, \ldots, x_n).$$

Carrying the expansion through all variables, we obtain the following result as a generalization of the standard-sum form of Eq. (15–22):

$$f(x_1, x_2, \ldots, x_n) = x_1 \cdot x_2 \cdot \cdots \cdot x_n \cdot f(1, 1, \ldots, 1)$$
$$+ x_1 \cdot x_2 \cdot \cdots \cdot x_n' \cdot f(1, 1, \ldots, 0)$$
$$+ \cdots$$
$$+ x_1 \cdot x_2' \cdot \cdots \cdot x_n' \cdot f(1, 0, \ldots, 0)$$
$$+ x_1' \cdot x_2' \cdot \cdots \cdot x_n' \cdot f(0, 0, \ldots, 0). \quad (15\text{-}26)$$

The dual of Eq. (15–26) may be similarly derived (see Problem 15–13) as a generalization of the standard product form of Eq. (15–22′).

Example 15-2 $f(x_1, x_2, x_3, x_4) = 1$ for the following combinations of variables:

x_1	x_2	x_3	x_4
0	1	0	1
1	1	0	0

and $f = 0$ for the rest. Each row corresponds to a product term of the standard sum: 0's are replaced by the complement of the corresponding variables and 1's by the corresponding variables themselves. By replacing:

(0 1 0 1)   with $(x_1' \cdot x_2 \cdot x_3' \cdot x_4)$

and

(1 1 0 0)   with $(x_1 \cdot x_2 \cdot x_3' \cdot x_4')$,

we obtain

$$f = x_1' \cdot x_2 \cdot x_3' \cdot x_4 + x_1 \cdot x_2 \cdot x_3' \cdot x_4'.$$

It is also possible to expand functions into final forms where only NAND or NOR operators will appear. For brevity we shall show the first canonical form for a function of two variables. (See Problem 15-14 for the second canonical form.) Writing f_{00}, f_{01}, f_{10} and f_{11} for $f(0,0)$, $f(0,1)$, $f(1,0)$, and $f(1,1)$, respectively, in Eq. (15-22), we first apply the DeMorgan theorem (Eq. 15-24) to each product term in Eq. (15-22):

$$(x_1 \cdot x_2 \cdot f_{11})' = \text{Na}(x_1, x_2, f_{11}),$$

or

$$x_1 \cdot x_2 \cdot f_{11} = [\text{Na}(x_1, x_2, f_{11})]',$$

where Na denotes the NAND operation. Therefore the first canonical form of Eq. (15-22) can be converted to

$$f(x_1, x_2) = [\text{Na}(x_1, x_2, f_{11})]' + [\text{Na}(x_1, x_2', f_{10})]'$$
$$+ [\text{Na}(x_1', x_2, f_{01})]' + [\text{Na}(x_1', x_2', f_{00})]'.$$

According to Eq. (15-24), the entire right side of this last equation is a global NAND operation on four NAND operators. Hence,

$$f_1(x_1, x_2)$$
$$= \text{Na}[\text{Na}(x_1, x_2, f_{11}), \text{Na}(x_1, x_2', f_{10}), \text{Na}(x_1', x_2, f_{01}), \text{Na}(x_1', x_2', f_{00})].$$
(15-27)

The dual of Eq. (15-27), to be derived in Problem 15-14, will have NOR's replacing the NAND's, and all subscripts on the f's complemented (that is, f_{11} becomes f_{00}, f_{10} becomes f_{01}, etc.). Since we have

$$\text{Na}(x_1, x_2, 0) = 1, \qquad \text{Na}(x_1, x_2, 1) = \text{Na}(x_1, x_2),$$

all NAND's for which $f_{ij} = 0$ drop out from the right side of Eq. (15-27).

Example 15-3 We shall derive the NAND-gate implementation of the function given in Example 15-2. Since we take only those NAND's for which f is 1,

$$f(x_1, x_2, x_3, x_4) = \text{Na}[\text{Na}(x_1', x_2, x_3', x_4), \text{Na}(x_1, x_2, x_3', x_4')].$$

The realization of this system via NAND gates is shown in Fig. 15-15.

The switching systems shown in Figs. 15-12 and 15-15 for branch-type and gate-type elements, respectively, are not necessarily the "best" final form for engineering realization, since they are directly based on the canonical forms (the standard sum and product) which in general are not the simplest. The complexity and cost of a switching system depends on the total number

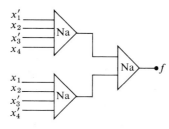

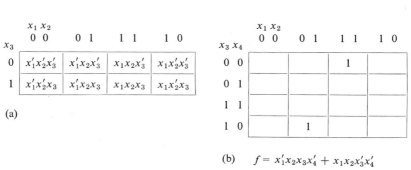

Fig. 15-15 Logic function realized by gate elements.

(a)

(b) $f = x_1'x_2x_3x_4' + x_1x_2x_3'x_4'$

Fig. 15-16 Karnaugh maps for three and four variables.

of literals (the number of variables, including repeated or complemented variables as new literals) and the total number of products or sums that appear in a switching function. In the design procedure it is generally desirable to reduce these numbers.

We shall present a design method for circuit simplification based upon the Karnaugh map [18] in the next section. The Karnaugh map is a graphical representation of a switching function and its variables, and in a sense is an extension of the Venn diagram to three- and four-variable systems. The format of the map is shown in Fig. 15-16 where the dots of the logical AND function are omitted for convenience. For a four-variable system where the Venn diagram approach fails due to complexity, the matrix structure of the Karnaugh map provides an excellent tool for design. Two-variable systems are so simple that the map is not needed. Complexity quickly mounts, however, when the number of variables is increased, with the Karnaugh map approaching a limit of practicality at five or six variables.

The three-variable map shown in Fig. 15-16(a), has column headings x_1x_2 and row headings x_3 for all possible combinations of the input variables. The eight cells within the map represent all possible states of the three input-variable combinations, each of which may yield an output of "0" or "1." Similarly, the map for four variables [part (b) of the figure] consists of sixteen cells for complete coverage of all possibilities. If, for example, a function is expressed in the first canonical form as:

$$f(x_1, x_2, x_3, x_4) = x_1'x_2x_3x_4' + x_1x_2x_3'x_4',$$
$$\quad\quad\quad\quad\quad (0\ 1\ 1\ 0)\quad\quad (1\ 1\ 0\ 0)$$

we write 1's in the corresponding cells, as shown in Fig. 15–16(b). If a function is expressed in the second canonical form, 0's will be "plotted" in the map.

The coding of the different columns and rows in Fig. 15–16(b) follows the Gray code, which is defined in Problem 15–6. By using this code we permit only one variable to change in going from a cell to any adjacent cell (not diagonally, however). Note that, defining adjacency of two cells as cells differing by only one variable, the upper right-hand cell is adjacent to both the upper left-hand and lower right-hand cells of Fig. 15–16(b). In fact, the whole top row is adjacent to the whole bottom row, etc. It is necessary to use this Gray code scheme when executing simplifications directly on the Karnaugh map, as explained in the next section.

A switching function *in its canonical form* can also be expressed numerically. For this purpose we adopt the following coding for literals:

For the first canonical form (standard sum), code uncomplemented literals as 1 and complemented literals as 0. Thus, $x_1' \cdot x_2 \cdot x_3 \cdot x_4'$ is coded as 0 1 1 0.

For the second canonical form (standard product), code complemented variables as 1 and uncomplemented variables as 0. Thus, $(x_1 + x_2' + x_3' + x_4)$ has the code 0 1 1 0.

The coded expression, taken as a binary number, is now converted into its decimal equivalent. For instance, a four-bit binary word $a_3a_2a_1a_0$ is equal to a decimal number N such that

$$N = a_3 2^3 + a_2 2^2 + a_1 2^1 + a_0 2^0 = 8a_3 + 4a_2 + 2a_1 + a_0,$$

where the a_i, $i = 0, 1, \ldots$, are either 1 or 0. A canonical expression of a switching function is then numerically represented as a sum or product of

$x_3 x_4$ \ $x_1 x_2$	0 0	0 1	1 1	1 0
0 0	0	4	12	8
0 1	1	5	13	9
1 1	3	7	15	11
1 0	2	6	14	10

$f(x_1, x_2, x_3, x_4) = $ standard sum
$\quad\quad\quad\quad\quad\quad = \sum_4 $ (decimal numbers)

or

$f(x_1, x_2, x_3, x_4) = $ standard product
$\quad\quad\quad\quad\quad\quad = \prod_4 $ (decimal numbers)

Fig. 15–17 Canonical form of the switching function and its numerical representation.

Table 15-5 MINTERMS AND MAXTERMS OF A THREE-VARIABLE SYSTEM

First canonical form (standard sum)			Second canonical form (standard product)		
Minterms m_i	Numerical representation		Maxterms M_j	Numerical representation	
	Binary	Decimal		Binary	Decimal
$m_0 = x_1' \cdot x_2' \cdot x_3'$	0 0 0	0	$M_7 = x_1' + x_2' + x_3'$	1 1 1	7
$m_1 = x_1' \cdot x_2' \cdot x_3$	0 0 1	1	$M_6 = x_1' + x_2' + x_3$	1 1 0	6
$m_2 = x_1' \cdot x_2 \cdot x_3'$	0 1 0	2	$M_5 = x_1' + x_2 + x_3'$	1 0 1	5
$m_3 = x_1' \cdot x_2 \cdot x_3$	0 1 1	3	$M_4 = x_1' + x_2 + x_3$	1 0 0	4
$m_4 = x_1 \cdot x_2' \cdot x_3'$	1 0 0	4	$M_3 = x_1 + x_2' + x_3'$	0 1 1	3
$m_5 = x_1 \cdot x_2' \cdot x_3$	1 0 1	5	$M_2 = x_1 + x_2' + x_3$	0 1 0	2
$m_6 = x_1 \cdot x_2 \cdot x_3'$	1 1 0	6	$M_1 = x_1 + x_2 + x_3'$	0 0 1	1
$m_7 = x_1 \cdot x_2 \cdot x_3$	1 1 1	7	$M_0 = x_1 + x_2 + x_3$	0 0 0	0

components expressed by decimal numbers, where it is understood that the functional output in each decimally numbered box (see Fig. 15-17) that appears in the canonical form is 1 for the standard-sum form and 0 for the standard-product form.

Example 15-4

If $f(x_1, x_2, x_3, x_4) = x_1' \cdot x_2 \cdot x_3 \cdot x_4' + x_1 \cdot x_2 \cdot x_3' \cdot x_4'$

$= (0\ 1\ 1\ 0) + (1\ 1\ 0\ 0) = 6 + 12$

$= \sum_4 (6, 12),$ enter 1's in boxes 6 and 12 of Fig. 15-17.

If $f(x_1, x_2, x_3, x_4) = (x_1 + x_2' + x_3' + x_4) \cdot (x_1' + x_2' + x_3 + x_4)$

$= (0\ 1\ 1\ 0) \cdot (1\ 1\ 0\ 0)$

$= 6 \cdot 12 = \prod_4 (6, 12),$ enter 0's in boxes 6 and 12.

The subscript 4 in the final expressions indicates the number of variables. The cells of the Karnaugh map are related to the decimal number representations as shown in Fig. 15-17 for four variable systems. Similar arguments hold for systems having more than four variables.

We conclude this section with some remarks on the canonical forms. It was shown by Eq. (15-26) and Problem 15-13 that product or summation terms in the canonical forms must include all literals in each term. These terms are called the minterms and maxterms, for the first and second canonical forms, respectively. All possible minterms and maxterms are listed in Table 15-5 for a three-variable system. The total number of possible terms of an

n-variable system is 2^n, and the following properties hold:

In the first canonical (or standard-sum) form, the sum of all possible minterms is 1,

$$\sum_{i=0}^{2^n-1} m_i = 1 . \tag{15-28}$$

This is because there must exist at least one m_k for which $m_k = 1$; otherwise the function is trivial.

In the second canonical (or standard-product) form, the product of all possible maxterms is 0,

$$\prod_{i=0}^{2^n-1} M_j = 0 . \tag{15-28'}$$

If this were not true all M_k would be 1, and the function would be trivial.

15-3 COMBINATIONAL SYSTEMS

A switching system whose output at any time is determined by the inputs at that time is called a *combinational system*. The short time delay involved in the operation of a logic element is ignored in such a system. A system is said to be *sequential* when it has memory, so that its output at any instant depends not only on the inputs at that instant, but also on previous inputs. In this case, as we shall see in the next section, time delays are of the utmost importance. While combinational systems play an essential role in the implementation of sequential systems, they have important applications in their own right. In this section we shall see how the design of combinational systems follows directly from an understanding of the fundamentals of switching algebra.

The main tasks we must face in combinational-system design involve the following problems: (1) interpretation of an assigned job in the form of a truth table or switching function; (2) manipulation and simplification of the switching function; (3) implementation of the final algebraic result by means of branch-type or gate-type elements; and (4) investigation of hazards. The first item is often the most important and difficult phase in engineering design where a thorough knowledge of a given process and its operation is required to account for all combinations of events that can possibly occur. The designer's problem is not limited to routine implementation. He must often be aware of such crucial factors as reliability, flexibility, fail-safe capability, maintainability, and many other diverse problems, including malfunction detection and diagnosis. However, we shall limit the discussions here to combinational systems design per se, that is, items (2) through (4). We begin with a very simple example to introduce the subject.

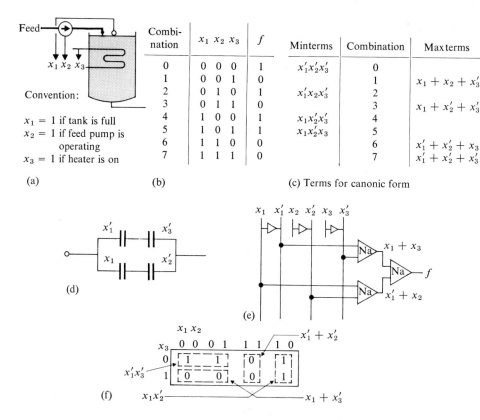

Fig. 15-18 Combinational system applied for an interlocking.

Example 15-5 Consider a plant in which a feed pump will operate when a tank is not full and a heater will operate when the tank is full (Fig. 15-18a). Taking the three logic variables as defined in the figure, we construct a truth table in Fig. 15-18(b) for all possible combinations of the variables. Let us suppose that the combinations 1, 3, 6, and 7 are to be avoided (coded $f = 0$), while the others are allowed to occur ($f = 1$). The switching function f may cause an interlocking action or may serve as a warning system.

Design of this system may begin either with the first canonical form, Eq. (15-26), or its dual, as determined in Prob. 15-13. We arbitrarily choose the former course, leaving the standard-product solution as an exercise (Prob. 15-15). Our first task is thus to write Eq. (15-26) in the standard-sum first canonical form, for the truth table of Fig. 15-18(b). Taking the left-side convention of Fig. 15-18(c), we collect minterms for which f is 1 and in which a zero of x_i is expressed by x_i'. (As an aid in the solution of Prob. 15-15, the right side of Fig. 15-18(c) lists the maxterms for which f is 0 and in which a zero of x_i is expressed by x_i.) From the minterms in Fig. 15-18(c):

$$f = \sum_{3} (0, 2, 4, 5) ,$$

or

$$f = x_1' \cdot x_2' \cdot x_3' + x_1' \cdot x_2 \cdot x_3' + x_1 \cdot x_2' \cdot x_3' + x_1 \cdot x_2' \cdot x_3 . \tag{15-29}$$

15-3 Combinational systems 721

First canonic form

$$f = \sum_{4} (3, 4, 5, 7, 11, 13, 15)$$

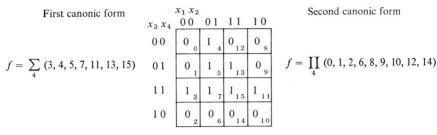

(a) Problem statement

Second canonic form

$$f = \prod_{4} (0, 1, 2, 6, 8, 9, 10, 12, 14)$$

0 1 0 0
0 1 0 1
0 1 0 –
$= x_1' x_2 x_3'$

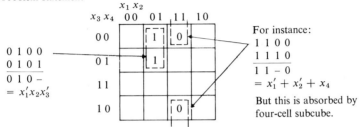

(b) Two-cell subcubes

For instance:
1 1 0 0
1 1 1 0
1 1 – 0
$= x_1' + x_2' + x_4$

But this is absorbed by four-cell subcube.

– 1 – 1 = $x_2 x_4$

– – 1 1 = $x_3 x_4$

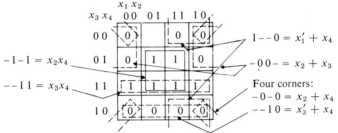

(c) Four-cell subcubes

$1 – – 0 = x_1' + x_4$

$– 0 0 – = x_2 + x_3$

Four corners:
$– 0 – 0 = x_2 + x_4$
$– – 1 0 = x_3' + x_4$

Letting $d = 1$:
$– – – 1 = x_4$

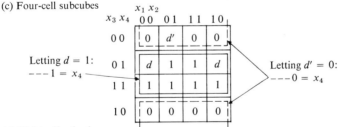

(d) Eight-cell subcubes
if cells labeled d
are "don't care" cells

Letting $d' = 0$:
$– – – 0 = x_4$

Fig. 15–19 Reduction using the Karnaugh map.

Applying the rule $(x + x') = 1$ and the distributive law, we can reduce Eq. (15-29) to

$$f = x_1' \cdot x_3' + x_1 \cdot x_2' . \qquad (15\text{-}30)$$

This function may be realized by four relay contacts, as shown in the branch-type circuit of Fig. 15–18(d). If gate-type-element implementation is desired, we apply Eq. (15-27) directly to the simplified first canonical form of Eq. (15-30) and get:

$$f = \text{Na} \left[\text{Na} \, (x_1', x_3'), \, \text{Na} \, (x_1, x_2') \right] . \qquad (15\text{-}31)$$

Note that the number of literals in Eq. (15-31), corresponding to the number of gate inputs required, is the same as the number of literals in Eq. (15-30), corresponding to the number of relay contacts required. The NAND gate circuit of Eq. (15-31) is shown in Fig. 15-18(e). NOR gate implementation of *this* function (see Prob. 15-15) results in the same number of required gates and negations. In general, this will not be true, and both canonical forms should be checked.

The manipulation of Eq. (15-29) into the reduced form of Eq. (15-30) was done algebraically. The algebraic approach will become difficult for more complex situations and a graphical approach would be more efficient. The Karnaugh map for our system is shown in Fig. 15-18(f). As indicated in the figure, the product terms in the reduced-sum form of Eq. (15-30) correspond to pairs of adjacent one-cells in the map. The adjacent zero-cells of the figure are for use in the second-canonical form solution of the problem.

The Karnaugh map provides a systematic approach to logical function simplification due to the property that *adjacent cells* (in rows, columns, and along the edges, as discussed in the previous section) differ by only one variable. Compare, for example, boxes 2 and 10 in Fig. 15-17, for which we have $x_1' \cdot x_2' \cdot x_3 \cdot x_4'$ (0 0 1 0) and $x_1 \cdot x_2' \cdot x_3 \cdot x_4'$ (1 0 1 0), respectively. For this property, the sequence of cells does not follow ordinary binary counting. For instance in Fig. 15-17 we note that $x_1 x_2$ and $x_3 x_4$ are both in the order 00 01 11 10 (that is, the Gray code, discussed previously). We shall present the map method by the following example.

Example 15-6 A switching function to be simplified is given in Fig. 15-19(a). Stated in numerical representation, the function is (putting 1 boxes in $\sum$ format, and 0's in $\prod$)

$$f = \sum_4 (3, 4, 5, 7, 11, 13, 15) = \prod_4 (0, 1, 2, 6, 8, 9, 10, 12, 14) . \qquad (15\text{-}32)$$

As shown in Fig. 15-19(b), a pair of adjacent cells can be combined into a "two-cell subcube" to eliminate the noncommon literal. The effect of eliminating this noncommon literal is to leave a simplified expression (either a minterm for the sum form or a maxterm for the product form) which completely specifies the original ungrouped function of the isolated cells. Thus, in the present case,

$$x_1' \cdot x_2 \cdot x_3' \cdot x_4' + x_1' \cdot x_2 \cdot x_3' \cdot x_4 = x_1' \cdot x_2 \cdot x_3' .$$

It is also possible to eliminate two literals by grouping four cells into a subcube, as shown in Fig. 15-19(c). The minterms or products [for the first-canonical-sum form (left-hand side of the figure)] which we obtain by carrying out *all* possible groupings of one-cells are termed the *prime implicants* of the given function. A prime implicant of a given function is defined to be a product f_i such that (1) $f_i = 1$ implies $f = 1$, but (2) this property will be lost if any literal is eliminated from f_i. The prime implicants of Eq. (15-32) are, from the left sides of Figs. 15-19(b) and (c):

$$f_1 = x_1' \cdot x_2 \cdot x_3' , \qquad f_2 = x_2 \cdot x_4 , \qquad \text{and} \qquad f_3 = x_3 \cdot x_4 , \qquad (15\text{-}33)$$

so that the function f is

$$f = f_1 + f_2 + f_3 .$$

The maxterms or sums [for the second-canonical-product form (right-hand side of the figure)] which we obtain by carrying out *all* possible groupings of zero-cells are termed the *prime implicates*. A prime implicate of a given function is defined to be a sum F_j such that (1) $F_j = 0$ implies $f = 0$, but (2) this property will be lost if any literal is eliminated from F_j. The prime implicates of Eq. (15-32) are, from the right side of Fig. 15-19(c),

$$F_1 = x_1' + x_4, \qquad F_2 = x_3' + x_4,$$
$$F_3 = x_2 + x_4, \qquad F_4 = x_2 + x_3, \tag{15-33}$$

so that the function f is

$$f = F_1 \cdot F_2 \cdot F_3 \cdot F_4.$$

Even though the results given by the two approaches are logically equivalent, the first f appears to be less complex to implement, involving one literal less and one term less than the second. Before jumping to this conclusion however, we should read ahead to Eq. (15-34)!

In general, there are four possible formats for the prime implicants (or implicates) of a four-variable system: (a) an implicant with four literals that corresponds to an isolated cell, (b) a two-cell subcube (3 literals), (c) a four-cell subcube (2 literals), and (d) an eight-cell subcube (one literal). For the last case, see Fig. 15-19(d).* Note that other combinations are not allowed by the logic of the situation.

The sum (first form) or product (second form) of *all* the prime implicants (or implicates) is *not* necessarily the simplest form of a given switching function. The final form, after reduction, is given by a set of prime implicants that is *necessary and sufficient* to represent all the cells for which $f = 1$ (first form) or $f = 0$ (second form). In our example all three prime implicants are required to give the first form. However, in the second form we note that the "territory" of the last four-cell subcube taken at the four corners completely overlaps with others (Fig. 15-19c). Therefore, all the cells for which $f = 0$ are completely represented by F_1, F_2, and F_4 in Eq. (15-33), hence F_3 is not needed. The final result for f in the second form is:

$$f = x_1'x_2x_3' + x_2x_4 + x_3x_4 = (x_1' + x_4)(x_3' + x_4)(x_2 + x_3), \tag{15-34}$$

so that in the case of Example 15-6 the second canonical form is to be preferred—it involves the same number of (NOR) gates and negations as the first (NAND) form implementation, but has one literal less and no triple-input gate. In general, the technique for determining the simplest form of a switching function is to search for those prime implicants (or implicates)

---

* If in some cells it is immaterial whether $f = 1$ (or $f = 0$) or not, these are called the "don't care" cells. The don't-care condition adds flexibility in simplification. For instance, don't-care cells, marked "d" in Fig. 15-19(d), will enable us to construct an eight-cell cube.

which are both as large as possible (in cube size) and yet overlap as little as possible. We shall see later, however, that this process of eliminating redundancy by looking for nonoverlapping prime implicants can lead to hazardous (false) circuit operation. In fact, when confronted with such hazards we shall have to *add* redundancy to eliminate them, which is exactly the opposite of our present emphasis on circuit simplification. For the moment, though, we shall continue our discussion of circuit simplification to formalize the above generalizations on simplification. We should note, however, that the prime implicant (implicate) form may not yield the simplest form, and further reduction will often be possible (for example, $x_1 \cdot x_2 \cdot (x_3 + x_4)$ is simpler to implement than $x_1 \cdot x_2 \cdot x_3 + x_1 \cdot x_2 \cdot x_4$).

The map method of detecting and choosing the prime implicants is most effective for three- and four-variable switching functions. Although it is still applicable (with slight modifications) when the number of variables is five or even six, perhaps six is the upper limit. The tabulation method of Quine and McCluskey [19, 20], on the other hand, places no limit on the number of variables. Also, it offers the possibility of computer solution. Its basic principle is identical to the map method. It consists of two distinct operations: (1) detection of a complete set of prime implicants using tabulation, and (2) selection of a minimal number of prime implicants which are necessary to represent the original function, using a simplification chart. We shall present the tabulation method by once more solving the preceding example.

Example 15-7 The system to be simplified is given by Eq. (15-32), restated:

$$f = \sum_4 (3, 4, 5, 7, 11, 13, 15) = \prod_4 (0, 1, 2, 6, 8, 9, 10, 12, 14).$$

Step 1 (Fig. 15-20a). Group the binary representations of the minterms in the standard sum (left-hand side of the figure, the first form) or maxterms in the standard product (right-hand side, the second form) according to the number of 1's (or 0's) they contain, so that all members of a group have the same number of 1's (or 0's). Begin with the group containing the lowest number of 1's (or 0's) and draw a line between each group.

Step 2. We now apply the rule of cancellation that corresponds to a two-cell subcube in the map method. Any two products (or sums) which differ from each other by *only one* variable x_i can be combined, and this common x_i can be eliminated. Such pairs can exist only in two adjacent groups of Fig. 15-20(a).

Starting with the first member in the top group [minterm number 4 on the left side in Fig. 15-20(a) and (b)], we compare it with *all* members in the next group. The sums 4 and 3 cannot combine because they differ in more than one literal, but sums 4 and 5 differ only in the last literal, and hence can be combined into 0 1 0 –, where the bar denotes an eliminated literal. When the new term, labeled "4, 5," is entered in the second table [Fig. 15-20(b), left], the terms 4 and 5 are checked off, as shown at the left of this table. The remainder of the table in Fig. 15-20(b) is filled out in the same fashion.

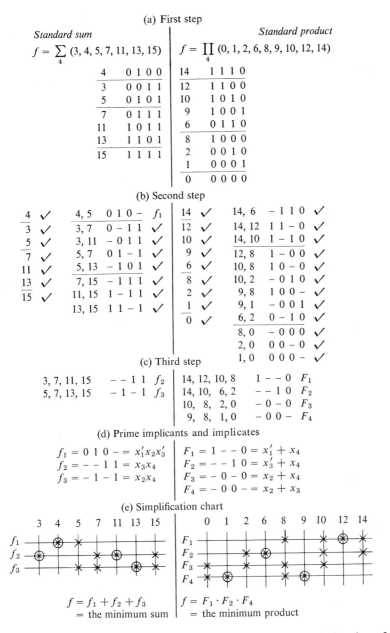

Fig. 15-20 The tabulation method to detect prime implicants and the chart for selection of necessary and sufficient implicants.

Step 3. This step is the same in its basic procedure as the preceding step. We go through an exhaustive comparison process for the terms listed in the second step, where a term can be compared with a member of the next (adjacent) group only when the two terms have bars at the same location. For instance, (3, 7) and (11, 15) combine into − − 1 1, and (3, 7), (11, 15), (3, 11), (7, 15) in the left-hand side of Fig. 15–20b will be checked off. The elimination process must be carried out as long as possible. In this example the third step is the last one (that is, no eight-cell subcubes are possible). The unchecked terms, labelled f_1, f_2, f_3 on the left, and F_1 through F_4 on the right of Fig. 15–20(b) and (c) and listed in (d), are the prime implicants and implicates for the first and second forms, respectively.

Step 4. The last step of simplification is performed on a chart (Fig. 15–20e), where a set of prime implicants (or implicates) necessary and sufficient to represent the given switching function is to be selected from the complete set listed in Fig. 15–20(d). In this example, it turns out that there is no redundancy in prime implicants when the first form is used; all three, f_1, f_2, and f_3, are needed. On the other hand, F_3 turns out to be a redundant implicate in the second form. To see how to identify this redundancy, let us focus our attention on the right-hand chart of Fig. 15–20(e). The top entry, 0, 1, 2, 6, ..., 14, represents the set of sums in the standard product $f = \prod_4 (0, 1, 2, 6, 8, 9, 10, 12, 14)$. The left-hand entry of the chart is the set of prime implicates, F_1 through F_4. According to part (c) of the figure, F_1 represents maxterms numbers 8, 10, 12, and 14. In the first row of the chart, we mark crosses at the intersections 8, 10, 12, 14. The remainder of the chart is completed in the same manner.

Each column is next inspected to see whether it has only one cross in it, or more than one. We then place a circle around a cross which stands alone in a column; for instance, the first row is circled at 12. This circle signifies that the prime implicate F_1 must be present in the final form of f to represent the maxterm 12. Circles are located in the rows for F_1, F_2, and F_4, meaning that these three must be present in the final form. By inspection of F_3 we find that all of its crosses are overlapping with those of other necessary prime implicates. Therefore there is no need for F_3. We thus obtain the final forms shown at the bottom of Fig. 15–20. The same result was given by Eq. (15–34). These final forms, called the *minimum sum or product*, turn out to be unique in our example. In some problems the end result is not unique. We have shown two complementary approaches side by side, for the map method and the tabulation method. In practice there is no necessity to work on the first and the second form simultaneously. However, it is advisable to check both forms of the final solution to determine which leads to less costly implementation.

While the above simplification technique will work well for single-output, two-level AND–OR logic, it is questionable whether such simplification approaches are feasible for multi-input, multi-output implementations with fan-in and fan-out constraints. Since all literals seldom represent equal cost, and, further, since redundancy is required anyhow for reliability and hazard-free operation, the effort involved in the saving of a few gates is sometimes not warranted.

A switching function reduced to a minimum sum or product can be implemented by either branch-type or gate-type elements. The principle of

15-3   Combinational systems   727

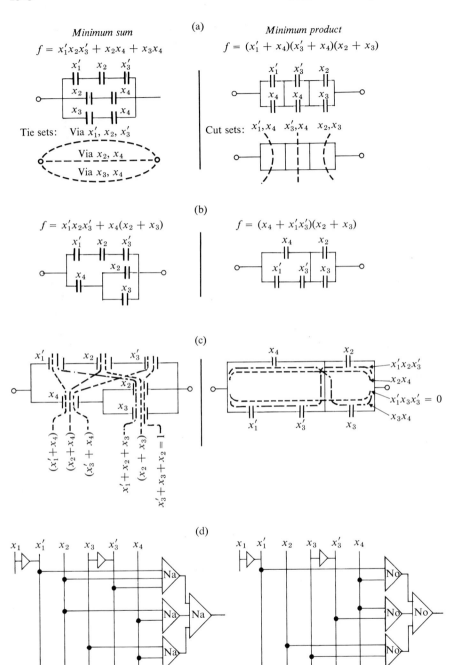

Fig. 15-21 Implementation of a reduced logic function.

realization, shown in Fig. 15–21 for the function of Eq. (15–34), is basically the same as in Example 15–5. However, a further simplification in the series-parallel structure of branch-type circuits is possible for the present system. This is shown in Fig. 15–21(b).

The dashed lines in Fig. 15–21(a) indicate the *tie sets* and *cut sets* for the first and second forms, respectively. For the first form (on the left), each term of the minimum sum is a tie set. A tie set is thus a minimum set of contacts, which, when closed, ensures that $f = 1$ regardless of the state of the other contacts. For the second form (right-hand side of the figure), each term of the minimum product is a cut set. A cut set is thus a minimum set of contacts which, when open, ensures that $f = 0$ regardless of the state of the other contacts.

Applying the concept of tie set or cut set, we can derive the minimum product of a function from its minimum sum, or vice versa. This is shown for our example in Fig. 15–21(c). To derive the minimum product from the minimum sum given as on the left-hand side of the figure, we consider all possible cuts shown by the dashed lines, and obtain the sums for the minimum product. $F_3 = (x_2 + x_4)$ is not included for the reason given in Fig. 15–20(e). Since $(x_2 + x_3)$ is already included, $(x_1' + x_2 + x_3)$ is redundant and is omitted. We also omit $(x_3 + x_3' + x_2)$ since, always being equal to 1, it is not a cut set. To derive the minimum sum starting from the minimum product on the right, we consider all possible tie sets as shown by dashed lines, and obtain the products for the minimum sum. It is not necessary to consider those tie sets which contain a variable and its complement, because, for instance, $x_1' \cdot (x_3 \cdot x_3') = 0$ and is not a tie set.

Figure 15–21(d) shows realization of the same logical function by gate-type elements. The first and second forms utilize the NAND and NOR gate, respectively. In the first form we obtain

$$f = (x_1' \cdot x_2 \cdot x_3') + (x_2 \cdot x_4) + (x_3 \cdot x_4)$$
$$= \text{Na}\,[\text{Na}\,(x_1', x_2, x_3'), \text{Na}\,(x_2, x_4), \text{Na}\,(x_2, x_4)],$$

where Na signifies the NAND operation, and we note two stages of the NAND operation. These stages are indicated by the two vertical lines of NAND operation in Fig. 15–21(d). If NOR gate operation is preferred, we convert f to f' by the DeMorgan theorem. Since the resulting f' is in the second form (see below), it can be realized by a two-stage NOR-gate circuit. In the second form,

$$f = (x_1' + x_4) \cdot (x_2' + x_4) \cdot (x_2 + x_3)$$
$$= \text{No}\,[\text{No}\,(x_1', x_4), \text{No}\,(x_2', x_4), \text{No}\,(x_2, x_3)],$$

where No signifies the NOR operation, and we again note two stages of operation.

15-3

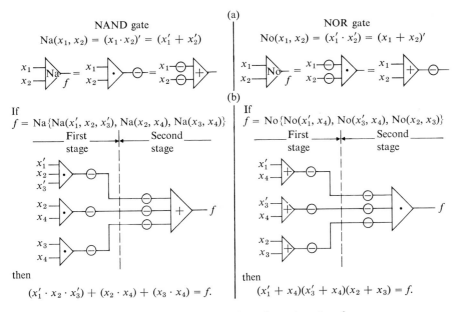

Fig. 15-22 Writing equations for gate networks.

In general, both NOR and NAND gates may be represented by either the first (product) or second (sum) form. These representations are shown in Fig. 15-22(a), where the symbols " $\cdot$ ", " $+$ ", " $-$ " represent product (AND), sum (OR) and complement (NOT), respectively. A system expressed in terms of NOR or NAND operations can be *analyzed* by means of these basic relations. For this purpose a multistage gate system is expressed in terms of alternating " $\cdot$ " and " $+$ " operations. As a result of this alternation, pairs of $-$'s will appear next to each other between stages. These pairs simply drop out by the rule of Eq. (15-10). An example of this approach is shown in Fig. 15-22(b), where the original logic function is regained in both canonical forms.

Design and analysis of both branch- and gate-type systems were shown in Figs. 15-21 and 15-22 for series-parallel structural patterns. Some complications arise when the structure is not series-parallel [1 through 3].

Throughout the preceding discussion, switching systems were considered to be ideal in the sense that well-defined logical information was assumed to exist everywhere in a system at all times. This assumption may fail when a variable x (and its complement), appearing at several locations in a system, changes state with a slight time difference between operation of x and x'. When this occurs, the equations

$$x + x' = 1, \tag{15-35}$$
$$x \cdot x' = 0 \tag{15-35'}$$

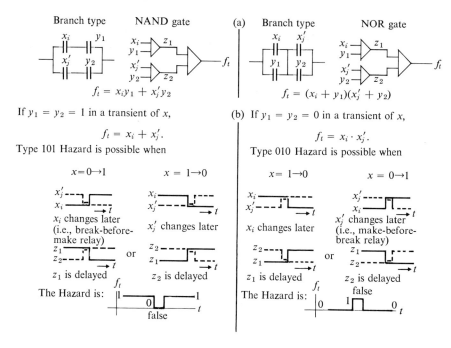

Fig. 15-23 Example of hazards.

may not hold during the short interval of the time difference. Consequently, a false value, $f = 0$ or 1, will momentarily appear, as shown in Fig. 15-23 for two simple examples. To investigate possible hazards in a switching system due to a variable x, we assign subscripts i and j to this variable that will change from x to x' with a slight time difference, while designating other switching variables by y_k, $k = 1, 2, \ldots$ The switching function for a hazardous transient will be designated by subscript t, as f_t.

There are two kinds of "static" hazards, type 1-0-1 and type 0-1-0, for the cases of momentary false values of 0 and 1, respectively. This hazard-type notation has not been standardized, and care should be exercized in interpreting reports in the literature where our 1-0-1 type of hazard is usually called type 1, and our 0-1-0 is called type 0. We feel that the three-digit depiction of the hazard graphically relates to the actual transient output shown at the bottom of Fig. 15-23. There are also "dynamic" hazards which can appear when a variable switches from 0 to 1, which we would call a type 0-1-0-1 dynamic hazard, or from 1 to 0, which we would call a type 1-0-1-0 dynamic hazard. The sequence of numbers in the notation exactly describes the time sequence of the output function f_t. We need not concern ourselves with dynamic hazards, however, since they may be avoided by using canonical-form expressions and eliminating static hazards (as explained below). There is still another form of hazard, called an essential hazard, which we shall refer to in the next section on sequential circuits.

15-3 Combinational systems

As shown in Fig. 15–23, the possibility of static hazards exists in both branch- and gate-type elements. In branch-type switching systems, static hazards are caused by one switch (x_i or x_j') closing or opening sooner than the other. In gate-type systems they are due to a time difference in the operation of the first stage gate (in electronic gates the delay is usually found within the gate elements, whereas in pneumatic–fluidic systems the delay generally occurs in connecting tubing between elements). The outputs of the first-stage gates are shown as z_1 and z_2 in the figure.

The hazard analysis of the simple examples shown in Fig. 15–23 may be generalized into a pair of rules on the possibility of static hazards due to a single variable x.

Type 1–0–1 hazards are possible if a switching function takes the form

$$f_t = [x_i + x_j' + h(y_1, y_2, \ldots)] \cdot g(x, y_1, y_2, \ldots), \tag{15-36}$$

and x changes while $g = 1$ and $h = 0$. The variable x in $g(x, y_1, y_2, \ldots)$ may involve both x_i and x_j; but in order for the hazard to occur g must remain 1. Figure 15–23 is a special case with $h \equiv 0$ and

$$g = (x_i + y_2) \cdot (x_j' + y_1) \cdot (y_1 + y_2),$$

where this form of the switching function is derived from the cut-set concept applied to the branch-type circuit on the left-hand side of Fig. 15–23(a).

Type 0–1–0 hazards are possible if a switching function takes the form

$$f_t = x_i \cdot x_j' \cdot g(y, y_2, \ldots) + h(x, y_1, y_2, \ldots), \tag{15-36'}$$

and x changes when $g = 1$ and $h = 0$. The variable x in $h(x, y_1, y_2, \ldots)$ may involve both x_i and x_j; but in order for the hazard to occur h must remain 0. Figure 15–23 is a special case of $g \equiv 1$ and

$$h = x_i y_2 + x_j' y_1 + y_1 y_2,$$

where this form of the switching function is derived from the tie-set concept applied to the branch-type circuit on the right-hand side of Fig. 15–23(a).

Static hazards will have no serious effect if a combinational switching system is applied to a manipulator that acts as a low-pass filter. In a sequential switching system, however, even a very short false value may trigger a false sequence.

To remove the hazard we provide a redundant path which is independent of the time-delayed variable being changed. The design of the redundant path is shown in Fig. 15–24 for our example system of the preceding figure. Since only one type of hazard is involved in each form, $(x \cdot x')$ will not appear in the first form; hence, only type 1–0–1 hazards are possible. Similarly, $(x + x')$ will not appear in the second form; hence only type 0–1–0 hazards are possible. From these considerations we deduce that it is necessary to

(a)

In the first form

$$f = xy_1 + x'y_2.$$

Type 101 Hazard occurs when

$$y_1 = y_2 = 1.$$

In the second form

$$f = (x + y_1)(x' + y_2).$$

Type 010 Hazard occurs when

$$y_1 = y_2 = 0.$$

(b)
To avoid the hazards

add a redundant term f_0:

$$f = xy_1 + x'y_2 + f_0,$$

where f_0 remains at 1 during hazardous transient:

$$f_0 = 1 \text{ for } x = 0 \leftrightarrows 1.$$

multiply by a redundant term F_0:

$$f = (x + y_1)(x' + y_2)F_0,$$

where F_0 remains at 0 during hazardous transient:

$$F_0 = 0 \text{ for } x = 1 \leftrightarrows 0.$$

(c)
The redundant term can be found:

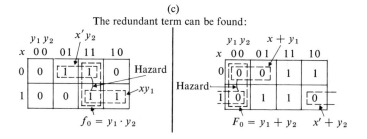

(d)
The hazard-free system is:

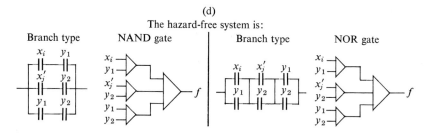

Fig. 15-24 Example of hazard compensation.

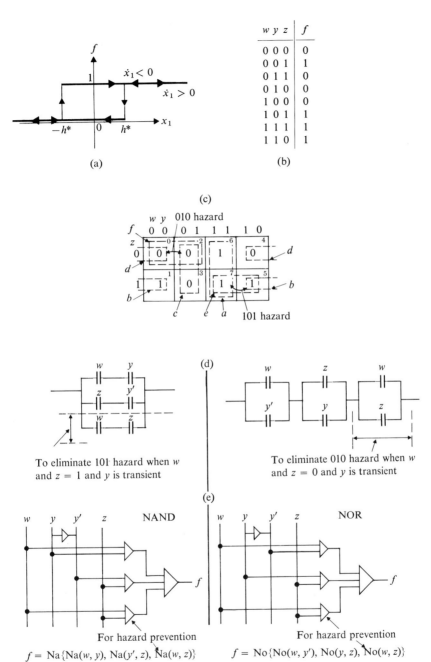

Fig. 15-25 Reverse hysteresis: (a) Fig. 12-6(b) (repeated); (b) truth table; (c) Karnaugh map for reverse hysteresis; (d) branch-type and (e) gate-type implementation.

remove only one type of hazard for a given system. The redundant term that will maintain a constant value while x is changing is spotted in the Karnaugh map in Fig. 15-24(c) as the prime implicant (or implicate) that joins two adjacent but nonoverlapping prime implicants (or implicates). The term is thus obtained by combining an adjacent pair of cells that correspond to x_i and x_j'. With this term f_0 or F_0 added, the hazards will be removed, because the term keeps $f_0 = 1$ or $F_0 = 0$ while x is changing.

Example 15-8 As a further illustration of the foregoing techniques let us consider implementation of the reverse hysteresis discussed in Example 12-6. The switching function to be generated is completely specified by Fig. 12-6(b), which is repeated here (Fig. 15-25a) for convenience. Examination of the figure reveals that six different line segments (that is, functions) will have to be independently specified. Since two variables yield only four combinations of output, we will thus need three variables to completely specify the required switching function. These variables are chosen as follows:

$$w = \begin{cases} 1, & \text{if } x_1 > 0, \\ 0, & \text{if } x_1 < 0; \end{cases}$$

$$y = \begin{cases} 1, & \text{if } |x_1| > h^*, \\ 0, & \text{if } |x_1| < h^*; \end{cases}$$

$$z = \begin{cases} 1, & \text{if } \dot{x}_1 > 0, \\ 0, & \text{if } \dot{x}_1 < 0. \end{cases}$$

For these variables, the truth table and Karnaugh map are as shown in Fig. 15-25(b) and (c), respectively. We first solve the problem by using prime implicants a, b, and e, where the last is necessary to eliminate the 1-0-1 hazard noted in Fig. 15-25(c). We thus get

$$f = \sum_3 (1, 5, 6, 7) = w \cdot y + y' \cdot z + w \cdot z ,$$

where the last term is the redundancy corresponding to prime implicant e. The contact and gate-type implementations of the required switching function are shown on the left-hand side of Fig. 15-25(d) and (e), respectively. The problem may also be solved by using prime implicates c, d, and f, where the last is required to eliminate the 0-1-0 hazard noted in Fig. 15-25(c). We thus get, as an alternative solution,

$$f = \prod_3 (0, 2, 3, 4) = (w + y') \cdot (z + y) \cdot (w + z) ,$$

where the last term is the redundancy corresponding to the prime implicate f. The contact and gate-type implementations of the required switching function, when expressed in the second form, are shown on the right-hand side in Fig. 15-25(d) and (e), respectively.

As a final note in this section on combinational circuits, we briefly review a recent article by Krigman [21] in which the concept of bond graphs (see Chapter 6) is utilized in circuit design. In Fig. 15-26(a), the logical function OR is represented by a 0 and the function AND by a 1. (For more than two inputs to an OR or AND, a junction may have more than three

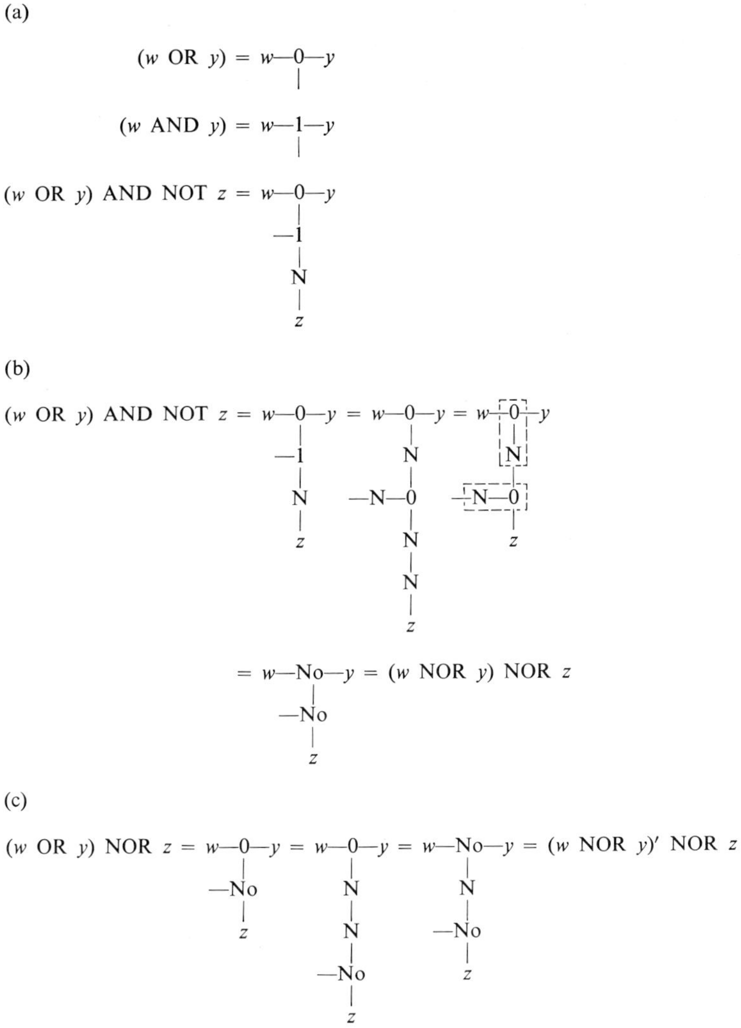

15-3   Combinational systems   735

(a)

$$(w \text{ OR } y) = w\text{—}0\text{—}y$$
$$\hspace{3cm}|$$

$$(w \text{ AND } y) = w\text{—}1\text{—}y$$
$$\hspace{3cm}|$$

$$(w \text{ OR } y) \text{ AND NOT } z = w\text{—}0\text{—}y$$
$$\hspace{5cm}|$$
$$\hspace{5cm}-1$$
$$\hspace{5cm}|$$
$$\hspace{5cm}N$$
$$\hspace{5cm}|$$
$$\hspace{5cm}z$$

(b)

$$(w \text{ OR } y) \text{ AND NOT } z = w\text{—}0\text{—}y = w\text{—}0\text{—}y = w\text{—}\overline{0}\text{—}y$$

with intermediate forms showing negations and junctions

$$= w\text{—}No\text{—}y = (w \text{ NOR } y) \text{ NOR } z$$
$$\hspace{2.5cm}|$$
$$\hspace{2.5cm}\text{—}No$$
$$\hspace{2.5cm}|$$
$$\hspace{2.5cm}z$$

(c)

$$(w \text{ OR } y) \text{ NOR } z = w\text{—}0\text{—}y = w\text{—}0\text{—}y = w\text{—}No\text{—}y = (w \text{ NOR } y)' \text{ NOR } z$$

Fig. 15-26 Bond graphs and switching logic.

bonds emanating from it.) NOT is represented by an "N" or by priming a variable. Each proposition represents a signal source, and every bond represents a fluid line or an electrical conductor. Each logical function (junction) will have three bonds, and each negation two. The single free bond in any graph is the switching-function output.

These graphs can be manipulated by applying the following two basic rules:

a) two negations in tandem are equivalent to a through bond (see Eq. 15-10), or

$$\text{—N—N—} \equiv \text{—},$$

b) AND junctions can be converted into OR junctions (or vice versa), according to the dual forms of DeMorgan's theorem [Eqs. (15-16) and (15-16′)]:

```
—N—1—N—≡—0—
     |        |
     N
     |

—N—0—N—≡—1—
     |        |
     N
     |
```

To convert any Boolean statement to OR–NOR logic, simply draw the bond graph, replace 1-junctions with negated 0-junctions, and remove all double negatives. A simple example is shown in Fig. 15-26(b), where N0 in the figure represents the NOR operation and N1 would represent the NAND operation. To convert a statement to all NOR logic, it will also be necessary to insert tandem negations into each bond that defines an OR operation. One N then combines with the 0 to form a NOR junction and the remaining N negates this junction (Fig. 15-26c). The referenced paper gives several other examples, including an application to a drill-press operation. While it presents no schemes that could not be effected by other available methods, it does offer a unified and simple bookkeeping system for switching-system design and simplification, where the final circuit layout may be drawn directly from the bond graph.

15-4 SEQUENTIAL SYSTEMS

In sequential switching systems the current output depends on past as well as present inputs. Therefore, in addition to "instantaneous" logical relations, time must also be taken into account to deal with input–output relations of the system. Since all signals take on only one of two values, 0 or 1, the time difference between the input to and the output from a system will be a pure delay (or dead time), Δt. In the analysis of a sequential system we introduce a delay element D as needed, and thus separate the delay effects from the instantaneous logical relations. (Recall that Boolean algebra completely ignores time as a variable.)

Sequential switching systems can be divided into two classes: asynchronous and synchronous. Sequential operation in the synchronous mode depends on clock pulses, whereas the clock is removed in an asynchronous system. Some of the basic principles of sequential-system design will be introduced in this section using the asynchronous mode. In the first example, however, we consider a synchronous system.

15-4   Sequential systems

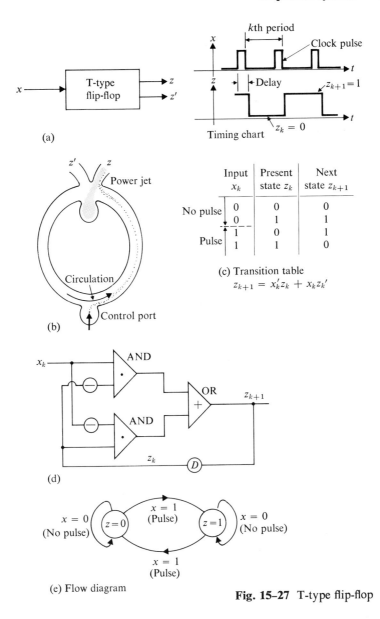

Fig. 15-27 T-type flip-flop

Example 15-9 The T-type (trigger-type) flip-flop, shown in Fig. 15-27, is one of the easiest types of memory element to analyze. The system can be either synchronous or asynchronous, but the input must be a pulse. The output changes state *after* each *pulse input*, as shown in the timing chart (Fig. 15-25a) for a synchronous mode of operation. The fluid amplifier shown in part (b) of the figure, known as a Warren loop, after its inventor, is a flip-flop of this type. It consists of a bistable type (also called a wall-attachment type) fluid amplifier and a feedback loop connecting the two side control ports. Suppose that the jet is attached (due to the Coanda effect) to the right-hand side wall of the output duct labeled z, thus making $z = 1$. Through aspiration, the jet will set up a counterclockwise circulation in the loop circuit. This flow provides sufficient bias so that the next input pulse from the control port follows the loop circulation and, in this case, impinges on the power jet from the right, thus switching the jet to the left-hand side duct z'. This also reverses the circulation flow in the loop. It could be said that memory of the current state is stored by the bias circulation in the loop. Clearly, successive input pulses will cause the output to trigger back and forth between z and z'.

The switching sequence is shown in the state transition table (Fig. 15-27c). By inspection of the table it is possible to express the relation for the next output, z_{k+1}, by the following difference equation:

$$z_{k+1} = x'_k \cdot z_k + x_k \cdot z'_k . \tag{15-37}$$

In a synchronous system the input–output relation can be expressed by a difference equation since the system operates according to a fixed period. The variables x_k and z_k in Eq. (15-37) are input and output in the kth period, respectively, where $x_k = 1$ signifies that a pulse exists in that period. Referring to the timing chart in Fig. 15-27(a), we see that the delay between z_k and z_{k+1} has to be larger than the width of the clock pulse, but smaller than the period of clock pulses. The logic circuit of Eq. (15-37) in terms of AND and OR gates and a delay element is shown in Fig. 15-27(d). In Example 15-12 we shall emphasize the importance of the delay element. In the present example, the delay may be nothing more than a length of air hose. The diagram shown in part (e) of the figure is called the flow diagram. Two equilibrium states (called the *stable* states) are shown by small circles, labeled 0 and 1. No change in state occurs so long as there is no pulse input (that is, $x = 0$). The state changes from one to the other when a pulse ($x = 1$) hits the system. These statements are symbolically indicated by the directed lines in Fig. 15-27(e).

If the output of one T-type flip-flop is used as the input to a second flip-flop, it is clear that the second will flip only when the first output goes from 1 to 0. But this occurs on every *second* input to the first flip-flop. The output of the second flip-flop thus switches at one-half the frequency of the input to the first. Obviously, we may continue the procedure by cascading T-type flip-flops to divide by 4, 8, 16, etc. (By using feedback we can construct circuits that divide by numbers other than powers of 2.) The value of such a scheme for developing counters should be readily apparent. To develop a four-bit binary counter, to count from 0 to 15 for example, we would need four cascaded T-type flip-flops where the output of the first represents the least significant bit and the output of the fourth the most significant.

Example 15-10 A different kind of flip-flop, called the RS-type (reset-set type), is shown in Fig. 15-28. A bistable fluid amplifier without a loop, shown in part (a) of the figure, exhibits a switching action of this type. The input (air flow) is admitted via control port S or R. Suppose that the power jet is attached to the right side wall when control inputs S and R are both zero, so that the right-hand output z is 1. A signal (air flow) introduced in the right control port (labelled R, meaning $R = 1$ is a reset signal) will switch the jet to the left-hand duct where z' now becomes 1. Now the jet attaches itself to the left side wall, achieving the memory function. Hence there will be no change in output state by 0's on both S and R until a 1-input is

15-4 Sequential systems **739**

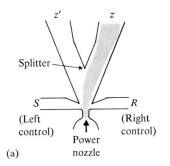

(a)

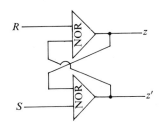

(b)

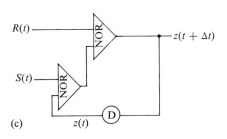

(c)

(d)

Transition Table

Input		Present state	Next state
R	S	$z(t)$	$z(t + \Delta t)$
d	0	0	0
0	1	0	1
0	d	1	1
1	0	1	0

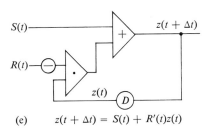

(e) $z(t + \Delta t) = S(t) + R'(t)z(t)$

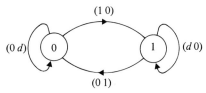

(f) Flow diagram for input pairs (SR)

Fig. 15-28 RS-type flip-flop.

applied via the *set* port S. Simultaneous 1 inputs via both R and S are not allowed in this type of flip-flop.

The RS-type flip-flop implemented with a pair of NOR-gate elements is shown in Fig. 15-28(b). Since the output of the NOR gate is 1 only if both its inputs are zero, the system has two possible stable states: either $z = 1$ or $z = 0$, when S and R are both zero. Suppose $z = 0$ and S goes to 1 (with $R = 0$), then z' drops to 0 and z will go to 1. When S goes back to 0, $z = 1$ will hold due to the crossover feedback, and thus the state will be stored until R is brought up to 1. As noted in the preceding paragraph, S and R are never allowed to be 1 at the same time; otherwise the complementary relation (z and z') between the two outputs will fail.

A concentrated delay element D is introduced in (c) of the figure for a better understanding of the sequential operation. Two possible states of the delay element exist:

a) an equilibrium (or "stable") condition, $z(t + \Delta t) = z(t)$, or
b) a transient condition (which is said to be an unstable state), $z(t + \Delta t) \neq z(t)$.

An unstable state will always change into a stable state.

We now display the property of the flip-flop in a transition table (Fig. 15–28d). The first and third rows of the table represent stable states. If S is kept at 0 and z is 0, then z remains at 0 regardless of the R-input. Therefore R in the first row is represented by a symbol d signifying that R can be either 1 or 0. It is now possible to express the asynchronous (or synchronous) operation of the RS-type flip-flop by the following equation:

$$z(t + \Delta t) = S(t) + R'(t) \cdot z(t) . \qquad (15\text{--}38)$$

A circuit equivalent to this expression is shown in Fig. 15–28(e). Its flow diagram is in part (f) of the figure, where the numbers in parentheses next to each path are a pair of causative SR-inputs, (S, R, in that order).

The procedure discussed in the preceding examples must be reversed when *designing* a sequential switching system for a prescribed job statement. A job statement may be given either by word description, a timing chart, or a sequence diagram. (In a sequence diagram only the order in which events occur is shown by drawing lines for the 1's of each variable, without indicating a relative time scale.) From the job statement, we then construct a primitive flow (that is, truth) table. This is done by starting with a given, arbitrary combination of input variables and associated outputs. The starting condition will be the initial condition if specified by the job statement. (Detailed application of these and the following steps is shown in Example 15–11.) The first combination is associated with the state ①. From here we go to state ② by changing one of the input variables according to the job statement: this process is continued until no further states can be established. The primitive flow table can be reduced to a final flow table by eliminating those similar states which occur in the same column (see Fig. 15–30c), lead to the same succeeding states, and have the same outputs. The difference between a flow table representing a logic circuit and a flow table representing a sequential circuit is that the logical flow table identifies the outputs only by the arrangement of the input variables. In a sequential circuit several states can occur in the same column with different outputs. In order to differentiate between the different rows in which these states appear, so-called secondary variables or internal variables must be introduced. They are used to code the different rows of the merged flow table. The secondary variables appear as outputs of a logic circuit depending on the inputs and the delayed secondary variables. This is achieved by establishing delayed feedback loops around the internal logic circuit. The output of the sequential circuit is still described by a pure logic circuit depending on the input variables and the secondary variables. If the number of input variables and the

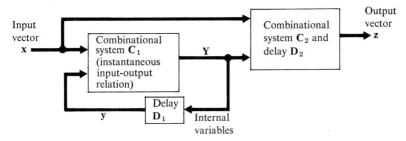

Fig. 15-29 Feedback configuration of sequential switching system.

number of rows in the merged flow table is not too large, the equations for the secondary variables and the output variables can be derived from a Karnaugh map representation.

In the typical configuration of a sequential system shown in Fig. 15-29, the vectors **x**, **Y**, **y**, and **z** are input, internal variables (before and after delay D_1), and output, respectively. In Fig. 15-29, the delay D_1 is usually incorporated into the feedback loop rather than into the feedforward path (following C_1) in order to increase the speed of the input-output loop. It is not necessary that the secondary variable appearing in the output matrix be delayed, because this part of the system consists of a pure *logical* circuit. However, the alternative arrangement is sometimes necessary to avoid essential hazards. (An essential hazard differs from a static or dynamic hazard in that it is not just a momentary false output, but rather a complete failure of the circuit to behave according to the job specification (see [1], pp. 595–599).

The two combinational sections are separated into instantaneous combinational matrices C_1, C_2 and delay matrices D_1 and D_2, respectively. By a Karnaugh map called the excitation matrix we design the combinational system C_1. Since **x** and **y** are inputs to C_1, and the vector variable **Y** is its output, entries in the Karnaugh map are elements of **Y**. Since C_2 is a separate combinational system, entries for **z** are not required in the excitation matrix. After completing the combinational-system design for C_1, we construct a second Karnaugh map for C_2. This map is called the output matrix, where the entries are **z**. Creating a primitive flow table, reducing it to a flow table, and then assigning a set of secondary variables constitute the crucial steps in sequential-switching system design. Due consideration of hazardous situations and attention to important details may be developed with experience. In the following example we shall show an outline of the design approach.

Example 15-11 *Design of the RS-type flip-flop which was analyzed in Example 15-10.* Let us suppose that the input–output relation indicated by the timing chart (Fig. 15-30a) is desired. We can find the stable states by picking out those sets of variables in the timing chart where, following an input change, the output variable achieves a constant state and remains there as long as the input doesn't change again. Circled

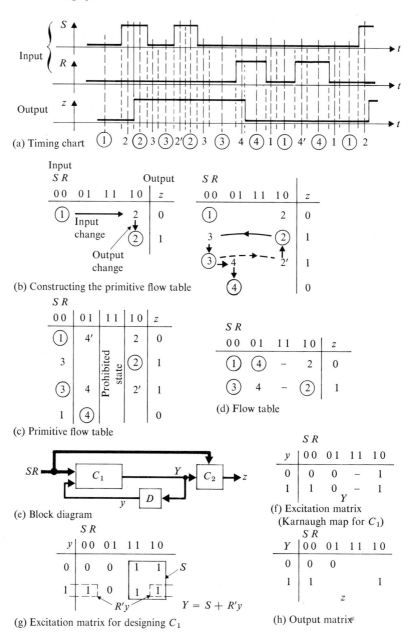

Fig. 15–30 Design of an RS-type flip-flop.

numbers are assigned to the stable states. (The choice of the starting state may influence the complexity of the end result [2], p. 204). To clarify the transition of states we make a table, as shown in Fig. 15-30(b). We begin with stable state ①, which is entered in the first row of the colume for $SR = 00$. At the right end of the table is the column for the output z, where a 0 is entered. In the timing chart (Fig. 15-30a) a transition via unstable state 2 to a new stable state ② is shown. The transition is caused by a change in input variable S. In the flow table (Fig. 15-30b) an input change is a horizontal movement from ① to 2. This is followed by a vertical movement due to a resulting output change, from 2 to ②. Although no change in the output occurs for a change of S from 1 to 0, the flow table is continued from ② via 3 to ③, as shown in the right-hand side of Fig. 15-30(b). Input S may change to 1 again without causing any change in z, as indicated by the transition from ③ via 2' to ②, where ② is a stable state already entered in the table. Output change from 1 to 0 can take place only when the reset signal R goes up to 1. This is the transition indicated by 4 and ④. A completed flow table, called the primitive flow table, is shown in Fig. 15-30(c). One stable state (circled entry) appears in each row of a primitive flow table.

We next simplify or merge the primitive flow table by combining the rows. The simplification rules are the following:

a) If all the sequences that can be reached from two states n_1 and n_2 are identical and if the outputs are the same, we call the states *equivalent states*. One of these two states can be eliminated.

b) Two lines of a primitive flow table can be combined if the following conditions are satisfied:

1. Both lines give identical outputs.
2. The stable states indicated in each line differ by only one input variable (R or S).
3. The preceding state of one of these states can be the other one only for a variation of R or S.

In a systematic study, the primitive flow table is first examined to determine if the simplification rule (a) can be applied. As an example, let us reconstruct the primitive flow table in Fig. 15-30(b). Starting with stable state ③ a change of S led us through 2' back to stable state ②. In a simple flow table like the one shown, such a transition is obvious; but for a more complex flow table it is safer to establish a new stable state for every new transition. If, instead of leading the sequence from ③ back to ②, we had created a stable state ⑤, simplification rule (a) would show that states ② and ⑤ are equivalent states and therefore state ⑤ would have been eliminated. Simplification rule (b) applies to rows ① and ④ as well as rows ② and ③ in Fig. 15-30(c). The resulting merged flow table is shown in Fig. 15-30(d). When two identical state numbers, one circled and the other uncircled, are combined, the resulting entry is circled. If a row contains a blank space, indicating that the corresponding combination of input variables cannot be reached from the stable state in this row, it is possible to introduce any number that facilitates simplification. Since the purpose of the flow table is to arrive at the design of the combinational system C_1 (Fig. 15-29) whose output is not the final output z, the entries for z are not needed in the flow table.

In general, it is known that 2^n rows of a merged flow table can be coded by $(2n - 1)$ secondary variables. Therefore, for our system, we need only one secondary variable, Y (Fig. 15-30e). To design a combinational system C_1 whose inputs are SR and y and whose output is Y, we construct a Karnaugh map (Fig. 15-30f) whose entries are output Y [which is one of the inputs to C_2 in part (e)]. Referring to the flow table [part (d) in the figure], we first enter stable states for which $y = Y$ (that is, by columns, in each row, we assign the y value to Y when that column entry is stable). Since Y and y are, respectively, the input and output of a delay element D, Y is a future value of y. We therefore enter a 1 in the cell that corresponds to unstable state 2 in the flow table, and 0 in the cell of unstable state 4. The Karnaugh map for combinational system C_1 is customarily called the *excitation matrix*. Our excitation matrix has two blank entries in the column for the prohibited input pair $SR = 11$. Hence we can write four different excitation functions depending on what values are chosen for the blank entries. Entering 1 in both blank cells (Fig. 15-30g) we obtain the switching function

$$Y = S + R'y. \qquad (15\text{-}39)$$

Since, in general, the output will be a function of the input and the internal variables, the Karnaugh map for the output (called the output matrix) has the same format as the excitation matrix, with entries for output z (or outputs). We enter the desired output z in the cells of the output matrix (Fig. 15-30h) that correspond to the stable entries of the merged flow table [① through ④ in Fig. 15-30(d)] and find that

$$Y = z \qquad (15\text{-}40)$$

will satisfy the output requirement. Recalling that Y signifies a future variable at $(t + \Delta t)$, and y a present value at t, we see that Eqs. (15-39) and (15-40) combined yield Eq. (15-38) which was derived earlier by analysis. It is now possible to implement the system in the form of Fig. 15-28(e) or (c).

The steps in the design procedure shown in the preceding example are summarized as follows: (a) Construct a primitive flow table from a problem statement (given in words, timing chart, or sequence diagram). All possibilities must be considered at this step (including don't-care conditions). We circle each stable state in the entry. (b) Reduce the primitive flow table to the merged flow table, and assign the internal variables (this step is sometimes referred to as "coding the row states"). (c) Write the excitation matrix and determine the switching function for the internal (or feedback) variable(s). (d) Write the output matrix referring to the flow table and determine the output switching function. (e) Implement the combinational systems and check possible hazards. Present-day design techniques of sequential switching systems are still at a stage where insight, experience, and skill are essential in each of these steps. Since it is impossible to cover the details without carefully going through a number of examples, we conclude this section by reviewing the outline of the design procedure by means of another very simple example.

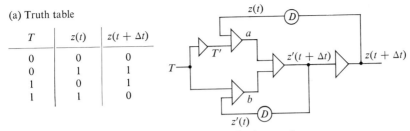

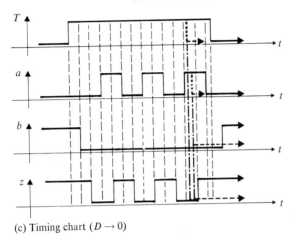

(c) Timing chart ($D \to 0$)

Fig. 15-31 T-type flip-flop operation for zero delay.

Example 15-12 *Design of the T-type flip-flop which was analyzed in Example 15-9.* Let us first determine the effect of deleting the delay in Fig. 15-27(d) while permitting asynchronous operation. Figure 15-31(a) shows the truth table for the desired relation between the presence (1) or absence (0) of a trigger pulse, T, and the present, $z(t)$, and next value, $z(t + \Delta t)$, of the output. From this table we write

$$z(t + \Delta t) = T' \cdot z(t) + T \cdot z(t)'$$
$$= [T + z(t)']' + [T' + z(t)]' = [(b + a)']' , \qquad (15\text{-}41)$$

where $b = [T + z(t)']'$ and $a = [T' + z(t)]'$. Equation (15-41) may be implemented with NOR gates as indicated in Fig. 15-31(b). It may be shown, by using Fig. 15-22(a), that the results of Figs. 15-31(b) and 15-27(d) are identical. To draw the timing chart, we introduce an input-signal change and trace it sequentially through the circuit. The timing chart for T, a, b, and z is shown in Fig. 15-31(c) for the case of the delay time approaching zero (but considering small delay time through each gate). The solid and dashed trajectories are two possible paths we shall consider. The first important thing we note is that the variables a and z oscillate (slightly out of phase, as shown) as long as T remains 1 (that is, the pulse stays on). Secondly, we observe that the final state of the output depends upon *when* the trigger pulse, T, is removed. For the case of the solid line, T goes down when a is down and z is

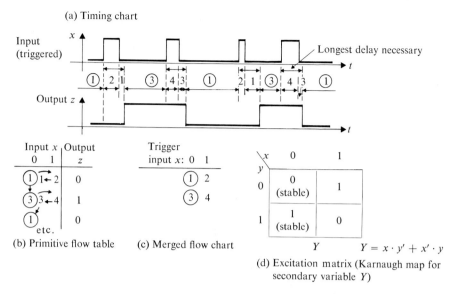

Fig. 15-32 Synthesis of the T-type flip-flop.

up. The final stable state of the output is then $z = 1$ and the output has *not* flipped due to the trigger. To get the flip-flop action of the output, the trigger pulse must be removed when a is up and z is down. Since this is not the desired trigger flip-flop operation, it is clear that the delay must play an important role in the flip-flop action.

We now consider the timing chart for the input-output relationship of the T-type flip-flop with the delay included. As shown in Fig. 15-32(a) we permit aperiodic trigger pulses of varying pulse lengths, but we do require that the delay time be longer than the longest pulse width, yet shorter than the shortest period of the trigger. In Fig. 15-32(b) we construct the primitive flow table by referring to the sequential steps in the timing chart. As in the previous example, we then determine the merged flow chart (Fig. 15-32c) which again has 2 $(=2^n)$ rows. Thus $n = 1$ and there must be $(2n - 1)$ or one secondary variable. We construct the excitation matrix for this secondary variable and determine

$$Y = x \cdot y' + x' \cdot y. \tag{15-42}$$

Recalling that y is the signal Y delayed, we call $Y = z_{k+1}$ and $y = z_k$ and note that Eq. (15-42) is thus exactly the same result we arrived at in the previous analysis (Eq. 15-37). The circuit may be implemented as indicated in Fig. 15-27(d).

As suggested by even these simple examples, the problem of sequential-circuit design can be complex. If we endeavor to combine flip-flops to design counters, for example, the complexity quickly mounts; yet this is precisely the problem to be solved in designing the sequential circuitry of a digital computer. To cope with such problems, and to understand essential hazard analysis, we recommend further reading ([1], Chapter 12, and [22], Chapters 5 and 6) and a good deal of practice. (See also [5], in Japanese, for a thorough

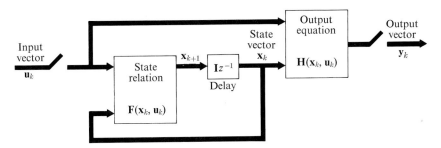

Fig. 15-33 Block diagram of a discrete-time system.

coverage of the field.) The intuition and artistry necessary to design and implement sequential-switching circuits can come only with experience.

15-5 THEORY OF FINITE-STATE SYSTEMS

The structure of a sequential switching system, shown in Fig. 15-29, closely resembles that of the discrete-time system with which we dealt in Chapters 1 through 14. General equations for the latter were

$$\text{state equation: } \mathbf{x}_{k+1} = \mathbf{F}(\mathbf{x}_k, \mathbf{u}_k) ,$$
$$\text{output equation: } \mathbf{y}_k = \mathbf{H}(\mathbf{x}_k, \mathbf{u}_k) ,$$
(15-43)

which are represented by the block diagram of Fig. 15-33. Despite the close resemblance of the two block diagrams, there exists a crucial difference in that the number of states a sequential system can assume is finite, whereas that of a discrete-time system is infinite because its state $\mathbf{x}_k$ is not discretized in state space. A system where the number of possible states is finite is said to be finite-state-determined. A binary-logic sequential system is, however, just one example of a finite-state system. We shall introduce here an abstract but simple formulation for stochastic finite-state systems (see Ref. 23 for more details).

Let us assume that a finite-state system has n possible states; $1, 2, \ldots, i, \ldots, n$. Thus $\mathbf{i}(k)$ is a state vector at the kth time point, where $k = 0, 1, 2, \ldots$ Note that n is not necessarily the dimension of the state vector. When the finite system is stochastic, we consider $\mathbf{i}(k)$ to occur with a probability $x_i(k)$. In other words, the *state probability* $x_i(k)$ is the probability of what the ith state will be at the kth instant. Note that the value of $x_i(k)$ is always nonnegative, and is less than (or equal to) one.

We next designate by $p_{ji}(k)$ the probability that the state $\mathbf{i}(k)$ will move into another state $\mathbf{j}(k + 1)$ at the next time point $(k + 1)$. This implies that the *transition probability* $p_{ji}(k)$ depends only on i, j, and k, regardless of how the ith state is reached. As noted earlier (see Sec. 13.5), a stochastic system with this property is called a Markov process. If a finite-state system is a Markov process, the state probability $x_j(k + 1)$ will be given in terms of

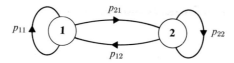

Fig. 15-34 Flow diagram of a two-state system.

all states and their transition probabilities into the jth state;

$$x_j(k+1) = \sum_{i=1}^{n} p_{ji}(k)x_i(k), \tag{15-44}$$

where, like x_i, p_{ji} must satisfy the following conditions:

$$p_{ji}(k) \geq 0 \quad \text{for all} \quad i, j, \text{ and } k,$$

$$\sum_{j=1}^{n} p_{ji}(k) = 1 \quad \text{for all} \quad i \text{ and } k. \tag{15-45}$$

Since Eq. (15-44) applies for all j ($j = 1, 2, \ldots, n$), the set of n equations can be expressed in the following vector form:

$$\mathbf{x}(k+1) = \mathbf{P}(k)\mathbf{x}(k). \tag{15-46}$$

$\mathbf{P}(k)$ in this expression has the transition probabilities $p_{ji}(k)$ as its elements; hence, it is a transition probability matrix. If the Markov process is stationary, the matrix will not depend on k, and thus,

$$\mathbf{x}(k+1) = \mathbf{P}\mathbf{x}(k). \tag{15-47}$$

Although the n-vector $\mathbf{x}$ is not a state vector but a state probability vector, and a finite-state system is always nonlinear, the last two equations have the same form as the state equation of a linear nth-order free system whose state vector is $\mathbf{x}$.

A flow diagram similar to the form of Fig. 15-28(f) serves as a graphical description of a Markov process for a finite-state system. As shown in Fig. 15-34 for a two-state system, it is a kind of signal flow graph where p_{ji} written along a directed line signifies the transition probability from the i state to the j state in one discrete time interval. Like Eq. (2-40), the z-domain expression of (15-47) is

$$z\mathbf{X}(z) - z\mathbf{x}(0) = \mathbf{P}\mathbf{X}(z),$$

so that

$$\mathbf{X}(z) = (z\mathbf{I} - \mathbf{P})^{-1}z\mathbf{x}(0). \tag{15-48}$$

The time-domain solution of (15-47) agrees (in its form) with the free response of a linear, stationary, discrete-time system ($\mathbf{u}_i = 0$ in Eq. (3-51)),

$$\mathbf{x}(k) = \mathscr{Z}^{-1}\mathbf{X}(z) = \mathbf{P}^k\mathbf{x}(0), \tag{15-49}$$

15-5   Theory of finite-state systems

whose modes are characterized by the eigenvalues of the system, which, in turn, are the roots of the characteristic equation

$$|z\mathbf{I} - \mathbf{P}| = 0. \tag{15-50}$$

The final value theorem in the z-domain,

$$\lim_{k \to \infty} \mathbf{x}(k) = \lim_{z \to 1} (z - 1)\mathbf{X}(z) \tag{15-51}$$

can be conveniently applied to the z-domain solution, (15-48), to obtain the final distribution of state probability.

Example 15-13 The z-domain solution of the two-state system, Fig. 15-34, is:

$$\mathbf{X}(z) = \begin{bmatrix} z - p_{11} & -p_{12} \\ -p_{21} & z - p_{22} \end{bmatrix}^{-1} z\mathbf{x}(0) = \frac{1}{\Delta} \begin{bmatrix} z - p_{22} & p_{12} \\ p_{21} & z - p_{11} \end{bmatrix} z\mathbf{x}(0),$$

where

$$\Delta = z^2 - (p_{11} + p_{22})z + (p_{11} + p_{22} - p_{21}p_{12}).$$

If, for instance,

$$p_{11} = 0.6, \qquad p_{21} = 0.4, \qquad p_{12} = 0.8, \qquad \text{and} \qquad p_{22} = 0.2,$$

the eigenvalues are 1 and -0.2, and the z-domain solution becomes

$$\mathbf{X}(z) = \frac{z}{(z-1)(z+0.2)} \begin{bmatrix} z - 0.2 & 0.8 \\ 0.4 & z - 0.6 \end{bmatrix} \mathbf{x}(0).$$

By Eq. (15-51) we find the final probability vector as

$$\mathbf{x}(\infty) = \lim_{z \to 1} \frac{z}{z + 0.2} \begin{bmatrix} z - 0.2 & 0.8 \\ 0.4 & z - 0.6 \end{bmatrix} \mathbf{x}(0) = \begin{bmatrix} \frac{2}{3}(x_1(0) + x_2(0)) \\ \frac{1}{3}(x_1(0) + x_2(0)) \end{bmatrix} = \begin{bmatrix} \frac{2}{3} \\ \frac{1}{3} \end{bmatrix},$$

where $x_1(0) + x_2(0) = 1$ is taken into account.

Example 15-14 An interesting special case of Fig. 15-34 arises when $p_{11} = 0$ and $p_{22} = 0$. By Eq. (15-45) the transition probabilities are $p_{21} = p_{12} = 1$, so that the system for this special case is deterministic. By intuition we see that the system will flip-flop periodically. The characteristic polynomial for this case is

$$\Delta = z^2 - 1,$$

and the final value of $\mathbf{x}$ is found to be

$$\mathbf{x}(\infty) = \begin{bmatrix} \frac{1}{2} \\ \frac{1}{2} \end{bmatrix}.$$

A system like this is said to be periodic.

Example 15-15 We consider another interesting special case of Fig. 15-34 where $p_{21} = 0$ but $p_{12} \neq 0$; for instance,

$$p_{11} = 1, \qquad p_{21} = 0, \qquad p_{12} = 0.8, \qquad \text{and} \qquad p_{22} = 0.2.$$

Since the state transition is directed only from **2** to **1**, it can be intuitively said that the final state will be **1** with 100% probability regardless of the initial state. The statement can be confirmed by a calculation similar to the above examples. A state like **1** in this example is generally called an absorbing state.

Example 15-16 A transient will eventually settle out if a system has an absorbing state. The mean or expected number of periods after which the transient will end can be computed by Eq. (13-4), restated:

$$E\{x\} = \mu = \int_a^b x p(x)\, dx\ ,$$

where x must be replaced by k for our problem so that

$$E\{k\} = \sum_{k=0}^{\infty} k p(k)\ . \tag{15-52}$$

In this expression, $p(k)$ is the probability that the absorbing state is entered for the first time at the kth instant, so that for absorbing state **1** it is:

$$p(k) = x_1(k) - x_1(k-1)\ .$$

For the system of Example 15-15 we have

$$\begin{bmatrix} x_1(k+1) \\ x_2(k+1) \end{bmatrix} = \begin{bmatrix} 1 & 0.8 \\ 0 & 0.2 \end{bmatrix} \begin{bmatrix} x_1(k) \\ x_2(k) \end{bmatrix},$$

so that

$$p(0) = x_1(0)\ , \qquad p(1) = 0.8 x_2(0)\ , \qquad p(2) = 0.8 \cdot 0.2 x_2(0)\ ,\ \ldots$$

and

$$\begin{aligned}
E\{k\} &= 0.8\{1 + 0.2 \cdot 2 + (0.2)^2 \cdot 3 + \cdots\} x_2(0) \\
&= 0.8\{1 + 0.2 + (0.2)^2 + \cdots \\
&\quad\quad + 0.2 + (0.2)^2 + \cdots \\
&\quad\quad\quad + (0.2)^2 + \cdots \\
&\quad\quad\quad\quad + \cdots\} x_2(0) \\
&= \frac{0.8}{(1-0.2)^2} x_2(0) = 1.25 x_2(0)\ .
\end{aligned}$$

The same answer can be obtained by the following (general) procedure. In the z-domain $p(k)$ is given by

$$P(z) = (1 - z^{-1}) X(z)\ ,$$

and, by definition of the transformation,

$$P(z) = \sum_{k=0}^{\infty} p(k) z^{-k}\ .$$

Differentiating both sides of this last expression with respect to z, we obtain

$$\frac{d}{dz} P(z) = -\sum_{k=0}^{\infty} k p(k) z^{-(k-1)}$$

or

$$\sum_{k=0}^{\infty} kp(k)z^{-k} = -z\frac{d}{dz}P(z),$$

in which the left side is given by $\mathscr{Z}kp(k)$. It therefore follows (by #1 of Table 1-2) that

$$\mathscr{Z}\left[\sum_{k=0}^{k} kp(k)\right] = -\frac{z^2}{z-1}\frac{d}{dz}P(z).$$

By the final-value theorem we finally obtain

$$\sum_{k=0}^{\infty} kp(k) = -\lim_{z \to 1} z^2 \frac{d}{dz}P(z).$$

In our problem we have

$$P(z) = x_1(0) + \frac{0.8}{z - 0.2}x_2(0), \qquad \frac{d}{dz}P(z) = \frac{-0.8}{(z - 0.2)^2}x_2(0);$$

thus

$$E\{k\} = \sum_{k=0}^{\infty} kp(k) = \lim_{z \to 1} \frac{0.8}{(z - 0.2)^2}x_2(0) = 1.25x_2(0).$$

To generalize the concept of the absorbing state which we saw in Example 15–15, we replace the states of that example by sets of states, where each set consists of a group of states. When this is done, an absorbing state will appear as a set, as shown in Fig. 15–35. The set is called *ergodic* when a state never leaves the set once it has entered. All states that do not belong to an ergodic set constitute a transient set. Since modes for the states in a transient set must eventually vanish, the magnitude of their eigenvalues must be less than one. On the other hand, nonzero final-state probabilities must exist for some states which belong to an ergodic set(s); hence, eigenvalues for these must involve one. For instance, the system of Example

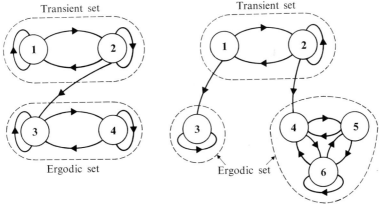

Fig. 15–35 Ergodic set versus transient set.

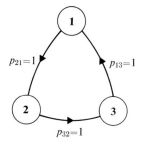

Fig. 15-36 An example of a periodic system.

15-13 is an ergodic set, and its eigenvalue of $z = 1$ is not accidental. Moreover, if an ergodic set is periodic, with period m, its characteristic equation will include a factor $(z^m - 1)$.

Example 15-17 The system of Example 15-14 is made three-state in Fig. 15-36, for which we have

$$\mathbf{P} = \begin{bmatrix} 0 & 0 & 1 \\ 1 & 0 & 0 \\ 0 & 1 & 0 \end{bmatrix},$$

so that

$$|z\mathbf{I} - \mathbf{P}| = z^3 - 1 .$$

In a further continuation of this abstract theory, cost associated with state and control of a stochastic nature are introduced, and this develops into an optimal-control problem based upon the principle of optimality [23].

REFERENCES

1. Caldwell, S. H., *Switching Circuits and Logical Design*. New York: J. Wiley, 1958.
2. Maley, G. A., Earle, J., *The Logic Design of Transistor Digital Computers*. Englewood Cliffs, N. J.. Prentice-Hall, 1963.
3. Torng, H. C., *Introduction to the Logical Design of Switching Systems*. Reading, Mass.: Addison-Wesley, 1964.
4. Harrison, M., *Introduction to Switching and Automata Theory*. New York: McGraw-Hill, 1965.
5. Mori, M., (ed.), *Sequential Controls Handbook* (in Japanese). Tokyo: Ohm-Sha, 1964.
6. "Special report on logic devices in digital control," *Control Engineering*, **14**, No. 1, January 1967, pp. 61-124.
7. Kirshner, J. M. (ed.), *Fluid Amplifiers*. New York: McGraw-Hill, 1966.
8. Humphrey, E. F., Tarumoto, D. H. (eds.), *Fluidics*. Boston: Fluid Amplifier Assoc., Inc., 1965.

9. "Fluidics Bibliography," *Hydraulics and Pneumatics*, **20**, No. 2, February 1967, pp. 65-74.
10. Glaettli, H. H., "Hydraulic logic; what's its potential?" *Control Engineering*. May 1961, p. 83.
11. Togino, K., Inone, K., "Universal fluid logic element," *Control Engineering*. May 1965, pp. 78-87.
12. Holbrook, E. L., "Pneumatic logic," *Control Engineering*. July 1961, pp. 104-108, August 1961, pp. 92-96; November 1961, pp. 110-113; February 1962, pp. 89-92; December 1962, pp. 85-88.
13. Boole, G., *The Mathematical Analysis of Logic*. Cambridge, England, 1847.
14. Shannon, C. E., "A symbolic analysis of relay and switching circuits," *Trans. AIEE*, **57**, 1938, pp. 713-723.
15. Richalet, J., "Calcul opérationnel booléen," *L'Onde Électrique*, **43**, October 1963, pp. 1003-1021.
16. Richalet, J., "Les systèmes discrets," *L'Onde Électrique*, **44**, October 1964, pp. 1011-1020.
17. Pfeiffer, P. E., *Sets, Events, and Switching*. New York: McGraw-Hill, 1964.
18. Karnaugh, M., "The map method for systems of combinational logic circuits," *Trans. AIEE, Communications and Electronics*, **72**, part 1, November 1953, pp. 593-599.
19. Quine, W. V., "On cores and prime implicants of truth functions," *Am. Math. Monthly*, **66**, No. 9, November 1959, pp. 755-760.
20. McCluskey, E. J., Jr., "Minimization of Boolean functions," *Bell System Tech. J.*, **35**, No. 6, November 1956, pp. 1417-1444.
21. Krigman, A., "Bond graphs for designing logic circuits," *Control Engineering*, **15**, No. 2, February 1968, pp. 91-92.
22. McCluskey, E. J., *Introduction to the Theory of Switching Circuits*. New York: McGraw-Hill, 1965.
23. Freeman, H., *Discrete-Time Systems*. New York: Wiley, 1965.

PROBLEMS

15-1 Using turbulence amplifiers, implement the following three functions:

a) NAND,

b) exclusive OR, and

c) set-reset flip-flop.

Write the sequential truth table for the last item.

15-2 Draw the Venn diagrams for the functions f_0 through f_7 of Table 15-4. Label the functions in terms of AND, OR, NOR, etc. How can the Venn diagrams for f_8 through f_{15} be determined from your results?

754 Switching systems

Logic block available	Function desired			
	AND	OR	NAND	NOR
AND	⊐AND⊢			
OR		⊐OR⊢		
NAND			⊐NAND⊢	
NOR				⊐NOR⊢

Fig. P15-3

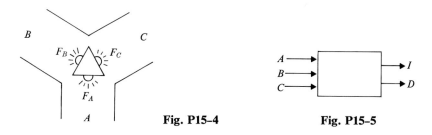

Fig. P15-4   **Fig. P15-5**

15-3 In the left-hand column of Fig. P15-3 are available logic blocks from which the desired functions of the remaining columns are to be constructed by simple negation. Indicating negation by small circles at input and/or output ends of the building blocks, fill in the remainder of the table.

15-4 Traffic lights at the intersection shown in Fig. P15-4 are governed by the following traffic conditions:

  i) a car approaching the intersection on A always has a green light;
  ii) a car approaching it on B has a green light only when no cars are approaching it on A;
  iii) a car approaching on C has a green light only when no cars are approaching on either A or B; and
  iv) in the absence of traffic all lights are red.

Considering the presence of an approaching car as a 1 and its absence as a 0, draw up a truth table for the lights F_A, F_B, and F_C (taking green as 1 and red as 0). Obtain the first canonical (sum of products) expressions for F_A, F_B, and F_C, and simplify them to single-term expressions.

15-5 The combinational circuit shown schematically in Fig. P15-5 has three inputs and two outputs. Output I is to equal 1 when any one and only one of the input variables is 1, and when all three inputs are 1. Output D is to equal 1 when any two

Gray			Binary			Decimal
a	b	c	x	y	z	
0	0	0	0	0	0	0
0	0	1	0	0	1	1
0	1	1	0	1	0	2
0	1	0	0	1	1	3
1	1	0	1	0	0	4
1	1	1	1	0	1	5
1	0	1	1	1	0	6
1	0	0	1	1	1	7

Fig. P15-6

x_3 \ $x_1 x_2$	0 0	0 1	1 1	1 0
0	1	0	0	1
1	1	0	1	1

Fig. P15-9

of the input variables are 1, and when all three inputs are 1. Draw the truth table for these switching functions and design a single-gate-type circuit that will produce both outputs.

15-6 Shown in Fig. P15-6 are the Gray and binary three-bit codes for the decimal numbers 0 through 7. Assuming the availability of the four logical building blocks of Problem 15-3, as well as negation and EXCLUSIVE OR, design a circuit that will convert a number in the Gray code (a, b, c) into its equivalent in the binary code (x, y, z). [*Hint:* This problem may be solved with three EXCLUSIVE OR's alone.]

15-7 Four thermostats w, x, y, and z are spaced evenly around a large room. Each thermostat is capable of two states, 0 and 1. A rough method of controlling the temperature in the room is to turn on the heat supply to the room when two or more of the thermostats indicate a 1. Taking heat supply ON as a 1, draw the Karnaugh map that describes this situation. Synthesize the function using NOR gates only.

15-8 A third-order state feedback, relay control system must have the relay switch in the ON position under any of the following four conditions: all three state variables positive; the first positive and the second and third negative; the first and third negative and the second positive; and the first and second negative and the third positive. Coding the positive-state variable condition as a 1, construct the Karnaugh map for this switching function (take ON = 1). Implement this function with a relay contact circuit using the sum-of-products form. Repeat, using the product-of-sums form.

15-9 For a system similar to the previous problem, a Karnaugh map is given as shown in Fig. P15-9. Give two simplified word statements describing this switching function, one based on prime implicants and the other on prime implicates. Implement each with a gate-type circuit.

15-10 Using the Karnaugh map method, simplify the following statements into one concise expression:

"Insurance policy No. 22 may be issued when the applicant satisfies any one of the following:

i) has been issued policy No. 19 and is a married male,

ii) has been issued policy No. 19 and is married and under 25 years old,

iii) has not been issued policy No. 19 and is a married female,

iv) is a male under 25 years old, and

v) is married and 25 or over."

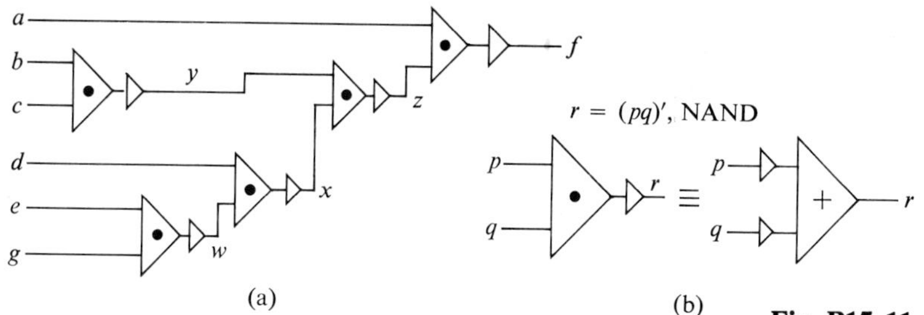

Fig. P15-11

15-11 The NAND gate circuit of Fig. P15-11(a) may be analyzed for the simplest form of the output function f in either of two ways:

 i) by writing f from inspection of the circuit and applying DeMorgan's theorem; or

 ii) by making use of Fig. P15-11(b) to redraw the circuit and then writing f by inspection.

The second method is a direct application of DeMorgan's theorem to the circuit diagram, in which the network is rearranged so that no NOT's appear to the right of any gate. Note that two NOT's on an uninterrupted line cancel each other. Show that the two methods give the same result.

15-12 Derive Eq. (15-22′) in the second canonical form by starting with Eq. (15-13′). Use the results to find an expression for f_9 of Table 15-4.

15-13 Generalize the results of the previous problem to an n-variable switching function and obtain an expression in the standard-product second canonical form that is equivalent to Eq. (15-26). Apply the results to find the switching function for

$$f(x_1, x_2, x_3, x_4) = 1$$

for all combinations of the four variables except the following two:

x_1	x_2	x_3	x_4
0	1	0	1
0	0	1	1

15-14 Starting with Eq. (15-22) and using the DeMorgan theorem, derive the equivalent to Eq. (15-27) in the second canonical form, using NOR gates. Apply the results to the function described at the end of the previous problem to obtain the NOR-gate implementation of the function.

15-15 Re-solve the example problem of Fig. 15-18 using the second canonical form of standard products. Draw the simplest possible relay and gate circuits to implement the given function, and check that they give the same results as shown in Fig. 15-18.

15-16 A switching function of three variables, each of which may take on a value of 0 or 1, is to be ON (or 1) only when the three variables are not the same. From the Karnaugh map of this function, choose a canonical form that will yield a gate-type circuit implementation with the fewest possible logical blocks.

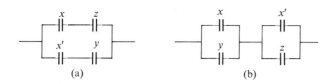

Fig. P15-18

$x_3\,x_4$	$x_1\,x_2$ 0 0	0 1	1 1	1 0
0 0	0	0	1	0
0 1	0	1	1	0
1 1	1	1	1	1
1 0	1	0	1	1

Fig. P15-19

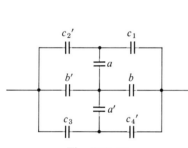

Fig. P15-20

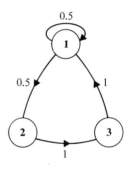

Fig. P15-22

15-17 If, in the previous problem, implementation is to be accomplished by using relay contacts, which canonical form is to be preferred? (Check for hazards!)

15-18 At first glance, the relay contact circuits of Fig. P15-18(a) and (b) appear to yield the same output, but with respect to hazard-free operation they differ. Analyze each circuit for hazards, and specify which should have make-before-break and which a break-before-make relay x for hazard-free operation.

15-19 Write the function represented by Fig. P15-19 first by using prime implicants (first canonical form) and then by using prime implicates (second canonical form). In either case redundancies are to be added as necessary to eliminate any possible hazards.

15-20 Does the relay contact circuit of Fig. P15-20 have any static hazards? Check the circuit for dynamic hazard (by drawing a timing chart) if $a = 0$, $b = 1$, and c goes from 0 to 1 in the order of the subscripts in the figure.

15-21 For the system of Example 15-13, show that one eigenvalue is always 1.

15-22 For the system of Fig. P15-22, obtain $\mathbf{x}(\infty)$.

APPENDIX

MATRIX ALGEBRA

The purpose of this appendix is to review the basic definitions and rules of matrix algebra which are necessary for an understanding of the main text. Most of the relations are shown in terms of simple examples (instead of their general forms) in such a way that a generalization to an arbitrary number of rows and columns is possible without additional information.

A-1 DEFINITIONS

1. A vector **x** in an n-dimensional space is expressed by a column with coordinates $x_1, x_2, \ldots, x_n$ as its elements. For $n = 2$,

$$\mathbf{x} = \begin{bmatrix} x_1 \\ x_2 \end{bmatrix}.$$

2. Switching of columns and rows is called *transposition*. It is designated by a prime, or a superscript T. A column vector **x**, when transposed, becomes a row vector.

Example

$$\mathbf{x} = \begin{bmatrix} x_1 \\ x_2 \end{bmatrix}; \quad \mathbf{x}' = [x_1 \ x_2].$$

$$\mathbf{c} = [c_1 \ c_2]; \quad \mathbf{c}^T = \begin{bmatrix} c_1 \\ c_2 \end{bmatrix}.$$

3. Addition and subtraction of vectors are defined for two or more vectors of the same form (either column or row) and the same dimension. Thus

$$\mathbf{x} \pm \mathbf{y} = \begin{bmatrix} x_1 \\ x_2 \end{bmatrix} \pm \begin{bmatrix} y_1 \\ y_2 \end{bmatrix} = \begin{bmatrix} x_1 \pm y_1 \\ x_2 \pm y_2 \end{bmatrix}, \tag{A-1}$$

$$\mathbf{c} \pm \mathbf{d} = [c_1 \ c_2] \pm [d_1 \ d_2] = [c_1 \pm d_1 \ c_2 \pm d_2]. \tag{A-2}$$

4. An array of $n \times m$ elements c_{ij} in n rows and m columns is called an $n \times m$ rectangular matrix when $n \neq m$. The matrix is equivalent to a column matrix where each element is a row vector, or a row matrix having column vectors as its elements.

Example 3×2 *matrix:*

$$\mathbf{C} = \begin{bmatrix} c_{11} & c_{12} \\ c_{21} & c_{22} \\ c_{31} & c_{32} \end{bmatrix} = \begin{bmatrix} \mathbf{c}_1 \\ \mathbf{c}_2 \\ \mathbf{c}_3 \end{bmatrix}, \quad \text{where} \quad \mathbf{c}_i = [c_{i1} \;\; c_{i2}], \quad i = 1, 2, 3.$$

Example 2×3 *matrix:*

$$\mathbf{D} = \begin{bmatrix} d_{11} & d_{12} & d_{13} \\ d_{21} & d_{22} & d_{23} \end{bmatrix} = [\mathbf{d}_1 \;\; \mathbf{d}_2 \;\; \mathbf{d}_3],$$

where

$$\mathbf{d}_j = \begin{bmatrix} d_{1j} \\ d_{2j} \end{bmatrix}, \quad j = 1, 2, 3.$$

Transposition. For the matrix **C** given above,

$$\mathbf{C}' = \begin{bmatrix} c_{11} & c_{21} & c_{31} \\ c_{12} & c_{22} & c_{32} \end{bmatrix} = [\mathbf{c}_1' \;\; \mathbf{c}_2' \;\; \mathbf{c}_3'].$$

5. A matrix is square when $n = m$. For instance,

$$\mathbf{A} = \begin{bmatrix} a_{11} & a_{12} \\ a_{21} & a_{22} \end{bmatrix}, \quad \mathbf{I} = \begin{bmatrix} 1 & 0 \\ 0 & 1 \end{bmatrix}, \quad \Lambda = \begin{bmatrix} \lambda_1 & 0 \\ 0 & \lambda_2 \end{bmatrix},$$

where **I** is the unit or identity matrix, always square. A square matrix in the form of Λ is called *diagonal*. Both **I** and Λ belong to a class of square matrices called *symmetric*; their elements are symmetric with respect to the diagonal,

$$a_{ij} = a_{ji}.$$

6. Matrices with the same number of rows and columns can be added or subtracted according to Eqs. (A–1) and (A–2). For example,

$$s\mathbf{I} - \mathbf{A} = \begin{bmatrix} s - a_{11} & -a_{12} \\ -a_{21} & s - a_{22} \end{bmatrix}, \tag{A-3}$$

where s is a common scalar factor that is involved in all elements of the matrix.

A-2 MULTIPLICATION

1. A *scalar* (or *dot*) *product* is defined for two vectors of the same dimension n where a row vector must be followed by a column vector. For instance,

$$\mathbf{bc} = [b_1 \;\; b_2] \begin{bmatrix} c_1 \\ c_2 \end{bmatrix} = b_1 c_1 + b_2 c_2. \tag{A-4}$$

Matrix algebra

In general, a scalar product (also called the *inner product*) is given by the sum of the products of all corresponding elements. The product is expressed by a pair of parentheses (sometimes triangular brackets are used):

$$\mathbf{a} = \begin{bmatrix} a_1 \\ a_2 \end{bmatrix}, \quad \mathbf{c} = \begin{bmatrix} c_1 \\ c_2 \end{bmatrix}, \quad (\mathbf{a}, \mathbf{c}) = \langle \mathbf{a}, \mathbf{c} \rangle = a_1 c_1 + a_2 c_2 . \quad (A\text{-}5)$$

We therefore have the following equality when both $\mathbf{a}$ and $\mathbf{c}$ are column vectors:

$$(\mathbf{a}, \mathbf{c}) = \mathbf{a}'\mathbf{c} = \mathbf{c}'\mathbf{a} . \quad (A\text{-}6)$$

2. The inner product of two vectors serves as the basis for defining the product of two matrices. When two square matrices are both 2×2, their product is given by

$$\mathbf{AB} = \begin{bmatrix} \mathbf{a}_1 \\ \mathbf{a}_2 \end{bmatrix} [\mathbf{b}_1 \ \mathbf{b}_2] = \begin{bmatrix} \mathbf{a}_1\mathbf{b}_1 & \mathbf{a}_1\mathbf{b}_2 \\ \mathbf{a}_2\mathbf{b}_1 & \mathbf{a}_2\mathbf{b}_2 \end{bmatrix}$$

$$= \begin{bmatrix} a_{11}b_{11} + a_{12}b_{21} & a_{11}b_{12} + a_{12}b_{22} \\ a_{21}b_{11} + a_{22}b_{21} & a_{21}b_{12} + a_{22}b_{22} \end{bmatrix}. \quad (A\text{-}7)$$

Note that the $\mathbf{a}_i$ are row vectors and the $\mathbf{b}_j$ column vectors, so that the inner products on the right-hand side of Eq. (A-7) are defined by Eq. (A-5). The column premultiplying a row operation of the middle step in Eq. (A-7) gives the result shown on the right according to the rule: the ijth element of the product is formed by multiplying the jth column of the row vector by the ith row of the column vector.

3. In general, a matrix product does not commute:

$$\mathbf{AB} \neq \mathbf{BA} . \quad (A\text{-}8)$$

For instance, if $\mathbf{b}$ follows $\mathbf{c}$ in Eq. (A-4), we have

$$\mathbf{cb} = \begin{bmatrix} c_1 \\ c_2 \end{bmatrix} [b_1 \ b_2] = \begin{bmatrix} c_1 b_1 & c_1 b_2 \\ c_2 b_1 & c_2 b_2 \end{bmatrix}; \quad (A\text{-}9)$$

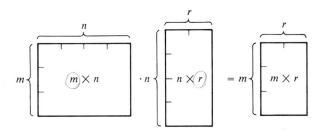

Fig. A-1 Product of rectangular matrices.

hence $\mathbf{bc} \neq \mathbf{cb}$. However, there are some exceptions. For instance,

$$\mathbf{AI} = \mathbf{IA} = \mathbf{A}. \tag{A-10}$$

4. The product of rectangular matrices is defined in terms of the pair shown in Fig. A-1. For two matrices to be conformable (that is, to have the dimensionality to permit multiplication) the number of columns in the first must be the same as the number of rows in the second. The product matrix will have the same number of rows as the first, and the same number of columns as the second. If, for instance, $n = 3$, $m = 2$, and $r = 1$, we have

$$\begin{bmatrix} c_{11} & c_{12} & c_{13} \\ c_{21} & c_{22} & c_{23} \end{bmatrix} \begin{bmatrix} x_1 \\ x_2 \\ x_3 \end{bmatrix} = \begin{bmatrix} y_1 \\ y_2 \end{bmatrix}. \tag{A-11}$$

5. Equation (A-11), $\mathbf{Cx} = \mathbf{y}$, is equivalent to the following pair of simultaneous scalar algebraic equations:

$$\begin{aligned} c_{11}x_1 + c_{12}x_2 + c_{13}x_3 &= y_1, \\ c_{21}x_1 + c_{22}x_2 + c_{23}x_3 &= y_2. \end{aligned} \tag{A-12}$$

Conversely, a set of simultaneous equations can be replaced by a matrix equation. For example, the following pair,

$$\frac{dx_1}{dt} = a_{11}x_1 + a_{12}x_2 + b_1 u, \qquad \frac{dx_2}{dt} = a_{21}x_1 + a_{22}x_2 + b_2 u,$$

can be combined into the matrix form,

$$\frac{d}{dt}\begin{bmatrix} x_1 \\ x_2 \end{bmatrix} = \begin{bmatrix} a_{11} & a_{12} \\ a_{21} & a_{22} \end{bmatrix} \begin{bmatrix} x_1 \\ x_2 \end{bmatrix} + \begin{bmatrix} b_1 \\ b_2 \end{bmatrix} u,$$

$$\frac{d\mathbf{x}}{dt} = \mathbf{Ax} + \mathbf{b}u.$$

6. The following relation holds for transposition of a matrix product:

$$(\mathbf{Ax})' = \mathbf{x}'\mathbf{A}'. \tag{A-13}$$

Example

$$(\mathbf{Ax})' = \left\{ \begin{bmatrix} a_{11} & a_{12} \\ a_{21} & a_{22} \end{bmatrix} \begin{bmatrix} x_1 \\ x_2 \end{bmatrix} \right\}' = \mathbf{x}'\mathbf{A}' = \begin{bmatrix} x_1 & x_2 \end{bmatrix} \begin{bmatrix} a_{11} & a_{21} \\ a_{12} & a_{22} \end{bmatrix}$$

$$= [(a_{11}x_1 + a_{12}x_2) \quad (a_{21}x_1 + a_{22}x_2)].$$

A-3 MATRIX INVERSION

1. The *inverse* of a matrix is defined when the matrix is square; it plays the role of division. An inverse matrix $\mathbf{M}^{-1}$ of a square matrix $\mathbf{M}$ is defined

by
$$\mathbf{M}\mathbf{M}^{-1} = \mathbf{M}^{-1}\mathbf{M} = \mathbf{I}.\tag{A-14}$$

The inverse exists if and only if

$$\det \mathbf{M} \neq 0.\tag{A-15}$$

2. If all diagonal elements of a diagonal matrix are other than zero, the inverse of this matrix is a diagonal matrix whose elements are the inverse of the original elements. For example:

$$\Lambda = \begin{bmatrix} \lambda_1 & 0 \\ 0 & \lambda_2 \end{bmatrix}, \quad \Lambda^{-1} = \begin{bmatrix} 1/\lambda_1 & 0 \\ 0 & 1/\lambda_2 \end{bmatrix},\tag{A-16}$$

where

$$\lambda_1 \lambda_2 \neq 0.$$

3. A set of inhomogeneous, first-order, algebraic equations can be solved using an inverse matrix. For instance, let

$$ax_1 + bx_2 = y_1, \quad cx_1 + dx_2 = y_2,$$

or

$$\begin{bmatrix} a & b \\ c & d \end{bmatrix} \begin{bmatrix} x_1 \\ x_2 \end{bmatrix} = \begin{bmatrix} y_1 \\ y_2 \end{bmatrix}.$$

Then the solution is given by

$$\begin{bmatrix} x_1 \\ x_2 \end{bmatrix} = \begin{bmatrix} a & b \\ c & d \end{bmatrix}^{-1} \begin{bmatrix} y_1 \\ y_2 \end{bmatrix},\tag{A-17}$$

where

$$\begin{bmatrix} a & b \\ c & d \end{bmatrix}^{-1} = \frac{1}{\Delta} \begin{bmatrix} d & -b \\ -c & a \end{bmatrix},\tag{A-18}$$

and

$$\Delta = \begin{vmatrix} a & b \\ c & d \end{vmatrix} = ad - bc \neq 0.\tag{A-19}$$

If, however, the equation is homogeneous (that is, $y_j = 0$ for all j), a nontrivial solution exists if and only if the determinant is zero,

$$\Delta = 0.$$

The solution in this case can be determined up to a proportionality factor.

For example, if
$$ax_1 + bx_2 = 0, \qquad cx_1 + dx_2 = 0,$$
then we must have
$$\frac{x_2}{x_1} = -\frac{a}{b} = -\frac{c}{d}.$$
Hence,
$$\Delta = ad - bc = 0,$$
and, with this condition satisfied,
$$x_1 = k, \qquad x_2 = -\frac{a}{b}k,$$
where k is an arbitrary constant.

4. The inverse matrix $\mathbf{M}^{-1}$ of an $n \times n$ matrix $\mathbf{M}$ (for which det $\mathbf{M} \neq 0$) is determined by the following computation rules:

(a) Compute the *cofactor* determinants D_{ij} where D_{ij} is a determinant obtained by striking off the ith row and the jth column of $\mathbf{M}$.

(b) Fix m_{ij} by
$$m_{ij} = (-1)^{i+j} D_{ij}.$$

(c) Determine the matrix such that
$$\mathbf{M}_{ij} = [m_{ij}]. \tag{A-20}$$

(d) The inverse matrix is now given by
$$\mathbf{M}^{-1} = \frac{\mathbf{M}'_{ij}}{\Delta}, \tag{A-21}$$
where
$$\Delta = \det \mathbf{M}.$$

5. *Example:* If
$$\mathbf{M} = \begin{bmatrix} a & b_1 & d_1 \\ b_2 & c & f_1 \\ d_2 & f_2 & e \end{bmatrix},$$
then
$$\mathbf{M}^{-1} = \frac{1}{\det \mathbf{M}} \begin{bmatrix} m_{11} & m_{21} & m_{31} \\ m_{12} & m_{22} & m_{32} \\ m_{13} & m_{23} & m_{33} \end{bmatrix},$$

where

$$m_{11} = \begin{vmatrix} c & f_1 \\ f_2 & e \end{vmatrix}, \quad m_{12} = -\begin{vmatrix} b_2 & f_1 \\ d_2 & e \end{vmatrix}, \quad m_{13} = \begin{vmatrix} b_2 & c \\ d_2 & f_2 \end{vmatrix},$$

$$m_{21} = -\begin{vmatrix} b_1 & d_1 \\ f_2 & e \end{vmatrix}, \quad m_{22} = \begin{vmatrix} a & d_1 \\ d_2 & e \end{vmatrix}, \quad m_{23} = -\begin{vmatrix} a & b_1 \\ d_2 & f_2 \end{vmatrix},$$

$$m_{31} = \begin{vmatrix} b_1 & d_1 \\ c & f_1 \end{vmatrix}, \quad m_{32} = -\begin{vmatrix} a & d_1 \\ b_2 & f_1 \end{vmatrix}, \quad m_{33} = \begin{vmatrix} a & b_1 \\ b_2 & c \end{vmatrix}.$$

(If **M** is symmetric, then $\mathbf{M}^{-1}$ is also symmetric.)

6. If **M** and $\mathbf{M}^{-1}$ are expressed in terms of row and column vectors, as

$$\mathbf{M} = \begin{bmatrix} \mathbf{m}'_1 \\ \mathbf{m}'_2 \\ \mathbf{m}'_3 \end{bmatrix}, \quad \mathbf{M}^{-1} = [\hat{\mathbf{m}}_1 \ \hat{\mathbf{m}}_2 \ \hat{\mathbf{m}}_3],$$

then, by virtue of the property $\mathbf{M}\mathbf{M}^{-1} = \mathbf{I}$, it follows that

$$\mathbf{m}'_i \hat{\mathbf{m}}_j = \begin{cases} 1 & \text{if } i = j, \\ 0 & \text{otherwise.} \end{cases}$$

In other words, the vectors $\hat{\mathbf{m}}'_i$ and $\hat{\mathbf{m}}_j$ are *orthogonal* to each other when $i \neq j$.

A-4 MINIMUM RIGHT-INVERSE

1. An inverse does not exist for a rectangular matrix. It is therefore impossible to solve a set of simultaneous equations when the number of unknowns exceeds the number of knowns. For instance, if x_1 and x_2 in the following equation are unknowns,

$$ax_1 + bx_2 = y,$$

or

$$\begin{bmatrix} a & b \\ 0 & 0 \end{bmatrix} \begin{bmatrix} x_1 \\ x_2 \end{bmatrix} = \begin{bmatrix} y \\ 0 \end{bmatrix}, \quad (A\text{-}22)$$

we cannot solve for

$$\begin{bmatrix} x_1 \\ x_2 \end{bmatrix} \text{ because } \begin{vmatrix} a & b \\ 0 & 0 \end{vmatrix} = 0.$$

2. The problem in its general form can be stated as follows: Given an m-vector **y** and an $m \times n$ matrix **F** such that (Fig. A–2a).

$$\mathbf{Fx} = \mathbf{y}, \quad (A\text{-}23)$$

Appendix

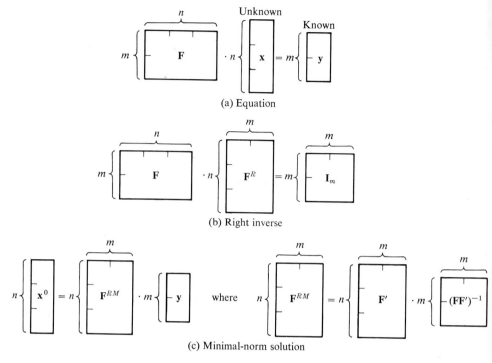

Fig. A-2 Minimum right inverse.

find an n-vector $\mathbf{x}$ which satisfies the equation, where $n > m$.

The "solution" to this problem is given by

$$\mathbf{x} = \mathbf{F}^R \mathbf{y} . \tag{A-24}$$

$\mathbf{F}^R$ in this expression is called the *right-inverse* of $\mathbf{F}$. It is an $n \times m$ matrix (Fig. A-2b) determined by

$$\mathbf{F}\mathbf{F}^R = \mathbf{I}_m , \tag{A-25}$$

where $\mathbf{I}_m$ is the $m \times m$ identity (or unit) matrix. $\mathbf{F}^R$ exists when the rank of $\mathbf{F}$ is m or greater. (The *rank* of a matrix is the order of the largest nonzero determinant that can be formed from the elements of the matrix.)

3. The "solution" (Eq. A-24), is not unique. If, for instance,

$$x_1 + 2x_2 = y = 10 , \tag{A-26}$$

then

$$\mathbf{F} = [1 \quad 2] , \tag{A-27}$$

and an infinite number of right inverses $\mathbf{F}^R$ exist along the straight line $P_1 P_3$

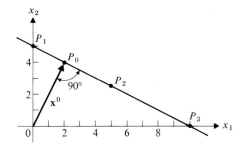

Fig. A-3 Example of the minimum norm solution.

shown in Fig. A-3. For example,

$$F^R = \begin{bmatrix} 0 \\ 0.5 \end{bmatrix}, \quad \text{or} \quad \begin{bmatrix} 0.5 \\ 0.25 \end{bmatrix}, \quad \text{or} \quad \begin{bmatrix} 1 \\ 0 \end{bmatrix}$$

for P_1, P_2, and P_3, respectively, in the figure.

4. There exists, however, a unique "solution" for which the *norm* (length) of **x** is minimal. This is indicated by point P_0 in the figure. Such a solution $\mathbf{x}^0$, called the *minimal solution*, is determined by the minimum right inverse $\mathbf{F}^{RM}$:

$$\mathbf{x}^0 = \mathbf{F}^{RM}\mathbf{y}, \tag{A-28}$$

where

$$\mathbf{F}^{RM} = \mathbf{F}'(\mathbf{FF}')^{-1}. \tag{A-29}$$

5. *Example*: For Eq. (A-27), we have

$$\mathbf{FF}' = \begin{bmatrix} 1 & 2 \end{bmatrix} \begin{bmatrix} 1 \\ 2 \end{bmatrix} = 5, \quad (\mathbf{FF}')^{-1} = \frac{1}{5};$$

thus

$$\mathbf{F}^{RM} = \frac{1}{5}\begin{bmatrix} 1 \\ 2 \end{bmatrix} = \begin{bmatrix} 0.2 \\ 0.4 \end{bmatrix}.$$

For $y = 10$, this gives $x_1 = 2$ and $x_2 = 4$, at point P_0 in Fig. A-3. Note that the vector $\mathbf{x}^0$ is orthogonal to the line P_1P.

6. Let us designate $(\mathbf{x}, \mathbf{x})$ by $||\mathbf{x}||$ and prove the minimal norm condition. We begin with the equality

$$||\mathbf{x}|| = ||\mathbf{x}^0 + (\mathbf{x} - \mathbf{x}^0)||$$
$$= ||\mathbf{x}^0|| + ||\mathbf{x} - \mathbf{x}^0|| + 2(\mathbf{x}^0)'(\mathbf{x} - \mathbf{x}^0), \tag{A-30}$$

where **x** is any solution to Eq. (A-23) and $\mathbf{x}^0$ is the hypothesized minimum norm solution. By Eqs. (A–23) and (A–28),

$$\mathbf{x}^0 = \mathbf{F}^{RM}\mathbf{y} = \mathbf{F}^{RM}\mathbf{F}\mathbf{x} ,$$

and

$$(\mathbf{x}^0)' = \mathbf{y}'(\mathbf{F}^{RM})' .$$

Also, we have

$$(\mathbf{x} - \mathbf{x}^0) = (\mathbf{I}_m - \mathbf{F}^{RM}\mathbf{F})\mathbf{x} ;$$

hence

$$\begin{aligned}(\mathbf{x}^0)'(\mathbf{x} - \mathbf{x}^0) &= \mathbf{y}'(\mathbf{F}^{RM})'(\mathbf{I}_m - \mathbf{F}^{RM}\mathbf{F})\mathbf{x} \\ &= \mathbf{y}'\{(\mathbf{F}^{RM})' - (\mathbf{F}^{RM})'\mathbf{F}^{RM}\mathbf{F}\}\mathbf{x} .\end{aligned} \quad (A\text{–}31)$$

By definition of the minimum right-inverse (Eq. A–29),

$$(\mathbf{F}^{RM})' = [(\mathbf{F}\mathbf{F}')^{-1}]'\mathbf{F} ;$$

thus

$$(\mathbf{F}^{RM})'\mathbf{F}^{RM}\mathbf{F} = [(\mathbf{F}\mathbf{F}')^{-1}]'\mathbf{F}\mathbf{F}'(\mathbf{F}\mathbf{F}')^{-1}\mathbf{F} = [(\mathbf{F}\mathbf{F}')^{-1}]'\mathbf{F} = (\mathbf{F}^{RM})' .$$

Substituting these conditions into Eq. (A–31), we find that

$$(\mathbf{x}^0)'(\mathbf{x} - \mathbf{x}^0) = 0 .$$

Therefore, Eq. (A–30) reduces to

$$||\mathbf{x}|| = ||\mathbf{x}^0|| + ||\mathbf{x} - \mathbf{x}^0|| . \quad (A\text{–}32)$$

Since all norms are positive definite (nonnegative), we conclude from the last equation that

$$||\mathbf{x}|| \geqslant ||\mathbf{x}^0|| ; \quad (A\text{–}33)$$

hence $\mathbf{x}^0$ is the minimal norm solution.

A-5 MINIMUM LEFT-INVERSE

1. We now consider the problem that is complementary to the preceding one: solution of systems of equations with more knowns than unknowns. That is, in the vector equation (Fig. A–4a)

$$\mathbf{H}\mathbf{u} = \mathbf{z} , \quad (A\text{–}34)$$

$\mathbf{z}$ is a known n-vector, and $\mathbf{u}$ is an unknown m-vector, where $n > m$.

2. As mentioned before, the inverse of $\mathbf{H}$ does not exist because it is an $n \times m$ rectangular matrix. In other words, since the number of unknowns is less than the number of knowns, an exact solution does not exist. For

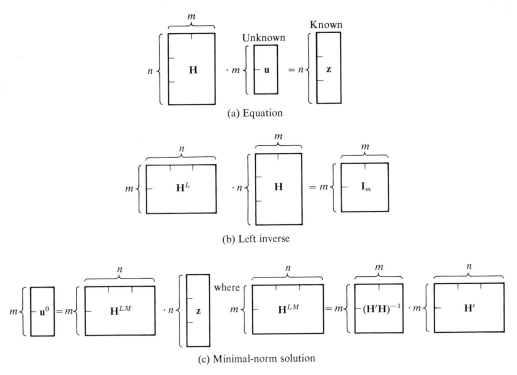

Fig. A-4 Minimum left inverse.

example, if

$$\mathbf{H} = \begin{bmatrix} 1 \\ 2 \end{bmatrix}, \quad \mathbf{z} = \begin{bmatrix} 10 \\ 4 \end{bmatrix},$$ (A-35)

then Eq. (A-34) implies either

$$u = 10, \quad \text{or} \quad u = 2,$$

and we cannot find a solution for u.

3. It is possible, however, to find an approximate solution in terms of the nonunique left-inverse of $\mathbf{H}$, $\mathbf{H}^L$, which is defined by

$$\mathbf{H}^L \mathbf{H} = \mathbf{I}_m.$$ (A-36)

Letting $\mathbf{H}^L = [h_1 \ h_2]$ for the system of Eqs. (A-35), we find that

$$h_1 + 2h_2 = 1.$$ (A-37)

Therefore $\mathbf{H}^L$ is given by all points on the line HH in Fig. A-5; hence $\mathbf{H}^L$ is not unique. The vector $\mathbf{H}$ is shown lying along the line PP. In other words, an infinite number of u's exist, none of which are solutions to the

Appendix

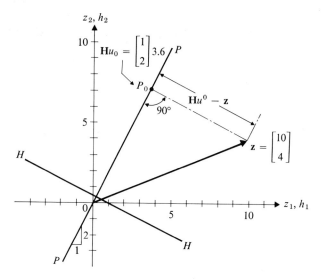

Fig. A-5 Examples of the minimum left inverse.

original problem:

$$\mathbf{u} = \mathbf{H}^L \mathbf{z}. \tag{A-38}$$

4. The "solution" for the minimum left-inverse $\mathbf{H}^{LM}$,

$$\mathbf{u}^0 = \mathbf{H}^{LM}\mathbf{z}, \tag{A-39}$$

is unique. $\mathbf{H}^{LM}$ exists when $\mathbf{H}$ has rank m or greater, and it is given by

$$\mathbf{H}^{LM} = (\mathbf{H}'\mathbf{H})^{-1}\mathbf{H}'. \tag{A-40}$$

We shall show in what follows that the distance between $\mathbf{Hu}$ and $\mathbf{z}$ is minimized for the "best" solution $\mathbf{u}^0$:

$$\|\mathbf{Hu} - \mathbf{z}\| \geqslant \|\mathbf{Hu}^0 - \mathbf{z}\|. \tag{A-41}$$

5. *Example*: The system of Eqs. (A-35):

$$\mathbf{H}'\mathbf{H} = (1)^2 + (2)^2 = 5,$$

$$\mathbf{H}^{LM} = (0.2)\begin{bmatrix} 1 & 2 \end{bmatrix} = \begin{bmatrix} 0.2 & 0.4 \end{bmatrix},$$

$$\mathbf{u}^0 = \begin{bmatrix} 0.2 & 0.4 \end{bmatrix}\begin{bmatrix} 10 \\ 4 \end{bmatrix} = 3.6.$$

$\mathbf{Hu}^0$ for this case is given by point P_0 in Fig. A-5. P_0, among all points on PP, is closest to the tip of $\mathbf{z}$. Note that $\mathbf{H}^{LM}$ lies at the intersection of the lines HH and PP.

6. To prove Eq. (A-41), we let $\mathbf{u}$ be any solution and $\mathbf{u}^0$ be the minimal norm solution and start with the equality:

$$\begin{aligned}
||\mathbf{Hu} - \mathbf{z}|| &= ||\mathbf{H}(\mathbf{u} - \mathbf{u}^0) + \mathbf{H}\mathbf{u}^0 - \mathbf{z}|| \\
&= ||\mathbf{H}(\mathbf{u} - \mathbf{u}^0) + (\mathbf{HH}^{LM} - \mathbf{I}_n)\mathbf{z}|| \\
&= ||\mathbf{H}(\mathbf{u} - \mathbf{u}^0)|| + ||(\mathbf{HH}^{LM} - \mathbf{I}_n)\mathbf{z}|| \\
&\quad + 2[\mathbf{H}(\mathbf{u} - \mathbf{u}^0)]'(\mathbf{HH}^{LM} - \mathbf{I}_n)\mathbf{z}.
\end{aligned} \quad (A-42)$$

The last term on the right-hand side of Eq. (A-42) vanishes:

$$\begin{aligned}
[\mathbf{H}(\mathbf{u} - \mathbf{u}^0)]'(\mathbf{HH}^{LM} - \mathbf{I}_n)\mathbf{z} &= (\mathbf{u} - \mathbf{u}^0)'\mathbf{H}'(\mathbf{HH}^{LM} - \mathbf{I}_n)\mathbf{z} \\
&= (\mathbf{u} - \mathbf{u}^0)'(\mathbf{H}'\mathbf{HH}^{LM} - \mathbf{H}')\mathbf{z} \\
&= (\mathbf{u} - \mathbf{u}^0)'\{\mathbf{H}' - \mathbf{H}'\}\mathbf{z} \\
&= 0.
\end{aligned}$$

Therefore, Eq. (A-42) reduces to

$$\begin{aligned}
||\mathbf{Hu} - \mathbf{z}|| &= ||\mathbf{H}(\mathbf{u} - \mathbf{u}^0)|| + ||\mathbf{HH}^{LM}\mathbf{z} - \mathbf{z}|| \\
&= ||\mathbf{H}(\mathbf{u} - \mathbf{u}^0)|| + ||\mathbf{H}\mathbf{u}^0 - \mathbf{z}||.
\end{aligned} \quad (A-43)$$

Since the first term on the right-hand side of this equation is nonnegative (that is, positive definite), the inequality condition (A-41) holds.

Note that the minimum pseudo-inverse, repeated twice, gives the original matrix:

$$(\mathbf{F}^{RM})^{LM} = \mathbf{F}, \quad (\mathbf{H}^{LM})^{RM} = \mathbf{H}. \quad (A-44)$$

MAJOR SYMBOLS

(The numbers in parentheses refer to chapter, and sections, for local use of symbols.)

- **A**   system's matrix in state equation
- $\mathbf{A}_J$ Jacobian matrix for nonlinear system (12)
- $A(s)$ polynomial in s that appears as denominator of transfer function (8)
- A area (of piston, free liquid level, etc.); scalar form of **A**
- a scalar form of **A**; element of matrix; sinusoidal amplitude (9); amplitude in the describing function (12)
- a_{ij} element of **A**
- a_i coefficient of $A(s)$
- a_h, a_c dimensionless parameters for heat exchanger (7-4)
- **B**   matrix for control vector in state equation
- $B(s)$ polynomial in s as numerator of transfer function (8)
- **b**   B-matrix as column vector (for scalar input)
- b scalar form of **B**; element of matrix; damper coefficient; feedback motion (5-5)
- b_j element of **b**
- **C**   matrix for state vector in output equation
- $\mathbf{C}_i$ coefficient matrix (3-1)
- C generalized capacitance; scalar form of **C**; constant
- C_i integration constant; coefficient
- **c**   C-matrix as row vector (for scalar output)
- c scalar form of **C**; element of matrix; sonic velocity; specific heat
- $c(t), C(s)$ controlled variable in the t- and s-domains, respectively (c_k, $C(z)$, respectively, for discrete-time systems)
- **D**   matrix for input vector in output equation
- d scalar form of **D**; diameter
- D derivative operator ($= d/dt$)
- $D(s)$ denominator polynomial in s
- $D(z)$ pulse transfer function for discrete-time control law
- det   determinant
- E quantity of energy; Young's modulus (7); expected value (13, 14)
- e base of natural logarithms ($= 2.71828$), element of **Y**
- $e(t), E(s)$ error signal in the t- and s-domains, respectively (e_k, $E(z)$, respectively, for discrete-time systems)
- **F**   vector function; rectangular matrix for minimum right-inverse ($=\mathbf{CT}$ in modal control); weighting matrix (10-5, 14); gain matrix (11-4)
- F force; probability distribution function (13-1)
- **f**   vector function in state equation; gain vector (11-3)
- f generalized flow variable (5, 6 and 7); various functions
- f_i elements of **F**

Major symbols

f_0 cost function (14)
$f(t)$, $F(s)$ function of time and its Laplace transform
$G(s)$ matrix transfer function
$G(t)$ optimum feedback control matrix (14) (G_k for discrete-time systems)
$G(s)$, $G(D)$ transfer function ($G(z)$ for pulse transfer function)
$g(t)$ unit-impulse response (weighting function) $= \mathscr{L}^{-1}[G(s)]$; scalar form of $\mathbf{G}(t)$ (g_k for discrete-time systems)
$\mathbf{H}$ vector function for the output equation; rectangular matrix for minimum left-inverse ($=\mathbf{T}^{-1}\mathbf{B}$ in modal control)
$\mathbf{H}(t)$ matrix function for $\mathbf{G}(t)$ ($\mathbf{H}_k$ for discrete-time systems) (14)
$H(s)$ transfer function of feedback element
H Hamiltonian (or inner product) in the maximum principle (14); heat of reaction (5-4)
h height; head; magnitude; film coefficient for heat transfer; hysteresis gap (12)
h_j scalar form for $\mathbf{H}_k$ (14-6)
$\mathbf{I}$ identity matrix
I generalized inductance
i integer number
$\mathscr{I}m$ imaginary part
J performance index (integrated cost function); moment of inertia (6-4)
j $=\sqrt{-1}$; integer number
$\mathbf{K}$ gain matrix in multivariable control systems
$\mathbf{K}(t)$ gain matrix for optimum filter ($\mathbf{K}_k$ for discrete-time systems) (13, 14)
K gain (of amplifier, in root locus, etc.)
$\mathbf{k}'$ component row vectors of $\mathbf{K}$
k gain; constant coefficient; spring constant; integer number of sampled data; thermal conductivity (5-2)
k_c gain of P-control
L dead time
$\mathscr{L}, \mathscr{L}^{-1}$ Laplace transformation and its inverse, respectively
l (total) length
$\mathbf{M}$ transfer matrix and other square matrices
$\mathbf{M}_J$ $(n+1)$ dimensional Jacobian matrix (14)
$\mathbf{M}_k$ covariance matrix of one-step prediction error ($\mathbf{M}(t)$ for continuous-time systems) (13, 14)
M magnitude of bang-bang control; supremum value of H (14); input–output magnitude ratio in frequency response; moment (7-1)
$m(t)$, $M(s)$ manipulated variable in feedback control system, in the t- and s-domains, respectively
m dimension of $\mathbf{y}$, order of $B(s)$, and other integer numbers; mass; $\int p\,dt$ (6)
m_k scalar form of $\mathbf{M}_k$ (13-4)
N integer number; noise intensity (13)
$N(a)$ describing function (12)
$N(s)$ numerator polynomial in s
n dimension of $\mathbf{x}$ (order of lumped-parameter system); order of $A(s)$, etc.
$\mathbf{P}$ system's matrix in difference equation of state
P power (5)

	P	power; period of oscillation; probability (13); etc.
	p	scalar form of **P**; generalized potential variable (5, 6 and 7); probability density function (13); transition probability (15-5)
$\mathbf{p}(t), p(t)$		particular solution (3)
	p_i	pole (eigenvalue)
	$p(x)$	first probability-density function (13)
	Q	matrix for input vector in difference equation of state; matrix to specify gradient in quadratic Lyapunov function
	Q	flowrate; heat quantity
	q	**Q** as a column vector (for scalar controlling input); vector for port variables (7)
	q	scalar form of **Q**; flowrate (5); $\int f\,dt$ (6)
	R	transfer matrix in partial differential equation for distributed-parameter systems (7); covariance matrix (13)
	R	generalized resistance; maximum slope of unit-step-input response (of monotone processes); reset input in flip-flop (15-4); gas constant (5-3)
	r, R	radius
$r(t), R(s)$		reference (or set point) input to feedback control system, in the t- and s-domains, respectively (r_k, $R(z)$, respectively, for discrete-time systems)
	r	dimension of input vector; ratio
	$\mathscr{R}e$	real part
	$\mathbf{S}(t)$	solution matrix
	S	switching function (12-5); set input to flip-flop (15); shear force (7-1)
$S(x_i, i)$		minimal value of performance index (14-6)
	s	parameter for Laplace transformation
	T	transformation matrix (modal matrix) for **x**
	T	time constant; sampling period; specified final time (of optimal control) (14)
T_i, T_d		I- and D-action time constants, respectively
	t	running time
	U	covariance matrix for $\mathbf{u}_k$ (13, 14)
$\mathbf{u}(t), \mathbf{U}(s)$		input vector, in the t- and s-domains, respectively ($\mathbf{u}_k$, $\mathbf{U}(z)$ for discrete-time systems)
$u(t), U(s)$		scalar input in the t- and s-domains, respectively (u_k, $U(z)$ for discrete-time systems)
	U	admissible range for **u**; overall heat transfer coefficient (5)
	V	covariance matrix for $\mathbf{v}_k$ (13, 14)
$\mathbf{v}(t), \mathbf{V}(s)$		disturbance vector in the t- and s-domains, respectively ($\mathbf{v}_k$, $\mathbf{V}(z)$ for discrete-time systems), measurement noise (13, 14)
$v(t), V(s)$		scalar form of $\mathbf{v}(t)$ and $\mathbf{V}(s)$, respectively (v_k, $V(z)$ for discrete-time systems)
	V	quadratic form; test function for stability; Lyapunov function; volume (5); element of **R** (7-3)
	v	specific volume; velocity; measurement noise (13-4; Problem 14-21)
	$\mathbf{v}^i$	ith eigenvector
	v_j^i	jth element of $\mathbf{v}^i$

Major symbols

W   matrix for vector partial differential equation of distributed-parameter systems; symmetric matrix for stability test; matrix weighting coefficients; covariance matrix for w_k (13, 14)

W   element of **R** (7–3)

w^i orthogonal vector to v^i (10–3)

$\mathbf{w}_k$ noise vector (13, 14)

w   flowrate or product of flowrate and specific heat of carrier; weighting factor

X   covariance matrix for $\mathbf{x}(t)$

$\mathbf{x}(t), \mathbf{X}(s)$ state vector in the *t*- and *s*-domains, respectively; $\mathbf{x}^*(t)$ for modal domain ($\mathbf{x}_k$, $\mathbf{X}(z)$ for discrete-time systems)

$\hat{\mathbf{x}}(t)$ optimal estimate of $\mathbf{x}(t)$

$\mathbf{X}(t)$ $(n + 1)$-dimensional state-vector (14)

$x_i(t)$ element of $\mathbf{x}(t)$ (state variable)

$x(t)$ scalar state variable

x length factor (5–1); random variable (13)

x_j Boolean variable (15)

$\mathbf{y}(t), \mathbf{Y}(s)$ output (or response) vector ($\mathbf{y}_k$, $\mathbf{Y}(z)$ for discrete-time systems)

$y_i(t)$ element of $\mathbf{y}(t)$

$y(t)$ scalar response

Y   symmetric matrix for generating a quadratic form

$Y(D)$ admittance [**Y** for admittance matrix (6–5)]

Y_k set of observations $y_0, y_1, \ldots, y_k$ or $\mathbf{y}_0, \mathbf{y}_1, \ldots, \mathbf{y}_k$ (13, 14)

y, *Y*   internal variable in sequential system (15)

$\mathbf{Z}_k$ covariance matrix for estimation error (13, 14)

Z_c characteristic impedance (7–2)

$Z(D)$ impedance [**Z** for impedance matrix (6–5)]

$\mathscr{Z}, \mathscr{Z}^{-1}$ *z*- and inverse *z*-transformation, respectively

z $= e^{sT}$, parameter in the *z*-domain; distance variable; output of sequential system (15)

z_j zero [roots of $B(s) = 0$]

$\mathbf{a}$ real part of v^i for complex eigenvalue

α real part of complex quantity; minimum slope of nonlinear gain (12)

$\boldsymbol{\beta}$ imaginary part of v^i for complex eigenvalue

β imaginary part of complex quantity; maximum slope of nonlinear gain (12); bulk modulus (5)

Γ propagation operator (7–2)

γ slope in isocline method

Δ determinant; step magnitude in quantization (*A/D* conversion); small difference; dead zone

$\delta(t)$ Dirac delta function

ε small quantity

ζ damping coefficient

η dummy variable; empirical coefficient; dimensionless distance (7)

Θ absolute temperature (5)

θ angle variable in polar coordinate; angle; temperature

κ gain and other constants

Λ A-matrix in the canonical form

λ quantity related to characteristic value
μ mean value
ν exponent in performance-index equation (14-3)
π = 3.141592653589793238462643383279
ρ radius (of ellipse); distance from origin in polar coordinates; mass density
σ real part in complex quantity (and complex eigenvalue); standard deviation
σ^2 variance
τ time (in correlation functions and dummy variable in convolution); dimensionless time
ϕ phase angle; temperature (6); correlation function in the time domain (Φ for the frequency domain)
$\boldsymbol{\Psi}$ $(n+1)$-dimensional covariant vector (14)
$\boldsymbol{\phi}$ n-dimensional covariant vector (14)
ψ_i element of $\boldsymbol{\Psi}$ or $\boldsymbol{\phi}$ (14)
Ω normalized circular frequency
Ω_s spinning velocity of gyro (6)
ω circular frequency [$P = 2\pi/\omega$, $f = \omega/2\pi$ = cycles per unit time]

INDEX

INDEX

A

Absolute error, integrated (*See* Integrated absolute error)
Absolute stability, 532
Absorbing state, 750, 751
Acceleration control, 689
Actuating error, 192, 317, 318
Adaptation of human operator, 198
Adaptive control, 468, 554–556
Addition, of matrices, 760
   of vectors, 759
Adjacent cells, 717, 722
Adjoint matrix, 43, 642
Adjoint system, 642
Adjoint vector (*See* Covariant vector)
Adjustable parameter, 142
Admissible range, 12, 625, 692, 775
Admittance, 165, 231, 268
Airplane, 202, 431, 555
Algebra of logic (*See* Switching algebra)
Algebraic equation, 42, 763
Algorithm, 103
Aliasing (*See* Noise aliasing)
Amplifier, 188, 252–254
Amplitude locus, 526
Amplitude modulation, 468
Amplitude ratio (*See* Magnitude ratio)
Amplitude ratio per cycle, 327
Analog computer, 19, 92–93, 370
Analog controller (*See* Controller, analog)
Analog filter, 474
Analog simulation, 93, 98, 370
Analog-to-digital converter, 468
Analogy, 162, 171, 220
Analyzer of modes, 433, 435, 447
AND, 701, 707, 710, 711, 734, 736
AND gate, 701
Angle condition of root locus, 332

Angular momentum, 243
Approximate model, 346, 372, 686–687
Argument, 359
Arrhenius law, 180
Associative law, 180, 713
Asymmetric variable, 287
Asymptotes, for Bode diagram, 364
   for root loci, 335
Asymptotic stability, 116, 137, 179, 320, 482
   in the large, 117, 512, 532, 539
   Theorem of, 121, 128, 133
Asynchronous system, 736
Ausman, J. S., 365
Autocatalytic reaction, 148, 186, 693
Autocorrelation function, 578, 579, 593
Automobile steering, 201
Automobile suspension, 219
Autonomous system, 115, 258, 653
Auxiliary vector (*See* Covariant vector)
Availability, 163

B

Backlash, 527, 573
Bacteria colony, 692
Baffle-nozzle mechanism, 188–189
Bakke, R. M., 554, 555
Band-limited signal, 449, 471–472, 614
Bandwidth, 391, 474, 687
Barbeyrac, J., 486
Bang-bang control, 12, 548, 630, 647
Basis vector, 88
Bass, R. W., 423
Batch chemical reaction, 184
Batch process, 552
Beam, bending of, 270–271, 287
Bellman, R. E., 624, 626
Bergen, A. R., 545
Bertram, J. E., 499

781

Bilateral coupling, 175, 207
Bilateral Laplace transformation, 587
Binary-coded data, 7
Binary counter, 738
Binary logic, 705 (*See also* Logic)
Binary signal, 582
Binary time series, 580
Binary white noise, 580
Binary word, 717
Biological system, 655
Biodynamics, 186
Biot number, 308
Bistable elements, 704
Black-box approach, 37, 197
Block diagrams, 8, 18, 39, 42
   analog controller, 192
   discrete-time control systems, 484
   distributed-parameter objects, 281, 292, 296
   finite-state systems, 747
   modal control systems, 433, 437, 448, 451
   multivariable-feedback control system, 414, 419
   optimal-feedback control systems, 660, 676
   optimum filter, 603, 607
   single-variable feedback control system, 318
   three-port systems, 249
   two-port systems, 222
   vector input, state and output, 41, 42, 51
Bode, H. W., 366
Bode diagrams, 361, 687
Bode theorem, 366
Bond, 207
Bond graph, 208, 212, 215, 220, 221, 242, 247, 258, 734–736
Boole, G., 699
Boolean algebra (*See* Switching algebra)
Boundary conditions, 297
Boundary input, 304
Boundary output, 304
Branch type elements, 701, 715, 731
Break contact, 700
Break frequency (*See* Corner frequency)

Brown, F. T., 269, 286
Bryson, A. E., 616
Bulk modulus, 156

C

Cadzow, J. A., 494
Calvert, J. F., 551
Cancellation (*See* Pole-zero cancellation)
Canonical product (*See* Standard product)
Canonical sum (*See* Standard sum)
Cantilever beam, 289
Cap symbol, 710
Capacitance, 156, 157, 163, 166
Capacitor, 163, 173, 208, 210
Capillary tube, 199
Car-following theory, 276
Carrier, 170, 176, 271, 290, 291–296
Cascade control, 419, 431
Cascade coupling (*See* Coupling, unilateral)
Causality (*also* Causal relation), 5, 8, 10, 18, 158, 208, 212, 214, 216, 280
Center, 97
Center of gravity of poles, 335
Central-limit theorem, 583
Central moment, 576
Centrifugal pump, 241
Chain of two-port elements, 226–236
Chain reaction, 182–183
Characteristic curves, 241, 702
Characteristic equation,
   continuous-time systems, 23, 43, 62, 65, 97, 137
   discrete-time systems, 52, 143, 481, 483, 749
   exponential modes, 63
   finite *vs.* infinite-order, 158
   ICR-system, 216
   linearized system, 533
   multivariable-feedback systems, 414, 415, 423, 426
   Nyquist theorem, 373–374
   open-loop *vs.* closed-loop, 423–426
   Routh test, 137

servomechanism, 327
single-variable feedback systems, 319–320, 332, 373
Characteristic impedance, 278
Characteristic polynomial (*See* Characteristic equation)
Characteristic roots (*See* Eigenvalues)
Characteristic variables, 280
Characteristics, 297
Chattering, 551, 569, 650
Chemical reactor, 179–182
Circle criterion, 546–548
Classical control theory, 4, 80–81, 317, 356, 412
Clock pulses, 736
Closed-loop control (*See* Feedback control)
Closed-loop frequency response, 391–398
Closed-loop poles, 319–320, 374, 421, 422, 427
Closed-loop response, 320, 374, 421, 482
Closed-loop transfer function, 320, 391, 484–485
Coanda effect, 254, 704, 738
Coasting, 639
Coding, 707, 717
Cofactor determinant (*See* Determinant, cofactor)
Colored noise, 594, 614
Column vector, 759
Combinational systems, 719–736, 741
Compensation, 420, 519
Compensation networks, 80, 384
Compensator, 555
Complementation, 701, 707 (*See also* Negation)
Complex eigenvalues (*See* Conjugate complex eigenvalues)
Complex frequency response, 388–389
Complex number, 358
Complex system, 7
Component concentration system, 175
Composition control, 468
Compressibility, 156, 169, 232
Computer (*See* Analog computer, Digital computer)
Condensing process, 274

Conditional expectation, 598
Conditional-probability density function, 602
Conditionally stable system, 381
Conduit (*See also* Pipeline), 157
Cone of attainability, 640–641, 644–645, 654
Conformal mapping, 372, 390
Conjugate complex eigenvalues, 63, 85, 96, 108, 321, 323, 403, 435, 483, 551, 552
Conjugate complex eigenvectors, 68
Conjugate complex poles (*See* Conjugate complex eigenvalues)
Conjugate complex roots (*See* Conjugate complex eigenvalues)
Conservative system, 121
Constitutive relation, 37, 159, 161, 163, 172, 210, 245
Constraints, 185, 347, 452, 625, 649, 657
Contacts, break, 700
  make, 700
Continuity condition, 156, 237
Continuous-time system, 7 ff
  as limiting case of discrete-time system, 593–594, 606, 669, 673, 677
Control accuracy, 319
Control actions (*See* D, I, P, PD, and PID actions)
Control algorithms
  minimum settling time, 486–490
  modal, 433–434, 437, 495–498
  optimal feedback, 673, 677
  position (*See* Position algorithm)
  vector feedback, 422, 429
  velocity (*See* Velocity algorithm)
Control computer, 463
Control laws (*See* Control actions, Control algorithms)
Control rod, 184
Control strategy, 624
Controllability, 8, 75–86, 100–101, 108, 349, 350, 355, 417–418, 428, 486
Controlled variable, 317, 413, 477
Controller, analog, 187–197, 467
  digital (*See* Direct Digital Control)
  multivariable, 413, 415

784　Index

Controller adjustment (*See* Controlling parameter tuning, Optimum tuning)
Controlling parameter tuning, 343–344, 479–480
Control object, 26, 317, 413, 452, 658, 668, 669, 677
Control perturbation, 640
Control quality (*See* Quality of control, Performance index)
Control sequence, 486, 492–494
Control valve, 11, 476, 527
Conveyor process, 274
Convolution (*See* Convolution integral)
Convolution integral, 15, 63, 329, 581, 587
Corner frequency, 362
Correlation functions (*See* Autocorrelation function, Cross-correlation function, Covariance matrix)
Costate vector (*See* Covariant vector)
Cost function (*See* Performance index)
Coulomb friction, 159, 171, 527
Counter, 738, 746
Counterelectromotive force, 240
Coupling, bilateral, 175
  unilateral, 175, 176, 252, 254
Coupling of feedback loops, 413
Coupling of multiple modes, 72
Coupling of processes, 178, 182
Covariance matrix, 592, 607, 676, 685
Covariant vector, 629, 636, 654, 669
Cox, J. B., 478
Critical damping, 327, 345
Critical size of population, 185
Critical switching boundary, 549
Critical trajectory, 551
Cross controller, 417
Cross-coupling effects in modal control, 444
Cross-correlation function, 578, 582, 593
Cross-power-density spectrum, 587
CR-system (*See* RC system)
Cup symbol, 710
Cutoff frequency, 474
Cut sets, 727, 731
Cybernetics, 4
Cylinder, 98, 127, 532

D

D-action, 336, 344, 386, 477
Damped natural frequency, 324, 331
Damped oscillation, 98, 322, 341, 388
Damped real frequency (*See* Damped natural frequency)
Damper, 119, 171
Damping coefficient (*See* Damping ratio)
Damping factor (*See* Damping ratio)
Damping ratio, 331, 359, 663
Dashpot (*See* Damper)
DC-gain, 10
DC-motor, 240
DDC (*See* Direct Digital Control systems)
DDC algorithms, 478
Deadbeat response, 345, 481
Dead time, 177, 267, 277, 301–302, 340–342, 367, 370, 383, 485, 514, 564, 700, 736
  approximation of, 355, 367, 370
  variable, 555
Dead zone, 510, 647
DeBolt, R. R., 478
Decade, 362
Decay rate, 345
Decayed oscillation (*See* Damped oscillation)
Decibels, 361
Decimal numbers, 717
Deconvolution, 329
Decouplability theorem, 416
Decoupling, 48, 66, 93, 99, 135, 415
Decoupling controller, 417
Degenerate system, 632
Delay element, 736 (*See also* Dead time)
Delay line, 277
Delay process (*See* Dead time)
Delayed integrator system, 341
DeMoivre's theorem, 371
DeMorgan theorem, 710, 714, 715, 736
Denominator polynomial, 44, 64, 137
Derivative action (*See* D-action)
Derivative kick, 477
Derivative operator, 18
Describing function, 524
Describing-function method, 523–534
Desired value, 318

Determinants, 44, 763
    cofactor, 764
Deterministic signal, 6, 575, 613, 676
Diagonal form of A-matrix, 61, 431
Diagonalization, 48, 65, 126, 279, 290
    of open-loop matrix transfer function, 415
Diagonalized domain (*See* Modal domain)
Diagonal matrix, 65, 760
Difference equation of dynamic systems, 18, 24, 50–52, 102–108, 302–303, 311
Differential equation of dynamic systems, 13, 35–38, 58–75
    partial (*See* Partial differential equation)
Differential gap, 512–513
Differential gear, 256–257
Differentiator, 37
Diffusion process, 170, 229, 236, 446–447
Digital computer, 17, 24, 277
    solution, 101, 103
Digital control algorithms, 476, 556
Digital filtering of noise, 475
Dilution process, 177, 371
Dimensionality of dynamic systems, 38
Diode gate, 701
Dirac delta function, 16, 64, 593
Direct Digital Control systems, 197, 457, 476, 541
Direct method of Lyapunov (*See* Lyapunov's direct method)
Discrete-time asymptotic stability theorem, 133, 144
Discrete-time maximum principle (*See* Maximum principle, discrete)
Discrete-time model of continuous-time system, 103–105
Discrete-time relay control system, 520–523
Discrete-time systems, 7, 17, 50, 101–109, 132, 457–499, 592, 603, 666, 669, 673, 677, 748
Discrete-time optimal control (*See* Optimal control, discrete-time)
Disjunctive function (*See* Exclusive OR)
Displacement-balance type controller, 190

Displacement feedback system, 685
Dissipation of energy (*See* Energy dissipation)
Distillation column, 276
Distinct real eigenvalues, 94, 134, 402
Distributed control, 448
Distributed-parameter systems, 6, 156, 236, 266–306, 340, 370, 448, 657
    root locus of, 340–342
Distributed input, 304–305, 447–448
Distributed output, 304–306, 448
Distributive law, 713
Disturbance input, 317, 323, 414, 419
Dither, 569, 650
Dominant branch of root locus, 336
Dominant conjugate complex poles, 321, 322, 331, 340, 344, 388, 393, 552
Dominant mode, 431
Dominant mode of oscillation (*See* Dominant conjugate complex poles)
Don't-care cell, 723
Double integrator system (*See* Inertial object)
Double oscillator, 99, 112
Double pole, 74, 95 (*See also* Multiple pole)
Draper, C. S., 558
Dry friction (*See* Coulomb friction)
Dual-mode control, 551
Dual relation, in C and I, 163
    in Riccati equation, 678
    in switching algebra, 707, 713
Dummy variable, 60, 64
Dynamic compensation, exact, 420
Dynamic optimization, 4, 560
Dynamic programming, 626–629, 645
Dynamic systems, 12, 162, 164
Dzung, L. S., 381

E

Earthquake simulator, 685
Eggleston, D. M., 651
Eigenvalues
    A-matrix (continuous-time system), 63, 66, 68, 71, 137
    ellipse, ellipsoid, hyperellipsoid, 126
    feedback control systems, 319

finite-state systems, 752
location in the s-plane, 331
P-matrix (discrete-time system), 106, 108, 134–135
RC system, 173
transmission line, 279
Eigenvectors, 67, 77, 82, 92, 100–101
Elastic shaft, 278, 282
Elastic vibration, 289
Elasticity of pipe, 232
Electrical system, 163–165
Electrical circuit, 214, 220
Electromagnetic relay, 699
Electromechanical transducer, 240
Ellipse, 97, 127–128, 403–404
Ellipsoid (*See* Hyperellipsoid)
Elliptic partial differential equation, 297
Encoder, 475
Encoding, 476
End state, 625, 651, 653, 670
End-to-end relation of a line, 279
Energetic system, 118, 206
Energy, 164
 potential, 118, 212
 kinetic, 118, 212
Energy converter, 236
Energy dissipation, 119, 163
Energy port, 207
Ensemble average, 578
Entropy, 236, 249
Environment, 153, 259
Epidemic process, 184–185, 205
Equilibrium state, 115
Equivalent state, 743
Ergodic hypothesis, 578
Ergodic set, 751, 752
Ergodic process, 578
Error, feedback control (*See* Actuating error)
 optimum prediction, 682
 state estimation, 598
 steady state (*See* Offset)
Essential hazards (*See* Hazards, essential)
Evans, W. R., 4, 332
Exact compensation, 420
Excitation, 700
Excitation matrix, 741, 744
Exclusive OR, 705–706, 711, 713

Existence of optimal control, 632
Exothermic reaction, 181
Expansion of logic function, 714
Expected value, 576
Exponential decay, 15, 180, 187, 211, 470
Exponential mode, 23, 63, 98
Exponential rise, 81, 180, 187
Exponential transfer function, 301
Extremal computer, 560

F

Fan-in, 726
Fan-out, 726
Fast mode, 92–93
Feedback, 254, 325, 741
Feedback amplifier, 254
Feedback compensation, 348
Feedback controls, 317–398, 413–416, 421–422, 430–434, 454, 657–666, 671–673, 675–690
Feedback control, ideal, 319
Feedback control systems, structure of, 317
Feedback-feedforward system, 420
Feedback element, 187, 318
Feedforward compensation (*See* Feedforward control)
Feedforward control, 348, 419–420, 468
Fermentation process, 186–187
Film coefficient, 170
Filters (*See* Noise filter, Optimum filter)
Final control element (*See* Control valve)
Final flow table, 740
Final value of error, 322 (*See also* Offset)
Final-value theorem, 158, 294, 323, 490, 749
Finite-order model (*See* Lumped model)
Finite-state systems, 747–752
Finite-time settling control, 482, 483–485, 486–494
First canonical form (*See* Standard sum)
First-level DDC, 468, 476
First method of Lyapunov (*See* Lyapunov's first method)

First-order lag (*See* First-order systems)
First-order systems, 12, 19, 35, 45, 162, 348, 480, 648, 678, 697
  with a delay, 488, 515, 518, 526
First probability density function, 576
Fission process, 183
Flapper–nozzle mechanism (*See* Baffle–nozzle mechanism)
Flip-flops
  RS-type, 738–739, 741–743
  T-type, 737, 745–746
Float, 262
Flow chart, 104, 603
Flow control, 468, 591
Flow diagram, 738, 740
Flow source, 210
Flow variable, 157, 162
Fluid amplifiers, proportional type, 197, 252–254
  wall-attachment type, 704, 737–738
Fluid inertia, 232
Fluidic devices, 700
Fluid-line, 269, 284, 289 (*See also* Pipeline)
Fluid machinery, 241
Fluid motor, 239
Focus, 96, 520
Force, 164, 171
Force-balance type controller, 190
Force-flow analogy, 171
Force-potential analogy, 220–221
Forced motion, 15, 100
Forced response (*See* Forced motion)
Forward-loop transfer function, 325
Four-terminal system, 164
Fourier coefficient, 447, 524
Fourier integral, 397
Fourier series, 397, 449, 468, 524
Fourier transform pair, 584
Fourier transformation, 584
Four-point difference algorithm, 477–478
Free response, 15, 748
Free system (*See* Autonomous system)
Frequency bandwidth (*See* Bandwidth)
Frequency domain, 372, 397, 524, 584
Frequency locus, 526
Frequency response, 4, 266, 356–408, 482

Frequency-response method, 4, 356–398
  Bode diagrams, 361
  Nichols chart, 393
  Nyquist diagrams, 359
Frequency-response trajectory, 403
Frequency spectrum, 584
Friction (*See* Coulomb friction, Viscous friction)
Fuller, A. T., 650

G

Gain, 10, 188, 190, 336, 344, 347, 382, 554
  time-varying optimal gain, 673, 677
  in frequency response (*See* Magnitude ratio)
Gain asymptotes, 364
Gain crossover, 382
Gain margin, 382, 394
Gas constant, 169
Gas-flow system, 164, 168–169
Gas-pressure system (*See* Pressure system)
Gate-type elements, 701, 715, 731
Gaussian distribution, 577, 582, 591
Gaussian white noise, 597, 675
Gaussian white-noise generator, 586
Gear train, 256
Genealogical process, 502
Gibbs, W., 4
Global stability (*See* Asymptotic stability in the large)
Goff, K. W., 480
Gould, L. A., 431, 446
Governor, 2, 3
Gradient, 637, 668, 670
Gradient vector, 120
Gray codes, 717, 722
Gustafson, R. D., 346
Gyrator, 242–243
Gyroscope, 243–244
Gyro stabilizer, 255

H

Half-adder, 705
Hamiltonian, 629, 636, 643, 651, 663, 667, 669

Hatanaka, H., 536
Hazards, 700, 719, 724
  dynamic, 730
  essential, 730, 741, 746
  static, 730
  type 1–0–1 (or Type 1), 730
  type 0–1–0 (or Type 0), 730
Hazard compensation, 732
Head, 165
Heart, 261
Heat balance, 175, 180, 184, 272
Heat capacitance, 170, 174
Heat conduction, 170, 236, 263, 269–270, 285, 446 (*See also* Diffusion process)
Heat exchanger, 174, 224, 272, 291–296, 328
  counter-flow, 276, 293
  parallel-flow, 275, 292
  tube-and-tank type, 267, 295–296
Heat flow, 169
Heat of reaction, 181
Heat-storage element (*See* Heat capacitance)
Heat transfer, 169
Heat-transfer coefficient, overall, 170 (*See also* Film coefficient)
Heat transport, 170
Heating coil, 272
Heating unit, 450
Hierarchy of control, 468
Highway traffic flow, 276–277
Hill-climbing control, 558–566
Hill-climbing technique, 347
History of control, 2
Ho, Yu-Chi, 657
Holder (*See* Zero-hold element)
Homogeneous equation, 65, 67, 630, 636, 763–764
Homogeneous solution, 58, 60
Human operator, 197
Hurwitz determinants, 114
Hybrid system of lumped- and distributed-parameters, 267, 295–296
Hydraulic actuator, 196, 205, 238, 685
  locking action of, 238–239
Hydraulic servo (*See* Hydraulic actuator)
Hydrodynamical time derivative, 275, 277, 299

Hyperbolic function, 157
Hyperbolic partial differential equation, 280, 297, 299
Hyperellipsoid, 119, 126
Hyperplane, 11, 249
Hypersphere, 119
Hypersurface, 11
Hysteresis, 512, 525

I

I-action, 3, 196, 344, 386, 483
IAE (*See* Integrated absolute error)
IAE criterion, 479
ISE (*See* Integrated squared error)
ITAE (Integral of time multiplied by absolute value of error), 347
ITSE (Integral of time multiplied by squared error), 347
Ideal control, 319, 433, 496, 498
Ideal element, 10, 158, 206
Ideal filter, 472
Ideal junction, 213
  one-junction, 213
  zero-junction, 213
Ideal measurement, 438, 496, 498
Identification, on-line, 556–558
  of gain, 554
  of processes, 329, 370, 372, 582, 587
  of state, 490
Identity matrix, 40, 760
Imaginary part, 358
Imaginary poles, 96
Impedance, 231, 268 (*See also* Series impedance)
  matching, 238, 255
Implementation of logic system, 712, 719, 727, 734
Impulse input, 16, 42
Impulse modulator, 469, 472
Impulse response (*See* Unit impulse response)
Impulse train, 25, 468
Incidental nonlinearity, 509, 527
Independent modal control, 435, 439, 449
Inductance, 156, 163, 171–172
Inductor, 163, 210

Inertia, 156
 fluid, 169
Inertial object, 625, 630–631, 637, 650, 662, 693, 694, 695
Initial condition (*See* Initial state)
Initial state, 17, 20, 42, 58, 59, 102, 212, 600, 655
Initial-value theorem, 158, 294
Initialization, 599, 662, 664, 672, 683
Inner product, 761
Input, 5, 35, 158, 208, 741
 vector, 38, 259
Input impedance, 255
Input–output cross-correlation function, 582
Input–output relation, 8, 36, 37, 43, 159, 188, 229, 249, 266, 324, 523, 581, 701, 703, 736, 738
 sinusoidal (*See* Frequency response)
Instability, 122, 180
Instrumentation, 436, 451
Integral windup (*See* Reset windup)
Integrated absolute error, 344, 346
Integrated microcircuits, 699
Integrated squared error, 130, 346, 651, 664 (*See also* Quadratic performance index)
Integrating control action (*See* I-action)
Integrator, 12, 173, 210, 692
Intentional nonlinearity, 509, 527
Interaction, energy, 154, 162, 206
 of control actions, 195
 of feedback loops, 413, 497
 of system, 257
Interlocking action, 720
Intermittent signal, 468
Internal state (*See* State)
Internal variable of sequential system, 740–741
Intersection, 710
Inventory control, 694
Invariant axis of rotation, 406
Inverse hysteresis, 518–520, 733–734
Inverse Laplace transformation, 20, 37, 61, 358
Inverse matrix, 43, 67, 424, 764
Inverse relation, 10
Inverse z-transformation, 108

Inverted pendulum, 110
Irrational transfer function, 370 (*See also* Exponential transfer function)
Irreversible causality, 11
Isocline, 90
Isocline method, 90–91, 511
Isothermal process, 169
Iterative chain, 226–236

J

Jacobian matrix, 537, 539, 632, 642, 667
James, H. M., 575
Johnson, G. W., 541
Joint probability density function, 577
Jordan canonical form, 48, 65, 67, 431
Jump resonance, 534
Junction, 736 (*See also* Ideal junction)
Jury, E. I., 108
Jury theorem, 144

K

Kalman, R. E., 499, 597
Kalman filter (*See* Optimum filter)
Karam, J., 372
Karnaugh map, 716, 721, 722, 734
Kinetic energy, 118, 212
Kirchhoff's law, 214
 loop law of Kirchhoff, 247
 node law of Kirchhoff, 247
Krasovskii, N. N., 539
Krigman, A., 734

L

Lag and delay system, 341
Lag network, 384, 387
Lagrange multiplier, 657
Landing problem, 637–639
Laplace domain (*See* s-domain)
Laplace transformation, 19, 41
 bilateral, 587
Laplace transformation pairs, 21, 22, 469
Law of mass action, 184
Lead–lag compensation, 348
Lead network, 348, 384–385
Leakage, 232
Leaking tank, 36, 54, 160

Learning process, 198
Left half plane (LHP), 142, 372
Left inverse, 769 (*See also* Minimal left inverse)
Length of vector, 89
Letov, A. M., 541
Level control, 352, 417, 468
Lever, 237
LHP (*See* Left half plane)
Li, Y. T., 558
Limit cycle, 514, 518, 523, 525, 529, 563
  stable (*See* Stable limit cycle)
  state-space description, 530–534
  symmetric, 517, 521
  unstable (*See* Unstable limit cycle)
Linear independence, 67, 89
Linear object (*See* Linear system)
Linear systems, 5, 127, 159, 637 ff
  time-varying linear system, 132, 607, 668
Linearization, 10, 13, 179, 182, 242, 250, 532
Linear superposition, 15, 284
Linear transformation, 65
Linkage, 256
Liquid-flow system, 155, 158, 165–168, 468
Liquid-level control (*See* Level control)
Liquid-storage tank, 160
Lissajous figure, 402–403
Literals, 715–716, 718, 723
Load current, 326
Load impedance, 281
Load upsets, 419
Local stability, 537
Locking action, 238
Log magnitude, 361–362
Logarithmic spiral, 97
Logic circuits (*See* Logic networks)
Logic elements, 699–706
Logic functions, 701, 710
  n-variable, 714
  single-variable, 708
  two-variable, 711
Logic networks, 519, 560, 741
Logic variable, 708
Logical product, 707 (*See also* AND)

Logical sum, 707 (*See also* OR)
Log magnitude diagram (*See* Bode diagram)
Low-pass filter, 473, 731
Lumped model, 156, 235 (*See also* Lumping)
Lumping, 5, 266, 302–304, 370
Lyapunov, M. A., 4, 114
Lyapunov's direct method, 114, 118
Lyapunov's first method, 536–537
Lyapunov's second method (*See* Lyapunov's direct method)
Lyapunov functions, 118, 120, 123, 134, 452, 453, 532, 535, 539, 541, 542
  discrete-time, 132

M

M-series, 582
Mach number, 309
Macroscopic model, 156
Magnitude of a complex number, 358
Magnitude ratio, 356, 359, 393
Major axis of ellipse, 404
Make-before-break relay, 700
Make contact, 700
Manifold, 403, 653
Manipulated variable, 317, 413, 422, 431, 485
Manipulator, 435, 450, 475
Manual control, 198
Manual-to-automatic transfer, 476
Mapping function, 372
Markov process, 608, 747–748
Mass, 171
Material balance, 180
Material exchange, 276
Mathematical modeling, 5, 6, 153, 155, 206, 267
Matrix, 759–760
  adjoint, 43
  diagonal, 65, 760
  Jordan canonical form, 71–72
  modified canonical form, 69
  null, 96
  product, 761–762
  pseudo-inverse, 69, 765–771
  Schwartz form, 72

sum, 760
transposition, 760
Matrix block diagram, 39
Matrix exponential, 58
Matrix inversion, 762–765
Matrix polynomial, 425
Matrix Riccati equation (*See* Riccati equation)
Matrix transfer function, 43, 64, 249, 413–414
Maximum principle, 629–630, 635–637, 645
 applications, 646–657
 discrete, 668
Maxterm, 718, 720, 723, 724
Maxwell, J. C., 3
McCluskey, E. J., Jr., 724
Mean value (*See* Expected value)
Measurement noise, 473, 599, 697
Mean squared error, 588, 598, 607
Mean squared quantization error, 475
Mechanical system, 164, 171–172
Memory, 719, 737, 738
Memory-type nonlinearity, 525
Merged flow chart, 746
Message, 591
Microorganisms, 186
Microscopic model, 156, 226
Microstructure, 227
Minimal (or minimum) left inverse, 444, 494, 496, 768–771
Minimal (or minimum) phase system, 366, 383
Minimal (or minimum) right inverse, 444, 452, 493, 496, 609, 765–768
Minimum principle, 630, 643
Minimum product, 726, 728
Minimum sum, 726, 728
Minimum-time settling control, 486
Minor axis of ellipse, 404
Minor feedback, 529
Minterm, 718, 720, 722, 724
Mixed system of lumped and distributed parameters (*See* Hybrid system)
Mixing tank, 175, 589
Mobility analogy, 172
Modal control, continuous-time lumped system, 430–446

continuous-time, distributed-parameter system, 446–452
 discrete–time, 495–498
Modal domain, 78, 81, 108, 431, 495
Modal matrix, 66, 107
Modal state space, 118
Modes, 63, 76–78, 92, 289, 321, 348, 419
Mode decoupling, 66
Mode-oriented design, 436
Mode-superposition method, 300
Modes of optimal control, 646–648
Modern control theory, 4
Modified canonical form, 69, 85, 97, 99, 108, 403, 431, 442
Modulation, 450
Modulator, 249–252, 469
Molar flow rate, 276
Mole balance, 177
Moment of inertia, 238
Momentum balance, 156
Mori, A. S., 431
Motion in state space, 87–101
M_p criterion, 396
Mullin, F. J., 486
Multiple poles, 43, 63, 71, 331, 435
Multiple roots (*See* Multiple poles)
Multiplexer, 468
Multiplexing, 468
Multiplication (*See* Product)
Multiport systems, 245–259
Multistage decision process, 628
Multivariable control system, 412–452, 491–499
Murray-Lasso, M. A., 431, 446

N

NAND, 703, 710, 711, 715
NAND gate, 703, 728
Natural causality (*See* Causality)
Natural frequency (*See* Damped natural frequency)
Negation, 701, 707
Negative definiteness, 119
Negative semidefiniteness, 119
Negative hysteresis (*See* Inverse hysteresis)
Nelson, W. L., 520, 523

Neutron, 182
   delayed, 183
   density, 180
Newton's law, of angular momentum, 243
   of cooling, 169
   of motion, 109, 171
Nichols, N. B., 269, 372, 575
Nichols chart, 393, 546
Node, 95, 520
Noise, 473, 610–615
Noise aliasing, 473–474
Noise filter, 473–475, 591
No-memory component (*See* Static element)
Nonautonomous system, 122, 653
Nonhomogeneous differential equation, 59
Nonideal control, 441, 443
Nonideal measurement, 439, 443
Noninteracting control, 416
Noninteracting coupling (*See* Coupling, unilateral)
Nonlinear distortion, 523, 527
Nonlinear process (*See* Nonlinear system)
Nonlinear system, 5, 40, 210, 229, 259, 509–566, 693
Nonlinear relation, 250
Nonlinearities, 159, 177, 179, 509, 527
Nonminimal phase system, 366–367, 383, 571
Nonstationary process (*See* Nonstationary system)
Nonstationary system, 6, 40, 177–178, 607, 668
Nontrivial solution, 67
NOR, 703, 710, 711, 715
NOR gate, 703, 704, 728, 739
Norm, 497, 609, 767
Normal distribution (*See* Gaussian distribution)
Normally closed contact (*See* Break contact)
Normally open contact (*See* Make contact)
NOT, 701, 707, 735

NOT gate, 702
Nuclear reactor, 179, 182–184
Null matrix, 96
Number of literals, 715–716
Numerator polynomial, 44, 64
Numerical expression of switching function, 717–718, 722
Numerical solution (*See* Digital computer solution)
Nyquist diagram (*See* Nyquist plot)
Nyquist plane, 526, 546
Nyquist plot, 359–360, 370, 379, 545
Nyquist stability criterion, 372–381
   modified, 377

O

Observability, 8, 75–86, 100–101, 108–109, 350, 355, 417–418, 489
Observable mode, 79–80
Offset, 324, 328, 388, 489
Oldenburger, R., 382
ONE, 699, 700–704, 707
One-port, 207, 209
On-off action, 512–514, 525, 526–527 (*See also* Bang-bang action)
On-off control, 518, 526
Open loop, 319, 382
Open-loop frequency response, 376, 393
Open-loop gain, 319, 686
Open-loop matrix transfer function, 414
Open-loop poles, 319, 332, 357, 373, 419, 422, 686
Open-loop transfer function, 319, 323, 332, 373, 414, 417
Open-loop zeros 319, 332, 373, 686
Operating point, 160, 242
Operating range, 159
Optimal control, 452–458, 491, 624–690, 752
Optimal feedback control, continuous time, 452–458, 657–666, 673
   discrete-time, 671–675
   stochastic, 675–685
Optimal prediction, 605, 682
Optimal trajectory, 456, 626–629, 633–635, 639

Optimalizing control, 558–559
Optimization, 4, 346, 468, 624
Optimum adjustments (*See* Optimum tuning)
Optimum filter, 597–618, 676, 677, 682, 698
Optimum sampling period, 480
Optimum tuning, 343
OR, 701, 707, 710, 711, 734, 736
OR gate, 701, 704
Order of system, 44, 155
Ordinary differential equation, 41, 156
Orifice resistance, 157, 168
Orthogonal coordinate transformation, 126
Orthogonality, 100, 126, 449, 765
Oscillation (*See* Damped oscillation, Undamped oscillation)
Oscillator, 99, 112, 379 (*See also* Undamped oscillator)
Oscillatory mode, 63, 98, 101
Oscillatory response, 331
Oscillatory stability limit, 533
Output, 5, 35, 36, 38, 158, 208, 741
Output equation, 38, 399, 747
Output impedance, 326
Output matrix, 744
Output vector (*See* Output)
Overloading, 159 (*See also* Saturation)
Overshoot, 322, 346

P

P-action, 3, 324, 329, 343–344, 386, 422, 432, 448
P-control (*See* P-action)
P-controller, 190
Parabolic partial differential equation, 297
Parallel connection, 700
Parameter optimization (*See* Optimization)
Parametric control, 655
Parametric change, 250–251, 324
Partial differential equations, 156, 267, 268, 296–297
Partial fraction expansion, 48, 106, 321, 357

Particular solution, 59
Passive feedback, 188
Paynter, H. M., 103, 213
PD action, 190
PD control (*See* PD action)
Peak-holding (*See* Static peak-holding)
Percolation, 271–272, 297–302
Percentage overshoot (*See* Overshoot)
Perfect induction, 708
Performance index, 7, 130, 135, 346, 492, 625, 635, 661, 663, 666, 669, 676, 677, 678, 685
  quadratic, 125–132, 133–136, 452, 539, 542, 658, 661, 668, 669
Periodic system, 749
Period of oscillation, 327, 381
Periodicity of sampled data, 470
Perturbation variables, 161
Perturbation in optimal control, 640
Perret, R., 560, 706
Phase angle, 359, 393
Phase crossover, 382
Phase margin, 382, 394
Phase-plane, 87, 514
Phase shift, 356, 359, 524–525
Phase space, 87
Phasor, 352, 359, 372
Phillips, R. S., 575
Physical realizability, of continuous-time transfer function, 46
  of exact compensation, 420
  of D-action, 195
  of ideal filter, 472
  of pulse transfer function, 484
Physically realizable system, 46, 319
PI action, 191, 343–344, 348, 476, 480
PI control (*See* PI action)
PID action, 193–195, 343–344, 388, 476
PID algorithm, 476
PID control law (*See* PID action)
Piecewise linear system, 510
Pipe elasticity, 232
Pipe friction, 232
Pipeline, 177, 200, 207, 278
Piston logic elements, 700, 705
Platoon, 277
Pneumatic controller, 3, 188–189

Pneumatic instruments, 188
Pneumatic line dynamics, 372
Pneumatic logic system, 705
Poincaré, H., 4
Poiseuille's law, 200
Polar plot (*See* Nyquist plot)
Poles (*See* Eigenvalues)
Pole cancellation, 80–81, 348–349, 388, 417, 485
Pole–zero cancellation (*See* Pole cancellation)
Polytropic process, 169
Pontryagin, L. S., 624, 629
Popov, V. M., 543
Popov criterion, 543–545
Popov line, 545
Popov locus, 545
Population, 185, 186, 502, 692
Port (*See* Energy port)
Posicast, 348, 485, 551–552
Positive definiteness, 119, 123
Positive semidefiniteness, 119
Position algorithm, 476
Potential source, 210, 214
Potential energy, 118, 212
Potential variable, 162
Powell, R. E., 478
Power, 162
Power-density spectrum, 474, 584, 585
Power transmission, 162, 207, 231
Precession, 243–244
Predictive display, 197
Predictor algorithm, 612
Pressure, 157, 165
Pressure indicator, 255
Pressure system, 173, 201
Pressure control, 468
Preview control, 198
Prime implicants, 722, 724, 725
Prime implicates, 723
Primitive flow table, 740, 743
Principal minors, 126
Principle of optimality, 626, 646, 679–680
Probability, 576
Probability density function, 576
Probability distribution function, 576
  of quantization error, 476
Probability theory, 575–576

Process control, 386, 419, 527, 528
Process dynamics, 554
Product, scalar, 760
  matrix, 761–762
Profos, P., 271
Propagation operator, 278
Proportional band, 190
Proportional control action (*See* P-action)
Proportional-plus-derivative control action (*See* PD-action)
Proportional-plus-integral control action (*See* PI-action)
Proportional-plus-integral-plus-derivative control action (*See* PID-action)
Proportional-plus-reset action (*See* PI-action)
Prototype elements, 226
Pseudoinverses (*See* Minimum left inverse, Minimum right inverse)
Pseudo-random signal, 582
P-source (*See* Potential source)
Pulse duration control, 483, 523
Pulse train, 468
Pulse transfer function, 52, 481, 483, 490
Pulse width control (*See* Pulse duration control)
Pump, 29

Q

Quadratic form, 125
Quadratic Lyapunov function, 125–132, 133–136, 539, 542
Quadratic performance index (*See* Performance index, quadratic)
Quality of control, 329
Quantization error, 475–476, 541–543
Quantization step(-size), 475, 542
Quantized signal, 7
Quarter-decay condition, 341
Quarter-decay response, 341–342, 344
Quasi-linear solution, 159
Quine, W. V., 724

R

Ramp input, 268, 590
Random sequence, 608

Random variable, 7, 575, 576, 591, 679
Rank, 766
Rankine cycle, 183
Radar system, 590, 594
Rate gyro, 245, 255
RC systems, 172–178, 210–211, 512–513
RCI circuit, 215
Reaction processes, 178–187
Real part, 358
Realizability (*See* Physical realizability)
Real poles, 323 (*See also* Distinct real eigenvalues)
Real zero, 323
Recursive algorithm, 600, 602, 683, 684
Redundancy, 724, 726
Reference input, 192, 318, 414, 477
Reflection coefficient, 283, 284
Regulator (*See* Controller)
Relay (*See* Electromagnetic relay)
Relay control action (*See also* Bang-bang control)
Relay circuit, 712
Repeated eigenvalue (*See* Multiple poles)
Reset action, 191
Reset-set type flip-flop (*See* Flip-flops, RS-type)
Reset windup, 194, 478
Residue at pole, 48, 62
Resistance, 163, 165, 169, 171, 199
  variable, 250
Resistor, 163
Resolvent, 424
Resonance peak, 393, 534
Response, 57, 283, 285 (*See also* Output)
Response characteristics, 77
Reverse hysteresis (*See* Inverse hysteresis)
Reverse reaction process, 147, 370
Reversible causality, 10
RHP (*See* Right half plane)
Riccati equation, for optimal control, 661, 662, 672, 673, 678, 688
  for optimum filter, 611–612, 677, 678
Right half plane, 138, 372
Right inverse, 766 (*See also* Minimal right inverse)
Ring voltage, 247
Rise time, 322

RMS value (Root mean square value), 597
Rocket system, 202–204
Root-locus method, 4, 330, 332–341, 482
  rules of, 335
Rosenbrock, H. H., 302, 431
Routh array, 114, 139, 141, 346, 377
Routh, E. J., 137
Routh test, 137–143, 329
Rouxel, R., 560
Row vector, 759
RS-type flip-flop (*See* Flip-flops, RS-type)

S

Saddle, 95
Sales strategy, 694
Salukvanze, M. E., 452
Sample space, 576
Sampled data, 469, 471–473
Sampled-data control, 4, 136, 457–499, 520
Sampled-data systems (*See* Discrete-time systems)
Sampled signal (*See* Sampled data)
Sampler, 51
Sampling interval (*See* Sampling period)
Sampling period, 17, 102, 468, 593
Sampling theorem, 471
Sapiro, M. A., 545
Saturation, 12, 193, 525, 647, 649
  velocity, 528–534
Saturation amplifier, 534
Scalar product, 760
Scatter, 577
Schlaefer, F. M., 431, 446
Schwartz form, 50, 72, 74, 84, 108, 176, 402
s-domain, 19, 25, 41, 61, 106, 159, 318
Second canonical form (*See* Standard product)
Second central moment, 577
Second-level control, 468
Second-order approximation, 346
Secondary variable, 740
Self-adaptive process, 198

796   Index

Self-excited oscillations, 523, 569 (*See also* Limit cycle)
Semi-infinite line, 284
Sensitivity, 324–325, 336, 347
Sequence diagram (*See* Timing chart)
Sequential control, 468
Sequential systems, 566, 582, 719, 731, 736–747
  asynchronous, 736
  synchronous, 736
Series compensation, 348
Series connection, 700
Series impedance, 165, 216, 268, 279
Series inductance, 169
Series–parallel structure, 729
Series resistance, 168, 251
Servomechanism, 4, 252, 325–326, 336, 386, 512, 534, 571, 594
Servomotor, 511
Set point (*See* Reference input)
Setpoint band, 478
Set-reset type flip-flop (*See* Flip-flops, RS-type)
Settling time, 322
Shaft, 207, 229
  elastic, 278
  frictionless elastic, 282
Shannon, C. E., 699, 706
Shaping filter, 594, 615, 666, 685, 687
Sharpness of Lyapunov function, 123, 540
Shifting theorem, 24
Shift operator, 25
Shunt admittance, 165, 216, 268, 279
Shunt admittance system, 217
Shunt capacitance, 168, 171, 216
Shut-down operation, 485
Signal-flow diagrams, scalar, 9, 18, 45, 47, 69, 85, 195
  vector, 39, 42, 52
Signal generator, 666, 685
Signal variables, 162
Sign function, 510
Simple harmonic motion, 212, 696
Simplification chart, 724
Simson, A. K., 254
Simulation, 229, 370, 611

Single time-constant system (*See* First-order system)
Single-loop control, 317, 479
Sinusoidal forcing, 100, 356, 400, 534
Sinusoidal signal, 357, 470
Sinusoidal steady state, 356–357, 399 (*See also* Frequency response)
Sliding friction (*See* Coulomb friction)
Slope of gain asymptotes, 364
Slow mode, 92–93, 435
Smith, O. J. M., 551
Smoothing, 589
Solution matrix, 58, 106, 110, 658, 664
Sonic velocity, 282
Source (*See* Flow source, Potential source)
Space eigenfunction, 447, 449, 450
Spectrum, 397
  power density, 584, 585
Speed governor (*See* Governor)
Speed of response, 386
Sphere (*See* Hypersphere)
Spirule, 333
s-plane, 330
Spring, 159, 171
Spring constant, 171
Spring–mass(damper) systems, 5, 118, 121, 218, 262, 552
Spurious component, 470
Square matrix, 760
Stable focus, 520
Stable limit cycle, 514, 530–531
Stable node, 520
Stable state, 738, 740
Stability, 142, 179, 319, 536–548
  in the large, 116, 512
  in the sense of Lyapunov, 116, 514, 563
Stability boundary, 532–534
Stability limit, 143, 337, 340, 341, 382, 531
Stability map, 329
Stability of motion, 117
Stack-type pneumatic instrument, 205
Staircase input pattern, 17, 26
Standard deviation, 577
Standard product, 709, 712, 718, 720
Standard sum, 709, 712, 714, 718, 720

Star, 96, 100, 438
Star current, 247
Start-up operation, 485
Start-up trajectory, 514
State, 35, 77, 210
State equation, 35, 38, 45, 48, 233, 234, 258, 399, 747
State estimation, 597, 610, 676
State model, 44, 229
State plane, 87 (*See also* Phase plane)
State probability, 747–748
State probability vector, 748
State space, 57, 77, 87, 97, 116
  statistical relationship in, 591–596
State-space formulation, 192, 532, 596
State-space frequency response, 398–408
State-space viewpoint, 4, 5, 421
State-transition matrix (*See* Solution matrix)
State-transition table, 738, 740
State variables, 35, 229, 347
State vector, 38, 300, 747
State-vector feedback, 322, 386, 486, 519, 671
Static accuracy, 322, 386
Static approximation, 9
Static characteristics, 248, 249
Static elements, 8, 10, 246
Static error, 3 (*See also* Offset)
Static friction (*See* Coulomb friction)
Static optimization (*See* Static peak holding)
Static peak holding, 4, 559–560
Static relation, 6, 163, 224, 238
Stationary systems, 6, 14, 40, 60, 155
Stationary random processes, 578
Statistical approach, 575
Statistical performance of control systems, 683–685
Steady-state error, 323, 324, 489 (*See also* Offset)
Steady-state relationship, 9
Step input, 15
Step-input response, 16, 343, 397
Stepping motor, 468, 479
Stirred-tank reactor (*See* Chemical reactor)
Stochastic optimal control, 676

Stochastic signal (*See* Stochastic variable)
Stochastic systems, 575–618, 675–690, 747
Stochastic variable, 575
Stored-program digital computer, 468
Storey, C., 302
Straight-line trajectory, 88
Strong condition of stability, 123
Structural model, 208
Subcube, 722
Subspace (*See* Manifold)
Subtraction, of matrix, 760
  of vector, 759
Superposition (*See* Linear superposition)
Supervisory control, 468
Supplementary vector (*See* Covariant vector)
Switching algebra, 706
Switching control, 12, 632 (*See also* Bang-bang control, On-off control)
Switching function, 510, 561, 630, 719 (*See also* Logic function)
Switching line, 512, 522, 551
Switching systems, 699–747
Sylvester condition, 125–126, 129, 540
Sylvester inequalities (*See* Sylvester condition)
Symmetric form (*See* Modified canonical form)
Symmetric limit cycle, 517
Symmetric matrix, 125, 247, 286, 454, 592, 760, 765
Symmetric variable, 287
Synchronous system, 736
Synthesizer for modal control, 433, 447
System model, 685
Systems, 153
  adjoint (*See* Adjoint system)
  continuous-time (*See* Continuous-time systems)
  discrete-time (*See* Discrete-time systems)
  finite-state (*See* Finite-state systems)
  linear (*See* Linear system)
  memoryless (*See* Static element)
  order of (*See* Dimensionality of dynamic systems)

798   Index

sampled data (*See* Discrete-time system)
time-varying (*See* Nonstationary system)
System boundary, 153
System output (*See* Output)
System response (*See* Output)

T

Tabulation method, 724–726
Tangential approximation, 160 (*See also* Linearization)
Tapered tank, 160
Target state (*See* End state)
Taylor-series expansion, 537, 641, 646
Temperature, 169
Temperature control, 328, 468, 755
Template for Bode diagram, 361–362
Terminal control, 693
Terminal state (*See* End state)
Thal-Larsen, H., 581
Thermal conductivity, 170
Thermal systems, 31, 164, 169–170, 337, 692
Thermometer, 8
Thermostats, 755
Third-level control, 468
Third-order systems, 98, 101
Three-dimensional trajectory, 98
Three-integrator system, 550–551
Three-port, 208, 213, 245–249
  input–output relation of, 248–249
  static characteristic of, 246, 248
Tie set, 728, 731
Time average, 578
Time constant, 37, 191, 211, 272, 331
Time delay (*See* Dead time)
Time-delay operator, 24, 52, 266
Time-domain, 41
Time-domain relation, 591
Time-domain solution, 63, 101, 102, 329
Time-domain specification, 322
Time-invariant system (*See* Stationary system)
Time moment, 346
Time-optimal control, 198, 549–551, 625, 628–629

Time-optimal positioning servo, 549
Time sharing, 468
Time-variant system (*See* Nonstationary system)
Time-varying system (*See* Nonstationary system)
Time-varying gain, 610, 660, 685
Timing chart, 738, 740
T-matrix (*See* Modal matrix)
Tolle, Max, 3
Torque, 164
Total solution, 60
Tracking error, 590, 664
Traffic lights, 754
Trajectory, 87, 98, 120, 129, 185, 403, 512, 626
  optimal (*See* Optimal trajectory)
Trajectory patterns, 94–98, 115
Transducer, 236
Transduction, ideal, 240
Transfer function(s)
  closed-loop, 320, 391
  exponential, 301, 370
  factored, 364
  irrational, 281, 286, 370
  open-loop, 319, 323, 332, 373, 414, 417
  matrix, 43, 64, 249, 414
  transcendental, 267
  unfactored, 365–366
Transfer matrix, 216–219, 221, 234, 237, 279, 300, 326
Transformer, 236–237
  ideal, 237, 238, 242
Transient set, 751
Transistor, 252
Transistor gate, 699, 702, 703, 707
Transition probability, 747–748
Transition probability matrix, 748
Transmission, 701
Transmission line, 226–236, 278–286
Transmission matrix (*See* Transfer matrix)
Transportation lag (*See* Dead time)
Transport motion, 289
Transposition, 759, 760
Transversality condition, 653, 654, 659, 671, 672
Truncation, 158

Truth table, 701, 708, 719
Tsypkin, J., 517
T-type flip-flop (*See* Flip-flop, T-type)
Turbulence, 159
Turbulence amplifier, 704, 753
Turns ratio, 237
Two-point boundary-value problem, 671
Two-port, 207
   input–output combinations of, 249
   prototype, 217, 226
   static behavior of, 241
Two-port chain, 225
Two-port elements, 216, 242
Two-port system, 236, 249
Two-tank system, 39, 105
Type 1 system, 323, 328
Type 0 system, 323

U

Unbounded response, 331
Unconditional expectation, 676
Uncontrollability (*See* Uncontrollable mode)
Uncontrollable mode, 79, 93, 349, 379, 386, 388, 437, 485, 666, 697
Undamped natural frequency, 324, 331
Undamped oscillator, 96, 99, 379, 381
Underdamped system, 327
Uniform line, 268
Unilateral coupling (*See* Coupling, unilateral)
Unilateral element, 187
Unilateral measurement, 328
Union, 710 (*See also* Logical sum)
Unit-circle stability criterion, 135, 144, 482
Unit impulse, 19, 37
Unit-impulse input response, 63, 329, 330–331, 581
Unit step, 19, 23, 37
Unit-step input response, 321, 327
Unit vector, 88
Unity feedback, 319, 414
Unnatural causality, 212
Unobservability (*See* Unobservable mode)
Unobservable mode, 79, 100, 350, 379, 436

Unobservable system, 350
Unstable limit cycle, 530, 532, 534
Unstable mode, 81
Unstable pole, 381
Unstable state, 740
Unstable systems, 331, 367
U-tube manometer, 201

V

Valve, 249 (*See also* Control valve)
Vapor condenser, 274
Variance, 577
Variational vector, 641
Vectors, 759
   basis, 88
   length of, 89
   linear dependence, 67
   linear independence, 89
   norm of, 116, 444
   normalized, 89
   orthogonal, 126
   scalar product of, 760
Vector equations (*See* Output equation, State equation)
Vector feedback control law, 421
Vector input, 38, 400
Vector output, 38, 400
Vehicle motion, 201
Velocity, 171
Velocity algorithm, 476, 480, 489, 497
Velocity-potential analogy, 171
Velocity saturation, 528–534
Venn diagram, 709–710, 711, 713, 716
Vibration testing system, 685
Viscous friction, 171

W

Warren loop, 737–738
Water tank, 13, 54, 199
Watt governor, 3
Wave propagation process, 277, 634–635
Wave reflection, 283
Wave-scattering variables, 280, 281, 283, 284
Weighting factor, 130, 347, 452, 658, 679
Weighting matrix, 457, 668, 670
White noise, 581, 586, 592

Wiener, N., 4
Wiener filter, 606
Wiper mechanism, 325
Word length, 475
Wykes, J. H., 431

X

x-y plotter, 93

Y

Y-system (*See* Shunt admittance system)
Yamashita, S., 519
Yore, E. E., 558
Young's modulus, 200, 271

Z

z-domain, 25, 51, 482, 749
Z-system (*See* Series impedance system)
z-transformation, 23, 25, 51
z-transformation pairs, 21, 22
Zalkind, C. S., 483
Zames, G., 546
ZERO, 699, 700–704, 707
Zero-hold element, 26, 51, 370, 473, 481
Zeros, 319
Ziegler, J. G., 343
Ziegler-Nichols approximation of process dynamics, 479
Ziegler-Nichols rule, 343–344, 383, 480